Small Office Building Handbook

Small Office Building Handbook:

Design for Reducing First Costs & Utility Costs

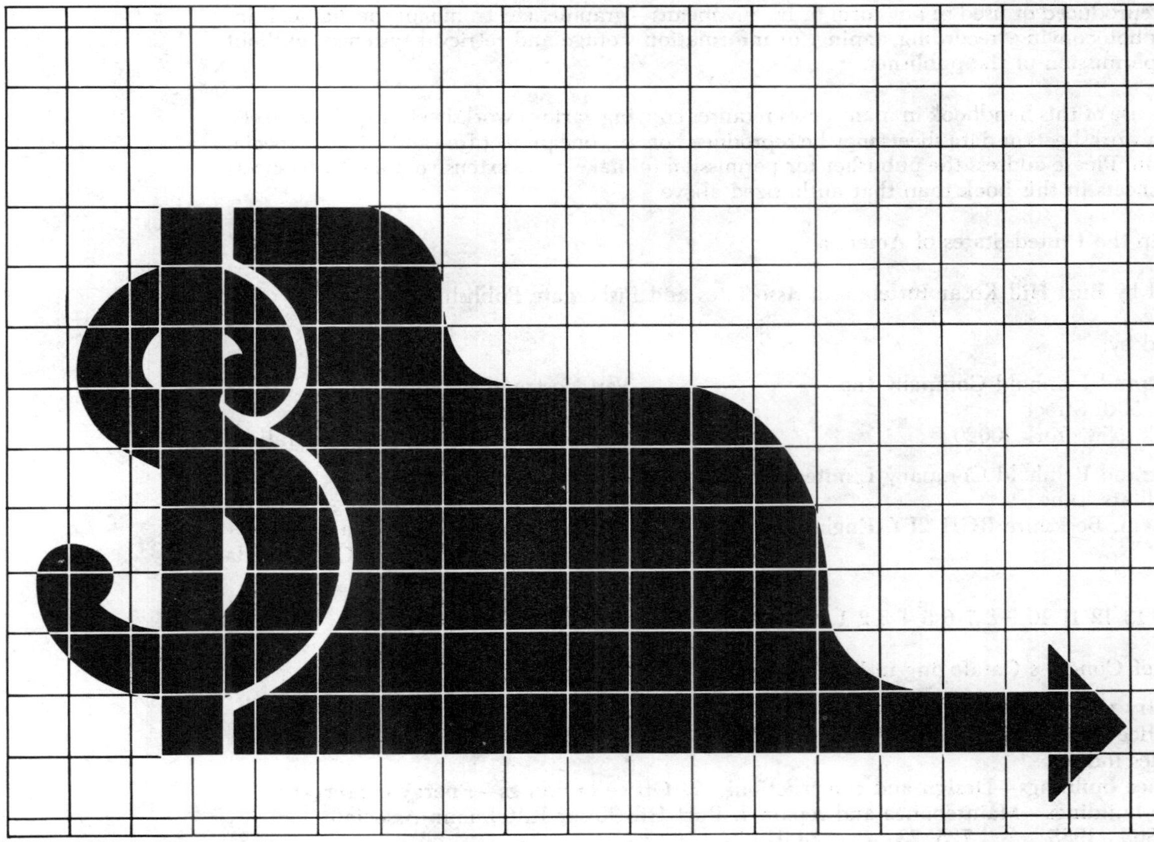

A project team headed by

BURT HILL KOSAR RITTELMANN ASSOCIATES

VNR VAN NOSTRAND REINHOLD COMPANY

Library of Congress Catalog Card Number 84-11917

ISBN 0-442-21126-0

Work supported by U.S. Department of Energy, Assistant Secretary for Conservation and Renewable Energy, Office of Building Energy Research and Development, under Prime Contract DE-AC06-76RLO 1830 with Battelle Memorial Institute, Pacific Northwest Laboratories, Richland, Washington.

Printed in the United States of America

Designed by Burt Hill Kosar Rittelmann Associates and Fishergate Publishing Company

Published by:

Van Nostrand Reinhold Company Inc.
135 West 50th Street
New York, New York 10020

Van Nostrand Reinhold Company Limited
Molly Millars Lane
Wokingham, Berkshire RG11 2PY, England

Van Nostrand Reinhold
480 La Trobe Street
Melbourne, Victoria 3000, Australia

Macmillan of Canada
Division of Gage Publishing Limited
164 Commander Boulevard
Agincourt, Ontario M1S 3C7, Canada

16 15 14 13 12 11 10 9 8 7 6 5 4 3 2 1

Library of Congress Cataloging in Publication Data

Main entry under title:
Small office building handbook.
Includes index.
1. Office buildings—Design and construction. 2. Office buildings—Energy conservation. 3. Office buildings—Maintenance and repair. I. Burt Hill Kosar Rittelmann Associates.
TH4311.S63 1985 725′.23 84-11917
ISBN 0-442-21126-0

Acknowledgments

PROJECT TEAM. The research and analyses for the Small Office Building Handbook were a team effort. The team was headed by Burt Hill Kosar Rittelmann Associates (BHKRA) and included Mueller Associates, Inc. (MAI); Control Data Corporation-Energy Technology Center (CDC); Hart, McMurphy & Parks (HMP); W. I. Whiddon & Associates (WIW); Scharf-Godfrey Associates (SG); and Daylighting/Energy Design Associates (DED).

SUPPORT AND GUIDANCE. The continuous support and thoughtful guidance of Stanley H. Pansky and Raymond W. Reilly at the Pacific Northwest Laboratories, operated for the Department of Energy by Battelle Memorial Institute, and of James L. Binkley, AIA, and Ted L. Kurkowski of the U.S. Department of Energy, Office of Buildings Energy Research and Development, have made the project possible.

MAJOR CONTRIBUTORS. The following persons made major contributions to the project:

John K. Holton, AIA, PE, Principal Researcher, Principal Author, Project Manager (BHKRA)
Theodore D. Swanson, PE, Principal Researcher, Author (MAI)
Ripudaman Gulati, PE, Principal Researcher (CDC)
G. Kimball Hart, Principal Author, Principal Editor (HMP)

PROJECT TEAM MEMBERS. The following persons were members of the project team:

Harry T. Gordon, AIA, Researcher, Partner-in-Charge (BHKRA)
William I. Whiddon, Researcher (WIW)
P. Richard Rittelmann, AIA, Researcher, Managing Partner (BHKRA)
James E. Moore, PE, Reseacher (MAI)
Paul C. Tseng, PE, Researcher (CDC)
Peter A. Lackler, Researcher (MAI)
Peter G. Rockwell, Researcher (BHKRA)
Ronald L. Overton, PE, Researcher (MAI)

Howard C. Huang, Graphic Design (BHKRA)
Barry L. Henderson, Researcher (MAI)
Paul W. Scanlon, PE, Researcher (BHKRA)
Alvin M. Sain, Researcher (BHKRA)
James E. Davy, Graphic Design (BHKRA)
Rene L. Carter, Administration/Text Production (BHKRA)
N. Scott Jones, Graphic Design (BHKRA)
Donald L. Anderson, Graphic Production (BHKRA)
Susan C. Holton, Editing/Graphics (BHKRA)

CONSULTANTS. The following persons were project consultants:
Howard T. Thomas, Cost Estimating, Mechanical (SG)
Joseph P. Jiacinto, Cost Estimating, Electrical (SG)
Joseph Sheets, Cost Estimating, Electrical (SG)
Benjamin H. Evans, Daylighting (DED)

ADVISORY PANEL. Important review and guidance was provided by two advisory groups. The Technical Advisory Panel was comprised of: Erv Bales; Harvey J. Bryan, AIA; Dr. Richard D. Cummings; Joseph J. Deringer, AIA; Harrison Fraker, AIA; Dr. Malcolm Lewis, PE; and Dr. Donald D. Ross, PE. The Building Owners and Managers Association representatives included: Maynard R. Hokanson, RPA, JD; Richard Allen, RPA; Russell V. Baltis, Jr., RPA; Robert L. Michaud, PE; and Dwayne A. Myers, RPA.

EDITORIAL GUIDANCE. Very helpful comment and guidance on editorial and publishing aspects was given by Morrie Helitzer of Helitzer Communications, Inc.

ORGANIZATIONS. Valuable basic research information on technical issues was provided by Lawrence Berkeley Laboratories, the California Energy Commission and the Tennessee Valley Authority. Critical review and comment of major importance to the project were provided by K. Michael McGrath and A. Joseph Van den Berg of the Edison Electric Institute and Louis A. Sarkes of the American Gas Association.

CLIMATE MAP. The climate region map was developed by Tony Olsen of Battelle/Pacific Northwest Laboratories.

Contents

SECTION III: ALTERNATIVE PROCESS

APPENDICES AND GLOSSARY

Introduction

In the decade since the energy crisis of 1973, a tremendous amount of research has been done to find ways of reducing energy use in buildings. For building owners, this translates into the possibility of significantly lower operating costs. For example, a well built and well maintained 40,000 square foot office building can use $60,000 less energy annually than it did just a few years ago. And that is with good design and improved equipment alone — nothing exotic.

This handbook focuses on energy saving strategies suitable for small office buildings (50,000 sq.ft. or less) and should be of interest to any building owner or manager concerned with energy costs. Although it is generally intended for use in designing new buildings, the handbook offers guidance applicable to buildings undergoing major rehabilitation. The strategies proposed are aimed at the great majority of buildings and are designed not to interfere with owner concerns about maintaining attractive rentable space.

This handbook starts from the premise that good current practice roughly matches *ASHRAE Standard 90-75 – Energy Conservation in New Building Design* and proceeds to demonstrate additional levels of savings. It lists dozens of cost effective strategies for your consideration. These recommendations are grouped so that you can easily pick the level of savings that best fits your current situation and interest. Several strategies cost virtually nothing to implement.

This handbook furthermore represents one of the first attempts to present guidance based on energy *cost* savings (in dollars) rather than on energy *use* savings (in Btus). This distinction is important because the principal energy problem in a building often differs significantly depending on whether one is looking at use or cost. This difference is caused by the wide range of fuel and electricity costs across the U.S. Calculation procedures have been developed for this handbook so that it can be used in the face of the growing complexity of utility rate structures. Enough data is given for a user to be able either to apply a local utility rate structure himself or to contact a local utility rate specialist and have that person compute an estimated annual fuel bill for the proposed project.

The underlying research for this project includes literally hundreds of computer simulation runs with state-of-the-art programs, including DOE 2.1b and a new computer version of the ASHRAE TC 4.7 method. This analysis has been cross-checked using professional judgment and the experience of a wide range of building experts, including a Technical Review Panel from the Building Owners and Managers Association.

Following this introduction is a summary of key findings from this project. It is a quick course for anyone considering using the handbook. Following the summary, a detailed sample problem illustrates precisely how this handbook is used to select cost effective energy saving strategies.

Section I of the handbook provides the reader with a complete decision-making process for selecting energy saving strategies in five climates covering the continental U.S. Section II contains detailed descriptive material explaining the first section. Finally, Section III provides an alternative design process that can be used in situations where the more simplified selector is not appropriate.

Summary

The purpose of this summary is to give the reader a quick introduction to the type of design guidance that will be found in this handbook. It is intended to give a preview of what is possible. For actually choosing and applying strategies presented here, the reader should use the main body of the book. As with the main text, this section is organized into several levels of savings so the reader can easily pick the actions that best fit a given project. A sample problem follows this section and gives a detailed introduction to the process of using the handbook.

LEVEL 1: Start Right

Matching good practice is the first step

The first level of energy cost savings can be achieved simply by following good current practice, which means at least meeting ASHRAE 90–75 Standards. The American Society of Heating, Refrigerating and Air-Conditioning Engineers first presented its energy standard 90 in 1975 (thus the name ASHRAE 90–75) and has continued to upgrade it ever since. It has been adopted by the building codes of the majority of states, and it has a very broad consensus across the design professions as the basis of good current practice—good, but not extreme in its efforts to reduce building energy use. However, there are still many buildings constructed every year that do not meet ASHRAE 90–75 Standards.

The checklist shown is a tremendous simplification of the key energy steps that must be taken to meet the ASHRAE 90–75 Standard.

Although abbreviated, this list (and the supporting tables) is a useful summary of good current practice. The strategy selector process—Section I—includes a much more detailed explanation of the assumptions about good current practice used in this analysis. This is called "LEVEL 1" because matching current practice is the first step necessary for anyone who is not already doing so. As you will see shortly, the costs of not meeting current practice can be significant.

COMPARE YOUR BUILDING TO THIS BASIC CHECKLIST	YES ☑
1. ADEQUATE INSULATION (SEE TABLE 1.)	☐
2. PROPER GLAZING (SEE TABLE 1.)	☐
3 NIGHTTIME TEMPERATURE SETBACK (55°HTG; 90°CLG).	☐
4. AIR AND TEMPERATURE CONTROLS (SEE TABLE 2.)	☐
5 REASONABLE LIGHTING LEVELS (2.25 W/SF. OR LESS + 0.75 W/SF EQUIP.	☐
6. GOOD HVAC CONTROLS (2° DEADBAND, LOW LEAK DAMPERS, TABLE 2.) . .	☐
7. REASONABLE VENTILATION RATES (.07 CFM/S.F OR LESS)	☐
8. REASONABLE WALL TIGHTNESS (INFILT.: OCC. HRS. PRESSURIZED; UN-OCC. HRS. - 1/3 - 1/2 ACH	☐

The map and tables on this page are used in conjunction with the checklist on the previous page to indicate selections of building materials and components that represent good current practice. It is only one way to meet ASHRAE 90–75 Standards. We believe, however, that it represents the way that most owners and builders can cost effectively meet the standard. Note that insulation levels and glazing characteristics (glazing % and glazing R value) are interrelated through the Overall Thermal Transfer Value (OTTV).

TABLE 2 - GOOD PRACTICE HVAC CONTROLS BY SYSTEM TYPE.

SYSTEM	TEMPERATURE CONTROL (IN ADD'N TO T'STAT)	OUTSIDE AIR CONTROL *
PTAC	ZONE	FIXED VOLUME
RT	OA RESET	ECONOMIZER
HP	ZONE	FIXED VOLUME
CV	OA RESET	ECONOMIZER
VAV	FIXED	ECONOMIZER
HP/HP	ZONE	ECONOMIZER
MZ	ZONE	ECONOMIZER
HYD. HP	ZONE	FIXED VOLUME

*ECONOMIZER VARIES BY CLIMATE (TABLE 1)

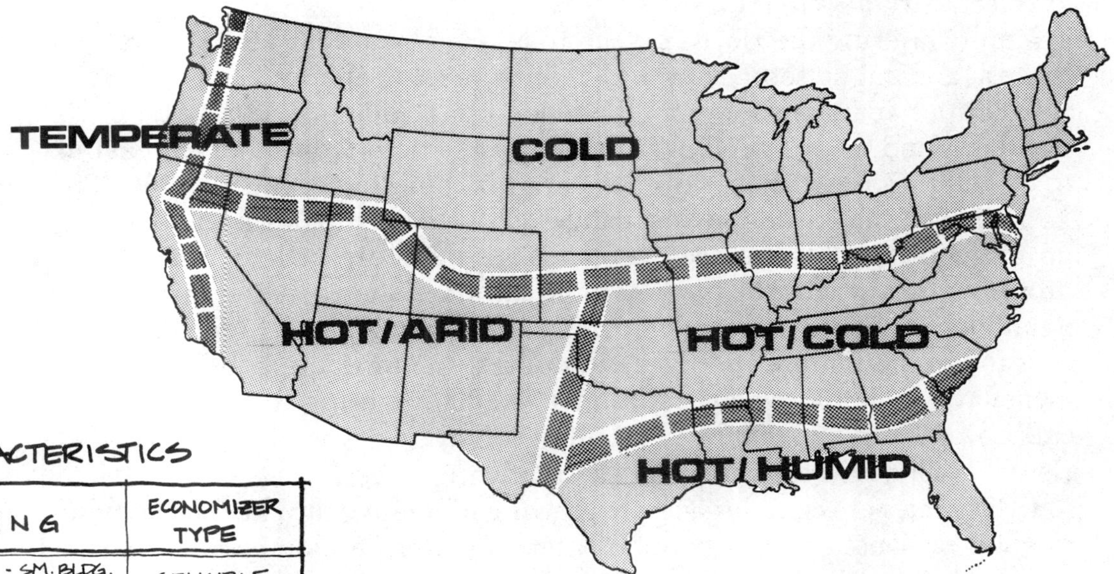

TABLE 1 - GOOD PRACTICE BUILDING CHARACTERISTICS

CLIMATE	ROOF INSULATION	WALL INSULATION	GLAZING	ECONOMIZER TYPE
COLD	R-14.3	3½" BATT	DOUBLE CLEAR - SM. BLDG. DOUBLE TINTED - OTHERS	SENSIBLE
TEMPERATE	R-8.3	3½" BATT	DOUBLE CLEAR - SM. BLDG. DOUBLE TINTED - OTHERS	ENTHALPY
HOT/COLD	R-8.3	3½" BATT	DOUBLE CLEAR - SM. BLDG. DOUBLE TINTED - OTHERS	ENTHALPY
HOT/HUMID	R-8.3	3/4" FOAM ON CONC. BLK. OR 3½" BATT	SINGLE TINTED OR REFLECTIVE - SM. DOUBLE TINTED OR SINGLE REFLECTIVE — OTHERS	NONE
HOT/ARID	R-8.3	3½" BATT	SINGLE TINTED - SM. BLDG. DOUBLE TINTED - OTHERS	ENTHALPY

Example

On the graph shown, the bar on the left represents a typical annual energy bill for a 40,000 sq. ft. office building in Washington, D.C. (an average climate with average utility costs). The building has minimal insulation, single clear glazing, no nighttime setback, no deadband on the thermostat, and a constant volume HVAC system with reheat. This is a poor building by current standards, calculated to cost its owner about $125,000 per year for electricity and heating fuel.

Simply altering the HVAC system to allow for outdoor air temperature control and night setback would reduce annual energy costs to about $90,000 per year, as shown by the second bar. Meeting ASHRAE 90–75 standards and today's good practice (Level 1) would reduce the annual bill to $67,000 – a total savings over the original building of roughly $60,000.

Obviously, the greatest savings can be achieved by bringing a bad building up to good current practice. Nevertheless, there are still significant improvements over current practice that can be made cost effectively. This handbook is about what you can still do after you have met today's standards to save even more energy – and dollars. In our example building, the owner was able to save an additional $12,000 by using the strategies we call Level 2 and $25,000 by combining the strategies we call Level 2 and Level 3. They are explained on the following pages.

Utility rates are complex and are growing more so each year. The analysis method presented in this handbook lets the user develop costs of energy savings in a realistic manner. Data are presented in both annual and monthly quantities for both consumption and demand and are further subdivided to distinguish heating use. This detailed data subdivision lets either the reader or a utility prepare an accurate projected utility bill based on most current utility rates.

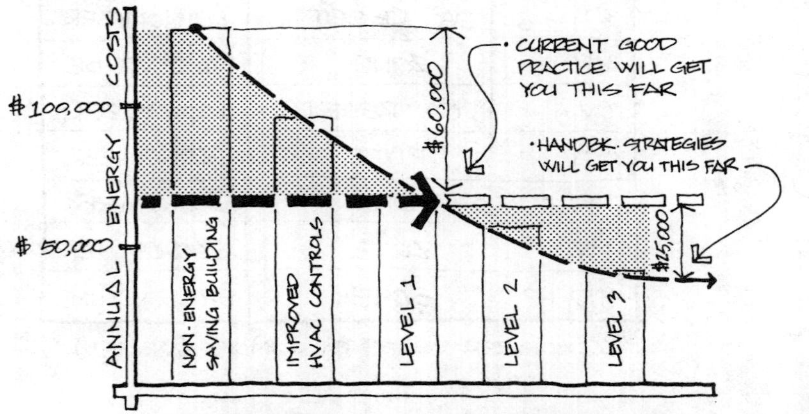

• EXAMPLE BASED ON SAMPLE CALCULATIONS FOR A TYPICAL 40,000 SQ. FT. OFFICE

Selecting the right HVAC system and the appropriate amount of glass can substantially reduce energy costs.

TABLE 3: GOOD EQUIPMENT SELECTIONS (equipment COPs should meet ASHRAE 90-75, 1980 standards)

Packaged Terminal AC (PTAC) - self-contained units, single zone, no economizer, electric.

Heat Pump (HP) - self-contained units, single zone, no economizer, electric.

Roof Top Packaged DX (RT) - roof mounted package DX, single zone, electric, gas or oil heat.

Multizone (MZ) - package unit, three-deck, (hot, cold, return bypass), constant volume, return air mixed with hot or cold air (no reheat), electric or gas (coils) heat.

Constant Volume (CV) - single duct supply, plenum return, constant volume, multiple zones, heat is electric duct coil or gas/oil hydronic base board.

Variable Air Volume (VAV) - single duct supply, plenum return, multiple zones, pressure activated VAV boxes, supply and return fans with variable inlet vanes and constant speed motor, heat is electric duct coil or gas/oil hydronic base board.

Split Heat Pump (HP/HP) - air to air heat pump, package units, one unit for core, one unit for perimeter, constant volume system, backup heat is electric duct coils.

Hydronic Heat Pump (Hyd HP) - multiple small air/water heat pumps on hydronic loop, load redistribution through loop, backup boiler (usually electric), cooling tower, no economizer.

BUILDING SIZE	HVAC SYSTEM	CLIMATE				
		COLD	TEMP.	HOT/COLD	HOT/HUMID	HOT/ARID
SMALL 4,000 - 10,000 SF	PTAC	●	●	●	●	●
	HP	●	●	●	●	●
	RT	●	●	●	●	●
MEDIUM 10,000 - 25,000 SF	MZ				●	●
	CV	●	●	●		
	VAV	●	●	●	●	●
	HP/HP	●	●	●		
	hyd HP				●	●
LARGE 25,000 - 50,000 SF	CV	●	●	●		
	VAV	●	●	●	●	●
	hyd HP	●	●	●	●	●

LEVEL 2: Make Initial Choices

The table 3 at left lists HVAC systems that represent good selections for each building size category. Dots within the matrix below further specify the climate zone where each system is appropriate. Any building that is not using one of these "good" systems is probably wasting money for its owner.

Selecting among the various "good" alternatives is more difficult and depends a great deal on the situation. First costs and annual operation costs are two major factors to consider. Within each building size category in the matrix below, systems are roughly listed with the lowest first cost system first, up to the highest costs system last. It is generally true, however, that the lower cost systems have higher annual energy costs. It is necessary to calculate the factors in this tradeoff based on your own local construction costs and utility rates, in order to make the proper system selection. Section I of the handbook provides a way to compute this important tradeoff.

Other key factors that must be considered by the designer in making a selection of HVAC system include: system flexibility, maintainability, fuel availability, manufacturer or dealer service, and previous experience. These important considerations are left to the best judgment of the design team and are not addressed in this handbook.

Windows, if properly selected and controlled, can save substantial energy and improve the quality of the office space at the same time. Reasonable amounts of window area (up to 30 to 40% of the wall surface) actually save energy when used in conjunction with daylight dimming controls. To control heat loss, double glazing is usually adequate even in the most northern climates. To control heat gain, either external or internal shading devices or modest tinting are usually the best solutions in all climates. However, shading devices require careful design, particularly for east and west windows.

LEVEL 3: Select Strategy Sets

After making Level 1 and 2 improvements, upgrading the lighting system is the single most important change you can make.

Install high efficiency ballasts in all fluorescent fixtures. High efficiency ballasts are improved conventional electromagnetic ballasts that use 8–16 watts less input to operate two standard lamps—a direct reduction in energy use of 10–15% over standard ballasts. These ballasts also operate with less power input than standard ballasts, reducing internal heat gains and, in turn, reducing cooling loads—a further reduction in energy costs.

These ballasts are widely available for both two-lamp and four-lamp fixtures. Some manufacturers have one-lamp and three-lamp models. Many fixtures can now be ordered from the manufacturer with high efficiency ballasts already installed.

A big advantage of this strategy is the reliability of the reduction in cooling load. Because ballasts are permanently installed, resulting load reductions can be used in downsizing HVAC equipment.

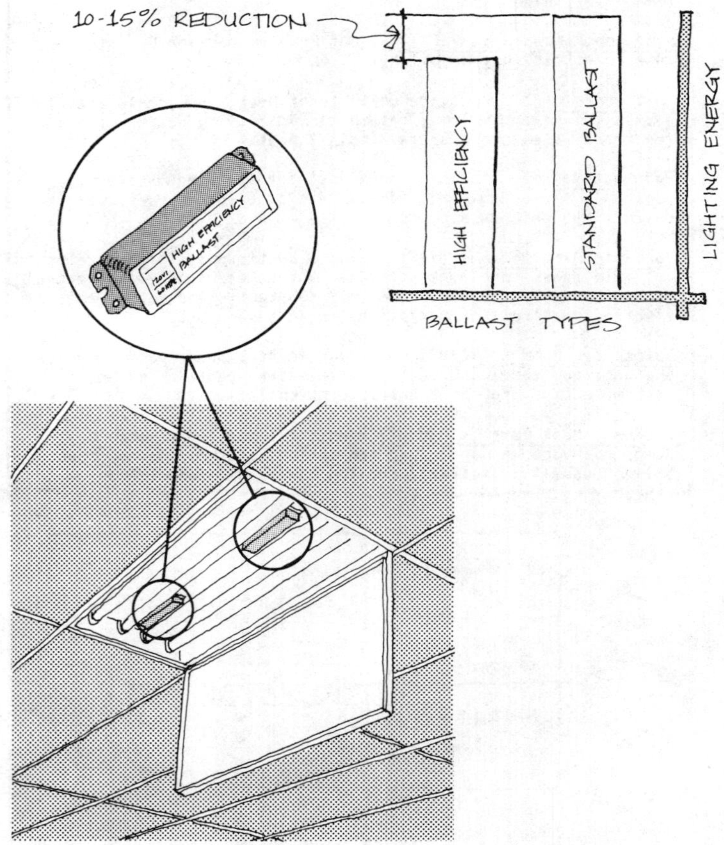

Downsize cooling equipment. As most HVAC system components are sized to meet peak cooling loads, a reduction in cooling load translates directly into the potential for a downsized system—including cooling plant, ducts, and fans. This can mean many thousands of dollars in first cost savings.

First cost savings are typically large enough not only to pay for the higher efficiency ballasts but to finance other energy improvements as well. Annual energy cost savings continue for the life of the building. These sizable savings will be partly offset by modest increases in heating costs in winter.

A few HVAC systems do not readily lend themselves to size changes in small increments. Such systems include PTAC and heat pump systems for small buildings and hydronic loop heat pumps on medium and large buildings. When selecting any of these systems the potential for cost savings due to downsizing will be lost.

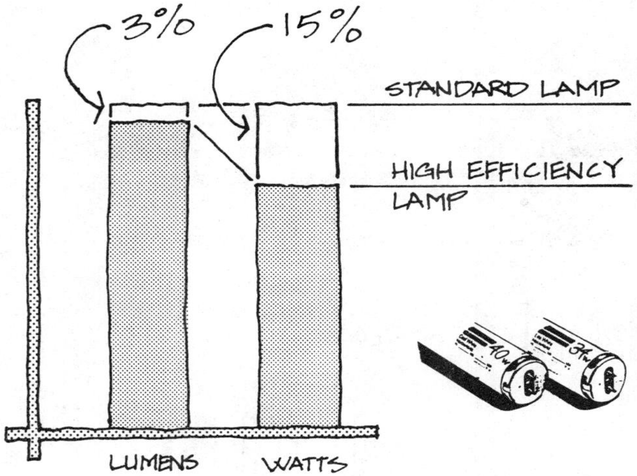

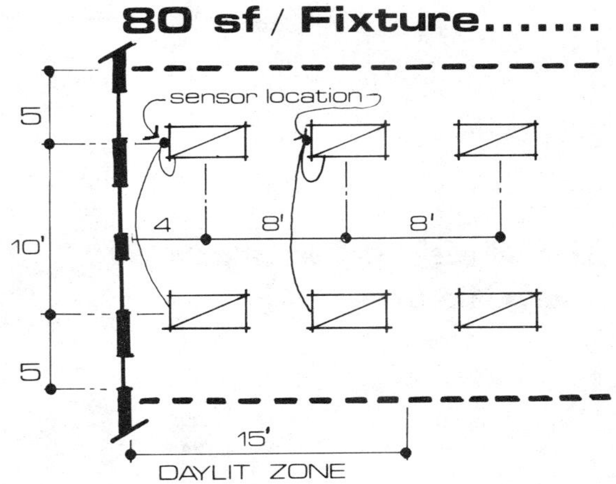

Strategies that often work . . . at some cost

Replace standard fluorescent lamps with high efficiency lamps. High efficiency lamps that provide essentially the same light output and useful life as standard lamps and that draw 34 watts instead of the usual 40 are available from several manufacturers (see diagram). With only a 3% reduction in lumen output, it is possible to reduce wattage used up to 15%.

Because there is no guarantee that low wattage lamps will be used by all subsequent tenants, downsizing HVAC equipment for the lower load is not recommended.

This strategy is a very practical addition to the two previous strategies. Downsizing of the HVAC system because of the use of the high efficiency ballasts can typically pay for both the ballasts and high efficiency lamps.

Install Automatic Daylight Dimmers. This strategy reduces both electric consumption and demand over the entire year. Daylight dimmers are particularly helpful in reducing energy costs because they dim unneeded lights on bright summer days, thus cutting both peak electric demand and consumption at the time of year when electric rates are highest in many utilities. Cooling costs will also be reduced.

It is recommended that only continuous (not step) dimmers be installed. Dimmers offer benefits in buildings with either clear or tinted glass (but not reflective glass) as long as glass area is at least 20% of exterior wall area. Optimum results are achieved in buildings with 30–40% glass area. Dimmers may control one or more fixtures up to a depth of 15 feet from the window wall (see sketch) and may be used with most standard or high efficiency ballasts, but usually not electronic ballasts. They are not recommended for use with high efficiency lamps, as reliable dimming of such lamps is questionable. Because of the potential for occupant intervention, reductions resulting from dimming should *not* be included in calculations aimed at downsizing HVAC equipment.

Replace standard fixtures with new high efficiency designs. Replacing standard *four*-lamp fixtures with *three*-lamp parabolic designs reduces lighting energy 25% while maintaining lighting levels. Including high efficiency ballasts and lamps can bring total savings in lighting up to 40%.

Nearly the same effect can be achieved using standard fixtures with high efficiency prismatic lenses. Reductions resulting from high efficiency fixtures and ballasts may in many cases be used to downsize HVAC equipment. This strategy is not recommended in combination with daylight dimmers because of high initial cost.

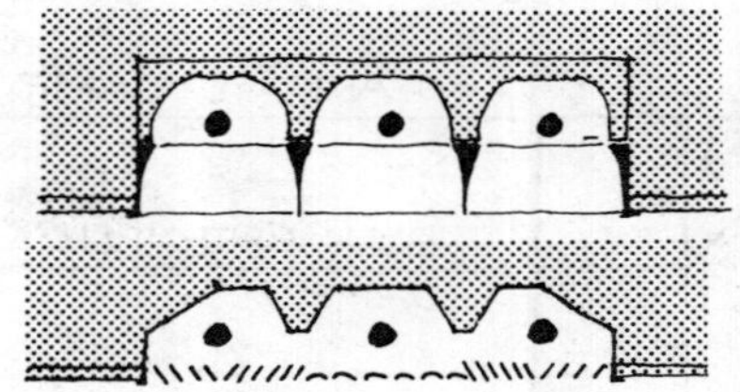

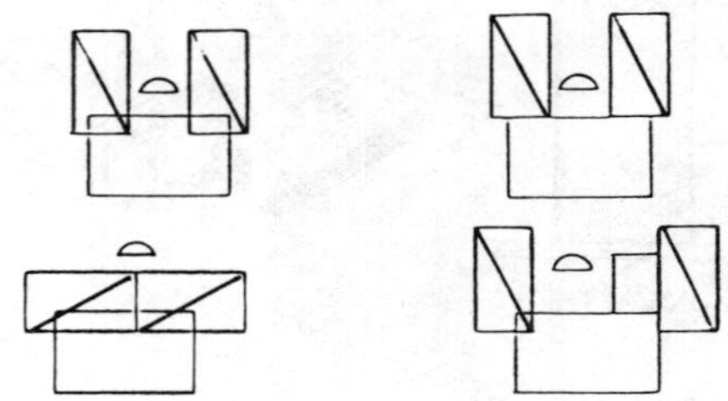

Design a non-uniform lighting system to match occupant needs. Carefully designing a lighting system of varying intensity—50 foot candles on work tasks, 30 foot candles in surrounding areas, and 10 foot candles in corridors, bathrooms, and core areas—will reduce the number of required fixtures and will cut both lighting and cooling costs. Downsizing HVAC equipment based on this strategy should be done only if there is reasonable assurance that lighting intensities will not be increased in the future.

Upgrade specified cooling equipment Coefficients of Performance. Upgrading from 1980 to 1984 ASHRAE Standards level will improve equipment performance an average of 10–15%.

Strategies for specific conditions

Combine skylights and daylight dimmers over large core areas in 1-story buildings. In areas that cannot be daylit from perimeter windows, use transluscent double glazed skylights in combination with daylight dimmers.

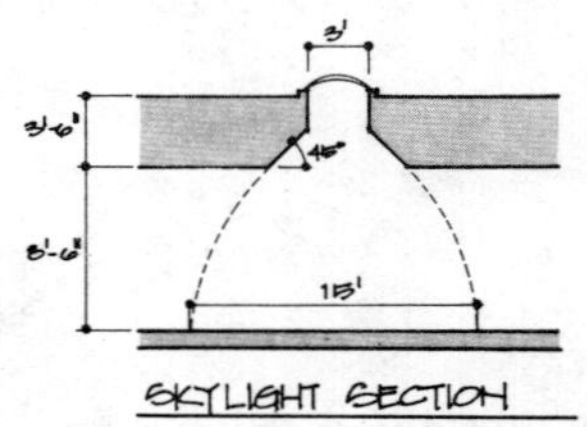

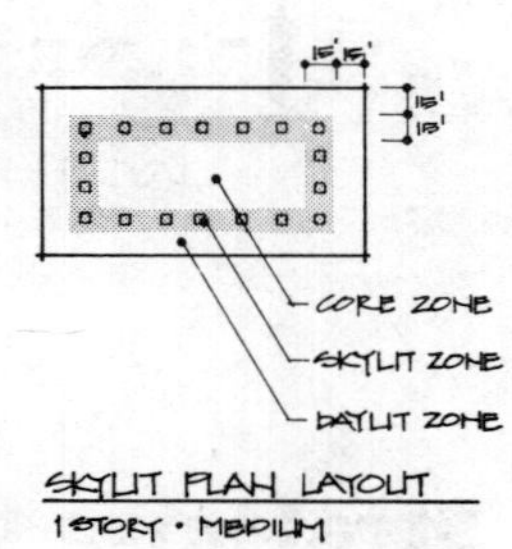

Consider high efficiency pulse combustion boilers to meet large heating loads. With annual efficiencies of 85% or better, fuel savings can offset modestly higher equipment cost.

Consider a combined evaporative cooler and DX system in hot/arid climates. Evaporative coolers may be cost effective in meeting the high cooling loads of hot arid climates. They should be used in combination with a downsized DX system to insure occupant comfort and reasonable equipment cost.

Additional strategies requiring careful design

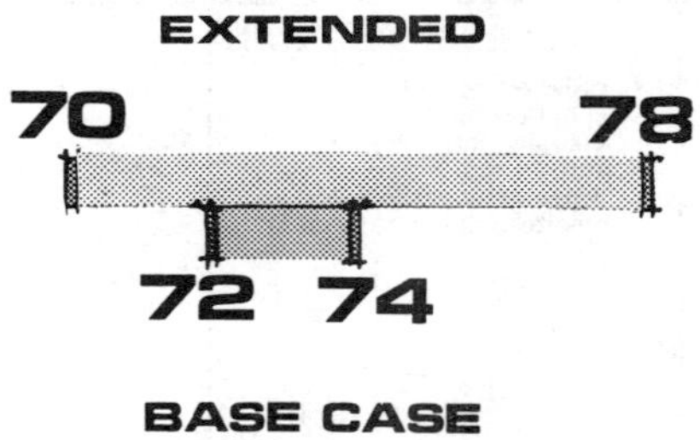

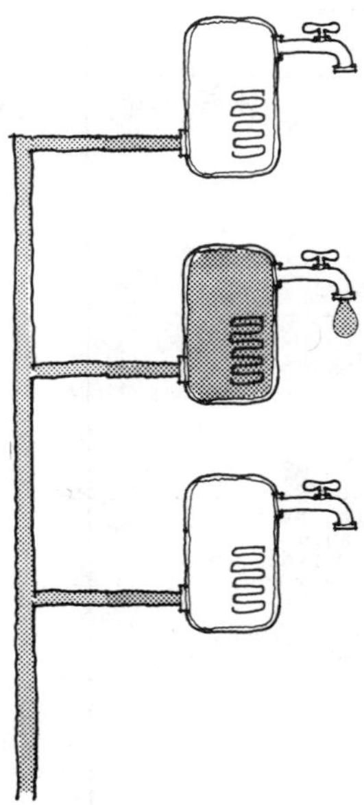

LEVEL 4: . . . Add More

The following strategies can further increase energy savings—often significantly so. They must be applied carefully, however. In most cases, special knowledge of the proposed strategy and experience with its application are necessary to insure proper design. Inappropriate applications can not only eliminate energy savings but also lead to serious occupant discomfort.

- **Increase thermostat deadband** during occupied periods from 2°F to 8°F (70 degrees for heating and 78 degrees for cooling).
- **Install a central energy monitoring and control system** to control energy consumption and demand with a computer and sensors to measure energy use, switch lights, and control HVAC equipment in the most efficient manner.
- **Install external shading devices** on windows in climate areas with high solar radiation during the cooling season.
- **Install point of use DHW heaters** in place of conventional single tank central system to eliminate the large heat losses from hot water distribution piping.
- **Change tinted glass to reflective glass** in climates with high cooling requirements if daylighting is not likely to be used as a strategy.
- **Design a building with high mass** (concrete or masonry construction) which in conjunction with night air flushing can lower cooling costs in appropriate climates.

LEVEL 5: Finish Right

All of the savings discussed in the previous four levels can be lost if a building is not properly built and maintained. Proper turnover and maintenance must be considered from the earliest stage of design. This checklist is a simple list of responsibilities for each member of the design-build team at each period in the process of delivering a building. These items are covered in detail in Section II of the text.

A CHECKLIST FOR SUCCESSFUL TURNOVER AND BUILDING OPERATION

	DESIGN	CONSTRUCTION	TURNOVER	BUILDING OPERATION
DESIGN TEAM (ARCHITECTS/ ENGINEERS)	• Document all design assumptions (1) • Estimate energy use by subsystem (2) **& For Large Buildings:** • Document all systems and control strategies (3) • Specify contents of O&M Manuals (4) • Specify HVAC balancing requirements (5) • Specify equipment protection and cleaning (6) • Specify on-site instruction (7) • Specify access doors, valves, test ports (8) • Specify piping system ID (9) • Specify price quotations on all long-term maintenance contracts (10)	• Prepare "as-built" drawings (11) • Observe construction process for compliance (12)	• Conduct "punch list" inspection (15) • Visually check for air leakage (15) • If possible, participate in on-site training session(s) (7)	
CONTRACTOR & SUB-CONTRACTOR	**For Small Buildings Only:** • Document all systems and control strategies (3) • Specify contents of O&M Manuals (4) • Specify HVAC balancing requirements (5) • Specify equipment protection and cleaning (6) • Specify on-site instruction (7) • Specify access doors, valves, test ports (8) • Specify piping system ID (9) • Specify price quotations on all long-term maintenance contracts (10)	• Prepare O&M Manual (4) • Perform final calibration of all controls (5) • Hire independent balancing agent or require of subs (5) • Protect and clean equipment (6) • Provide all access doors, valves, and test ports (8) • Place piping system ID tags (9) • Prepare "as-built" drawings (11) • Use HVAC system to "flush" construction odors (13)	• Deliver O&M Manual (4) • Conduct on-site instruction and training (7)	• Honor warranty and call back as required (16)
EQUIPMENT MANUFACT.		• Support calibration of controls (5) • Provide all special tools required (14)	• Support O&M Manual preparation (4) • Support on-site instruction and training (7)	
OWNER & BLDG. STAFF	• Check all key design assumptions (1)	• Periodically check compliance on-site (12)	• Attend on-site instruction and training sessions (7)	• Monitor energy consumption against estimates (2) • Periodically test equipment efficiency (17) • Periodically sample and test water (18) • Inspect and calibrate controls (19) • Monitor occupant complains (20)

Sample Problem

This section presents one completed sample problem using the "Selector Process" that is at the heart of this handbook. All of the forms necessary to complete an analysis for one of your proposed buildings are contained in Section I, "Selector Process," or Appendix A "Worksheets." We have selected one set of those forms and will use them to illustrate how we would select a set of energy-conservation strategies for a proposed 40,000 sq. ft. office building in Washington, D.C.

This handbook is divided into five climate regions so that all analysis and recommendations might more closely match specific climate conditions. Climate regions have been developed by statistical analysis of weather data to be representative of the entire U.S. These five climate regions are used in current ASHRAE Standards work and are felt to be representative of the principal climate differences that are important in office building energy design. Before proceeding into the first level of energy saving analysis, it is necessary to select the appropriate climate region for your project. If your site is near the dividing line between two regions, you may wish to compare the climate characteristics of both regions and select the closest match. After you have identified the right set of forms to use based on your climate region, the process consists of five levels as follows:

1. START RIGHT

The first level of energy savings can be achieved by making sure that your proposed small office building comes up to good current practice. This handbook starts with the assumption that current practice meets or exceeds ASHRAE Standard 90–75. These assumptions are spelled out in detail in Level 1 documentation. Start right by making sure your building meets these minimums.

2. MAKE INITIAL CHOICES

Level 2 energy savings can be achieved by wisely selecting basic building characteristics such as amount of glazing, HVAC system type, and heating fuel. A simple decision tree is used to assist you in comparing options based on actual energy costs in your area. If you are not familiar with your utility rate schedules, this section also provides guidance on working with your local utilities to obtain accurate energy cost estimates.

3. SELECT STRATEGY SET

Level 3 energy savings are achieved by applying a set of lighting, mechanical, and daylighting strategies to improve upon the design already achieved in Levels 1 and 2. Strategy sets are presented in increasing levels of aggressiveness in saving energy—starting with modest no-cost options and ending with daylight strategies. First cost estimates and a computation process are provided so you can estimate the cost effectiveness of each strategy set for your building before making a selection.

4. . . . ADD MORE

Level 4 presents strategies which are either unique to your given climate or which require a strong commitment to energy conservation, such as increased manual switching of lights or a wide temperature deadband. The design team is encouraged to investigate further those which may be appropriate. Criteria for selection and economic analysis are left to you.

5. FINISH RIGHT

The final level of energy savings is achieved when the building, both constructed and operating according to design, is turned over to the owner and maintenance staff. This section presents a checklist of responsibilities for all members of the design and construction teams to insure that building turnover and operation will be successful. Each item on the checklist is explained in detail, including the time at which it must occur in the design/build sequence.

SECTION I

SELECTOR PROCESS

CLIMATE REGIONS
IDENTIFY REGION
COLD CLIMATE
TEMPERATE CLIMATE
HOT/COLD CLIMATE
HOT/HUMID CLIMATE
HOT/ARID CLIMATE

In the example problem that follows, actual pages from Section I and Appendix A have been reduced to serve as illustrations. Where appropriate, the pages are filled in by hand to demonstrate how you would actually proceed through a problem. Additional explanatory material is presented in the margins to assist you. Technical issues are not addressed in this sample problem. That information is provided in the main text—principally in Section II.

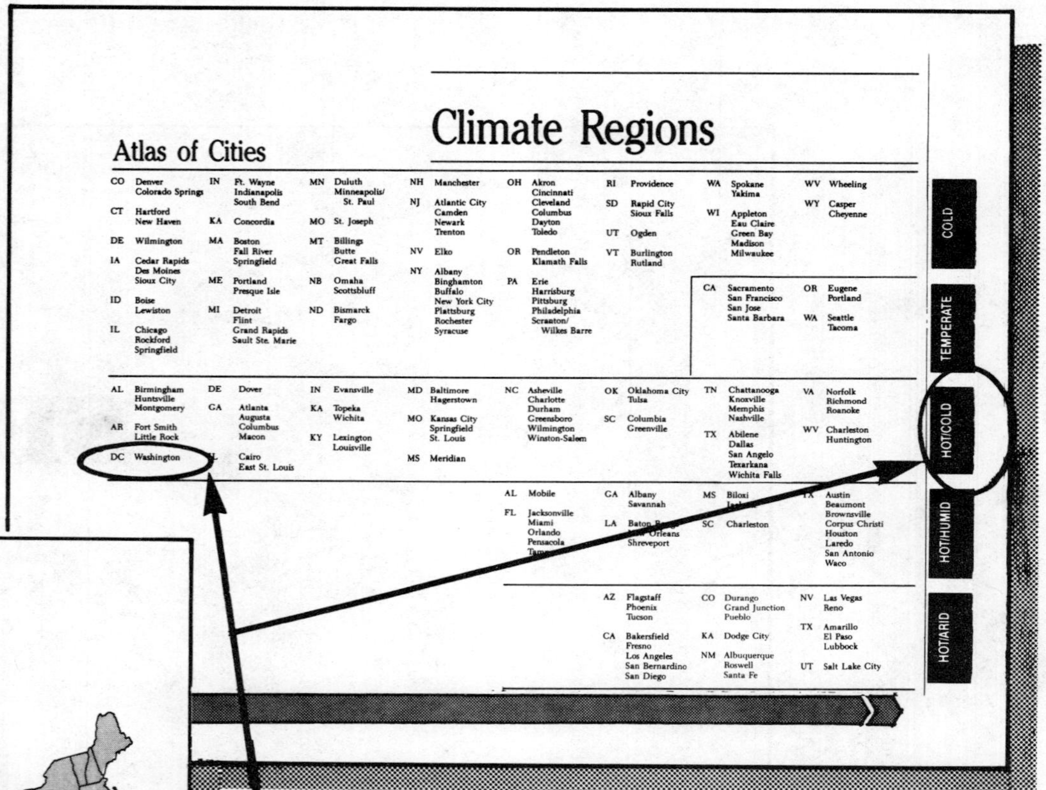

Climate Regions

Atlas of Cities

COLD

CO	Denver, Colorado Springs	IN	Ft. Wayne, Indianapolis, South Bend	MN	Duluth, Minneapolis/ St. Paul	NH	Manchester	OH	Akron, Cincinnati, Cleveland, Columbus, Dayton, Toledo	RI	Providence	WA	Spokane, Yakima	WV	Wheeling
CT	Hartford, New Haven	KA	Concordia	MO	St. Joseph	NJ	Atlantic City, Camden, Newark, Trenton	OR	Pendleton, Klamath Falls	SD	Rapid City, Sioux Falls	WI	Appleton, Eau Claire, Green Bay, Madison, Milwaukee	WY	Casper, Cheyenne
DE	Wilmington	MA	Boston, Fall River, Springfield	MT	Billings, Butte, Great Falls	NV	Elko	PA	Erie, Harrisburg, Pittsburg, Philadelphia, Scranton/ Wilkes Barre	UT	Ogden				
IA	Cedar Rapids, Des Moines, Sioux City	ME	Portland, Presque Isle	NB	Omaha, Scottsbluff	NY	Albany, Binghamton, Buffalo, New York City, Plattsburg, Rochester, Syracuse			VT	Burlington, Rutland				
ID	Boise, Lewiston	MI	Detroit, Flint, Grand Rapids, Sault Ste. Marie	ND	Bismarck, Fargo										
IL	Chicago, Rockford, Springfield														

TEMPERATE

CA	Sacramento, San Francisco, San Jose, Santa Barbara	OR	Eugene, Portland
		WA	Seattle, Tacoma

HOT/COLD

AL	Birmingham, Huntsville, Montgomery	DE	Dover	IN	Evansville	MD	Baltimore, Hagerstown	NC	Asheville, Charlotte, Durham, Greensboro, Wilmington, Winston-Salem	OK	Oklahoma City, Tulsa	TN	Chattanooga, Knoxville, Memphis, Nashville	VA	Norfolk, Richmond, Roanoke
AR	Fort Smith, Little Rock	GA	Atlanta, Augusta, Columbus, Macon	KA	Topeka, Wichita	MO	Kansas City, Springfield, St. Louis			SC	Columbia, Greenville	TX	Abilene, Dallas, San Angelo, Texarkana, Wichita Falls	WV	Charleston, Huntington
DC	Washington	IL	Cairo, East St. Louis	KY	Lexington, Louisville	MS	Meridian								

HOT/HUMID

AL	Mobile	GA	Albany, Savannah	MS	Biloxi, [illegible]	TX	Austin, Beaumont, Brownsville, Corpus Christi, Houston, Laredo, San Antonio, Waco
FL	Jacksonville, Miami, Orlando, Pensacola, [illegible]	LA	Baton [illegible], [illegible] Orleans, Shreveport	SC	Charleston		

HOT/ARID

AZ	Flagstaff, Phoenix, Tucson	CO	Durango, Grand Junction, Pueblo	NV	Las Vegas, Reno
CA	Bakersfield, Fresno, Los Angeles, San Bernardino, San Diego	KA	Dodge City	TX	Amarillo, El Paso, Lubbock
		NM	Albuquerque, Roswell, Santa Fe	UT	Salt Lake City

All five climate regions are illustrated on the map below. If your proposed building falls on the edge of two climate regions, you may wish to do the analysis for both regions to see which better fits your own experience in the area.

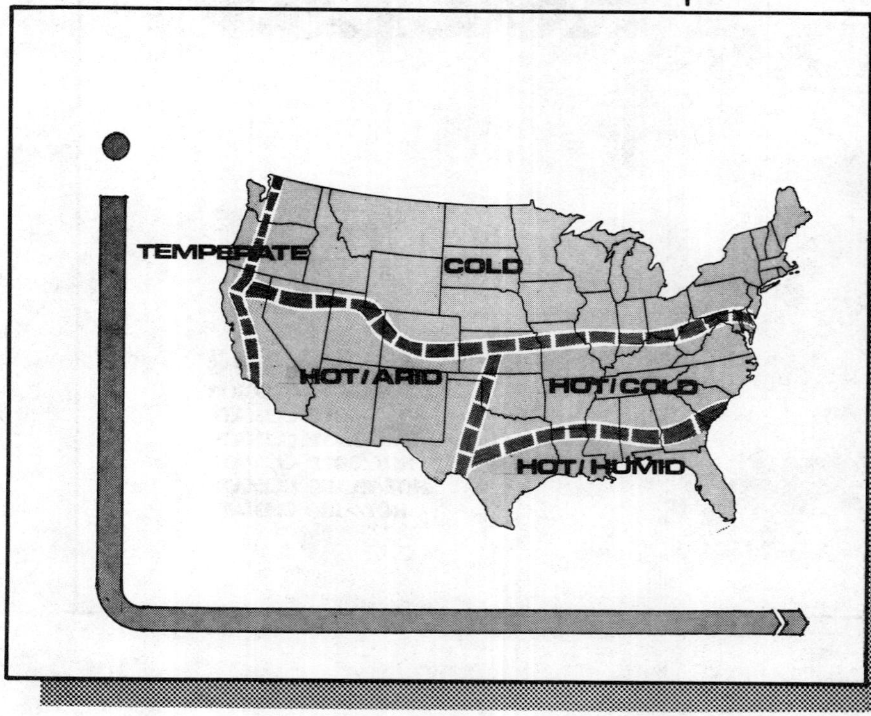

Find your city and note the relevant climate region.

The entire selection process is climate specific. Therefore, you need to select the climate region and related forms that match the location of the building for which you are preparing an analysis. To facilitate this selection, an atlas of major cities arranged by climate region is provided. For our sample problems, the city is Washington, D.C. The climate region, therefore, is "Hot/Cold." Both are indicated on the atlas above. If your city is not listed, consult the map at left.

Confirm that you are in the right region and proceed.

Hot/Cold Climate

Identify Region

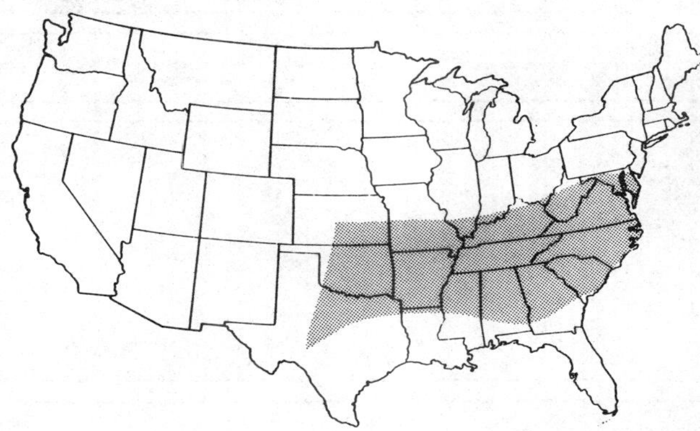

CITIES IN REGION

AL	Birmingham Huntsville Montgomery	MS	Meridian
AR	Fort Smith Little Rock	NC	Asheville Charlotte Durham Greensboro Wilmington Winston-Salem
DC	Washington		
DE	Dover	OK	Oklahoma City Tulsa
GA	Atlanta Augusta Columbus Macon	SC	Columbia Greenville
IL	Cairo East St. Louis	TN	Chattanooga Knoxville Memphis Nashville
IN	Evansville	TX	Abilene Dallas San Angelo Texarkana Wichita Falls
KA	Topeka Wichita		
KY	Lexington Louisville	VA	Norfolk Richmond Roanoke
MD	Baltimore Hagerstown	WV	Charleston Huntington
MO	Kansas City Springfield St. Louis		

BEGIN YOUR 5 LEVELS OF STRATEGY SELECTION

Note: See Section II for a full discussion of the five climate regions used in this manual.

The cover page for the *Hot/Cold* climate region clearly indicates the area contained within the region and again lists selected cities. Use this page to confirm that you have selected the correct climate for your analysis.

Check your design against current good practice and common design assumptions for all buildings.

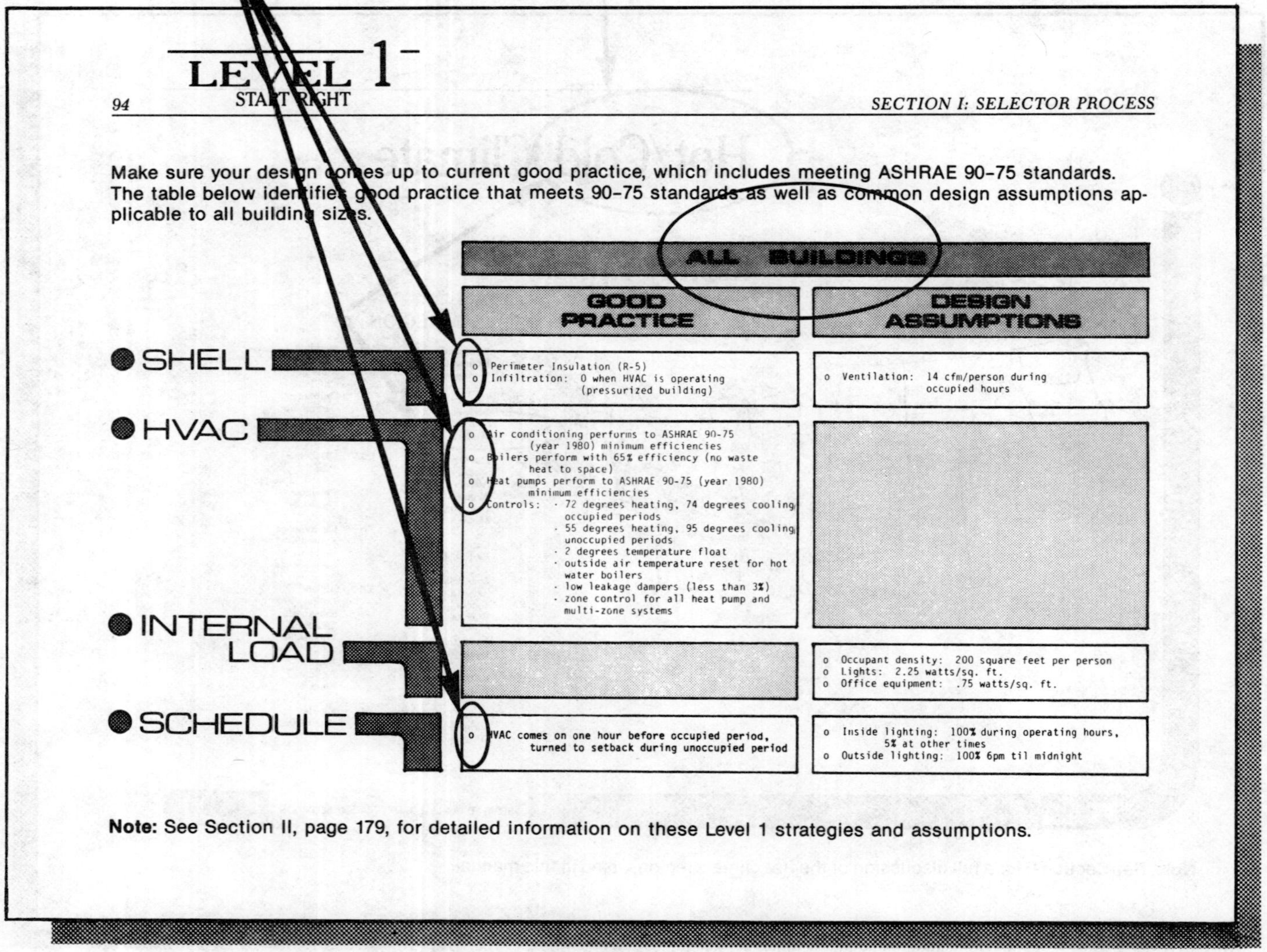

LEVEL 1 START RIGHT

94 SECTION I: SELECTOR PROCESS

Make sure your design comes up to current good practice, which includes meeting ASHRAE 90-75 standards. The table below identifies good practice that meets 90-75 standards as well as common design assumptions applicable to all building sizes.

ALL BUILDINGS	GOOD PRACTICE	DESIGN ASSUMPTIONS
SHELL	o Perimeter Insulation (R-5) o Infiltration: 0 when HVAC is operating (pressurized building)	o Ventilation: 14 cfm/person during occupied hours
HVAC	o Air conditioning performs to ASHRAE 90-75 (year 1980) minimum efficiencies o Boilers perform with 65% efficiency (no waste heat to space) o Heat pumps perform to ASHRAE 90-75 (year 1980) minimum efficiencies o Controls: · 72 degrees heating, 74 degrees cooling occupied periods · 55 degrees heating, 95 degrees cooling unoccupied periods · 2 degrees temperature float · outside air temperature reset for hot water boilers · low leakage dampers (less than 3%) · zone control for all heat pump and multi-zone systems	
INTERNAL LOAD		o Occupant density: 200 square feet per person o Lights: 2.25 watts/sq. ft. o Office equipment: .75 watts/sq. ft.
SCHEDULE	o HVAC comes on one hour before occupied period, turned to setback during unoccupied period	o Inside lighting: 100% during operating hours, 5% at other times o Outside lighting: 100% 6pm til midnight

Note: See Section II, page 179, for detailed information on these Level 1 strategies and assumptions.

Check your design against our interpretation of good current practice that meets ASHRAE 90-75. If your building measures up, proceed. If it doesn't, you may want to consider additional analysis at this point. Our research shows that any building that doesn't meet these guidelines can be wasting significant amounts of energy—and dollars.

Some aspects of good practice and design assumptions vary by building size: match them to your case.

Note: Two HVAC system options selected to investigate.

HOT/COLD CLIMATE 95

Some aspects of ASHRAE 90–75 and current **good practice** vary according to building size and HVAC system type. Make sure your design comes up to these criteria. You may want to come back to this table as you explore HVAC options in Level 2.

		BUILDING SIZE								
		SMALL 2000-10000 sq ft			MEDIUM 10000-25000 sq ft			LARGE 25000-50000 sq ft		
SHELL	ROOF INSULATION	R-8.3			R-8.3			R-8.3		
	WALL INSULATION	R-7.5			R-7.5			R-7.5		
	GLAZING	DOUBLE GLAZED WITH CLEAR GLASS			DOUBLE GLAZED WITH TINTED GLASS			DOUBLE GLAZED WITH TINTED GLASS		
	AIR CHANGES	1/2 AC/HR UNOCCUPIED			1/3 AC/HR UNOCCUPIED			1/3 AC/HR UNOCCUPIED		
HVAC		PACKAGED TERMINAL AIR CONDITIONERS	ROOFTOP DX	HEAT PUMP	CONSTANT VOLUME	VARIABLE AIR VOLUME	SPLIT HEAT PUMP	CONSTANT VOLUME	VARIABLE AIR VOLUME	HYDRONIC HEAT PUMP
	TEMPERATURE CONTROL	ZONE	OA RESET	ZONE	OA RESET	FIXED	ZONE	OA RESET	FIXED	ZONE
	AIR CONTROL	FIXED VOLUME	ENTHALPY ECONOMIZER	FIXED VOLUME	ENTHALPY ECONOMIZER	ENTHALPY ECONOMIZER	ENTHALPY ECONOMIZER	ENTHALPY ECONOMIZER	ENTHALPY ECONOMIZER	FIXED VOLUME

Some common **design assumptions** also vary according to building size and no. of stories.

INT. LOAD		1 STORY	1 STORY	2 STORY	2 STORY	4 STORY ①
	ELEVATOR & OUTDOOR LIGHTS	1345 WATTS	4035 WATTS	4535 WATTS	7860 WATTS	9360 WATTS
SCHEDULE	OCCUPANCY	80% 8am - 6pm Monday - Friday	80% 8am - 6pm Monday - Friday		80% 8am - 6pm: M-F 100% 9am - 1pm: SAT	

GO TO THE LEVEL 2 PAGE SPECIFIC TO YOUR BUILDING SIZE & NO. OF STORIES.

Note ① For 3 or 5 story structures, use "4-story" classification.

This step is the same as the previous one; however, the list of criteria for good practice varies according to building size.

Note: that under "HVAC," there are several options. Each of these options we believe represents good practice. You may wish to choose more than one option at this point and compare the results of each choice in the following Level 2 analysis. For our sample problem, we have chosen two options: a "Constant Volume" system and a "Variable Air Volume" system. We also had the option of choosing either a 2-story or 4-story building shape. In our sample problem, the client made that selection for us. We are limited to a 4-story design. Finally, be sure to note that from here on the analysis is specific for building size and number of stories. Choose Level 2 analysis forms to match the selections you made here, i.e.: "Large 4-Story."

Note: Choose Level 2 pages based on building size and number of stories.

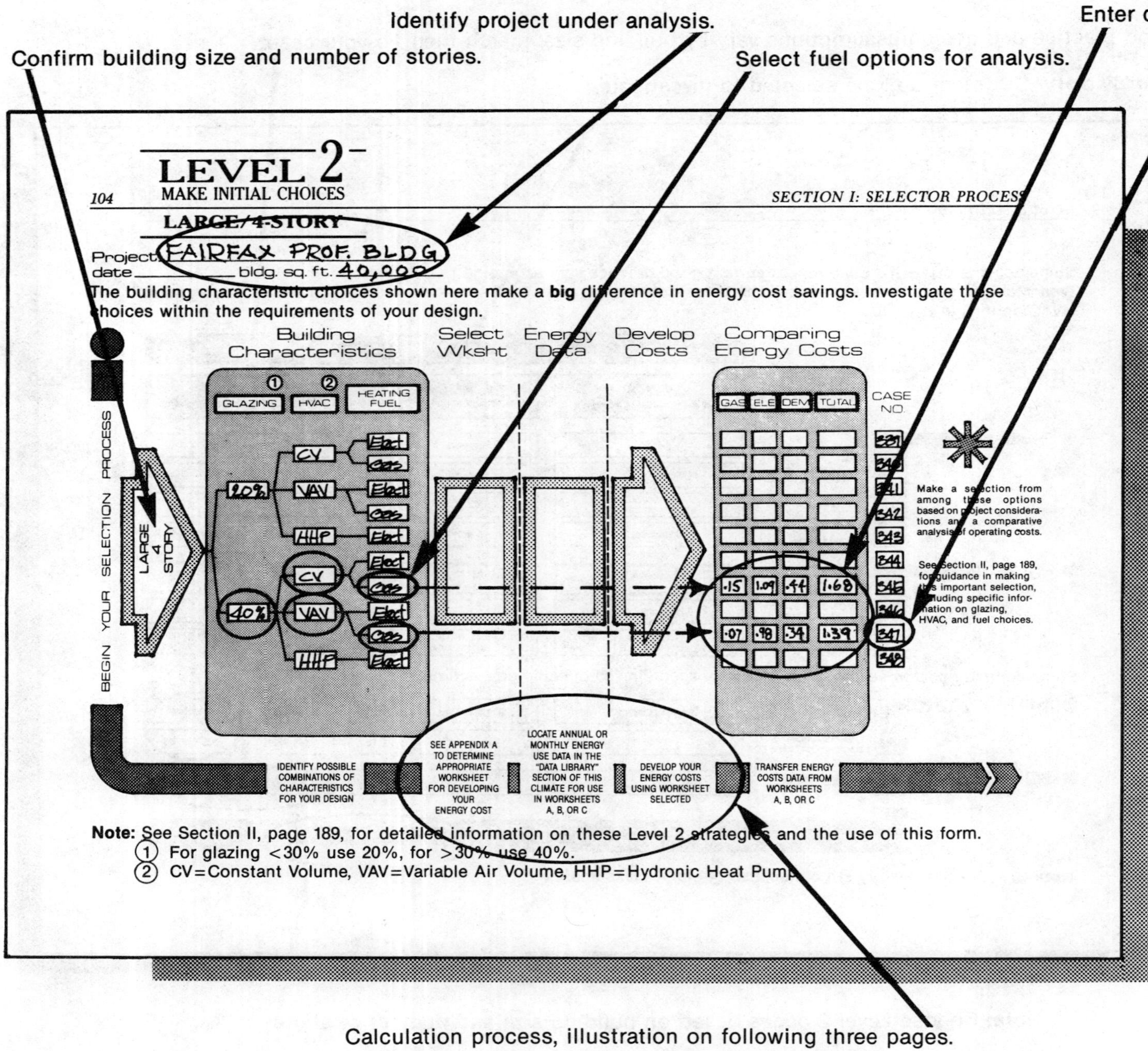

Before beginning Level 2, confirm that you have selected the correct worksheet, conforming to the size and number of stories decided upon in Level 1. For your own records, enter the name, date, and building square footage in the upper left hand corner of the page. Starting through the selector process (from left to right), you will note that level 2 presents you with three choices: glazing area, HVAC system type, and heating fuel. For the sample problem, we have chosen only the 40% glazing option. We want to compare a Constant Volume (CV) system against a Variable Air Volume (VAV) system, so we have circled both. We are not particularly interested in the Hydronic Heat Pump (HHP) at this point. We have gas at the site, so it is our heating fuel of choice. These choices together mean that we will investigate Case. No. 345 and Case No. 347. These case numbers will be used later to locate data in the data library. The balance of the page leads you through a simple analysis process, the results of which are entered in the blank boxes provided. The numbers shown here have been developed using Worksheet "A." The process is illustrated over the next three pages. Once analysis is complete, a single case is selected to carry forward to Level 3. For our sample problem, we selected Case No. 347.

Select the worksheet that best fits your situation.

UTILITY RATE CHARACTERISTICS

Select Worksheet A, B, or C for energy cost analysis with the aid of these guidelines (note that more than one utility rate may apply to a project). You may wish to contact your utility to clarify any of the guidelines below.

A Annual Analysis

IF UTILITY RATES:

1. Have relatively uniform demand *and* consumption charges throughout the year.

(You do the analysis)

B Monthly Analysis

IF UTILITY RATES:

1. Have relatively uniform demand charges throughout the year with no "ratchet" charges.
2. Have consumption rates with only seasonal changes.

(You do the analysis)

C Form Letter To Utility

IF UTILITY RATES:

1. Have complicated "ratchet" clauses for demand.
2. Have consumption rates which depend on level of demand.
3. Have seasonal changes in demand and consumption rates.
4. Have separate heating and general service rates that require many calculations

and

Your HVAC system can be wired and metered to use these rates—(including VC/elec., VAV/elec., MZ/elec., or RT-DX/elec.).

(Utility does the analysis)

Following selection of building characteristics, select a worksheet for the balance of the analysis. An instructional page from Appendix A, shown at right, is used to guide your selection. We have chosen to use Worksheet "A," first because we want to do a simple rate calculation ourselves and second, because we believe our local rate schedule is uniform enough to be characterized on an annual basis. The other worksheets and instructions are contained in Appendix A.

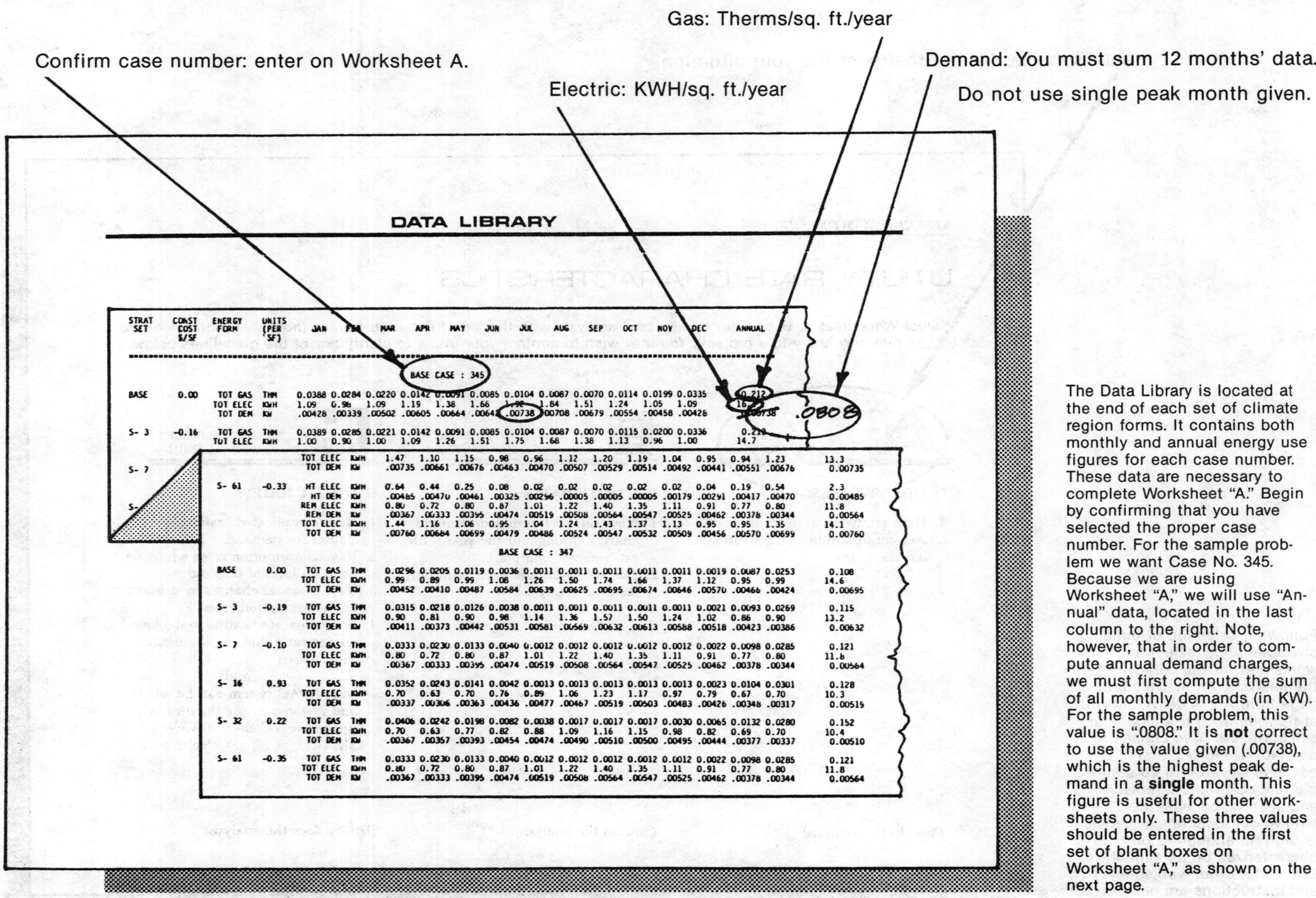

STRAT SET	CONST COST $/SF	ENERGY FORM	UNITS (PER SF)	JAN	FEB	MAR	APR	MAY	JUN	JUL	AUG	SEP	OCT	NOV	DEC	ANNUAL
							BASE CASE : 345									
BASE	0.00	TOT GAS	THM	0.0388	0.0284	0.0220	0.0142	0.0091	0.0085	0.0104	0.0087	0.0070	0.0114	0.0199	0.0335	0.212
		TOT ELEC	KWH	1.09	0.96	1.09	1.19	1.38	1.66	1.9[illegible]	1.84	1.51	1.24	1.05	1.09	16.[illegible]
		TOT DEM	KW	.00428	.00339	.00502	.00605	.00664	.0064[illegible]	.00738	.00708	.00679	.00554	.00458	.00428	[illegible]0738
S- 3	-0.16	TOT GAS	THM	0.0389	0.0285	0.0221	0.0142	0.0091	0.0085	0.0104	0.0087	0.0070	0.0115	0.0200	0.0336	0.21[illegible]
		TOT ELEC	KWH	1.00	0.90	1.00	1.09	1.26	1.51	1.75	1.68	1.38	1.13	0.96	1.00	14.7
S- 7		TOT ELEC	KWH	1.47	1.10	1.15	0.98	0.96	1.12	1.20	1.19	1.04	0.95	0.94	1.23	13.3
		TOT DEM	KW	.00735	.00661	.00676	.00463	.00470	.00507	.00529	.00514	.00492	.00441	.00551	.00676	0.00735
S- 61	-0.33	HT ELEC	KWH	0.64	0.44	0.25	0.08	0.02	0.02	0.02	0.02	0.02	0.04	0.19	0.54	2.3
		HT DEM	KW	.00485	.00470	.00461	.00325	.00296	.00005	.00005	.00005	.00179	.00291	.00417	.00470	0.00485
		REM ELEC	KWH	0.80	0.72	0.80	0.87	1.01	1.22	1.40	1.35	1.11	0.91	0.77	0.80	11.8
		REM DEM	KW	.00367	.00333	.00395	.00474	.00519	.00508	.00564	.00547	.00525	.00462	.00378	.00344	0.00564
		TOT ELEC	KWH	1.44	1.16	1.06	0.95	1.04	1.24	1.43	1.37	1.13	0.95	0.95	1.35	14.1
		TOT DEM	KW	.00760	.00684	.00699	.00479	.00486	.00524	.00547	.00532	.00509	.00456	.00570	.00699	0.00760
							BASE CASE : 347									
BASE	0.00	TOT GAS	THM	0.0296	0.0205	0.0119	0.0036	0.0011	0.0011	0.0011	0.0011	0.0011	0.0019	0.0087	0.0253	0.108
		TOT ELEC	KWH	0.99	0.89	0.99	1.08	1.26	1.50	1.74	1.66	1.37	1.12	0.95	0.99	14.6
		TOT DEM	KW	.00452	.00410	.00487	.00584	.00639	.00625	.00695	.00674	.00646	.00570	.00466	.00424	0.00695
S- 3	-0.19	TOT GAS	THM	0.0315	0.0218	0.0126	0.0038	0.0011	0.0011	0.0011	0.0011	0.0011	0.0021	0.0093	0.0269	0.115
		TOT ELEC	KWH	0.90	0.81	0.90	0.98	1.14	1.36	1.57	1.50	1.24	1.02	0.86	0.90	13.2
		TOT DEM	KW	.00411	.00373	.00442	.00531	.00581	.00569	.00632	.00613	.00588	.00518	.00423	.00386	0.00632
S- 7	-0.10	TOT GAS	THM	0.0333	0.0230	0.0133	0.0040	0.0012	0.0012	0.0012	0.0012	0.0012	0.0022	0.0098	0.0285	0.121
		TOT ELEC	KWH	0.80	0.72	0.80	0.87	1.01	1.22	1.40	1.35	1.11	0.91	0.77	0.80	11.6
		TOT DEM	KW	.00367	.00333	.00395	.00474	.00519	.00508	.00564	.00547	.00525	.00462	.00378	.00344	0.00564
S- 16	0.93	TOT GAS	THM	0.0352	0.0243	0.0141	0.0042	0.0013	0.0013	0.0013	0.0013	0.0013	0.0023	0.0104	0.0301	0.128
		TOT ELEC	KWH	0.70	0.63	0.70	0.76	0.89	1.06	1.23	1.17	0.97	0.79	0.67	0.70	10.3
		TOT DEM	KW	.00337	.00306	.00363	.00436	.00477	.00467	.00519	.00503	.00483	.00426	.00348	.00317	0.00515
S- 32	0.22	TOT GAS	THM	0.0406	0.0242	0.0198	0.0082	0.0038	0.0017	0.0017	0.0017	0.0030	0.0065	0.0132	0.0280	0.152
		TOT ELEC	KWH	0.70	0.63	0.77	0.82	0.88	1.09	1.16	1.15	0.98	0.82	0.69	0.70	10.4
		TOT DEM	KW	.00367	.00357	.00393	.00454	.00474	.00490	.00510	.00500	.00495	.00444	.00377	.00337	0.00510
S- 61	-0.35	TOT GAS	THM	0.0333	0.0230	0.0133	0.0040	0.0012	0.0012	0.0012	0.0012	0.0012	0.0022	0.0098	0.0285	0.121
		TOT ELEC	KWH	0.80	0.72	0.80	0.87	1.01	1.22	1.40	1.35	1.11	0.91	0.77	0.80	11.8
		TOT DEM	KW	.00367	.00333	.00395	.00474	.00519	.00508	.00564	.00547	.00525	.00462	.00378	.00344	0.00564

The Data Library is located at the end of each set of climate region forms. It contains both monthly and annual energy use figures for each case number. These data are necessary to complete Worksheet "A." Begin by confirming that you have selected the proper case number. For the sample problem we want Case No. 345. Because we are using Worksheet "A," we will use "Annual" data, located in the last column to the right. Note, however, that in order to compute annual demand charges, we must first compute the sum of all monthly demands (in KW). For the sample problem, this value is ".0808." It is **not** correct to use the value given (.00738), which is the highest peak demand in a **single** month. This figure is useful for other worksheets only. These three values should be entered in the first set of blank boxes on Worksheet "A," as shown on the next page.

Begin by entering the case number and data from the Data Library (we have shown Case No. 345). Next, enter the number of building square feet. The sample problem is 40,000 sq. ft. Compute annual total energy use by simple multiplication. Based on your own knowledge or research, enter your local utility rates for gas (in therms), electric consumption (in KWH), and electric demand (in KW). By simple multiplication, compute annual energy costs for each. Finally, divide by building square footage to get energy costs per sq. ft. Enter all four figures on the Level 2 worksheet. It is necessary to multiply energy use by building size the first time so that energy rates are applied to total building energy use. In most utilities, the amount of total use is significant in determining the rate category into which your building will fall. For the sake of comparison, the total building bill must be again reduced by dividing by the building square footage to get the energy cost per square foot. This process is repeated for all cases under analysis, with a Worksheet "A" completed for each. Following this analysis, answers are compared and, with the help of professional judgment, a selection of one best option is made and carried to Level 3. For the sample problem, we selected Case No. 347.

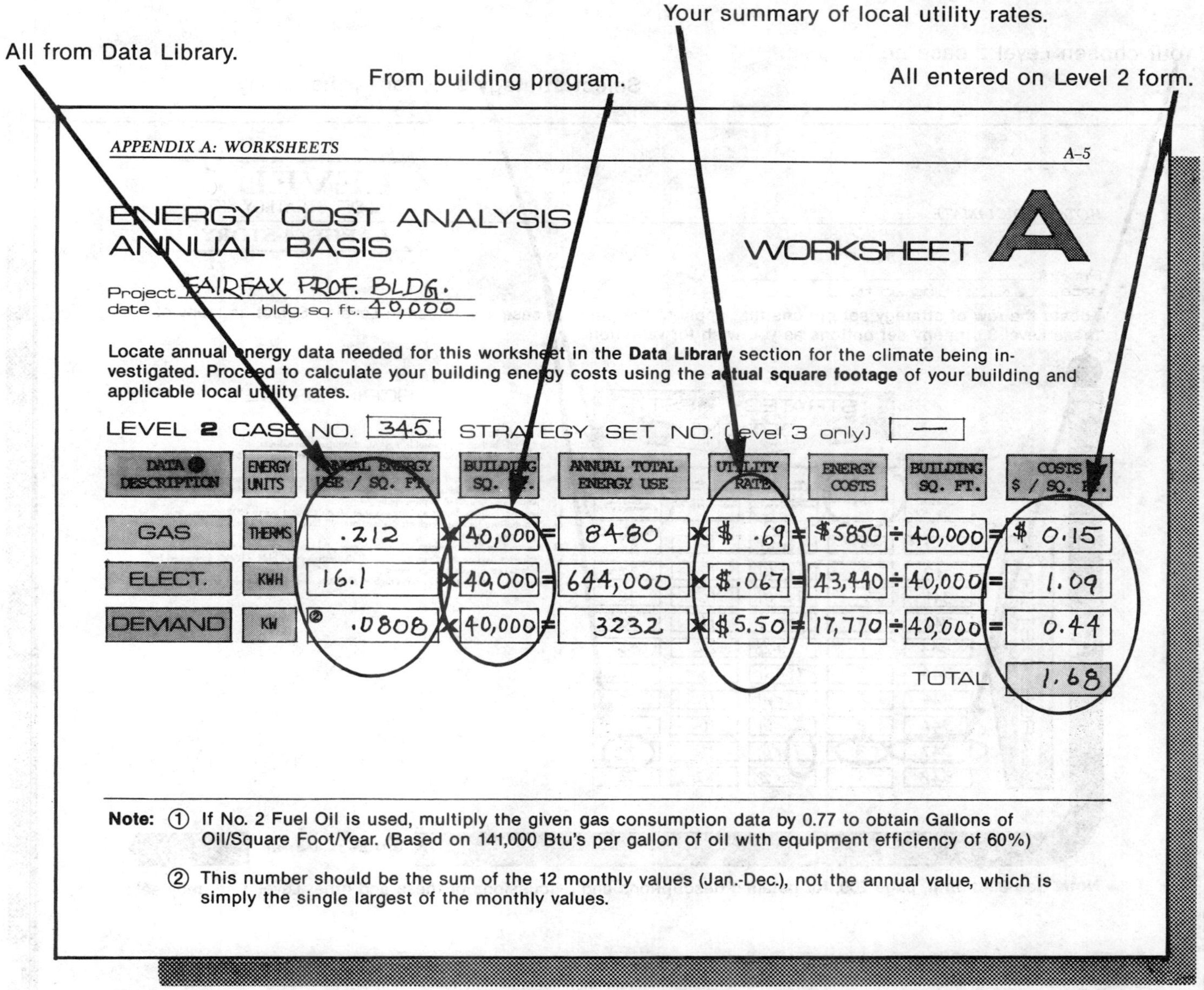

APPENDIX A: WORKSHEETS A–5

ENERGY COST ANALYSIS
ANNUAL BASIS

WORKSHEET A

Project FAIRFAX PROF. BLDG.
date ______ bldg. sq. ft. 40,000

Locate annual energy data needed for this worksheet in the **Data Library** section for the climate being investigated. Proceed to calculate your building energy costs using the **actual square footage** of your building and applicable local utility rates.

LEVEL **2** CASE NO. 345 STRATEGY SET NO. (Level 3 only) —

DATA ① DESCRIPTION	ENERGY UNITS	ANNUAL ENERGY USE / SQ. FT.		BUILDING SQ. FT.		ANNUAL TOTAL ENERGY USE		UTILITY RATE		ENERGY COSTS		BUILDING SQ. FT.		COSTS $ / SQ. FT.
GAS	THERMS	.212	×	40,000	=	8480	×	$ .69	=	$5850	÷	40,000	=	$ 0.15
ELECT.	KWH	16.1	×	40,000	=	644,000	×	$.067	=	43,440	÷	40,000	=	1.09
DEMAND	KW	② .0808	×	40,000	=	3232	×	$5.50	=	17,770	÷	40,000	=	0.44
												TOTAL		1.68

Note: ① If No. 2 Fuel Oil is used, multiply the given gas consumption data by 0.77 to obtain Gallons of Oil/Square Foot/Year. (Based on 141,000 Btu's per gallon of oil with equipment efficiency of 60%)

② This number should be the sum of the 12 monthly values (Jan.-Dec.), not the annual value, which is simply the single largest of the monthly values.

Your chosen Level 2 case no.

Select strategy sets for further analysis.

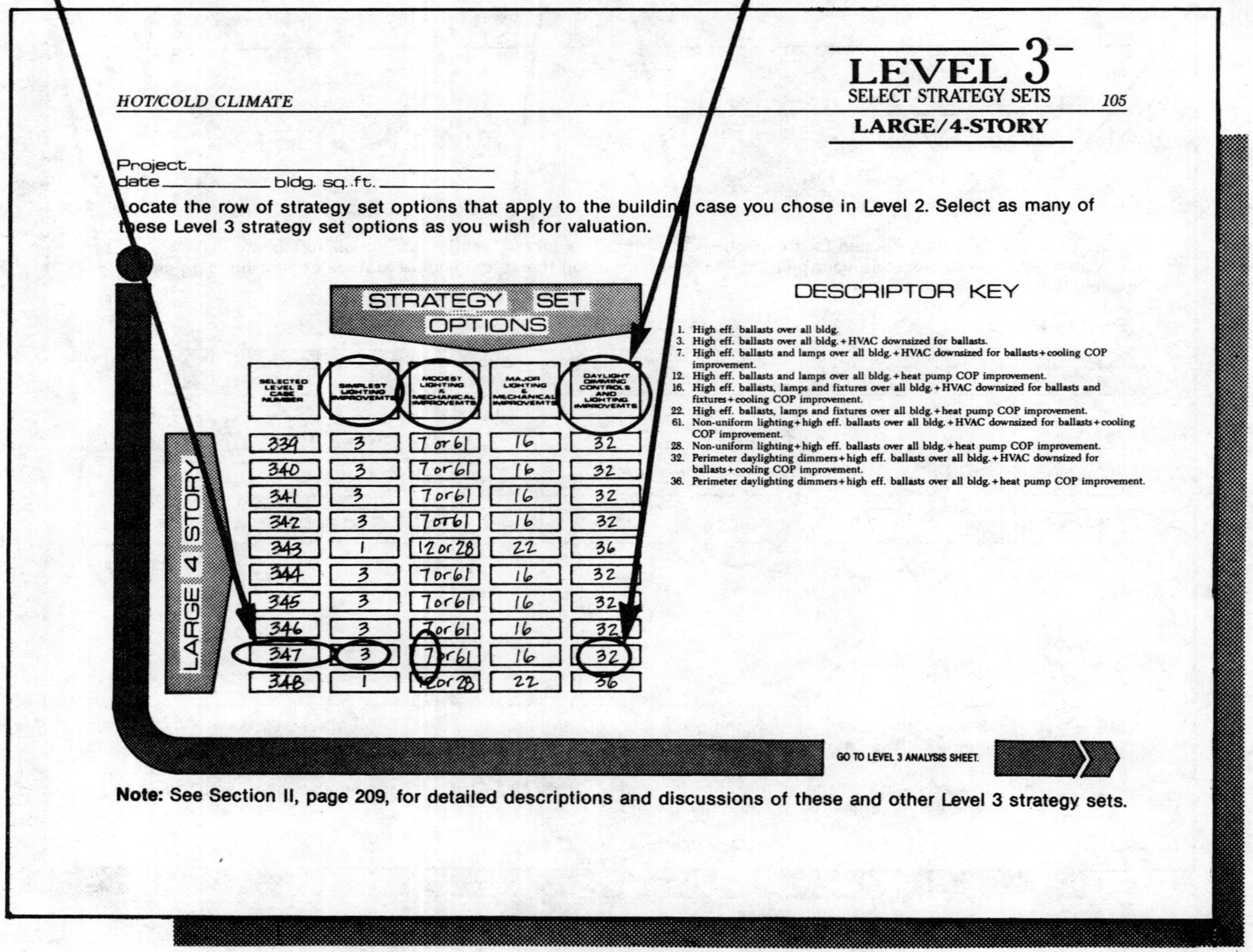

HOT/COLD CLIMATE

LEVEL 3

SELECT STRATEGY SETS 105

LARGE/4-STORY

Project________________
date________ bldg. sq. ft.________

Locate the row of strategy set options that apply to the building case you chose in Level 2. Select as many of these Level 3 strategy set options as you wish for valuation.

LARGE 4 STORY

STRATEGY SET OPTIONS

SELECTED LEVEL 2 CASE NUMBER	SIMPLEST LIGHTING IMPROVEMTS	MODEST LIGHTING & MECHANICAL IMPROVEMTS	MAJOR LIGHTING & MECHANICAL IMPROVEMTS	DAYLIGHT DIMMING CONTROLS AND LIGHTING IMPROVEMTS
339	3	7 or 61	16	32
340	3	7 or 61	16	32
341	3	7 or 61	16	32
342	3	7 or 61	16	32
343	1	12 or 28	22	36
344	3	7 or 61	16	32
345	3	7 or 61	16	32
346	3	7 or 61	16	32
347	3	7 or 61	16	32
348	1	12 or 28	22	36

DESCRIPTOR KEY

1. High eff. ballasts over all bldg.
3. High eff. ballasts over all bldg. + HVAC downsized for ballasts.
7. High eff. ballasts and lamps over all bldg. + HVAC downsized for ballasts + cooling COP improvement.
12. High eff. ballasts and lamps over all bldg. + heat pump COP improvement.
16. High eff. ballasts, lamps and fixtures over all bldg. + HVAC downsized for ballasts and fixtures + cooling COP improvement.
22. High eff. ballasts, lamps and fixtures over all bldg. + heat pump COP improvement.
61. Non-uniform lighting + high eff. ballasts over all bldg. + HVAC downsized for ballasts + cooling COP improvement.
28. Non-uniform lighting + high eff. ballasts over all bldg. + heat pump COP improvement.
32. Perimeter daylighting dimmers + high eff. ballasts over all bldg. + HVAC downsized for ballasts + cooling COP improvement.
36. Perimeter daylighting dimmers + high eff. ballasts over all bldg. + heat pump COP improvement.

GO TO LEVEL 3 ANALYSIS SHEET.

Note: See Section II, page 209, for detailed descriptions and discussions of these and other Level 3 strategy sets.

The first step of Level 3 analysis is to select one or more strategy sets for analysis. As shown on the worksheet at left, each Level 2 case has a number of strategy set options. In general, as you move from left to right, the amount of energy that can be saved annually increases. First costs may go up, but not consistently. Analysis is required to know with certainty. For our sample problem, we intend to investigate three strategy sets (3, 7, and 32), representing the full range of options possible. These strategy sets are discussed in detail in Section II, page 209.

Separate data sets exist for each strategy set.

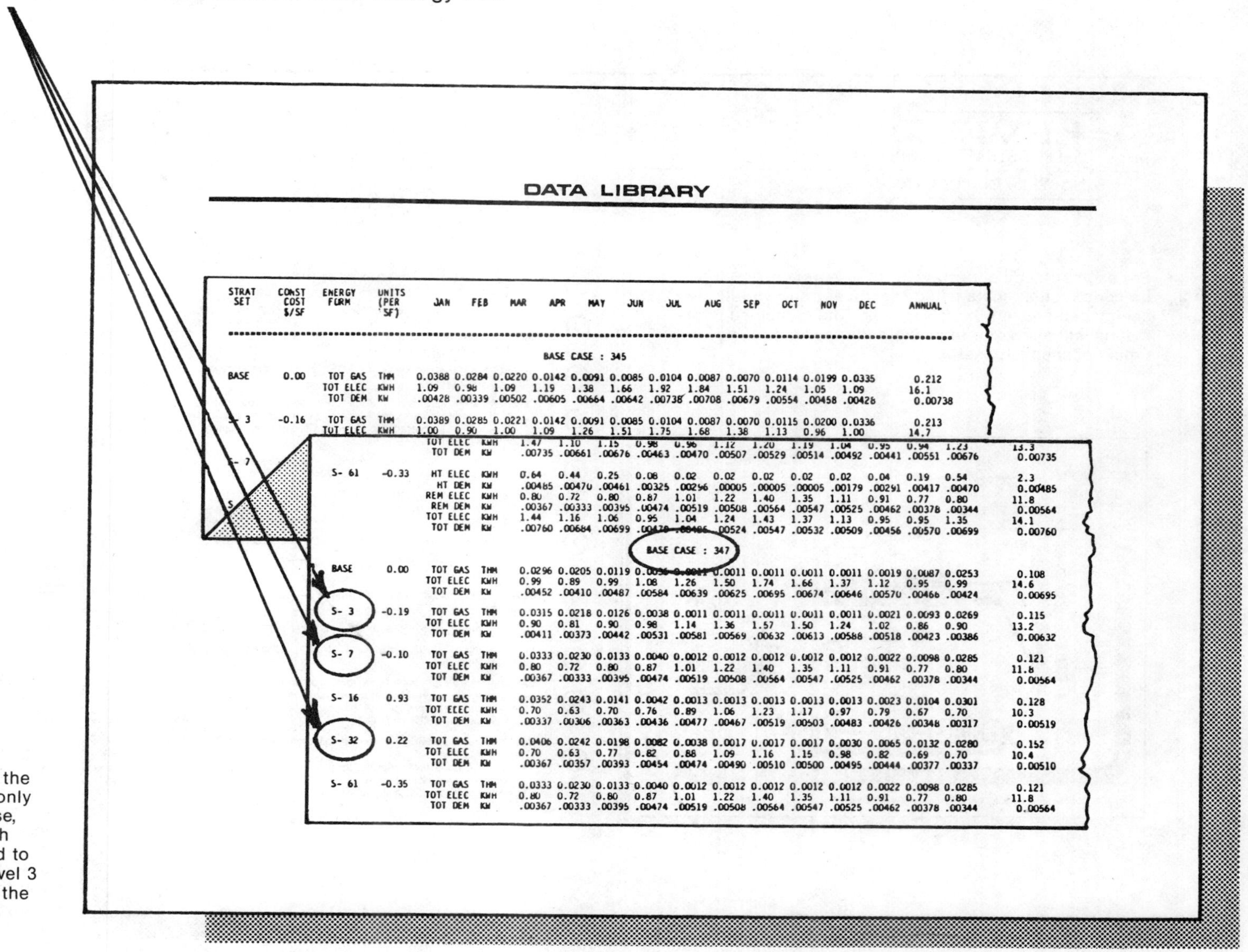

Observe that each case in the Data Library contains not only a data set for the base case, but also a data set for each strategy set as it is applied to the base case. In doing Level 3 analysis, be sure to select the proper data set for each strategy set under study.

LEVEL 3

SELECT STRATEGY SETS

ANALYSIS/INSTRUCTIONS

For each strategy set you want to investigate, develop its construction costs using the upper route and its energy costs using the lower route. The combined information will allow you to select a strategy set for optimum economic benefit.

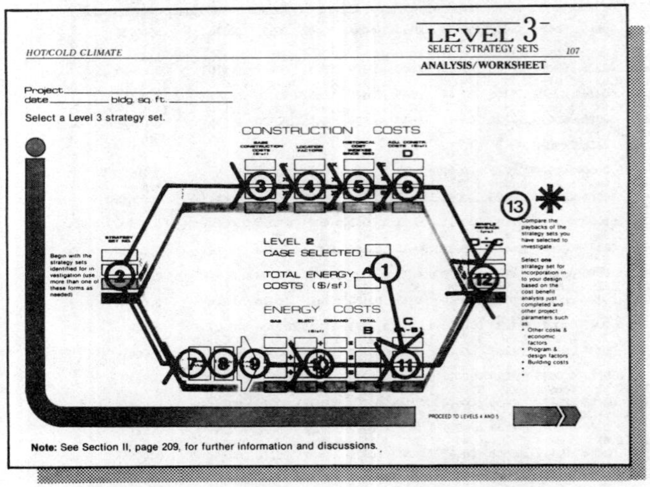

1. Enter the Level 2 case selected and its total energy costs as developed in Level 2.
2. As many as four strategy sets for any *one* Level 2 case can be analyzed on each sheet. You will need to use more sheets to analyze more than four strategy sets or to analyze strategy sets for another Level 2 case.
3. Obtain *base construction costs* ($/SF) for each strategy set from the Data Library.
4. Obtain *location factors* from commonly used cost estimating handbooks. Additional multipliers may be entered to account for above or below average construction costs due to specific project conditions.
5. Obtain *historical cost indexes* (factors since 1983) from commonly used cost estimating handbooks.
6. Multiply 3, 4, and 5 to obtain *adjusted construction costs* ($/SF). User may wish to use more refined construction cost data as supplied by a cost estimator for the specific project.
7. Use the same worksheet you selected for use in Level 2.
8. Locate energy use data for the selected strategy sets in the Data Library.
9. Develop energy costs for each strategy set using the worksheet selected.
10. Transfer energy cost information from worksheets A, B, or C.
11. Find the energy savings from each strategy set by subtracting each "B" from "A."
12. Calculate *simple payback* by dividing "D" (initial costs) by "C" (yearly savings). Alternative economic analysis techniques other than simple payback may be employed using information from columns "D" and "C."
13. Cost factors to consider in making a decision on these strategy sets include:
    - Negative construction costs (first cost savings)
    - Positive construction costs
    - Annual energy cost savings

The second step of Level 3 analysis compares construction costs and energy costs in order to compute a simple payback for each strategy set under consideration. The process is not complex, although there are several variables that must be considered. A special sheet of instructions is provided—shown at left. Use these specific instructions to guide you through the Level 3 analysis. A few highlights are shown on the next page.

Start by entering the numbers of those strategies selected for analysis in the boxes provided at the far left of this form. You may then proceed along either "track" suggested as long as both are done. To compute "Energy Costs," the lower "track," you will need to do another Worksheet "A" (or B or C if you have chosen them). Our computations on Worksheet "A" for the sample problem follow on the next two pages. Results from Worksheet "A" are entered in the boxes provided—as shown. To compute "Construction Costs," the upper "track," start with the base construction cost of the strategy set. This is found in the second column of the Data Library, as shown on the next page. Two additional numbers, the "Location Factor" and the "Historical Cost Index," you will have to develop on your own, using a current cost estimating handbook and your own judgment. Some easy arithmetic will produce for you the simple payback for each strategy set under analysis. Each user will have his or her own interpretation of financial indicators. For our sample problem, we are impressed that we can get up to $.22 savings per square foot or $8,000 ($.22×40,000 sq. ft.) per year at no increase in first cost (set #7). We would have to stop and think a minute before spending $.21 a sq. ft. to save only another $.08 per sq. ft. per year (set #32). For our conservative client, we will probably select strategy set No. 7.

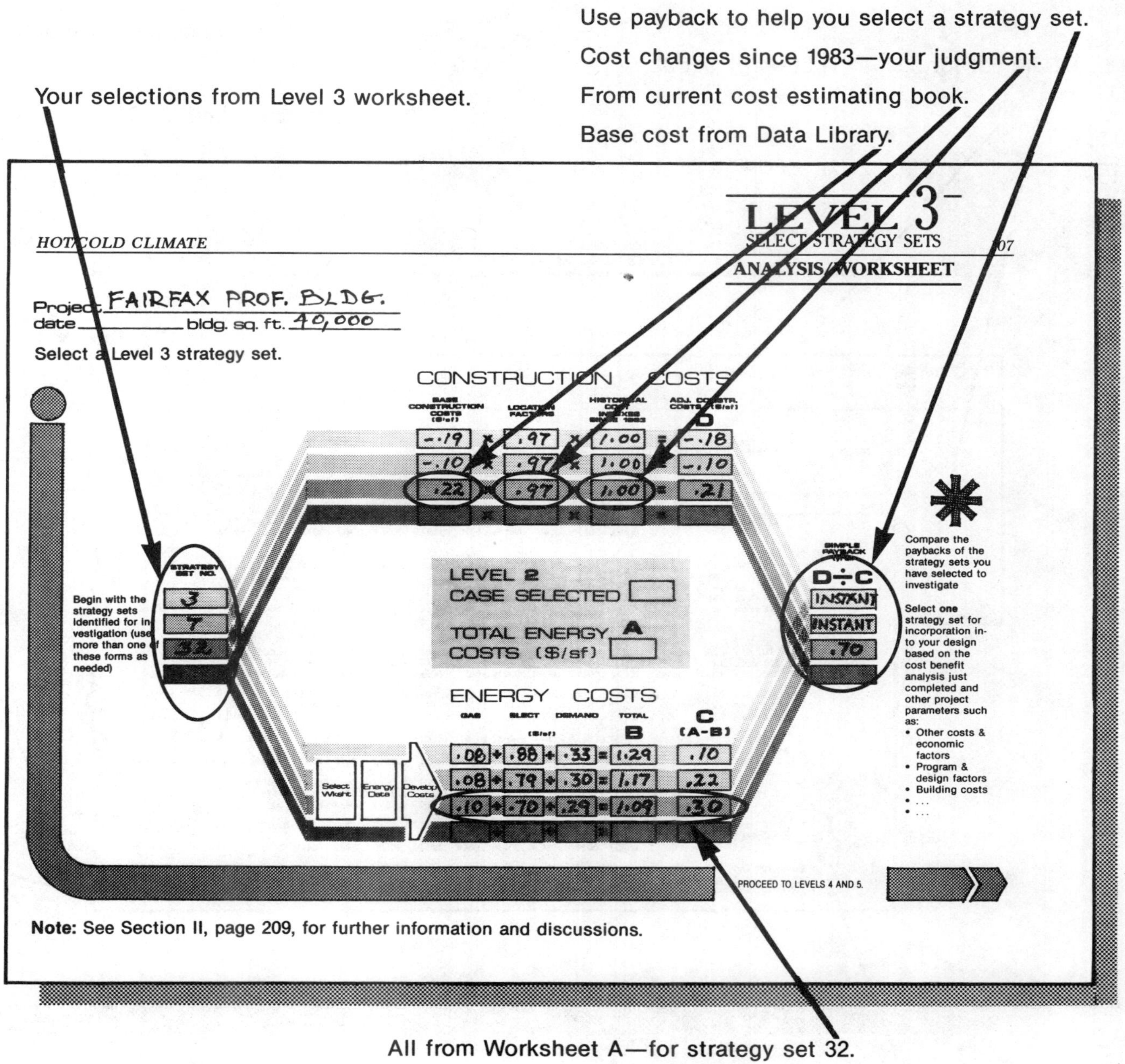

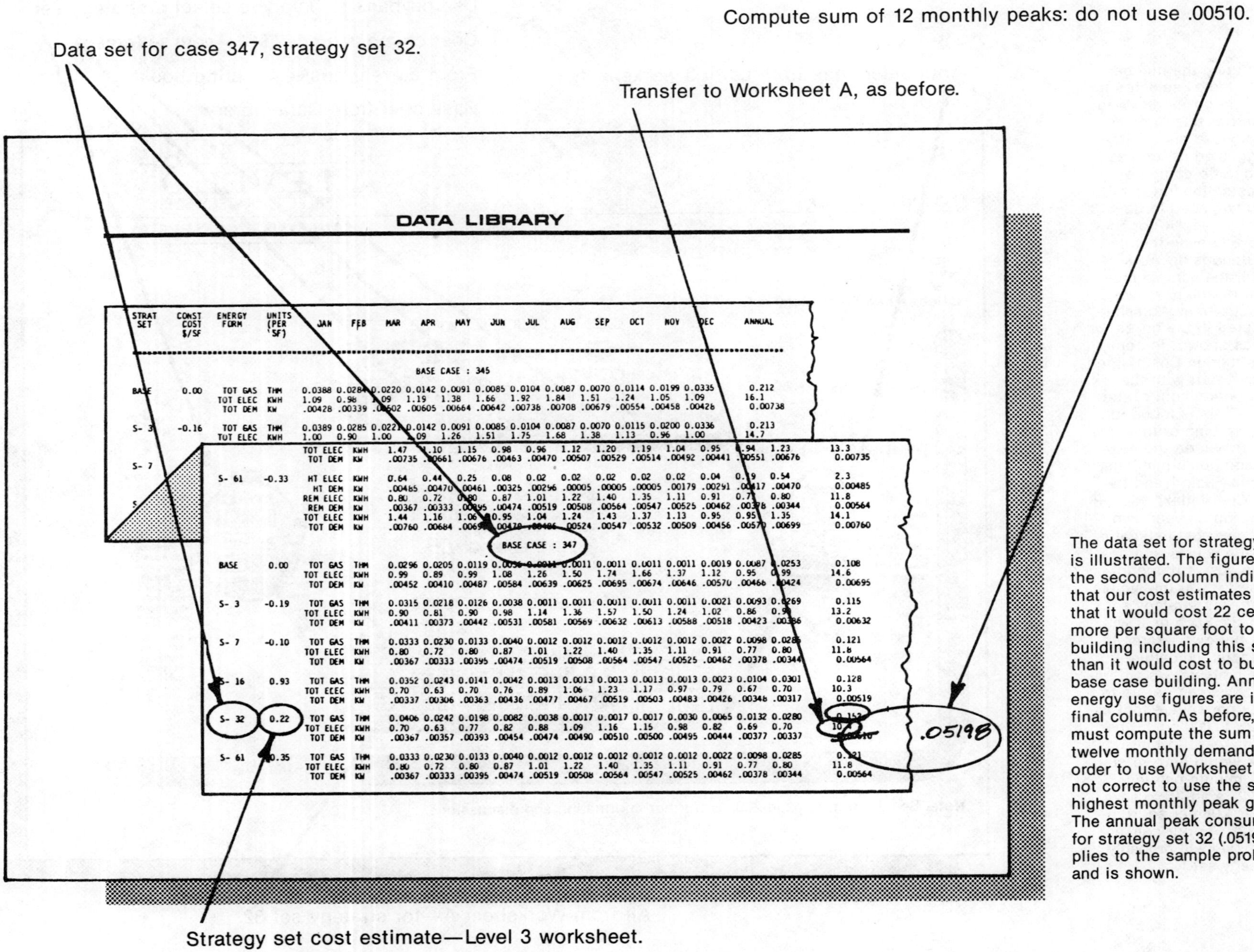

The data set for strategy set 32 is illustrated. The figure ".22" in the second column indicates that our cost estimates show that it would cost 22 cents more per square foot to build a building including this strategy than it would cost to build the base case building. Annual energy use figures are in the final column. As before, you must compute the sum of all twelve monthly demands in order to use Worksheet "A." It is not correct to use the simple highest monthly peak given. The annual peak consumption for strategy set 32 (.05198) applies to the sample problem and is shown.

Transfer to Level 3 worksheet.

Note: Data set for strategy set 32 in case 347.

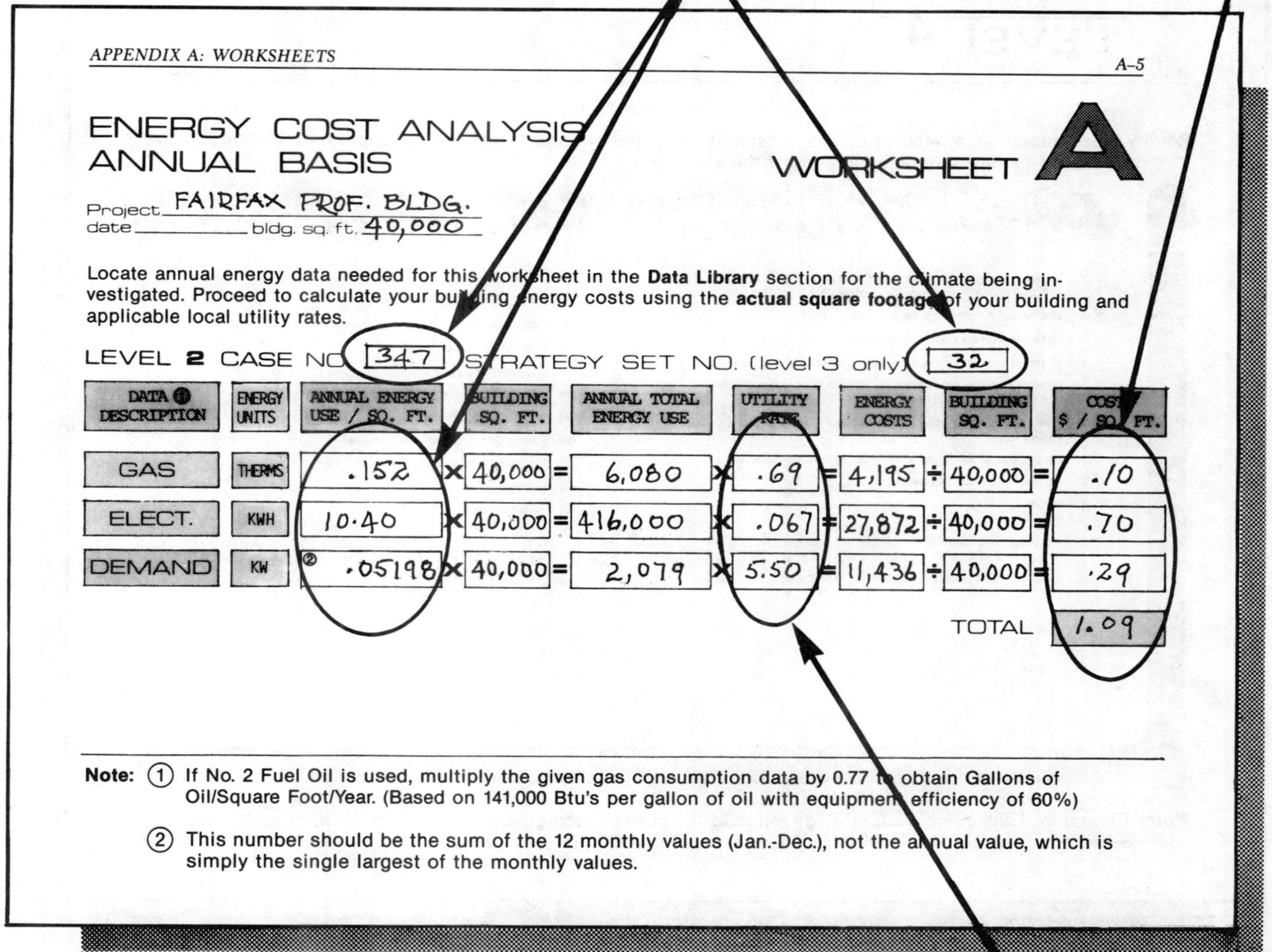
APPENDIX A: WORKSHEETS A-5

ENERGY COST ANALYSIS
ANNUAL BASIS

WORKSHEET A

Project FAIRFAX PROF. BLDG.
date ______ bldg. sq. ft. 40,000

Locate annual energy data needed for this worksheet in the **Data Library** section for the climate being investigated. Proceed to calculate your building energy costs using the **actual square footage** of your building and applicable local utility rates.

LEVEL 2 CASE NO. 347 STRATEGY SET NO. (level 3 only) 32

DATA ① DESCRIPTION	ENERGY UNITS	ANNUAL ENERGY USE / SQ. FT.		BUILDING SQ. FT.		ANNUAL TOTAL ENERGY USE		UTILITY RATE		ENERGY COSTS		BUILDING SQ. FT.		COST $ / SQ. FT.
GAS	THERMS	.152	×	40,000	=	6,080	×	.69	=	4,195	÷	40,000	=	.10
ELECT.	KWH	10.40	×	40,000	=	416,000	×	.067	=	27,872	÷	40,000	=	.70
DEMAND	KW	② .05198	×	40,000	=	2,079	×	5.50	=	11,436	÷	40,000	=	.29
												TOTAL		1.09

Note: ① If No. 2 Fuel Oil is used, multiply the given gas consumption data by 0.77 to obtain Gallons of Oil/Square Foot/Year. (Based on 141,000 Btu's per gallon of oil with equipment efficiency of 60%)

② This number should be the sum of the 12 monthly values (Jan.-Dec.), not the annual value, which is simply the single largest of the monthly values.

From your knowledge of utility rates, as before.

Worksheet "A" is computed as before, BUT with new data for each strategy set. As you can see on the worksheet illustrated, this is case number 347 and strategy set number 32. It is important to always enter the strategy set number in the box provided to insure proper identification of the data in the future. Utility rates are the same as those used previously. Results of the calculation are entered on the lower "track" of the Level 3 analysis worksheet.

LEVEL 4
... ADD MORE

108 SECTION I: SELECTOR PROCESS

Review these additional strategies for the possibility of further savings. Check the description in Section II, page 235, carefully for guidance on adopting any of the strategies.

A SELECT ON THE BASIS OF TECHNICAL JUDGEMENT

1. Increase thermostat deadband for occupied periods from 2°F to 8°F (70°F heating–78°F cooling)
2. ✓ Design for efficient space planning
3. Optimize building orientation
4. Minimize building height to reduce impact of glazed area
5. Reduce shading coefficient to minimally acceptable levels (not with daylighting)
6. Reduce glazing to minimally acceptable levels (not with daylighting)
7. Landscape to minimize wind induced infiltration effect
12. ✓ Increase manual light switching
13. Install operable windows (best for buildings without economizer cycle)
15. Install energy monitoring and control system (medium and large buildings only)
16. Locate outdoor cooling equipment in the shade

B VALUE MAY BE DETERMINED BY SIMPLE CALCULATION

19. High efficiency motors and pumps
20. Multiple speed compressors
21. Variable or two speed circulating fan
22. Electric resistance heaters
23. Desuperheater or stand-alone heat pump for DHW
24. ✓ Point-of-use electric domestic hot water heaters
25. External shading
26. Solar domestic hot water heater
29. Demand limiter
30. Absorption chiller-heater
31. Cogeneration

Note: The number preceding each strategy refers to the corresponding paragraph in Section II, page 235. See Section II, page 235, for further information and discussion of each strategy.

Additional savings can be achieved by applying strategies suggested in Level 4. Unfortunately, the analysis required to determine the appropriateness or cost effectiveness of these strategies is beyond the scope of this book. We offer this list of suggestions for those who wish to pursue even greater energy savings. As usual, these should be applied only after careful analysis and with sound professional judgment.

The final step in any energy analysis is to do the best you can to see that your good conserving ideas are carried out in actual practice. Level 5 in the handbook provides a checklist of actions that should be taken by members of the design team to insure proper building turnover and operation. Portions of the guidelines are further subdivided as being most applicable at either the small or large end of our building size range 4,000–50,000 sq. ft. The list was carefully examined and a number of actions were marked for use in the project. Owner, contractors and equipment suppliers will be contacted about their role in the turnover process.

HOT/COLD CLIMATE LEVEL 5 – FINISH RIGHT *109*

Review this checklist and adopt the procedures that are appropriate to your project. These actions are **important** if you are to achieve and maintain the savings from the selected Level 1–4 strategies.

A CHECKLIST FOR SUCCESSFUL TURNOVER AND BUILDING OPERATION

	DESIGN	CONSTRUCTION	TURNOVER	BUILDING OPERATION
DESIGN TEAM (ARCHITECTS/ ENGINEERS)	✓ • Document all design assumptions (1) • Estimate energy use by sub-system (2) **& For Large Buildings:** ✓ • Document all systems and control strategies (3) ✓ • Specify contents of O&M Manuals (4) ✓ • Specify HVAC balancing requirements (5) ✓ • Specify equipment protection and cleaning (6) ✓ • Specify on-site instruction (7) ✓ • Specify access doors, valves, test ports (8) ✓ • Specify piping system ID (9) • Specify price quotations on all long-term maintenance contracts (10)	✓ • Prepare "as-built" drawings (11) ✓ • Observe construction process for compliance (12)	✓ • Conduct "punch list" inspection (15) ✓ • Visually check for air leakage (15) • If possible, participate in on-site training session(s) (7)	
CONTRACTOR & SUB-CONTRACTOR	**For Small Buildings Only:** • Document all systems and control strategies (3) • Specify contents of O&M Manuals (4) • Specify HVAC balancing requirements (5) • Specify equipment protection and cleaning (6) • Specify on-site instruction (7) • Specify access doors, valves, test ports (8) • Specify piping system ID (9) • Specify price quotations on all long-term maintenance contracts (10)	✓ • Prepare O&M Manual (4) • Perform final calibration of all controls (5) ✓ • Hire independent balancing agent or require of subs (5) ✓ • Protect and clean equipment (6) ✓ • Provide all access doors, valves, and test ports (8) ✓ • Place piping system ID tags (9) ✓ • Prepare "as-built" drawings (11) • Use HVAC system to "flush" construction odors (13)	✓ • Deliver O&M Manual (4) ✓ • Conduct on-site instruction and training (7)	✓ • Honor warranty and call back as required (16)
EQUIPMENT MANUFACT.		• Support calibration of controls (5) • Provide all special tools required (14)	✓ • Support O&M Manual preparation (4) ✓ • Support on-site instruction and training (7)	
OWNER & BLDG. STAFF	✓ • Check all key design assumptions (1)	✓ • Periodically check compliance on-site (12)	✓ • Attend on-site instruction and training sessions (7)	• Monitor energy consumption against estimates (2) ✓ • Periodically test equipment efficiency (17) • Periodically sample and test water (18) ✓ • Inspect and calibrate controls (19) ✓ • Monitor occupant complains (20)

Note: See Section II, page 255, for more information and discussion. Numbers shown in ( ) refer to the numbered paragraphs in Section II, page 255.

SECTION I

SELECTOR PROCESS

CLIMATE REGIONS
COLD CLIMATE
TEMPERATE CLIMATE
HOT/COLD CLIMATE
HOT/HUMID CLIMATE
HOT/ARID CLIMATE

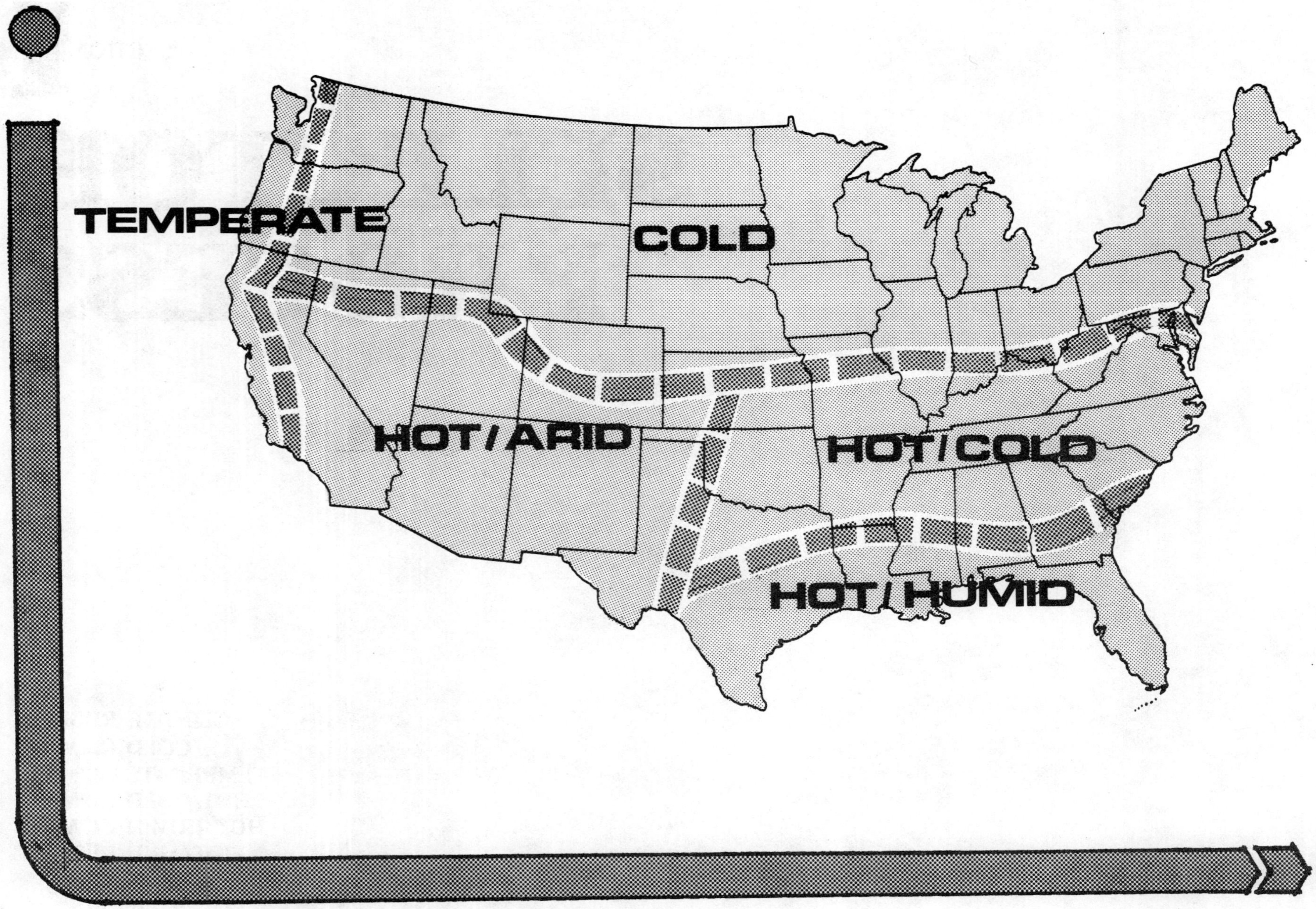
TEMPERATE
COLD
HOT / ARID
HOT / COLD
HOT / HUMID

Atlas of Cities

Climate Regions

COLD

CO	Denver, Colorado Springs
CT	Hartford, New Haven
DE	Wilmington
IA	Cedar Rapids, Des Moines, Sioux City
ID	Boise, Lewiston
IL	Chicago, Rockford, Springfield
IN	Ft. Wayne, Indianapolis, South Bend
KA	Concordia
MA	Boston, Fall River, Springfield
ME	Portland, Presque Isle
MI	Detroit, Flint, Grand Rapids, Sault Ste. Marie
MN	Duluth, Minneapolis/St. Paul
MO	St. Joseph
MT	Billings, Butte, Great Falls
NB	Omaha, Scottsbluff
ND	Bismarck, Fargo
NH	Manchester
NJ	Atlantic City, Camden, Newark, Trenton
NV	Elko
NY	Albany, Binghamton, Buffalo, New York City, Plattsburg, Rochester, Syracuse
OH	Akron, Cincinnati, Cleveland, Columbus, Dayton, Toledo
OR	Pendleton, Klamath Falls
PA	Erie, Harrisburg, Pittsburg, Philadelphia, Scranton/Wilkes Barre
RI	Providence
SD	Rapid City, Sioux Falls
UT	Ogden
VT	Burlington, Rutland
WA	Spokane, Yakima
WI	Appleton, Eau Claire, Green Bay, Madison, Milwaukee
WV	Wheeling
WY	Casper, Cheyenne

TEMPERATE

CA	Sacramento, San Francisco, San Jose, Santa Barbara
OR	Eugene, Portland
WA	Seattle, Tacoma

HOT/COLD

AL	Birmingham, Huntsville, Montgomery
AR	Fort Smith, Little Rock
DC	Washington
DE	Dover
GA	Atlanta, Augusta, Columbus, Macon
IL	Cairo, East St. Louis
IN	Evansville
KA	Topeka, Wichita
KY	Lexington, Louisville
MD	Baltimore, Hagerstown
MO	Kansas City, Springfield, St. Louis
MS	Meridian
NC	Asheville, Charlotte, Durham, Greensboro, Wilmington, Winston-Salem
OK	Oklahoma City, Tulsa
SC	Columbia, Greenville
TN	Chattanooga, Knoxville, Memphis, Nashville
TX	Abilene, Dallas, San Angelo, Texarkana, Wichita Falls
VA	Norfolk, Richmond, Roanoke
WV	Charleston, Huntington

HOT/HUMID

AL	Mobile
FL	Jacksonville, Miami, Orlando, Pensacola, Tampa
GA	Albany, Savannah
LA	Baton Rouge, New Orleans, Shreveport
MS	Biloxi, Jackson
SC	Charleston
TX	Austin, Beaumont, Brownsville, Corpus Christi, Houston, Laredo, San Antonio, Waco

HOT/ARID

AZ	Flagstaff, Phoenix, Tucson
CA	Bakersfield, Fresno, Los Angeles, San Bernardino, San Diego
CO	Durango, Grand Junction, Pueblo
KA	Dodge City
NM	Albuquerque, Roswell, Santa Fe
NV	Las Vegas, Reno
TX	Amarillo, El Paso, Lubbock
UT	Salt Lake City

Cold Climate

Identify Region

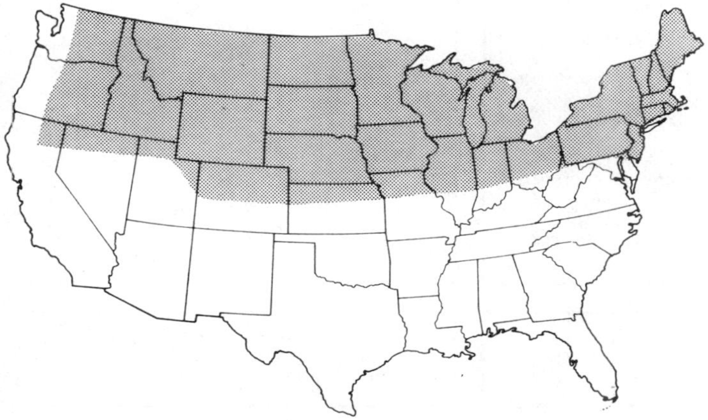

CITIES IN REGION

CO	Denver Colorado Springs
CT	Hartford New Haven
DE	Wilmington
IA	Cedar Rapids Des Moines Sioux City
ID	Boise Lewiston
IL	Chicago Rockford Springfield
IN	Ft. Wayne Indianapolis South Bend
KA	Concordia
MA	Boston Fall River Springfield
ME	Portland Presque Isle
MI	Detroit Flint Grand Rapids Sault Ste. Marie
MN	Duluth Minneapolis/ St. Paul
MO	St. Joseph
MT	Billings Butte Great Falls
NB	Omaha Scottsbluff
ND	Bismarck Fargo
NH	Manchester
NJ	Atlantic City Camden Newark Trenton
NV	Elko
NY	Albany Binghamton Buffalo New York City Plattsburg Rochester Syracuse
OH	Akron Cincinnati Cleveland Columbus Dayton Toledo
OR	Pendleton Klamath Falls
PA	Erie Harrisburg Pittsburg Philadelphia Scranton/ Wilkes Barre
RI	Providence
SD	Rapid City Sioux Falls
UT	Ogden
VT	Burlington Rutland
WA	Spokane Yakima
WI	Appleton Eau Claire Green Bay Madison Milwaukee
WV	Wheeling
WY	Casper Cheyenne

BEGIN YOUR 5 LEVELS OF STRATEGY SELECTION

Note: See Section II for a full discussion of the five climate regions used in this manual.

Make sure your design comes up to current good practice, which includes meeting ASHRAE 90-75 Standards. The table below identifies good practice that meets 90-75 Standards as well as common design assumptions applicable to all building sizes.

	ALL BUILDINGS	
	GOOD PRACTICE	**DESIGN ASSUMPTIONS**
SHELL	o Perimeter Insulation (R-5) o Infiltration: 0 when HVAC is operating (pressurized building)	o Ventilation: 14 cfm/person during occupied hours
HVAC	o Air conditioning performs to ASHRAE 90-75 (year 1980) minimum efficiencies o Boilers perform with 65% efficiency (no waste heat to space) o Heat pumps perform to ASHRAE 90-75 (year 1980) minimum efficiencies o Controls: · 72 degrees heating, 74 degrees cooling occupied periods · 55 degrees heating, 95 degrees cooling unoccupied periods · 2 degrees temperature float · outside air temperature reset for hot water boilers · low leakage dampers (less than 3%) · zone control for all heat pump and multi-zone systems	
INTERNAL LOAD		o Occupant density: 200 square feet per person o Lights: 2.25 watts/sq. ft. o Office equipment: .75 watts/sq. ft.
SCHEDULE	o HVAC comes on one hour before occupied period, turned to setback during unoccupied period	o Inside lighting: 100% during operating hours, 5% at other times o Outside lighting: 100% 6pm til midnight

Note: See Section II, page 179, for detailed information on these Level 1 strategies and assumptions.

Some aspects of ASHRAE 90-75 and current **good practice** vary according to building size and HVAC system type. Make sure your design comes up to these criteria. You may want to come back to this table as you explore HVAC options in Level 2.

		BUILDING SIZE								
		SMALL 2000-10000 sq ft			MEDIUM 10000-25000 sq ft			LARGE 25000-50000 sq ft		
SHELL	ROOF INSULATION	R-14.3			R-14.3			R-14.3		
	WALL INSULATION	R-7.5			R-7.5			R-7.5		
	GLAZING	DOUBLE GLAZED WITH CLEAR GLASS			DOUBLE GLAZED WITH TINTED GLASS			DOUBLE GLAZED WITH TINTED GLASS		
	AIR CHANGES	1/2 AC/HR UNOCCUPIED			1/3 AC/HR UNOCCUPIED			1/3 AC/HR UNOCCUPIED		
HVAC		PACKAGED TERMINAL AIR CONDITIONERS	ROOFTOP DX	HEAT PUMP	CONSTANT VOLUME	VARIABLE AIR VOLUME	SPLIT HEAT PUMP	CONSTANT VOLUME	VARIABLE AIR VOLUME	HYDRONIC HEAT PUMP
	TEMPERATURE CONTROL	ZONE	OA RESET	ZONE	OA RESET	FIXED	ZONE	OA RESET	FIXED	ZONE
	AIR CONTROL	FIXED VOLUME	SENSIBLE ECONOMIZER	FIXED VOLUME	SENSIBLE ECONOMIZER	SENSIBLE ECONOMIZER	SENSIBLE ECONOMIZER	SENSIBLE ECONOMIZER	SENSIBLE ECONOMIZER	FIXED VOLUME

Some common **design assumptions** also vary according to building size and no. of stories.

		SMALL	MEDIUM		LARGE	
INT. LOAD		1 STORY	1 STORY	2 STORY	2 STORY	4 STORY ①
	ELEVATOR & OUTDOOR LIGHTS	1345 WATTS	4035 WATTS	4535 WATTS	7860 WATTS	9360 WATTS
SCHEDULE	OCCUPANCY	80% 8am - 6pm Monday - Friday	80% 8am - 6pm Monday - Friday		80% 8am - 6pm: M-F 100% 9am - 1pm: SAT	

GO TO LEVEL-2, PAGE SPECIFIC TO YOUR BUILDING SIZE & NO. OF STORIES.

Note ① For 3 or 5 story structures, use "4-story" classification.

LEVEL 2
MAKE INITIAL CHOICES
SMALL/1-STORY

Project ______________________
date ____________ bldg. sq. ft. ____________

The building characteristic choices shown here make a **big** difference in energy cost savings. Investigate these choices within the requirements of your design.

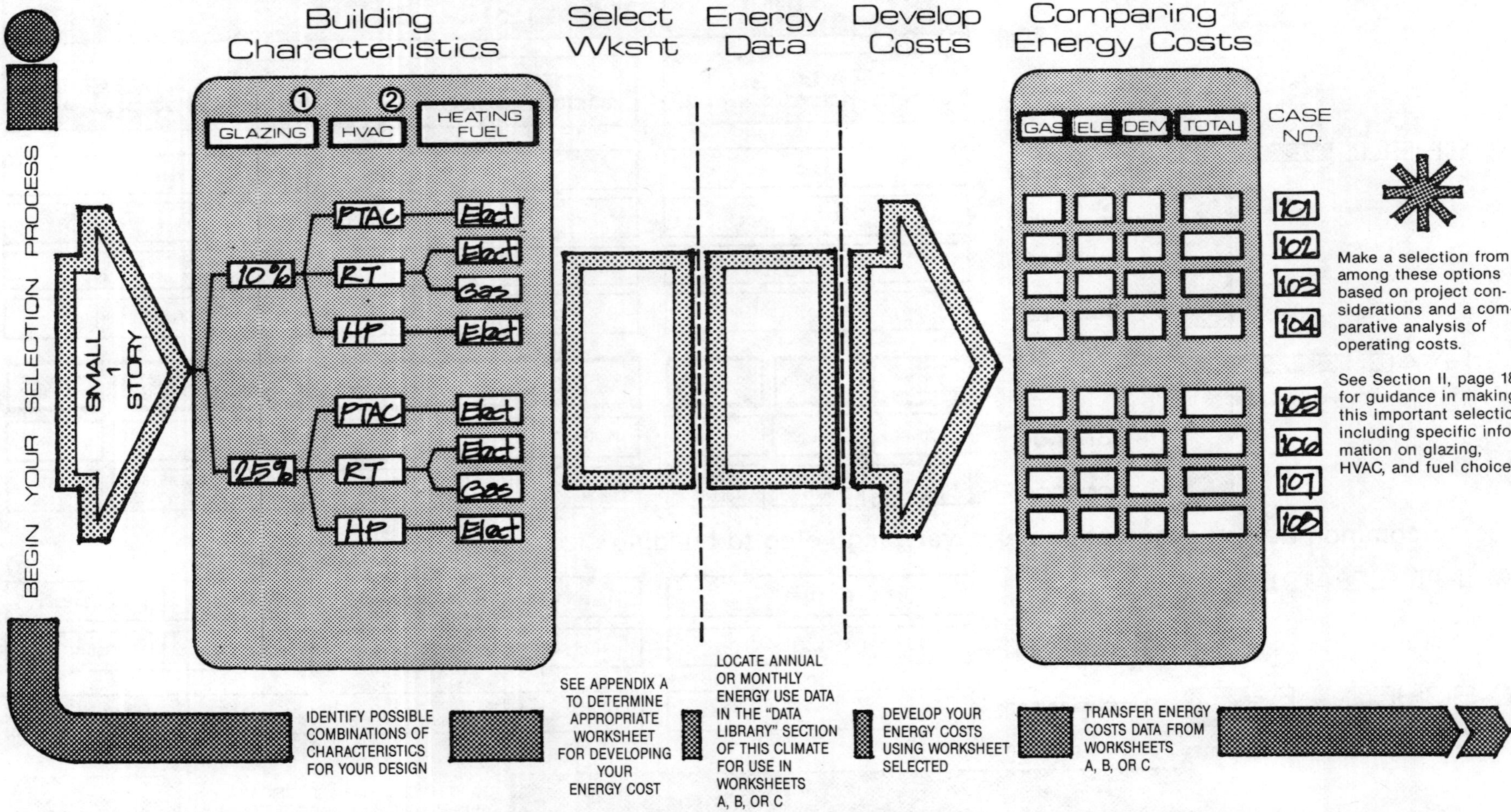

Make a selection from among these options based on project considerations and a comparative analysis of operating costs.

See Section II, page 189, for guidance in making this important selection, including specific information on glazing, HVAC, and fuel choices.

Note: See Section II, page 189, for detailed information on these Level 2 strategies and the use of this form.

① For glazing <17% use 10%, for >17% use 25%.

② PTAC = Packaged Terminal A/C, RT = Rooftop DX, HP = Packaged Terminal Heat Pump

Project______________________________
date____________ bldg. sq. ft.____________

Locate the row of strategy set options that apply to the building case you chose in Level 2. Select as many of these Level 3 strategy sets options as you wish for valuation.

STRATEGY SET OPTIONS

SMALL 1 STORY

SELECTED LEVEL 2 CASE NUMBER	SIMPLEST LIGHTING IMPROVEMTS	MODEST LIGHTING & MECHANICAL IMPROVEMTS	MAJOR LIGHTING & MECHANICAL IMPROVEMTS	DAYLIGHT DIMMING CONTROLS AND LIGHTING IMPROVEMTS
101	1	5 or 24	14	
102	3	6 or 60	15	
103	3	6 or 60	15	
104	1	12 or 28	22	
105	1	5 or 24	14	30
106	3	6 or 60	15	31
107	3	6 or 60	15	31
108	1	12 or 28	22	36

DESCRIPTOR KEY

1. High eff. ballasts over all bldg.
3. High eff. ballasts over all bldg. + HVAC downsized for ballasts.
5. High eff. ballasts + high eff. fluorescent lamps over all bldg.
6. High eff. ballasts and lamps over all building + HVAC downsized for ballasts.
12. High eff. ballasts and lamps over all bldg + heat pump COP improvement.
14. High eff. ballasts and lamps + high eff. fixtures over all bldg.
15. High eff. ballasts, lamps and fixtures over all bldg. + HVAC downsized for ballasts and fixtures.
22. High eff. ballasts, lamps and fixtures over all bldg. + heat pump COP improvement.
24. Non-uniform lighting + high eff. ballasts over all bldg.
60. Non-uniform lighting + high eff. ballasts over all bldg. + HVAC downsized for ballasts.
28. Non-uniform lighting + high eff. ballasts over all bldg. + heat pump COP improvement.
30. Perimeter daylighting dimmers + high eff. ballasts over all bldg.
31. Perimeter daylighting dimmers + high eff. ballasts over all bldg. + HVAC downsized for ballasts.
36. Perimeter daylighting dimmers + high eff. ballasts over all bldg. + heat pump COP improvement.

GO TO LEVEL 3 ANALYSIS SHEET.

Note: See Section II, page 209, for detailed descriptions and discussions of these and other Level 3 strategy sets.

LEVEL 2

MAKE INITIAL CHOICES

MEDIUM/1-STORY

Project ______________________
date __________ bldg. sq. ft. __________

The building characteristic choices shown here make a **big** difference in energy cost savings. Investigate these choices within the requirements of your design.

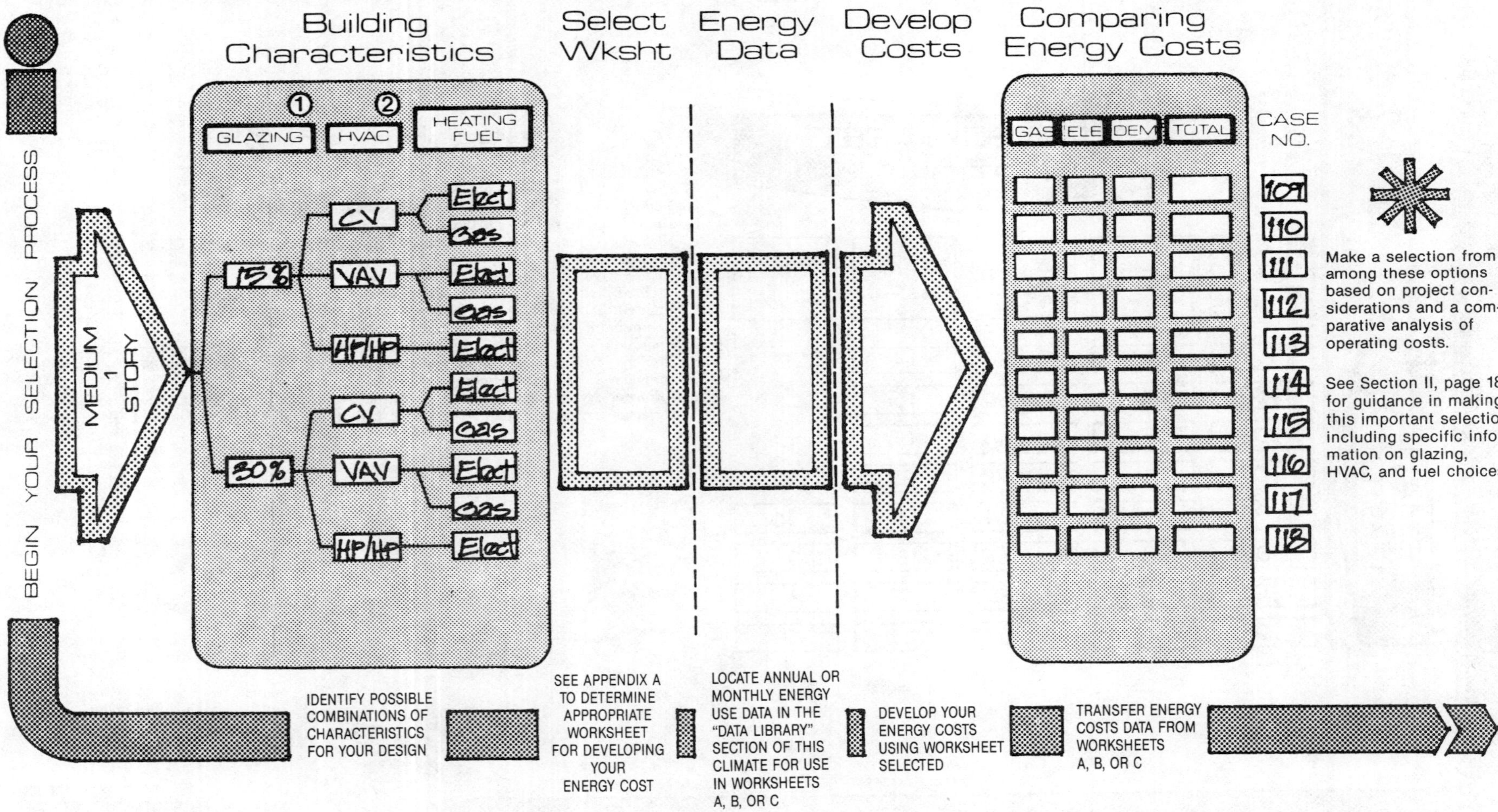

Make a selection from among these options based on project considerations and a comparative analysis of operating costs.

See Section II, page 189, for guidance in making this important selection, including specific information on glazing, HVAC, and fuel choices.

Note: See Section II, page 189, for detailed information on these Level 2 strategies and the use of this form.

① For glazing <23% use 15%, for >23% use 30%.

② CV=Constant Volume, VAV=Variable Air Volume, HP/HP=Split System Heat Pump

MEDIUM/1-STORY

Project ____________________
date __________ bldg. sq. ft. __________

Locate the row of strategy set options that apply to the building case you chose in Level 2. Select as many of these Level 3 strategy set options as you wish for valuation.

STRATEGY SET OPTIONS

MEDIUM 1 STORY

SELECTED LEVEL 2 CASE NUMBER	SIMPLEST LIGHTING IMPROVEMTS	MODEST LIGHTING & MECHANICAL IMPROVEMTS	MAJOR LIGHTING & MECHANICAL IMPROVEMTS	DAYLIGHT DIMMING CONTROLS AND LIGHTING IMPROVEMTS	SKYLIGHT'G, DAYLIGHT DIMMING CONTROLS & LIGHTING IMPROVEMTS
109	3	6 or 60	15	31 or 44 or 39	49
110	3	6 or 60	19	34 or 47 or 42	52
111	3	6 or 60	15	31 or 44 or 39	49
112	3	6 or 60	19	34 or 47 or 42	52
113	3	9 or 62	18	33 or 46 or 41	51
114	3	6 or 60	15	31 or 44 or 39	49
115	3	6 or 60	19	34 or 47 or 42	52
116	3	6 or 60	15	31 or 44 or 39	49
117	3	6 or 60	19	34 or 47 or 42	52
118	3	9 or 62	18	33 or 46 or 41	51

DESCRIPTOR KEY

3. High eff. ballasts over all bldg. + HVAC downsized for ballasts.
6. High eff. ballasts and lamps over all building + HVAC downsized for ballasts.
9. High eff. ballasts and lamps over all bldg. + HVAC downsized for ballasts + heat pump COP improvement.
15. High eff. ballasts, lamps and fixtures over all bldg. + HVAC downsized for ballasts and fixtures.
18. High eff. ballasts, lamps and fixtures over all bldg. + HVAC downsized for ballasts and fixtures + heat pump COP improvement.
19. High eff. ballasts, lamps and fixtures over all bldg. + HVAC downsized for ballasts and fixtures + high eff. boilers.
60. Non-uniform lighting + high eff. ballasts over all bldg. + HVAC downsized for ballasts.
62. Non-uniform lighting + high eff. ballasts over all bldg. + HVAC downsized for ballasts + heat pump COP improvement.
31. Perimeter daylighting dimmers + high eff. ballasts over all bldg. + HVAC downsized for ballasts.
33. Perimeter daylighting dimmers + high eff. ballasts over all bldg. + HVAC downsized for ballasts + heat pump COP improvement.
34. Perimeter daylighting dimmers + high eff. ballasts over all bldg. + HVAC downsized for ballasts + high eff. boilers.
39. Perimeter daylighting dimmers + high eff. ballasts over all bldg. + high eff. lamps and fixtures in core + HVAC downsized for ballasts and fixtures.
41. Perimeter daylighting dimmers + high eff. ballasts over all bldg. + high eff. lamps and fixtures in core + HVAC downsized for ballasts and fixtures + heat pump COP improvement.
42. Perimeter daylighting dimmers + high eff. ballasts over all bldg. + high eff. lamps and fixtures in core + HVAC downsized for ballasts and fixtures + high eff. boilers.
44. Perimeter daylighting dimmers + high eff. ballasts over all bldg. + high eff. lamps in core + HVAC downsized for ballasts.
46. Perimeter daylighting dimmers + high eff. ballasts over all bldg. + high eff. lamps in core + HVAC downsized for ballasts + heat pump COP improvement.
47. Perimeter daylighting dimmers + high eff. ballasts over all bldg. + high eff. lamps in core + HVAC downsized for ballasts + high eff. boilers.
49. Perimeter daylighting dimmers + skylights with dimmers + high eff. ballasts over all bldg. + HVAC downsized for ballasts.
51. Perimeter daylighting dimmers + skylights with dimmers + high eff. ballasts over all bldg. + HVAC downsized for ballasts + heat pump COP improvement.
52. Perimeter daylighting dimmers + skylights with dimmers + high eff. ballasts over all bldg. + HVAC downsized for ballasts + high eff. boilers.

GO TO LEVEL 3 ANALYSIS SHEET.

Note: See Section II, page 209, for detailed descriptions and discussions of these and other Level 3 strategy sets.

LEVEL 2
MAKE INITIAL CHOICES

MEDIUM/2-STORY

Project ______________________
date __________ bldg. sq. ft. __________

The building characteristic choices shown here make a **big** difference in energy cost savings. Investigate these choices within the requirements of your design.

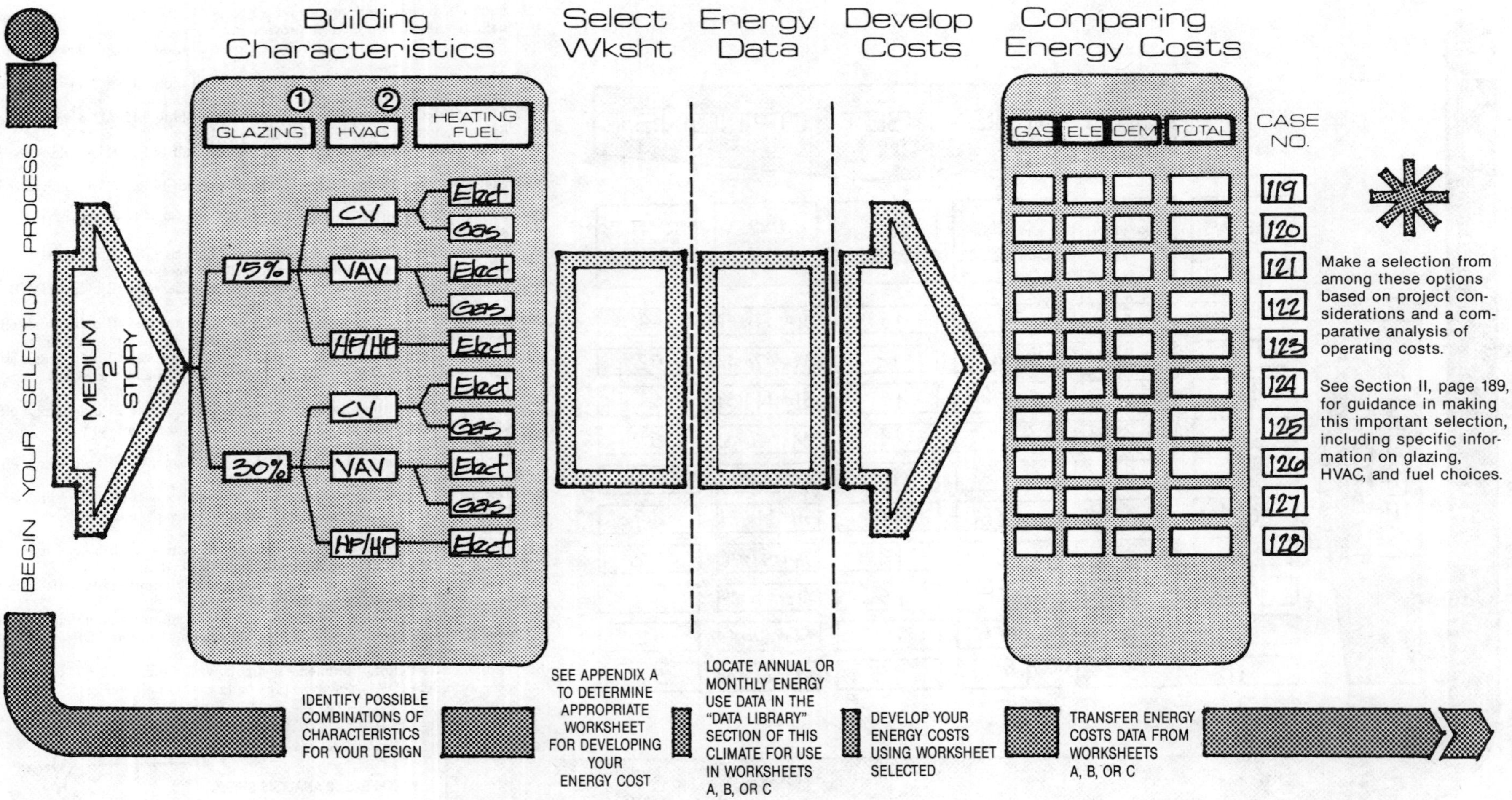

Note: See Section II, page 189, for detailed information on these Level 2 strategies and the use of this form.

① For glazing <23% use 15%, for >23% use 30%.

② CV=Constant Volume, VAV=Variable Air Volume, HP/HP=Split System Heat Pump.

Project ______________________
date __________ bldg. sq. ft. __________

Locate the row of strategy set options that apply to the building case you chose in Level 2. Select as many of these Level 3 strategy set options as you wish for valuation.

MEDIUM 2 STORY

STRATEGY SET OPTIONS

SELECTED LEVEL 2 CASE NUMBER	SIMPLEST LIGHTING IMPROVEMTS	MODEST LIGHTING & MECHANICAL IMPROVEMTS	MAJOR LIGHTING & MECHANICAL IMPROVEMTS	DAYLIGHT DIMMING CONTROLS AND LIGHTING IMPROVEMTS
119	3	6 or 60	15	31
120	3	6 or 60	19	34
121	3	6 or 60	15	31
122	3	6 or 60	19	34
123	3	9 or 62	18	33
124	3	6 or 60	15	31
125	3	6 or 60	19	34
126	3	6 or 60	15	31
127	3	6 or 60	19	34
128	3	9 or 62	18	33

DESCRIPTOR KEY

3. High eff. ballasts over all bldg. + HVAC downsized for ballasts.
6. High eff. ballasts and lamps over all building + HVAC downsized for ballasts.
9. High eff. ballasts and lamps over all bldg. + HVAC downsized for ballasts + heat pump COP improvement.
15. High eff. ballasts, lamps and fixtures over all bldg. + HVAC downsized for ballasts and fixtures.
18. High eff. ballasts, lamps and fixtures over all bldg. + HVAC downsized for ballasts and fixtures + heat pump COP improvement.
19. High eff. ballasts, lamps and fixtures over all bldg. + HVAC downsized for ballasts and fixtures + high eff. boilers.
60. Non-uniform lighting + high eff. ballasts over all bldg. + HVAC downsized for ballasts.
62. Non-uniform lighting + high eff. ballasts over all bldg. + HVAC downsized for ballasts + heat pump COP improvement.
31. Perimeter daylighting dimmers + high eff. ballasts over all bldg. + HVAC downsized for ballasts.
33. Perimeter daylighting dimmers + high eff. ballasts over all bldg. + HVAC downsized for ballasts + heat pump COP improvement.
34. Perimeter daylighting dimmers + high eff. ballasts over all bldg. + HVAC downsized for ballasts + high eff. boilers.

GO TO LEVEL 3 ANALYSIS SHEET.

Note: See Section II, page 209, for detailed descriptions and discussions of these and other Level 3 strategy sets.

LEVEL 2
MAKE INITIAL CHOICES
LARGE/2-STORY

Project ______________________
date ____________ bldg. sq. ft. ____________

The building characteristic choices shown here make a **big** difference in energy cost savings. Investigate these choices within the requirements of your design.

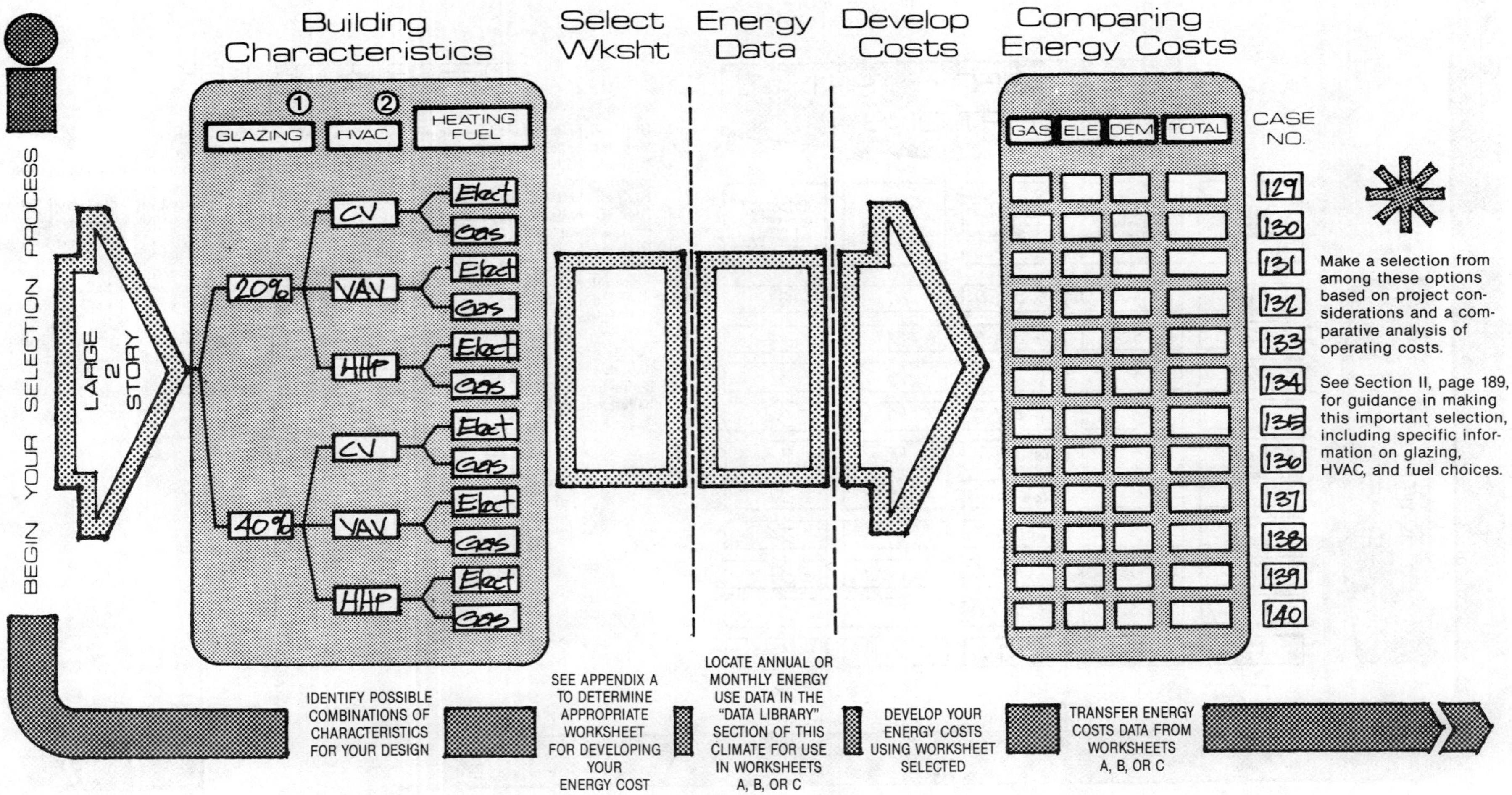

Note: See Section II, page 189, for detailed information on these Level 2 strategies and the use of this form.

① For glazing <30% use 20%, for >30% use 40%.

② CV=Constant Volume, VAV=Variable Air Volume, HHP=Hydronic Heat Pump.

Project ____________________

date __________ bldg. sq. ft. __________

Locate the row of strategy set options that apply to the building case you chose in Level 2. Select as many of these Level 3 strategy set options as you wish for valuation.

STRATEGY SET OPTIONS

LARGE 2 STORY

SELECTED LEVEL 2 CASE NUMBER	SIMPLEST LIGHTING IMPROVEMTS	MODEST LIGHTING & MECHANICAL IMPROVEMTS	MAJOR LIGHTING & MECHANICAL IMPROVEMTS	DAYLIGHT DIMMING CONTROLS AND LIGHTING IMPROVEMTS
129	3	6 or 60	15	31 or 44 or 39
130	3	6 or 60	15	31 or 44 or 39
131	3	6 or 60	15	31 or 44 or 39
132	3	6 or 60	15	31 or 44 or 39
133	1	12 or 28	22	36 or 48 or 43
134	1	12 or 28	22	36 or 48 or 43
135	3	6 or 60	15	31 or 44 or 39
136	3	6 or 60	15	31 or 44 or 39
137	3	6 or 60	15	31 or 44 or 39
138	3	6 or 60	15	31 or 44 or 39
139	1	12 or 28	22	36 or 48 or 43
140	1	12 or 28	22	36 or 48 or 43

DESCRIPTOR KEY

1. High eff. ballasts over all bldg.
3. High eff. ballasts over all bldg.+HVAC downsized for ballasts.
6. High eff. ballasts and lamps over all building+HVAC downsized for ballasts.
12. High eff. ballasts and lamps over all bldg+heat pump COP improvement.
15. High eff. ballasts, lamps and fixtures over all bldg.+HVAC downsized for ballasts and fixtures.
22. High eff. ballasts, lamps and fixtures over all bldg.+heat pump COP improvement.
60. Non-uniform lighting+high eff. ballasts over all bldg.+HVAC downsized for ballasts.
28. Non-uniform lighting+high eff. ballasts over all bldg.+heat pump COP improvement.
31. Perimeter daylighting dimmers+high eff. ballasts over all bldg.+HVAC downsized for ballasts.
36. Perimeter daylighting dimmers+high eff. ballasts over all bldg.+heat pump COP improvement.
39. Perimeter daylighting dimmers+high eff. ballasts over all bldg.+high eff. lamps and fixtures in core+HVAC downsized for ballasts and fixtures.
43. Perimeter daylighting dimmers+high eff. ballasts over all bldg.+high eff. lamps and fixtures in core+heat pump COP improvement.
44. Perimeter daylighting dimmers+high eff. ballasts over all bldg.+high eff. lamps in core+HVAC downsized for ballasts.
48. Perimeter daylighting dimmers+high eff. ballasts over all bldg.+high eff. lamps in core+heat pump COP improvement.

GO TO LEVEL 3 ANALYSIS SHEET.

Note: See Section II, page 209, for detailed descriptions and discussions of these and other Level 3 strategy sets.

LEVEL 2
MAKE INITIAL CHOICES
LARGE/4-STORY

Project ______________________
date __________ bldg. sq. ft. __________

The building characteristic choices shown here make a **big** difference in energy cost savings. Investigate these choices within the requirements of your design.

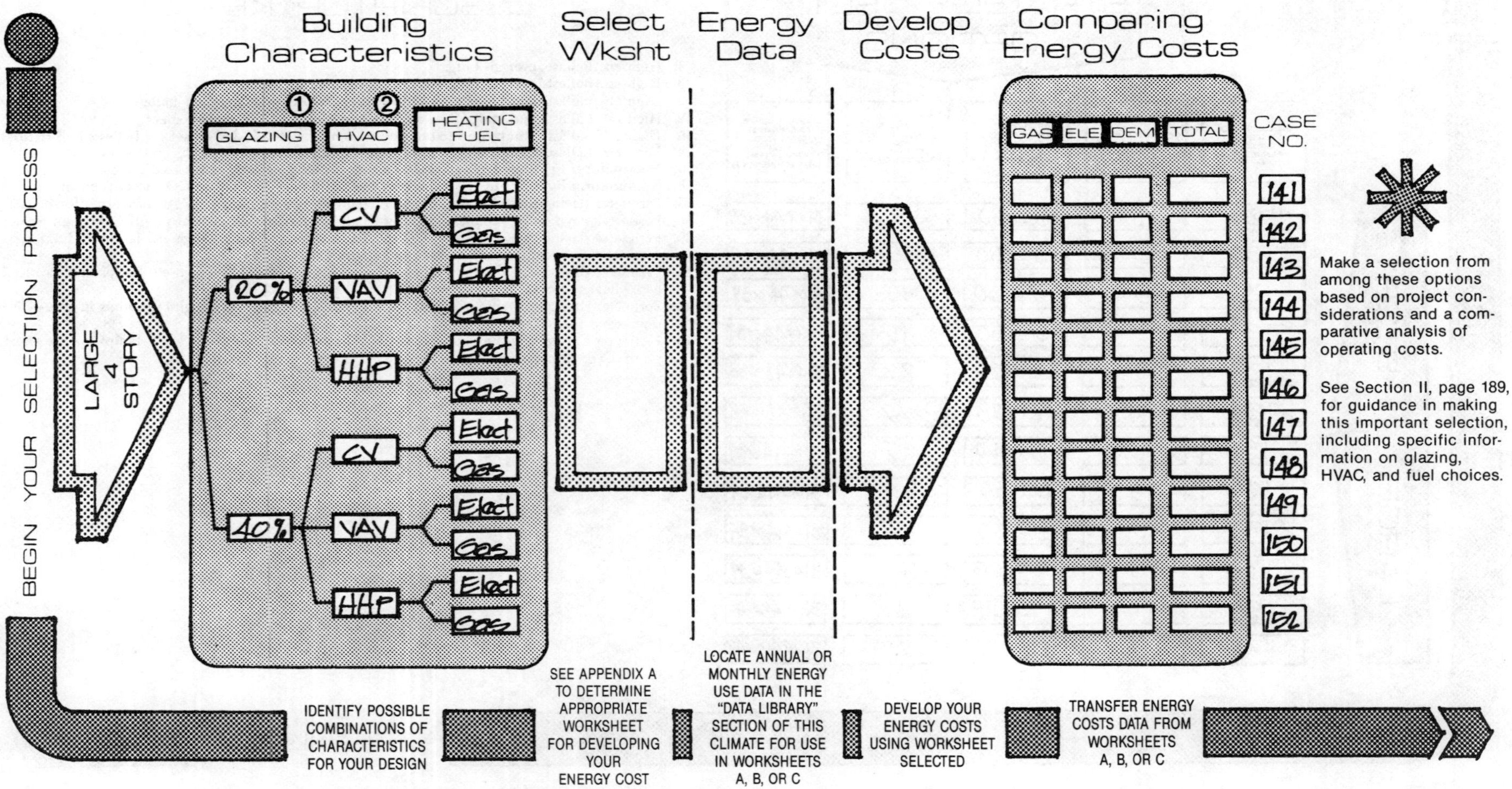

Note: See Section II, page 189, for detailed information on these Level 2 strategies and the use of this form.

① For glazing $<30\%$ use 20%, for $>30\%$ use 40%.

② CV=Constant Volume, VAV=Variable Air Volume, HHP=Hydronic Heat Pump.

Project ____________________
date __________ bldg. sq. ft. __________

Locate the row of strategy set options that apply to the building case you chose in Level 2. Select as many of these Level 3 strategy set options as you wish for valuation.

STRATEGY SET OPTIONS

LARGE 4 STORY

SELECTED LEVEL 2 CASE NUMBER	SIMPLEST LIGHTING IMPROVEMTS	MODEST LIGHTING & MECHANICAL IMPROVEMTS	MAJOR LIGHTING & MECHANICAL IMPROVEMTS	DAYLIGHT DIMMING CONTROLS AND LIGHTING IMPROVEMTS
141	3	6 or 60	15	31
142	3	6 or 60	15	31
143	3	6 or 60	15	31
144	3	6 or 60	15	31
145	1	12 or 28	22	36
146	1	12 or 28	22	36
147	3	6 or 60	15	31
148	3	6 or 60	15	31
149	3	6 or 60	15	31
150	3	6 or 60	15	31
151	1	12 or 28	22	36
152	1	12 or 28	22	36

DESCRIPTOR KEY

1. High eff. ballasts over all bldg.
3. High eff. ballasts over all bldg.+HVAC downsized for ballasts.
6. High eff. ballasts and lamps over all building+HVAC downsized for ballasts.
12. High eff. ballasts and lamps over all bldg+heat pump COP improvement.
15. High eff. ballasts, lamps and fixtures over all bldg.+HVAC downsized for ballasts and fixtures.
22. High eff. ballasts, lamps and fixtures over all bldg.+heat pump COP improvement.
60. Non-uniform lighting+high eff. ballasts over all bldg.+HVAC downsized for ballasts.
28. Non-uniform lighting+high eff. ballasts over all bldg.+heat pump COP improvement.
31. Perimeter daylighting dimmers+high eff. ballasts over all bldg.+HVAC downsized for ballasts.
36. Perimeter daylighting dimmers+high eff. ballasts over all bldg.+heat pump COP improvement.

GO TO LEVEL 3 ANALYSIS SHEET.

Note: See Section II, page 209, for detailed descriptions and discussions of these and other Level 3 strategy sets.

ANALYSIS/INSTRUCTIONS

For each strategy set you want to investigate, develop its construction costs using the upper route and its energy costs using the lower route. The combined information will allow you to select a strategy set for optimum economic benefit.

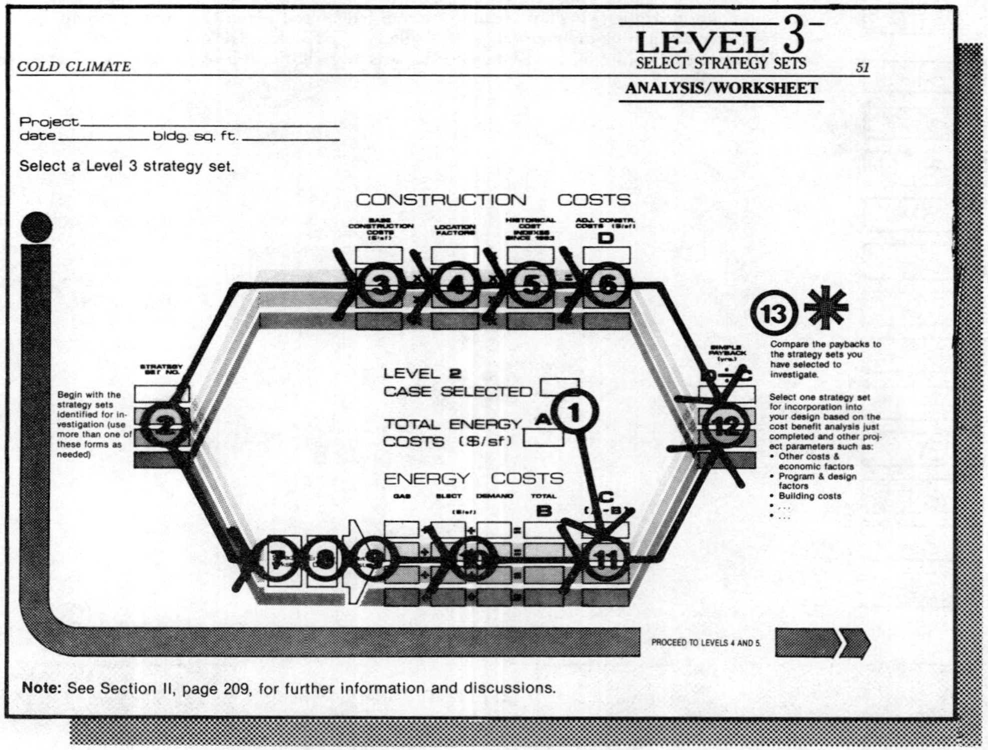

(1) Enter the Level 2 case selected and its total energy costs as developed in Level 2.

(2) As many as four strategy sets for any *one* Level 2 case can be analyzed on each sheet. You will need to use more sheets to analyze more than four strategy sets or to analyze strategy sets for another Level 2 case.

(3) Obtain *base construction costs* ($/SF) for each strategy set from the Data Library.

(4) Obtain *location factors* from commonly used cost estimating handbooks. Additional multipliers may be entered to account for above or below average construction costs due to specific project conditions.

(5) Obtain *historical cost indexes* (factors since 1983) from commonly used cost estimating handbooks.

(6) Multiply 3, 4, and 5 to obtain *adjusted construction costs* ($/SF). User may wish to use more refined construction cost data as supplied by a cost estimator for the specific project.

(7) Use the same worksheet you selected for use in Level 2.

(8) Locate energy use data for the selected strategy sets in the Data Library.

(9) Develop energy costs for each strategy set using the worksheet selected.

(10) Transfer energy cost information from worksheets A, B, or C.

(11) Find the energy savings from each strategy set by subtracting each "B" from "A."

(12) Calculate *simple payback* by dividing "D" (initial costs) by "C" (yearly savings). Alternative economic analysis techniques other than simple payback may be employed using information from columns "D" and "C."

(13) Cost factors to consider in making a decision on these strategy sets include:
- Negative construction costs (first cost savings)
- Positive construction costs
- Annual energy cost savings

ANALYSIS/WORKSHEET

Project__________________________________
date______________ bldg. sq. ft. ______________

Select a Level 3 strategy set.

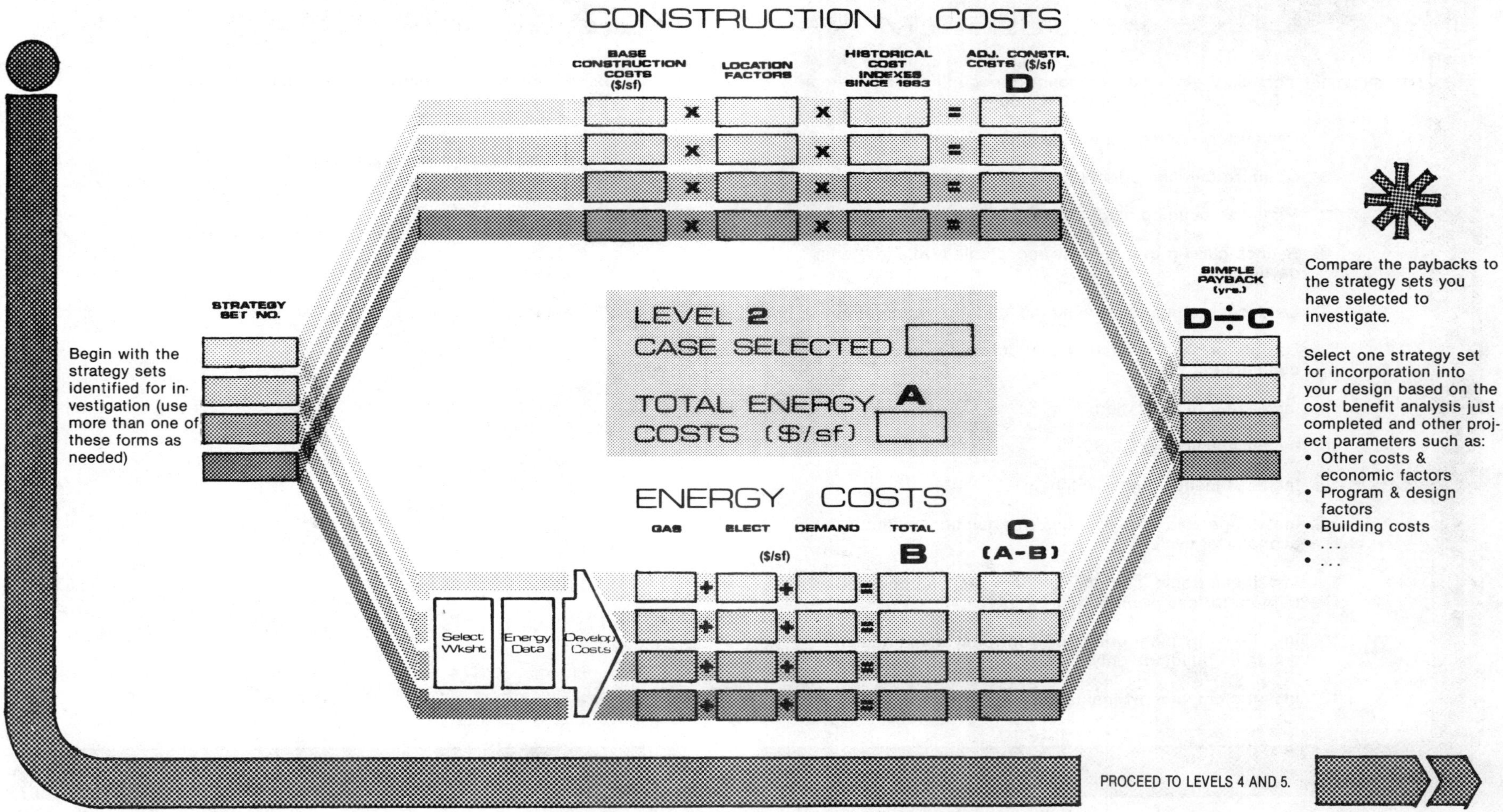

Note: See Section II, page 209, for further information and discussions.

Review these additional strategies for the possibility of further savings. Check the description in Section II, page 235, carefully for guidance on adopting any of the strategies.

A SELECT ON THE BASIS OF TECHNICAL JUDGEMENT

1. Increase thermostat deadband for occupied periods from 2°F to 8°F (70°F heating-78°F cooling)
2. Design for efficient space planning
3. Optimize building orientation
4. Mimimize building height to reduce impact of glazed area
6. Reduce glazing to minimally acceptable levels (not with daylighting)
7. Landscape to minimize wind induced infiltration effect
8. Design for passive heating (particularly for small buildings)
9. Increase roof insulation
10. Increase wall insulation
12. Increase manual light switching
13. Install operable windows (best for buildings without economizer cycle)
14. Install electronic ignition, flue dampers, and powered burners for gas or oil boilers
15. Install energy monitoring and control system (medium and large buildings only)
17. Install night window insulation

B VALUE MAY BE DETERMINED BY SIMPLE CALCULATION

19. High efficiency motors and pumps
20. Multiple speed compressors
21. Variable or two speed circulating fan
23. Desuperheater or stand-alone heat pump for DHW
24. Point-of-use electric domestic hot water heaters
25. External shading
26. Solar domestic hot water heater
28. Hot water storage
29. Demand limiter
31. Cogeneration

Note: The number preceding each strategy refers to the corresponding paragraph in Section II, page 235. See Section II, page 235, for further information and discussion of each strategy.

Review this checklist and adopt the procedures that are appropriate to your project. These actions are **important** if you are to achieve and maintain the savings from the selected Level 1-4 strategies.

A CHECKLIST FOR SUCCESSFUL TURNOVER AND BUILDING OPERATION

	DESIGN	CONSTRUCTION	TURNOVER	BUILDING OPERATION
DESIGN TEAM (ARCHITECTS/ ENGINEERS)	• Document all design assumptions (1) • Estimate energy use by sub-system (2) **& For Large Buildings:** • Document all systems and control strategies (3) • Specify contents of O&M Manuals (4) • Specify HVAC balancing requirements (5) • Specify equipment protection and cleaning (6) • Specify on-site instruction (7) • Specify access doors, valves, test ports (8) • Specify piping system ID (9) • Specify price quotations on all long-term maintenance contracts (10)	• Prepare "as-built" drawings (11) • Observe construction process for compliance (12)	• Conduct "punch list" inspection (15) • Visually check for air leakage (15) • If possible, participate in on-site training session(s) (7)	
CONTRACTOR & SUB-CONTRACTOR	**For Small Buildings Only:** • Document all systems and control strategies (3) • Specify contents of O&M Manuals (4) • Specify HVAC balancing requirements (5) • Specify equipment protection and cleaning (6) • Specify on-site instruction (7) • Specify access doors, valves, test ports (8) • Specify piping system ID (9) • Specify price quotations on all long-term maintenance contracts (10)	• Prepare O&M Manual (4) • Perform final calibration of all controls (5) • Hire independent balancing agent or require of subs (5) • Protect and clean equipment (6) • Provide all access doors, valves, and test ports (8) • Place piping system ID tags (9) • Prepare "as-built" drawings (11) • Use HVAC system to "flush" construction odors (13)	• Deliver O&M Manual (4) • Conduct on-site instruction and training (7)	• Honor warranty and call back as required (16)
EQUIPMENT MANUFACT.		• Support calibration of controls (5) • Provide all special tools required (14)	• Support O&M Manual preparation (4) • Support on-site instruction and training (7)	
OWNER & BLDG. STAFF	• Check all key design assumptions (1)	• Periodically check compliance on-site (12)	• Attend on-site instruction and training sessions (7)	• Monitor energy consumption against estimates (2) • Periodically test equipment efficiency (17) • Periodically sample and test water (18) • Inspect and calibrate controls (19) • Monitor occupant complains (20)

Note: See Section II, page 255, for more information and discussion. Numbers shown in ( ) refer to the numbered paragraphs in Section II, page 255.

1. Locate data by Base Case number.
2. Locate monthly or annual data for Base Case or Strategy Set.

Use monthly data for Worksheets B or C

Use annual data for Worksheet A

BASE CASE 326

STRAT SET	CONST COST $/SF	ENERGY FORM	UNITS (PER SF)	JAN	FEB	MAR	APR	MAY	JUN	JUL	AUG	SEP	OCT	NOV	DEC	ANNUAL
BASE	0.00	HT ELEC	KWH	0.76	0.53	0.31	0.09	0.03	0.03	0.03	0.03	0.03	0.05	0.23	0.65	2.8
		HT DEM	KW	.00605	.00587	.00575	.00405	.00369	.00006	.00006	.00006	.00224	.00363	.00520	.00587	0.00605
		REM ELEC	KWH	0.97	0.87	0.97	1.06	1.23	1.47	1.70	1.63	1.34	1.10	0.93	0.97	14.3
		REM DEM	KW	.00504	.00457	.00543	.00651	.00713	.00698	.00775	.00752	.00721	.00635	.00519	.00473	0.00775
		TOT ELEC	KWH	1.74	1.40	1.28	1.15	1.26	1.50	1.73	1.66	1.37	1.15	1.15	1.63	17.1
		TOT DEM	KW	.00930	.00828	.00828	.00698	.00716	.00763	.00781	.00763	.00744	.00670	.00679	.00837	0.00930
S- 3	-0.24	HT ELEC	KWH	0.81	0.56	0.32	0.10	0.03	0.03	0.03	0.03	0.03	0.05	0.24	0.69	3.0
		HT DEM	KW	.00605	.00587	.00575	.00405	.00369	.00006	.00006	.00006	.00224	.00363	.00520	.00587	0.00605
		REM ELEC	KWH	0.88	0.79	0.88	0.96	1.12	1.34	1.55	1.48	1.22	1.00	0.85	0.88	13.0
		REM DEM	KW	.00464	.00421	.00500	.00600	.00657	.00643	.00714	.00693	.00664	.00585	.00478	.00436	0.00714
S-																

BASE CASE 327

STRAT SET	CONST COST $/SF	ENERGY FORM	UNITS (PER SF)	JAN	FEB	MAR	APR	MAY	JUN	JUL	AUG	SEP	OCT	NOV	DEC	ANNUAL
BASE	0.00	TOT GAS	THM	0.0402	0.0277	0.0161	0.0048	0.0015	0.0015	0.0015	0.0015	0.0015	0.0026	0.0118	0.0343	0.146
		TOT ELEC	KWH	0.97	0.87	0.97	1.06	1.23	1.47	1.70	1.63	1.34	1.10	0.93	0.97	14.3
		TOT DEM	KW	.00504	.00457	.00543	.00651	.00713	.00698	.00775	.00752	.00721	.00635	.00519	.00473	0.00775
S- 3	-0.24	TOT GAS	THM	0.0426	0.0295	0.0171	0.0051	0.0015	0.0015	0.0015	0.0015	0.0015	0.0028	0.0126	0.0364	0.155
		TOT ELEC	KWH	0.88	0.79	0.88	0.96	1.12	1.34	1.55	1.48	1.22	1.00	0.85	0.88	13.0
		TOT DEM	KW	.00464	.00421	.00500	.00600	.00657	.00643	.00714	.00693	.00664	.00585	.00478	.00436	0.00714
S- 7	-0.17	TOT GAS	THM	0.0451	0.0312	0.0180	0.0054	0.0016	0.0016	0.0016	0.0016	0.0016	0.0030	0.0133	0.0385	0.164
		TOT ELEC	KWH	0.79	0.71	0.79	0.86	1.00	1.19	1.38	1.32	1.09	0.89	0.75	0.79	11.6
		TOT DEM	KW	.00406	.00369	.00438	.00525	.00575	.00563	.00625	.00606	.00581	.00513	.00419	.00381	0.00625
S- 16	0.40	TOT GAS	THM	0.0476	0.0329	0.0190	0.0057	0.0017	0.0017	0.0017	0.0017	0.0017	0.0031	0.0140	0.0407	0.173
		TOT ELEC	KWH	0.69	0.62	0.69	0.75	0.87	1.04	1.20	1.15	0.95	0.78	0.66	0.69	10.1
		TOT DEM	KW	.00370	.00336	.00398	.00478	.00523	.00512	.00569	.00552	.00529	.00467	.00381	.00347	0.00569
S- 25	0.38	TOT GAS	THM	0.0451	0.0312	0.0180	0.0054	0.0016	0.0016	0.0016	0.0016	0.0016	0.0030	0.0133	0.0385	0.164
		TOT ELEC	KWH	0.79	0.71	0.79	0.86	1.00	1.19	1.38	1.32	1.09	0.89	0.75	0.79	11.6
		TOT DEM	KW	.00406	.00369	.00438	.00525	.00575	.00563	.00625	.00606	.00581	.00513	.00419	.00381	0.00625
S- 32	0.09	TOT GAS	THM	0.0433	0.0258	0.0211	0.0087	0.0041	0.0018	0.0018	0.0018	0.0032	0.0070	0.0141	0.0298	0.162
		TOT ELEC	KWH	0.66	0.62	0.75	0.81	0.87	1.07	1.14	1.13	0.96	0.81	0.67	0.66	10.2
		TOT DEM	KW	.00414	.00403	.00443	.00512	.00535	.00552	.00575	.00564	.00558	.00500	.00426	.00380	0.00575

HT ELEC = Electric consumption for space and water heating.

REM ELEC = Remaining electric consumption (not used for heating).

TOT ELEC = Total electric consumption.

HT DEM = Peak electrical demand for space and water heating.

REM DEM = Peak electrical demand for remaining load (not heating).

TOT DEM = Total peak electrical demand.

GAS = Total gas consumption.

(Annual figures are sums of monthly consumption values and are yearly maximums of monthly peak demand values.)

STRAT SET	CONST COST $/SF	ENERGY FORM	UNITS (PER SF)	JAN	FEB	MAR	APR	MAY	JUN	JUL	AUG	SEP	OCT	NOV	DEC	ANNUAL
BASE CASE : 101																
BASE	0.00	TOT ELEC	KWH	2.26	1.60	1.49	1.25	1.32	1.36	1.51	1.49	1.32	1.29	1.44	2.06	18.4
		TOT DEM	KW	.00950	.00684	.00694	.00446	.00352	.00446	.00456	.00484	.00437	.00475	.00570	.00864	0.00950
S- 1	0.09	TOT ELEC	KWH	2.15	1.52	1.42	1.19	1.26	1.30	1.44	1.42	1.26	1.23	1.37	1.96	17.5
		TOT DEM	KW	.00854	.00615	.00623	.00401	.00316	.00401	.00410	.00436	.00393	.00427	.00512	.00777	0.00854
S- 5	0.21	TOT ELEC	KWH	2.05	1.45	1.35	1.14	1.20	1.24	1.37	1.35	1.20	1.17	1.30	1.87	16.7
		TOT DEM	KW	.00800	.00576	.00584	.00376	.00296	.00376	.00384	.00408	.00368	.00400	.00480	.00728	0.00800
S- 14	1.52	TOT ELEC	KWH	1.92	1.36	1.26	1.06	1.12	1.15	1.28	1.26	1.12	1.09	1.22	1.75	15.6
		TOT DEM	KW	.00788	.00567	.00575	.00370	.00292	.00370	.00378	.00402	.00362	.00394	.00473	.00717	0.00788
S- 24	-0.04	TOT ELEC	KWH	2.05	1.45	1.35	1.14	1.20	1.24	1.37	1.35	1.20	1.17	1.30	1.87	16.7
		TOT DEM	KW	.00800	.00576	.00584	.00376	.00296	.00376	.00384	.00408	.00368	.00400	.00480	.00728	0.00800
BASE CASE : 102																
BASE	0.00	TOT ELEC	KWH	2.17	1.75	1.57	1.15	1.08	1.20	1.32	1.29	1.07	1.14	1.46	1.98	17.2
		TOT DEM	KW	.00950	.00674	.00674	.00427	.00247	.00266	.00276	.00304	.00276	.00513	.00560	.00903	0.00950
S- 3	-0.02	TOT ELEC	KWH	2.08	1.68	1.50	1.11	1.04	1.15	1.27	1.24	1.02	1.09	1.40	1.90	16.5
		TOT DEM	KW	.00854	.00606	.00606	.00384	.00222	.00239	.00248	.00273	.00248	.00461	.00504	.00811	0.00854
S- 6	0.10	TOT ELEC	KWH	1.99	1.61	1.44	1.06	1.00	1.11	1.22	1.19	0.98	1.04	1.34	1.82	15.8
		TOT DEM	KW	.00800	.00568	.00568	.00360	.00208	.00224	.00232	.00256	.00232	.00432	.00472	.00760	0.00800
S- 15	1.28	TOT ELEC	KWH	1.88	1.52	1.36	1.00	0.94	1.04	1.15	1.12	0.92	0.98	1.27	1.71	14.9
		TOT DEM	KW	.00787	.00559	.00559	.00354	.00205	.00220	.00228	.00252	.00228	.00425	.00464	.00748	0.00787
S- 60	-0.15	TOT ELEC	KWH	1.99	1.61	1.44	1.06	1.00	1.11	1.22	1.19	0.98	1.04	1.34	1.82	15.8
		TOT DEM	KW	.00800	.00568	.00568	.00360	.00208	.00224	.00232	.00256	.00232	.00432	.00472	.00760	0.00800
BASE CASE : 103																
BASE	0.00	TOT GAS	THM	0.0569	0.0432	0.0317	0.0152	0.0066	0.0047	0.0037	0.0034	0.0047	0.0123	0.0304	0.0493	0.262
		TOT ELEC	KWH	0.88	0.79	0.89	0.88	1.04	1.23	1.40	1.37	1.07	0.95	0.82	0.88	12.2
		TOT DEM	KW	.00346	.00346	.00346	.00335	.00432	.00464	.00481	.00540	.00481	.00470	.00346	.00346	0.00540
S- 3	-0.02	TOT GAS	THM	0.0599	0.0455	0.0334	0.0160	0.0069	0.0050	0.0039	0.0036	0.0050	0.0130	0.0320	0.0519	0.276
		TOT ELEC	KWH	0.81	0.73	0.82	0.81	0.95	1.13	1.29	1.25	0.99	0.87	0.75	0.81	11.2
		TOT DEM	KW	.00305	.00305	.00305	.00296	.00382	.00410	.00425	.00477	.00425	.00415	.00305	.00305	0.00477
S- 6	0.10	TOT GAS	THM	0.0629	0.0479	0.0351	0.0168	0.0073	0.0052	0.0041	0.0038	0.0052	0.0136	0.0336	0.0545	0.290
		TOT ELEC	KWH	0.74	0.67	0.75	0.74	0.88	1.04	1.18	1.15	0.91	0.80	0.69	0.74	10.3
		TOT DEM	KW	.00286	.00286	.00286	.00277	.00358	.00384	.00398	.00447	.00398	.00389	.00286	.00286	0.00447
S- 15	1.27	TOT GAS	THM	0.0668	0.0508	0.0373	0.0179	0.0077	0.0055	0.0043	0.0040	0.0055	0.0145	0.0357	0.0579	0.308
		TOT ELEC	KWH	0.65	0.59	0.66	0.65	0.77	0.92	1.04	1.02	0.80	0.71	0.61	0.65	9.1
		TOT DEM	KW	.00262	.00262	.00262	.00254	.00328	.00353	.00365	.00410	.00365	.00357	.00262	.00262	0.00410
S- 60	-0.15	TOT GAS	THM	0.0629	0.0479	0.0351	0.0168	0.0073	0.0052	0.0041	0.0038	0.0052	0.0136	0.0336	0.0545	0.290
		TOT ELEC	KWH	0.74	0.67	0.75	0.74	0.88	1.04	1.18	1.15	0.91	0.80	0.69	0.74	10.3
		TOT DEM	KW	.00286	.00286	.00286	.00277	.00358	.00384	.00398	.00447	.00398	.00389	.00286	.00286	0.00447
BASE CASE : 104																
BASE	0.00	TOT ELEC	KWH	2.12	1.50	1.39	1.17	1.24	1.27	1.41	1.39	1.24	1.20	1.34	1.93	17.2
		TOT DEM	KW	.00950	.00665	.00684	.00370	.00323	.00380	.00399	.00437	.00389	.00399	.00418	.00741	0.00950
S- 1	0.09	TOT ELEC	KWH	1.99	1.41	1.31	1.10	1.17	1.20	1.33	1.31	1.17	1.13	1.26	1.81	16.2
		TOT DEM	KW	.00854	.00598	.00615	.00333	.00290	.00342	.00359	.00393	.00350	.00359	.00376	.00666	0.00854
S- 12	0.22	TOT ELEC	KWH	1.78	1.26	1.17	0.99	1.04	1.07	1.19	1.17	1.04	1.01	1.13	1.62	14.5
		TOT DEM	KW	.00800	.00560	.00576	.00312	.00272	.00320	.00336	.00368	.00328	.00336	.00352	.00624	0.00800
S- 22	1.54	TOT ELEC	KWH	1.65	1.17	1.09	0.91	0.96	0.99	1.10	1.09	0.96	0.94	1.05	1.50	13.4
		TOT DEM	KW	.00787	.00551	.00567	.00307	.00268	.00315	.00331	.00362	.00323	.00331	.00346	.00614	0.00787
S- 28	-0.03	TOT ELEC	KWH	1.78	1.26	1.17	0.99	1.04	1.07	1.19	1.17	1.04	1.01	1.13	1.62	14.5
		TOT DEM	KW	.00800	.00560	.00576	.00312	.00272	.00320	.00336	.00368	.00328	.00336	.00352	.00624	0.00800
BASE CASE : 105																
BASE	0.00	TOT ELEC	KWH	2.40	1.70	1.58	1.33	1.40	1.44	1.60	1.58	1.40	1.37	1.52	2.18	19.5
		TOT DEM	KW	.00950	.00684	.00694	.00446	.00352	.00446	.00456	.00484	.00437	.00475	.00570	.00864	0.00950
S- 1	0.09	TOT ELEC	KWH	2.28	1.61	1.50	1.26	1.33	1.37	1.52	1.50	1.33	1.29	1.44	2.07	18.5
		TOT DEM	KW	.00854	.00615	.00623	.00401	.00316	.00401	.00410	.00436	.00393	.00427	.00512	.00777	0.00854
S- 5	0.22	TOT ELEC	KWH	2.18	1.54	1.43	1.20	1.27	1.31	1.45	1.43	1.27	1.24	1.38	1.98	17.7
		TOT DEM	KW	.00800	.00576	.00584	.00376	.00296	.00376	.00384	.00408	.00368	.00400	.00480	.00728	0.00800
S- 14	1.53	TOT ELEC	KWH	2.03	1.44	1.34	1.12	1.19	1.22	1.35	1.34	1.19	1.15	1.29	1.85	16.5
		TOT DEM	KW	.00788	.00567	.00575	.00370	.00292	.00370	.00378	.00402	.00362	.00394	.00473	.00717	0.00788
S- 24	-0.03	TOT ELEC	KWH	2.18	1.54	1.43	1.20	1.27	1.31	1.45	1.43	1.27	1.24	1.38	1.98	17.7
		TOT DEM	KW	.00800	.00576	.00584	.00376	.00296	.00376	.00384	.00408	.00368	.00400	.00480	.00728	0.00800
S- 30	0.69	TOT ELEC	KWH	2.13	1.70	1.42	0.81	0.72	0.91	0.95	0.97	0.82	0.80	1.15	1.99	14.2
		TOT DEM	KW	.00975	.00702	.00712	.00458	.00361	.00458	.00468	.00497	.00448	.00488	.00585	.00887	0.00975
BASE CASE : 106																
BASE	0.00	TOT ELEC	KWH	2.26	1.83	1.63	1.20	1.13	1.25	1.38	1.34	1.11	1.18	1.52	2.06	17.9
		TOT DEM	KW	.00950	.00674	.00674	.00427	.00247	.00266	.00276	.00304	.00276	.00513	.00560	.00903	0.00950
S- 3	-0.05	TOT ELEC	KWH	2.15	1.74	1.56	1.15	1.08	1.20	1.32	1.28	1.06	1.13	1.45	1.97	17.1
		TOT DEM	KW	.00854	.00606	.00606	.00384	.00222	.00239	.00248	.00273	.00248	.00461	.00504	.00811	0.00854

STRAT SET	CONST COST $/SF	ENERGY FORM	UNITS (PER SF)	JAN	FEB	MAR	APR	MAY	JUN	JUL	AUG	SEP	OCT	NOV	DEC	ANNUAL
BASE CASE : 106 (CONTINUED)																
S- 6	0.08	TOT ELEC	KWH	2.07	1.67	1.49	1.10	1.03	1.15	1.26	1.23	1.02	1.08	1.39	1.89	16.4
		TOT DEM	KW	.00800	.00568	.00568	.00360	.00208	.00224	.00232	.00256	.00232	.00432	.00472	.00760	0.00800
S- 15	1.25	TOT ELEC	KWH	1.95	1.58	1.41	1.04	0.98	1.09	1.19	1.16	0.96	1.02	1.32	1.78	15.5
		TOT DEM	KW	.00782	.00555	.00555	.00352	.00203	.00219	.00227	.00250	.00227	.00422	.00461	.00743	0.00782
S- 31	0.56	TOT ELEC	KWH	2.40	1.95	1.62	0.89	0.67	0.92	1.00	1.01	0.78	0.84	1.34	2.18	15.6
		TOT DEM	KW	.00975	.00692	.00692	.00439	.00254	.00273	.00283	.00312	.00283	.00527	.00575	.00926	0.00975
S- 60	-0.17	TOT ELEC	KWH	2.07	1.67	1.49	1.10	1.03	1.15	1.26	1.23	1.02	1.08	1.39	1.89	16.4
		TOT DEM	KW	.00800	.00568	.00568	.00360	.00208	.00224	.00232	.00256	.00232	.00432	.00472	.00760	0.00800
BASE CASE : 107																
BASE	0.00	TOT GAS	THM	0.0571	0.0434	0.0318	0.0153	0.0066	0.0047	0.0037	0.0034	0.0047	0.0124	0.0305	0.0494	0.263
		TOT ELEC	KWH	0.93	0.84	0.94	0.93	1.10	1.30	1.48	1.44	1.14	1.01	0.86	0.93	12.9
		TOT DEM	KW	.00346	.00346	.00346	.00335	.00432	.00464	.00481	.00540	.00481	.00470	.00346	.00346	0.00540
S- 3	-0.05	TOT GAS	THM	0.0601	0.0457	0.0335	0.0161	0.0069	0.0050	0.0039	0.0036	0.0050	0.0130	0.0321	0.0521	0.277
		TOT ELEC	KWH	0.85	0.77	0.86	0.85	1.00	1.19	1.36	1.32	1.04	0.92	0.79	0.85	11.8
		TOT DEM	KW	.00305	.00305	.00305	.00296	.00382	.00410	.00425	.00477	.00425	.00415	.00305	.00305	0.00477
S- 6	0.08	TOT GAS	THM	0.0631	0.0480	0.0352	0.0169	0.0073	0.0052	0.0041	0.0038	0.0052	0.0137	0.0338	0.0547	0.291
		TOT ELEC	KWH	0.78	0.71	0.80	0.78	0.93	1.10	1.25	1.22	0.96	0.85	0.73	0.78	10.9
		TOT DEM	KW	.00286	.00286	.00286	.00277	.00358	.00384	.00398	.00447	.00398	.00389	.00286	.00286	0.00447
S- 15	1.24	TOT GAS	THM	0.0671	0.0510	0.0374	0.0179	0.0077	0.0056	0.0043	0.0040	0.0056	0.0145	0.0358	0.0581	0.309
		TOT ELEC	KWH	0.69	0.62	0.70	0.69	0.82	0.97	1.10	1.07	0.84	0.75	0.64	0.69	9.6
		TOT DEM	KW	.00262	.00262	.00262	.00254	.00328	.00353	.00365	.00410	.00365	.00357	.00262	.00262	0.00410
S- 31	0.55	TOT GAS	THM	0.0768	0.0608	0.0448	0.0147	0.0045	0.0026	0.0022	0.0022	0.0035	0.0102	0.0330	0.0659	0.320
		TOT ELEC	KWH	0.77	0.69	0.73	0.66	0.66	0.96	1.03	1.05	0.78	0.70	0.73	0.80	9.5
		TOT DEM	KW	.00320	.00320	.00320	.00310	.00400	.00430	.00445	.00500	.00445	.00435	.00320	.00320	0.00500
S- 60	-0.17	TOT GAS	THM	0.0631	0.0480	0.0352	0.0169	0.0073	0.0052	0.0041	0.0038	0.0052	0.0137	0.0338	0.0547	0.291
		TOT ELEC	KWH	0.78	0.71	0.80	0.78	0.93	1.10	1.25	1.22	0.96	0.85	0.73	0.78	10.9
		TOT DEM	KW	.00286	.00286	.00286	.00277	.00358	.00384	.00398	.00447	.00398	.00389	.00286	.00286	0.00447
BASE CASE : 108																
BASE	0.00	TOT ELEC	KWH	2.25	1.59	1.48	1.24	1.32	1.35	1.50	1.48	1.32	1.28	1.43	2.05	18.3
		TOT DEM	KW	.00950	.00665	.00684	.00370	.00323	.00380	.00399	.00437	.00389	.00399	.00418	.00741	0.00950
S- 1	0.09	TOT ELEC	KWH	2.12	1.50	1.39	1.17	1.24	1.27	1.41	1.39	1.24	1.20	1.34	1.93	17.2
		TOT DEM	KW	.00854	.00598	.00615	.00333	.00290	.00342	.00359	.00393	.00350	.00359	.00376	.00666	0.00854
S- 12	0.24	TOT ELEC	KWH	1.89	1.34	1.25	1.05	1.11	1.14	1.26	1.25	1.11	1.08	1.20	1.72	15.4
		TOT DEM	KW	.00788	.00552	.00567	.00307	.00268	.00315	.00331	.00362	.00323	.00331	.00347	.00615	0.00788
S- 22	1.56	TOT ELEC	KWH	1.75	1.24	1.15	0.97	1.02	1.05	1.16	1.15	1.02	0.99	1.11	1.59	14.2
		TOT DEM	KW	.00787	.00551	.00567	.00307	.00268	.00315	.00331	.00362	.00323	.00331	.00346	.00614	0.00787
S- 28	-0.01	TOT ELEC	KWH	1.89	1.34	1.25	1.05	1.11	1.14	1.26	1.25	1.11	1.08	1.20	1.72	15.4
		TOT DEM	KW	.00788	.00552	.00567	.00307	.00268	.00315	.00331	.00362	.00323	.00331	.00347	.00615	0.00788
S- 36	0.72	TOT ELEC	KWH	1.71	1.23	1.14	0.77	0.65	0.82	0.85	0.87	0.75	0.74	1.02	1.54	12.1
		TOT DEM	KW	.00950	.00665	.00684	.00370	.00323	.00380	.00399	.00437	.00389	.00399	.00418	.00741	0.00950
BASE CASE : 109																
BASE	0.00	HT ELEC	KWH	1.05	0.80	0.59	0.28	0.12	0.09	0.07	0.06	0.09	0.23	0.56	0.91	4.8
		HT DEM	KW	.00908	.00863	.00772	.00445	.00309	.00209	.00072	.00114	.00281	.00463	.00645	.00908	0.00908
		REM ELEC	KWH	0.94	0.84	0.95	0.94	1.11	1.31	1.50	1.46	1.14	1.01	0.87	0.94	13.0
		REM DEM	KW	.00380	.00380	.00373	.00485	.00563	.00629	.00648	.00655	.00616	.00583	.00380	.00380	0.00655
		TOT ELEC	KWH	1.99	1.64	1.53	1.22	1.23	1.40	1.56	1.52	1.23	1.24	1.43	1.85	17.8
		TOT DEM	KW	.01288	.01069	.00902	.00670	.00657	.00734	.00708	.00721	.00683	.00708	.00863	.01288	0.01288
S- 3	-0.11	HT ELEC	KWH	1.10	0.84	0.61	0.29	0.13	0.09	0.07	0.07	0.09	0.24	0.59	0.95	5.1
		HT DEM	KW	.00908	.00863	.00772	.00445	.00309	.00209	.00072	.00114	.00281	.00463	.00645	.00908	0.00908
		REM ELEC	KWH	0.85	0.77	0.86	0.85	1.00	1.19	1.36	1.32	1.04	0.92	0.79	0.85	11.8
		REM DEM	KW	.00349	.00349	.00343	.00445	.00518	.00578	.00596	.00602	.00566	.00536	.00349	.00349	0.00602
		TOT ELEC	KWH	1.95	1.60	1.47	1.14	1.13	1.28	1.43	1.39	1.13	1.16	1.38	1.80	16.9
		TOT DEM	KW	.01240	.01029	.00868	.00645	.00632	.00707	.00682	.00694	.00657	.00682	.00831	.01240	0.01240
S- 6	-0.03	HT ELEC	KWH	1.17	0.89	0.65	0.31	0.13	0.10	0.08	0.07	0.10	0.25	0.62	1.01	5.4
		HT DEM	KW	.00908	.00863	.00772	.00445	.00309	.00209	.00072	.00114	.00281	.00463	.00645	.00908	0.00908
		REM ELEC	KWH	0.78	0.70	0.79	0.78	0.92	1.09	1.24	1.21	0.95	0.84	0.72	0.78	10.8
		REM DEM	KW	.00322	.00322	.00316	.00411	.00477	.00533	.00549	.00555	.00522	.00494	.00322	.00322	0.00555
		TOT ELEC	KWH	1.94	1.59	1.44	1.09	1.05	1.19	1.32	1.28	1.05	1.09	1.35	1.79	16.2
		TOT DEM	KW	.01190	.00988	.00833	.00619	.00607	.00678	.00655	.00666	.00631	.00655	.00797	.01190	0.01190
S- 15	1.19	HT ELEC	KWH	1.22	0.93	0.68	0.33	0.14	0.10	0.08	0.07	0.10	0.26	0.65	1.06	5.6
		HT DEM	KW	.00908	.00863	.00772	.00445	.00309	.00209	.00072	.00114	.00281	.00463	.00645	.00908	0.00908
		REM ELEC	KWH	0.68	0.62	0.69	0.68	0.81	0.96	1.09	1.07	0.84	0.74	0.64	0.68	9.5
		REM DEM	KW	.00292	.00292	.00287	.00372	.00433	.00483	.00498	.00503	.00473	.00448	.00292	.00292	0.00503
		TOT ELEC	KWH	1.90	1.55	1.37	1.01	0.95	1.06	1.17	1.14	0.94	1.01	1.29	1.74	15.1
		TOT DEM	KW	.01140	.00958	.00809	.00581	.00547	.00604	.00581	.00604	.00559	.00616	.00764	.01140	0.01140

(HT ELEC=ELECTRIC CONSUMPTION FOR SPACE AND WATER HEATING. REM ELEC=REMAINING ELECTRIC CONSUMPTION (NOT USED FOR HEATING). TOT ELEC=TOTAL ELECTRIC CONSUMPTION. HT DEM=PEAK ELECTRIC DEMAND FOR SPACE AND WATER HEATING. REM DEM=PEAK ELECTRIC DEMAND FOR REMAINING LOAD (NOT HEATING). TOT DEM=TOTAL PEAK ELECTRIC DEMAND. GAS=TOTAL GAS CONSUMPTION. ANNUAL FIGURES ARE SUMS OF MONTHLY CONSUMPTION VALUES OR ARE YEARLY MAXIMUMS OF MONTHLY PEAK DEMAND VALUES.)

STRAT SET	CONST COST $/SF	ENERGY FORM	UNITS (PER SF)	JAN	FEB	MAR	APR	MAY	JUN	JUL	AUG	SEP	OCT	NOV	DEC	ANNUAL
								BASE CASE : 109 (CONTINUED)								
S- 31	0.24	HT ELEC	KWH	1.02	0.81	0.77	0.51	0.33	0.24	0.19	0.19	0.27	0.42	0.68	0.95	6.4
		HT DEM	KW	.00908	.00863	.00772	.00445	.00309	.00209	.00072	.00114	.00281	.00463	.00645	.00908	0.00908
		REM ELEC	KWH	0.78	0.68	0.79	0.76	0.80	1.10	1.18	1.20	0.92	0.80	0.72	0.80	10.5
		REM DEM	KW	.00319	.00319	.00313	.00407	.00473	.00528	.00545	.00550	.00517	.00489	.00319	.00319	0.00550
		TOT ELEC	KWH	1.80	1.49	1.55	1.27	1.12	1.34	1.36	1.39	1.20	1.22	1.40	1.75	16.9
		TOT DEM	KW	.01240	.01042	.00880	.00632	.00595	.00657	.00632	.00657	.00608	.00670	.00831	.01240	0.01240
S- 39	0.95	HT ELEC	KWH	1.09	0.87	0.82	0.55	0.35	0.25	0.20	0.20	0.29	0.45	0.72	1.02	6.8
		HT DEM	KW	.00908	.00863	.00772	.00445	.00309	.00209	.00072	.00114	.00281	.00463	.00645	.00908	0.00908
		REM ELEC	KWH	0.71	0.63	0.72	0.69	0.73	1.01	1.08	1.10	0.85	0.73	0.67	0.73	9.6
		REM DEM	KW	.00300	.00300	.00295	.00383	.00445	.00497	.00513	.00518	.00487	.00461	.00300	.00300	0.00518
		TOT ELEC	KWH	1.81	1.49	1.54	1.24	1.08	1.27	1.28	1.31	1.14	1.18	1.39	1.75	16.5
		TOT DEM	KW	.01200	.01008	.00852	.00612	.00576	.00636	.00612	.00636	.00588	.00648	.00804	.01200	0.01200
S- 44	0.29	HT ELEC	KWH	1.05	0.83	0.78	0.52	0.33	0.24	0.19	0.20	0.28	0.43	0.69	0.97	6.5
		HT DEM	KW	.00908	.00863	.00772	.00445	.00309	.00209	.00072	.00114	.00281	.00463	.00645	.00908	0.00908
		REM ELEC	KWH	0.75	0.66	0.76	0.73	0.77	1.06	1.13	1.15	0.89	0.77	0.70	0.77	10.1
		REM DEM	KW	.00310	.00310	.00305	.00396	.00460	.00514	.00530	.00535	.00503	.00476	.00310	.00310	0.00535
		TOT ELEC	KWH	1.79	1.49	1.54	1.25	1.10	1.30	1.32	1.35	1.17	1.20	1.39	1.74	16.6
		TOT DEM	KW	.01220	.01025	.00866	.00622	.00586	.00647	.00622	.00647	.00598	.00659	.00817	.01220	0.01220
S- 49	1.55	HT ELEC	KWH	1.07	0.85	0.80	0.54	0.34	0.25	0.19	0.20	0.29	0.44	0.71	1.00	6.7
		HT DEM	KW	.00908	.00863	.00772	.00445	.00309	.00209	.00072	.00114	.00281	.00463	.00645	.00908	0.00908
		REM ELEC	KWH	0.78	0.68	0.79	0.76	0.80	1.10	1.18	1.20	0.92	0.80	0.72	0.80	10.5
		REM DEM	KW	.00319	.00319	.00313	.00407	.00473	.00528	.00545	.00550	.00517	.00489	.00319	.00319	0.00550
		TOT ELEC	KWH	1.85	1.53	1.59	1.29	1.14	1.35	1.37	1.40	1.21	1.24	1.43	1.80	17.2
		TOT DEM	KW	.01240	.01042	.00880	.00632	.00595	.00657	.00632	.00657	.00608	.00670	.00831	.01240	0.01240
S- 60	-0.28	HT ELEC	KWH	1.17	0.89	0.65	0.31	0.13	0.10	0.08	0.07	0.10	0.25	0.62	1.01	5.4
		HT DEM	KW	.00908	.00863	.00772	.00445	.00309	.00209	.00072	.00114	.00281	.00463	.00645	.00908	0.00908
		REM ELEC	KWH	0.78	0.70	0.79	0.78	0.92	1.09	1.24	1.21	0.95	0.84	0.72	0.78	10.8
		REM DEM	KW	.00322	.00322	.00316	.00411	.00477	.00533	.00549	.00555	.00522	.00494	.00322	.00322	0.00555
		TOT ELEC	KWH	1.94	1.59	1.44	1.09	1.05	1.19	1.32	1.28	1.05	1.09	1.35	1.79	16.2
		TOT DEM	KW	.01190	.00988	.00833	.00619	.00607	.00678	.00655	.00666	.00631	.00655	.00797	.01190	0.01190
								BASE CASE : 110								
BASE	0.00	TOT GAS	THM	0.0551	0.0419	0.0307	0.0147	0.0063	0.0046	0.0036	0.0033	0.0046	0.0119	0.0295	0.0478	0.254
		TOT ELEC	KWH	0.94	0.84	0.95	0.94	1.11	1.31	1.50	1.46	1.14	1.01	0.87	0.94	13.0
		TOT DEM	KW	.00380	.00380	.00373	.00485	.00563	.00629	.00648	.00655	.00616	.00583	.00380	.00380	0.00655
S- 3	-0.12	TOT GAS	THM	0.0577	0.0439	0.0322	0.0154	0.0066	0.0048	0.0037	0.0035	0.0048	0.0125	0.0309	0.0500	0.266
		TOT ELEC	KWH	0.85	0.77	0.86	0.85	1.00	1.19	1.36	1.32	1.04	0.92	0.79	0.85	11.8
		TOT DEM	KW	.00349	.00349	.00343	.00445	.00518	.00578	.00596	.00602	.00566	.00536	.00349	.00349	0.00602
S- 6	-0.03	TOT GAS	THM	0.0612	0.0465	0.0341	0.0164	0.0071	0.0051	0.0039	0.0037	0.0051	0.0133	0.0327	0.0530	0.282
		TOT ELEC	KWH	0.78	0.70	0.79	0.78	0.92	1.09	1.24	1.21	0.95	0.84	0.72	0.78	10.8
		TOT DEM	KW	.00322	.00322	.00316	.00411	.00477	.00533	.00549	.00555	.00522	.00494	.00322	.00322	0.00555
S- 19	1.44	TOT GAS	THM	0.0521	0.0396	0.0290	0.0139	0.0060	0.0043	0.0034	0.0031	0.0043	0.0113	0.0278	0.0451	0.240
		TOT ELEC	KWH	0.68	0.62	0.69	0.68	0.81	0.96	1.09	1.07	0.84	0.74	0.64	0.68	9.5
		TOT DEM	KW	.00292	.00292	.00287	.00372	.00433	.00483	.00498	.00503	.00473	.00448	.00292	.00292	0.00503
S- 34	0.49	TOT GAS	THM	0.0536	0.0425	0.0402	0.0268	0.0171	0.0124	0.0097	0.0100	0.0144	0.0221	0.0355	0.0499	0.335
		TOT ELEC	KWH	0.78	0.68	0.79	0.76	0.80	1.10	1.18	1.20	0.92	0.80	0.72	0.80	10.5
		TOT DEM	KW	.00319	.00319	.00313	.00407	.00473	.00528	.00545	.00550	.00517	.00489	.00319	.00319	0.00550
S- 42	1.20	TOT GAS	THM	0.0574	0.0456	0.0431	0.0287	0.0183	0.0133	0.0104	0.0108	0.0154	0.0237	0.0381	0.0535	0.359
		TOT ELEC	KWH	0.71	0.63	0.72	0.69	0.73	1.01	1.08	1.10	0.85	0.73	0.67	0.73	9.6
		TOT DEM	KW	.00300	.00300	.00295	.00383	.00445	.00497	.00513	.00518	.00487	.00461	.00300	.00300	0.00518
S- 47	0.54	TOT GAS	THM	0.0549	0.0436	0.0412	0.0274	0.0175	0.0127	0.0099	0.0103	0.0147	0.0226	0.0364	0.0511	0.343
		TOT ELEC	KWH	0.75	0.66	0.76	0.73	0.77	1.06	1.13	1.15	0.89	0.77	0.70	0.77	10.1
		TOT DEM	KW	.00310	.00310	.00305	.00396	.00460	.00514	.00530	.00535	.00503	.00476	.00310	.00310	0.00535
S- 52	1.80	TOT GAS	THM	0.0563	0.0447	0.0422	0.0282	0.0180	0.0130	0.0102	0.0106	0.0151	0.0232	0.0373	0.0524	0.352
		TOT ELEC	KWH	0.78	0.68	0.79	0.76	0.80	1.10	1.18	1.20	0.92	0.80	0.72	0.80	10.5
		TOT DEM	KW	.00319	.00319	.00313	.00407	.00473	.00528	.00545	.00550	.00517	.00489	.00319	.00319	0.00550
S- 60	-0.28	TOT GAS	THM	0.0612	0.0465	0.0341	0.0164	0.0071	0.0051	0.0039	0.0037	0.0051	0.0133	0.0327	0.0530	0.282
		TOT ELEC	KWH	0.78	0.70	0.79	0.78	0.92	1.09	1.24	1.21	0.95	0.84	0.72	0.78	10.8
		TOT DEM	KW	.00322	.00322	.00316	.00411	.00477	.00533	.00549	.00555	.00522	.00494	.00322	.00322	0.00555
								BASE CASE : 111								
BASE	0.00	HT ELEC	KWH	1.07	0.82	0.60	0.29	0.12	0.09	0.07	0.06	0.09	0.23	0.57	0.93	4.9
		HT DEM	KW	.00580	.00580	.00580	.00418	.00232	.00104	.00031	.00054	.00238	.00487	.00580	.00580	0.00580
		REM ELEC	KWH	0.90	0.81	0.91	0.90	1.06	1.26	1.44	1.40	1.10	0.98	0.84	0.90	12.5
		REM DEM	KW	.00354	.00354	.00348	.00425	.00502	.00555	.00572	.00590	.00525	.00490	.00354	.00354	0.00590
		TOT ELEC	KWH	1.97	1.63	1.51	1.19	1.19	1.35	1.51	1.46	1.19	1.21	1.41	1.83	17.5
		TOT DEM	KW	.00885	.00876	.00876	.00487	.00558	.00602	.00619	.00628	.00575	.00540	.00858	.00885	0.00885
S- 3	-0.13	HT ELEC	KWH	1.12	0.85	0.62	0.30	0.13	0.09	0.07	0.07	0.09	0.24	0.60	0.97	5.1
		HT DEM	KW	.00580	.00580	.00580	.00418	.00232	.00104	.00031	.00054	.00238	.00487	.00580	.00580	0.00580
		REM ELEC	KWH	0.81	0.73	0.82	0.81	0.95	1.13	1.29	1.25	0.99	0.87	0.75	0.81	11.2
		REM DEM	KW	.00324	.00324	.00319	.00389	.00459	.00508	.00524	.00540	.00481	.00448	.00324	.00324	0.00540
		TOT ELEC	KWH	1.92	1.58	1.44	1.10	1.08	1.22	1.36	1.32	1.08	1.12	1.35	1.77	16.3
		TOT DEM	KW	.00850	.00842	.00842	.00468	.00536	.00578	.00595	.00604	.00553	.00519	.00825	.00850	0.00850
S- 6	-0.05	HT ELEC	KWH	1.16	0.88	0.64	0.31	0.13	0.10	0.07	0.07	0.10	0.25	0.62	1.00	5.3
		HT DEM	KW	.00580	.00580	.00580	.00418	.00232	.00104	.00031	.00054	.00238	.00487	.00580	.00580	0.00580
		REM ELEC	KWH	0.73	0.66	0.74	0.73	0.86	1.02	1.16	1.13	0.89	0.79	0.68	0.73	10.1
		REM DEM	KW	.00300	.00300	.00295	.00360	.00425	.00470	.00485	.00500	.00445	.00415	.00300	.00300	0.00500
		TOT ELEC	KWH	1.88	1.54	1.38	1.04	0.99	1.12	1.24	1.20	0.98	1.04	1.29	1.73	15.4
		TOT DEM	KW	.00820	.00812	.00779	.00467	.00394	.00443	.00467	.00476	.00443	.00533	.00779	.00820	0.00820

STRAT SET	CONST COST $/SF	ENERGY FORM	UNITS (PER SF)	JAN	FEB	MAR	APR	MAY	JUN	JUL	AUG	SEP	OCT	NOV	DEC	ANNUAL
								BASE CASE : 111 (CONTINUED)								
S- 15	1.07	HT ELEC	KWH	1.21	0.92	0.67	0.32	0.14	0.10	0.08	0.07	0.10	0.26	0.64	1.05	5.6
		HT DEM	KW	.00580	.00580	.00580	.00418	.00232	.00104	.00031	.00054	.00238	.00487	.00580	.00580	0.00580
		REM ELEC	KWH	0.64	0.58	0.65	0.64	0.75	0.90	1.02	0.99	0.78	0.69	0.59	0.64	8.9
		REM DEM	KW	.00271	.00271	.00267	.00325	.00384	.00425	.00438	.00452	.00402	.00375	.00271	.00271	0.00452
		TOT ELEC	KWH	1.85	1.49	1.32	0.96	0.89	1.00	1.10	1.07	0.88	0.95	1.24	1.68	14.4
		TOT DEM	KW	.00785	.00777	.00746	.00447	.00377	.00424	.00447	.00455	.00424	.00510	.00746	.00785	0.00785
S- 31	0.22	HT ELEC	KWH	1.43	1.01	0.78	0.33	0.13	0.05	0.03	0.04	0.10	0.24	0.60	1.19	6.0
		HT DEM	KW	.00580	.00580	.00580	.00418	.00232	.00104	.00031	.00054	.00238	.00487	.00580	.00580	0.00580
		REM ELEC	KWH	0.69	0.62	0.69	0.71	0.74	0.97	1.04	1.07	0.86	0.74	0.66	0.69	9.5
		REM DEM	KW	.00288	.00288	.00283	.00346	.00408	.00451	.00466	.00480	.00427	.00398	.00288	.00288	0.00480
		TOT ELEC	KWH	2.12	1.63	1.47	1.04	0.86	1.01	1.07	1.11	0.96	0.98	1.26	1.89	15.4
		TOT DEM	KW	.00842	.00834	.00800	.00480	.00404	.00455	.00480	.00488	.00455	.00547	.00800	.00842	0.00842
S- 39	0.88	HT ELEC	KWH	1.62	1.15	0.88	0.37	0.14	0.05	0.03	0.04	0.11	0.28	0.68	1.35	6.8
		HT DEM	KW	.00580	.00580	.00580	.00418	.00232	.00104	.00031	.00054	.00238	.00487	.00580	.00580	0.00580
		REM ELEC	KWH	0.61	0.54	0.61	0.63	0.65	0.85	0.92	0.95	0.76	0.65	0.59	0.61	8.4
		REM DEM	KW	.00259	.00259	.00255	.00311	.00367	.00406	.00419	.00432	.00384	.00359	.00259	.00259	0.00432
		TOT ELEC	KWH	2.24	1.70	1.49	1.00	0.80	0.91	0.96	0.99	0.87	0.93	1.26	1.97	15.1
		TOT DEM	KW	.00805	.00797	.00765	.00459	.00386	.00435	.00459	.00467	.00435	.00523	.00765	.00805	0.00805
S- 44	0.26	HT ELEC	KWH	1.51	1.07	0.82	0.35	0.13	0.05	0.03	0.04	0.10	0.26	0.63	1.26	6.3
		HT DEM	KW	.00580	.00580	.00580	.00418	.00232	.00104	.00031	.00054	.00238	.00487	.00580	.00580	0.00580
		REM ELEC	KWH	0.65	0.58	0.65	0.67	0.69	0.91	0.98	1.01	0.81	0.69	0.62	0.65	8.9
		REM DEM	KW	.00273	.00273	.00268	.00328	.00387	.00428	.00441	.00455	.00405	.00378	.00273	.00273	0.00455
		TOT ELEC	KWH	2.16	1.65	1.47	1.01	0.83	0.96	1.01	1.04	0.91	0.95	1.25	1.91	15.2
		TOT DEM	KW	.00825	.00817	.00784	.00470	.00396	.00446	.00470	.00478	.00446	.00536	.00784	.00825	0.00825
S- 49	1.53	HT ELEC	KWH	1.51	1.07	0.82	0.35	0.13	0.05	0.03	0.04	0.10	0.26	0.63	1.26	6.3
		HT DEM	KW	.00580	.00580	.00580	.00418	.00232	.00104	.00031	.00054	.00238	.00487	.00580	.00580	0.00580
		REM ELEC	KWH	0.64	0.57	0.64	0.65	0.68	0.89	0.96	0.98	0.79	0.68	0.61	0.64	8.7
		REM DEM	KW	.00288	.00288	.00283	.00346	.00408	.00451	.00466	.00480	.00427	.00398	.00288	.00288	0.00480
		TOT ELEC	KWH	2.15	1.64	1.45	1.00	0.81	0.94	0.99	1.02	0.89	0.94	1.24	1.89	15.0
		TOT DEM	KW	.00842	.00834	.00800	.00480	.00404	.00455	.00480	.00488	.00455	.00547	.00800	.00842	0.00842
S- 60	-0.30	HT ELEC	KWH	1.16	0.88	0.64	0.31	0.13	0.10	0.07	0.07	0.10	0.25	0.62	1.00	5.3
		HT DEM	KW	.00580	.00580	.00580	.00418	.00232	.00104	.00031	.00054	.00238	.00487	.00580	.00580	0.00580
		REM ELEC	KWH	0.73	0.66	0.74	0.73	0.86	1.02	1.16	1.13	0.89	0.79	0.68	0.73	10.1
		REM DEM	KW	.00300	.00300	.00295	.00360	.00425	.00470	.00485	.00500	.00445	.00415	.00300	.00300	0.00500
		TOT ELEC	KWH	1.88	1.54	1.38	1.04	0.99	1.12	1.24	1.20	0.98	1.04	1.29	1.73	15.4
		TOT DEM	KW	.00820	.00812	.00779	.00467	.00394	.00443	.00467	.00476	.00443	.00533	.00779	.00820	0.00820
								BASE CASE : 112								
BASE	0.00	TOT GAS	THM	0.0564	0.0429	0.0315	0.0151	0.0065	0.0047	0.0036	0.0034	0.0047	0.0122	0.0302	0.0489	0.260
		TOT ELEC	KWH	0.90	0.81	0.91	0.90	1.06	1.26	1.44	1.40	1.10	0.98	0.84	0.90	12.5
		TOT DEM	KW	.00360	.00360	.00354	.00431	.00490	.00549	.00572	.00590	.00543	.00496	.00360	.00389	0.00590
S- 3	-0.14	TOT GAS	THM	0.0586	0.0446	0.0327	0.0157	0.0068	0.0049	0.0038	0.0035	0.0049	0.0127	0.0313	0.0508	0.270
		TOT ELEC	KWH	0.81	0.73	0.82	0.81	0.95	1.13	1.29	1.25	0.99	0.87	0.75	0.81	11.2
		TOT DEM	KW	.00329	.00329	.00324	.00394	.00448	.00502	.00524	.00540	.00497	.00454	.00329	.00356	0.00540
S- 6	-0.06	TOT GAS	THM	0.0608	0.0462	0.0339	0.0162	0.0070	0.0050	0.0039	0.0036	0.0050	0.0132	0.0325	0.0526	0.280
		TOT ELEC	KWH	0.73	0.66	0.74	0.73	0.86	1.02	1.16	1.13	0.89	0.79	0.68	0.73	10.1
		TOT DEM	KW	.00305	.00305	.00300	.00365	.00415	.00465	.00485	.00500	.00460	.00420	.00305	.00330	0.00500
S- 19	1.31	TOT GAS	THM	0.0514	0.0391	0.0287	0.0137	0.0059	0.0043	0.0033	0.0031	0.0043	0.0111	0.0275	0.0446	0.237
		TOT ELEC	KWH	0.64	0.58	0.65	0.64	0.75	0.90	1.02	0.99	0.78	0.69	0.59	0.64	8.9
		TOT DEM	KW	.00276	.00276	.00271	.00330	.00375	.00420	.00438	.00452	.00416	.00380	.00276	.00298	0.00452
S- 34	0.47	TOT GAS	THM	0.0751	0.0532	0.0407	0.0172	0.0066	0.0025	0.0016	0.0019	0.0050	0.0128	0.0313	0.0626	0.313
		TOT ELEC	KWH	0.69	0.62	0.69	0.71	0.74	0.97	1.04	1.07	0.86	0.74	0.66	0.69	9.5
		TOT DEM	KW	.00293	.00293	.00288	.00350	.00398	.00446	.00466	.00480	.00442	.00403	.00293	.00317	0.00480
S- 42	1.12	TOT GAS	THM	0.0852	0.0604	0.0461	0.0195	0.0075	0.0028	0.0018	0.0021	0.0057	0.0146	0.0355	0.0710	0.355
		TOT ELEC	KWH	0.61	0.54	0.61	0.63	0.65	0.85	0.92	0.95	0.76	0.65	0.59	0.61	8.4
		TOT DEM	KW	.00264	.00264	.00259	.00315	.00359	.00402	.00419	.00432	.00397	.00363	.00264	.00285	0.00432
S- 47	0.51	TOT GAS	THM	0.0794	0.0563	0.0430	0.0182	0.0070	0.0026	0.0017	0.0020	0.0053	0.0136	0.0331	0.0662	0.331
		TOT ELEC	KWH	0.65	0.58	0.65	0.67	0.69	0.91	0.98	1.01	0.81	0.69	0.62	0.65	8.9
		TOT DEM	KW	.00278	.00278	.00273	.00332	.00378	.00423	.00441	.00455	.00419	.00382	.00278	.00300	0.00455
S- 52	1.78	TOT GAS	THM	0.0792	0.0561	0.0429	0.0181	0.0069	0.0026	0.0016	0.0020	0.0053	0.0135	0.0330	0.0660	0.330
		TOT ELEC	KWH	0.64	0.57	0.64	0.65	0.68	0.89	0.96	0.98	0.79	0.68	0.61	0.64	8.7
		TOT DEM	KW	.00293	.00293	.00288	.00350	.00398	.00446	.00466	.00480	.00442	.00403	.00293	.00317	0.00480
S- 60	-0.30	TOT GAS	THM	0.0608	0.0462	0.0339	0.0162	0.0070	0.0050	0.0039	0.0036	0.0050	0.0132	0.0325	0.0526	0.280
		TOT ELEC	KWH	0.73	0.66	0.74	0.73	0.86	1.02	1.16	1.13	0.89	0.79	0.68	0.73	10.1
		TOT DEM	KW	.00305	.00305	.00300	.00365	.00415	.00465	.00485	.00500	.00460	.00420	.00305	.00330	0.00500
								BASE CASE : 113								
BASE	0.00	TOT ELEC	KWH	1.77	1.39	1.25	0.98	1.12	1.37	1.50	1.44	1.09	1.06	1.17	1.66	15.8
		TOT DEM	KW	.00414	.00371	.00396	.00426	.00538	.00599	.00612	.00618	.00470	.00433	.00297	.00420	0.00618
S- 3	-0.20	TOT ELEC	KWH	1.67	1.31	1.18	0.92	1.06	1.30	1.42	1.36	1.03	1.00	1.10	1.56	14.9
		TOT DEM	KW	.00413	.00370	.00394	.00425	.00536	.00598	.00610	.00616	.00468	.00431	.00296	.00419	0.00616
S- 9	-0.12	TOT ELEC	KWH	1.53	1.21	1.08	0.85	0.97	1.19	1.30	1.25	0.95	0.92	1.01	1.44	13.7
		TOT DEM	KW	.00362	.00325	.00346	.00373	.00471	.00525	.00536	.00541	.00411	.00379	.00260	.00368	0.00541
S- 18	1.14	TOT ELEC	KWH	1.43	1.13	1.01	0.79	0.91	1.11	1.22	1.16	0.88	0.86	0.95	1.34	12.8
		TOT DEM	KW	.00326	.00292	.00311	.00335	.00423	.00471	.00481	.00486	.00369	.00340	.00233	.00330	0.00486

(HT ELEC=ELECTRIC CONSUMPTION FOR SPACE AND WATER HEATING. REM ELEC=REMAINING ELECTRIC CONSUMPTION (NOT USED FOR HEATING). TOT ELEC=TOTAL ELECTRIC CONSUMPTION. HT DEM=PEAK ELECTRIC DEMAND FOR SPACE AND WATER HEATING. REM DEM=PEAK ELECTRIC DEMAND FOR REMAINING LOAD (NOT HEATING). TOT DEM=TOTAL PEAK ELECTRIC DEMAND. GAS=TOTAL GAS CONSUMPTION. ANNUAL FIGURES ARE SUMS OF MONTHLY CONSUMPTION VALUES OR ARE YEARLY MAXIMUMS OF MONTHLY PEAK DEMAND VALUES.)

STRAT SET	CONST COST $/SF	ENERGY FORM	UNITS (PER SF)	JAN	FEB	MAR	APR	MAY	JUN	JUL	AUG	SEP	OCT	NOV	DEC	ANNUAL
								BASE CASE : 113 (CONTINUED)								
S- 33	0.15	TOT ELEC	KWH	1.57	1.23	1.11	0.87	0.99	1.22	1.33	1.27	0.97	0.94	1.04	1.47	14.0
		TOT DEM	KW	.00411	.00368	.00393	.00424	.00534	.00596	.00608	.00614	.00467	.00430	.00295	.00418	0.00614
S- 41	0.88	TOT ELEC	KWH	1.53	1.21	1.08	0.85	0.97	1.19	1.30	1.25	0.95	0.92	1.01	1.44	13.7
		TOT DEM	KW	.00392	.00351	.00374	.00404	.00509	.00567	.00579	.00585	.00445	.00409	.00281	.00398	0.00585
S- 46	0.20	TOT ELEC	KWH	1.55	1.21	1.09	0.86	0.98	1.20	1.31	1.26	0.95	0.92	1.02	1.45	13.8
		TOT DEM	KW	.00403	.00361	.00385	.00415	.00523	.00583	.00595	.00601	.00457	.00421	.00288	.00409	0.00601
S- 51	1.46	TOT ELEC	KWH	1.52	1.20	1.07	0.84	0.97	1.18	1.29	1.24	0.94	0.91	1.01	1.43	13.6
		TOT DEM	KW	.00411	.00368	.00393	.00424	.00534	.00596	.00608	.00614	.00467	.00430	.00295	.00418	0.00614
S- 62	-0.37	TOT ELEC	KWH	1.53	1.21	1.08	0.85	0.97	1.19	1.30	1.25	0.95	0.92	1.01	1.44	13.7
		TOT DEM	KW	.00362	.00325	.00346	.00373	.00471	.00525	.00536	.00541	.00411	.00379	.00260	.00368	0.00541
								BASE CASE : 114								
BASE	0.00	HT ELEC	KWH	1.24	0.95	0.69	0.33	0.14	0.10	0.08	0.07	0.10	0.27	0.66	1.08	5.7
		HT DEM	KW	.00908	.00863	.00772	.00445	.00309	.00209	.00072	.00114	.00281	.00463	.00645	.00908	0.00908
		REM ELEC	KWH	0.98	0.88	0.99	0.98	1.16	1.37	1.56	1.52	1.20	1.06	0.91	0.98	13.6
		REM DEM	KW	.00428	.00428	.00421	.00546	.00635	.00708	.00731	.00738	.00694	.00657	.00428	.00428	0.00738
		TOT ELEC	KWH	2.22	1.83	1.69	1.31	1.30	1.48	1.64	1.60	1.30	1.33	1.58	2.06	19.3
		TOT DEM	KW	.01390	.01293	.01056	.00806	.00806	.00917	.00903	.00903	.00834	.00862	.01043	.01390	0.01390
S- 3	-0.15	HT ELEC	KWH	1.30	0.99	0.73	0.35	0.15	0.11	0.08	0.08	0.11	0.28	0.70	1.13	6.0
		HT DEM	KW	.00908	.00863	.00772	.00445	.00309	.00209	.00072	.00114	.00281	.00463	.00645	.00908	0.00908
		REM ELEC	KWH	0.89	0.80	0.90	0.89	1.05	1.24	1.41	1.38	1.08	0.96	0.82	0.89	12.3
		REM DEM	KW	.00341	.00341	.00335	.00435	.00506	.00564	.00582	.00588	.00553	.00523	.00341	.00341	0.00588
		TOT ELEC	KWH	2.19	1.79	1.62	1.23	1.20	1.35	1.50	1.46	1.19	1.24	1.52	2.01	18.3
		TOT DEM	KW	.01260	.01172	.00958	.00731	.00731	.00832	.00819	.00819	.00756	.00781	.00945	.01260	0.01260
S- 6	-0.06	HT ELEC	KWH	1.38	1.05	0.77	0.37	0.16	0.11	0.09	0.08	0.11	0.30	0.74	1.20	6.4
		HT DEM	KW	.00908	.00863	.00772	.00445	.00309	.00209	.00072	.00114	.00281	.00463	.00645	.00908	0.00908
		REM ELEC	KWH	0.81	0.73	0.82	0.81	0.96	1.14	1.30	1.27	0.99	0.88	0.76	0.81	11.3
		REM DEM	KW	.00331	.00331	.00325	.00422	.00490	.00547	.00564	.00570	.00536	.00507	.00331	.00331	0.00570
		TOT ELEC	KWH	2.19	1.78	1.59	1.18	1.12	1.26	1.39	1.35	1.11	1.18	1.49	2.01	17.7
		TOT DEM	KW	.01270	.01181	.00965	.00737	.00737	.00838	.00825	.00825	.00762	.00787	.00953	.01270	0.01270
S- 15	1.16	HT ELEC	KWH	1.44	1.10	0.80	0.39	0.17	0.12	0.09	0.09	0.12	0.31	0.77	1.25	6.7
		HT DEM	KW	.00908	.00863	.00772	.00445	.00309	.00209	.00072	.00114	.00281	.00463	.00645	.00908	0.00908
		REM ELEC	KWH	0.72	0.65	0.73	0.72	0.85	1.00	1.14	1.11	0.88	0.78	0.67	0.72	10.0
		REM DEM	KW	.00316	.00316	.00311	.00403	.00469	.00523	.00540	.00545	.00512	.00485	.00316	.00316	0.00545
		TOT ELEC	KWH	2.16	1.74	1.53	1.10	1.01	1.12	1.24	1.20	1.00	1.09	1.44	1.97	16.6
		TOT DEM	KW	.01160	.01137	.00951	.00661	.00638	.00708	.00696	.00696	.00650	.00696	.00893	.01160	0.01160
S- 31	0.21	HT ELEC	KWH	1.05	0.84	0.79	0.53	0.34	0.24	0.19	0.20	0.28	0.43	0.70	0.98	6.6
		HT DEM	KW	.00908	.00863	.00772	.00445	.00309	.00209	.00072	.00114	.00281	.00463	.00645	.00908	0.00908
		REM ELEC	KWH	0.80	0.70	0.81	0.78	0.82	1.13	1.21	1.23	0.95	0.82	0.75	0.82	10.8
		REM DEM	KW	.00360	.00360	.00353	.00459	.00533	.00595	.00614	.00620	.00583	.00552	.00360	.00360	0.00620
		TOT ELEC	KWH	1.85	1.54	1.60	1.30	1.16	1.38	1.40	1.43	1.23	1.26	1.44	1.80	17.4
		TOT DEM	KW	.01260	.01235	.01033	.00718	.00693	.00769	.00756	.00756	.00706	.00756	.00970	.01260	0.01260
S- 39	0.92	HT ELEC	KWH	1.12	0.89	0.84	0.56	0.36	0.26	0.20	0.21	0.30	0.46	0.74	1.04	7.0
		HT DEM	KW	.00908	.00863	.00772	.00445	.00309	.00209	.00072	.00114	.00281	.00463	.00645	.00908	0.00908
		REM ELEC	KWH	0.75	0.66	0.76	0.73	0.77	1.06	1.13	1.15	0.89	0.77	0.70	0.77	10.1
		REM DEM	KW	.00344	.00344	.00338	.00439	.00510	.00569	.00587	.00593	.00557	.00528	.00344	.00344	0.00593
		TOT ELEC	KWH	1.86	1.54	1.60	1.29	1.12	1.32	1.33	1.36	1.19	1.23	1.44	1.81	17.1
		TOT DEM	KW	.01220	.01196	.01000	.00695	.00671	.00744	.00732	.00732	.00683	.00732	.00939	.01220	0.01220
S- 44	0.25	HT ELEC	KWH	1.08	0.86	0.81	0.54	0.35	0.25	0.20	0.20	0.29	0.45	0.72	1.01	6.8
		HT DEM	KW	.00908	.00863	.00772	.00445	.00309	.00209	.00072	.00114	.00281	.00463	.00645	.00908	0.00908
		REM ELEC	KWH	0.78	0.68	0.79	0.76	0.80	1.10	1.18	1.20	0.92	0.80	0.72	0.80	10.5
		REM DEM	KW	.00353	.00353	.00347	.00450	.00523	.00584	.00602	.00608	.00572	.00541	.00353	.00353	0.00608
		TOT ELEC	KWH	1.86	1.54	1.60	1.30	1.14	1.35	1.37	1.40	1.22	1.24	1.44	1.81	17.3
		TOT DEM	KW	.01240	.01215	.01017	.00707	.00682	.00756	.00744	.00744	.00694	.00744	.00955	.01240	0.01240
S- 49	1.52	HT ELEC	KWH	1.12	0.89	0.84	0.56	0.36	0.26	0.20	0.21	0.30	0.46	0.74	1.04	7.0
		HT DEM	KW	.00908	.00863	.00772	.00445	.00309	.00209	.00072	.00114	.00281	.00463	.00645	.00908	0.00908
		REM ELEC	KWH	0.73	0.65	0.74	0.71	0.75	1.04	1.11	1.13	0.87	0.75	0.69	0.75	9.9
		REM DEM	KW	.00360	.00360	.00353	.00459	.00533	.00595	.00614	.00620	.00583	.00552	.00360	.00360	0.00620
		TOT ELEC	KWH	1.85	1.53	1.58	1.27	1.11	1.30	1.31	1.34	1.17	1.22	1.43	1.79	16.9
		TOT DEM	KW	.01260	.01235	.01033	.00718	.00693	.00769	.00756	.00756	.00706	.00756	.00970	.01260	0.01260
S- 60	-0.31	HT ELEC	KWH	1.38	1.05	0.77	0.37	0.16	0.11	0.09	0.08	0.11	0.30	0.74	1.20	6.4
		HT DEM	KW	.00908	.00863	.00772	.00445	.00309	.00209	.00072	.00114	.00281	.00463	.00645	.00908	0.00908
		REM ELEC	KWH	0.81	0.73	0.82	0.81	0.96	1.14	1.30	1.27	0.99	0.88	0.76	0.81	11.3
		REM DEM	KW	.00331	.00331	.00325	.00422	.00490	.00547	.00564	.00570	.00536	.00507	.00331	.00331	0.00570
		TOT ELEC	KWH	2.19	1.78	1.59	1.18	1.12	1.26	1.39	1.35	1.11	1.18	1.49	2.01	17.7
		TOT DEM	KW	.01270	.01181	.00965	.00737	.00737	.00838	.00825	.00825	.00762	.00787	.00953	.01270	0.01270
								BASE CASE : 115								
BASE	0.00	TOT GAS	THM	0.0653	0.0497	0.0364	0.0175	0.0075	0.0054	0.0042	0.0039	0.0054	0.0141	0.0349	0.0566	0.301
		TOT ELEC	KWH	0.98	0.88	0.99	0.98	1.16	1.37	1.56	1.52	1.20	1.06	0.91	0.98	13.6
		TOT DEM	KW	.00428	.00428	.00421	.00546	.00635	.00708	.00731	.00738	.00694	.00657	.00428	.00428	0.00738
S- 3	-0.15	TOT GAS	THM	0.0684	0.0520	0.0381	0.0183	0.0079	0.0057	0.0044	0.0041	0.0057	0.0148	0.0365	0.0592	0.315
		TOT ELEC	KWH	0.89	0.80	0.90	0.89	1.05	1.24	1.41	1.38	1.08	0.96	0.82	0.89	12.3
		TOT DEM	KW	.00341	.00341	.00335	.00435	.00506	.00564	.00582	.00588	.00553	.00523	.00341	.00341	0.00588
S- 6	-0.06	TOT GAS	THM	0.0725	0.0551	0.0404	0.0194	0.0083	0.0060	0.0047	0.0043	0.0060	0.0157	0.0387	0.0628	0.334
		TOT ELEC	KWH	0.81	0.73	0.82	0.81	0.96	1.14	1.30	1.27	0.99	0.88	0.76	0.81	11.3
		TOT DEM	KW	.00331	.00331	.00325	.00422	.00490	.00547	.00564	.00570	.00536	.00507	.00331	.00331	0.00570
S- 19	1.41	TOT GAS	THM	0.0616	0.0469	0.0344	0.0165	0.0071	0.0051	0.0040	0.0037	0.0051	0.0133	0.0329	0.0534	0.284
		TOT ELEC	KWH	0.72	0.65	0.73	0.72	0.85	1.00	1.14	1.11	0.88	0.78	0.67	0.72	10.0
		TOT DEM	KW	.00316	.00316	.00311	.00403	.00469	.00523	.00540	.00545	.00512	.00485	.00316	.00316	0.00545
								BASE CASE : 115 (CONTINUED)								
S- 34	0.46	TOT GAS	THM	0.0554	0.0439	0.0415	0.0277	0.0176	0.0128	0.0100	0.0104	0.0149	0.0228	0.0367	0.0516	0.346
		TOT ELEC	KWH	0.80	0.70	0.81	0.78	0.82	1.13	1.21	1.23	0.95	0.82	0.75	0.82	10.8
		TOT DEM	KW	.00360	.00360	.00353	.00459	.00533	.00595	.00614	.00620	.00583	.00552	.00360	.00360	0.00620
S- 42	1.17	TOT GAS	THM	0.0587	0.0466	0.0440	0.0294	0.0187	0.0136	0.0106	0.0110	0.0158	0.0242	0.0389	0.0547	0.367
		TOT ELEC	KWH	0.75	0.66	0.76	0.73	0.77	1.06	1.13	1.15	0.89	0.77	0.70	0.77	10.1
		TOT DEM	KW	.00344	.00344	.00338	.00439	.00510	.00569	.00587	.00593	.00557	.00528	.00344	.00344	0.00593
S- 47	0.51	TOT GAS	THM	0.0568	0.0451	0.0426	0.0284	0.0181	0.0131	0.0103	0.0106	0.0153	0.0234	0.0376	0.0529	0.355
		TOT ELEC	KWH	0.78	0.68	0.79	0.76	0.80	1.10	1.18	1.20	0.92	0.80	0.72	0.80	10.5
		TOT DEM	KW	.00353	.00353	.00347	.00450	.00523	.00584	.00602	.00608	.00572	.00541	.00353	.00353	0.00608
S- 52	1.77	TOT GAS	THM	0.0587	0.0466	0.0440	0.0294	0.0187	0.0136	0.0106	0.0110	0.0158	0.0242	0.0389	0.0547	0.367
		TOT ELEC	KWH	0.73	0.65	0.74	0.71	0.75	1.04	1.11	1.13	0.87	0.75	0.69	0.75	9.9
		TOT DEM	KW	.00360	.00360	.00353	.00459	.00533	.00595	.00614	.00620	.00583	.00552	.00360	.00360	0.00620
S- 60	-0.31	TOT GAS	THM	0.0725	0.0551	0.0404	0.0194	0.0083	0.0060	0.0047	0.0043	0.0060	0.0157	0.0387	0.0628	0.334
		TOT ELEC	KWH	0.81	0.73	0.82	0.81	0.96	1.14	1.30	1.27	0.99	0.88	0.76	0.81	11.3
		TOT DEM	KW	.00331	.00331	.00325	.00422	.00490	.00547	.00564	.00570	.00536	.00507	.00331	.00331	0.00570
								BASE CASE : 116								
BASE	0.00	HT ELEC	KWH	1.22	0.93	0.68	0.33	0.14	0.10	0.08	0.07	0.10	0.26	0.65	1.06	5.6
		HT DEM	KW	.00580	.00580	.00580	.00418	.00232	.00104	.00031	.00054	.00238	.00487	.00580	.00580	0.00580
		REM ELEC	KWH	0.92	0.83	0.93	0.92	1.09	1.29	1.47	1.43	1.13	1.00	0.86	0.92	12.8
		REM DEM	KW	.00369	.00369	.00363	.00443	.00523	.00578	.00597	.00615	.00547	.00510	.00369	.00369	0.00615
		TOT ELEC	KWH	2.14	1.76	1.61	1.25	1.23	1.39	1.55	1.51	1.23	1.26	1.51	1.98	18.4
		TOT DEM	KW	.00850	.00842	.00842	.00468	.00536	.00578	.00595	.00604	.00553	.00519	.00825	.00850	0.00850
S- 3	-0.15	HT ELEC	KWH	1.27	0.97	0.71	0.34	0.15	0.11	0.08	0.08	0.11	0.27	0.68	1.10	5.8
		HT DEM	KW	.00580	.00580	.00580	.00418	.00232	.00104	.00031	.00054	.00238	.00487	.00580	.00580	0.00580
		REM ELEC	KWH	0.82	0.74	0.83	0.82	0.97	1.15	1.31	1.28	1.00	0.89	0.76	0.82	11.4
		REM DEM	KW	.00294	.00294	.00289	.00353	.00417	.00461	.00475	.00490	.00436	.00407	.00294	.00294	0.00490
		TOT ELEC	KWH	2.09	1.71	1.54	1.16	1.12	1.26	1.39	1.35	1.11	1.16	1.44	1.92	17.3
		TOT DEM	KW	.00850	.00842	.00842	.00468	.00536	.00578	.00595	.00604	.00553	.00519	.00825	.00850	0.00850
S- 6	-0.07	HT ELEC	KWH	1.32	1.00	0.73	0.35	0.15	0.11	0.08	0.08	0.11	0.28	0.70	1.14	6.1
		HT DEM	KW	.00580	.00580	.00580	.00418	.00232	.00104	.00031	.00054	.00238	.00487	.00580	.00580	0.00580
		REM ELEC	KWH	0.75	0.68	0.76	0.75	0.88	1.05	1.20	1.16	0.92	0.81	0.70	0.75	10.4
		REM DEM	KW	.00285	.00285	.00280	.00342	.00404	.00446	.00461	.00475	.00423	.00394	.00285	.00285	0.00475
		TOT ELEC	KWH	2.06	1.68	1.49	1.10	1.04	1.16	1.28	1.24	1.02	1.10	1.40	1.89	16.5
		TOT DEM	KW	.00820	.00812	.00779	.00467	.00394	.00443	.00467	.00476	.00443	.00533	.00779	.00820	0.00820
S- 15	1.02	HT ELEC	KWH	1.37	1.04	0.76	0.37	0.16	0.11	0.09	0.08	0.11	0.30	0.73	1.19	6.3
		HT DEM	KW	.00580	.00580	.00580	.00418	.00232	.00104	.00031	.00054	.00238	.00487	.00580	.00580	0.00580
		REM ELEC	KWH	0.65	0.59	0.66	0.65	0.77	0.92	1.05	1.02	0.80	0.71	0.61	0.65	9.1
		REM DEM	KW	.00273	.00273	.00268	.00328	.00387	.00428	.00441	.00455	.00405	.00378	.00273	.00273	0.00455
		TOT ELEC	KWH	2.02	1.63	1.43	1.02	0.93	1.03	1.13	1.10	0.91	1.01	1.34	1.84	15.4
		TOT DEM	KW	.00785	.00777	.00746	.00447	.00377	.00424	.00447	.00455	.00424	.00510	.00746	.00785	0.00785
S- 31	0.20	HT ELEC	KWH	1.84	1.30	0.99	0.42	0.16	0.06	0.04	0.05	0.12	0.31	0.76	1.53	7.7
		HT DEM	KW	.00580	.00580	.00580	.00418	.00232	.00104	.00031	.00054	.00238	.00487	.00580	.00580	0.00580
		REM ELEC	KWH	0.70	0.62	0.70	0.72	0.75	0.98	1.05	1.08	0.87	0.75	0.67	0.70	9.6
		REM DEM	KW	.00294	.00294	.00289	.00353	.00417	.00461	.00475	.00490	.00436	.00407	.00294	.00294	0.00490
		TOT ELEC	KWH	2.54	1.92	1.69	1.14	0.91	1.04	1.09	1.13	0.99	1.06	1.44	2.23	17.2
		TOT DEM	KW	.00845	.00837	.00803	.00482	.00406	.00456	.00482	.00490	.00456	.00549	.00803	.00845	0.00845
S- 39	0.85	HT ELEC	KWH	2.00	1.42	1.08	0.46	0.18	0.07	0.04	0.05	0.13	0.34	0.83	1.67	8.3
		HT DEM	KW	.00580	.00580	.00580	.00418	.00232	.00104	.00031	.00054	.00238	.00487	.00580	.00580	0.00580
		REM ELEC	KWH	0.64	0.57	0.64	0.66	0.68	0.89	0.96	0.99	0.80	0.68	0.61	0.64	8.7
		REM DEM	KW	.00273	.00273	.00268	.00328	.00387	.00428	.00441	.00455	.00405	.00378	.00273	.00273	0.00455
		TOT ELEC	KWH	2.64	1.99	1.72	1.11	0.86	0.96	1.00	1.04	0.93	1.02	1.45	2.31	17.1
		TOT DEM	KW	.00805	.00797	.00765	.00459	.00386	.00435	.00459	.00467	.00435	.00523	.00765	.00805	0.00805
S- 44	0.24	HT ELEC	KWH	1.90	1.35	1.03	0.44	0.17	0.06	0.04	0.05	0.13	0.32	0.79	1.58	7.9
		HT DEM	KW	.00580	.00580	.00580	.00418	.00232	.00104	.00031	.00054	.00238	.00487	.00580	.00580	0.00580
		REM ELEC	KWH	0.67	0.60	0.67	0.69	0.72	0.94	1.02	1.04	0.84	0.72	0.65	0.67	9.2
		REM DEM	KW	.00285	.00285	.00280	.00342	.00404	.00446	.00461	.00475	.00423	.00394	.00285	.00285	0.00475
		TOT ELEC	KWH	2.58	1.95	1.70	1.13	0.89	1.01	1.06	1.09	0.97	1.05	1.44	2.26	17.2
		TOT DEM	KW	.00827	.00819	.00786	.00471	.00397	.00447	.00471	.00480	.00447	.00538	.00786	.00827	0.00827
S- 49	1.51	HT ELEC	KWH	1.93	1.37	1.05	0.44	0.17	0.06	0.04	0.05	0.13	0.33	0.81	1.61	8.1
		HT DEM	KW	.00580	.00580	.00580	.00418	.00232	.00104	.00031	.00054	.00236	.00487	.00580	.00580	0.00580
		REM ELEC	KWH	0.64	0.57	0.64	0.66	0.69	0.90	0.97	1.00	0.80	0.69	0.62	0.64	8.8
		REM DEM	KW	.00294	.00294	.00289	.00353	.00417	.00461	.00475	.00490	.00436	.00407	.00294	.00294	0.00490
		TOT ELEC	KWH	2.58	1.94	1.69	1.10	0.86	0.96	1.01	1.04	0.93	1.02	1.42	2.26	16.9
		TOT DEM	KW	.00845	.00837	.00803	.00482	.00406	.00456	.00482	.00490	.00456	.00549	.00803	.00845	0.00845
S- 60	-0.32	HT ELEC	KWH	1.32	1.00	0.73	0.35	0.15	0.11	0.08	0.08	0.11	0.28	0.70	1.14	6.1
		HT DEM	KW	.00580	.00580	.00580	.00418	.00232	.00104	.00031	.00054	.00238	.00487	.00580	.00580	0.00580
		REM ELEC	KWH	0.75	0.68	0.76	0.75	0.88	1.05	1.20	1.16	0.92	0.81	0.70	0.75	10.4
		REM DEM	KW	.00285	.00285	.00280	.00342	.00404	.00446	.00461	.00475	.00423	.00394	.00285	.00285	0.00475
		TOT ELEC	KWH	2.06	1.68	1.49	1.10	1.04	1.16	1.28	1.24	1.02	1.10	1.40	1.89	16.5
		TOT DEM	KW	.00820	.00812	.00779	.00467	.00394	.00443	.00467	.00476	.00443	.00533	.00779	.00820	0.00820

(HT ELEC=ELECTRIC CONSUMPTION FOR SPACE AND WATER HEATING. REM ELEC=REMAINING ELECTRIC CONSUMPTION (NOT USED FOR HEATING). TOT ELEC=TOTAL ELECTRIC CONSUMPTION. HT DEM=PEAK ELECTRIC DEMAND FOR SPACE AND WATER HEATING. REM DEM=PEAK ELECTRIC DEMAND FOR REMAINING LOAD (NOT HEATING). TOT DEM=TOTAL PEAK ELECTRIC DEMAND. GAS=TOTAL GAS CONSUMPTION. ANNUAL FIGURES ARE SUMS OF MONTHLY CONSUMPTION VALUES OR ARE YEARLY MAXIMUMS OF MONTHLY PEAK DEMAND VALUES.)

STRAT SET	CONST COST $/SF	ENERGY FORM	UNITS (PER SF)	JAN	FEB	MAR	APR	MAY	JUN	JUL	AUG	SEP	OCT	NOV	DEC	ANNUAL
BASE CASE : 117																
BASE	0.00	TOT GAS	THM	0.0640	0.0487	0.0357	0.0171	0.0074	0.0053	0.0041	0.0038	0.0053	0.0139	0.0342	0.0555	0.295
		TOT ELEC	KWH	0.92	0.83	0.93	0.92	1.09	1.29	1.47	1.43	1.13	1.00	0.86	0.92	12.8
		TOT DEM	KW	.00375	.00375	.00369	.00449	.00510	.00572	.00597	.00615	.00566	.00517	.00375	.00406	0.00615
S- 3	-0.16	TOT GAS	THM	0.0666	0.0507	0.0371	0.0178	0.0077	0.0055	0.0043	0.0040	0.0055	0.0144	0.0356	0.0577	0.307
		TOT ELEC	KWH	0.82	0.74	0.83	0.82	0.97	1.15	1.31	1.28	1.00	0.89	0.76	0.82	11.4
		TOT DEM	KW	.00299	.00299	.00294	.00358	.00407	.00456	.00475	.00490	.00451	.00412	.00299	.00323	0.00490
S- 6	-0.08	TOT GAS	THM	0.0690	0.0525	0.0385	0.0184	0.0080	0.0057	0.0045	0.0041	0.0057	0.0149	0.0369	0.0598	0.318
		TOT ELEC	KWH	0.75	0.68	0.76	0.75	0.88	1.05	1.20	1.16	0.92	0.81	0.70	0.75	10.4
		TOT DEM	KW	.00290	.00290	.00285	.00347	.00394	.00442	.00461	.00475	.00437	.00399	.00290	.00314	0.00475
S- 19	1.26	TOT GAS	THM	0.0584	0.0444	0.0325	0.0156	0.0067	0.0048	0.0038	0.0035	0.0048	0.0126	0.0312	0.0506	0.269
		TOT ELEC	KWH	0.65	0.59	0.66	0.65	0.77	0.92	1.05	1.02	0.80	0.71	0.61	0.65	9.1
		TOT DEM	KW	.00278	.00278	.00273	.00332	.00378	.00423	.00441	.00455	.00419	.00382	.00278	.00300	0.00455
S- 34	0.44	TOT GAS	THM	0.0965	0.0683	0.0523	0.0221	0.0084	0.0032	0.0020	0.0024	0.0064	0.0165	0.0402	0.0804	0.402
		TOT ELEC	KWH	0.70	0.62	0.70	0.72	0.75	0.98	1.05	1.08	0.87	0.75	0.67	0.70	9.6
		TOT DEM	KW	.00299	.00299	.00294	.00358	.00407	.00456	.00475	.00490	.00451	.00412	.00299	.00323	0.00490
S- 42	1.09	TOT GAS	THM	0.1051	0.0745	0.0569	0.0241	0.0092	0.0035	0.0022	0.0026	0.0070	0.0180	0.0438	0.0876	0.438
		TOT ELEC	KWH	0.64	0.57	0.64	0.66	0.68	0.89	0.96	0.99	0.80	0.68	0.61	0.64	8.7
		TOT DEM	KW	.00278	.00278	.00273	.00332	.00378	.00423	.00441	.00455	.00419	.00382	.00278	.00300	0.00455
S- 47	0.49	TOT GAS	THM	0.0998	0.0707	0.0541	0.0229	0.0087	0.0033	0.0021	0.0025	0.0067	0.0171	0.0416	0.0832	0.416
		TOT ELEC	KWH	0.67	0.60	0.67	0.69	0.72	0.94	1.02	1.04	0.84	0.72	0.65	0.67	9.2
		TOT DEM	KW	.00290	.00290	.00285	.00347	.00394	.00442	.00461	.00475	.00437	.00399	.00290	.00314	0.00475
S- 52	1.76	TOT GAS	THM	0.1015	0.0719	0.0550	0.0233	0.0089	0.0034	0.0021	0.0025	0.0068	0.0173	0.0423	0.0846	0.423
		TOT ELEC	KWH	0.64	0.57	0.64	0.66	0.69	0.90	0.97	1.00	0.80	0.69	0.62	0.64	8.8
		TOT DEM	KW	.00299	.00299	.00294	.00358	.00407	.00456	.00475	.00490	.00451	.00412	.00299	.00323	0.00490
S- 60	-0.33	TOT GAS	THM	0.0690	0.0525	0.0385	0.0184	0.0080	0.0057	0.0045	0.0041	0.0057	0.0149	0.0369	0.0598	0.318
		TOT ELEC	KWH	0.75	0.68	0.76	0.75	0.88	1.05	1.20	1.16	0.92	0.81	0.70	0.75	10.4
		TOT DEM	KW	.00290	.00290	.00285	.00347	.00394	.00442	.00461	.00475	.00437	.00399	.00290	.00314	0.00475
BASE CASE : 118																
BASE	0.00	TOT ELEC	KWH	1.87	1.47	1.32	1.04	1.19	1.45	1.59	1.52	1.15	1.12	1.24	1.75	16.7
		TOT DEM	KW	.00490	.00439	.00468	.00505	.00637	.00710	.00725	.00732	.00556	.00512	.00351	.00498	0.00732
S- 3	-0.24	TOT ELEC	KWH	1.76	1.38	1.24	0.97	1.11	1.37	1.49	1.43	1.08	1.05	1.16	1.65	15.7
		TOT DEM	KW	.00489	.00438	.00467	.00504	.00635	.00708	.00723	.00730	.00555	.00511	.00350	.00496	0.00730
S- 9	-0.15	TOT ELEC	KWH	1.62	1.28	1.15	0.90	1.03	1.26	1.38	1.32	1.00	0.97	1.07	1.52	14.5
		TOT DEM	KW	.00429	.00385	.00410	.00442	.00558	.00622	.00635	.00641	.00487	.00449	.00308	.00436	0.00641
S- 18	1.10	TOT ELEC	KWH	1.51	1.19	1.07	0.84	0.96	1.17	1.28	1.23	0.93	0.90	1.00	1.42	13.5
		TOT DEM	KW	.00386	.00346	.00369	.00397	.00501	.00559	.00570	.00576	.00438	.00403	.00276	.00392	0.00576
S- 33	0.12	TOT ELEC	KWH	1.75	1.37	1.23	0.97	1.11	1.36	1.48	1.42	1.08	1.05	1.15	1.64	15.6
		TOT DEM	KW	.00488	.00437	.00466	.00502	.00633	.00706	.00721	.00728	.00553	.00510	.00349	.00495	0.00728
S- 41	0.85	TOT ELEC	KWH	1.72	1.36	1.22	0.95	1.09	1.34	1.46	1.40	1.06	1.03	1.14	1.62	15.4
		TOT DEM	KW	.00464	.00416	.00444	.00478	.00603	.00672	.00686	.00693	.00527	.00485	.00333	.00471	0.00693
S- 46	0.16	TOT ELEC	KWH	1.74	1.36	1.22	0.96	1.10	1.35	1.47	1.41	1.07	1.04	1.15	1.63	15.5
		TOT DEM	KW	.00477	.00427	.00456	.00491	.00619	.00691	.00705	.00712	.00541	.00498	.00342	.00484	0.00712
S- 51	1.43	TOT ELEC	KWH	1.70	1.34	1.20	0.94	1.08	1.32	1.44	1.38	1.05	1.02	1.12	1.60	15.2
		TOT DEM	KW	.00488	.00437	.00466	.00502	.00633	.00706	.00721	.00728	.00553	.00510	.00349	.00495	0.00728
S- 62	-0.40	TOT ELEC	KWH	1.62	1.28	1.15	0.90	1.03	1.26	1.38	1.32	1.00	0.97	1.07	1.52	14.5
		TOT DEM	KW	.00429	.00385	.00410	.00442	.00558	.00622	.00635	.00641	.00487	.00449	.00308	.00436	0.00641
BASE CASE : 119																
BASE	0.00	HT ELEC	KWH	0.89	0.68	0.50	0.24	0.10	0.07	0.06	0.05	0.07	0.19	0.48	0.77	4.1
		HT DEM	KW	.00908	.00863	.00772	.00445	.00309	.00209	.00072	.00114	.00281	.00463	.00645	.00908	0.00908
		REM ELEC	KWH	0.94	0.84	0.95	0.94	1.11	1.31	1.50	1.46	1.14	1.01	0.87	0.94	13.0
		REM DEM	KW	.00386	.00386	.00379	.00492	.00572	.00638	.00658	.00665	.00625	.00592	.00386	.00386	0.00665
		TOT ELEC	KWH	1.83	1.52	1.45	1.17	1.21	1.39	1.55	1.51	1.22	1.21	1.35	1.71	17.1
		TOT DEM	KW	.01290	.01071	.00903	.00671	.00658	.00735	.00710	.00722	.00684	.00710	.00864	.01290	0.01290
S- 3	-0.12	HT ELEC	KWH	0.94	0.71	0.52	0.25	0.11	0.08	0.06	0.06	0.08	0.20	0.50	0.81	4.3
		HT DEM	KW	.00908	.00863	.00772	.00445	.00309	.00209	.00072	.00114	.00281	.00463	.00645	.00908	0.00908
		REM ELEC	KWH	0.85	0.77	0.86	0.85	1.00	1.19	1.36	1.32	1.04	0.92	0.79	0.85	11.8
		REM DEM	KW	.00355	.00355	.00349	.00453	.00526	.00588	.00606	.00612	.00575	.00545	.00355	.00355	0.00612
		TOT ELEC	KWH	1.78	1.48	1.38	1.10	1.11	1.27	1.42	1.38	1.12	1.12	1.29	1.66	16.1
		TOT DEM	KW	.01240	.01029	.00868	.00645	.00632	.00707	.00682	.00694	.00657	.00682	.00831	.01240	0.01240
S- 6	-0.04	HT ELEC	KWH	0.99	0.75	0.55	0.27	0.11	0.08	0.06	0.06	0.08	0.21	0.53	0.86	4.6
		HT DEM	KW	.00908	.00863	.00772	.00445	.00309	.00209	.00072	.00114	.00281	.00463	.00645	.00908	0.00908
		REM ELEC	KWH	0.78	0.70	0.79	0.78	0.92	1.09	1.24	1.21	0.95	0.84	0.72	0.78	10.8
		REM DEM	KW	.00327	.00327	.00321	.00417	.00484	.00540	.00557	.00563	.00529	.00501	.00327	.00327	0.00563
		TOT ELEC	KWH	1.77	1.46	1.34	1.04	1.03	1.17	1.31	1.27	1.03	1.06	1.25	1.64	15.4
		TOT DEM	KW	.01190	.00988	.00833	.00619	.00607	.00678	.00655	.00666	.00631	.00655	.00797	.01190	0.01190
S- 15	1.18	HT ELEC	KWH	1.04	0.79	0.58	0.28	0.12	0.09	0.07	0.06	0.09	0.22	0.55	0.90	4.8
		HT DEM	KW	.00908	.00863	.00772	.00445	.00309	.00209	.00072	.00114	.00281	.00463	.00645	.00908	0.00908
		REM ELEC	KWH	0.68	0.62	0.69	0.68	0.81	0.96	1.09	1.07	0.84	0.74	0.64	0.68	9.5
		REM DEM	KW	.00296	.00296	.00291	.00377	.00439	.00490	.00505	.00510	.00479	.00454	.00296	.00296	0.00510
		TOT ELEC	KWH	1.72	1.41	1.27	0.96	0.93	1.05	1.16	1.13	0.92	0.97	1.19	1.58	14.3
		TOT DEM	KW	.01140	.00946	.00798	.00593	.00581	.00650	.00627	.00638	.00604	.00627	.00764	.01140	0.01140
BASE CASE : 119 (CONTINUED)																
S- 31	0.32	HT ELEC	KWH	0.88	0.69	0.66	0.44	0.28	0.20	0.16	0.16	0.24	0.36	0.58	0.82	5.5
		HT DEM	KW	.00908	.00863	.00772	.00445	.00309	.00209	.00072	.00114	.00281	.00463	.00645	.00908	0.00908
		REM ELEC	KWH	0.77	0.68	0.78	0.75	0.79	1.09	1.16	1.19	0.92	0.79	0.72	0.79	10.4
		REM DEM	KW	.00321	.00321	.00315	.00409	.00476	.00531	.00547	.00553	.00520	.00492	.00321	.00321	0.00553
		TOT ELEC	KWH	1.64	1.37	1.44	1.19	1.07	1.29	1.32	1.35	1.15	1.15	1.30	1.61	15.9
		TOT DEM	KW	.01245	.01033	.00872	.00647	.00635	.00710	.00685	.00697	.00660	.00685	.00834	.01245	0.01245
S- 60	-0.29	HT ELEC	KWH	0.99	0.75	0.55	0.27	0.11	0.08	0.06	0.06	0.08	0.21	0.53	0.86	4.6
		HT DEM	KW	.00908	.00863	.00772	.00445	.00309	.00209	.00072	.00114	.00281	.00463	.00645	.00908	0.00908
		REM ELEC	KWH	0.78	0.70	0.79	0.78	0.92	1.09	1.24	1.21	0.95	0.84	0.72	0.78	10.8
		REM DEM	KW	.00327	.00327	.00321	.00417	.00484	.00540	.00557	.00563	.00529	.00501	.00327	.00327	0.00563
		TOT ELEC	KWH	1.77	1.46	1.34	1.04	1.03	1.17	1.31	1.27	1.03	1.06	1.25	1.64	15.4
		TOT DEM	KW	.01190	.00988	.00833	.00619	.00607	.00678	.00655	.00666	.00631	.00655	.00797	.01190	0.01190
BASE CASE : 120																
BASE	0.00	TOT GAS	THM	0.0469	0.0356	0.0261	0.0125	0.0054	0.0039	0.0030	0.0028	0.0039	0.0102	0.0251	0.0406	0.216
		TOT ELEC	KWH	0.94	0.84	0.95	0.94	1.11	1.31	1.50	1.46	1.14	1.01	0.87	0.94	13.0
		TOT DEM	KW	.00386	.00386	.00379	.00492	.00572	.00638	.00658	.00665	.00625	.00592	.00386	.00386	0.00665
S- 3	-0.13	TOT GAS	THM	0.0490	0.0373	0.0273	0.0131	0.0057	0.0041	0.0032	0.0029	0.0041	0.0106	0.0262	0.0425	0.226
		TOT ELEC	KWH	0.85	0.77	0.86	0.85	1.00	1.19	1.36	1.32	1.04	0.92	0.79	0.85	11.8
		TOT DEM	KW	.00355	.00355	.00349	.00453	.00526	.00588	.00606	.00612	.00575	.00545	.00355	.00355	0.00612
S- 6	-0.05	TOT GAS	THM	0.0521	0.0396	0.0290	0.0139	0.0060	0.0043	0.0034	0.0031	0.0043	0.0113	0.0278	0.0451	0.240
		TOT ELEC	KWH	0.78	0.70	0.79	0.78	0.92	1.09	1.24	1.21	0.95	0.84	0.72	0.78	10.8
		TOT DEM	KW	.00327	.00327	.00321	.00417	.00484	.00540	.00557	.00563	.00529	.00501	.00327	.00327	0.00563
S- 19	1.43	TOT GAS	THM	0.0443	0.0337	0.0247	0.0118	0.0051	0.0037	0.0029	0.0027	0.0037	0.0096	0.0237	0.0384	0.204
		TOT ELEC	KWH	0.68	0.62	0.69	0.68	0.81	0.96	1.09	1.07	0.84	0.74	0.64	0.68	9.5
		TOT DEM	KW	.00296	.00296	.00291	.00377	.00439	.00490	.00505	.00510	.00479	.00454	.00296	.00296	0.00510
S- 34	0.58	TOT GAS	THM	0.0459	0.0364	0.0344	0.0230	0.0146	0.0106	0.0083	0.0086	0.0123	0.0189	0.0304	0.0428	0.287
		TOT ELEC	KWH	0.77	0.68	0.78	0.75	0.79	1.09	1.16	1.19	0.92	0.79	0.72	0.79	10.4
		TOT DEM	KW	.00321	.00321	.00315	.00409	.00476	.00531	.00547	.00553	.00520	.00492	.00321	.00321	0.00553
S- 60	-0.39	TOT GAS	THM	0.0521	0.0396	0.0290	0.0139	0.0060	0.0043	0.0034	0.0031	0.0043	0.0113	0.0278	0.0451	0.240
		TOT ELEC	KWH	0.78	0.70	0.79	0.78	0.92	1.09	1.24	1.21	0.95	0.84	0.72	0.78	10.8
		TOT DEM	KW	.00327	.00327	.00321	.00417	.00484	.00540	.00557	.00563	.00529	.00501	.00327	.00327	0.00563
BASE CASE : 121																
BASE	0.00	HT ELEC	KWH	0.94	0.72	0.53	0.25	0.11	0.08	0.06	0.06	0.08	0.20	0.50	0.82	4.3
		HT DEM	KW	.00580	.00580	.00580	.00418	.00232	.00104	.00031	.00054	.00238	.00487	.00580	.00580	0.00580
		REM ELEC	KWH	0.93	0.84	0.94	0.93	1.10	1.30	1.48	1.44	1.14	1.01	0.86	0.93	12.9
		REM DEM	KW	.00365	.00365	.00359	.00438	.00517	.00572	.00590	.00608	.00541	.00505	.00365	.00365	0.00608
		TOT ELEC	KWH	1.87	1.55	1.47	1.18	1.21	1.38	1.54	1.50	1.21	1.21	1.37	1.74	17.2
		TOT DEM	KW	.00885	.00876	.00876	.00487	.00558	.00602	.00619	.00628	.00575	.00540	.00858	.00885	0.00885
S- 3	-0.16	HT ELEC	KWH	0.98	0.75	0.55	0.26	0.11	0.08	0.06	0.06	0.08	0.21	0.52	0.85	4.5
		HT DEM	KW	.00580	.00580	.00580	.00418	.00232	.00104	.00031	.00054	.00238	.00487	.00580	.00580	0.00580
		REM ELEC	KWH	0.83	0.75	0.84	0.83	0.98	1.16	1.32	1.29	1.01	0.90	0.77	0.83	11.5
		REM DEM	KW	.00335	.00335	.00330	.00402	.00475	.00525	.00542	.00559	.00498	.00464	.00335	.00335	0.00559
		TOT ELEC	KWH	1.81	1.49	1.39	1.09	1.09	1.24	1.39	1.35	1.09	1.11	.1.29	1.68	16.0
		TOT DEM	KW	.00850	.00842	.00842	.00468	.00536	.00578	.00595	.00604	.00553	.00519	.00825	.00850	0.00850
S- 6	-0.07	HT ELEC	KWH	1.02	0.77	0.57	0.27	0.12	0.08	0.07	0.06	0.08	0.22	0.54	0.88	4.7
		HT DEM	KW	.00580	.00580	.00580	.00418	.00232	.00104	.00031	.00054	.00238	.00487	.00580	.00580	0.00580
		REM ELEC	KWH	0.76	0.68	0.77	0.76	0.89	1.06	1.21	1.18	0.92	0.82	0.70	0.76	10.5
		REM DEM	KW	.00309	.00309	.00304	.00371	.00438	.00484	.00500	.00515	.00458	.00427	.00309	.00309	0.00515
		TOT ELEC	KWH	1.77	1.46	1.33	1.03	1.01	1.14	1.27	1.24	1.01	1.04	1.25	1.64	15.2
		TOT DEM	KW	.00820	.00812	.00779	.00467	.00394	.00443	.00467	.00476	.00443	.00533	.00779	.00820	0.00820
S- 15	1.02	HT ELEC	KWH	1.06	0.81	0.59	0.28	0.12	0.09	0.07	0.06	0.09	0.23	0.57	0.92	4.9
		HT DEM	KW	.00580	.00580	.00580	.00418	.00232	.00104	.00031	.00054	.00238	.00487	.00580	.00580	0.00580
		REM ELEC	KWH	0.66	0.60	0.67	0.66	0.78	0.93	1.05	1.03	0.81	0.71	0.61	0.66	9.2
		REM DEM	KW	.00280	.00280	.00275	.00336	.00396	.00438	.00452	.00466	.00415	.00387	.00280	.00280	0.00466
		TOT ELEC	KWH	1.72	1.40	1.26	0.94	0.90	1.01	1.12	1.09	0.89	0.94	1.18	1.58	14.0
		TOT DEM	KW	.00785	.00777	.00746	.00447	.00377	.00424	.00447	.00455	.00424	.00510	.00746	.00785	0.00785
S- 31	0.29	HT ELEC	KWH	1.41	1.00	0.76	0.32	0.12	0.05	0.03	0.04	0,09	0.24	0.59	1.17	5.9
		HT DEM	KW	.00580	.00580	.00580	.00418	.00232	.00104	.00031	.00054	.00238	.00487	.00580	.00580	0.00580
		REM ELEC	KWH	0.72	0.64	0.72	0.74	0.76	1.00	1.08	1.11	0.89	0.76	0.69	0.72	9.8
		REM DEM	KW	.00293	.00293	.00288	.00351	.00415	.00459	.00473	.00488	.00434	.00405	.00293	.00293	0.00488
		TOT ELEC	KWH	2.12	1.63	1.48	1.06	0.89	1.05	1.11	1.14	0.99	1.01	1.27	1.89	15.7
		TOT DEM	KW	.00847	.00839	.00805	.00483	.00407	.00457	.00483	.00491	.00457	.00551	.00805	.00847	0.00847
S- 60	-0.32	HT ELEC	KWH	1.02	0.77	0.57	0.27	0.12	0.08	0.07	0.06	0.08	0.22	0.54	0.88	4.7
		HT DEM	KW	.00580	.00580	.00580	.00418	.00232	.00104	.00031	.00054	.00238	.00487	.00580	.00580	0.00580
		REM ELEC	KWH	0.76	0.68	0.77	0.76	0.89	1.06	1.21	1.18	0.92	0.82	0.70	0.76	10.5
		REM DEM	KW	.00309	.00309	.00304	.00371	.00438	.00484	.00500	.00515	.00458	.00427	.00309	.00309	0.00515
		TOT ELEC	KWH	1.77	1.46	1.33	1.03	1.01	1.14	1.27	1.24	1.01	1.04	1.25	1.64	15.2
		TOT DEM	KW	.00820	.00812	.00779	.00467	.00394	.00443	.00467	.00476	.00443	.00533	.00779	.00820	0.00820
BASE CASE : 122																
BASE	0.00	TOT GAS	THM	0.0495	0.0376	0.0276	0.0132	0.0057	0.0041	0.0032	0.0030	0.0041	0.0107	0.0264	0.0429	0.228
		TOT ELEC	KWH	0.93	0.84	0.94	0.93	1.10	1.30	1.48	1.44	1.14	1.01	0.86	0.93	12.9
		TOT DEM	KW	.00371	.00371	.00365	.00444	.00505	.00565	.00590	.00608	.00559	.00511	.00371	.00401	0.00608
S- 3	-0.16	TOT GAS	THM	0.0514	0.0391	0.0287	0.0137	0.0059	0.0043	0.0033	0.0031	0.0043	0.0111	0.0275	0.0446	0.237
		TOT ELEC	KWH	0.83	0.75	0.84	0.83	0.98	1.16	1.32	1.29	1.01	0.90	0.77	0.83	11.5
		TOT DEM	KW	.00341	.00341	.00335	.00408	.00464	.00520	.00542	.00559	.00514	.00470	.00341	.00369	0.00559

(HT ELEC=ELECTRIC CONSUMPTION FOR SPACE AND WATER HEATING. REM ELEC=REMAINING ELECTRIC CONSUMPTION (NOT USED FOR HEATING). TOT ELEC=TOTAL ELECTRIC CONSUMPTION. HT DEM=PEAK ELECTRIC DEMAND FOR SPACE AND WATER HEATING. REM DEM=PEAK ELECTRIC DEMAND FOR REMAINING LOAD (NOT HEATING). TOT DEM=TOTAL PEAK ELECTRIC DEMAND. GAS=TOTAL GAS CONSUMPTION. ANNUAL FIGURES ARE SUMS OF MONTHLY CONSUMPTION VALUES OR ARE YEARLY MAXIMUMS OF MONTHLY PEAK DEMAND VALUES.)

STRAT SET	CONST COST $/SF	ENERGY FORM	UNITS (PER SF)	JAN	FEB	MAR	APR	MAY	JUN	JUL	AUG	SEP	OCT	NOV	DEC	ANNUAL
BASE CASE : 122 (CONTINUED)																
S- 6	-0.08	TOT GAS	THM	0.0534	0.0406	0.0298	0.0143	0.0062	0.0044	0.0034	0.0032	0.0044	0.0116	0.0285	0.0462	0.246
		TOT ELEC	KWH	0.76	0.68	0.77	0.76	0.89	1.06	1.21	1.18	0.92	0.82	0.70	0.76	10.5
		TOT DEM	KW	.00314	.00314	.00309	.00376	.00427	.00479	.00500	.00515	.00474	.00433	.00314	.00340	0.00515
S- 19	1.26	TOT GAS	THM	0.0451	0.0343	0.0252	0.0121	0.0052	0.0037	0.0029	0.0027	0.0037	0.0098	0.0241	0.0391	0.208
		TOT ELEC	KWH	0.66	0.60	0.67	0.66	0.78	0.93	1.05	1.03	0.81	0.71	0.61	0.66	9.2
		TOT DEM	KW	.00284	.00284	.00280	.00340	.00387	.00433	.00452	.00466	.00429	.00391	.00284	.00308	0.00466
S- 34	0.54	TOT GAS	THM	0.0739	0.0524	0.0400	0.0169	0.0065	0.0025	0.0015	0.0018	0.0049	0.0126	0.0308	0.0616	0.308
		TOT ELEC	KWH	0.72	0.64	0.72	0.74	0.76	1.00	1.08	1.11	0.89	0.76	0.69	0.72	9.8
		TOT DEM	KW	.00298	.00298	.00293	.00356	.00405	.00454	.00473	.00488	.00449	.00410	.00298	.00322	0.00488
S- 60	-0.33	TOT GAS	THM	0.0534	0.0406	0.0298	0.0143	0.0062	0.0044	0.0034	0.0032	0.0044	0.0116	0.0285	0.0462	0.246
		TOT ELEC	KWH	0.76	0.68	0.77	0.76	0.89	1.06	1.21	1.18	0.92	0.82	0.70	0.76	10.5
		TOT DEM	KW	.00314	.00314	.00309	.00376	.00427	.00479	.00500	.00515	.00474	.00433	.00314	.00340	0.00515
BASE CASE : 123																
BASE	0.00	TOT ELEC	KWH	1.76	1.38	1.24	0.97	1.11	1.37	1.49	1.43	1.08	1.05	1.16	1.65	15.7
		TOT DEM	KW	.00472	.00422	.00451	.00486	.00612	.00683	.00697	.00704	.00535	.00493	.00338	.00479	0.00704
S- 3	-0.23	TOT ELEC	KWH	1.66	1.30	1.17	0.92	1.05	1.29	1.41	1.35	1.02	0.99	1.10	1.55	14.8
		TOT DEM	KW	.00470	.00421	.00449	.00484	.00611	.00681	.00695	.00702	.00534	.00491	.00337	.00477	0.00702
S- 9	-0.14	TOT ELEC	KWH	1.53	1.21	1.08	0.85	0.97	1.19	1.30	1.25	0.95	0.92	1.01	1.44	13.7
		TOT DEM	KW	.00413	.00370	.00395	.00426	.00537	.00598	.00611	.00617	.00469	.00432	.00296	.00420	0.00617
S- 18	1.11	TOT ELEC	KWH	1.42	1.12	1.00	0.79	0.90	1.10	1.21	1.16	0.88	0.85	0.94	1.33	12.7
		TOT DEM	KW	.00371	.00332	.00355	.00382	.00482	.00537	.00548	.00554	.00421	.00388	.00266	.00377	0.00554
S- 33	0.22	TOT ELEC	KWH	1.60	1.26	1.13	0.89	1.02	1.24	1.36	1.30	0.99	0.96	1.06	1.50	14.3
		TOT DEM	KW	.00452	.00404	.00431	.00465	.00586	.00654	.00667	.00674	.00512	.00472	.00324	.00458	0.00674
S- 62	-0.39	TOT ELEC	KWH	1.53	1.21	1.08	0.85	0.97	1.19	1.30	1.25	0.95	0.92	1.01	1.44	13.7
		TOT DEM	KW	.00413	.00370	.00395	.00426	.00537	.00598	.00611	.00617	.00469	.00432	.00296	.00420	0.00617
BASE CASE : 124																
BASE	0.00	HT ELEC	KWH	1.13	0.86	0.63	0.30	0.13	0.09	0.07	0.07	0.09	0.25	0.61	0.98	5.2
		HT DEM	KW	.00908	.00863	.00772	.00445	.00309	.00209	.00072	.00114	.00281	.00463	.00645	.00908	0.00908
		REM ELEC	KWH	1.00	0.90	1.01	1.00	1.18	1.40	1.60	1.56	1.22	1.08	0.93	1.00	13.9
		REM DEM	KW	.00450	.00450	.00442	.00574	.00667	.00744	.00767	.00775	.00728	.00690	.00450	.00450	0.00775
		TOT ELEC	KWH	2.13	1.76	1.65	1.30	1.31	1.50	1.67	1.62	1.32	1.33	1.54	1.98	19.1
		TOT DEM	KW	.01315	.01223	.00999	.00763	.00763	.00868	.00855	.00855	.00789	.00815	.00986	.01315	0.01315
S- 3	-0.15	HT ELEC	KWH	1.19	0.90	0.66	0.32	0.14	0.10	0.08	0.07	0.10	0.26	0.63	1.03	5.5
		HT DEM	KW	.00908	.00863	.00772	.00445	.00309	.00209	.00072	.00114	.00281	.00463	.00645	.00908	0.00908
		REM ELEC	KWH	0.91	0.82	0.92	0.91	1.07	1.27	1.45	1.41	1.11	0.98	0.84	0.91	12.6
		REM DEM	KW	.00358	.00358	.00352	.00457	.00531	.00592	.00611	.00617	.00580	.00549	.00358	.00358	0.00617
		TOT ELEC	KWH	2.09	1.72	1.58	1.22	1.21	1.37	1.53	1.48	1.21	1.24	1.48	1.94	18.1
		TOT DEM	KW	.01260	.01172	.00958	.00731	.00731	.00832	.00819	.00819	.00756	.00781	.00945	.01260	0.01260
S- 6	-0.06	HT ELEC	KWH	1.26	0.96	0.70	0.34	0.14	0.10	0.08	0.08	0.10	0.27	0.67	1.09	5.8
		HT DEM	KW	.00908	.00863	.00772	.00445	.00309	.00209	.00072	.00114	.00281	.00463	.00645	.00908	0.00908
		REM ELEC	KWH	0.84	0.75	0.85	0.84	0.99	1.17	1.33	1.30	1.02	0.90	0.78	0.84	11.6
		REM DEM	KW	.00347	.00347	.00341	.00443	.00515	.00575	.00593	.00599	.00563	.00533	.00347	.00347	0.00599
		TOT ELEC	KWH	2.09	1.71	1.55	1.17	1.13	1.28	1.42	1.37	1.13	1.18	1.45	1.92	17.4
		TOT DEM	KW	.01220	.01135	.00927	.00708	.00708	.00805	.00793	.00793	.00732	.00756	.00915	.01220	0.01220
S- 19	1.15	HT ELEC	KWH	1.32	1.00	0.73	0.35	0.15	0.11	0.08	0.08	0.11	0.28	0.70	1.14	6.1
		HT DEM	KW	.00908	.00863	.00772	.00445	.00309	.00209	.00072	.00114	.00281	.00463	.00645	.00908	0.00908
		REM ELEC	KWH	0.73	0.66	0.74	0.73	0.87	1.03	1.17	1.14	0.90	0.80	0.68	0.73	10.2
		REM DEM	KW	.00332	.00332	.00327	.00424	.00493	.00550	.00567	.00573	.00539	.00510	.00332	.00332	0.00573
		TOT ELEC	KWH	2.05	1.66	1.48	1.09	1.02	1.14	1.26	1.22	1.01	1.08	1.39	1.87	16.3
		TOT DEM	KW	.01170	.01147	.00959	.00667	.00644	.00714	.00702	.00702	.00655	.00702	.00901	.01170	0.01170
S- 34	0.31	HT ELEC	KWH	0.97	0.77	0.72	0.48	0.31	0.22	0.18	0.18	0.26	0.40	0.64	0.90	6.0
		HT DEM	KW	.00908	.00863	.00772	.00445	.00309	.00209	.00072	.00114	.00281	.00463	.00645	.00908	0.00908
		REM ELEC	KWH	0.80	0.70	0.81	0.78	0.82	1.13	1.21	1.23	0.95	0.82	0.75	0.82	10.8
		REM DEM	KW	.00371	.00371	.00365	.00474	.00550	.00614	.00634	.00640	.00602	.00570	.00371	.00371	0.00640
		TOT ELEC	KWH	1.77	1.47	1.53	1.26	1.13	1.36	1.38	1.41	1.21	1.22	1.39	1.72	16.8
		TOT DEM	KW	.01275	.01249	.01046	.00727	.00701	.00778	.00765	.00765	.00714	.00765	.00982	.01275	0.01275
S- 60	-0.31	HT ELEC	KWH	1.26	0.96	0.70	0.34	0.14	0.10	0.08	0.08	0.10	0.27	0.67	1.09	5.8
		HT DEM	KW	.00908	.00863	.00772	.00445	.00309	.00209	.00072	.00114	.00281	.00463	.00645	.00908	0.00908
		REM ELEC	KWH	0.84	0.75	0.85	0.84	0.99	1.17	1.33	1.30	1.02	0.90	0.78	0.84	11.6
		REM DEM	KW	.00347	.00347	.00341	.00443	.00515	.00575	.00593	.00599	.00563	.00533	.00347	.00347	0.00599
		TOT ELEC	KWH	2.09	1.71	1.55	1.17	1.13	1.28	1.42	1.37	1.13	1.18	1.45	1.92	17.4
		TOT DEM	KW	.01220	.01135	.00927	.00708	.00708	.00805	.00793	.00793	.00732	.00756	.00915	.01220	0.01220
BASE CASE : 125																
BASE	0.00	TOT GAS	THM	0.0595	0.0452	0.0332	0.0159	0.0069	0.0049	0.0038	0.0036	0.0049	0.0129	0.0318	0.0515	0.274
		TOT ELEC	KWH	1.00	0.90	1.01	1.00	1.18	1.40	1.60	1.56	1.22	1.08	0.93	1.00	13.9
		TOT DEM	KW	.00450	.00450	.00442	.00574	.00667	.00744	.00767	.00775	.00728	.00690	.00450	.00450	0.00775
S- 3	-0.15	TOT GAS	THM	0.0623	0.0474	0.0347	0.0166	0.0072	0.0052	0.0040	0.0037	0.0052	0.0135	0.0333	0.0540	0.287
		TOT ELEC	KWH	0.91	0.82	0.92	0.91	1.07	1.27	1.45	1.41	1.11	0.98	0.84	0.91	12.6
		TOT DEM	KW	.00358	.00358	.00352	.00457	.00531	.00592	.00611	.00617	.00580	.00549	.00358	.00358	0.00617
S- 6	-0.07	TOT GAS	THM	0.0660	0.0502	0.0368	0.0176	0.0076	0.0055	0.0043	0.0040	0.0055	0.0143	0.0353	0.0572	0.304
		TOT ELEC	KWH	0.84	0.75	0.85	0.84	0.99	1.17	1.33	1.30	1.02	0.90	0.78	0.84	11.6
		TOT DEM	KW	.00347	.00347	.00341	.00443	.00515	.00575	.00593	.00599	.00563	.00533	.00347	.00347	0.00599
S- 19	1.40	TOT GAS	THM	0.0560	0.0426	0.0312	0.0150	0.0065	0.0046	0.0036	0.0034	0.0046	0.0121	0.0299	0.0485	0.258
		TOT ELEC	KWH	0.73	0.66	0.74	0.73	0.87	1.03	1.17	1.14	0.90	0.80	0.68	0.73	10.2
		TOT DEM	KW	.00332	.00332	.00327	.00424	.00493	.00550	.00567	.00573	.00539	.00510	.00332	.00332	0.00573

STRAT SET	CONST COST $/SF	ENERGY FORM	UNITS (PER SF)	JAN	FEB	MAR	APR	MAY	JUN	JUL	AUG	SEP	OCT	NOV	DEC	ANNUAL
BASE CASE : 125 (CONTINUED)																
S- 34	0.56	TOT GAS	THM	0.0507	0.0403	0.0380	0.0254	0.0162	0.0117	0.0092	0.0095	0.0136	0.0209	0.0336	0.0472	0.317
		TOT ELEC	KWH	0.80	0.70	0.81	0.78	0.82	1.13	1.21	1.23	0.95	0.82	0.75	0.82	10.8
		TOT DEM	KW	.00371	.00371	.00365	.00474	.00550	.00614	.00634	.00640	.00602	.00570	.00371	.00371	0.00640
S- 60	-0.31	TOT GAS	THM	0.0660	0.0502	0.0368	0.0176	0.0076	0.0055	0.0043	0.0040	0.0055	0.0143	0.0353	0.0572	0.304
		TOT ELEC	KWH	0.84	0.75	0.85	0.84	0.99	1.17	1.33	1.30	1.02	0.90	0.78	0.84	11.6
		TOT DEM	KW	.00347	.00347	.00341	.00443	.00515	.00575	.00593	.00599	.00563	.00533	.00347	.00347	0.00599
BASE CASE : 126																
BASE	0.00	HT ELEC	KWH	1.13	0.86	0.63	0.30	0.13	0.09	0.07	0.07	0.09	0.24	0.60	0.98	5.2
		HT DEM	KW	.00580	.00580	.00580	.00418	.00232	.00104	.00031	.00054	.00238	.00487	.00580	.00580	0.00580
		REM ELEC	KWH	0.96	0.87	0.98	0.96	1.14	1.35	1.54	1.50	1.18	1.05	0.90	0.96	13.4
		REM DEM	KW	.00391	.00391	.00385	.00469	.00554	.00613	.00632	.00652	.00580	.00541	.00391	.00391	0.00652
		TOT ELEC	KWH	2.09	1.73	1.61	1.27	1.27	1.45	1.61	1.57	1.27	1.29	1.50	1.94	18.6
		TOT DEM	KW	.00895	.00886	.00886	.00492	.00564	.00609	.00627	.00635	.00582	.00546	.00868	.00895	0.00895
S- 3	-0.19	HT ELEC	KWH	1.17	0.89	0.65	0.31	0.14	0.10	0.08	0.07	0.10	0.25	0.63	1.02	5.4
		HT DEM	KW	.00580	.00580	.00580	.00418	.00232	.00104	.00031	.00054	.00238	.00487	.00580	.00580	0.00580
		REM ELEC	KWH	0.86	0.78	0.88	0.86	1.02	1.21	1.38	1.34	1.06	0.94	0.80	0.86	12.0
		REM DEM	KW	.00311	.00311	.00306	.00374	.00441	.00488	.00503	.00519	.00462	.00431	.00311	.00311	0.00519
		TOT ELEC	KWH	2.04	1.67	1.53	1.18	1.16	1.31	1.46	1.41	1.15	1.19	1.43	1.88	17.4
		TOT DEM	KW	.00860	.00851	.00851	.00473	.00542	.00585	.00602	.00611	.00559	.00525	.00834	.00860	0.00860
S- 6	-0.11	HT ELEC	KWH	1.22	0.92	0.68	0.32	0.14	0.10	0.08	0.07	0.10	0.26	0.65	1.05	5.6
		HT DEM	KW	.00580	.00580	.00580	.00418	.00232	.00104	.00031	.00054	.00238	.00487	.00580	.00580	0.00580
		REM ELEC	KWH	0.78	0.71	0.80	0.78	0.93	1.10	1.25	1.22	0.96	0.85	0.73	0.78	10.9
		REM DEM	KW	.00302	.00302	.00297	.00363	.00428	.00474	.00489	.00504	.00449	.00418	.00302	.00302	0.00504
		TOT ELEC	KWH	2.00	1.63	1.47	1.11	1.07	1.20	1.33	1.29	1.06	1.11	1.38	1.84	16.5
		TOT DEM	KW	.00829	.00821	.00788	.00473	.00398	.00448	.00473	.00481	.00448	.00539	.00788	.00829	0.00829
S- 15	0.95	HT ELEC	KWH	1.27	0.97	0.71	0.34	0.15	0.11	0.08	0.08	0.11	0.27	0.68	1.10	5.8
		HT DEM	KW	.00580	.00580	.00580	.00418	.00232	.00104	.00031	.00054	.00238	.00487	.00580	.00580	0.00580
		REM ELEC	KWH	0.68	0.62	0.69	0.68	0.81	0.96	1.09	1.07	0.84	0.74	0.64	0.68	9.5
		REM DEM	KW	.00290	.00290	.00285	.00348	.00411	.00454	.00469	.00483	.00430	.00401	.00290	.00290	0.00483
		TOT ELEC	KWH	1.95	1.58	1.40	1.02	0.95	1.07	1.18	1.14	0.94	1.02	1.32	1.78	15.4
		TOT DEM	KW	.00794	.00786	.00754	.00453	.00381	.00429	.00453	.00461	.00429	.00516	.00754	.00794	0.00794
S- 31	0.26	HT ELEC	KWH	1.33	0.94	0.72	0.30	0.12	0.04	0.03	0.03	0.09	0.23	0.55	1.11	5.5
		HT DEM	KW	.00580	.00580	.00580	.00418	.00232	.00104	.00031	.00054	.00238	.00487	.00580	.00580	0.00580
		REM ELEC	KWH	0.72	0.64	0.72	0.74	0.76	1.00	1.08	1.11	0.89	0.76	0.69	0.72	9.8
		REM DEM	KW	.00293	.00293	.00288	.00351	.00415	.00459	.00473	.00488	.00434	.00405	.00293	.00293	0.00488
		TOT ELEC	KWH	2.04	1.58	1.43	1.04	0.88	1.04	1.11	1.14	0.98	0.99	1.24	1.82	15.3
		TOT DEM	KW	.00845	.00837	.00803	.00482	.00406	.00456	.00482	.00490	.00456	.00549	.00803	.00845	0.00845
S- 60	-0.35	HT ELEC	KWH	1.22	0.92	0.68	0.32	0.14	0.10	0.08	0.07	0.10	0.26	0.65	1.05	5.6
		HT DEM	KW	.00580	.00580	.00580	.00418	.00232	.00104	.00031	.00054	.00238	.00487	.00580	.00580	0.00580
		REM ELEC	KWH	0.78	0.71	0.80	0.78	0.93	1.10	1.25	1.22	0.96	0.85	0.73	0.78	10.9
		REM DEM	KW	.00302	.00302	.00297	.00363	.00428	.00474	.00489	.00504	.00449	.00418	.00302	.00302	0.00504
		TOT ELEC	KWH	2.00	1.63	1.47	1.11	1.07	1.20	1.33	1.29	1.06	1.11	1.38	1.84	16.5
		TOT DEM	KW	.00829	.00821	.00788	.00473	.00398	.00448	.00473	.00481	.00448	.00539	.00788	.00829	0.00829
BASE CASE : 127																
BASE	0.00	TOT GAS	THM	0.0592	0.0450	0.0330	0.0158	0.0068	0.0049	0.0038	0.0035	0.0049	0.0128	0.0317	0.0513	0.273
		TOT ELEC	KWH	0.96	0.87	0.98	0.96	1.14	1.35	1.54	1.50	1.18	1.05	0.90	0.96	13.4
		TOT DEM	KW	.00398	.00398	.00391	.00476	.00541	.00606	.00632	.00652	.00600	.00548	.00398	.00430	0.00652
S- 3	-0.20	TOT GAS	THM	0.0616	0.0469	0.0344	0.0165	0.0071	0.0051	0.0040	0.0037	0.0051	0.0133	0.0329	0.0534	0.284
		TOT ELEC	KWH	0.86	0.78	0.88	0.86	1.02	1.21	1.38	1.34	1.06	0.94	0.80	0.86	12.0
		TOT DEM	KW	.00317	.00317	.00311	.00379	.00431	.00483	.00503	.00519	.00477	.00436	.00317	.00343	0.00519
S- 6	-0.12	TOT GAS	THM	0.0638	0.0485	0.0356	0.0171	0.0074	0.0053	0.0041	0.0038	0.0053	0.0138	0.0341	0.0553	0.294
		TOT ELEC	KWH	0.78	0.71	0.80	0.78	0.93	1.10	1.25	1.22	0.96	0.85	0.73	0.78	10.9
		TOT DEM	KW	.00307	.00307	.00302	.00368	.00418	.00469	.00489	.00504	.00464	.00423	.00307	.00333	0.00504
S- 19	1.19	TOT GAS	THM	0.0540	0.0411	0.0301	0.0144	0.0062	0.0045	0.0035	0.0032	0.0045	0.0117	0.0289	0.0468	0.249
		TOT ELEC	KWH	0.68	0.62	0.69	0.68	0.81	0.96	1.09	1.07	0.84	0.74	0.64	0.68	9.5
		TOT DEM	KW	.00295	.00295	.00290	.00353	.00401	.00449	.00469	.00483	.00444	.00406	.00295	.00319	0.00483
S- 34	0.51	TOT GAS	THM	0.0696	0.0493	0.0377	0.0159	0.0061	0.0023	0.0014	0.0017	0.0046	0.0119	0.0290	0.0580	0.290
		TOT ELEC	KWH	0.72	0.64	0.72	0.74	0.76	1.00	1.08	1.11	0.89	0.76	0.69	0.72	9.8
		TOT DEM	KW	.00298	.00298	.00293	.00356	.00405	.00454	.00473	.00488	.00449	.00410	.00298	.00322	0.00488
S- 60	-0.36	TOT GAS	THM	0.0638	0.0485	0.0356	0.0171	0.0074	0.0053	0.0041	0.0038	0.0053	0.0138	0.0341	0.0553	0.294
		TOT ELEC	KWH	0.78	0.71	0.80	0.78	0.93	1.10	1.25	1.22	0.96	0.85	0.73	0.78	10.9
		TOT DEM	KW	.00307	.00307	.00302	.00368	.00418	.00469	.00489	.00504	.00464	.00423	.00307	.00333	0.00504
BASE CASE : 128																
BASE	0.00	TOT ELEC	KWH	1.89	1.49	1.34	1.05	1.20	1.47	1.61	1.54	1.17	1.13	1.25	1.77	16.9
		TOT DEM	KW	.00508	.00455	.00485	.00523	.00659	.00735	.00750	.00758	.00576	.00531	.00364	.00515	0.00758
S- 3	-0.29	TOT ELEC	KWH	1.78	1.40	1.26	0.99	1.13	1.38	1.51	1.45	1.10	1.07	1.18	1.67	15.9
		TOT DEM	KW	.00507	.00454	.00484	.00522	.00658	.00733	.00748	.00756	.00575	.00529	.00363	.00514	0.00756
S- 9	-0.19	TOT ELEC	KWH	1.65	1.29	1.16	0.91	1.04	1.28	1.40	1.34	1.01	0.98	1.09	1.54	14.7
		TOT DEM	KW	.00445	.00398	.00425	.00458	.00578	.00644	.00657	.00664	.00505	.00465	.00319	.00452	0.00664
S- 18	1.06	TOT ELEC	KWH	1.53	1.21	1.08	0.85	0.97	1.19	1.30	1.25	0.95	0.92	1.01	1.44	13.7
		TOT DEM	KW	.00399	.00358	.00381	.00411	.00519	.00578	.00590	.00596	.00453	.00417	.00286	.00405	0.00596

(HT ELEC=ELECTRIC CONSUMPTION FOR SPACE AND WATER HEATING. REM ELEC=REMAINING ELECTRIC CONSUMPTION (NOT USED FOR HEATING). TOT ELEC=TOTAL ELECTRIC CONSUMPTION. HT DEM=PEAK ELECTRIC DEMAND FOR SPACE AND WATER HEATING. REM DEM=PEAK ELECTRIC DEMAND FOR REMAINING LOAD (NOT HEATING). TOT DEM=TOTAL PEAK ELECTRIC DEMAND. GAS=TOTAL GAS CONSUMPTION. ANNUAL FIGURES ARE SUMS OF MONTHLY CONSUMPTION VALUES OR ARE YEARLY MAXIMUMS OF MONTHLY PEAK DEMAND VALUES.)

STRAT SET	CONST COST $/SF	ENERGY FORM	UNITS (PER SF)	JAN	FEB	MAR	APR	MAY	JUN	JUL	AUG	SEP	OCT	NOV	DEC	ANNUAL
BASE CASE : 128 (CONTINUED)																
S- 33	0.17	TOT ELEC	KWH	1.56	1.22	1.10	0.86	0.99	1.21	1.32	1.26	0.96	0.93	1.03	1.46	13.9
		TOT DEM	KW	.00480	.00430	.00458	.00494	.00623	.00695	.00709	.00716	.00544	.00501	.00344	.00487	0.00716
S- 62	-0.44	TOT ELEC	KWH	1.65	1.29	1.16	0.91	1.04	1.28	1.40	1.34	1.01	0.98	1.09	1.54	14.7
		TOT DEM	KW	.00445	.00398	.00425	.00458	.00578	.00644	.00657	.00664	.00505	.00465	.00319	.00452	0.00664
BASE CASE : 129																
BASE	0.00	HT ELEC	KWH	0.72	0.55	0.40	0.19	0.08	0.06	0.05	0.04	0.06	0.16	0.39	0.63	3.3
		HT DEM	KW	.00908	.00863	.00772	.00445	.00309	.00209	.00072	.00114	.00261	.00463	.00645	.00908	0.00908
		REM ELEC	KWH	0.98	0.88	0.99	0.98	1.16	1.37	1.56	1.52	1.20	1.06	0.91	0.98	13.6
		REM DEM	KW	.00380	.00380	.00373	.00485	.00563	.00629	.00648	.00655	.00616	.00583	.00380	.00380	0.00655
		TOT ELEC	KWH	1.70	1.43	1.40	1.17	1.24	1.43	1.61	1.57	1.26	1.22	1.30	1.61	16.9
		TOT DEM	KW	.01288	.01069	.00902	.00670	.00657	.00734	.00708	.00721	.00683	.00708	.00863	.01288	0.01288
S- 3	-0.11	HT ELEC	KWH	0.76	0.58	0.42	0.20	0.09	0.06	0.05	0.05	0.06	0.16	0.40	0.66	3.5
		HT DEM	KW	.00908	.00863	.00772	.00445	.00309	.00209	.00072	.00114	.00281	.00463	.00645	.00908	0.00908
		REM ELEC	KWH	0.89	0.80	0.90	0.89	1.05	1.24	1.41	1.38	1.08	0.96	0.82	0.89	12.3
		REM DEM	KW	.00349	.00349	.00343	.00445	.00518	.00578	.00596	.00602	.00566	.00536	.00349	.00349	0.00602
		TOT ELEC	KWH	1.64	1.38	1.32	1.09	1.13	1.31	1.46	1.42	1.15	1.12	1.23	1.54	15.8
		TOT DEM	KW	.01238	.01028	.00867	.00644	.00631	.00706	.00681	.00693	.00656	.00681	.00829	.01238	0.01238
S- 6	-0.03	HT ELEC	KWH	0.80	0.61	0.45	0.21	0.09	0.07	0.05	0.05	0.07	0.17	0.43	0.70	3.7
		HT DEM	KW	.00908	.00863	.00772	.00445	.00309	.00209	.00072	.00114	.00281	.00463	.00645	.00908	0.00908
		REM ELEC	KWH	0.81	0.73	0.82	0.81	0.96	1.14	1.30	1.27	0.99	0.88	0.76	0.81	11.3
		REM DEM	KW	.00322	.00322	.00316	.00411	.00477	.00533	.00549	.00555	.00522	.00494	.00322	.00322	0.00555
		TOT ELEC	KWH	1.62	1.35	1.27	1.03	1.05	1.21	1.35	1.31	1.06	1.06	1.19	1.51	15.0
		TOT DEM	KW	.01193	.00990	.00835	.00620	.00608	.00680	.00656	.00668	.00632	.00656	.00799	.01193	0.01193
S- 15	1.20	HT ELEC	KWH	0.84	0.64	0.47	0.22	0.10	0.07	0.05	0.05	0.07	0.18	0.45	0.73	3.9
		HT DEM	KW	.00908	.00863	.00772	.00445	.00309	.00209	.00072	.00114	.00281	.00463	.00645	.00908	0.00908
		REM ELEC	KWH	0.72	0.65	0.73	0.72	0.85	1.00	1.14	1.11	0.88	0.78	0.67	0.72	10.0
		REM DEM	KW	.00292	.00292	.00287	.00372	.00433	.00483	.00498	.00503	.00473	.00448	.00292	.00292	0.00503
		TOT ELEC	KWH	1.56	1.29	1.19	0.94	0.94	1.07	1.20	1.16	0.95	0.96	1.12	1.44	13.8
		TOT DEM	KW	.01143	.00960	.00812	.00583	.00549	.00606	.00583	.00606	.00560	.00617	.00766	.01143	0.01143
S- 31	0.20	HT ELEC	KWH	0.70	0.56	0.53	0.35	0.22	0.16	0.13	0.13	0.19	0.29	0.47	0.65	4.4
		HT DEM	KW	.00908	.00863	.00772	.00445	.00309	.00209	.00072	.00114	.00281	.00463	.00645	.00908	0.00908
		REM ELEC	KWH	0.81	0.71	0.83	0.79	0.84	1.15	1.23	1.25	0.97	0.84	0.76	0.84	11.0
		REM DEM	KW	.00319	.00319	.00313	.00407	.00473	.00528	.00545	.00550	.00517	.00489	.00319	.00319	0.00550
		TOT ELEC	KWH	1.52	1.27	1.35	1.14	1.06	1.32	1.36	1.39	1.16	1.13	1.22	1.49	15.4
		TOT DEM	KW	.01240	.01042	.00880	.00632	.00595	.00657	.00632	.00657	.00608	.00670	.00831	.01240	0.01240
S- 39	1.00	HT ELEC	KWH	0.75	0.60	0.56	0.38	0.24	0.17	0.14	0.14	0.20	0.31	0.50	0.70	4.7
		HT DEM	KW	.00908	.00863	.00772	.00445	.00309	.00209	.00072	.00114	.00281	.00463	.00645	.00908	0.00908
		REM ELEC	KWH	0.75	0.66	0.76	0.73	0.77	1.06	1.13	1.15	0.89	0.77	0.70	0.77	10.1
		REM DEM	KW	.00300	.00300	.00295	.00383	.00445	.00497	.00513	.00518	.00487	.00461	.00300	.00300	0.00518
		TOT ELEC	KWH	1.50	1.25	1.32	1.10	1.01	1.23	1.27	1.29	1.09	1.08	1.20	1.47	14.8
		TOT DEM	KW	.01200	.01008	.00852	.00612	.00576	.00636	.00612	.00636	.00588	.00648	.00804	.01200	0.01200
S- 44	0.23	HT ELEC	KWH	0.72	0.57	0.54	0.36	0.23	0.17	0.13	0.13	0.19	0.30	0.48	0.67	4.5
		HT DEM	KW	.00908	.00863	.00772	.00445	.00309	.00209	.00072	.00114	.00281	.00463	.00645	.00908	0.00908
		REM ELEC	KWH	0.78	0.69	0.80	0.76	0.81	1.11	1.19	1.21	0.93	0.81	0.73	0.81	10.6
		REM DEM	KW	.00310	.00310	.00305	.00396	.00460	.00514	.00530	.00535	.00503	.00476	.00310	.00310	0.00535
		TOT ELEC	KWH	1.50	1.26	1.34	1.12	1.04	1.28	1.32	1.34	1.13	1.10	1.21	1.48	15.1
		TOT DEM	KW	.01220	.01025	.00866	.00622	.00586	.00647	.00622	.00647	.00598	.00659	.00817	.01220	0.01220
S- 60	-0.28	HT ELEC	KWH	0.80	0.61	0.45	0.21	0.09	0.07	0.05	0.05	0.07	0.17	0.43	0.70	3.7
		HT DEM	KW	.00908	.00863	.00772	.00445	.00309	.00209	.00072	.00114	.00281	.00463	.00645	.00908	0.00908
		REM ELEC	KWH	0.81	0.73	0.82	0.81	0.96	1.14	1.30	1.27	0.99	0.88	0.76	0.81	11.3
		REM DEM	KW	.00322	.00322	.00316	.00411	.00477	.00533	.00549	.00555	.00522	.00494	.00322	.00322	0.00555
		TOT ELEC	KWH	1.62	1.35	1.27	1.03	1.05	1.21	1.35	1.31	1.06	1.06	1.19	1.51	15.0
		TOT DEM	KW	.01193	.00990	.00835	.00620	.00608	.00680	.00656	.00668	.00632	.00656	.00799	.01193	0.01193
BASE CASE : 130																
BASE	0.00	TOT GAS	THM	0.0380	0.0289	0.0212	0.0101	0.0044	0.0031	0.0025	0.0023	0.0031	0.0082	0.0203	0.0329	0.175
		TOT ELEC	KWH	0.98	0.88	0.99	0.98	1.16	1.37	1.56	1.52	1.20	1.06	0.91	0.98	13.6
		TOT DEM	KW	.00380	.00380	.00373	.00485	.00563	.00629	.00648	.00655	.00616	.00583	.00380	.00380	0.00655
S- 3	-0.12	TOT GAS	THM	0.0397	0.0302	0.0221	0.0106	0.0046	0.0033	0.0026	0.0024	0.0033	0.0086	0.0212	0.0344	0.183
		TOT ELEC	KWH	0.89	0.80	0.90	0.89	1.05	1.24	1.41	1.38	1.08	0.96	0.82	0.89	12.3
		TOT DEM	KW	.00349	.00349	.00343	.00445	.00518	.00578	.00596	.00602	.00566	.00536	.00349	.00349	0.00602
S- 6	-0.03	TOT GAS	THM	0.0421	0.0320	0.0235	0.0113	0.0049	0.0035	0.0027	0.0025	0.0035	0.0091	0.0225	0.0365	0.194
		TOT ELEC	KWH	0.81	0.73	0.82	0.81	0.96	1.14	1.30	1.27	0.99	0.88	0.76	0.81	11.3
		TOT DEM	KW	.00322	.00322	.00316	.00411	.00477	.00533	.00549	.00555	.00522	.00494	.00322	.00322	0.00555
S- 15	1.19	TOT GAS	THM	0.0441	0.0335	0.0246	0.0118	0.0051	0.0037	0.0028	0.0026	0.0037	0.0095	0.0235	0.0382	0.203
		TOT ELEC	KWH	0.72	0.65	0.73	0.72	0.85	1.00	1.14	1.11	0.88	0.78	0.67	0.72	10.0
		TOT DEM	KW	.00292	.00292	.00287	.00372	.00433	.00483	.00498	.00503	.00473	.00448	.00292	.00292	0.00503
S- 31	0.20	TOT GAS	THM	0.0336	0.0267	0.0252	0.0168	0.0107	0.0078	0.0061	0.0063	0.0090	0.0139	0.0223	0.0313	0.210
		TOT ELEC	KWH	0.81	0.71	0.83	0.79	0.84	1.15	1.23	1.25	0.97	0.84	0.76	0.84	11.0
		TOT DEM	KW	.00319	.00319	.00313	.00407	.00473	.00528	.00545	.00550	.00517	.00489	.00319	.00319	0.00550
S- 39	0.99	TOT GAS	THM	0.0395	0.0314	0.0296	0.0198	0.0126	0.0091	0.0072	0.0074	0.0106	0.0163	0.0262	0.0368	0.247
		TOT ELEC	KWH	0.75	0.66	0.76	0.73	0.77	1.06	1.13	1.15	0.89	0.77	0.70	0.77	10.1
		TOT DEM	KW	.00300	.00300	.00295	.00383	.00445	.00497	.00513	.00518	.00487	.00461	.00300	.00300	0.00518
S- 44	0.22	TOT GAS	THM	0.0378	0.0300	0.0283	0.0189	0.0120	0.0087	0.0068	0.0071	0.0101	0.0156	0.0250	0.0352	0.236
		TOT ELEC	KWH	0.78	0.69	0.80	0.76	0.81	1.11	1.19	1.21	0.93	0.81	0.73	0.81	10.6
		TOT DEM	KW	.00310	.00310	.00305	.00396	.00460	.00514	.00530	.00535	.00503	.00476	.00310	.00310	0.00535
S- 60	-0.28	TOT GAS	THM	0.0421	0.0320	0.0235	0.0113	0.0049	0.0035	0.0027	0.0025	0.0035	0.0091	0.0225	0.0365	0.194
		TOT ELEC	KWH	0.81	0.73	0.82	0.81	0.96	1.14	1.30	1.27	0.99	0.88	0.76	0.81	11.3
		TOT DEM	KW	.00322	.00322	.00316	.00411	.00477	.00533	.00549	.00555	.00522	.00494	.00322	.00322	0.00555

STRAT SET	CONST COST $/SF	ENERGY FORM	UNITS (PER SF)	JAN	FEB	MAR	APR	MAY	JUN	JUL	AUG	SEP	OCT	NOV	DEC	ANNUAL
BASE CASE : 131																
BASE	0.00	HT ELEC	KWH	0.74	0.56	0.41	0.20	0.09	0.06	0.05	0.04	0.06	0.16	0.40	0.64	3.4
		HT DEM	KW	.00580	.00580	.00580	.00418	.00232	.00104	.00031	.00054	.00238	.00487	.00580	.00580	0.00580
		REM ELEC	KWH	0.95	0.86	0.96	0.95	1.12	1.33	1.52	1.48	1.16	1.03	0.88	0.95	13.2
		REM DEM	KW	.00354	.00354	.00348	.00425	.00502	.00555	.00572	.00590	.00525	.00490	.00354	.00354	0.00590
		TOT ELEC	KWH	1.69	1.42	1.38	1.15	1.21	1.39	1.57	1.52	1.22	1.19	1.28	1.59	16.6
		TOT DEM	KW	.00885	.00876	.00876	.00487	.00558	.00602	.00619	.00628	.00575	.00540	.00858	.00885	0.00885
S- 3	-0.15	HT ELEC	KWH	0.77	0.58	0.43	0.21	0.09	0.06	0.05	0.05	0.06	0.17	0.41	0.67	3.5
		HT DEM	KW	.00580	.00580	.00580	.00418	.00232	.00104	.00031	.00054	.00238	.00487	.00580	.00580	0.00580
		REM ELEC	KWH	0.85	0.77	0.86	0.85	1.00	1.19	1.36	1.32	1.04	0.92	0.79	0.85	11.8
		REM DEM	KW	.00325	.00325	.00320	.00390	.00461	.00509	.00526	.00542	.00482	.00450	.00325	.00325	0.00542
		TOT ELEC	KWH	1.62	1.35	1.29	1.05	1.09	1.26	1.41	1.37	1.10	1.09	1.20	1.52	15.3
		TOT DEM	KW	.00850	.00842	.00842	.00468	.00536	.00578	.00595	.00604	.00553	.00519	.00825	.00850	0.00850
S- 6	-0.07	HT ELEC	KWH	0.80	0.61	0.45	0.21	0.09	0.07	0.05	0.05	0.07	0.17	0.43	0.69	3.7
		HT DEM	KW	.00580	.00580	.00580	.00418	.00232	.00104	.00031	.00054	.00238	.00487	.00580	.00580	0.00580
		REM ELEC	KWH	0.77	0.70	0.78	0.77	0.91	1.08	1.23	1.20	0.94	0.83	0.72	0.77	10.7
		REM DEM	KW	.00300	.00300	.00295	.00360	.00425	.00470	.00485	.00500	.00445	.00415	.00300	.00300	0.00500
		TOT ELEC	KWH	1.57	1.30	1.23	0.98	1.00	1.15	1.28	1.25	1.01	1.01	1.14	1.46	14.4
		TOT DEM	KW	.00820	.00812	.00779	.00467	.00394	.00443	.00467	.00476	.00443	.00533	.00779	.00820	0.00820
S- 15	1.01	HT ELEC	KWH	0.83	0.63	0.46	0.22	0.10	0.07	0.05	0.05	0.07	0.18	0.44	0.72	3.8
		HT DEM	KW	.00580	.00580	.00580	.00418	.00232	.00104	.00031	.00054	.00238	.00487	.00580	.00580	0.00580
		REM ELEC	KWH	0.67	0.61	0.68	0.67	0.80	0.95	1.08	1.05	0.82	0.73	0.63	0.67	9.4
		REM DEM	KW	.00271	.00271	.00267	.00325	.00384	.00425	.00438	.00452	.00402	.00375	.00271	.00271	0.00452
		TOT ELEC	KWH	1.51	1.24	1.15	0.90	0.89	1.02	1.13	1.10	0.89	0.91	1.07	1.39	13.2
		TOT DEM	KW	.00785	.00777	.00746	.00447	.00377	.00424	.00447	.00455	.00424	.00510	.00746	.00785	0.00785
S- 31	0.16	HT ELEC	KWH	0.99	0.70	0.53	0.23	0.09	0.03	0.02	0.02	0.07	0.17	0.41	0.82	4.1
		HT DEM	KW	.00580	.00580	.00580	.00418	.00232	.00104	.00031	.00054	.00238	.00487	.00580	.00580	0.00580
		REM ELEC	KWH	0.73	0.65	0.73	0.75	0.78	1.02	1.10	1.13	0.91	0.78	0.70	0.73	10.0
		REM DEM	KW	.00288	.00288	.00283	.00346	.00408	.00451	.00466	.00480	.00427	.00398	.00288	.00288	0.00480
		TOT ELEC	KWH	1.72	1.35	1.26	0.98	0.87	1.05	1.12	1.15	0.98	0.95	1.11	1.55	14.1
		TOT DEM	KW	.00842	.00834	.00800	.00480	.00404	.00455	.00480	.00488	.00455	.00547	.00800	.00842	0.00842
S- 39	0.88	HT ELEC	KWH	1.12	0.79	0.61	0.26	0.10	0.04	0.02	0.03	0.07	0.19	0.47	0.93	4.7
		HT DEM	KW	.00580	.00580	.00580	.00418	.00232	.00104	.00031	.00054	.00238	.00487	.00580	.00580	0.00580
		REM ELEC	KWH	0.65	0.58	0.65	0.66	0.69	0.90	0.97	1.00	0.81	0.69	0.62	0.65	8.9
		REM DEM	KW	.00259	.00259	.00255	.00311	.00367	.00406	.00419	.00432	.00384	.00359	.00259	.00259	0.00432
		TOT ELEC	KWH	1.76	1.37	1.25	0.92	0.79	0.94	1.00	1.03	0.88	0.88	1.09	1.58	13.5
		TOT DEM	KW	.00805	.00797	.00765	.00459	.00386	.00435	.00459	.00467	.00435	.00523	.00765	.00805	0.00805
S- 44	0.18	HT ELEC	KWH	1.04	0.74	0.57	0.24	0.09	0.03	0.02	0.03	0.07	0.18	0.44	0.87	4.3
		HT DEM	KW	.00580	.00580	.00580	.00418	.00232	.00104	.00031	.00054	.00238	.00487	.00580	.00580	0.00580
		REM ELEC	KWH	0.69	0.61	0.69	0.70	0.73	0.96	1.03	1.06	0.86	0.73	0.66	0.69	9.4
		REM DEM	KW	.00273	.00273	.00268	.00328	.00387	.00428	.00441	.00455	.00405	.00378	.00273	.00273	0.00455
		TOT ELEC	KWH	1.73	1.35	1.25	0.94	0.82	0.99	1.06	1.09	0.92	0.91	1.09	1.56	13.7
		TOT DEM	KW	.00825	.00817	.00784	.00470	.00396	.00446	.00470	.00478	.00446	.00536	.00784	.00825	0.00825
S- 60	-0.32	HT ELEC	KWH	0.80	0.61	0.45	0.21	0.09	0.07	0.05	0.05	0.07	0.17	0.43	0.69	3.7
		HT DEM	KW	.00580	.00580	.00580	.00418	.00232	.00104	.00031	.00054	.00238	.00487	.00580	.00580	0.00580
		REM ELEC	KWH	0.77	0.70	0.78	0.77	0.91	1.08	1.23	1.20	0.94	0.83	0.72	0.77	10.7
		REM DEM	KW	.00300	.00300	.00295	.00360	.00425	.00470	.00485	.00500	.00445	.00415	.00300	.00300	0.00500
		TOT ELEC	KWH	1.57	1.30	1.23	0.98	1.00	1.15	1.28	1.25	1.01	1.01	1.14	1.46	14.4
		TOT DEM	KW	.00820	.00812	.00779	.00467	.00394	.00443	.00467	.00476	.00443	.00533	.00779	.00820	0.00820
BASE CASE : 132																
BASE	0.00	TOT GAS	THM	0.0398	0.0295	0.0217	0.0104	0.0045	0.0032	0.0025	0.0023	0.0032	0.0084	0.0208	0.0337	0.179
		TOT ELEC	KWH	0.95	0.86	0.96	0.95	1.12	1.33	1.52	1.48	1.16	1.03	0.88	0.95	13.2
		TOT DEM	KW	.00360	.00360	.00354	.00431	.00490	.00549	.00572	.00590	.00543	.00496	.00360	.00389	0.00590
S- 3	-0.17	TOT GAS	THM	0.0404	0.0307	0.0225	0.0108	0.0047	0.0033	0.0026	0.0024	0.0033	0.0087	0.0216	0.0350	0.186
		TOT ELEC	KWH	0.85	0.77	0.86	0.85	1.00	1.19	1.36	1.32	1.04	0.92	0.79	0.85	11.8
		TOT DEM	KW	.00331	.00331	.00325	.00396	.00450	.00504	.00526	.00542	.00499	.00455	.00331	.00358	0.00542
S- 6	-0.09	TOT GAS	THM	0.0419	0.0318	0.0234	0.0112	0.0048	0.0035	0.0027	0.0025	0.0035	0.0091	0.0224	0.0363	0.193
		TOT ELEC	KWH	0.77	0.70	0.78	0.77	0.91	1.08	1.23	1.20	0.94	0.83	0.72	0.77	10.7
		TOT DEM	KW	.00305	.00305	.00300	.00365	.00415	.00465	.00485	.00500	.00460	.00420	.00305	.00330	0.00500
S- 15	0.98	TOT GAS	THM	0.0436	0.0332	0.0243	0.0117	0.0050	0.0036	0.0028	0.0026	0.0036	0.0094	0.0233	0.0378	0.201
		TOT ELEC	KWH	0.67	0.61	0.68	0.67	0.80	0.95	1.08	1.05	0.82	0.73	0.63	0.67	9.4
		TOT DEM	KW	.00276	.00276	.00271	.00330	.00375	.00420	.00438	.00452	.00416	.00380	.00276	.00298	0.00452
S- 31	0.14	TOT GAS	THM	0.0523	0.0371	0.0283	0.0120	0.0046	0.0017	0.0011	0.0013	0.0035	0.0089	0.0218	0.0436	0.218
		TOT ELEC	KWH	0.73	0.65	0.73	0.75	0.78	1.02	1.10	1.13	0.91	0.78	0.70	0.73	10.0
		TOT DEM	KW	.00293	.00293	.00288	.00350	.00398	.00446	.00466	.00480	.00442	.00403	.00293	.00317	0.00480
S- 39	0.86	TOT GAS	THM	0.0588	0.0417	0.0318	0.0135	0.0051	0.0020	0.0012	0.0015	0.0039	0.0100	0.0245	0.0490	0.245
		TOT ELEC	KWH	0.65	0.58	0.65	0.66	0.69	0.90	0.97	1.00	0.81	0.69	0.62	0.65	8.9
		TOT DEM	KW	.00264	.00264	.00259	.00315	.00359	.00402	.00419	.00432	.00397	.00363	.00264	.00285	0.00432
S- 44	0.17	TOT GAS	THM	0.0547	0.0388	0.0296	0.0125	0.0048	0.0018	0.0011	0.0014	0.0036	0.0093	0.0228	0.0456	0.228
		TOT ELEC	KWH	0.69	0.61	0.69	0.71	0.73	0.96	1.03	1.06	0.86	0.73	0.66	0.69	9.4
		TOT DEM	KW	.00278	.00278	.00273	.00332	.00378	.00423	.00441	.00455	.00419	.00382	.00278	.00300	0.00455
S- 60	-0.34	TOT GAS	THM	0.0419	0.0318	0.0234	0.0112	0.0048	0.0035	0.0027	0.0025	0.0035	0.0091	0.0224	0.0363	0.193
		TOT ELEC	KWH	0.77	0.70	0.78	0.77	0.91	1.08	1.23	1.20	0.94	0.83	0.72	0.77	10.7
		TOT DEM	KW	.00305	.00305	.00300	.00365	.00415	.00465	.00485	.00500	.00460	.00420	.00305	.00330	0.00500

(HT ELEC=ELECTRIC CONSUMPTION FOR SPACE AND WATER HEATING. REM ELEC=REMAINING ELECTRIC CONSUMPTION (NOT USED FOR HEATING). TOT ELEC=TOTAL ELECTRIC CONSUMPTION. HT DEM=PEAK ELECTRIC DEMAND FOR SPACE AND WATER HEATING. REM DEM=PEAK ELECTRIC DEMAND FOR REMAINING LOAD (NOT HEATING). TOT DEM=TOTAL PEAK ELECTRIC DEMAND. GAS=TOTAL GAS CONSUMPTION. ANNUAL FIGURES ARE SUMS OF MONTHLY CONSUMPTION VALUES OR ARE YEARLY MAXIMUMS OF MONTHLY PEAK DEMAND VALUES.)

STRAT SET	CONST COST $/SF	ENERGY FORM	UNITS (PER SF)	JAN	FEB	MAR	APR	MAY	JUN	JUL	AUG	SEP	OCT	NOV	DEC	ANNUAL
BASE CASE : 133																
BASE	0.00	TOT ELEC	KWH	1.12	1.52	1.38	1.09	1.09	1.28	1.34	1.34	1.17	1.09	1.12	1.71	16.0
		TOT DEM	KW	.00853	.00785	.00751	.00392	.00444	.00478	.00495	.00503	.00469	.00444	.00717	.00827	0.00853
S- 1	0.09	TOT ELEC	KWH	1.68	1.40	1.26	1.00	1.00	1.18	1.23	1.23	1.07	1.00	1.03	1.57	14.7
		TOT DEM	KW	.00827	.00761	.00728	.00380	.00430	.00463	.00480	.00488	.00455	.00430	.00695	.00802	0.00827
S- 12	0.19	TOT ELEC	KWH	1.50	1.25	1.14	0.90	0.90	1.06	1.11	1.11	0.96	0.90	0.92	1.41	13.2
		TOT DEM	KW	.00785	.00722	.00691	.00361	.00408	.00440	.00455	.00463	.00432	.00408	.00659	.00761	0.00785
S- 22	1.50	TOT ELEC	KWH	1.38	1.15	1.04	0.82	0.82	0.97	1.02	1.02	0.88	0.82	0.85	1.29	12.1
		TOT DEM	KW	.00759	.00698	.00668	.00349	.00395	.00425	.00440	.00448	.00417	.00395	.00638	.00736	0.00759
S- 28	-0.06	TOT ELEC	KWH	1.50	1.25	1.14	0.90	0.90	1.06	1.11	1.11	0.96	0.90	0.92	1.41	13.2
		TOT DEM	KW	.00785	.00722	.00691	.00361	.00408	.00440	.00455	.00463	.00432	.00408	.00659	.00761	0.00785
S- 36	0.42	TOT ELEC	KWH	1.55	1.29	1.17	0.92	0.92	1.09	1.14	1.14	0.99	0.92	0.95	1.46	13.6
		TOT DEM	KW	.00793	.00730	.00698	.00365	.00412	.00444	.00460	.00468	.00436	.00412	.00666	.00769	0.00793
S- 43	1.26	TOT ELEC	KWH	1.42	1.19	1.08	0.85	0.85	1.00	1.05	1.05	0.91	0.85	0.88	1.34	12.5
		TOT DEM	KW	.00765	.00704	.00673	.00352	.00398	.00428	.00444	.00451	.00421	.00398	.00643	.00742	0.00765
S- 48	0.45	TOT ELEC	KWH	1.49	1.24	1.13	0.89	0.89	1.05	1.10	1.10	0.96	0.89	0.92	1.40	13.1
		TOT DEM	KW	.00781	.00719	.00687	.00359	.00406	.00437	.00453	.00461	.00430	.00406	.00656	.00758	0.00781
BASE CASE : 134																
BASE	0.00	TOT GAS	THM	0.0272	0.0200	0.0104	0.0012	0.0011	0.0012	0.0012	0.0012	0.0012	0.0012	0.0037	0.0213	0.091
		TOT ELEC	KWH	1.35	1.12	1.15	1.04	1.04	1.28	1.35	1.35	1.15	1.07	1.04	1.28	14.2
		TOT DEM	KW	.00625	.00550	.00506	.00425	.00481	.00519	.00538	.00544	.00513	.00487	.00475	.00594	0.00625
S- 1	0.09	TOT GAS	THM	0.0339	0.0249	0.0130	0.0015	0.0014	0.0015	0.0015	0.0015	0.0015	0.0015	0.0046	0.0266	0.113
		TOT ELEC	KWH	1.19	0.99	1.01	0.91	0.91	1.13	1.19	1.19	1.01	0.94	0.91	1.13	12.5
		TOT DEM	KW	.00603	.00531	.00488	.00410	.00464	.00500	.00519	.00525	.00494	.00470	.00458	.00573	0.00603
S- 12	0.19	TOT GAS	THM	0.0390	0.0286	0.0150	0.0017	0.0016	0.0017	0.0017	0.0017	0.0017	0.0017	0.0053	0.0305	0.130
		TOT ELEC	KWH	1.02	0.85	0.87	0.78	0.78	0.96	1.02	1.02	0.87	0.80	0.78	0.96	10.7
		TOT DEM	KW	.00575	.00506	.00466	.00391	.00443	.00477	.00495	.00500	.00471	.00448	.00437	.00546	0.00575
S- 22	1.50	TOT GAS	THM	0.0492	0.0361	0.0189	0.0021	0.0021	0.0021	0.0021	0.0021	0.0021	0.0021	0.0066	0.0385	0.164
		TOT ELEC	KWH	0.86	0.71	0.73	0.66	0.66	0.81	0.86	0.86	0.73	0.68	0.66	0.81	9.0
		TOT DEM	KW	.00556	.00489	.00450	.00378	.00428	.00461	.00478	.00484	.00456	.00434	.00423	.00528	0.00556
S- 28	-0.06	TOT GAS	THM	0.0390	0.0286	0.0150	0.0017	0.0016	0.0017	0.0017	0.0017	0.0017	0.0017	0.0053	0.0305	0.130
		TOT ELEC	KWH	1.02	0.85	0.87	0.78	0.78	0.96	1.02	1.02	0.87	0.80	0.78	0.96	10.7
		TOT DEM	KW	.00575	.00506	.00466	.00391	.00443	.00477	.00495	.00500	.00471	.00448	.00437	.00546	0.00575
S- 36	0.42	TOT GAS	THM	0.0426	0.0312	0.0163	0.0018	0.0018	0.0018	0.0018	0.0018	0.0018	0.0018	0.0058	0.0334	0.142
		TOT ELEC	KWH	1.04	0.86	0.88	0.80	0.80	0.98	1.04	1.04	0.88	0.82	0.80	0.98	10.9
		TOT DEM	KW	.00520	.00458	.00421	.00354	.00400	.00432	.00447	.00452	.00426	.00406	.00395	.00494	0.00520
S- 43	1.26	TOT GAS	THM	0.0534	0.0392	0.0205	0.0023	0.0022	0.0023	0.0023	0.0023	0.0023	0.0023	0.0072	0.0418	0.178
		TOT ELEC	KWH	0.95	0.79	0.81	0.73	0.73	0.90	0.95	0.95	0.81	0.75	0.73	0.90	10.0
		TOT DEM	KW	.00500	.00440	.00405	.00340	.00385	.00415	.00430	.00435	.00410	.00390	.00380	.00475	0.00500
S- 48	0.45	TOT GAS	THM	0.0474	0.0348	0.0182	0.0021	0.0020	0.0021	0.0021	0.0021	0.0021	0.0021	0.0064	0.0371	0.158
		TOT ELEC	KWH	1.00	0.83	0.85	0.77	0.77	0.95	1.00	1.00	0.85	0.79	0.77	0.95	10.5
		TOT DEM	KW	.00510	.00449	.00413	.00347	.00393	.00423	.00439	.00444	.00418	.00398	.00388	.00484	0.00510
BASE CASE : 135																
BASE	0.00	HT ELEC	KWH	0.99	0.75	0.55	0.27	0.11	0.08	0.06	0.06	0.08	0.21	0.53	0.86	4.6
		HT DEM	KW	.00908	.00863	.00772	.00445	.00309	.00209	.00072	.00114	.00281	.00463	.00645	.00908	0.00908
		REM ELEC	KWH	1.03	0.93	1.04	1.03	1.22	1.44	1.64	1.60	1.26	1.12	0.96	1.03	14.3
		REM DEM	KW	.00428	.00428	.00421	.00546	.00635	.00708	.00731	.00738	.00694	.00657	.00428	.00428	0.00738
		TOT ELEC	KWH	2.02	1.68	1.60	1.29	1.33	1.53	1.71	1.66	1.34	1.33	1.49	1.89	18.9
		TOT DEM	KW	.01310	.01218	.00996	.00760	.00760	.00865	.00851	.00851	.00786	.00812	.00982	.01310	0.01310
S- 3	-0.14	HT ELEC	KWH	1.04	0.79	0.58	0.28	0.12	0.09	0.07	0.06	0.09	0.22	0.55	0.90	4.8
		HT DEM	KW	.00908	.00863	.00772	.00445	.00309	.00209	.00072	.00114	.00281	.00463	.00645	.00908	0.00908
		REM ELEC	KWH	0.93	0.84	0.94	0.93	1.10	1.30	1.48	1.44	1.14	1.01	0.86	0.93	12.9
		REM DEM	KW	.00341	.00341	.00335	.00435	.00506	.00564	.00582	.00588	.00553	.00523	.00341	.00341	0.00588
		TOT ELEC	KWH	1.97	1.63	1.52	1.21	1.22	1.39	1.55	1.51	1.22	1.23	1.42	1.83	17.7
		TOT DEM	KW	.01260	.01172	.00958	.00731	.00731	.00832	.00819	.00819	.00756	.00781	.00945	.01260	0.01260
S- 6	-0.05	HT ELEC	KWH	1.10	0.84	0.61	0.29	0.13	0.09	0.07	0.07	0.09	0.24	0.59	0.95	5.1
		HT DEM	KW	.00908	.00863	.00772	.00445	.00309	.00209	.00072	.00114	.00281	.00463	.00645	.00908	0.00908
		REM ELEC	KWH	0.86	0.77	0.87	0.86	1.01	1.20	1.37	1.33	1.05	0.93	0.80	0.86	11.9
		REM DEM	KW	.00331	.00331	.00325	.00422	.00490	.00547	.00564	.00570	.00536	.00507	.00331	.00331	0.00570
		TOT ELEC	KWH	1.96	1.61	1.48	1.15	1.14	1.29	1.44	1.40	1.14	1.17	1.39	1.81	17.0
		TOT DEM	KW	.01210	.01125	.00920	.00702	.00702	.00799	.00786	.00786	.00726	.00750	.00908	.01210	0.01210
S- 15	1.16	HT ELEC	KWH	1.15	0.87	0.64	0.31	0.13	0.10	0.07	0.07	0.10	0.25	0.61	1.00	5.3
		HT DEM	KW	.00908	.00863	.00772	.00445	.00309	.00209	.00072	.00114	.00281	.00463	.00645	.00908	0.00908
		REM ELEC	KWH	0.76	0.68	0.77	0.76	0.89	1.06	1.21	1.18	0.92	0.82	0.70	0.76	10.5
		REM DEM	KW	.00316	.00316	.00311	.00403	.00469	.00523	.00540	.00545	.00512	.00485	.00316	.00316	0.00545
		TOT ELEC	KWH	1.91	1.56	1.41	1.06	1.02	1.16	1.28	1.24	1.02	1.07	1.32	1.75	15.8
		TOT DEM	KW	.01160	.01137	.00951	.00661	.00638	.00708	.00696	.00696	.00650	.00696	.00893	.01160	0.01160
S- 31	0.17	HT ELEC	KWH	0.84	0.67	0.63	0.42	0.27	0.19	0.15	0.16	0.23	0.35	0.56	0.78	5.3
		HT DEM	KW	.00908	.00863	.00772	.00445	.00309	.00209	.00072	.00114	.00281	.00463	.00645	.00908	0.00908
		REM ELEC	KWH	0.84	0.74	0.86	0.82	0.87	1.20	1.28	1.30	1.00	0.87	0.79	0.87	11.4
		REM DEM	KW	.00360	.00360	.00353	.00459	.00533	.00595	.00614	.00620	.00583	.00552	.00360	.00360	0.00620
		TOT ELEC	KWH	1.68	1.41	1.49	1.24	1.13	1.39	1.43	1.46	1.23	1.21	1.34	1.65	16.6
		TOT DEM	KW	.01260	.01235	.01033	.00718	.00693	.00769	.00756	.00756	.00706	.00756	.00970	.01260	0.01260
BASE CASE : 135 (CONTINUED)																
S- 39	0.96	HT ELEC	KWH	0.89	0.71	0.67	0.45	0.28	0.21	0.16	0.17	0.24	0.37	0.59	0.83	5.6
		HT DEM	KW	.00908	.00863	.00772	.00445	.00309	.00209	.00072	.00114	.00281	.00463	.00645	.00908	0.00908
		REM ELEC	KWH	0.78	0.69	0.80	0.76	0.81	1.11	1.19	1.21	0.93	0.81	0.73	0.81	10.6
		REM DEM	KW	.00344	.00344	.00338	.00439	.00510	.00569	.00587	.00593	.00557	.00528	.00344	.00344	0.00593
		TOT ELEC	KWH	1.68	1.40	1.46	1.21	1.09	1.32	1.35	1.38	1.17	1.17	1.32	1.64	16.2
		TOT DEM	KW	.01220	.01196	.01000	.00695	.00671	.00744	.00732	.00732	.00683	.00732	.00939	.01220	0.01220
S- 44	0.20	HT ELEC	KWH	0.86	0.69	0.65	0.43	0.28	0.20	0.16	0.16	0.23	0.36	0.57	0.80	5.4
		HT DEM	KW	.00908	.00863	.00772	.00445	.00309	.00209	.00072	.00114	.00281	.00463	.00645	.00908	0.00908
		REM ELEC	KWH	0.81	0.71	0.83	0.79	0.84	1.15	1.23	1.25	0.97	0.84	0.76	0.84	11.0
		REM DEM	KW	.00353	.00353	.00347	.00450	.00523	.00584	.00602	.00608	.00572	.00541	.00353	.00353	0.00608
		TOT ELEC	KWH	1.68	1.40	1.47	1.22	1.11	1.35	1.39	1.42	1.20	1.19	1.33	1.64	16.4
		TOT DEM	KW	.01240	.01215	.01017	.00707	.00682	.00756	.00744	.00744	.00694	.00744	.00955	.01240	0.01240
S- 60	-0.30	HT ELEC	KWH	1.10	0.84	0.61	0.29	0.13	0.09	0.07	0.07	0.09	0.24	0.59	0.95	5.1
		HT DEM	KW	.00908	.00863	.00772	.00445	.00309	.00209	.00072	.00114	.00281	.00463	.00645	.00908	0.00908
		REM ELEC	KWH	0.86	0.77	0.87	0.86	1.01	1.20	1.37	1.33	1.05	0.93	0.80	0.86	11.9
		REM DEM	KW	.00331	.00331	.00325	.00422	.00490	.00547	.00564	.00570	.00536	.00507	.00331	.00331	0.00570
		TOT ELEC	KWH	1.96	1.61	1.48	1.15	1.14	1.29	1.44	1.40	1.14	1.17	1.39	1.81	17.0
		TOT DEM	KW	.01210	.01125	.00920	.00702	.00702	.00799	.00786	.00786	.00726	.00750	.00908	.01210	0.01210
BASE CASE : 136																
BASE	0.00	TOT GAS	THM	0.0521	0.0396	0.0290	0.0139	0.0060	0.0043	0.0034	0.0031	0.0043	0.0113	0.0278	0.0451	0.240
		TOT ELEC	KWH	1.03	0.93	1.04	1.03	1.22	1.44	1.64	1.60	1.26	1.12	0.96	1.03	14.3
		TOT DEM	KW	.00428	.00428	.00421	.00546	.00635	.00708	.00731	.00738	.00694	.00657	.00428	.00428	0.00738
S- 3	-0.15	TOT GAS	THM	0.0545	0.0414	0.0304	0.0146	0.0063	0.0045	0.0035	0.0033	0.0045	0.0118	0.0291	0.0472	0.251
		TOT ELEC	KWH	0.93	0.84	0.94	0.93	1.10	1.30	1.48	1.44	1.14	1.01	0.86	0.93	12.9
		TOT DEM	KW	.00341	.00341	.00335	.00435	.00506	.00564	.00582	.00588	.00553	.00523	.00341	.00341	0.00588
S- 6	-0.06	TOT GAS	THM	0.0577	0.0439	0.0322	0.0154	0.0066	0.0048	0.0037	0.0035	0.0048	0.0125	0.0309	0.0500	0.266
		TOT ELEC	KWH	0.86	0.77	0.87	0.86	1.01	1.20	1.37	1.33	1.05	0.93	0.80	0.86	11.9
		TOT DEM	KW	.00331	.00331	.00325	.00422	.00490	.00547	.00564	.00570	.00536	.00507	.00331	.00331	0.00570
S- 15	1.15	TOT GAS	THM	0.0603	0.0459	0.0336	0.0161	0.0070	0.0050	0.0039	0.0036	0.0050	0.0131	0.0322	0.0523	0.278
		TOT ELEC	KWH	0.76	0.68	0.77	0.76	0.89	1.06	1.21	1.18	0.92	0.82	0.70	0.76	10.5
		TOT DEM	KW	.00316	.00316	.00311	.00403	.00469	.00523	.00540	.00545	.00512	.00485	.00316	.00316	0.00545
S- 31	0.17	TOT GAS	THM	0.0442	0.0351	0.0331	0.0221	0.0141	0.0102	0.0080	0.0083	0.0119	0.0182	0.0293	0.0411	0.276
		TOT ELEC	KWH	0.84	0.74	0.86	0.82	0.87	1.20	1.28	1.30	1.00	0.87	0.79	0.87	11.4
		TOT DEM	KW	.00360	.00360	.00353	.00459	.00533	.00595	.00614	.00620	.00583	.00552	.00360	.00360	0.00620
S- 39	0.96	TOT GAS	THM	0.0467	0.0371	0.0350	0.0234	0.0149	0.0108	0.0085	0.0088	0.0126	0.0193	0.0310	0.0435	0.292
		TOT ELEC	KWH	0.78	0.69	0.80	0.76	0.81	1.11	1.19	1.21	0.93	0.81	0.73	0.81	10.6
		TOT DEM	KW	.00344	.00344	.00338	.00439	.00510	.00569	.00587	.00593	.00557	.00528	.00344	.00344	0.00593
S- 44	0.20	TOT GAS	THM	0.0454	0.0361	0.0341	0.0227	0.0145	0.0105	0.0082	0.0085	0.0122	0.0187	0.0301	0.0423	0.284
		TOT ELEC	KWH	0.81	0.71	0.83	0.79	0.84	1.15	1.23	1.25	0.97	0.84	0.76	0.84	11.0
		TOT DEM	KW	.00353	.00353	.00347	.00450	.00523	.00584	.00602	.00608	.00572	.00541	.00353	.00353	0.00608
S- 60	-0.31	TOT GAS	THM	0.0577	0.0439	0.0322	0.0154	0.0066	0.0048	0.0037	0.0035	0.0048	0.0125	0.0309	0.0500	0.266
		TOT ELEC	KWH	0.86	0.77	0.87	0.86	1.01	1.20	1.37	1.33	1.05	0.93	0.80	0.86	11.9
		TOT DEM	KW	.00331	.00331	.00325	.00422	.00490	.00547	.00564	.00570	.00536	.00507	.00331	.00331	0.00570
BASE CASE : 137																
BASE	0.00	HT ELEC	KWH	0.91	0.69	0.51	0.24	0.11	0.08	0.06	0.05	0.08	0.20	0.49	0.79	4.2
		HT DEM	KW	.00580	.00580	.00580	.00418	.00232	.00104	.00031	.00054	.00238	.00487	.00580	.00580	0.00580
		REM ELEC	KWH	0.97	0.88	0.99	0.97	1.15	1.36	1.55	1.51	1.19	1.05	0.90	0.97	13.5
		REM DEM	KW	.00369	.00369	.00363	.00443	.00523	.00578	.00597	.00615	.00547	.00510	.00369	.00369	0.00615
		TOT ELEC	KWH	1.89	1.57	1.49	1.22	1.25	1.44	1.61	1.57	1.26	1.25	1.39	1.76	17.7
		TOT DEM	KW	.00885	.00876	.00876	.00487	.00558	.00602	.00619	.00628	.00575	.00540	.00858	.00885	0.00885
S- 3	-0.18	HT ELEC	KWH	0.95	0.72	0.53	0.25	0.11	0.08	0.06	0.06	0.08	0.21	0.51	0.82	4.4
		HT DEM	KW	.00580	.00580	.00580	.00418	.00232	.00104	.00031	.00054	.00238	.00487	.00580	.00580	0.00580
		REM ELEC	KWH	0.87	0.79	0.88	0.87	1.03	1.22	1.39	1.36	1.06	0.94	0.81	0.87	12.1
		REM DEM	KW	.00294	.00294	.00289	.00353	.00417	.00461	.00475	.00490	.00436	.00407	.00294	.00294	0.00490
		TOT ELEC	KWH	1.82	1.51	1.41	1.13	1.14	1.30	1.45	1.41	1.14	1.15	1.32	1.69	16.5
		TOT DEM	KW	.00850	.00842	.00842	.00468	.00536	.00578	.00595	.00604	.00553	.00519	.00825	.00850	0.00850
S- 6	-0.10	HT ELEC	KWH	0.98	0.75	0.55	0.26	0.11	0.08	0.06	0.06	0.08	0.21	0.53	0.85	4.5
		HT DEM	KW	.00580	.00580	.00580	.00418	.00232	.00104	.00031	.00054	.00238	.00487	.00580	.00580	0.00580
		REM ELEC	KWH	0.78	0.71	0.80	0.78	0.93	1.10	1.25	1.22	0.96	0.85	0.73	0.78	10.9
		REM DEM	KW	.00285	.00285	.00280	.00342	.00404	.00446	.00461	.00475	.00423	.00394	.00285	.00285	0.00475
		TOT ELEC	KWH	1.77	1.46	1.34	1.05	1.04	1.18	1.32	1.28	1.04	1.06	1.26	1.64	15.4
		TOT DEM	KW	.00820	.00812	.00779	.00467	.00394	.00443	.00467	.00476	.00443	.00533	.00779	.00820	0.00820
S- 15	0.95	HT ELEC	KWH	1.02	0.78	0.57	0.27	0.12	0.08	0.07	0.06	0.08	0.22	0.55	0.89	4.7
		HT DEM	KW	.00580	.00580	.00580	.00418	.00232	.00104	.00031	.00054	.00238	.00487	.00580	.00580	0.00580
		REM ELEC	KWH	0.69	0.62	0.70	0.69	0.81	0.97	1.10	1.07	0.84	0.75	0.64	0.69	9.6
		REM DEM	KW	.00273	.00273	.00268	.00328	.00387	.00428	.00441	.00455	.00405	.00378	.00273	.00273	0.00455
		TOT ELEC	KWH	1.71	1.40	1.27	0.96	0.93	1.05	1.17	1.13	0.93	0.97	1.19	1.58	14.3
		TOT DEM	KW	.00785	.00777	.00746	.00447	.00377	.00424	.00447	.00455	.00424	.00510	.00746	.00785	0.00785
S- 31	0.13	HT ELEC	KWH	1.38	0.97	0.74	0.32	0.12	0.05	0.03	0.03	0.09	0.23	0.57	1.15	5.7
		HT DEM	KW	.00580	.00580	.00580	.00418	.00232	.00104	.00031	.00054	.00238	.00487	.00580	.00580	0.00580
		REM ELEC	KWH	0.74	0.66	0.74	0.76	0.79	1.03	1.11	1.14	0.92	0.79	0.71	0.74	10.1
		REM DEM	KW	.00294	.00294	.00289	.00353	.00417	.00461	.00475	.00490	.00436	.00407	.00294	.00294	0.00490
		TOT ELEC	KWH	2.11	1.63	1.48	1.07	0.91	1.08	1.14	1.18	1.01	1.02	1.28	1.88	15.8
		TOT DEM	KW	.00845	.00837	.00803	.00482	.00406	.00456	.00482	.00490	.00456	.00549	.00803	.00845	0.00845

(HT ELEC=ELECTRIC CONSUMPTION FOR SPACE AND WATER HEATING. REM ELEC=REMAINING ELECTRIC CONSUMPTION (NOT USED FOR HEATING). TOT ELEC=TOTAL ELECTRIC CONSUMPTION. HT DEM=PEAK ELECTRIC DEMAND FOR SPACE AND WATER HEATING. REM DEM=PEAK ELECTRIC DEMAND FOR REMAINING LOAD (NOT HEATING). TOT DEM=TOTAL PEAK ELECTRIC DEMAND. GAS=TOTAL GAS CONSUMPTION. ANNUAL FIGURES ARE SUMS OF MONTHLY CONSUMPTION VALUES OR ARE YEARLY MAXIMUMS OF MONTHLY PEAK DEMAND VALUES.)

STRAT SET	CONST COST $/SF	ENERGY FORM	UNITS (PER SF)	JAN	FEB	MAR	APR	MAY	JUN	JUL	AUG	SEP	OCT	NOV	DEC	ANNUAL
BASE CASE : 137 (CONTINUED)																
S- 39	0.84	HT ELEC	KWH	1.50	1.06	0.81	0.34	0.13	0.05	0.03	0.04	0.10	0.26	0.63	1.25	6.3
		HT DEM	KW	.00580	.00580	.00580	.00418	.00232	.00104	.00031	.00054	.00238	.00487	.00580	.00580	0.00580
		REM ELEC	KWH	0.67	0.60	0.67	0.69	0.72	0.94	1.01	1.04	0.84	0.72	0.65	0.67	9.2
		REM DEM	KW	.00273	.00273	.00268	.00328	.00387	.00428	.00441	.00455	.00405	.00378	.00273	.00273	0.00455
		TOT ELEC	KWH	2.17	1.66	1.49	1.04	0.85	0.99	1.05	1.08	0.94	0.98	1.27	1.92	15.5
		TOT DEM	KW	.00805	.00797	.00765	.00459	.00386	.00435	.00459	.00467	.00435	.00523	.00765	.00805	0.00805
S- 44	0.15	HT ELEC	KWH	1.42	1.01	0.77	0.33	0.12	0.05	0.03	0.04	0.09	0.24	0.59	1.19	5.9
		HT DEM	KW	.00580	.00580	.00580	.00418	.00232	.00104	.00031	.00054	.00238	.00487	.00580	.00580	0.00580
		REM ELEC	KWH	0.71	0.63	0.71	0.73	0.76	0.99	1.07	1.10	0.89	0.76	0.68	0.71	9.7
		REM DEM	KW	.00285	.00285	.00280	.00342	.00404	.00446	.00461	.00475	.00423	.00394	.00285	.00285	0.00475
		TOT ELEC	KWH	2.13	1.64	1.48	1.06	0.88	1.04	1.10	1.14	0.98	1.00	1.27	1.90	15.7
		TOT DEM	KW	.00827	.00819	.00786	.00471	.00397	.00447	.00471	.00480	.00447	.00538	.00786	.00827	0.00827
S- 60	-0.35	HT ELEC	KWH	0.98	0.75	0.55	0.26	0.11	0.08	0.06	0.06	0.08	0.21	0.53	0.85	4.5
		HT DEM	KW	.00580	.00580	.00580	.00418	.00232	.00104	.00031	.00054	.00238	.00487	.00580	.00580	0.00580
		REM ELEC	KWH	0.78	0.71	0.80	0.78	0.93	1.10	1.25	1.22	0.96	0.85	0.73	0.78	10.9
		REM DEM	KW	.00285	.00285	.00280	.00342	.00404	.00446	.00461	.00475	.00423	.00394	.00285	.00285	0.00475
		TOT ELEC	KWH	1.77	1.46	1.34	1.05	1.04	1.18	1.32	1.28	1.04	1.06	1.26	1.64	15.4
		TOT DEM	KW	.00820	.00812	.00779	.00467	.00394	.00443	.00467	.00476	.00443	.00533	.00779	.00820	0.00820
BASE CASE : 138																
BASE	0.00	TOT GAS	THM	0.0480	0.0365	0.0267	0.0128	0.0055	0.0040	0.0031	0.0029	0.0040	0.0104	0.0256	0.0415	0.221
		TOT ELEC	KWH	0.97	0.88	0.99	0.97	1.15	1.36	1.55	1.51	1.19	1.05	0.90	0.97	13.5
		TOT DEM	KW	.00375	.00375	.00369	.00449	.00510	.00572	.00597	.00615	.00566	.00517	.00375	.00406	0.00615
S- 3	-0.20	TOT GAS	THM	0.0499	0.0380	0.0278	0.0133	0.0058	0.0041	0.0032	0.0030	0.0041	0.0108	0.0267	0.0432	0.230
		TOT ELEC	KWH	0.87	0.79	0.88	0.87	1.03	1.22	1.39	1.36	1.06	0.94	0.81	0.87	12.1
		TOT DEM	KW	.00299	.00299	.00294	.00358	.00407	.00456	.00475	.00490	.00451	.00412	.00299	.00323	0.00490
S- 6	-0.12	TOT GAS	THM	0.0516	0.0393	0.0288	0.0138	0.0060	0.0043	0.0033	0.0031	0.0043	0.0112	0.0276	0.0447	0.238
		TOT ELEC	KWH	0.78	0.71	0.80	0.78	0.93	1.10	1.25	1.22	0.96	0.85	0.73	0.78	10.9
		TOT DEM	KW	.00290	.00290	.00285	.00347	.00394	.00442	.00461	.00475	.00437	.00399	.00290	.00314	0.00475
S- 15	0.92	TOT GAS	THM	0.0538	0.0409	0.0300	0.0144	0.0062	0.0045	0.0035	0.0032	0.0045	0.0117	0.0288	0.0466	0.248
		TOT ELEC	KWH	0.69	0.62	0.70	0.69	0.81	0.97	1.10	1.07	0.84	0.75	0.64	0.69	9.6
		TOT DEM	KW	.00278	.00278	.00273	.00332	.00378	.00423	.00441	.00455	.00419	.00382	.00278	.00300	0.00455
S- 31	0.11	TOT GAS	THM	0.0722	0.0512	0.0391	0.0166	0.0063	0.0024	0.0015	0.0018	0.0048	0.0123	0.0301	0.0602	0.301
		TOT ELEC	KWH	0.74	0.66	0.74	0.76	0.79	1.03	1.11	1.14	0.92	0.79	0.71	0.74	10.1
		TOT DEM	KW	.00299	.00299	.00294	.00358	.00407	.00456	.00475	.00490	.00451	.00412	.00299	.00323	0.00490
S- 39	0.81	TOT GAS	THM	0.0787	0.0558	0.0426	0.0180	0.0069	0.0026	0.0016	0.0020	0.0052	0.0134	0.0328	0.0656	0.328
		TOT ELEC	KWH	0.67	0.60	0.67	0.69	0.72	0.94	1.01	1.04	0.84	0.72	0.65	0.67	9.2
		TOT DEM	KW	.00278	.00278	.00273	.00332	.00378	.00423	.00441	.00455	.00419	.00382	.00278	.00300	0.00455
S- 44	0.14	TOT GAS	THM	0.0746	0.0529	0.0404	0.0171	0.0065	0.0025	0.0016	0.0019	0.0050	0.0128	0.0311	0.0622	0.311
		TOT ELEC	KWH	0.71	0.63	0.71	0.73	0.76	0.99	1.07	1.10	0.89	0.76	0.68	0.71	9.7
		TOT DEM	KW	.00290	.00290	.00285	.00347	.00394	.00442	.00461	.00475	.00437	.00399	.00290	.00314	0.00475
S- 60	-0.37	TOT GAS	THM	0.0516	0.0393	0.0288	0.0138	0.0060	0.0043	0.0033	0.0031	0.0043	0.0112	0.0276	0.0447	0.238
		TOT ELEC	KWH	0.78	0.71	0.80	0.78	0.93	1.10	1.25	1.22	0.96	0.85	0.73	0.78	10.9
		TOT DEM	KW	.00290	.00290	.00285	.00347	.00394	.00442	.00461	.00475	.00437	.00399	.00290	.00314	0.00475
BASE CASE : 139																
BASE	0.00	TOT ELEC	KWH	2.17	1.80	1.63	1.29	1.29	1.52	1.60	1.60	1.39	1.29	1.33	2.03	19.0
		TOT DEM	KW	.00753	.00693	.00663	.00346	.00392	.00422	.00437	.00444	.00414	.00392	.00633	.00730	0.00753
S- 1	0.09	TOT ELEC	KWH	2.00	1.66	1.51	1.19	1.19	1.40	1.47	1.47	1.28	1.19	1.23	1.87	17.5
		TOT DEM	KW	.00625	.00575	.00550	.00288	.00325	.00350	.00362	.00369	.00344	.00325	.00525	.00606	0.00625
S- 12	0.20	TOT ELEC	KWH	1.80	1.50	1.36	1.07	1.07	1.26	1.33	1.33	1.15	1.07	1.11	1.69	15.8
		TOT DEM	KW	.00587	.00540	.00517	.00270	.00305	.00329	.00340	.00346	.00323	.00305	.00493	.00569	0.00587
S- 22	1.51	TOT ELEC	KWH	1.64	1.37	1.24	0.98	0.98	1.15	1.21	1.21	1.05	0.98	1.01	1.54	14.4
		TOT DEM	KW	.00560	.00515	.00493	.00258	.00291	.00314	.00325	.00330	.00308	.00291	.00470	.00543	0.00560
S- 28	-0.05	TOT ELEC	KWH	1.80	1.50	1.36	1.07	1.07	1.26	1.33	1.33	1.15	1.07	1.11	1.69	15.8
		TOT DEM	KW	.00587	.00540	.00517	.00270	.00305	.00329	.00340	.00346	.00323	.00305	.00493	.00569	0.00587
S- 36	0.43	TOT ELEC	KWH	1.63	1.36	1.23	0.97	0.97	1.14	1.20	1.20	1.04	0.97	1.00	1.53	14.3
		TOT DEM	KW	.00815	.00750	.00717	.00375	.00424	.00456	.00473	.00481	.00448	.00424	.00685	.00791	0.00815
S- 43	1.27	TOT ELEC	KWH	1.53	1.27	1.15	0.91	0.91	1.07	1.13	1.13	0.98	0.91	0.94	1.43	13.4
		TOT DEM	KW	.00790	.00727	.00695	.00363	.00411	.00442	.00458	.00466	.00435	.00411	.00664	.00766	0.00790
S- 48	0.46	TOT ELEC	KWH	1.57	1.31	1.19	0.94	0.94	1.10	1.16	1.16	1.01	0.94	0.97	1.48	13.8
		TOT DEM	KW	.00803	.00739	.00707	.00369	.00418	.00450	.00466	.00474	.00442	.00418	.00675	.00779	0.00803
BASE CASE : 140																
BASE	0.00	TOT GAS	THM	0.0319	0.0234	0.0122	0.0014	0.0013	0.0014	0.0014	0.0014	0.0014	0.0014	0.0043	0.0250	0.106
		TOT ELEC	KWH	1.62	1.34	1.38	1.24	1.24	1.53	1.62	1.62	1.38	1.28	1.24	1.53	17.0
		TOT DEM	KW	.00683	.00601	.00553	.00464	.00526	.00567	.00587	.00594	.00560	.00533	.00519	.00649	0.00683
S- 1	0.09	TOT GAS	THM	0.0387	0.0284	0.0148	0.0017	0.0016	0.0017	0.0017	0.0017	0.0017	0.0017	0.0052	0.0303	0.129
		TOT ELEC	KWH	1.42	1.19	1.22	1.10	1.10	1.35	1.42	1.42	1.22	1.13	1.10	1.35	15.0
		TOT DEM	KW	.00567	.00499	.00459	.00386	.00437	.00471	.00488	.00493	.00465	.00442	.00431	.00539	0.00567
S- 12	0.20	TOT GAS	THM	0.0453	0.0332	0.0174	0.0020	0.0019	0.0020	0.0020	0.0020	0.0020	0.0020	0.0061	0.0355	0.151
		TOT ELEC	KWH	1.23	1.02	1.04	0.94	0.94	1.16	1.23	1.23	1.04	0.97	0.94	1.16	12.9
		TOT DEM	KW	.00533	.00469	.00432	.00362	.00410	.00442	.00458	.00464	.00437	.00416	.00405	.00506	0.00533
S- 22	1.51	TOT GAS	THM	0.0585	0.0429	0.0224	0.0025	0.0024	0.0025	0.0025	0.0025	0.0025	0.0025	0.0079	0.0458	0.195
		TOT ELEC	KWH	1.02	0.85	0.87	0.78	0.78	0.96	1.02	1.02	0.87	0.80	0.78	0.96	10.7
		TOT DEM	KW	.00508	.00447	.00411	.00345	.00391	.00422	.00437	.00442	.00417	.00396	.00386	.00483	0.00508

STRAT SET	CONST COST $/SF	ENERGY FORM	UNITS (PER SF)	JAN	FEB	MAR	APR	MAY	JUN	JUL	AUG	SEP	OCT	NOV	DEC	ANNUAL
BASE CASE : 140 (CONTINUED)																
S- 28	-0.05	TOT GAS	THM	0.0453	0.0332	0.0174	0.0020	0.0019	0.0020	0.0020	0.0020	0.0020	0.0020	0.0061	0.0355	0.151
		TOT ELEC	KWH	1.23	1.02	1.04	0.94	0.94	1.16	1.23	1.23	1.04	0.97	0.94	1.16	12.9
		TOT DEM	KW	.00533	.00469	.00432	.00362	.00410	.00442	.00458	.00464	.00437	.00416	.00405	.00506	0.00533
S- 36	0.43	TOT GAS	THM	0.0512	0.0376	0.0196	0.0022	0.0021	0.0022	0.0022	0.0022	0.0022	0.0022	0.0069	0.0401	0.171
		TOT ELEC	KWH	1.05	0.88	0.90	0.81	0.81	1.00	1.05	1.05	0.90	0.83	0.81	1.00	11.1
		TOT DEM	KW	.00545	.00480	.00441	.00371	.00420	.00452	.00469	.00474	.00447	.00425	.00414	.00518	0.00545
S- 43	1.27	TOT GAS	THM	0.0597	0.0438	0.0229	0.0026	0.0025	0.0026	0.0026	0.0026	0.0026	0.0026	0.0081	0.0468	0.199
		TOT ELEC	KWH	0.91	0.76	0.78	0.70	0.70	0.86	0.91	0.91	0.78	0.72	0.70	0.86	9.6
		TOT DEM	KW	.00528	.00465	.00428	.00359	.00407	.00438	.00454	.00459	.00433	.00412	.00401	.00502	0.00528
S- 48	0.46	TOT GAS	THM	0.0561	0.0411	0.0215	0.0024	0.0023	0.0024	0.0024	0.0024	0.0024	0.0024	0.0076	0.0439	0.187
		TOT ELEC	KWH	0.97	0.81	0.83	0.74	0.74	0.92	0.97	0.97	0.83	0.77	0.74	0.92	10.2
		TOT DEM	KW	.00537	.00473	.00435	.00365	.00413	.00446	.00462	.00467	.00440	.00419	.00408	.00510	0.00537
BASE CASE : 141																
BASE	0.00	HT ELEC	KWH	0.78	0.59	0.43	0.21	0.09	0.06	0.05	0.05	0.06	0.17	0.42	0.67	3.6
		HT DEM	KW	.00908	.00863	.00772	.00445	.00309	.00209	.00072	.00114	.00281	.00463	.00645	.00908	0.00908
		REM ELEC	KWH	0.99	0.90	1.01	0.99	1.17	1.39	1.59	1.55	1.21	1.08	0.92	0.99	13.8
		REM DEM	KW	.00386	.00386	.00379	.00492	.00572	.00638	.00658	.00665	.00625	.00592	.00386	.00386	0.00665
		TOT ELEC	KWH	1.77	1.49	1.44	1.20	1.26	1.46	1.64	1.59	1.28	1.24	1.34	1.67	17.4
		TOT DEM	KW	.01288	.01069	.00902	.00670	.00657	.00734	.00708	.00721	.00683	.00708	.00863	.01288	0.01288
S- 3	-0.12	HT ELEC	KWH	0.81	0.62	0.45	0.22	0.09	0.07	0.05	0.05	0.07	0.18	0.44	0.70	3.8
		HT DEM	KW	.00908	.00863	.00772	.00445	.00309	.00209	.00072	.00114	.00281	.00463	.00645	.00908	0.00908
		REM ELEC	KWH	0.90	0.81	0.91	0.90	1.06	1.26	1.44	1.40	1.10	0.98	0.84	0.90	12.5
		REM DEM	KW	.00355	.00355	.00349	.00453	.00526	.00588	.00606	.00612	.00575	.00545	.00355	.00355	0.00612
		TOT ELEC	KWH	1.71	1.43	1.37	1.12	1.16	1.33	1.49	1.45	1.17	1.15	1.27	1.61	16.3
		TOT DEM	KW	.01240	.01029	.00868	.00645	.00632	.00707	.00682	.00694	.00657	.00682	.00831	.01240	0.01240
S- 6	-0.03	HT ELEC	KWH	0.86	0.65	0.48	0.23	0.10	0.07	0.06	0.05	0.07	0.19	0.46	0.74	4.0
		HT DEM	KW	.00908	.00863	.00772	.00445	.00309	.00209	.00072	.00114	.00281	.00463	.00645	.00908	0.00908
		REM ELEC	KWH	0.83	0.75	0.84	0.83	0.98	1.16	1.32	1.29	1.01	0.90	0.77	0.83	11.5
		REM DEM	KW	.00327	.00327	.00321	.00417	.00484	.00540	.00557	.00563	.00529	.00501	.00327	.00327	0.00563
		TOT ELEC	KWH	1.69	1.40	1.32	1.06	1.08	1.23	1.38	1.34	1.08	1.08	1.23	1.57	15.5
		TOT DEM	KW	.01190	.00988	.00833	.00619	.00607	.00678	.00655	.00666	.00631	.00655	.00797	.01190	0.01190
S- 15	1.19	HT ELEC	KWH	0.90	0.68	0.50	0.24	0.10	0.07	0.06	0.05	0.07	0.20	0.48	0.78	4.2
		HT DEM	KW	.00908	.00863	.00772	.00445	.00309	.00209	.00072	.00114	.00281	.00463	.00645	.00908	0.00908
		REM ELEC	KWH	0.73	0.66	0.74	0.73	0.86	1.02	1.16	1.13	0.89	0.79	0.68	0.73	10.1
		REM DEM	KW	.00296	.00296	.00291	.00377	.00439	.00490	.00505	.00510	.00479	.00454	.00296	.00296	0.00510
		TOT ELEC	KWH	1.63	1.34	1.24	0.97	0.96	1.09	1.22	1.19	0.96	0.98	1.16	1.51	14.3
		TOT DEM	KW	.01140	.00958	.00809	.00581	.00547	.00604	.00581	.00604	.00559	.00616	.00764	.01140	0.01140
S- 31	0.28	HT ELEC	KWH	0.76	0.60	0.57	0.38	0.24	0.18	0.14	0.14	0.20	0.31	0.50	0.71	4.8
		HT DEM	KW	.00908	.00863	.00772	.00445	.00309	.00209	.00072	.00114	.00281	.00463	.00645	.00908	0.00908
		REM ELEC	KWH	0.81	0.71	0.83	0.79	0.84	1.15	1.23	1.25	0.97	0.84	0.76	0.84	11.0
		REM DEM	KW	.00321	.00321	.00315	.00409	.00476	.00531	.00547	.00553	.00520	.00492	.00321	.00321	0.00553
		TOT ELEC	KWH	1.57	1.32	1.39	1.17	1.08	1.33	1.37	1.40	1.17	1.15	1.26	1.54	15.8
		TOT DEM	KW	.01245	.01046	.00884	.00635	.00598	.00660	.00635	.00660	.00610	.00672	.00834	.01245	0.01245
S- 60	-0.28	HT ELEC	KWH	0.86	0.65	0.48	0.23	0.10	0.07	0.06	0.05	0.07	0.19	0.46	0.74	4.0
		HT DEM	KW	.00908	.00863	.00772	.00445	.00309	.00209	.00072	.00114	.00281	.00463	.00645	.00908	0.00908
		REM ELEC	KWH	0.83	0.75	0.84	0.83	0.98	1.16	1.32	1.29	1.01	0.90	0.77	0.83	11.5
		REM DEM	KW	.00327	.00327	.00321	.00417	.00484	.00540	.00557	.00563	.00529	.00501	.00327	.00327	0.00563
		TOT ELEC	KWH	1.69	1.40	1.32	1.06	1.08	1.23	1.38	1.34	1.08	1.08	1.23	1.57	15.5
		TOT DEM	KW	.01190	.00988	.00833	.00619	.00607	.00678	.00655	.00666	.00631	.00655	.00797	.01190	0.01190
BASE CASE : 142																
BASE	0.00	TOT GAS	THM	0.0408	0.0310	0.0227	0.0109	0.0047	0.0034	0.0026	0.0024	0.0034	0.0088	0.0218	0.0353	0.188
		TOT ELEC	KWH	0.99	0.90	1.01	0.99	1.17	1.39	1.59	1.55	1.21	1.08	0.92	0.99	13.8
		TOT DEM	KW	.00386	.00386	.00379	.00492	.00572	.00638	.00658	.00665	.00625	.00592	.00386	.00386	0.00665
S- 3	-0.12	TOT GAS	THM	0.0427	0.0325	0.0238	0.0114	0.0049	0.0035	0.0028	0.0026	0.0035	0.0093	0.0229	0.0370	0.197
		TOT ELEC	KWH	0.90	0.81	0.91	0.90	1.06	1.26	1.44	1.40	1.10	0.98	0.84	0.90	12.5
		TOT DEM	KW	.00355	.00355	.00349	.00453	.00526	.00588	.00606	.00612	.00575	.00545	.00355	.00355	0.00612
S- 6	-0.04	TOT GAS	THM	0.0451	0.0343	0.0252	0.0121	0.0052	0.0037	0.0029	0.0027	0.0037	0.0098	0.0241	0.0391	0.208
		TOT ELEC	KWH	0.83	0.75	0.84	0.83	0.98	1.16	1.32	1.29	1.01	0.90	0.77	0.83	11.5
		TOT DEM	KW	.00327	.00327	.00321	.00417	.00484	.00540	.00557	.00563	.00529	.00501	.00327	.00327	0.00563
S- 15	1.19	TOT GAS	THM	0.0473	0.0360	0.0264	0.0126	0.0055	0.0039	0.0031	0.0028	0.0039	0.0102	0.0253	0.0410	0.218
		TOT ELEC	KWH	0.73	0.66	0.74	0.73	0.86	1.02	1.16	1.13	0.89	0.79	0.68	0.73	10.1
		TOT DEM	KW	.00296	.00296	.00291	.00377	.00439	.00490	.00505	.00510	.00479	.00454	.00296	.00296	0.00510
S- 31	0.28	TOT GAS	THM	0.0398	0.0316	0.0299	0.0199	0.0127	0.0092	0.0072	0.0075	0.0107	0.0164	0.0264	0.0371	0.249
		TOT ELEC	KWH	0.81	0.71	0.83	0.79	0.84	1.15	1.23	1.25	0.97	0.84	0.76	0.84	11.0
		TOT DEM	KW	.00321	.00321	.00315	.00409	.00476	.00531	.00547	.00553	.00520	.00492	.00321	.00321	0.00553
S- 60	-0.29	TOT GAS	THM	0.0451	0.0343	0.0252	0.0121	0.0052	0.0037	0.0029	0.0027	0.0037	0.0098	0.0241	0.0391	0.208
		TOT ELEC	KWH	0.83	0.75	0.84	0.83	0.98	1.16	1.32	1.29	1.01	0.90	0.77	0.83	11.5
		TOT DEM	KW	.00327	.00327	.00321	.00417	.00484	.00540	.00557	.00563	.00529	.00501	.00327	.00327	0.00563
BASE CASE : 143																
BASE	0.00	HT ELEC	KWH	0.75	0.57	0.42	0.20	0.09	0.06	0.05	0.05	0.06	0.16	0.40	0.65	3.5
		HT DEM	KW	.00580	.00580	.00580	.00418	.00232	.00104	.00031	.00054	.00238	.00487	.00580	.00580	0.00580
		REM ELEC	KWH	0.96	0.86	0.97	0.96	1.13	1.34	1.53	1.49	1.17	1.04	0.89	0.96	13.3
		REM DEM	KW	.00365	.00365	.00359	.00438	.00517	.00572	.00590	.00608	.00541	.00505	.00365	.00365	0.00608
		TOT ELEC	KWH	1.71	1.44	1.39	1.16	1.22	1.41	1.58	1.53	1.23	1.20	1.29	1.61	16.8
		TOT DEM	KW	.00885	.00876	.00876	.00487	.00558	.00602	.00619	.00628	.00575	.00540	.00858	.00885	0.00885

(HT ELEC=ELECTRIC CONSUMPTION FOR SPACE AND WATER HEATING. REM ELEC=REMAINING ELECTRIC CONSUMPTION (NOT USED FOR HEATING). TOT ELEC=TOTAL ELECTRIC CONSUMPTION. HT DEM=PEAK ELECTRIC DEMAND FOR SPACE AND WATER HEATING. REM DEM=PEAK ELECTRIC DEMAND FOR REMAINING LOAD (NOT HEATING). TOT DEM=TOTAL PEAK ELECTRIC DEMAND. GAS=TOTAL GAS CONSUMPTION. ANNUAL FIGURES ARE SUMS OF MONTHLY CONSUMPTION VALUES OR ARE YEARLY MAXIMUMS OF MONTHLY PEAK DEMAND VALUES.)

BASE CASE : 143

STRAT SET	CONST COST $/SF	ENERGY FORM	UNITS (PER SF)	JAN	FEB	MAR	APR	MAY	JUN	JUL	AUG	SEP	OCT	NOV	DEC	ANNUAL
BASE	0.00	HT ELEC	KWH	0.75	0.57	0.42	0.20	0.09	0.06	0.05	0.05	0.06	0.16	0.40	0.65	3.5
		HT DEM	KW	.00580	.00580	.00580	.00418	.00232	.00104	.00031	.00054	.00238	.00487	.00580	.00580	0.00580
		REM ELEC	KWH	0.96	0.86	0.97	0.96	1.13	1.34	1.53	1.49	1.17	1.04	0.89	0.96	13.3
		REM DEM	KW	.00365	.00365	.00359	.00438	.00517	.00572	.00590	.00608	.00541	.00505	.00365	.00365	0.00608
		TOT ELEC	KWH	1.71	1.44	1.39	1.16	1.22	1.41	1.58	1.53	1.23	1.20	1.29	1.61	16.8
		TOT DEM	KW	.00885	.00876	.00876	.00487	.00558	.00602	.00619	.00628	.00575	.00540	.00858	.00885	0.00885
S- 3	-0.16	HT ELEC	KWH	0.78	0.59	0.44	0.21	0.09	0.06	0.05	0.05	0.06	0.17	0.42	0.68	3.6
		HT DEM	KW	.00580	.00580	.00580	.00418	.00232	.00104	.00031	.00054	.00238	.00487	.00580	.00580	0.00580
		REM ELEC	KWH	0.86	0.77	0.87	0.86	1.01	1.20	1.37	1.33	1.05	0.93	0.80	0.86	11.9
		REM DEM	KW	.00335	.00335	.00330	.00402	.00475	.00525	.00542	.00559	.00498	.00464	.00335	.00335	0.00559
		TOT ELEC	KWH	1.64	1.37	1.30	1.07	1.10	1.27	1.42	1.38	1.11	1.10	1.21	1.53	15.5
		TOT DEM	KW	.00850	.00842	.00842	.00468	.00536	.00578	.00595	.00604	.00553	.00519	.00825	.00850	0.00850
S- 6	-0.08	HT ELEC	KWH	0.81	0.62	0.45	0.22	0.09	0.07	0.05	0.05	0.07	0.18	0.43	0.70	3.7
		HT DEM	KW	.00580	.00580	.00580	.00418	.00232	.00104	.00031	.00054	.00238	.00487	.00580	.00580	0.00580
		REM ELEC	KWH	0.78	0.70	0.79	0.78	0.92	1.09	1.24	1.21	0.95	0.84	0.72	0.78	10.8
		REM DEM	KW	.00309	.00309	.00304	.00371	.00438	.00484	.00500	.00515	.00458	.00427	.00309	.00309	0.00515
		TOT ELEC	KWH	1.59	1.32	1.24	0.99	1.01	1.16	1.29	1.26	1.02	1.02	1.16	1.48	14.5
		TOT DEM	KW	.00820	.00812	.00779	.00467	.00394	.00443	.00467	.00476	.00443	.00533	.00779	.00820	0.00820
S- 15	0.99	HT ELEC	KWH	0.84	0.64	0.47	0.23	0.10	0.07	0.05	0.05	0.07	0.18	0.45	0.73	3.9
		HT DEM	KW	.00580	.00580	.00580	.00418	.00232	.00104	.00031	.00054	.00238	.00487	.00580	.00580	0.00580
		REM ELEC	KWH	0.68	0.61	0.69	0.68	0.80	0.95	1.09	1.06	0.83	0.74	0.63	0.68	9.4
		REM DEM	KW	.00280	.00280	.00275	.00336	.00396	.00438	.00452	.00466	.00415	.00387	.00280	.00280	0.00466
		TOT ELEC	KWH	1.52	1.26	1.16	0.91	0.90	1.02	1.14	1.11	0.90	0.92	1.08	1.41	13.3
		TOT DEM	KW	.00785	.00777	.00746	.00447	.00377	.00424	.00447	.00455	.00424	.00510	.00746	.00785	0.00785
S- 31	0.23	HT ELEC	KWH	1.13	0.80	0.61	0.26	0.10	0.04	0.02	0.03	0.08	0.19	0.47	0.94	4.7
		HT DEM	KW	.00580	.00580	.00580	.00418	.00232	.00104	.00031	.00054	.00238	.00487	.00580	.00580	0.00580
		REM ELEC	KWH	0.74	0.66	0.74	0.76	0.79	1.03	1.11	1.14	0.92	0.79	0.71	0.74	10.1
		REM DEM	KW	.00293	.00293	.00288	.00351	.00415	.00459	.00473	.00488	.00434	.00405	.00293	.00293	0.00488
		TOT ELEC	KWH	1.86	1.45	1.35	1.02	0.89	1.07	1.13	1.17	0.99	0.98	1.18	1.68	14.8
		TOT DEM	KW	.00847	.00839	.00805	.00483	.00407	.00457	.00483	.00491	.00457	.00551	.00805	.00847	0.00847
S- 60	-0.33	HT ELEC	KWH	0.81	0.62	0.45	0.22	0.09	0.07	0.05	0.05	0.07	0.18	0.43	0.70	3.7
		HT DEM	KW	.00580	.00580	.00580	.00418	.00232	.00104	.00031	.00054	.00238	.00487	.00580	.00580	0.00580
		REM ELEC	KWH	0.78	0.70	0.79	0.78	0.92	1.09	1.24	1.21	0.95	0.84	0.72	0.78	10.8
		REM DEM	KW	.00309	.00309	.00304	.00371	.00438	.00484	.00500	.00515	.00458	.00427	.00309	.00309	0.00515
		TOT ELEC	KWH	1.59	1.32	1.24	0.99	1.01	1.16	1.29	1.26	1.02	1.02	1.16	1.48	14.5
		TOT DEM	KW	.00820	.00812	.00779	.00467	.00394	.00443	.00467	.00476	.00443	.00533	.00779	.00820	0.00820

BASE CASE : 144

STRAT SET	CONST COST $/SF	ENERGY FORM	UNITS (PER SF)	JAN	FEB	MAR	APR	MAY	JUN	JUL	AUG	SEP	OCT	NOV	DEC	ANNUAL
BASE	0.00	TOT GAS	THM	0.0395	0.0300	0.0220	0.0106	0.0046	0.0033	0.0025	0.0024	0.0033	0.0086	0.0211	0.0342	0.182
		TOT ELEC	KWH	0.96	0.86	0.97	0.96	1.13	1.34	1.53	1.49	1.17	1.04	0.89	0.96	13.3
		TOT DEM	KW	.00371	.00371	.00365	.00444	.00505	.00565	.00590	.00608	.00559	.00511	.00371	.00401	0.00608
S- 3	-0.18	TOT GAS	THM	0.0410	0.0312	0.0229	0.0110	0.0047	0.0034	0.0026	0.0025	0.0034	0.0089	0.0219	0.0355	0.189
		TOT ELEC	KWH	0.86	0.77	0.87	0.86	1.01	1.20	1.37	1.33	1.05	0.93	0.80	0.86	11.9
		TOT DEM	KW	.00351	.00351	.00345	.00420	.00477	.00535	.00558	.00575	.00529	.00483	.00351	.00380	0.00575
S- 6	-0.10	TOT GAS	THM	0.0425	0.0323	0.0237	0.0114	0.0049	0.0035	0.0027	0.0025	0.0035	0.0092	0.0227	0.0368	0.196
		TOT ELEC	KWH	0.78	0.70	0.79	0.78	0.92	1.09	1.24	1.21	0.95	0.84	0.72	0.78	10.8
		TOT DEM	KW	.00320	.00320	.00314	.00383	.00435	.00487	.00508	.00524	.00482	.00440	.00320	.00346	0.00524
S- 15	0.95	TOT GAS	THM	0.0443	0.0337	0.0247	0.0118	0.0051	0.0037	0.0029	0.0027	0.0037	0.0096	0.0237	0.0384	0.204
		TOT ELEC	KWH	0.68	0.61	0.69	0.68	0.80	0.95	1.09	1.06	0.83	0.74	0.63	0.68	9.4
		TOT DEM	KW	.00288	.00288	.00283	.00345	.00392	.00439	.00458	.00472	.00434	.00396	.00288	.00312	0.00472
S- 31	0.22	TOT GAS	THM	0.0590	0.0418	0.0320	0.0135	0.0052	0.0020	0.0012	0.0015	0.0039	0.0101	0.0246	0.0492	0.246
		TOT ELEC	KWH	0.74	0.66	0.74	0.76	0.79	1.03	1.11	1.14	0.92	0.79	0.71	0.74	10.1
		TOT DEM	KW	.00298	.00298	.00293	.00356	.00405	.00454	.00473	.00488	.00449	.00410	.00298	.00322	0.00488
S- 60	-0.35	TOT GAS	THM	0.0425	0.0323	0.0237	0.0114	0.0049	0.0035	0.0027	0.0025	0.0035	0.0092	0.0227	0.0368	0.196
		TOT ELEC	KWH	0.78	0.70	0.79	0.78	0.92	1.09	1.24	1.21	0.95	0.84	0.72	0.78	10.8
		TOT DEM	KW	.00320	.00320	.00314	.00383	.00435	.00487	.00508	.00524	.00482	.00440	.00320	.00346	0.00524

BASE CASE : 145

STRAT SET	CONST COST $/SF	ENERGY FORM	UNITS (PER SF)	JAN	FEB	MAR	APR	MAY	JUN	JUL	AUG	SEP	OCT	NOV	DEC	ANNUAL
BASE	0.00	TOT ELEC	KWH	1.89	1.58	1.43	1.13	1.13	1.33	1.39	1.39	1.21	1.13	1.16	1.78	16.6
		TOT DEM	KW	.00985	.00906	.00867	.00453	.00512	.00552	.00571	.00581	.00542	.00512	.00827	.00955	0.00985
S- 1	0.09	TOT ELEC	KWH	1.73	1.44	1.31	1.03	1.03	1.22	1.28	1.28	1.11	1.03	1.06	1.63	15.2
		TOT DEM	KW	.00877	.00807	.00772	.00403	.00456	.00491	.00509	.00517	.00482	.00456	.00737	.00851	0.00877
S- 12	0.19	TOT ELEC	KWH	1.55	1.29	1.17	0.92	0.92	1.09	1.14	1.14	0.99	0.92	0.95	1.46	13.6
		TOT DEM	KW	.00837	.00770	.00737	.00385	.00435	.00469	.00485	.00494	.00460	.00435	.00703	.00812	0.00837
S- 22	1.51	TOT ELEC	KWH	1.40	1.17	1.06	0.84	0.84	0.98	1.03	1.03	0.90	0.84	0.86	1.32	12.3
		TOT DEM	KW	.00798	.00734	.00702	.00367	.00415	.00447	.00463	.00471	.00439	.00415	.00670	.00774	0.00798
S- 28	-0.06	TOT ELEC	KWH	1.55	1.29	1.17	0.92	0.92	1.09	1.14	1.14	0.99	0.92	0.95	1.46	13.6
		TOT DEM	KW	.00837	.00770	.00737	.00385	.00435	.00469	.00485	.00494	.00460	.00435	.00703	.00812	0.00837
S- 36	0.51	TOT ELEC	KWH	1.48	1.24	1.12	0.88	0.88	1.04	1.09	1.09	0.95	0.88	0.91	1.39	13.0
		TOT DEM	KW	.00920	.00846	.00810	.00423	.00478	.00515	.00534	.00543	.00506	.00478	.00773	.00892	0.00920

BASE CASE : 146

STRAT SET	CONST COST $/SF	ENERGY FORM	UNITS (PER SF)	JAN	FEB	MAR	APR	MAY	JUN	JUL	AUG	SEP	OCT	NOV	DEC	ANNUAL
BASE	0.00	TOT GAS	THM	0.0286	0.0210	0.0110	0.0012	0.0012	0.0012	0.0012	0.0012	0.0012	0.0012	0.0039	0.0224	0.095
		TOT ELEC	KWH	1.41	1.17	1.20	1.08	1.08	1.33	1.41	1.41	1.20	1.11	1.08	1.33	14.8
		TOT DEM	KW	.00515	.00453	.00417	.00350	.00397	.00427	.00443	.00448	.00422	.00402	.00391	.00489	0.00515
S- 1	0.09	TOT GAS	THM	0.0342	0.0251	0.0131	0.0015	0.0014	0.0015	0.0015	0.0015	0.0015	0.0015	0.0046	0.0268	0.114
		TOT ELEC	KWH	1.24	1.03	1.05	0.95	0.95	1.17	1.24	1.24	1.05	0.98	0.95	1.17	13.0
		TOT DEM	KW	.00458	.00403	.00371	.00311	.00353	.00380	.00394	.00398	.00376	.00357	.00348	.00435	0.00458

BASE CASE : 146 (CONTINUED)

STRAT SET	CONST COST $/SF	ENERGY FORM	UNITS (PER SF)	JAN	FEB	MAR	APR	MAY	JUN	JUL	AUG	SEP	OCT	NOV	DEC	ANNUAL
S- 12	0.19	TOT GAS	THM	0.0387	0.0284	0.0148	0.0017	0.0016	0.0017	0.0017	0.0017	0.0017	0.0017	0.0052	0.0303	0.129
		TOT ELEC	KWH	1.05	0.88	0.90	0.81	0.81	1.00	1.05	1.05	0.90	0.83	0.81	1.00	11.1
		TOT DEM	KW	.00438	.00385	.00355	.00298	.00337	.00364	.00377	.00381	.00359	.00342	.00333	.00416	0.00438
S- 22	1.51	TOT GAS	THM	0.0486	0.0356	0.0186	0.0021	0.0020	0.0021	0.0021	0.0021	0.0021	0.0021	0.0066	0.0381	0.162
		TOT ELEC	KWH	0.87	0.73	0.75	0.67	0.67	0.83	0.87	0.87	0.75	0.69	0.67	0.83	9.2
		TOT DEM	KW	.00381	.00335	.00309	.00259	.00293	.00316	.00328	.00331	.00312	.00297	.00290	.00362	0.00381
S- 28	-0.06	TOT GAS	THM	0.0387	0.0284	0.0148	0.0017	0.0016	0.0017	0.0017	0.0017	0.0017	0.0017	0.0052	0.0303	0.129
		TOT ELEC	KWH	1.05	0.88	0.90	0.81	0.81	1.00	1.05	1.05	0.90	0.83	0.81	1.00	11.1
		TOT DEM	KW	.00438	.00385	.00355	.00298	.00337	.00364	.00377	.00381	.00359	.00342	.00333	.00416	0.00438
S- 36	0.51	TOT GAS	THM	0.0444	0.0326	0.0170	0.0019	0.0019	0.0019	0.0019	0.0019	0.0019	0.0019	0.0060	0.0348	0.148
		TOT ELEC	KWH	0.97	0.81	0.83	0.74	0.74	0.92	0.97	0.97	0.83	0.77	0.74	0.92	10.2
		TOT DEM	KW	.00515	.00453	.00417	.00350	.00397	.00427	.00443	.00448	.00422	.00402	.00391	.00489	0.00515

BASE CASE : 147

STRAT SET	CONST COST $/SF	ENERGY FORM	UNITS (PER SF)	JAN	FEB	MAR	APR	MAY	JUN	JUL	AUG	SEP	OCT	NOV	DEC	ANNUAL
BASE	0.00	HT ELEC	KWH	1.15	0.87	0.64	0.31	0.13	0.10	0.07	0.07	0.10	0.25	0.61	1.00	5.3
		HT DEM	KW	.00908	.00863	.00772	.00445	.00309	.00209	.00072	.00114	.00281	.00463	.00645	.00908	0.00908
		REM ELEC	KWH	1.07	0.97	1.09	1.07	1.27	1.50	1.71	1.67	1.31	1.16	1.00	1.07	14.9
		REM DEM	KW	.00450	.00450	.00442	.00574	.00667	.00744	.00767	.00775	.00728	.00690	.00450	.00450	0.00775
		TOT ELEC	KWH	2.22	1.84	1.73	1.38	1.40	1.60	1.79	1.74	1.41	1.41	1.61	2.07	20.2
		TOT DEM	KW	.01315	.01223	.00999	.00763	.00763	.00868	.00855	.00855	.00789	.00815	.00986	.01315	0.01315
S- 3	-0.16	HT ELEC	KWH	1.20	0.91	0.67	0.32	0.14	0.10	0.08	0.07	0.10	0.26	0.64	1.04	5.5
		HT DEM	KW	.00908	.00863	.00772	.00445	.00309	.00209	.00072	.00114	.00281	.00463	.00645	.00908	0.00908
		REM ELEC	KWH	0.97	0.88	0.99	0.97	1.15	1.36	1.55	1.51	1.19	1.05	0.90	0.97	13.5
		REM DEM	KW	.00358	.00358	.00352	.00457	.00531	.00592	.00611	.00617	.00580	.00549	.00358	.00358	0.00617
		TOT ELEC	KWH	2.17	1.79	1.66	1.29	1.29	1.46	1.63	1.58	1.29	1.31	1.55	2.01	19.0
		TOT DEM	KW	.01260	.01172	.00958	.00731	.00731	.00832	.00819	.00819	.00756	.00781	.00945	.01260	0.01260
S- 6	-0.07	HT ELEC	KWH	1.27	0.97	0.71	0.34	0.15	0.11	0.08	0.08	0.11	0.28	0.68	1.10	5.9
		HT DEM	KW	.00908	.00863	.00772	.00445	.00309	.00209	.00072	.00114	.00281	.00463	.00645	.00908	0.00908
		REM ELEC	KWH	0.89	0.81	0.91	0.89	1.05	1.25	1.43	1.39	1.09	0.97	0.83	0.89	12.4
		REM DEM	KW	.00347	.00347	.00341	.00443	.00515	.00575	.00593	.00599	.00563	.00533	.00347	.00347	0.00599
		TOT ELEC	KWH	2.17	1.77	1.62	1.23	1.20	1.36	1.51	1.47	1.20	1.24	1.51	2.00	18.3
		TOT DEM	KW	.01220	.01135	.00927	.00708	.00708	.00805	.00793	.00793	.00732	.00756	.00915	.01220	0.01220
S- 15	1.13	HT ELEC	KWH	1.33	1.01	0.74	0.36	0.15	0.11	0.09	0.08	0.11	0.29	0.71	1.16	6.2
		HT DEM	KW	.00908	.00863	.00772	.00445	.00309	.00209	.00072	.00114	.00281	.00463	.00645	.00908	0.00908
		REM ELEC	KWH	0.78	0.71	0.80	0.78	0.93	1.10	1.25	1.22	0.96	0.85	0.73	0.78	10.9
		REM DEM	KW	.00332	.00332	.00327	.00424	.00493	.00550	.00567	.00573	.00539	.00510	.00332	.00332	0.00573
		TOT ELEC	KWH	2.12	1.72	1.54	1.14	1.08	1.21	1.34	1.30	1.07	1.14	1.44	1.94	17.0
		TOT DEM	KW	.01170	.01147	.00959	.00667	.00644	.00714	.00702	.00702	.00655	.00702	.00901	.01170	0.01170
S- 31	0.25	HT ELEC	KWH	0.98	0.78	0.74	0.49	0.31	0.23	0.18	0.18	0.26	0.40	0.65	0.91	6.1
		HT DEM	KW	.00908	.00863	.00772	.00445	.00309	.00209	.00072	.00114	.00281	.00463	.00645	.00908	0.00908
		REM ELEC	KWH	0.86	0.75	0.87	0.84	0.88	1.22	1.30	1.32	1.02	0.88	0.80	0.88	11.6
		REM DEM	KW	.00371	.00371	.00365	.00474	.00550	.00614	.00634	.00640	.00602	.00570	.00371	.00371	0.00640
		TOT ELEC	KWH	1.84	1.53	1.61	1.33	1.19	1.44	1.48	1.51	1.28	1.29	1.45	1.79	17.7
		TOT DEM	KW	.01275	.01249	.01046	.00727	.00701	.00778	.00765	.00765	.00714	.00765	.00982	.01275	0.01275
S- 60	-0.32	HT ELEC	KWH	1.27	0.97	0.71	0.34	0.15	0.11	0.08	0.08	0.11	0.28	0.68	1.10	5.9
		HT DEM	KW	.00908	.00863	.00772	.00445	.00309	.00209	.00072	.00114	.00281	.00463	.00645	.00908	0.00908
		REM ELEC	KWH	0.89	0.81	0.91	0.89	1.05	1.25	1.43	1.39	1.09	0.97	0.83	0.89	12.4
		REM DEM	KW	.00347	.00347	.00341	.00443	.00515	.00575	.00593	.00599	.00563	.00533	.00347	.00347	0.00599
		TOT ELEC	KWH	2.17	1.77	1.62	1.23	1.20	1.36	1.51	1.47	1.20	1.24	1.51	2.00	18.3
		TOT DEM	KW	.01220	.01135	.00927	.00708	.00708	.00805	.00793	.00793	.00732	.00756	.00915	.01220	0.01220

BASE CASE : 148

STRAT SET	CONST COST $/SF	ENERGY FORM	UNITS (PER SF)	JAN	FEB	MAR	APR	MAY	JUN	JUL	AUG	SEP	OCT	NOV	DEC	ANNUAL
BASE	0.00	TOT GAS	THM	0.0603	0.0459	0.0336	0.0161	0.0070	0.0050	0.0039	0.0036	0.0050	0.0131	0.0322	0.0523	0.278
		TOT ELEC	KWH	1.07	0.97	1.09	1.07	1.27	1.50	1.71	1.67	1.31	1.16	1.00	1.07	14.9
		TOT DEM	KW	.00450	.00450	.00442	.00574	.00667	.00744	.00767	.00775	.00728	.00690	.00450	.00450	0.00775
S- 3	-0.17	TOT GAS	THM	0.0631	0.0480	0.0352	0.0169	0.0073	0.0052	0.0041	0.0038	0.0052	0.0137	0.0338	0.0547	0.291
		TOT ELEC	KWH	0.97	0.88	0.99	0.97	1.15	1.36	1.55	1.51	1.19	1.05	0.90	0.97	13.5
		TOT DEM	KW	.00393	.00393	.00386	.00501	.00582	.00650	.00670	.00677	.00636	.00603	.00393	.00393	0.00677
S- 6	-0.08	TOT GAS	THM	0.0668	0.0508	0.0373	0.0179	0.0077	0.0055	0.0043	0.0040	0.0055	0.0145	0.0357	0.0579	0.308
		TOT ELEC	KWH	0.89	0.81	0.91	0.89	1.05	1.25	1.43	1.39	1.09	0.97	0.83	0.89	12.4
		TOT DEM	KW	.00347	.00347	.00341	.00443	.00515	.00575	.00593	.00599	.00563	.00533	.00347	.00347	0.00599
S- 15	1.13	TOT GAS	THM	0.0701	0.0533	0.0391	0.0187	0.0081	0.0058	0.0045	0.0042	0.0058	0.0152	0.0375	0.0607	0.323
		TOT ELEC	KWH	0.78	0.71	0.80	0.78	0.93	1.10	1.25	1.22	0.96	0.85	0.73	0.78	10.9
		TOT DEM	KW	.00332	.00332	.00327	.00424	.00493	.00550	.00567	.00573	.00539	.00510	.00332	.00332	0.00573
S- 31	0.24	TOT GAS	THM	0.0515	0.0409	0.0386	0.0258	0.0164	0.0119	0.0093	0.0097	0.0138	0.0213	0.0341	0.0480	0.322
		TOT ELEC	KWH	0.86	0.75	0.87	0.84	0.88	1.22	1.30	1.32	1.02	0.88	0.80	0.88	11.6
		TOT DEM	KW	.00371	.00371	.00365	.00474	.00550	.00614	.00634	.00640	.00602	.00570	.00371	.00371	0.00640
S- 60	-0.32	TOT GAS	THM	0.0668	0.0508	0.0373	0.0179	0.0077	0.0055	0.0043	0.0040	0.0055	0.0145	0.0357	0.0579	0.308
		TOT ELEC	KWH	0.89	0.81	0.91	0.89	1.05	1.25	1.43	1.39	1.09	0.97	0.83	0.89	12.4
		TOT DEM	KW	.00347	.00347	.00341	.00443	.00515	.00575	.00593	.00599	.00563	.00533	.00347	.00347	0.00599

BASE CASE : 149

STRAT SET	CONST COST $/SF	ENERGY FORM	UNITS (PER SF)	JAN	FEB	MAR	APR	MAY	JUN	JUL	AUG	SEP	OCT	NOV	DEC	ANNUAL
BASE	0.00	HT ELEC	KWH	0.99	0.75	0.55	0.26	0.11	0.08	0.06	0.06	0.08	0.21	0.53	0.86	4.6
		HT DEM	KW	.00580	.00580	.00580	.00418	.00232	.00104	.00031	.00054	.00238	.00487	.00580	.00580	0.00580
		REM ELEC	KWH	1.00	0.90	1.01	1.00	1.18	1.40	1.60	1.56	1.22	1.08	0.93	1.00	13.9
		REM DEM	KW	.00391	.00391	.00385	.00469	.00554	.00613	.00632	.00652	.00580	.00541	.00391	.00391	0.00652
		TOT ELEC	KWH	1.99	1.65	1.57	1.26	1.30	1.49	1.66	1.62	1.31	1.30	1.46	1.86	18.5
		TOT DEM	KW	.00895	.00886	.00886	.00492	.00564	.00609	.00627	.00635	.00582	.00546	.00868	.00895	0.00895

(HT ELEC=ELECTRIC CONSUMPTION FOR SPACE AND WATER HEATING. REM ELEC=REMAINING ELECTRIC CONSUMPTION (NOT USED FOR HEATING). TOT ELEC=TOTAL ELECTRIC CONSUMPTION. HT DEM=PEAK ELECTRIC DEMAND FOR SPACE AND WATER HEATING. REM DEM=PEAK ELECTRIC DEMAND FOR REMAINING LOAD (NOT HEATING). TOT DEM=TOTAL PEAK ELECTRIC DEMAND. GAS=TOTAL GAS CONSUMPTION. ANNUAL FIGURES ARE SUMS OF MONTHLY CONSUMPTION VALUES OR ARE YEARLY MAXIMUMS OF MONTHLY PEAK DEMAND VALUES.)

STRAT SET	CONST COST $/SF	ENERGY FORM	UNITS (PER SF)	JAN	FEB	MAR	APR	MAY	JUN	JUL	AUG	SEP	OCT	NOV	DEC	ANNUAL
BASE CASE : 149 (CONTINUED)																
S- 3	-0.20	HT ELEC	KWH	1.02	0.78	0.57	0.27	0.12	0.08	0.07	0.06	0.08	0.22	0.55	0.89	4.7
		HT DEM	KW	.00580	.00580	.00580	.00418	.00232	.00104	.00031	.00054	.00238	.00487	.00580	.00580	0.00580
		REM ELEC	KWH	0.89	0.81	0.91	0.89	1.05	1.25	1.43	1.39	1.09	0.97	0.83	0.89	12.4
		REM DEM	KW	.00311	.00311	.00306	.00374	.00441	.00488	.00503	.00519	.00462	.00431	.00311	.00311	0.00519
		TOT ELEC	KWH	1.92	1.58	1.48	1.17	1.17	1.34	1.49	1.45	1.18	1.19	1.38	1.78	17.1
		TOT DEM	KW	.00860	.00851	.00851	.00473	.00542	.00585	.00602	.00611	.00559	.00525	.00834	.00860	0.00860
S- 6	-0.12	HT ELEC	KWH	1.07	0.81	0.60	0.29	0.12	0.09	0.07	0.06	0.09	0.23	0.57	0.92	4.9
		HT DEM	KW	.00580	.00580	.00580	.00418	.00232	.00104	.00031	.00054	.00238	.00487	.00580	.00580	0.00580
		REM ELEC	KWH	0.81	0.73	0.82	0.81	0.96	1.14	1.30	1.27	0.99	0.88	0.76	0.81	11.3
		REM DEM	KW	.00302	.00302	.00297	.00363	.00428	.00474	.00489	.00504	.00449	.00418	.00302	.00302	0.00504
		TOT ELEC	KWH	1.88	1.55	1.42	1.10	1.08	1.23	1.37	1.33	1.08	1.11	1.33	1.74	16.2
		TOT DEM	KW	.00829	.00821	.00788	.00473	.00398	.00448	.00473	.00481	.00448	.00539	.00788	.00829	0.00829
S- 15	0.91	HT ELEC	KWH	1.11	0.84	0.62	0.30	0.13	0.09	0.07	0.07	0.09	0.24	0.59	0.96	5.1
		HT DEM	KW	.00580	.00580	.00580	.00418	.00232	.00104	.00031	.00054	.00238	.00487	.00580	.00580	0.00580
		REM ELEC	KWH	0.71	0.64	0.72	0.71	0.84	1.00	1.14	1.11	0.87	0.77	0.66	0.71	9.9
		REM DEM	KW	.00290	.00290	.00285	.00348	.00411	.00454	.00469	.00483	.00430	.00401	.00290	.00290	0.00483
		TOT ELEC	KWH	1.82	1.48	1.34	1.01	0.97	1.09	1.21	1.17	0.96	1.01	1.25	1.67	15.0
		TOT DEM	KW	.00794	.00786	.00754	.00453	.00381	.00429	.00453	.00461	.00429	.00516	.00754	.00794	0.00794
S- 31	0.20	HT ELEC	KWH	1.22	0.87	0.66	0.28	0.11	0.04	0.03	0.03	0.08	0.21	0.51	1.02	5.1
		HT DEM	KW	.00580	.00580	.00580	.00418	.00232	.00104	.00031	.00054	.00238	.00487	.00580	.00580	0.00580
		REM ELEC	KWH	0.72	0.64	0.72	0.74	0.77	1.00	1.08	1.11	0.89	0.77	0.69	0.72	9.8
		REM DEM	KW	.00293	.00293	.00288	.00351	.00415	.00459	.00473	.00488	.00434	.00405	.00293	.00293	0.00488
		TOT ELEC	KWH	1.94	1.50	1.38	1.02	0.87	1.04	1.10	1.14	0.97	0.97	1.20	1.74	14.9
		TOT DEM	KW	.00845	.00837	.00803	.00482	.00406	.00456	.00482	.00490	.00456	.00549	.00803	.00845	0.00845
S- 60	-0.37	HT ELEC	KWH	1.07	0.81	0.60	0.29	0.12	0.09	0.07	0.06	0.09	0.23	0.57	0.92	4.9
		HT DEM	KW	.00580	.00580	.00580	.00418	.00232	.00104	.00031	.00054	.00238	.00487	.00580	.00580	0.00580
		REM ELEC	KWH	0.81	0.73	0.82	0.81	0.96	1.14	1.30	1.27	0.99	0.88	0.76	0.81	11.3
		REM DEM	KW	.00302	.00302	.00297	.00363	.00428	.00474	.00489	.00504	.00449	.00418	.00302	.00302	0.00504
		TOT ELEC	KWH	1.88	1.55	1.42	1.10	1.08	1.23	1.37	1.33	1.08	1.11	1.33	1.74	16.2
		TOT DEM	KW	.00829	.00821	.00788	.00473	.00398	.00448	.00473	.00481	.00448	.00539	.00788	.00829	0.00829
BASE CASE : 150																
BASE	0.00	TOT GAS	THM	0.0519	0.0394	0.0289	0.0139	0.0060	0.0043	0.0033	0.0031	0.0043	0.0112	0.0277	0.0449	0.239
		TOT ELEC	KWH	1.00	0.90	1.01	1.00	1.18	1.40	1.60	1.56	1.22	1.08	0.93	1.00	13.9
		TOT DEM	KW	.00398	.00398	.00391	.00476	.00541	.00606	.00632	.00652	.00600	.00548	.00398	.00430	0.00652
S- 3	-0.22	TOT GAS	THM	0.0538	0.0409	0.0300	0.0144	0.0062	0.0045	0.0035	0.0032	0.0045	0.0117	0.0288	0.0466	0.248
		TOT ELEC	KWH	0.89	0.81	0.91	0.89	1.05	1.25	1.43	1.39	1.09	0.97	0.83	0.89	12.4
		TOT DEM	KW	.00317	.00317	.00311	.00379	.00431	.00483	.00503	.00519	.00477	.00436	.00317	.00343	0.00519
S- 6	-0.14	TOT GAS	THM	0.0560	0.0426	0.0312	0.0150	0.0065	0.0046	0.0036	0.0034	0.0046	0.0121	0.0299	0.0485	0.258
		TOT ELEC	KWH	0.81	0.73	0.82	0.81	0.96	1.14	1.30	1.27	0.99	0.88	0.76	0.81	11.3
		TOT DEM	KW	.00307	.00307	.00302	.00368	.00418	.00469	.00489	.00504	.00464	.00423	.00307	.00333	0.00504
S- 15	0.88	TOT GAS	THM	0.0582	0.0442	0.0324	0.0155	0.0067	0.0048	0.0038	0.0035	0.0048	0.0126	0.0311	0.0504	0.268
		TOT ELEC	KWH	0.71	0.64	0.72	0.71	0.84	1.00	1.14	1.11	0.87	0.77	0.66	0.71	9.9
		TOT DEM	KW	.00295	.00295	.00290	.00353	.00401	.00449	.00469	.00483	.00444	.00406	.00295	.00319	0.00483
S- 31	0.18	TOT GAS	THM	0.0643	0.0456	0.0348	0.0147	0.0056	0.0021	0.0013	0.0016	0.0043	0.0110	0.0268	0.0536	0.268
		TOT ELEC	KWH	0.72	0.64	0.72	0.74	0.77	1.00	1.08	1.11	0.89	0.77	0.69	0.72	9.8
		TOT DEM	KW	.00298	.00298	.00293	.00356	.00405	.00454	.00473	.00488	.00449	.00410	.00298	.00322	0.00488
S- 60	-0.39	TOT GAS	THM	0.0560	0.0426	0.0312	0.0150	0.0065	0.0046	0.0036	0.0034	0.0046	0.0121	0.0299	0.0485	0.258
		TOT ELEC	KWH	0.81	0.73	0.82	0.81	0.96	1.14	1.30	1.27	0.99	0.88	0.76	0.81	11.3
		TOT DEM	KW	.00307	.00307	.00302	.00368	.00418	.00469	.00489	.00504	.00464	.00423	.00307	.00333	0.00504
BASE CASE : 151																
BASE	0.00	TOT ELEC	KWH	2.09	1.74	1.57	1.24	1.24	1.46	1.54	1.54	1.34	1.24	1.28	1.96	18.3
		TOT DEM	KW	.00890	.00819	.00783	.00409	.00463	.00498	.00516	.00525	.00490	.00463	.00748	.00863	0.00890
S- 1	0.09	TOT ELEC	KWH	1.94	1.62	1.46	1.16	1.16	1.36	1.43	1.43	1.24	1.16	1.19	1.82	17.0
		TOT DEM	KW	.00846	.00778	.00744	.00389	.00440	.00474	.00491	.00499	.00465	.00440	.00711	.00821	0.00846
S- 12	0.20	TOT ELEC	KWH	1.74	1.45	1.32	1.04	1.04	1.22	1.29	1.29	1.12	1.04	1.07	1.64	15.3
		TOT DEM	KW	.00748	.00688	.00658	.00344	.00389	.00419	.00434	.00441	.00411	.00389	.00628	.00726	0.00748
S- 22	1.52	TOT ELEC	KWH	1.58	1.32	1.20	0.95	0.95	1.11	1.17	1.17	1.01	0.95	0.97	1.49	13.9
		TOT DEM	KW	.00712	.00655	.00627	.00328	.00370	.00399	.00413	.00420	.00392	.00370	.00598	.00691	0.00712
S- 28	-0.05	TOT ELEC	KWH	1.74	1.45	1.32	1.04	1.04	1.22	1.29	1.29	1.12	1.04	1.07	1.64	15.3
		TOT DEM	KW	.00748	.00688	.00658	.00344	.00389	.00419	.00434	.00441	.00411	.00389	.00628	.00726	0.00748
S- 36	0.52	TOT ELEC	KWH	1.61	1.34	1.21	0.96	0.96	1.13	1.18	1.18	1.03	0.96	0.99	1.51	14.1
		TOT DEM	KW	.00970	.00892	.00854	.00446	.00504	.00543	.00563	.00572	.00534	.00504	.00815	.00941	0.00970
BASE CASE : 152																
BASE	0.00	TOT GAS	THM	0.0272	0.0200	0.0104	0.0012	0.0011	0.0012	0.0012	0.0012	0.0012	0.0012	0.0037	0.0213	0.091
		TOT ELEC	KWH	1.58	1.31	1.34	1.21	1.21	1.49	1.58	1.58	1.34	1.25	1.21	1.49	16.6
		TOT DEM	KW	.00560	.00493	.00454	.00381	.00431	.00465	.00482	.00487	.00459	.00437	.00426	.00532	0.00560
S- 1	0.09	TOT GAS	THM	0.0312	0.0229	0.0120	0.0014	0.0013	0.0014	0.0014	0.0014	0.0014	0.0014	0.0042	0.0244	0.104
		TOT ELEC	KWH	1.42	1.19	1.22	1.10	1.10	1.35	1.42	1.42	1.22	1.13	1.10	1.35	15.0
		TOT DEM	KW	.00530	.00466	.00429	.00360	.00408	.00440	.00456	.00461	.00435	.00413	.00403	.00503	0.00530
S- 12	0.20	TOT GAS	THM	0.0345	0.0253	0.0132	0.0015	0.0014	0.0015	0.0015	0.0015	0.0015	0.0015	0.0047	0.0270	0.115
		TOT ELEC	KWH	1.24	1.03	1.06	0.96	0.96	1.18	1.24	1.24	1.06	0.98	0.96	1.18	13.1
		TOT DEM	KW	.00470	.00414	.00381	.00320	.00362	.00390	.00404	.00409	.00385	.00367	.00357	.00446	0.00470
S- 22	1.52	TOT GAS	THM	0.0408	0.0299	0.0156	0.0018	0.0017	0.0018	0.0018	0.0018	0.0018	0.0018	0.0055	0.0320	0.136
		TOT ELEC	KWH	1.07	0.89	0.92	0.82	0.82	1.02	1.07	1.07	0.92	0.85	0.82	1.02	11.3
		TOT DEM	KW	.00450	.00396	.00365	.00306	.00346	.00374	.00387	.00392	.00369	.00351	.00342	.00427	0.00450
BASE CASE : 152 (CONTINUED)																
S- 28	-0.05	TOT GAS	THM	0.0345	0.0253	0.0132	0.0015	0.0014	0.0015	0.0015	0.0015	0.0015	0.0015	0.0047	0.0270	0.115
		TOT ELEC	KWH	1.24	1.03	1.06	0.96	0.96	1.18	1.24	1.24	1.06	0.98	0.96	1.18	13.1
		TOT DEM	KW	.00470	.00414	.00381	.00320	.00362	.00390	.00404	.00409	.00385	.00367	.00357	.00446	0.00470
S- 36	0.52	TOT GAS	THM	0.0558	0.0409	0.0214	0.0024	0.0023	0.0024	0.0024	0.0024	0.0024	0.0024	0.0075	0.0437	0.186
		TOT ELEC	KWH	1.00	0.83	0.85	0.77	0.77	0.95	1.00	1.00	0.85	0.79	0.77	0.95	10.5
		TOT DEM	KW	.00560	.00493	.00454	.00381	.00431	.00465	.00482	.00487	.00459	.00437	.00426	.00532	0.00560

(HT ELEC=ELECTRIC CONSUMPTION FOR SPACE AND WATER HEATING. REM ELEC=REMAINING ELECTRIC CONSUMPTION (NOT USED FOR HEATING). TOT ELEC=TOTAL ELECTRIC CONSUMPTION. HT DEM=PEAK ELECTRIC DEMAND FOR SPACE AND WATER HEATING. REM DEM=PEAK ELECTRIC DEMAND FOR REMAINING LOAD (NOT HEATING). TOT DEM=TOTAL PEAK ELECTRIC DEMAND. GAS=TOTAL GAS CONSUMPTION. ANNUAL FIGURES ARE SUMS OF MONTHLY CONSUMPTION VALUES OR ARE YEARLY MAXIMUMS OF MONTHLY PEAK DEMAND VALUES.)

Temperate Climate

Identify Region

CITIES IN REGION

CA	Sacramento San Francisco San Jose Santa Barbara
OR	Eugene Portland
WA	Seattle Tacoma

BEGIN YOUR 5 LEVELS OF STRATEGY SELECTION.

Note: See Section II for a full discussion of the five climate regions used in this manual.

Make sure your design comes up to current good practice, which includes meeting ASHRAE 90-75 Standards. The table below identifies good practice that meets 90-75 Standards as well as common design assumptions applicable to all building sizes.

	ALL BUILDINGS	
	GOOD PRACTICE	**DESIGN ASSUMPTIONS**
SHELL	o Perimeter Insulation (R-5) o Infiltration: 0 when HVAC is operating (pressurized building)	o Ventilation: 14 cfm/person during occupied hours
HVAC	o Air conditioning performs to ASHRAE 90-75 (year 1980) minimum efficiencies o Boilers perform with 65% efficiency (no waste heat to space) o Heat pumps perform to ASHRAE 90-75 (year 1980) minimum efficiencies o Controls: · 72 degrees heating, 74 degrees cooling occupied periods · 55 degrees heating, 95 degrees cooling unoccupied periods · 2 degrees temperature float · outside air temperature reset for hot water boilers · low leakage dampers (less than 3%) · zone control for all heat pump and multi-zone systems	
INTERNAL LOAD		o Occupant density: 200 square feet per person o Lights: 2.25 watts/sq. ft. o Office equipment: .75 watts/sq. ft.
SCHEDULE	o HVAC comes on one hour before occupied period, turned to setback during unoccupied period	o Inside lighting: 100% during operating hours, 5% at other times o Outside lighting: 100% 6pm til midnight

Note: See Section II, page 179, for detailed information on these Level 1 strategies and assumptions.

Some aspects of ASHRAE 90-75 and current **good practice** vary according to building size and HVAC system type. Make sure your design comes up to these criteria. You may want to come back to this table as you explore HVAC options in Level 2.

		BUILDING SIZE								
		SMALL 2000-10000 sq ft			MEDIUM 10000-25000 sq ft			LARGE 25000-50000 sq ft		
SHELL	ROOF INSULATION	R-8.3			R-8.3			R-8.3		
	WALL INSULATION	R-7.5			R-7.5			R-7.5		
	GLAZING	DOUBLE GLAZED WITH CLEAR GLASS			DOUBLE GLAZED WITH TINTED GLASS			DOUBLE GLAZED WITH TINTED GLASS		
	AIR CHANGES	1/2 AC/HR UNOCCUPIED			1/3 AC/HR UNOCCUPIED			1/3 AC/HR UNOCCUPIED		
HVAC		PACKAGED TERMINAL AIR CONDITIONERS	ROOFTOP DX	HEAT PUMP	CONSTANT VOLUME	VARIABLE AIR VOLUME	SPLIT HEAT PUMP	CONSTANT VOLUME	VARIABLE AIR VOLUME	HYDRONIC HEAT PUMP
	TEMPERATURE CONTROL	ZONE	OA RESET	ZONE	OA RESET	FIXED	ZONE	OA RESET	FIXED	ZONE
	AIR CONTROL	FIXED VOLUME	ENTHALPY ECONOMIZER	FIXED VOLUME	ENTHALPY ECONOMIZER	ENTHALPY ECONOMIZER	ENTHALPY ECONOMIZER	ENTHALPY ECONOMIZER	ENTHALPY ECONOMIZER	FIXED VOLUME

Some common **design assumptions** also vary according to building size and no. of stories.

		SMALL	MEDIUM		LARGE	
		1 STORY	1 STORY	2 STORY	2 STORY	4 STORY ①
INT. LOAD	ELEVATOR & OUTDOOR LIGHTS	1345 WATTS	4035 WATTS	4535 WATTS	7860 WATTS	9360 WATTS
SCHEDULE	OCCUPANCY	80% 8am - 6pm Monday - Friday	80% 8am - 6pm Monday - Friday		80% 8am - 6pm: M-F 100% 9am - 1pm: SAT	

GO TO THE LEVEL 2 PAGE SPECIFIC TO YOUR BUILDING SIZE & NO. OF STORIES.

Note ① For 3 or 5 story structures, use "4-story" classification.

LEVEL 2
MAKE INITIAL CHOICES
SMALL/1-STORY

Project____________________
date__________ bldg. sq. ft.__________

The building characteristic choices shown here make a **big** difference in energy cost savings. Investigate these choices within the requirements of your design.

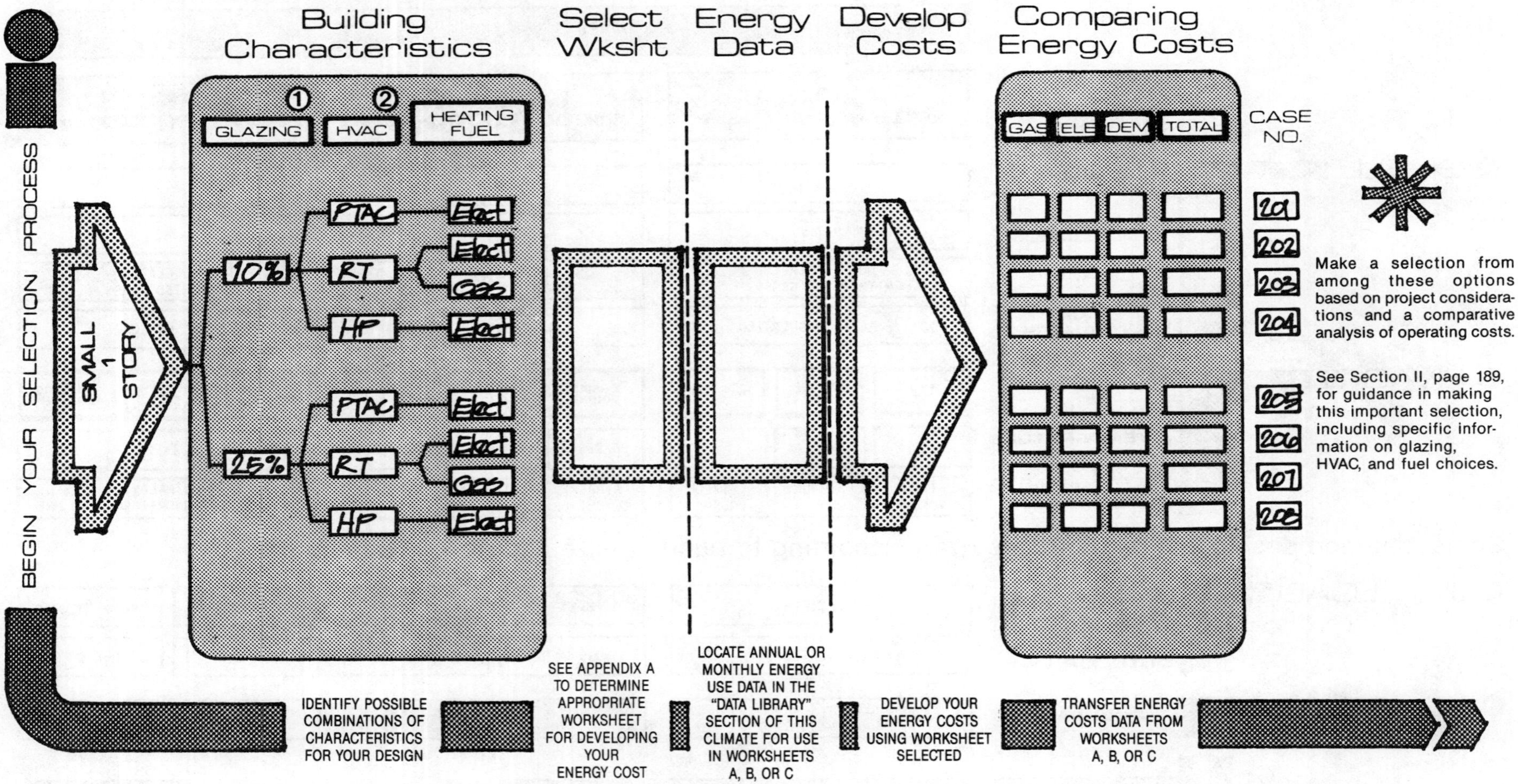

Note: See Section II, page 189, for detailed information on these Level 2 strategies and the use of this form.
① For glazing <17% use 10%, for >17% use 25%.
② PTAC = Packaged Terminal A/C, RT = Rooftop DX, HP = Packaged Terminal Heat Pump.

SMALL/1-STORY

Project________________________________
date______________ bldg. sq. ft. ______________

Locate the row of strategy set options that apply to the building case you chose in Level 2. Select as many of these Level 3 strategy set options as you wish for valuation.

STRATEGY SET OPTIONS

SMALL 1 STORY

SELECTED LEVEL 2 CASE NUMBER	SIMPLEST LIGHTING IMPROVEMTS	MODEST LIGHTING & MECHANICAL IMPROVEMTS	MAJOR LIGHTING & MECHANICAL IMPROVEMTS	DAYLIGHT DIMMING CONTROLS AND LIGHTING IMPROVEMTS
201	1	10 or 25	14	
202	3	6 or 60	15	
203	3	6 or 60	15	
204	1	12 or 28	22	
205	1	10 or 25	20	30
206	3	6 or 60	15	31
207	3	6 or 60	15	31
208	1	12 or 28	22	36

DESCRIPTOR KEY

1. High eff. ballasts over all bldg.
3. High eff. ballasts over all bldg.+HVAC downsized for ballasts.
6. High eff. ballasts and lamps over all building+HVAC downsized for ballasts.
10. High eff. ballasts and lamps over all bldg.+cooling COP improvement.
12. High eff. ballasts and lamps over all bldg.+heat pump COP improvement.
14. High eff. ballasts and lamps+high eff. fixtures over all bldg.
15. High eff. ballasts, lamps and fixtures over all bldg.+HVAC downsized for ballasts and fixtures.
20. High eff. ballasts, lamps and fixtures over all bldg.+cooling COP improvement.
22. High eff. ballasts, lamps and fixtures over all bldg.+heat pump COP improvement.
60. Non-uniform lighting+high eff. ballasts over all bldg.+HVAC downsized for ballasts.
25. Non-uniform lighting+high eff. ballasts over all bldg.+cooling COP improvement.
28. Non-uniform lighting+high eff. ballasts over all bldg.+heat pump COP improvement.
30. Perimeter daylighting dimmers+high eff. ballasts over all bldg.
31. Perimeter daylighting dimmers+high eff. ballasts over all bldg.+HVAC downsized for ballasts.
36. Perimeter daylighting dimmers+high eff. ballasts over all bldg.+heat pump COP improvement.

GO TO LEVEL 3 ANALYSIS SHEET.

Note: See Section II, page 209, for detailed descriptions and discussions of these and other Level 3 strategy sets.

MEDIUM/1-STORY

Project________________________________
date____________ bldg. sq. ft.______________

The building characteristic choices shown here make a **big** difference in energy cost savings. Investigate these choices within the requirements of your design.

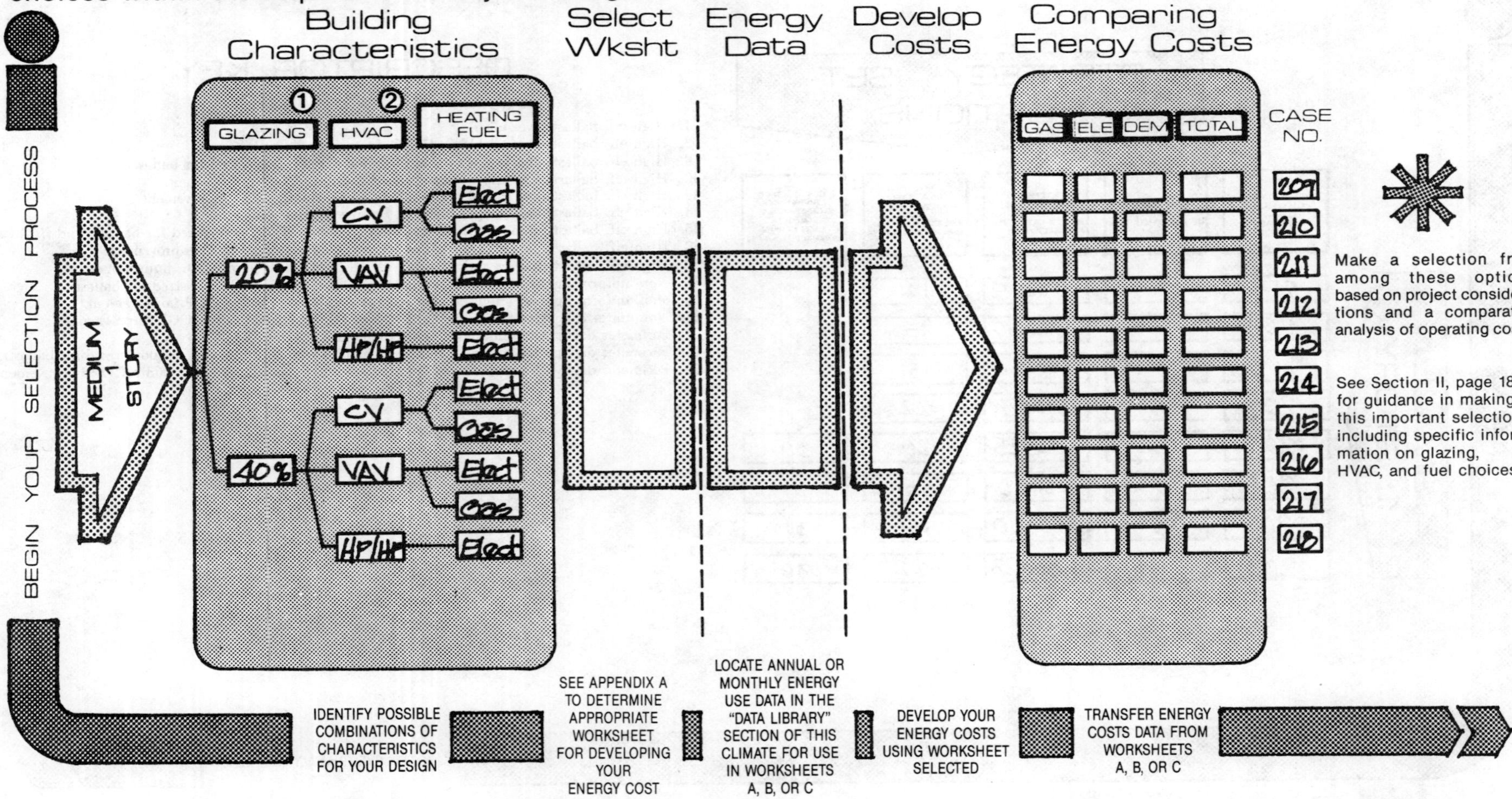

Make a selection from among these options based on project considerations and a comparative analysis of operating costs.

See Section II, page 189, for guidance in making this important selection, including specific information on glazing, HVAC, and fuel choices.

Note: See Section II, page 189, for detailed information on these Level 2 strategies and the use of this form.

① For glazing <30% use 20%, for >30% use 40%.

② CV=Constant Volume, VAV=Variable Air Volume, HP/HP=Split System Heat Pump.

Project ______________________
date __________ bldg. sq. ft. __________

Locate the row of strategy set options that apply to the building case you chose in Level 2. Select as many of these Level 3 strategy set options as you wish for valuation.

STRATEGY SET OPTIONS

MEDIUM 1 STORY

SELECTED LEVEL 2 CASE NUMBER	SIMPLEST LIGHTING IMPROVEMTS	MODEST LIGHTING & MECHANICAL IMPROVEMTS	MAJOR LIGHTING & MECHANICAL IMPROVEMTS	DAYLIGHT DIMMING CONTROLS AND LIGHTING IMPROVEMTS	SKYLIGHT'G, DAYLIGHT DIMMING CONTROLS & LIGHTING IMPROVEMTS
209	3	6 or 60	15	31 or 44 or 39	49
210	3	6 or 60	15	31 or 44 or 39	49
211	3	6 or 60	15	31 or 44 or 39	49
212	3	6 or 60	15	31 or 44 or 39	49
213	3	6 or 60	15	31 or 44 or 39	49
214	3	6 or 60	15	31 or 44 or 39	49
215	3	6 or 60	15	31 or 44 or 39	49
216	3	6 or 60	15	31 or 44 or 39	49
217	3	6 or 60	15	31 or 44 or 39	49
218	3	6 or 60	15	31 or 44 or 39	49

DESCRIPTOR KEY

3. High eff. ballasts over all bldg.+HVAC downsized for ballasts.
6. High eff. ballasts and lamps over all building+HVAC downsized for ballasts.
15. High eff. ballasts, lamps and fixtures over all bldg.+HVAC downsized for ballasts and fixtures.
60. Non-uniform lighting+high eff. ballasts over all bldg.+HVAC downsized for ballasts.
31. Perimeter daylighting dimmers + high eff. ballasts over all bldg. + HVAC downsized for ballasts.
39. Perimeter daylighting dimmers+high eff. ballasts over all bldg.+high eff. lamps and fixtures in core+HVAC downsized for ballasts and fixtures.
44. Perimeter daylighting dimmers+high eff. ballasts over all bldg.+high eff. lamps in core+HVAC downsized for ballasts.
49. Perimeter daylighting dimmers+skylights with dimmers+high eff. ballasts over all bldg.+HVAC downsized for ballasts.

GO TO LEVEL 3 ANALYSIS SHEET.

Note: See Section II, page 209, for detailed descriptions and discussions of these and other Level 3 strategy sets.

LEVEL 2 – MAKE INITIAL CHOICES

MEDIUM/2-STORY

Project ______________________

date ____________ bldg. sq. ft. ____________

The building characteristic choices shown here make a **big** difference in energy cost savings. Investigate these choices within the requirements of your design.

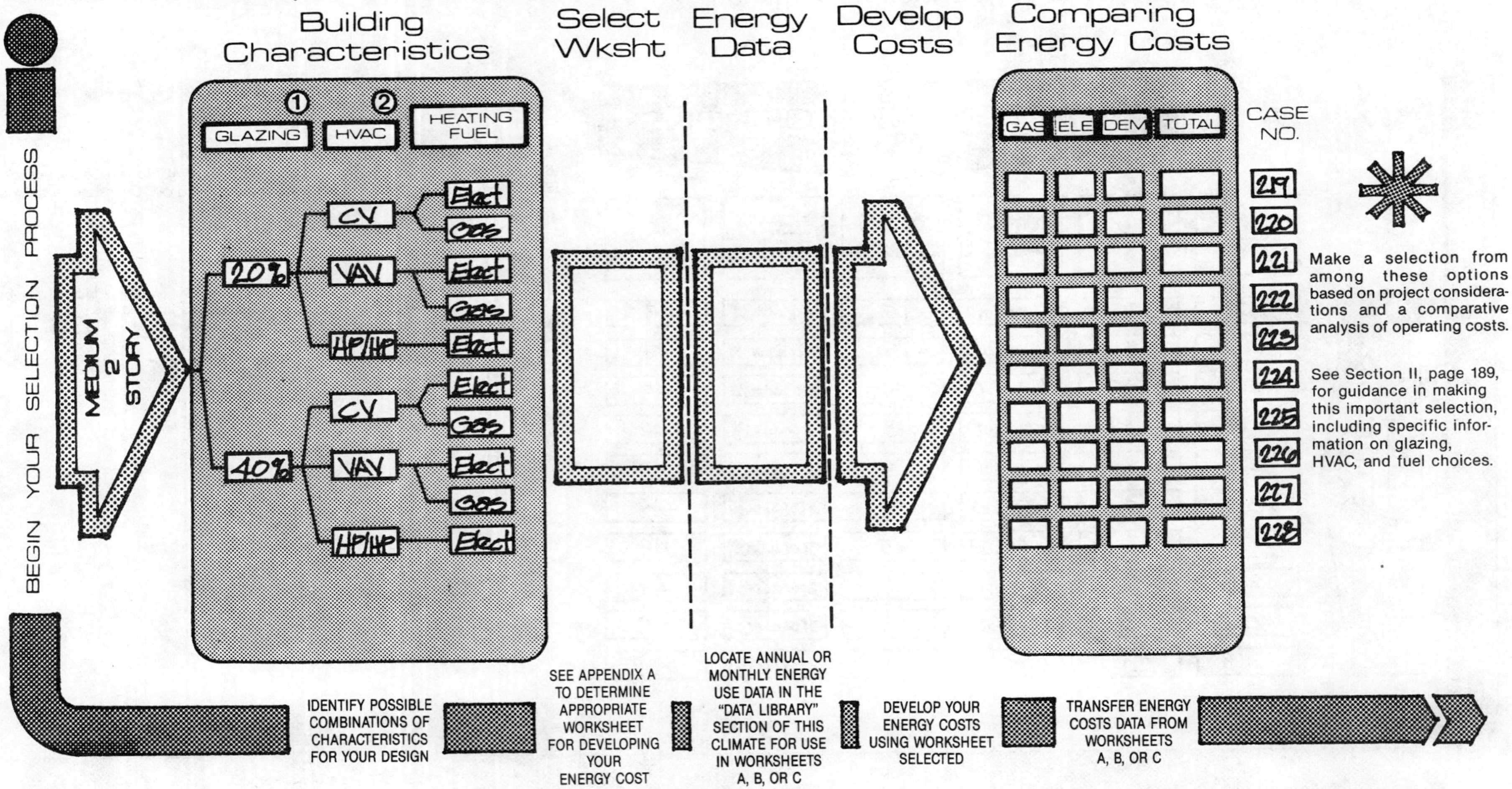

Note: See Section II, page 189, for detailed information on these Level 2 strategies and the use of this form.

① For glazing <30% use 20%, for >30% use 40%.

② CV=Constant Volume, VAV=Variable Air Volume, HP/HP=Split System Heat Pump.

Project ______________________
date __________ bldg. sq. ft. __________

Locate the row of strategy set options that apply to the building case you chose in Level 2. Select as many of these Level 3 strategy set options as you wish for valuation.

STRATEGY SET OPTIONS

MEDIUM 2 STORY

SELECTED LEVEL 2 CASE NUMBER	SIMPLEST LIGHTING IMPROVEMTS	MODEST LIGHTING & MECHANICAL IMPROVEMTS	MAJOR LIGHTING & MECHANICAL IMPROVEMTS	DAYLIGHT DIMMING CONTROLS AND LIGHTING IMPROVEMTS
219	3	6 or 60	15	31
220	3	6 or 60	15	31
221	3	6 or 60	15	31
222	3	6 or 60	15	31
223	3	6 or 60	15	31
224	3	6 or 60	15	31
225	3	6 or 60	15	31
226	3	6 or 60	15	31
227	3	6 or 60	15	31
228	3	6 or 60	15	31

DESCRIPTOR KEY

3. High eff. ballasts over all bldg.+HVAC downsized for ballasts.
6. High eff. ballasts and lamps over all building+HVAC downsized for ballasts.
15. High eff. ballasts, lamps and fixtures over all bldg.+HVAC downsized for ballasts and fixtures.
60. Non-uniform lighting+high eff. ballasts over all bldg.+HVAC downsized for ballasts.
31. Perimeter daylighting dimmers+high eff. ballasts over all bldg.+HVAC downsized for ballasts.

GO TO LEVEL 3 ANALYSIS SHEET.

Note: See Section II, page 209, for detailed descriptions and discussions of these and other Level 3 strategy sets.

LEVEL 2

MAKE INITIAL CHOICES

LARGE/2-STORY

Project ______________________
date __________ bldg. sq. ft. __________

The building characteristic choices shown here make a **big** difference in energy cost savings. Investigate these choices within the requirements of your design.

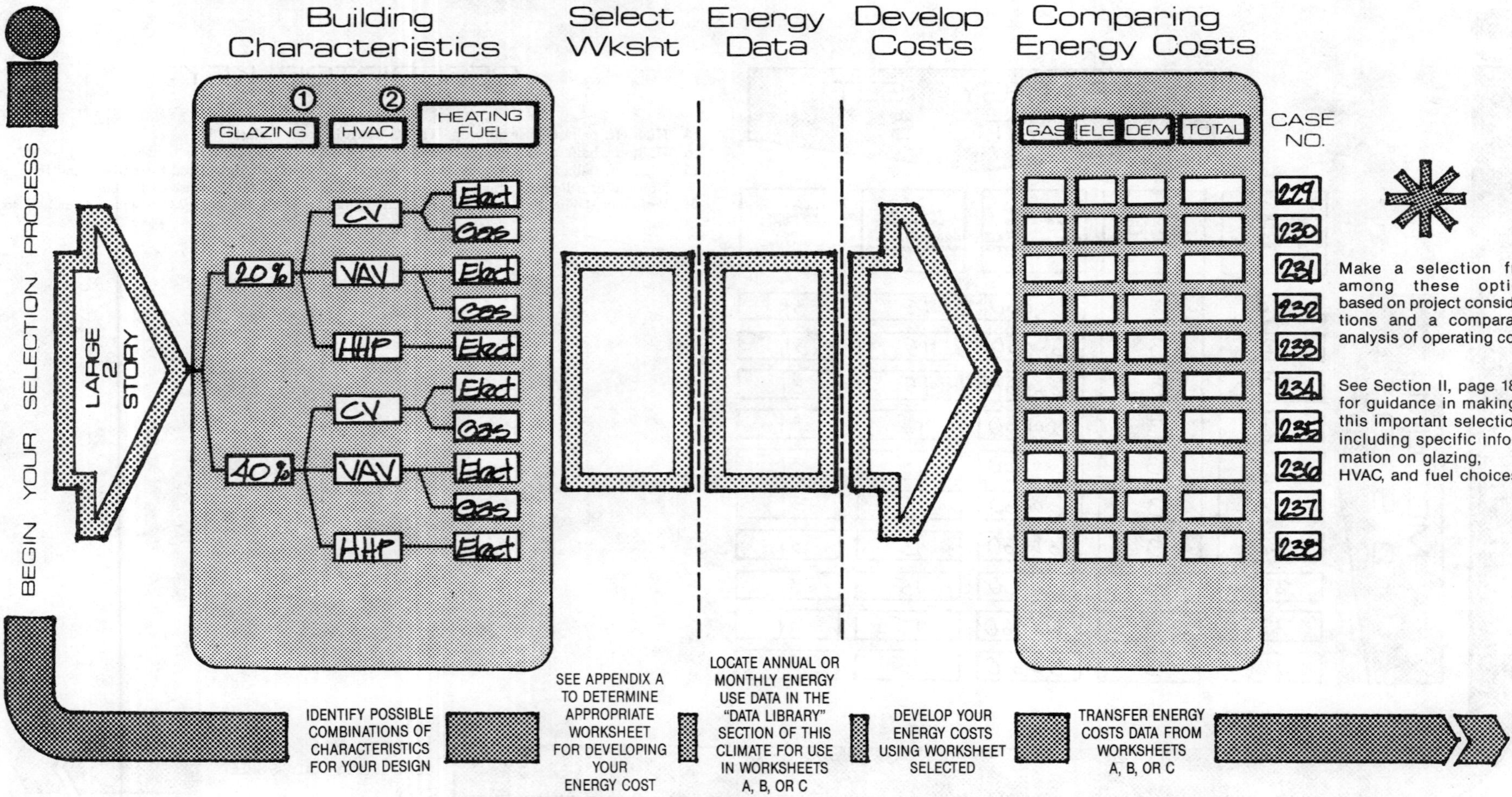

Make a selection from among these options based on project considerations and a comparative analysis of operating costs.

See Section II, page 189, for guidance in making this important selection, including specific information on glazing, HVAC, and fuel choices.

Note: See Section II, page 189, for detailed information on these Level 2 strategies and the use of this form.
① For glazing <30% use 20%, for >30% use 40%.
② CV=Constant Volume, VAV=Variable Air Volume, HHP=Hydronic Heat Pump.

LARGE/2-STORY

Project____________________________
date______________ bldg. sq. ft.______________

Locate the row of strategy set options that apply to the building case you chose in Level 2. Select as many of these Level 3 strategy set options as you wish for valuation.

STRATEGY SET OPTIONS

LARGE 2 STORY

SELECTED LEVEL 2 CASE NUMBER	SIMPLEST LIGHTING IMPROVEMTS	MODEST LIGHTING & MECHANICAL IMPROVEMTS	MAJOR LIGHTING & MECHANICAL IMPROVEMTS	DAYLIGHT DIMMING CONTROLS AND LIGHTING IMPROVEMTS
229	3	6 or 60	15	31 or 44 or 39
230	3	6 or 60	15	31 or 44 or 39
231	3	6 or 60	15	31 or 44 or 39
232	3	6 or 60	15	31 or 44 or 39
233	1	5 or 24	14	30 or 37 or 38
234	3	6 or 60	15	31 or 44 or 39
235	3	6 or 60	15	31 or 44 or 39
236	3	6 or 60	15	31 or 44 or 39
237	3	6 or 60	15	31 or 44 or 39
238	1	5 or 24	14	30 or 37 or 38

DESCRIPTOR KEY

1. High eff. ballasts over all bldg.
3. High eff. ballasts over all bldg.+HVAC downsized for ballasts.
5. High eff. ballasts+high eff. fluorescent lamps over all bldg.
6. High eff. ballasts and lamps over all building+HVAC downsized for ballasts.
14. High eff. ballasts and lamps+high eff. fixtures over all bldg.
15. High eff. ballasts, lamps and fixtures over all bldg.+HVAC downsized for ballasts and fixtures.
24. Non-uniform lighting+high eff. ballasts over all bldg.
60. Non-uniform lighting+high eff. ballasts over all bldg.+HVAC downsized for ballasts.
30. Perimeter daylighting dimmers+high eff. ballasts over all bldg.
31. Perimeter daylighting dimmers+high eff. ballasts over all bldg.+HVAC downsized for ballasts.
37. Perimeter daylighting dimmers+high eff. ballasts over all bldg.+high eff. lamps in core.
38. Perimeter daylighting dimmers+high eff. ballasts over all bldg.+high eff. lamps and fixtures in core.
39. Perimeter daylighting dimmers+high eff. ballasts over all bldg.+high eff. lamps and fixtures in core+HVAC downsized for ballasts and fixtures.
44. Perimeter daylighting dimmers+high eff. ballasts over all bldg.+high eff. lamps in core+HVAC downsized for ballasts.

GO TO LEVEL 3 ANALYSIS SHEET.

Note: See Section II, page 209, for detailed descriptions and discussions of these and other Level 3 strategy sets.

LEVEL 2
MAKE INITIAL CHOICES
LARGE/4-STORY

Project ______________________

date __________ bldg. sq. ft. __________

The building characteristic choices shown here make a **big** difference in energy cost savings. Investigate these choices within the requirements of your design.

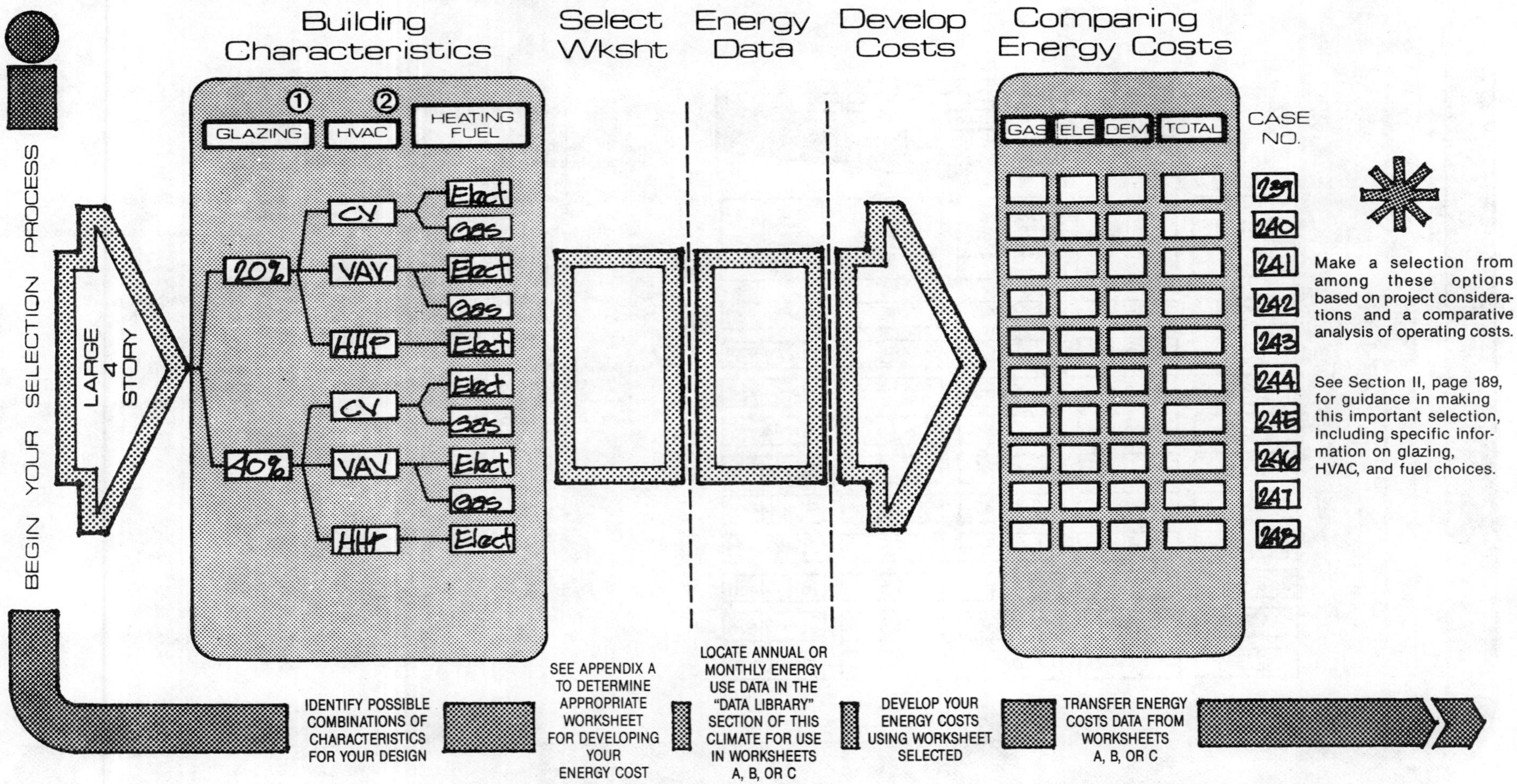

Note: See Section II, page 189, for detailed information on these Level 2 strategies and the use of this form.

① For glazing <30% use 20%, for >30% use 40%.

② CV=Constant Volume, VAV=Variable Air Volume, HHP=Hydronic Heat Pump.

Project______________________
date__________ bldg. sq. ft.__________

Locate the row of strategy set options that apply to the building case you chose in Level 2. Select as many of these Level 3 strategy set options as you wish for valuation.

LARGE 4 STORY

STRATEGY SET OPTIONS

SELECTED LEVEL 2 CASE NUMBER	SIMPLEST LIGHTING IMPROVEMTS	MODEST LIGHTING & MECHANICAL IMPROVEMTS	MAJOR LIGHTING & MECHANICAL IMPROVEMTS	DAYLIGHT DIMMING CONTROLS AND LIGHTING IMPROVEMTS
239	3	6 or 60	15	31
240	3	6 or 60	15	31
241	3	6 or 60	15	31
242	3	6 or 60	15	31
243	1	5 or 24	14	30
244	3	6 or 60	15	31
245	3	6 or 60	15	31
246	3	6 or 60	15	31
247	3	6 or 60	15	31
248	1	5 or 24	14	30

DESCRIPTOR KEY

1. High eff. ballasts over all bldg.
3. High eff. ballasts over all bldg. + HVAC downsized for ballasts.
5. High eff. ballasts + high eff. fluorescent lamps over all bldg.
6. High eff. ballasts and lamps over all building + HVAC downsized for ballasts.
14. High eff. ballasts and lamps + high eff. fixtures over all bldg.
15. High eff. ballasts, lamps and fixtures over all bldg. + HVAC downsized for ballasts and fixtures.
24. Non-uniform lighting + high eff. ballasts over all bldg.
60. Non-uniform lighting + high eff. ballasts over all bldg. + HVAC downsized for ballasts.
30. Perimeter daylighting dimmers + high eff. ballasts over all bldg.
31. Perimeter daylighting dimmers + high eff. ballasts over all bldg. + HVAC downsized for ballasts.

GO TO LEVEL 3 ANALYSIS SHEET.

Note: See Section II, page 209, for detailed descriptions and discussions of these and other Level 3 strategy sets.

LEVEL 3 –
SELECT STRATEGY SETS
ANALYSIS/INSTRUCTIONS

For each strategy set you want to investigate, develop its construction costs using the upper route and its energy costs using the lower route. The combined information will allow you to select a strategy set for optimum economic benefit.

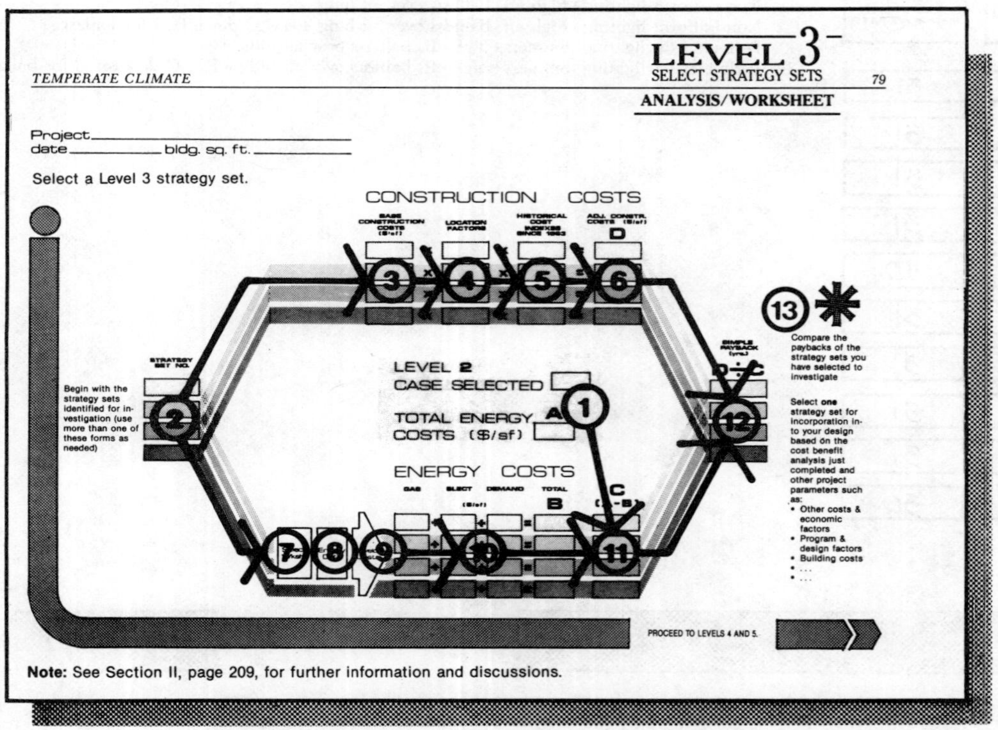

(1) Enter the Level 2 case selected and its total energy costs as developed in Level 2.

(2) As many as four strategy sets for any *one* Level 2 case can be analyzed on each sheet. You will need to use more sheets to analyze more than four strategy sets or to analyze strategy sets for another Level 2 case.

(3) Obtain *base construction costs* ($/SF) for each strategy set from the Data Library.

(4) Obtain *location factors* from commonly used cost estimating handbooks. Additional multipliers may be entered to account for above or below average construction costs due to specific project conditions.

(5) Obtain *historical cost indexes* (factors since 1983) from commonly used cost estimating handbooks.

(6) Multiply 3, 4, and 5 to obtain *adjusted construction costs* ($/SF). User may wish to use more refined construction cost data as supplied by a cost estimator for the specific project.

(7) Use the same worksheet you selected for use in Level 2.

(8) Locate energy use data for the selected strategy sets in the Data Library.

(9) Develop energy costs for each strategy set using the worksheet selected.

(10) Transfer energy cost information from worksheets A, B, or C.

(11) Find the energy savings from each strategy set by subtracting each "B" from "A."

(12) Calculate *simple payback* by dividing "D" (initial costs) by "C" (yearly savings). Alternative economic analysis techniques other than simple payback may be employed using information from columns "D" and "C."

(13) Cost factors to consider in making a decision on these strategy sets include:
- Negative construction costs (first cost savings)
- Positive construction costs
- Annual energy cost savings

ANALYSIS/WORKSHEET

Project______________________________
date____________ bldg. sq. ft.____________

Select a Level 3 strategy set.

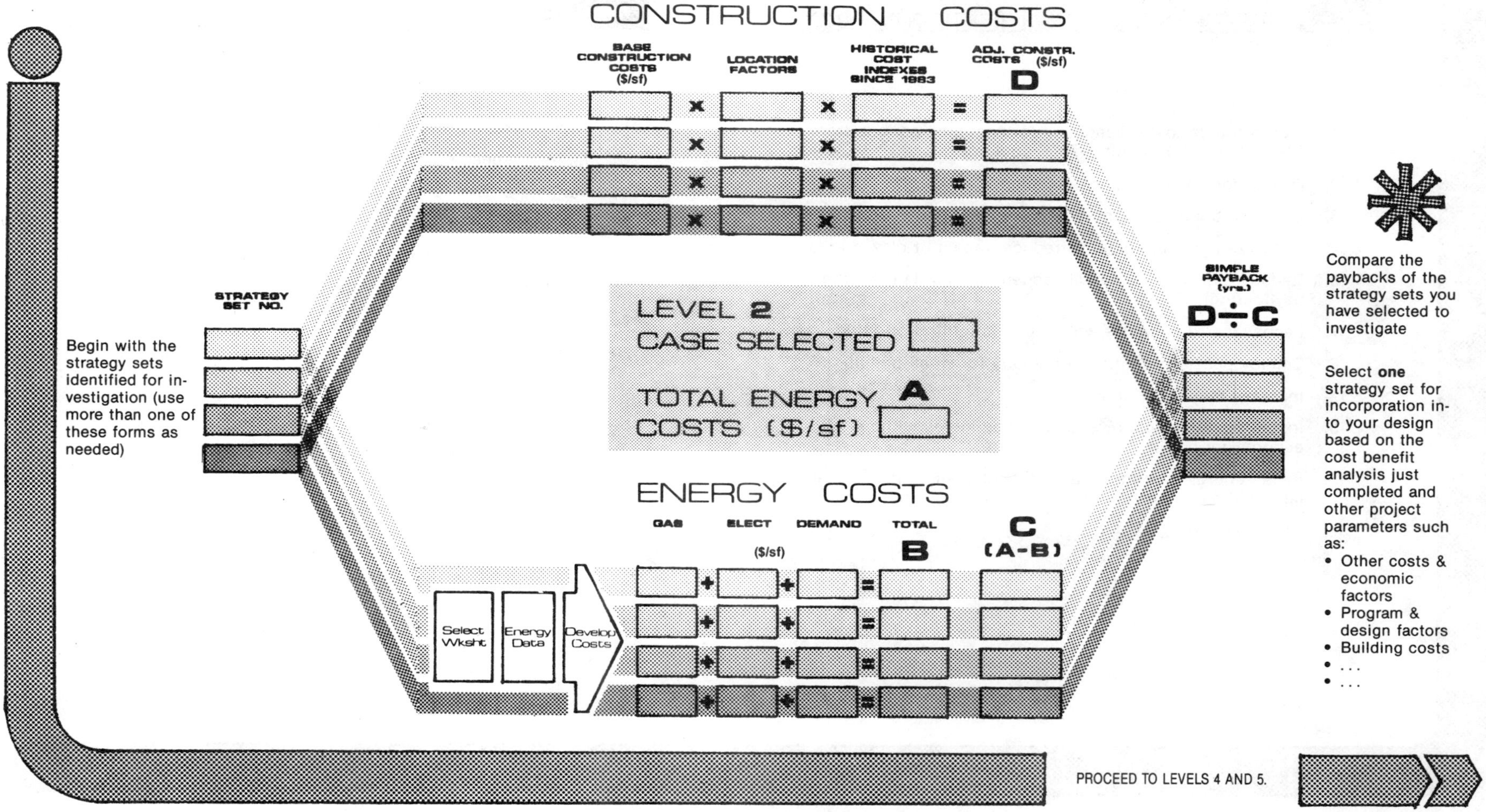

Note: See Section II, page 209, for further information and discussions.

Review these additional strategies for the possibility of further savings. Check the description in Section II, page 235, carefully for guidance on adopting any of the strategies.

A SELECT ON THE BASIS OF TECHNICAL JUDGEMENT

1. Increase thermostat deadband for occupied periods from 2°F to 8°F (70°F heating–78°F cooling)
2. Design for efficient space planning
3. Optimize building orientation
4. Minimize building height to reduce impact of glazed area
6. Reduce glazing to minimally acceptable levels (not with daylighting)
7. Landscape to minimize wind induced infiltration effect
8. Employ passive heating (particularly for small buildings)
12. Increase manual light switching
13. Install operable windows (best for buildings without economizer cycle)
15. Install energy monitoring and control system (medium and large buildings only)

B VALUE MAY BE DETERMINED BY SIMPLE CALCULATION

19. High efficiency motors and pumps
20. Multiple speed compressors
21. Variable or two speed circulating fan
22. Electric resistance heaters
23. Desuperheater or stand-alone heat pump for DHW
24. Point-of-use electric domestic hot water heaters
25. External shading
29. Demand limiter
31. Cogeneration

Note: The number preceding each strategy refers to the corresponding paragraph in Section II, page 235. See Section II, page 235, for further information and discussion of each strategy.

Review this checklist and adopt the procedures that are appropriate to your project. These actions are **important** if you are to achieve and maintain the savings from the selected Level 1–4 strategies.

A CHECKLIST FOR SUCCESSFUL TURNOVER AND BUILDING OPERATION

	DESIGN	CONSTRUCTION	TURNOVER	BUILDING OPERATION
DESIGN TEAM (ARCHITECTS/ ENGINEERS)	• Document all design assumptions (1) • Estimate energy use by sub-system (2) **& For Large Buildings:** • Document all systems and control strategies (3) • Specify contents of O&M Manuals (4) • Specify HVAC balancing requirements (5) • Specify equipment protection and cleaning (6) • Specify on-site instruction (7) • Specify access doors, valves, test ports (8) • Specify piping system ID (9) • Specify price quotations on all long-term maintenance contracts (10)	• Prepare "as-built" drawings (11) • Observe construction process for compliance (12)	• Conduct "punch list" inspection (15) • Visually check for air leakage (15) • If possible, participate in on-site training session(s) (7)	
CONTRACTOR & SUB-CONTRACTOR	**For Small Buildings Only:** • Document all systems and control strategies (3) • Specify contents of O&M Manuals (4) • Specify HVAC balancing requirements (5) • Specify equipment protection and cleaning (6) • Specify on-site instruction (7) • Specify access doors, valves, test ports (8) • Specify piping system ID (9) • Specify price quotations on all long-term maintenance contracts (10)	• Prepare O&M Manual (4) • Perform final calibration of all controls (5) • Hire independent balancing agent or require of subs (5) • Protect and clean equipment (6) • Provide all access doors, valves, and test ports (8) • Place piping system ID tags (9) • Prepare "as-built" drawings (11) • Use HVAC system to "flush" construction odors (13)	• Deliver O&M Manual (4) • Conduct on-site instruction and training (7)	• Honor warranty and call back as required (16)
EQUIPMENT MANUFACT.		• Support calibration of controls (5) • Provide all special tools required (14)	• Support O&M Manual preparation (4) • Support on-site instruction and training (7)	
OWNER & BLDG. STAFF	• Check all key design assumptions (1)	• Periodically check compliance on-site (12)	• Attend on-site instruction and training sessions (7)	• Monitor energy consumption against estimates (2) • Periodically test equipment efficiency (17) • Periodically sample and test water (18) • Inspect and calibrate controls (19) • Monitor occupant complains (20)

Note: See Section II, page 255, for more information and discussion. Numbers shown in ( ) refer to the numbered paragraphs in Section II, page 255.

1. Locate data by Base Case number.
2. Locate monthly or annual data for Base Case or Strategy Set.

Use monthly data for Worksheets B or C

Use annual data for Worksheet A

BASE CASE 326

STRAT SET	CONST COST $/SF	ENERGY FORM	UNITS (PER SF)	JAN	FEB	MAR	APR	MAY	JUN	JUL	AUG	SEP	OCT	NOV	DEC	ANNUAL
BASE	0.00	HT ELEC	KWH	0.76	0.53	0.31	0.09	0.03	0.03	0.03	0.03	0.03	0.05	0.23	0.65	2.8
		HT DEM	KW	.00605	.00587	.00575	.00405	.00369	.00006	.00006	.00006	.00224	.00363	.00520	.00587	0.00605
		REM ELEC	KWH	0.97	0.87	0.97	1.06	1.23	1.47	1.70	1.63	1.34	1.10	0.93	0.97	14.3
		REM DEM	KW	.00504	.00457	.00543	.00651	.00713	.00698	.00775	.00752	.00721	.00635	.00519	.00473	0.00775
		TOT ELEC	KWH	1.74	1.40	1.28	1.15	1.26	1.50	1.73	1.66	1.37	1.15	1.15	1.63	17.1
		TOT DEM	KW	.00930	.00828	.00828	.00698	.00716	.00763	.00781	.00763	.00744	.00670	.00679	.00837	0.00930
S- 3	-0.24	HT ELEC	KWH	0.81	0.56	0.32	0.10	0.03	0.03	0.03	0.03	0.03	0.05	0.24	0.69	3.0
		HT DEM	KW	.00605	.00587	.00575	.00405	.00369	.00006	.00006	.00006	.00224	.00363	.00520	.00587	0.00605
		REM ELEC	KWH	0.88	0.79	0.88	0.96	1.12	1.34	1.55	1.48	1.22	1.00	0.85	0.88	13.0
		REM DEM	KW	.00464	.00421	.00500	.00600	.00657	.00643	.00714	.00693	.00664	.00585	.00478	.00436	0.00714

BASE CASE 327

STRAT SET	CONST COST $/SF	ENERGY FORM	UNITS (PER SF)	JAN	FEB	MAR	APR	MAY	JUN	JUL	AUG	SEP	OCT	NOV	DEC	ANNUAL
BASE	0.00	TOT GAS	THM	0.0402	0.0277	0.0161	0.0048	0.0015	0.0015	0.0015	0.0015	0.0015	0.0026	0.0118	0.0343	0.146
		TOT ELEC	KWH	0.97	0.87	0.97	1.06	1.23	1.47	1.70	1.63	1.34	1.10	0.93	0.97	14.3
		TOT DEM	KW	.00504	.00457	.00543	.00651	.00713	.00698	.00775	.00752	.00721	.00635	.00519	.00473	0.00775
S- 3	-0.24	TOT GAS	THM	0.0426	0.0295	0.0171	0.0051	0.0015	0.0015	0.0015	0.0015	0.0015	0.0028	0.0126	0.0364	0.155
		TOT ELEC	KWH	0.88	0.79	0.88	0.96	1.12	1.34	1.55	1.48	1.22	1.00	0.85	0.88	13.0
		TOT DEM	KW	.00464	.00421	.00500	.00600	.00657	.00643	.00714	.00693	.00664	.00585	.00478	.00436	0.00714
S- 7	-0.17	TOT GAS	THM	0.0451	0.0312	0.0180	0.0054	0.0016	0.0016	0.0016	0.0016	0.0016	0.0030	0.0133	0.0385	0.164
		TOT ELEC	KWH	0.79	0.71	0.79	0.86	1.00	1.19	1.38	1.32	1.09	0.89	0.75	0.79	11.6
		TOT DEM	KW	.00406	.00369	.00438	.00525	.00575	.00563	.00625	.00606	.00581	.00513	.00419	.00381	0.00625
S- 16	0.40	TOT GAS	THM	0.0476	0.0329	0.0190	0.0057	0.0017	0.0017	0.0017	0.0017	0.0017	0.0031	0.0140	0.0407	0.173
		TOT ELEC	KWH	0.69	0.62	0.69	0.75	0.87	1.04	1.20	1.15	0.95	0.78	0.66	0.69	10.1
		TOT DEM	KW	.00370	.00336	.00398	.00478	.00523	.00512	.00569	.00552	.00529	.00467	.00381	.00347	0.00569
S- 25	0.38	TOT GAS	THM	0.0451	0.0312	0.0180	0.0054	0.0016	0.0016	0.0016	0.0016	0.0016	0.0030	0.0133	0.0385	0.164
		TOT ELEC	KWH	0.79	0.71	0.79	0.86	1.00	1.19	1.38	1.32	1.09	0.89	0.75	0.79	11.6
		TOT DEM	KW	.00406	.00369	.00438	.00525	.00575	.00563	.00625	.00606	.00581	.00513	.00419	.00381	0.00625
S- 32	0.09	TOT GAS	THM	0.0433	0.0258	0.0211	0.0087	0.0041	0.0018	0.0018	0.0018	0.0032	0.0070	0.0141	0.0298	0.162
		TOT ELEC	KWH	0.68	0.62	0.75	0.81	0.87	1.07	1.14	1.13	0.96	0.81	0.67	0.68	10.2
		TOT DEM	KW	.00414	.00403	.00443	.00512	.00535	.00552	.00575	.00564	.00558	.00500	.00426	.00380	0.00575

HT ELEC = Electric consumption for space and water heating.

REM ELEC = Remaining electric consumption (not used for heating).

TOT ELEC = Total electric consumption.

HT DEM = Peak electrical demand for space and water heating.

REM DEM = Peak electrical demand for remaining load (not heating).

TOT DEM = Total peak electrical demand.

GAS = Total gas consumption.

(Annual figures are sums of monthly consumption values and are yearly maximums of monthly peak demand values.)

STRAT SET	CONST COST $/SF	ENERGY FORM	UNITS (PER SF)	JAN	FEB	MAR	APR	MAY	JUN	JUL	AUG	SEP	OCT	NOV	DEC	ANNUAL
BASE CASE : 201																
BASE	0.00	TOT ELEC	KWH	1.59	1.12	1.22	1.16	1.21	1.24	1.42	1.33	1.18	1.13	1.24	1.45	15.3
		TOT DEM	KW	.00612	.00490	.00477	.00508	.00490	.00594	.00581	.00594	.00532	.00471	.00471	.00526	0.00612
S- 1	0.09	TOT ELEC	KWH	1.47	1.03	1.13	1.07	1.11	1.14	1.31	1.23	1.09	1.04	1.14	1.34	14.1
		TOT DEM	KW	.00566	.00453	.00441	.00470	.00453	.00549	.00538	.00549	.00492	.00436	.00436	.00487	0.00566
S- 10	0.14	TOT ELEC	KWH	1.36	0.96	1.05	1.00	1.03	1.06	1.22	1.14	1.01	0.97	1.06	1.24	13.1
		TOT DEM	KW	.00528	.00422	.00412	.00438	.00422	.00512	.00502	.00512	.00459	.00407	.00407	.00454	0.00528
S- 14	1.45	TOT ELEC	KWH	1.25	0.88	0.96	0.91	0.95	0.97	1.12	1.04	0.92	0.89	0.97	1.14	12.0
		TOT DEM	KW	.00500	.00400	.00390	.00415	.00400	.00485	.00475	.00485	.00435	.00385	.00385	.00430	0.00500
S- 25	-0.11	TOT ELEC	KWH	1.36	0.96	1.05	1.00	1.03	1.06	1.22	1.14	1.01	0.97	1.06	1.24	13.1
		TOT DEM	KW	.00528	.00422	.00412	.00438	.00422	.00512	.00502	.00512	.00459	.00407	.00407	.00454	0.00528
BASE CASE : 202																
BASE	0.00	TOT ELEC	KWH	1.51	1.07	1.13	0.96	0.92	0.94	1.09	1.07	0.92	0.90	1.12	1.37	13.0
		TOT DEM	KW	.00970	.00825	.00805	.00621	.00475	.00572	.00563	.00572	.00514	.00601	.00834	.00883	0.00970
S- 3	-0.05	TOT ELEC	KWH	1.40	0.99	1.05	0.90	0.86	0.87	1.02	0.99	0.86	0.83	1.04	1.27	12.1
		TOT DEM	KW	.00894	.00760	.00742	.00572	.00438	.00527	.00519	.00527	.00474	.00554	.00769	.00814	0.00894
S- 6	0.28	TOT ELEC	KWH	1.32	0.93	0.99	0.84	0.81	0.82	0.96	0.93	0.81	0.79	0.98	1.20	11.4
		TOT DEM	KW	.00832	.00707	.00691	.00532	.00408	.00491	.00483	.00491	.00441	.00516	.00716	.00757	0.00832
S- 15	1.24	TOT ELEC	KWH	1.23	0.87	0.92	0.78	0.75	0.76	0.89	0.87	0.75	0.73	0.91	1.11	10.6
		TOT DEM	KW	.00790	.00672	.00656	.00506	.00387	.00466	.00458	.00466	.00419	.00490	.00679	.00719	0.00790
S- 60	0.03	TOT ELEC	KWH	1.32	0.93	0.99	0.84	0.81	0.82	0.96	0.93	0.81	0.79	0.98	1.20	11.4
		TOT DEM	KW	.00832	.00707	.00691	.00532	.00408	.00491	.00483	.00491	.00441	.00516	.00716	.00757	0.00832
BASE CASE : 203																
BASE	0.00	TOT GAS	THM	0.0263	0.0131	0.0120	0.0065	0.0022	0.0010	0.0009	0.0010	0.0010	0.0026	0.0131	0.0212	0.101
		TOT ELEC	KWH	0.88	0.78	0.88	0.85	0.93	0.98	1.14	1.11	0.97	0.89	0.83	0.87	11.1
		TOT DEM	KW	.00320	.00320	.00429	.00472	.00456	.00543	.00527	.00532	.00483	.00434	.00358	.00320	0.00543
S- 3	-0.05	TOT GAS	THM	0.0278	0.0139	0.0127	0.0068	0.0024	0.0011	0.0010	0.0011	0.0011	0.0028	0.0139	0.0225	0.107
		TOT ELEC	KWH	0.80	0.71	0.80	0.78	0.85	0.89	1.04	1.01	0.88	0.81	0.76	0.79	10.1
		TOT DEM	KW	.00297	.00297	.00398	.00438	.00423	.00504	.00489	.00494	.00449	.00403	.00333	.00297	0.00504
S- 6	0.00	TOT GAS	THM	0.0299	0.0149	0.0137	0.0074	0.0025	0.0012	0.0010	0.0012	0.0012	0.0030	0.0149	0.0241	0.115
		TOT ELEC	KWH	0.73	0.64	0.73	0.71	0.77	0.81	0.95	0.92	0.80	0.74	0.69	0.72	9.2
		TOT DEM	KW	.00279	.00279	.00374	.00412	.00397	.00473	.00459	.00464	.00421	.00378	.00312	.00279	0.00473
S- 15	1.24	TOT GAS	THM	0.0333	0.0166	0.0152	0.0082	0.0028	0.0013	0.0012	0.0013	0.0013	0.0033	0.0166	0.0269	0.128
		TOT ELEC	KWH	0.64	0.57	0.64	0.63	0.68	0.72	0.84	0.81	0.71	0.65	0.61	0.63	8.1
		TOT DEM	KW	.00255	.00255	.00341	.00376	.00363	.00432	.00419	.00423	.00384	.00346	.00285	.00255	0.00432
S- 60	0.03	TOT GAS	THM	0.0299	0.0149	0.0137	0.0074	0.0025	0.0012	0.0010	0.0012	0.0012	0.0030	0.0149	0.0241	0.115
		TOT ELEC	KWH	0.73	0.64	0.73	0.71	0.77	0.81	0.95	0.92	0.80	0.74	0.69	0.72	9.2
		TOT DEM	KW	.00279	.00279	.00374	.00412	.00397	.00473	.00459	.00464	.00421	.00378	.00312	.00279	0.00473
BASE CASE : 204																
BASE	0.00	TOT ELEC	KWH	1.37	1.08	1.18	1.10	1.12	1.23	1.45	1.28	1.18	1.14	1.17	1.30	14.6
		TOT DEM	KW	.00428	.00390	.00428	.00461	.00450	.00542	.00531	.00537	.00488	.00434	.00401	.00379	0.00542
S- 1	0.09	TOT ELEC	KWH	1.26	0.99	1.09	1.01	1.03	1.13	1.33	1.18	1.09	1.05	1.07	1.19	13.4
		TOT DEM	KW	.00401	.00365	.00401	.00431	.00421	.00507	.00497	.00502	.00456	.00406	.00375	.00355	0.00507
S- 12	0.14	TOT ELEC	KWH	1.16	0.91	1.00	0.92	0.95	1.03	1.22	1.08	1.00	0.96	0.98	1.09	12.3
		TOT DEM	KW	.00374	.00341	.00374	.00402	.00393	.00473	.00464	.00468	.00426	.00378	.00350	.00331	0.00473
S- 22	1.45	TOT ELEC	KWH	1.04	0.82	0.90	0.83	0.85	0.93	1.10	0.98	0.90	0.87	0.89	0.99	11.1
		TOT DEM	KW	.00340	.00310	.00340	.00366	.00357	.00430	.00421	.00426	.00387	.00344	.00318	.00301	0.00430
S- 28	-0.11	TOT ELEC	KWH	1.16	0.91	1.00	0.92	0.95	1.03	1.22	1.08	1.00	0.96	0.98	1.09	12.3
		TOT DEM	KW	.00374	.00341	.00374	.00402	.00393	.00473	.00464	.00468	.00426	.00378	.00350	.00331	0.00473
BASE CASE : 205																
BASE	0.00	TOT ELEC	KWH	1.66	1.17	1.28	1.22	1.26	1.30	1.49	1.39	1.23	1.18	1.30	1.52	16.0
		TOT DEM	KW	.00697	.00558	.00544	.00579	.00558	.00676	.00662	.00676	.00606	.00537	.00537	.00599	0.00697
S- 1	0.09	TOT ELEC	KWH	1.53	1.07	1.18	1.12	1.16	1.19	1.37	1.28	1.13	1.09	1.19	1.40	14.7
		TOT DEM	KW	.00645	.00516	.00503	.00535	.00516	.00626	.00613	.00626	.00561	.00497	.00497	.00555	0.00645
S- 10	0.14	TOT ELEC	KWH	1.42	1.00	1.10	1.04	1.08	1.11	1.27	1.19	1.05	1.01	1.11	1.30	13.7
		TOT DEM	KW	.00601	.00511	.00469	.00379	.00294	.00343	.00337	.00343	.00331	.00337	.00505	.00529	0.00601
S- 20	1.45	TOT ELEC	KWH	1.30	0.91	1.00	0.95	0.99	1.01	1.16	1.09	0.96	0.93	1.01	1.19	12.5
		TOT DEM	KW	.00570	.00485	.00445	.00359	.00279	.00325	.00319	.00325	.00314	.00319	.00479	.00502	0.00570
S- 25	-0.11	TOT ELEC	KWH	1.42	1.00	1.10	1.04	1.08	1.11	1.27	1.19	1.05	1.01	1.11	1.30	13.7
		TOT DEM	KW	.00601	.00511	.00469	.00379	.00294	.00343	.00337	.00343	.00331	.00337	.00505	.00529	0.00601
S- 30	0.61	TOT ELEC	KWH	1.48	1.10	1.13	0.85	0.76	0.85	0.85	0.89	0.84	0.83	1.08	1.32	12.0
		TOT DEM	KW	.00675	.00574	.00526	.00425	.00331	.00385	.00378	.00385	.00371	.00378	.00567	.00594	0.00675
BASE CASE : 206																
BASE	0.00	TOT ELEC	KWH	1.54	1.09	1.16	0.98	0.94	0.96	1.12	1.09	0.94	0.92	1.14	1.40	13.3
		TOT DEM	KW	.01100	.00935	.00913	.00704	.00539	.00649	.00638	.00649	.00583	.00682	.00946	.01001	0.01100
S- 3	-0.09	TOT ELEC	KWH	1.43	1.01	1.07	0.91	0.87	0.89	1.03	1.01	0.87	0.85	1.06	1.29	12.3
		TOT DEM	KW	.01018	.00865	.00845	.00652	.00499	.00601	.00590	.00601	.00540	.00631	.00875	.00926	0.01018
BASE CASE : 206 (CONTINUED)																
S- 6	-0.04	TOT ELEC	KWH	1.32	0.93	0.99	0.84	0.81	0.82	0.96	0.93	0.81	0.79	0.98	1.20	11.4
		TOT DEM	KW	.00948	.00825	.00872	.00597	.00446	.00398	.00370	.00379	.00417	.00607	.00891	.00901	0.00948
S- 15	1.18	TOT ELEC	KWH	1.23	0.87	0.92	0.78	0.75	0.76	0.89	0.87	0.75	0.73	0.91	1.11	10.6
		TOT DEM	KW	.00900	.00783	.00828	.00567	.00423	.00378	.00351	.00360	.00396	.00576	.00846	.00855	0.00900
S- 31	0.44	TOT ELEC	KWH	1.27	0.95	0.97	0.73	0.65	0.73	0.73	0.76	0.72	0.71	0.93	1.13	10.3
		TOT DEM	KW	.01065	.00927	.00980	.00671	.00501	.00447	.00415	.00426	.00469	.00682	.01001	.01012	0.01065
S- 60	0.06	TOT ELEC	KWH	1.32	0.93	0.99	0.84	0.81	0.82	0.96	0.93	0.81	0.79	0.98	1.20	11.4
		TOT DEM	KW	.00948	.00825	.00872	.00597	.00446	.00398	.00370	.00379	.00417	.00607	.00891	.00901	0.00948
BASE CASE : 207																
BASE	0.00	TOT GAS	THM	0.0225	0.0112	0.0103	0.0055	0.0019	0.0009	0.0008	0.0009	0.0009	0.0022	0.0112	0.0182	0.087
		TOT ELEC	KWH	0.92	0.81	0.92	0.89	0.97	1.02	1.19	1.16	1.01	0.93	0.87	0.90	11.6
		TOT DEM	KW	.00365	.00365	.00489	.00539	.00520	.00619	.00600	.00607	.00551	.00495	.00409	.00365	0.00619
S- 3	-0.09	TOT GAS	THM	0.0239	0.0120	0.0109	0.0059	0.0020	0.0009	0.0008	0.0009	0.0009	0.0024	0.0120	0.0193	0.092
		TOT ELEC	KWH	0.83	0.74	0.83	0.81	0.88	0.92	1.08	1.05	0.91	0.84	0.79	0.82	10.5
		TOT DEM	KW	.00339	.00339	.00454	.00500	.00483	.00575	.00558	.00564	.00512	.00460	.00380	.00339	0.00575
S- 6	-0.04	TOT GAS	THM	0.0257	0.0128	0.0117	0.0063	0.0022	0.0010	0.0009	0.0010	0.0010	0.0026	0.0128	0.0207	0.099
		TOT ELEC	KWH	0.76	0.67	0.76	0.74	0.81	0.85	0.99	0.96	0.84	0.77	0.72	0.75	9.6
		TOT DEM	KW	.00318	.00318	.00426	.00469	.00453	.00539	.00523	.00528	.00480	.00431	.00356	.00318	0.00539
S- 15	1.14	TOT GAS	THM	0.0286	0.0143	0.0131	0.0070	0.0024	0.0011	0.0010	0.0011	0.0011	0.0029	0.0143	0.0231	0.110
		TOT ELEC	KWH	0.67	0.60	0.67	0.65	0.71	0.75	0.88	0.85	0.74	0.68	0.64	0.66	8.5
		TOT DEM	KW	.00291	.00291	.00389	.00429	.00414	.00493	.00478	.00483	.00439	.00394	.00325	.00291	0.00493
S- 31	0.44	TOT GAS	THM	0.0313	0.0196	0.0172	0.0082	0.0026	0.0013	0.0001	0.0007	0.0019	0.0053	0.0182	0.0238	0.130
		TOT ELEC	KWH	1.06	0.79	0.81	0.61	0.54	0.61	0.61	0.64	0.60	0.59	0.77	0.95	8.6
		TOT DEM	KW	.00236	.00236	.00316	.00348	.00336	.00400	.00388	.00392	.00356	.00320	.00264	.00236	0.00400
S- 60	0.07	TOT GAS	THM	0.0257	0.0128	0.0117	0.0063	0.0022	0.0010	0.0009	0.0010	0.0010	0.0026	0.0128	0.0207	0.099
		TOT ELEC	KWH	0.76	0.67	0.76	0.74	0.81	0.85	0.99	0.96	0.84	0.77	0.72	0.75	9.6
		TOT DEM	KW	.00318	.00318	.00426	.00469	.00453	.00539	.00523	.00528	.00480	.00431	.00356	.00318	0.00539
BASE CASE : 208																
BASE	0.00	TOT ELEC	KWH	1.45	1.14	1.25	1.16	1.19	1.29	1.52	1.36	1.25	1.20	1.23	1.37	15.4
		TOT DEM	KW	.00494	.00450	.00494	.00531	.00519	.00625	.00613	.00619	.00562	.00500	.00463	.00438	0.00625
S- 1	0.09	TOT ELEC	KWH	1.33	1.04	1.14	1.06	1.09	1.18	1.40	1.24	1.14	1.10	1.13	1.25	14.1
		TOT DEM	KW	.00457	.00416	.00457	.00491	.00480	.00578	.00566	.00572	.00520	.00462	.00428	.00405	0.00578
S- 12	0.14	TOT ELEC	KWH	1.22	0.96	1.05	0.98	1.00	1.09	1.29	1.14	1.05	1.01	1.04	1.16	13.0
		TOT DEM	KW	.00539	.00458	.00442	.00361	.00350	.00404	.00393	.00404	.00399	.00399	.00447	.00458	0.00539
S- 22	1.45	TOT ELEC	KWH	1.10	0.87	0.95	0.88	0.90	0.98	1.16	1.03	0.95	0.91	0.94	1.04	11.7
		TOT DEM	KW	.00491	.00417	.00403	.00329	.00319	.00368	.00358	.00368	.00363	.00363	.00408	.00417	0.00491
S- 28	-0.11	TOT ELEC	KWH	1.22	0.96	1.05	0.98	1.00	1.09	1.29	1.14	1.05	1.01	1.04	1.16	13.0
		TOT DEM	KW	.00539	.00458	.00442	.00361	.00350	.00404	.00393	.00404	.00399	.00399	.00447	.00458	0.00539
S- 36	0.61	TOT ELEC	KWH	1.33	0.99	1.02	0.77	0.68	0.77	0.77	0.80	0.76	0.75	0.97	1.19	10.8
		TOT DEM	KW	.00475	.00404	.00389	.00318	.00309	.00356	.00347	.00356	.00352	.00352	.00394	.00404	0.00475
BASE CASE : 209																
BASE	0.00	HT ELEC	KWH	0.70	0.44	0.41	0.27	0.12	0.06	0.00	0.02	0.09	0.26	0.48	0.61	3.5
		HT DEM	KW	.00608	.00578	.00547	.00468	.00359	.00298	.00146	.00249	.00292	.00474	.00590	.00553	0.00608
		REM ELEC	KWH	0.95	0.84	0.95	0.92	1.01	1.06	1.24	1.20	1.04	0.96	0.90	0.94	12.0
		REM DEM	KW	.00354	.00354	.00474	.00522	.00504	.00600	.00582	.00588	.00534	.00480	.00396	.00354	0.00600
		TOT ELEC	KWH	1.64	1.28	1.36	1.19	1.13	1.11	1.24	1.22	1.13	1.22	1.38	1.54	15.5
		TOT DEM	KW	.00988	.00988	.00988	.00909	.00741	.00840	.00781	.00781	.00721	.00958	.00988	.00988	0.00988
S- 3	-0.13	HT ELEC	KWH	0.74	0.47	0.43	0.28	0.13	0.06	0.00	0.02	0.09	0.28	0.50	0.64	3.6
		HT DEM	KW	.00608	.00578	.00547	.00468	.00359	.00298	.00146	.00249	.00292	.00474	.00590	.00553	0.00608
		REM ELEC	KWH	0.81	0.72	0.81	0.79	0.87	0.91	1.06	1.03	0.90	0.82	0.77	0.80	10.3
		REM DEM	KW	.00326	.00326	.00437	.00481	.00465	.00553	.00536	.00542	.00492	.00442	.00365	.00326	0.00553
		TOT ELEC	KWH	1.55	1.19	1.25	1.08	1.00	0.97	1.06	1.05	0.99	1.10	1.27	1.44	13.9
		TOT DEM	KW	.00950	.00950	.00950	.00874	.00713	.00807	.00751	.00751	.00694	.00922	.00950	.00950	0.00950
S- 6	-0.12	HT ELEC	KWH	0.80	0.51	0.47	0.31	0.14	0.07	0.00	0.02	0.10	0.30	0.55	0.70	4.0
		HT DEM	KW	.00608	.00578	.00547	.00468	.00359	.00298	.00146	.00249	.00292	.00474	.00590	.00553	0.00608
		REM ELEC	KWH	0.75	0.66	0.75	0.73	0.79	0.83	0.97	0.94	0.82	0.76	0.71	0.74	9.4
		REM DEM	KW	.00294	.00294	.00393	.00433	.00418	.00498	.00483	.00488	.00443	.00398	.00329	.00294	0.00498
		TOT ELEC	KWH	1.55	1.17	1.22	1.04	0.94	0.90	0.97	0.96	0.92	1.06	1.25	1.43	13.4
		TOT DEM	KW	.00885	.00885	.00885	.00814	.00664	.00752	.00699	.00699	.00646	.00858	.00885	.00885	0.00885
S- 15	1.13	HT ELEC	KWH	0.85	0.54	0.50	0.33	0.15	0.07	0.00	0.02	0.10	0.32	0.58	0.74	4.2
		HT DEM	KW	.00608	.00578	.00547	.00468	.00359	.00298	.00146	.00249	.00292	.00474	.00590	.00553	0.00608
		REM ELEC	KWH	0.65	0.58	0.65	0.63	0.69	0.73	0.85	0.82	0.72	0.66	0.62	0.64	8.2
		REM DEM	KW	.00261	.00261	.00350	.00385	.00372	.00443	.00430	.00434	.00394	.00354	.00292	.00261	0.00443
		TOT ELEC	KWH	1.50	1.11	1.15	0.96	0.84	0.80	0.85	0.84	0.82	0.98	1.20	1.38	12.4
		TOT DEM	KW	.00850	.00689	.00680	.00587	.00502	.00553	.00519	.00519	.00493	.00629	.00706	.00689	0.00850

(HT ELEC=ELECTRIC CONSUMPTION FOR SPACE AND WATER HEATING. REM ELEC=REMAINING ELECTRIC CONSUMPTION (NOT USED FOR HEATING). TOT ELEC=TOTAL ELECTRIC CONSUMPTION. HT DEM=PEAK ELECTRIC DEMAND FOR SPACE AND WATER HEATING. REM DEM=PEAK ELECTRIC DEMAND FOR REMAINING LOAD (NOT HEATING). TOT DEM=TOTAL PEAK ELECTRIC DEMAND. GAS=TOTAL GAS CONSUMPTION. ANNUAL FIGURES ARE SUMS OF MONTHLY CONSUMPTION VALUES OR ARE YEARLY MAXIMUMS OF MONTHLY PEAK DEMAND VALUES.)

STRAT SET	CONST COST $/SF	ENERGY FORM	UNITS (PER SF)	JAN	FEB	MAR	APR	MAY	JUN	JUL	AUG	SEP	OCT	NOV	DEC	ANNUAL
BASE CASE : 209 (CONTINUED)																
S- 31	0.15	HT ELEC	KWH	0.57	0.45	0.48	0.37	0.26	0.24	0.15	0.19	0.24	0.36	0.47	0.52	4.3
		HT DEM	KW	.00608	.00578	.00547	.00468	.00359	.00298	.00146	.00249	.00292	.00474	.00590	.00553	0.00608
		REM ELEC	KWH	0.78	0.71	0.82	0.77	0.74	0.89	1.00	0.98	0.87	0.75	0.76	0.81	9.9
		REM DEM	KW	.00325	.00325	.00395	.00430	.00410	.00500	.00480	.00485	.00435	.00425	.00325	.00325	0.00500
		TOT ELEC	KWH	1.35	1.16	1.30	1.14	1.00	1.13	1.15	1.17	1.11	1.11	1.23	1.33	14.2
		TOT DEM	KW	.00948	.00768	.00758	.00654	.00559	.00616	.00578	.00578	.00550	.00702	.00787	.00768	0.00948
S- 39	0.91	HT ELEC	KWH	0.64	0.50	0.54	0.41	0.29	0.27	0.17	0.21	0.27	0.40	0.53	0.57	4.8
		HT DEM	KW	.00608	.00578	.00547	.00468	.00359	.00298	.00146	.00249	.00292	.00474	.00590	.00553	0.00608
		REM ELEC	KWH	0.71	0.65	0.74	0.70	0.67	0.81	0.91	0.89	0.79	0.68	0.69	0.74	9.0
		REM DEM	KW	.00304	.00304	.00370	.00402	.00384	.00468	.00449	.00454	.00407	.00398	.00304	.00304	0.00468
		TOT ELEC	KWH	1.34	1.15	1.28	1.11	0.96	1.07	1.08	1.10	1.06	1.08	1.22	1.31	13.8
		TOT DEM	KW	.00910	.00737	.00728	.00628	.00537	.00591	.00555	.00555	.00528	.00673	.00755	.00737	0.00910
S- 44	0.19	HT ELEC	KWH	0.60	0.48	0.51	0.39	0.27	0.25	0.16	0.20	0.25	0.38	0.50	0.54	4.5
		HT DEM	KW	.00608	.00578	.00547	.00468	.00359	.00298	.00146	.00249	.00292	.00474	.00590	.00553	0.00608
		REM ELEC	KWH	0.75	0.68	0.78	0.74	0.71	0.85	0.95	0.93	0.83	0.72	0.73	0.77	9.4
		REM DEM	KW	.00317	.00317	.00386	.00420	.00400	.00488	.00468	.00473	.00425	.00415	.00317	.00317	0.00488
		TOT ELEC	KWH	1.35	1.16	1.29	1.13	0.98	1.10	1.12	1.13	1.08	1.09	1.23	1.32	14.0
		TOT DEM	KW	.00930	.00753	.00744	.00642	.00549	.00604	.00567	.00567	.00539	.00688	.00772	.00763	0.00930
S- 49	1.46	HT ELEC	KWH	0.55	0.44	0.47	0.36	0.25	0.23	0.15	0.18	0.23	0.35	0.46	0.50	4.2
		HT DEM	KW	.00608	.00578	.00547	.00468	.00359	.00298	.00146	.00249	.00292	.00474	.00590	.00553	0.00608
		REM ELEC	KWH	0.76	0.69	0.80	0.75	0.72	0.86	0.97	0.95	0.84	0.73	0.74	0.79	9.6
		REM DEM	KW	.00325	.00325	.00395	.00430	.00410	.00500	.00480	.00485	.00435	.00425	.00325	.00325	0.00500
		TOT ELEC	KWH	1.31	1.13	1.26	1.11	0.97	1.10	1.12	1.13	1.08	1.07	1.20	1.29	13.8
		TOT DEM	KW	.00948	.00768	.00758	.00654	.00559	.00616	.00578	.00578	.00550	.00702	.00787	.00768	0.00948
S- 60	-0.33	HT ELEC	KWH	0.80	0.51	0.47	0.31	0.14	0.07	0.00	0.02	0.10	0.30	0.55	0.70	4.0
		HT DEM	KW	.00608	.00578	.00547	.00468	.00359	.00298	.00146	.00249	.00292	.00474	.00590	.00553	0.00608
		REM ELEC	KWH	0.75	0.66	0.75	0.73	0.79	0.83	0.97	0.94	0.82	0.76	0.71	0.74	9.4
		REM DEM	KW	.00294	.00294	.00393	.00433	.00418	.00498	.00483	.00488	.00443	.00398	.00329	.00294	0.00498
		TOT ELEC	KWH	1.55	1.17	1.22	1.04	0.94	0.90	0.97	0.96	0.92	1.06	1.25	1.43	13.4
		TOT DEM	KW	.00885	.00885	.00885	.00814	.00664	.00752	.00699	.00699	.00646	.00858	.00885	.00885	0.00885
BASE CASE : 210																
BASE	0.00	TOT GAS	THM	0.0366	0.0232	0.0215	0.0141	0.0065	0.0031	0.0000	0.0009	0.0045	0.0138	0.0250	0.0319	0.181
		TOT ELEC	KWH	0.95	0.84	0.95	0.92	1.01	1.06	1.24	1.20	1.04	0.96	0.90	0.94	12.0
		TOT DEM	KW	.00354	.00354	.00474	.00522	.00504	.00600	.00582	.00588	.00534	.00480	.00396	.00354	0.00600
S- 3	-0.14	TOT GAS	THM	0.0386	0.0244	0.0227	0.0149	0.0069	0.0032	0.0000	0.0010	0.0048	0.0145	0.0264	0.0336	0.191
		TOT ELEC	KWH	0.81	0.72	0.81	0.79	0.87	0.91	1.06	1.03	0.90	0.82	0.77	0.80	10.3
		TOT DEM	KW	.00326	.00326	.00437	.00481	.00465	.00553	.00536	.00542	.00492	.00442	.00365	.00326	0.00553
S- 6	-0.09	TOT GAS	THM	0.0420	0.0266	0.0248	0.0162	0.0075	0.0035	0.0000	0.0010	0.0052	0.0158	0.0287	0.0366	0.208
		TOT ELEC	KWH	0.75	0.66	0.75	0.73	0.79	0.83	0.97	0.94	0.82	0.76	0.71	0.74	9.4
		TOT DEM	KW	.00294	.00294	.00393	.00433	.00418	.00498	.00483	.00488	.00443	.00398	.00329	.00294	0.00498
S- 15	1.13	TOT GAS	THM	0.0444	0.0282	0.0262	0.0172	0.0079	0.0037	0.0000	0.0011	0.0055	0.0167	0.0304	0.0387	0.220
		TOT ELEC	KWH	0.65	0.58	0.65	0.63	0.69	0.73	0.85	0.82	0.72	0.66	0.62	0.64	8.2
		TOT DEM	KW	.00261	.00261	.00350	.00385	.00372	.00443	.00430	.00434	.00394	.00354	.00292	.00261	0.00443
S- 31	0.14	TOT GAS	THM	0.0301	0.0237	0.0253	0.0194	0.0136	0.0127	0.0081	0.0099	0.0127	0.0188	0.0249	0.0271	0.226
		TOT ELEC	KWH	0.78	0.71	0.82	0.77	0.74	0.89	1.00	0.98	0.87	0.75	0.76	0.81	9.9
		TOT DEM	KW	.00325	.00325	.00395	.00430	.00410	.00500	.00480	.00485	.00435	.00425	.00325	.00325	0.00500
S- 39	0.90	TOT GAS	THM	0.0334	0.0264	0.0281	0.0216	0.0151	0.0141	0.0090	0.0110	0.0141	0.0208	0.0276	0.0301	0.251
		TOT ELEC	KWH	0.71	0.65	0.74	0.70	0.67	0.81	0.91	0.89	0.79	0.68	0.69	0.74	9.0
		TOT DEM	KW	.00304	.00304	.00370	.00402	.00384	.00468	.00449	.00454	.00407	.00398	.00304	.00304	0.00468
S- 44	0.19	TOT GAS	THM	0.0317	0.0250	0.0267	0.0205	0.0143	0.0133	0.0086	0.0105	0.0133	0.0198	0.0262	0.0286	0.238
		TOT ELEC	KWH	0.75	0.68	0.78	0.74	0.71	0.85	0.95	0.93	0.83	0.72	0.73	0.77	9.4
		TOT DEM	KW	.00317	.00317	.00386	.00420	.00400	.00488	.00468	.00473	.00425	.00415	.00317	.00317	0.00488
S- 49	1.46	TOT GAS	THM	0.0291	0.0230	0.0245	0.0188	0.0131	0.0123	0.0079	0.0096	0.0123	0.0182	0.0241	0.0263	0.219
		TOT ELEC	KWH	0.76	0.69	0.80	0.75	0.72	0.86	0.97	0.95	0.84	0.73	0.74	0.79	9.6
		TOT DEM	KW	.00325	.00325	.00395	.00430	.00410	.00500	.00480	.00485	.00435	.00425	.00325	.00325	0.00500
S- 60	-0.34	TOT GAS	THM	0.0420	0.0266	0.0248	0.0162	0.0075	0.0035	0.0000	0.0010	0.0052	0.0158	0.0287	0.0366	0.208
		TOT ELEC	KWH	0.75	0.66	0.75	0.73	0.79	0.83	0.97	0.94	0.82	0.76	0.71	0.74	9.4
		TOT DEM	KW	.00294	.00294	.00393	.00433	.00418	.00498	.00483	.00488	.00443	.00398	.00329	.00294	0.00498
BASE CASE : 211																
BASE	0.00	HT ELEC	KWH	0.59	0.37	0.35	0.23	0.10	0.05	0.00	0.01	0.07	0.22	0.40	0.51	2.9
		HT DEM	KW	.00395	.00371	.00371	.00340	.00265	.00194	.00071	.00142	.00197	.00332	.00379	.00379	0.00395
		REM ELEC	KWH	0.91	0.81	0.91	0.89	0.97	1.01	1.18	1.15	1.00	0.92	0.86	0.90	11.5
		REM DEM	KW	.00330	.00330	.00416	.00464	.00442	.00528	.00528	.00533	.00501	.00410	.00389	.00325	0.00533
		TOT ELEC	KWH	1.50	1.18	1.25	1.11	1.07	1.06	1.18	1.16	1.07	1.14	1.26	1.41	14.4
		TOT DEM	KW	.00675	.00607	.00601	.00506	.00439	.00520	.00513	.00520	.00472	.00526	.00621	.00614	0.00675
S- 3	-0.08	HT ELEC	KWH	0.63	0.40	0.37	0.24	0.11	0.05	0.00	0.02	0.08	0.24	0.43	0.55	3.1
		HT DEM	KW	.00395	.00371	.00371	.00340	.00265	.00194	.00071	.00142	.00197	.00332	.00379	.00379	0.00395
		REM ELEC	KWH	0.77	0.68	0.77	0.75	0.82	0.86	1.01	0.98	0.85	0.78	0.73	0.76	9.8
		REM DEM	KW	.00304	.00304	.00382	.00426	.00407	.00485	.00485	.00490	.00461	.00377	.00358	.00299	0.00490
		TOT ELEC	KWH	1.41	1.09	1.15	1.00	0.93	0.91	1.01	0.99	0.93	1.02	1.17	1.32	12.9
		TOT DEM	KW	.00653	.00588	.00581	.00490	.00424	.00503	.00496	.00503	.00457	.00509	.00601	.00594	0.00653
S- 6	-0.03	HT ELEC	KWH	0.68	0.43	0.40	0.26	0.12	0.06	0.00	0.02	0.08	0.25	0.46	0.59	3.3
		HT DEM	KW	.00395	.00371	.00371	.00340	.00265	.00194	.00071	.00142	.00197	.00332	.00379	.00379	0.00395
		REM ELEC	KWH	0.70	0.62	0.70	0.68	0.74	0.78	0.91	0.88	0.77	0.71	0.66	0.69	8.9
		REM DEM	KW	.00275	.00275	.00346	.00385	.00368	.00439	.00439	.00443	.00416	.00341	.00323	.00270	0.00443
		TOT ELEC	KWH	1.38	1.05	1.10	0.94	0.86	0.84	0.91	0.90	0.85	0.96	1.13	1.28	12.2
		TOT DEM	KW	.00608	.00547	.00541	.00456	.00395	.00468	.00462	.00468	.00426	.00474	.00559	.00553	0.00608

STRAT SET	CONST COST $/SF	ENERGY FORM	UNITS (PER SF)	JAN	FEB	MAR	APR	MAY	JUN	JUL	AUG	SEP	OCT	NOV	DEC	ANNUAL
BASE CASE : 211 (CONTINUED)																
S- 15	0.56	HT ELEC	KWH	0.72	0.46	0.43	0.28	0.13	0.06	0.00	0.02	0.09	0.27	0.49	0.63	3.6
		HT DEM	KW	.00395	.00371	.00371	.00340	.00265	.00194	.00071	.00142	.00197	.00332	.00379	.00379	0.00395
		REM ELEC	KWH	0.61	0.54	0.61	0.59	0.65	0.68	0.79	0.77	0.67	0.62	0.58	0.60	7.7
		REM DEM	KW	.00244	.00244	.00307	.00342	.00326	.00389	.00389	.00393	.00369	.00303	.00287	.00240	0.00393
		TOT ELEC	KWH	1.33	1.00	1.03	0.87	0.77	0.74	0.79	0.79	0.76	0.89	1.07	1.23	11.3
		TOT DEM	KW	.00590	.00531	.00525	.00442	.00383	.00454	.00448	.00454	.00413	.00460	.00543	.00537	0.00590
S- 31	0.24	HT ELEC	KWH	0.80	0.54	0.53	0.31	0.15	0.11	0.05	0.08	0.13	0.27	0.56	0.60	4.1
		HT DEM	KW	.00395	.00371	.00371	.00340	.00265	.00194	.00071	.00142	.00197	.00332	.00379	.00379	0.00395
		REM ELEC	KWH	0.70	0.63	0.72	0.69	0.72	0.83	0.92	0.92	0.82	0.71	0.68	0.72	9.1
		REM DEM	KW	.00293	.00288	.00334	.00347	.00347	.00418	.00410	.00414	.00368	.00368	.00313	.00301	0.00418
		TOT ELEC	KWH	1.50	1.17	1.25	1.00	0.87	0.95	0.98	1.01	0.94	0.98	1.24	1.33	13.2
		TOT DEM	KW	.00635	.00578	.00565	.00476	.00381	.00432	.00425	.00419	.00387	.00508	.00584	.00584	0.00635
S- 39	0.68	HT ELEC	KWH	0.98	0.66	0.64	0.38	0.19	0.14	0.07	0.10	0.16	0.33	0.68	0.73	5.0
		HT DEM	KW	.00395	.00371	.00371	.00340	.00265	.00194	.00071	.00142	.00197	.00332	.00379	.00379	0.00395
		REM ELEC	KWH	0.62	0.56	0.65	0.61	0.64	0.74	0.82	0.82	0.73	0.63	0.61	0.65	8.1
		REM DEM	KW	.00265	.00261	.00302	.00314	.00314	.00378	.00370	.00374	.00333	.00333	.00283	.00272	0.00378
		TOT ELEC	KWH	1.60	1.22	1.29	1.00	0.82	0.88	0.89	0.92	0.88	0.96	1.28	1.37	13.1
		TOT DEM	KW	.00610	.00555	.00543	.00458	.00366	.00415	.00409	.00403	.00372	.00488	.00561	.00561	0.00610
S- 44	0.28	HT ELEC	KWH	0.88	0.59	0.58	0.35	0.17	0.12	0.06	0.09	0.14	0.30	0.61	0.66	4.5
		HT DEM	KW	.00395	.00371	.00371	.00340	.00265	.00194	.00071	.00142	.00197	.00332	.00379	.00379	0.00395
		REM ELEC	KWH	0.66	0.59	0.68	0.65	0.68	0.79	0.87	0.87	0.77	0.67	0.64	0.68	8.6
		REM DEM	KW	.00280	.00276	.00320	.00332	.00332	.00400	.00392	.00396	.00352	.00352	.00300	.00288	0.00400
		TOT ELEC	KWH	1.54	1.19	1.27	1.00	0.84	0.91	0.93	0.96	0.91	0.96	1.25	1.34	13.1
		TOT DEM	KW	.00625	.00569	.00556	.00469	.00375	.00425	.00419	.00413	.00381	.00500	.00575	.00575	0.00625
S- 49	1.55	HT ELEC	KWH	0.78	0.53	0.51	0.31	0.15	0.11	0.05	0.08	0.12	0.26	0.54	0.58	4.0
		HT DEM	KW	.00395	.00371	.00371	.00340	.00265	.00194	.00071	.00142	.00197	.00332	.00379	.00379	0.00395
		REM ELEC	KWH	0.68	0.61	0.70	0.67	0.69	0.81	0.90	0.90	0.79	0.69	0.66	0.70	8.8
		REM DEM	KW	.00293	.00288	.00334	.00347	.00347	.00418	.00410	.00414	.00368	.00368	.00313	.00301	0.00418
		TOT ELEC	KWH	1.46	1.13	1.22	0.97	0.84	0.92	0.95	0.98	0.92	0.95	1.20	1.29	12.8
		TOT DEM	KW	.00635	.00578	.00565	.00476	.00381	.00432	.00425	.00419	.00387	.00508	.00584	.00584	0.00635
S- 60	-0.28	HT ELEC	KWH	0.68	0.43	0.40	0.26	0.12	0.06	0.00	0.02	0.08	0.25	0.46	0.59	3.3
		HT DEM	KW	.00395	.00371	.00371	.00340	.00265	.00194	.00071	.00142	.00197	.00332	.00379	.00379	0.00395
		REM ELEC	KWH	0.70	0.62	0.70	0.68	0.74	0.78	0.91	0.88	0.77	0.71	0.66	0.69	8.9
		REM DEM	KW	.00275	.00275	.00346	.00385	.00368	.00439	.00439	.00443	.00416	.00341	.00323	.00270	0.00443
		TOT ELEC	KWH	1.38	1.05	1.10	0.94	0.86	0.84	0.91	0.90	0.85	0.96	1.13	1.28	12.2
		TOT DEM	KW	.00608	.00547	.00541	.00456	.00395	.00468	.00462	.00468	.00426	.00474	.00559	.00553	0.00608
BASE CASE : 212																
BASE	0.00	TOT GAS	THM	0.0309	0.0196	0.0182	0.0119	0.0055	0.0026	0.0000	0.0008	0.0038	0.0116	0.0211	0.0269	0.153
		TOT ELEC	KWH	0.91	0.81	0.91	0.89	0.97	1.01	1.18	1.15	1.00	0.92	0.86	0.90	11.5
		TOT DEM	KW	.00330	.00330	.00416	.00464	.00442	.00528	.00528	.00533	.00501	.00410	.00389	.00325	0.00533
S- 3	-0.09	TOT GAS	THM	0.0333	0.0211	0.0196	0.0129	0.0059	0.0028	0.0000	0.0008	0.0041	0.0125	0.0228	0.0290	0.165
		TOT ELEC	KWH	0.77	0.68	0.77	0.75	0.82	0.86	1.01	0.98	0.85	0.78	0.73	0.76	9.8
		TOT DEM	KW	.00304	.00304	.00382	.00426	.00407	.00485	.00485	.00490	.00461	.00377	.00358	.00299	0.00490
S- 6	-0.04	TOT GAS	THM	0.0356	0.0225	0.0209	0.0137	0.0063	0.0030	0.0000	0.0009	0.0044	0.0134	0.0243	0.0310	0.176
		TOT ELEC	KWH	0.70	0.62	0.70	0.68	0.74	0.78	0.91	0.88	0.77	0.71	0.66	0.69	8.9
		TOT DEM	KW	.00275	.00275	.00346	.00385	.00368	.00439	.00439	.00443	.00416	.00341	.00323	.00270	0.00443
S- 15	0.53	TOT GAS	THM	0.0380	0.0241	0.0224	0.0147	0.0068	0.0032	0.0000	0.0009	0.0047	0.0143	0.0259	0.0331	0.188
		TOT ELEC	KWH	0.61	0.54	0.61	0.59	0.65	0.68	0.79	0.77	0.67	0.62	0.58	0.60	7.7
		TOT DEM	KW	.00244	.00244	.00307	.00342	.00326	.00389	.00389	.00393	.00369	.00303	.00287	.00240	0.00393
S- 31	0.23	TOT GAS	THM	0.0421	0.0284	0.0278	0.0165	0.0080	0.0059	0.0028	0.0043	0.0067	0.0141	0.0293	0.0315	0.217
		TOT ELEC	KWH	0.70	0.63	0.72	0.69	0.72	0.83	0.92	0.92	0.82	0.71	0.68	0.72	9.1
		TOT DEM	KW	.00293	.00293	.00334	.00347	.00347	.00418	.00410	.00414	.00368	.00368	.00313	.00301	0.00418
S- 39	0.66	TOT GAS	THM	0.0512	0.0346	0.0338	0.0201	0.0098	0.0071	0.0034	0.0053	0.0082	0.0172	0.0356	0.0383	0.264
		TOT ELEC	KWH	0.62	0.56	0.65	0.61	0.64	0.74	0.82	0.82	0.73	0.63	0.61	0.65	8.1
		TOT DEM	KW	.00265	.00265	.00302	.00314	.00314	.00378	.00370	.00374	.00333	.00333	.00283	.00272	0.00378
S- 44	0.28	TOT GAS	THM	0.0462	0.0312	0.0305	0.0181	0.0088	0.0064	0.0031	0.0048	0.0074	0.0155	0.0321	0.0345	0.238
		TOT ELEC	KWH	0.66	0.59	0.68	0.65	0.68	0.79	0.87	0.87	0.77	0.67	0.64	0.68	8.6
		TOT DEM	KW	.00280	.00280	.00320	.00332	.00332	.00400	.00392	.00396	.00352	.00352	.00300	.00288	0.00400
S- 49	1.54	TOT GAS	THM	0.0407	0.0275	0.0269	0.0160	0.0078	0.0057	0.0027	0.0042	0.0065	0.0136	0.0284	0.0304	0.210
		TOT ELEC	KWH	0.68	0.61	0.70	0.67	0.69	0.81	0.90	0.90	0.79	0.69	0.66	0.70	8.8
		TOT DEM	KW	.00293	.00293	.00334	.00347	.00347	.00418	.00410	.00414	.00368	.00368	.00313	.00301	0.00418
S- 60	-0.29	TOT GAS	THM	0.0356	0.0225	0.0209	0.0137	0.0063	0.0030	0.0000	0.0009	0.0044	0.0134	0.0243	0.0310	0.176
		TOT ELEC	KWH	0.70	0.62	0.70	0.68	0.74	0.78	0.91	0.88	0.77	0.71	0.66	0.69	8.9
		TOT DEM	KW	.00275	.00275	.00346	.00385	.00368	.00439	.00439	.00443	.00416	.00341	.00323	.00270	0.00443
BASE CASE : 213																
BASE	0.00	TOT ELEC	KWH	1.19	0.94	1.03	0.95	0.98	1.07	1.26	1.12	1.03	0.99	1.02	1.13	12.7
		TOT DEM	KW	.00367	.00367	.00491	.00541	.00522	.00622	.00603	.00610	.00554	.00498	.00411	.00367	0.00622
S- 3	-0.08	TOT ELEC	KWH	1.10	0.87	0.95	0.88	0.90	0.98	1.16	1.03	0.95	0.91	0.94	1.04	11.7
		TOT DEM	KW	.00343	.00343	.00460	.00506	.00489	.00582	.00565	.00570	.00518	.00466	.00384	.00343	0.00582
S- 6	-0.03	TOT ELEC	KWH	1.02	0.81	0.88	0.82	0.84	0.92	1.08	0.96	0.88	0.85	0.87	0.97	10.9
		TOT DEM	KW	.00320	.00320	.00428	.00472	.00455	.00542	.00526	.00531	.00482	.00434	.00358	.00320	0.00542
S- 15	1.11	TOT ELEC	KWH	0.93	0.73	0.80	0.74	0.76	0.83	0.98	0.87	0.80	0.77	0.79	0.88	9.9
		TOT DEM	KW	.00302	.00302	.00404	.00445	.00430	.00512	.00497	.00502	.00456	.00410	.00338	.00302	0.00512

(HT ELEC=ELECTRIC CONSUMPTION FOR SPACE AND WATER HEATING. REM ELEC=REMAINING ELECTRIC CONSUMPTION (NOT USED FOR HEATING). TOT ELEC=TOTAL ELECTRIC CONSUMPTION. HT DEM=PEAK ELECTRIC DEMAND FOR SPACE AND WATER HEATING. REM DEM=PEAK ELECTRIC DEMAND FOR REMAINING LOAD (NOT HEATING). TOT DEM=TOTAL PEAK ELECTRIC DEMAND. GAS=TOTAL GAS CONSUMPTION. ANNUAL FIGURES ARE SUMS OF MONTHLY CONSUMPTION VALUES OR ARE YEARLY MAXIMUMS OF MONTHLY PEAK DEMAND VALUES.)

STRAT SET	CONST COST $/SF	ENERGY FORM	UNITS (PER SF)	JAN	FEB	MAR	APR	MAY	JUN	JUL	AUG	SEP	OCT	NOV	DEC	ANNUAL
BASE CASE : 213 (CONTINUED)																
S- 31	0.24	TOT ELEC	KWH	1.09	0.86	0.94	0.87	0.89	0.97	1.15	1.02	0.94	0.90	0.93	1.03	11.6
		TOT DEM	KW	.00345	.00345	.00462	.00509	.00491	.00585	.00567	.00573	.00521	.00468	.00386	.00345	0.00585
S- 39	0.92	TOT ELEC	KWH	1.08	0.85	0.93	0.86	0.89	0.97	1.14	1.01	0.93	0.90	0.92	1.02	11.5
		TOT DEM	KW	.00332	.00332	.00444	.00489	.00472	.00562	.00545	.00551	.00500	.00450	.00371	.00332	0.00562
S- 44	0.29	TOT ELEC	KWH	1.08	0.85	0.93	0.86	0.89	0.97	1.14	1.01	0.93	0.90	0.92	1.02	11.5
		TOT DEM	KW	.00340	.00340	.00455	.00501	.00484	.00576	.00559	.00564	.00513	.00461	.00380	.00340	0.00576
S- 49	1.55	TOT ELEC	KWH	1.06	0.84	0.92	0.85	0.87	0.95	1.12	0.99	0.92	0.88	0.90	1.01	11.3
		TOT DEM	KW	.00345	.00345	.00462	.00509	.00491	.00585	.00567	.00573	.00521	.00468	.00386	.00345	0.00585
S- 60	-0.28	TOT ELEC	KWH	1.02	0.81	0.88	0.82	0.84	0.92	1.08	0.96	0.88	0.85	0.87	0.97	10.9
		TOT DEM	KW	.00320	.00320	.00428	.00472	.00455	.00542	.00526	.00531	.00482	.00434	.00358	.00320	0.00542
BASE CASE : 214																
BASE	0.00	HT ELEC	KWH	0.88	0.56	0.52	0.34	0.16	0.07	0.00	0.02	0.11	0.33	0.60	0.76	4.3
		HT DEM	KW	.00608	.00608	.00602	.00517	.00395	.00340	.00188	.00298	.00359	.00462	.00608	.00608	0.00608
		REM ELEC	KWH	0.99	0.88	0.99	0.96	1.05	1.10	1.29	1.25	1.09	1.00	0.94	0.98	12.5
		REM DEM	KW	.00392	.00392	.00525	.00579	.00559	.00665	.00645	.00652	.00592	.00532	.00439	.00392	0.00665
		TOT ELEC	KWH	1.86	1.43	1.50	1.30	1.21	1.17	1.29	1.27	1.20	1.33	1.54	1.74	16.8
		TOT DEM	KW	.01008	.00907	.00887	.00766	.00685	.00796	.00726	.00726	.00665	.00827	.00978	.00927	0.01008
S- 3	-0.17	HT ELEC	KWH	0.93	0.59	0.55	0.36	0.17	0.08	0.00	0.02	0.11	0.35	0.63	0.81	4.6
		HT DEM	KW	.00608	.00608	.00602	.00517	.00395	.00340	.00188	.00298	.00359	.00462	.00608	.00608	0.00608
		REM ELEC	KWH	0.85	0.75	0.85	0.82	0.90	0.94	1.10	1.07	0.93	0.86	0.80	0.83	10.7
		REM DEM	KW	.00362	.00362	.00484	.00533	.00515	.00613	.00595	.00601	.00546	.00490	.00405	.00362	0.00613
		TOT ELEC	KWH	1.77	1.34	1.39	1.18	1.06	1.02	1.10	1.09	1.05	1.20	1.44	1.64	15.3
		TOT DEM	KW	.01005	.00905	.00884	.00764	.00683	.00794	.00724	.00724	.00663	.00824	.00975	.00925	0.01005
S- 6	-0.12	HT ELEC	KWH	1.01	0.64	0.59	0.39	0.18	0.08	0.00	0.02	0.12	0.38	0.69	0.88	5.0
		HT DEM	KW	.00608	.00608	.00602	.00517	.00395	.00340	.00188	.00298	.00359	.00462	.00608	.00608	0.00608
		REM ELEC	KWH	0.78	0.69	0.78	0.76	0.83	0.87	1.01	0.98	0.86	0.79	0.74	0.77	9.8
		REM DEM	KW	.00326	.00326	.00437	.00481	.00465	.00553	.00536	.00542	.00492	.00442	.00365	.00326	0.00553
		TOT ELEC	KWH	1.78	1.33	1.37	1.15	1.01	0.95	1.01	1.01	0.98	1.17	1.43	1.64	14.8
		TOT DEM	KW	.00930	.00837	.00818	.00707	.00632	.00735	.00670	.00670	.00614	.00763	.00902	.00856	0.00930
S- 15	1.09	HT ELEC	KWH	1.07	0.68	0.63	0.41	0.19	0.09	0.00	0.03	0.13	0.40	0.73	0.93	5.3
		HT DEM	KW	.00608	.00608	.00602	.00517	.00395	.00340	.00188	.00298	.00359	.00462	.00608	.00608	0.00608
		REM ELEC	KWH	0.68	0.60	0.68	0.66	0.72	0.76	0.88	0.86	0.75	0.69	0.64	0.67	8.6
		REM DEM	KW	.00318	.00318	.00387	.00421	.00402	.00490	.00470	.00475	.00426	.00417	.00318	.00318	0.00490
		TOT ELEC	KWH	1.75	1.28	1.31	1.07	0.91	0.85	0.88	0.89	0.88	1.09	1.38	1.60	13.9
		TOT DEM	KW	.00878	.00817	.00799	.00641	.00562	.00650	.00597	.00606	.00571	.00711	.00860	.00852	0.00878
S- 31	0.15	HT ELEC	KWH	0.69	0.54	0.58	0.45	0.31	0.29	0.19	0.23	0.29	0.43	0.57	0.62	5.2
		HT DEM	KW	.00608	.00608	.00602	.00517	.00395	.00340	.00188	.00298	.00359	.00462	.00608	.00608	0.00608
		REM ELEC	KWH	0.79	0.72	0.83	0.78	0.75	0.90	1.01	0.99	0.88	0.76	0.77	0.82	10.0
		REM DEM	KW	.00359	.00359	.00437	.00476	.00453	.00553	.00531	.00536	.00481	.00470	.00359	.00359	0.00553
		TOT ELEC	KWH	1.48	1.26	1.41	1.23	1.06	1.19	1.20	1.22	1.17	1.19	1.34	1.44	15.2
		TOT DEM	KW	.00970	.00902	.00883	.00708	.00621	.00718	.00660	.00669	.00630	.00786	.00951	.00941	0.00970
S- 39	0.86	HT ELEC	KWH	0.74	0.58	0.62	0.48	0.33	0.31	0.20	0.24	0.31	0.46	0.61	0.66	5.5
		HT DEM	KW	.00608	.00608	.00602	.00517	.00395	.00340	.00188	.00298	.00359	.00462	.00608	.00608	0.00608
		REM ELEC	KWH	0.73	0.66	0.77	0.72	0.69	0.83	0.93	0.91	0.81	0.70	0.71	0.76	9.2
		REM DEM	KW	.00340	.00340	.00413	.00450	.00429	.00523	.00502	.00507	.00455	.00445	.00340	.00340	0.00523
		TOT ELEC	KWH	1.46	1.24	1.38	1.19	1.02	1.14	1.13	1.16	1.12	1.16	1.32	1.42	14.8
		TOT DEM	KW	.00930	.00865	.00846	.00679	.00595	.00688	.00632	.00642	.00604	.00753	.00911	.00902	0.00930
S- 44	0.19	HT ELEC	KWH	0.71	0.56	0.60	0.46	0.32	0.30	0.19	0.23	0.30	0.44	0.59	0.64	5.3
		HT DEM	KW	.00608	.00608	.00602	.00517	.00395	.00340	.00188	.00298	.00359	.00462	.00608	.00608	0.00608
		REM ELEC	KWH	0.77	0.70	0.81	0.76	0.73	0.88	0.99	0.97	0.86	0.74	0.75	0.80	9.8
		REM DEM	KW	.00350	.00350	.00425	.00463	.00441	.00538	.00516	.00522	.00468	.00457	.00350	.00350	0.00538
		TOT ELEC	KWH	1.48	1.26	1.41	1.22	1.05	1.18	1.18	1.20	1.16	1.18	1.34	1.44	15.1
		TOT DEM	KW	.00953	.00886	.00867	.00696	.00610	.00705	.00648	.00658	.00619	.00772	.00934	.00924	0.00953
S- 49	1.46	HT ELEC	KWH	0.67	0.53	0.56	0.43	0.30	0.28	0.18	0.22	0.28	0.42	0.55	0.60	5.0
		HT DEM	KW	.00608	.00608	.00602	.00517	.00395	.00340	.00188	.00298	.00359	.00462	.00608	.00608	0.00608
		REM ELEC	KWH	0.77	0.70	0.81	0.76	0.73	0.87	0.98	0.96	0.85	0.74	0.75	0.80	9.7
		REM DEM	KW	.00359	.00359	.00437	.00476	.00453	.00553	.00531	.00536	.00481	.00470	.00359	.00359	0.00553
		TOT ELEC	KWH	1.44	1.23	1.37	1.19	1.03	1.15	1.16	1.18	1.14	1.15	1.30	1.40	14.7
		TOT DEM	KW	.00970	.00902	.00883	.00708	.00621	.00718	.00660	.00669	.00630	.00786	.00951	.00941	0.00970
S- 60	-0.37	HT ELEC	KWH	1.01	0.64	0.59	0.39	0.18	0.08	0.00	0.02	0.12	0.38	0.69	0.88	5.0
		HT DEM	KW	.00608	.00608	.00602	.00517	.00395	.00340	.00188	.00298	.00359	.00462	.00608	.00608	0.00608
		REM ELEC	KWH	0.78	0.69	0.78	0.76	0.83	0.87	1.01	0.98	0.86	0.79	0.74	0.77	9.8
		REM DEM	KW	.00326	.00326	.00437	.00481	.00465	.00553	.00536	.00542	.00492	.00442	.00365	.00326	0.00553
		TOT ELEC	KWH	1.78	1.33	1.37	1.15	1.01	0.95	1.01	1.01	0.98	1.17	1.43	1.64	14.8
		TOT DEM	KW	.00930	.00837	.00818	.00707	.00632	.00735	.00670	.00670	.00614	.00763	.00902	.00856	0.00930
BASE CASE : 215																
BASE	0.00	TOT GAS	THM	0.0461	0.0292	0.0271	0.0178	0.0082	0.0039	0.0000	0.0011	0.0057	0.0173	0.0315	0.0401	0.228
		TOT ELEC	KWH	0.99	0.88	0.99	0.96	1.05	1.10	1.29	1.25	1.09	1.00	0.94	0.98	12.5
		TOT DEM	KW	.00392	.00392	.00525	.00579	.00559	.00665	.00645	.00652	.00592	.00532	.00439	.00392	0.00665
S- 3	-0.18	TOT GAS	THM	0.0487	0.0308	0.0287	0.0188	0.0087	0.0041	0.0000	0.0012	0.0060	0.0183	0.0333	0.0424	0.241
		TOT ELEC	KWH	0.85	0.75	0.85	0.82	0.90	0.94	1.10	1.07	0.93	0.86	0.80	0.83	10.7
		TOT DEM	KW	.00362	.00362	.00484	.00533	.00515	.00613	.00595	.00601	.00546	.00490	.00405	.00362	0.00613
S- 6	-0.13	TOT GAS	THM	0.0529	0.0335	0.0312	0.0204	0.0094	0.0045	0.0000	0.0013	0.0066	0.0199	0.0362	0.0461	0.262
		TOT ELEC	KWH	0.78	0.69	0.78	0.76	0.83	0.87	1.01	0.98	0.86	0.79	0.74	0.77	9.8
		TOT DEM	KW	.00326	.00326	.00437	.00481	.00465	.00553	.00536	.00542	.00492	.00442	.00365	.00326	0.00553
S- 15	1.08	TOT GAS	THM	0.0562	0.0356	0.0331	0.0217	0.0100	0.0047	0.0000	0.0014	0.0070	0.0211	0.0384	0.0489	0.278
		TOT ELEC	KWH	0.68	0.60	0.68	0.66	0.72	0.76	0.88	0.86	0.75	0.69	0.64	0.67	8.6
		TOT DEM	KW	.00289	.00289	.00387	.00426	.00412	.00490	.00475	.00480	.00436	.00392	.00323	.00289	0.00490

STRAT SET	CONST COST $/SF	ENERGY FORM	UNITS (PER SF)	JAN	FEB	MAR	APR	MAY	JUN	JUL	AUG	SEP	OCT	NOV	DEC	ANNUAL
BASE CASE : 215 (CONTINUED)																
S- 31	0.14	TOT GAS	THM	0.0362	0.0286	0.0305	0.0234	0.0163	0.0152	0.0098	0.0120	0.0152	0.0226	0.0299	0.0326	0.272
		TOT ELEC	KWH	0.79	0.72	0.83	0.78	0.75	0.90	1.01	0.99	0.88	0.76	0.77	0.82	10.0
		TOT DEM	KW	.00359	.00359	.00437	.00476	.00453	.00553	.00531	.00536	.00481	.00470	.00359	.00359	0.00553
S- 39	0.85	TOT GAS	THM	0.0386	0.0304	0.0325	0.0249	0.0174	0.0162	0.0104	0.0128	0.0162	0.0241	0.0319	0.0348	0.290
		TOT ELEC	KWH	0.73	0.66	0.77	0.72	0.69	0.83	0.93	0.91	0.81	0.70	0.71	0.76	9.2
		TOT DEM	KW	.00340	.00340	.00413	.00450	.00429	.00523	.00502	.00507	.00455	.00445	.00340	.00340	0.00523
S- 44	0.19	TOT GAS	THM	0.0372	0.0294	0.0314	0.0241	0.0168	0.0157	0.0101	0.0123	0.0157	0.0232	0.0308	0.0336	0.280
		TOT ELEC	KWH	0.77	0.70	0.81	0.76	0.73	0.88	0.99	0.97	0.86	0.74	0.75	0.80	9.8
		TOT DEM	KW	.00350	.00350	.00425	.00463	.00441	.00538	.00516	.00522	.00468	.00457	.00350	.00350	0.00538
S- 49	1.46	TOT GAS	THM	0.0351	0.0277	0.0296	0.0227	0.0158	0.0148	0.0095	0.0116	0.0148	0.0219	0.0290	0.0317	0.264
		TOT ELEC	KWH	0.77	0.70	0.81	0.76	0.73	0.87	0.98	0.96	0.85	0.74	0.75	0.80	9.7
		TOT DEM	KW	.00359	.00359	.00437	.00476	.00453	.00553	.00531	.00536	.00481	.00470	.00359	.00359	0.00553
S- 60	-0.37	TOT GAS	THM	0.0529	0.0335	0.0312	0.0204	0.0094	0.0045	0.0000	0.0013	0.0066	0.0199	0.0362	0.0461	0.262
		TOT ELEC	KWH	0.78	0.69	0.78	0.76	0.83	0.87	1.01	0.98	0.86	0.79	0.74	0.77	9.8
		TOT DEM	KW	.00326	.00326	.00437	.00481	.00465	.00553	.00536	.00542	.00492	.00442	.00365	.00326	0.00553
BASE CASE : 216																
BASE	0.00	HT ELEC	KWH	0.70	0.44	0.41	0.27	0.12	0.06	0.00	0.02	0.09	0.26	0.48	0.61	3.5
		HT DEM	KW	.00435	.00435	.00435	.00435	.00313	.00218	.00070	.00165	.00252	.00435	.00435	.00435	0.00435
		REM ELEC	KWH	0.92	0.82	0.92	0.90	0.98	1.03	1.21	1.17	1.02	0.94	0.88	0.91	11.7
		REM DEM	KW	.00355	.00355	.00447	.00499	.00476	.00567	.00567	.00573	.00539	.00441	.00418	.00350	0.00573
		TOT ELEC	KWH	1.62	1.26	1.33	1.17	1.11	1.09	1.21	1.19	1.10	1.20	1.35	1.52	15.1
		TOT DEM	KW	.00750	.00743	.00743	.00645	.00487	.00578	.00578	.00585	.00548	.00705	.00743	.00743	0.00750
S- 3	-0.08	HT ELEC	KWH	0.75	0.47	0.44	0.29	0.13	0.06	0.00	0.02	0.09	0.28	0.51	0.65	3.7
		HT DEM	KW	.00435	.00435	.00435	.00435	.00313	.00218	.00070	.00165	.00252	.00435	.00435	.00435	0.00435
		REM ELEC	KWH	0.79	0.70	0.79	0.77	0.84	0.88	1.02	1.00	0.87	0.80	0.75	0.78	10.0
		REM DEM	KW	.00327	.00327	.00412	.00459	.00438	.00523	.00523	.00528	.00496	.00407	.00385	.00322	0.00528
		TOT ELEC	KWH	1.54	1.17	1.23	1.06	0.97	0.94	1.02	1.01	0.96	1.08	1.26	1.43	13.7
		TOT DEM	KW	.00746	.00739	.00739	.00642	.00485	.00574	.00574	.00582	.00545	.00701	.00739	.00739	0.00746
S- 6	-0.03	HT ELEC	KWH	0.80	0.51	0.47	0.31	0.14	0.07	0.00	0.02	0.10	0.30	0.55	0.70	4.0
		HT DEM	KW	.00435	.00435	.00435	.00435	.00313	.00218	.00070	.00165	.00252	.00435	.00435	.00435	0.00435
		REM ELEC	KWH	0.71	0.63	0.71	0.69	0.76	0.79	0.93	0.90	0.78	0.72	0.68	0.70	9.0
		REM DEM	KW	.00295	.00295	.00370	.00413	.00394	.00470	.00470	.00475	.00446	.00366	.00347	.00290	0.00475
		TOT ELEC	KWH	1.51	1.14	1.18	1.00	0.90	0.86	0.93	0.92	0.88	1.02	1.22	1.40	13.0
		TOT DEM	KW	.00673	.00666	.00666	.00579	.00437	.00518	.00518	.00525	.00491	.00633	.00666	.00666	0.00673
S- 15	0.56	HT ELEC	KWH	0.85	0.54	0.50	0.33	0.15	0.07	0.00	0.02	0.11	0.32	0.58	0.74	4.2
		HT DEM	KW	.00435	.00435	.00435	.00435	.00313	.00218	.00070	.00165	.00252	.00435	.00435	.00435	0.00435
		REM ELEC	KWH	0.62	0.55	0.62	0.60	0.66	0.69	0.81	0.78	0.68	0.63	0.59	0.61	7.8
		REM DEM	KW	.00262	.00262	.00330	.00368	.00351	.00419	.00419	.00423	.00398	.00326	.00309	.00258	0.00423
		TOT ELEC	KWH	1.47	1.09	1.12	0.93	0.81	0.76	0.81	0.80	0.79	0.95	1.17	1.35	12.1
		TOT DEM	KW	.00653	.00646	.00646	.00562	.00424	.00503	.00503	.00509	.00477	.00614	.00646	.00646	0.00653
S- 31	0.24	HT ELEC	KWH	0.67	0.45	0.44	0.26	0.13	0.09	0.04	0.07	0.11	0.22	0.47	0.50	3.5
		HT DEM	KW	.00435	.00435	.00435	.00435	.00313	.00218	.00070	.00165	.00252	.00435	.00435	.00435	0.00435
		REM ELEC	KWH	0.68	0.61	0.71	0.67	0.70	0.82	0.90	0.90	0.80	0.69	0.67	0.71	8.9
		REM DEM	KW	.00296	.00292	.00338	.00351	.00351	.00423	.00415	.00419	.00372	.00372	.00317	.00305	0.00423
		TOT ELEC	KWH	1.35	1.06	1.15	0.94	0.83	0.91	0.95	0.97	0.91	0.92	1.13	1.21	12.3
		TOT DEM	KW	.00710	.00703	.00703	.00596	.00369	.00433	.00426	.00433	.00383	.00653	.00703	.00703	0.00710
S- 39	0.68	HT ELEC	KWH	0.78	0.52	0.51	0.30	0.15	0.11	0.05	0.08	0.12	0.26	0.54	0.58	4.0
		HT DEM	KW	.00395	.00395	.00395	.00395	.00284	.00197	.00063	.00150	.00229	.00395	.00395	.00395	0.00395
		REM ELEC	KWH	0.61	0.55	0.64	0.60	0.63	0.73	0.81	0.81	0.71	0.62	0.60	0.64	7.9
		REM DEM	KW	.00270	.00266	.00308	.00320	.00320	.00385	.00377	.00381	.00339	.00339	.00289	.00277	0.00385
		TOT ELEC	KWH	1.39	1.07	1.15	0.91	0.78	0.84	0.86	0.89	0.84	0.88	1.14	1.22	11.9
		TOT DEM	KW	.00683	.00676	.00676	.00574	.00355	.00417	.00410	.00417	.00369	.00628	.00676	.00676	0.00683
S- 44	0.28	HT ELEC	KWH	0.72	0.49	0.48	0.28	0.14	0.10	0.05	0.07	0.12	0.24	0.50	0.54	3.7
		HT DEM	KW	.00395	.00395	.00395	.00395	.00284	.00197	.00063	.00150	.00229	.00395	.00395	.00395	0.00395
		REM ELEC	KWH	0.65	0.58	0.68	0.64	0.67	0.78	0.86	0.86	0.76	0.66	0.63	0.68	8.4
		REM DEM	KW	.00283	.00279	.00324	.00336	.00336	.00405	.00397	.00401	.00356	.00356	.00304	.00292	0.00405
		TOT ELEC	KWH	1.37	1.07	1.15	0.92	0.80	0.88	0.91	0.94	0.87	0.90	1.14	1.21	12.2
		TOT DEM	KW	.00698	.00691	.00691	.00586	.00363	.00426	.00419	.00426	.00377	.00642	.00691	.00691	0.00698
S- 49	1.55	HT ELEC	KWH	0.65	0.44	0.43	0.25	0.12	0.09	0.04	0.07	0.10	0.22	0.45	0.49	3.3
		HT DEM	KW	.00435	.00435	.00435	.00435	.00313	.00218	.00070	.00165	.00252	.00435	.00435	.00435	0.00435
		REM ELEC	KWH	0.66	0.59	0.69	0.65	0.68	0.79	0.88	0.88	0.77	0.67	0.65	0.69	8.6
		REM DEM	KW	.00296	.00292	.00338	.00351	.00351	.00423	.00415	.00419	.00372	.00372	.00317	.00305	0.00423
		TOT ELEC	KWH	1.31	1.03	1.12	0.91	0.80	0.88	0.92	0.94	0.88	0.89	1.10	1.17	12.0
		TOT DEM	KW	.00710	.00703	.00703	.00596	.00369	.00433	.00426	.00433	.00383	.00653	.00703	.00703	0.00710
S- 60	-0.28	HT ELEC	KWH	0.80	0.51	0.47	0.31	0.14	0.07	0.00	0.02	0.10	0.30	0.55	0.70	4.0
		HT DEM	KW	.00435	.00435	.00435	.00435	.00313	.00218	.00070	.00165	.00252	.00435	.00435	.00435	0.00435
		REM ELEC	KWH	0.71	0.63	0.71	0.69	0.76	0.79	0.93	0.90	0.78	0.72	0.68	0.70	9.0
		REM DEM	KW	.00295	.00295	.00370	.00413	.00394	.00470	.00470	.00475	.00446	.00366	.00347	.00290	0.00475
		TOT ELEC	KWH	1.51	1.14	1.18	1.00	0.90	0.86	0.93	0.92	0.88	1.02	1.22	1.40	13.0
		TOT DEM	KW	.00673	.00666	.00666	.00579	.00437	.00518	.00518	.00525	.00491	.00633	.00666	.00666	0.00673

(HT ELEC=ELECTRIC CONSUMPTION FOR SPACE AND WATER HEATING. REM ELEC=REMAINING ELECTRIC CONSUMPTION (NOT USED FOR HEATING). TOT ELEC=TOTAL ELECTRIC CONSUMPTION. HT DEM=PEAK ELECTRIC DEMAND FOR SPACE AND WATER HEATING. REM DEM=PEAK ELECTRIC DEMAND FOR REMAINING LOAD (NOT HEATING). TOT DEM=TOTAL PEAK ELECTRIC DEMAND. GAS=TOTAL GAS CONSUMPTION. ANNUAL FIGURES ARE SUMS OF MONTHLY CONSUMPTION VALUES OR ARE YEARLY MAXIMUMS OF MONTHLY PEAK DEMAND VALUES.)

STRAT SET	CONST COST $/SF	ENERGY FORM	UNITS (PER SF)	JAN	FEB	MAR	APR	MAY	JUN	JUL	AUG	SEP	OCT	NOV	DEC	ANNUAL
BASE CASE : 217																
BASE	0.00	TOT GAS	THM	0.0366	0.0232	0.0215	0.0141	0.0065	0.0031	0.0000	0.0009	0.0045	0.0138	0.0250	0.0319	0.181
		TOT ELEC	KWH	0.92	0.82	0.92	0.90	0.98	1.03	1.21	1.17	1.02	0.94	0.88	0.91	11.7
		TOT DEM	KW	.00355	.00355	.00447	.00499	.00476	.00567	.00567	.00573	.00539	.00441	.00418	.00350	0.00573
S- 3	-0.09	TOT GAS	THM	0.0394	0.0250	0.0232	0.0152	0.0070	0.0033	0.0000	0.0010	0.0049	0.0148	0.0269	0.0343	0.195
		TOT ELEC	KWH	0.79	0.70	0.79	0.77	0.84	0.88	1.02	1.00	0.87	0.80	0.75	0.78	10.0
		TOT DEM	KW	.00327	.00327	.00412	.00459	.00438	.00523	.00523	.00528	.00496	.00407	.00385	.00322	0.00528
S- 6	-0.04	TOT GAS	THM	0.0420	0.0266	0.0248	0.0162	0.0075	0.0035	0.0000	0.0010	0.0052	0.0158	0.0287	0.0366	0.208
		TOT ELEC	KWH	0.71	0.63	0.71	0.69	0.76	0.79	0.93	0.90	0.78	0.72	0.68	0.70	9.0
		TOT DEM	KW	.00295	.00295	.00370	.00413	.00394	.00470	.00470	.00475	.00446	.00366	.00347	.00290	0.00475
S- 15	0.53	TOT GAS	THM	0.0448	0.0284	0.0264	0.0173	0.0080	0.0038	0.0000	0.0011	0.0055	0.0169	0.0306	0.0391	0.222
		TOT ELEC	KWH	0.62	0.55	0.62	0.60	0.66	0.69	0.81	0.78	0.68	0.63	0.59	0.61	7.8
		TOT DEM	KW	.00262	.00262	.00330	.00368	.00351	.00419	.00419	.00423	.00398	.00326	.00309	.00258	0.00423
S- 31	0.23	TOT GAS	THM	0.0351	0.0237	0.0232	0.0138	0.0067	0.0049	0.0024	0.0036	0.0056	0.0118	0.0244	0.0262	0.181
		TOT ELEC	KWH	0.68	0.61	0.71	0.67	0.70	0.82	0.90	0.90	0.80	0.69	0.67	0.71	8.9
		TOT DEM	KW	.00296	.00296	.00338	.00351	.00351	.00423	.00415	.00419	.00372	.00372	.00317	.00305	0.00423
S- 39	0.66	TOT GAS	THM	0.0407	0.0275	0.0269	0.0160	0.0078	0.0057	0.0027	0.0042	0.0065	0.0136	0.0284	0.0304	0.210
		TOT ELEC	KWH	0.61	0.55	0.64	0.60	0.63	0.73	0.81	0.81	0.71	0.62	0.60	0.64	7.9
		TOT DEM	KW	.00270	.00270	.00308	.00320	.00320	.00385	.00377	.00381	.00339	.00339	.00289	.00277	0.00385
S- 44	0.28	TOT GAS	THM	0.0378	0.0255	0.0250	0.0148	0.0072	0.0053	0.0025	0.0039	0.0060	0.0127	0.0263	0.0283	0.195
		TOT ELEC	KWH	0.65	0.58	0.68	0.64	0.67	0.78	0.86	0.86	0.76	0.66	0.63	0.68	8.4
		TOT DEM	KW	.00283	.00283	.00324	.00336	.00336	.00405	.00397	.00401	.00356	.00356	.00304	.00292	0.00405
S- 49	1.54	TOT GAS	THM	0.0341	0.0231	0.0225	0.0134	0.0065	0.0048	0.0023	0.0035	0.0055	0.0114	0.0238	0.0255	0.176
		TOT ELEC	KWH	0.66	0.59	0.69	0.65	0.68	0.79	0.88	0.88	0.77	0.67	0.65	0.69	8.6
		TOT DEM	KW	.00296	.00296	.00338	.00351	.00351	.00423	.00415	.00419	.00372	.00372	.00317	.00305	0.00423
S- 60	-0.29	TOT GAS	THM	0.0420	0.0266	0.0248	0.0162	0.0075	0.0035	0.0000	0.0010	0.0052	0.0158	0.0287	0.0366	0.208
		TOT ELEC	KWH	0.71	0.63	0.71	0.69	0.76	0.79	0.93	0.90	0.78	0.72	0.68	0.70	9.0
		TOT DEM	KW	.00295	.00295	.00370	.00413	.00394	.00470	.00470	.00475	.00446	.00366	.00347	.00290	0.00475
BASE CASE : 218																
BASE	0.00	TOT ELEC	KWH	1.26	0.99	1.09	1.01	1.03	1.13	1.33	1.18	1.09	1.05	1.07	1.19	13.4
		TOT DEM	KW	.00427	.00427	.00572	.00630	.00608	.00724	.00702	.00710	.00644	.00579	.00478	.00427	0.00724
S- 3	-0.12	TOT ELEC	KWH	1.16	0.91	1.00	0.92	0.95	1.03	1.22	1.08	1.00	0.96	0.98	1.09	12.3
		TOT DEM	KW	.00399	.00399	.00535	.00589	.00569	.00677	.00657	.00663	.00603	.00542	.00447	.00399	0.00677
S- 6	-0.07	TOT ELEC	KWH	1.08	0.85	0.93	0.86	0.89	0.97	1.14	1.01	0.93	0.90	0.92	1.02	11.5
		TOT DEM	KW	.00372	.00372	.00498	.00549	.00530	.00631	.00612	.00618	.00562	.00505	.00416	.00372	0.00631
S- 15	1.04	TOT ELEC	KWH	0.98	0.77	0.84	0.78	0.80	0.87	1.03	0.92	0.84	0.81	0.83	0.93	10.4
		TOT DEM	KW	.00352	.00352	.00472	.00519	.00501	.00597	.00579	.00585	.00531	.00478	.00394	.00352	0.00597
S- 31	0.20	TOT ELEC	KWH	1.02	0.81	0.88	0.82	0.84	0.92	1.08	0.96	0.88	0.85	0.87	0.97	10.9
		TOT DEM	KW	.00404	.00404	.00541	.00596	.00575	.00685	.00664	.00671	.00610	.00548	.00452	.00404	0.00685
S- 39	0.86	TOT ELEC	KWH	1.00	0.78	0.86	0.80	0.82	0.89	1.05	0.93	0.86	0.83	0.85	0.94	10.6
		TOT DEM	KW	.00389	.00389	.00521	.00573	.00554	.00659	.00639	.00646	.00587	.00527	.00435	.00389	0.00659
S- 44	0.29	TOT ELEC	KWH	1.02	0.80	0.87	0.81	0.83	0.91	1.07	0.95	0.87	0.84	0.86	0.96	10.8
		TOT DEM	KW	.00398	.00398	.00532	.00586	.00566	.00674	.00654	.00661	.00600	.00539	.00445	.00398	0.00674
S- 49	1.51	TOT ELEC	KWH	0.99	0.78	0.85	0.79	0.81	0.88	1.04	0.92	0.85	0.82	0.84	0.93	10.5
		TOT DEM	KW	.00404	.00404	.00541	.00596	.00575	.00685	.00664	.00671	.00610	.00548	.00452	.00404	0.00685
S- 60	-0.32	TOT ELEC	KWH	1.08	0.85	0.93	0.86	0.89	0.97	1.14	1.01	0.93	0.90	0.92	1.02	11.5
		TOT DEM	KW	.00372	.00372	.00498	.00549	.00530	.00631	.00612	.00618	.00562	.00505	.00416	.00372	0.00631
BASE CASE : 219																
BASE	0.00	HT ELEC	KWH	0.57	0.36	0.33	0.22	0.10	0.05	0.00	0.01	0.07	0.21	0.39	0.49	2.8
		HT DEM	KW	.00608	.00578	.00547	.00468	.00359	.00298	.00146	.00249	.00292	.00474	.00590	.00553	0.00608
		REM ELEC	KWH	0.95	0.84	0.95	0.92	1.01	1.06	1.24	1.20	1.04	0.96	0.90	0.94	12.0
		REM DEM	KW	.00354	.00354	.00474	.00522	.00504	.00600	.00582	.00588	.00534	.00480	.00396	.00354	0.00600
		TOT ELEC	KWH	1.51	1.20	1.28	1.14	1.11	1.10	1.24	1.21	1.11	1.17	1.29	1.43	14.8
		TOT DEM	KW	.00988	.00988	.00988	.00909	.00741	.00840	.00781	.00781	.00721	.00958	.00988	.00988	0.00988
S- 3	-0.12	HT ELEC	KWH	0.60	0.38	0.35	0.23	0.11	0.05	0.00	0.01	0.07	0.22	0.41	0.52	3.0
		HT DEM	KW	.00608	.00578	.00547	.00468	.00359	.00298	.00146	.00249	.00292	.00474	.00590	.00553	0.00608
		REM ELEC	KWH	0.81	0.72	0.81	0.79	0.87	0.91	1.06	1.03	0.90	0.82	0.77	0.80	10.3
		REM DEM	KW	.00326	.00326	.00437	.00481	.00465	.00553	.00536	.00542	.00492	.00442	.00365	.00326	0.00553
		TOT ELEC	KWH	1.41	1.10	1.16	1.02	0.97	0.96	1.06	1.04	0.97	1.05	1.18	1.32	13.3
		TOT DEM	KW	.00950	.00950	.00950	.00874	.00713	.00807	.00751	.00751	.00694	.00922	.00950	.00950	0.00950
S- 6	-0.07	HT ELEC	KWH	0.65	0.41	0.38	0.25	0.12	0.05	0.00	0.02	0.08	0.24	0.44	0.57	3.2
		HT DEM	KW	.00608	.00578	.00547	.00468	.00359	.00298	.00146	.00249	.00292	.00474	.00590	.00553	0.00608
		REM ELEC	KWH	0.75	0.66	0.75	0.73	0.79	0.83	0.97	0.94	0.82	0.76	0.71	0.74	9.4
		REM DEM	KW	.00294	.00294	.00393	.00433	.00418	.00498	.00483	.00488	.00443	.00398	.00329	.00294	0.00498
		TOT ELEC	KWH	1.40	1.07	1.13	0.98	0.91	0.89	0.97	0.96	0.90	1.00	1.15	1.30	12.7
		TOT DEM	KW	.00885	.00885	.00885	.00814	.00664	.00752	.00699	.00699	.00646	.00858	.00885	.00885	0.00885
S- 15	1.14	HT ELEC	KWH	0.69	0.44	0.41	0.27	0.12	0.06	0.00	0.02	0.09	0.26	0.47	0.60	3.4
		HT DEM	KW	.00608	.00578	.00547	.00468	.00359	.00298	.00146	.00249	.00292	.00474	.00590	.00553	0.00608
		REM ELEC	KWH	0.65	0.58	0.65	0.63	0.69	0.73	0.85	0.82	0.72	0.66	0.62	0.64	8.2
		REM DEM	KW	.00261	.00261	.00350	.00385	.00372	.00443	.00430	.00434	.00394	.00354	.00292	.00261	0.00443
		TOT ELEC	KWH	1.34	1.01	1.06	0.90	0.81	0.78	0.85	0.84	0.80	0.92	1.09	1.24	11.6
		TOT DEM	KW	.00850	.00689	.00680	.00587	.00502	.00553	.00519	.00519	.00493	.00629	.00706	.00689	0.00850
BASE CASE : 219 (CONTINUED)																
S- 31	0.29	HT ELEC	KWH	0.46	0.36	0.39	0.30	0.21	0.19	0.12	0.15	0.19	0.29	0.38	0.42	3.5
		HT DEM	KW	.00608	.00578	.00547	.00468	.00359	.00298	.00146	.00249	.00292	.00474	.00590	.00553	0.00608
		REM ELEC	KWH	0.75	0.68	0.78	0.74	0.71	0.85	0.95	0.93	0.83	0.72	0.73	0.77	9.4
		REM DEM	KW	.00325	.00325	.00395	.00430	.00410	.00500	.00480	.00485	.00435	.00425	.00325	.00325	0.00500
		TOT ELEC	KWH	1.21	1.04	1.17	1.03	0.92	1.04	1.08	1.09	1.03	1.01	1.11	1.19	12.9
		TOT DEM	KW	.00955	.00774	.00764	.00659	.00563	.00621	.00583	.00583	.00554	.00707	.00793	.00774	0.00955
S- 60	-0.32	HT ELEC	KWH	0.65	0.41	0.38	0.25	0.12	0.05	0.00	0.02	0.08	0.24	0.44	0.57	3.2
		HT DEM	KW	.00608	.00578	.00547	.00468	.00359	.00298	.00146	.00249	.00292	.00474	.00590	.00553	0.00608
		REM ELEC	KWH	0.75	0.66	0.75	0.73	0.79	0.83	0.97	0.94	0.82	0.76	0.71	0.74	9.4
		REM DEM	KW	.00294	.00294	.00393	.00433	.00418	.00498	.00483	.00488	.00443	.00398	.00329	.00294	0.00498
		TOT ELEC	KWH	1.40	1.07	1.13	0.98	0.91	0.89	0.97	0.96	0.90	1.00	1.15	1.30	12.7
		TOT DEM	KW	.00885	.00885	.00885	.00814	.00664	.00752	.00699	.00699	.00646	.00858	.00885	.00885	0.00885
BASE CASE : 220																
BASE	0.00	TOT GAS	THM	0.0297	0.0188	0.0175	0.0115	0.0053	0.0025	0.0000	0.0007	0.0037	0.0112	0.0203	0.0259	0.147
		TOT ELEC	KWH	0.95	0.84	0.95	0.92	1.01	1.06	1.24	1.20	1.04	0.96	0.90	0.94	12.0
		TOT DEM	KW	.00354	.00354	.00474	.00522	.00504	.00600	.00582	.00588	.00534	.00480	.00396	.00354	0.00600
S- 3	-0.13	TOT GAS	THM	0.0313	0.0198	0.0184	0.0121	0.0056	0.0026	0.0000	0.0008	0.0039	0.0118	0.0214	0.0273	0.155
		TOT ELEC	KWH	0.81	0.72	0.81	0.79	0.87	0.91	1.06	1.03	0.90	0.82	0.77	0.80	10.3
		TOT DEM	KW	.00326	.00326	.00437	.00481	.00465	.00553	.00536	.00542	.00492	.00442	.00365	.00326	0.00553
S- 6	-0.08	TOT GAS	THM	0.0341	0.0216	0.0201	0.0132	0.0061	0.0029	0.0000	0.0008	0.0042	0.0128	0.0233	0.0297	0.169
		TOT ELEC	KWH	0.75	0.66	0.75	0.73	0.79	0.83	0.97	0.94	0.82	0.76	0.71	0.74	9.4
		TOT DEM	KW	.00294	.00294	.00393	.00433	.00418	.00498	.00483	.00488	.00443	.00398	.00329	.00294	0.00498
S- 15	1.14	TOT GAS	THM	0.0362	0.0229	0.0213	0.0140	0.0064	0.0030	0.0000	0.0009	0.0045	0.0136	0.0247	0.0315	0.179
		TOT ELEC	KWH	0.65	0.58	0.65	0.63	0.69	0.73	0.85	0.82	0.72	0.66	0.62	0.64	8.2
		TOT DEM	KW	.00261	.00261	.00350	.00385	.00372	.00443	.00430	.00434	.00394	.00354	.00292	.00261	0.00443
S- 31	0.28	TOT GAS	THM	0.0242	0.0191	0.0204	0.0157	0.0109	0.0102	0.0066	0.0080	0.0102	0.0151	0.0200	0.0218	0.182
		TOT ELEC	KWH	0.75	0.68	0.78	0.74	0.71	0.85	0.95	0.93	0.83	0.72	0.73	0.77	9.4
		TOT DEM	KW	.00325	.00325	.00395	.00430	.00410	.00500	.00480	.00485	.00435	.00425	.00325	.00325	0.00500
S- 60	-0.33	TOT GAS	THM	0.0341	0.0216	0.0201	0.0132	0.0061	0.0029	0.0000	0.0008	0.0042	0.0128	0.0233	0.0297	0.169
		TOT ELEC	KWH	0.75	0.66	0.75	0.73	0.79	0.83	0.97	0.94	0.82	0.76	0.71	0.74	9.4
		TOT DEM	KW	.00294	.00294	.00393	.00433	.00418	.00498	.00483	.00488	.00443	.00398	.00329	.00294	0.00498
BASE CASE : 221																
BASE	-0.08	HT ELEC	KWH	0.46	0.29	0.27	0.18	0.08	0.04	0.00	0.01	0.06	0.17	0.32	0.40	2.3
		HT DEM	KW	.00435	.00409	.00409	.00374	.00291	.00213	.00078	.00157	.00218	.00365	.00418	.00418	0.00435
		REM ELEC	KWH	0.92	0.81	0.92	0.89	0.97	1.02	1.19	1.16	1.01	0.93	0.87	0.90	11.6
		REM DEM	KW	.00343	.00343	.00431	.00481	.00459	.00547	.00547	.00553	.00520	.00426	.00404	.00337	0.00553
		TOT ELEC	KWH	1.38	1.11	1.19	1.07	1.06	1.06	1.19	1.17	1.07	1.10	1.19	1.31	13.9
		TOT DEM	KW	.00748	.00673	.00666	.00561	.00486	.00576	.00568	.00576	.00524	.00583	.00688	.00681	0.00748
S- 3	-0.08	HT ELEC	KWH	0.50	0.32	0.30	0.19	0.09	0.04	0.00	0.01	0.06	0.19	0.34	0.44	2.5
		HT DEM	KW	.00435	.00409	.00409	.00374	.00291	.00213	.00078	.00157	.00218	.00365	.00418	.00418	0.00435
		REM ELEC	KWH	0.78	0.69	0.78	0.76	0.83	0.87	1.02	0.99	0.86	0.79	0.74	0.77	9.9
		REM DEM	KW	.00315	.00315	.00396	.00442	.00422	.00503	.00503	.00508	.00478	.00391	.00371	.00310	0.00508
		TOT ELEC	KWH	1.28	1.01	1.07	0.95	0.92	0.91	1.02	1.00	0.92	0.98	1.08	1.21	12.4
		TOT DEM	KW	.00724	.00652	.00644	.00543	.00471	.00557	.00550	.00557	.00507	.00565	.00666	.00659	0.00724
S- 6	-0.03	HT ELEC	KWH	0.54	0.34	0.32	0.21	0.10	0.05	0.00	0.01	0.07	0.20	0.37	0.47	2.7
		HT DEM	KW	.00435	.00409	.00409	.00374	.00291	.00213	.00078	.00157	.00218	.00365	.00418	.00418	0.00435
		REM ELEC	KWH	0.70	0.62	0.70	0.69	0.75	0.78	0.92	0.89	0.78	0.71	0.67	0.70	8.9
		REM DEM	KW	.00285	.00285	.00359	.00400	.00382	.00455	.00455	.00460	.00432	.00354	.00336	.00281	0.00460
		TOT ELEC	KWH	1.24	0.96	1.02	0.89	0.84	0.83	0.92	0.91	0.84	0.91	1.03	1.16	11.6
		TOT DEM	KW	.00674	.00607	.00600	.00506	.00438	.00519	.00512	.00519	.00472	.00526	.00620	.00613	0.00674
S- 15	0.55	HT ELEC	KWH	0.57	0.36	0.34	0.22	0.10	0.05	0.00	0.01	0.07	0.22	0.39	0.50	2.8
		HT DEM	KW	.00435	.00409	.00409	.00374	.00291	.00213	.00078	.00157	.00218	.00365	.00418	.00418	0.00435
		REM ELEC	KWH	0.61	0.54	0.61	0.60	0.65	0.68	0.80	0.78	0.68	0.62	0.58	0.61	7.8
		REM DEM	KW	.00253	.00253	.00318	.00355	.00339	.00404	.00404	.00408	.00384	.00314	.00298	.00249	0.00408
		TOT ELEC	KWH	1.19	0.91	0.95	0.82	0.75	0.73	0.80	0.79	0.75	0.84	0.97	1.11	10.6
		TOT DEM	KW	.00654	.00589	.00582	.00490	.00425	.00504	.00497	.00504	.00458	.00510	.00602	.00595	0.00654
S- 31	0.33	HT ELEC	KWH	0.72	0.49	0.48	0.28	0.14	0.10	0.05	0.07	0.12	0.24	0.50	0.54	3.7
		HT DEM	KW	.00435	.00409	.00409	.00374	.00291	.00213	.00078	.00157	.00218	.00365	.00418	.00418	0.00435
		REM ELEC	KWH	0.69	0.62	0.71	0.68	0.70	0.82	0.91	0.91	0.80	0.70	0.67	0.71	8.9
		REM DEM	KW	.00300	.00295	.00342	.00355	.00355	.00428	.00419	.00424	.00377	.00377	.00321	.00308	0.00428
		TOT ELEC	KWH	1.41	1.10	1.19	0.96	0.84	0.92	0.96	0.98	0.92	0.94	1.17	1.25	12.6
		TOT DEM	KW	.00715	.00651	.00636	.00536	.00429	.00486	.00479	.00472	.00436	.00572	.00658	.00658	0.00715
S- 60	-0.28	HT ELEC	KWH	0.54	0.34	0.32	0.21	0.10	0.05	0.00	0.01	0.07	0.20	0.37	0.47	2.7
		HT DEM	KW	.00435	.00409	.00409	.00374	.00291	.00213	.00078	.00157	.00218	.00365	.00418	.00418	0.00435
		REM ELEC	KWH	0.70	0.62	0.70	0.69	0.75	0.78	0.92	0.89	0.78	0.71	0.67	0.70	8.9
		REM DEM	KW	.00285	.00285	.00359	.00400	.00382	.00455	.00455	.00460	.00432	.00354	.00336	.00281	0.00460
		TOT ELEC	KWH	1.24	0.96	1.02	0.89	0.84	0.83	0.92	0.91	0.84	0.91	1.03	1.16	11.6
		TOT DEM	KW	.00674	.00607	.00600	.00506	.00438	.00519	.00512	.00519	.00472	.00526	.00620	.00613	0.00674
BASE CASE : 222																
BASE	0.00	TOT GAS	THM	0.0244	0.0155	0.0144	0.0094	0.0044	0.0021	0.0000	0.0006	0.0030	0.0092	0.0167	0.0213	0.121
		TOT ELEC	KWH	0.92	0.81	0.92	0.89	0.97	1.02	1.19	1.16	1.01	0.93	0.87	0.90	11.6
		TOT DEM	KW	.00343	.00343	.00431	.00481	.00459	.00547	.00547	.00553	.00520	.00426	.00404	.00337	0.00553
S- 3	-0.09	TOT GAS	THM	0.0263	0.0166	0.0155	0.0101	0.0047	0.0022	0.0000	0.0007	0.0033	0.0099	0.0179	0.0229	0.130
		TOT ELEC	KWH	0.78	0.69	0.78	0.76	0.83	0.87	1.02	0.99	0.86	0.79	0.74	0.77	9.9
		TOT DEM	KW	.00315	.00315	.00396	.00442	.00422	.00503	.00503	.00508	.00478	.00391	.00371	.00310	0.00508

(HT ELEC=ELECTRIC CONSUMPTION FOR SPACE AND WATER HEATING. REM ELEC=REMAINING ELECTRIC CONSUMPTION (NOT USED FOR HEATING). TOT ELEC=TOTAL ELECTRIC CONSUMPTION. HT DEM=PEAK ELECTRIC DEMAND FOR SPACE AND WATER HEATING. REM DEM=PEAK ELECTRIC DEMAND FOR REMAINING LOAD (NOT HEATING). TOT DEM=TOTAL PEAK ELECTRIC DEMAND. GAS=TOTAL GAS CONSUMPTION. ANNUAL FIGURES ARE SUMS OF MONTHLY CONSUMPTION VALUES OR ARE YEARLY MAXIMUMS OF MONTHLY PEAK DEMAND VALUES.)

STRAT SET	CONST COST $/SF	ENERGY FORM	UNITS (PER SF)	JAN	FEB	MAR	APR	MAY	JUN	JUL	AUG	SEP	OCT	NOV	DEC	ANNUAL
BASE CASE : 222 (CONTINUED)																
S- 6	-0.04	TOT GAS	THM	0.0281	0.0178	0.0165	0.0108	0.0050	0.0024	0.0000	0.0007	0.0035	0.0106	0.0192	0.0245	0.139
		TOT ELEC	KWH	0.70	0.62	0.70	0.69	0.75	0.78	0.92	0.89	0.78	0.71	0.67	0.70	8.9
		TOT DEM	KW	.00285	.00285	.00359	.00400	.00382	.00455	.00455	.00460	.00432	.00354	.00336	.00281	0.00460
S- 15	0.52	TOT GAS	THM	0.0301	0.0191	0.0177	0.0116	0.0054	0.0025	0.0000	0.0007	0.0037	0.0113	0.0206	0.0262	0.149
		TOT ELEC	KWH	0.61	0.54	0.61	0.60	0.65	0.68	0.80	0.78	0.68	0.62	0.58	0.61	7.8
		TOT DEM	KW	.00253	.00253	.00318	.00355	.00339	.00404	.00404	.00408	.00384	.00314	.00298	.00249	0.00408
S- 31	0.33	TOT GAS	THM	0.0380	0.0257	0.0251	0.0149	0.0073	0.0053	0.0025	0.0039	0.0061	0.0127	0.0265	0.0284	0.196
		TOT ELEC	KWH	0.69	0.62	0.71	0.68	0.70	0.82	0.91	0.91	0.80	0.70	0.67	0.71	8.9
		TOT DEM	KW	.00300	.00300	.00342	.00355	.00355	.00428	.00419	.00424	.00377	.00377	.00321	.00308	0.00428
S- 60	-0.29	TOT GAS	THM	0.0281	0.0178	0.0165	0.0108	0.0050	0.0024	0.0000	0.0007	0.0035	0.0106	0.0192	0.0245	0.139
		TOT ELEC	KWH	0.70	0.62	0.70	0.69	0.75	0.78	0.92	0.89	0.78	0.71	0.67	0.70	8.9
		TOT DEM	KW	.00285	.00285	.00359	.00400	.00382	.00455	.00455	.00460	.00432	.00354	.00336	.00281	0.00460
BASE CASE : 223																
BASE	0.00	TOT ELEC	KWH	1.19	0.94	1.03	0.95	0.98	1.07	1.26	1.12	1.03	0.99	1.02	1.13	12.7
		TOT DEM	KW	.00379	.00379	.00507	.00559	.00539	.00642	.00623	.00629	.00571	.00514	.00424	.00379	0.00642
S- 3	-0.09	TOT ELEC	KWH	1.10	0.87	0.95	0.88	0.90	0.98	1.16	1.03	0.95	0.91	0.94	1.04	11.7
		TOT DEM	KW	.00355	.00355	.00475	.00523	.00505	.00601	.00583	.00589	.00535	.00481	.00397	.00355	0.00601
S- 6	-0.04	TOT ELEC	KWH	1.02	0.81	0.88	0.82	0.84	0.92	1.08	0.96	0.88	0.85	0.87	0.97	10.9
		TOT DEM	KW	.00330	.00330	.00442	.00487	.00470	.00560	.00543	.00549	.00498	.00448	.00370	.00330	0.00560
S- 15	1.09	TOT ELEC	KWH	0.93	0.73	0.80	0.74	0.76	0.83	0.98	0.87	0.80	0.77	0.79	0.88	9.9
		TOT DEM	KW	.00312	.00312	.00418	.00460	.00444	.00529	.00513	.00518	.00471	.00423	.00349	.00312	0.00529
S- 31	0.32	TOT ELEC	KWH	1.09	0.86	0.94	0.87	0.89	0.97	1.15	1.02	0.94	0.90	0.93	1.03	11.6
		TOT DEM	KW	.00362	.00362	.00485	.00534	.00516	.00614	.00596	.00602	.00546	.00491	.00405	.00362	0.00614
S- 60	-0.29	TOT ELEC	KWH	1.02	0.81	0.88	0.82	0.84	0.92	1.08	0.96	0.88	0.85	0.87	0.97	10.9
		TOT DEM	KW	.00330	.00330	.00442	.00487	.00470	.00560	.00543	.00549	.00498	.00448	.00370	.00330	0.00560
BASE CASE : 224																
BASE	0.00	HT ELEC	KWH	0.84	0.53	0.50	0.33	0.15	0.07	0.00	0.02	0.10	0.32	0.58	0.73	4.2
		HT DEM	KW	.00608	.00608	.00602	.00517	.00395	.00340	.00188	.00298	.00359	.00462	.00608	.00608	0.00608
		REM ELEC	KWH	1.01	0.90	1.01	0.99	1.08	1.13	1.32	1.28	1.11	1.02	0.96	1.00	12.8
		REM DEM	KW	.00412	.00412	.00551	.00607	.00586	.00698	.00677	.00684	.00621	.00558	.00461	.00412	0.00698
		TOT ELEC	KWH	1.85	1.43	1.51	1.31	1.23	1.20	1.32	1.30	1.22	1.34	1.54	1.73	17.0
		TOT DEM	KW	.01015	.00913	.00893	.00771	.00690	.00802	.00731	.00731	.00670	.00832	.00985	.00934	0.01015
S- 3	-0.18	HT ELEC	KWH	0.89	0.56	0.52	0.34	0.16	0.07	0.00	0.02	0.11	0.33	0.61	0.77	4.4
		HT DEM	KW	.00608	.00608	.00602	.00517	.00395	.00340	.00188	.00298	.00359	.00462	.00608	.00608	0.00608
		REM ELEC	KWH	0.86	0.76	0.86	0.84	0.92	0.96	1.12	1.09	0.95	0.87	0.82	0.85	10.9
		REM DEM	KW	.00379	.00379	.00508	.00559	.00540	.00643	.00624	.00630	.00572	.00514	.00424	.00379	0.00643
		TOT ELEC	KWH	1.75	1.33	1.38	1.18	1.07	1.03	1.12	1.11	1.06	1.21	1.42	1.62	15.3
		TOT DEM	KW	.01012	.00911	.00891	.00769	.00688	.00799	.00729	.00729	.00668	.00830	.00982	.00931	0.01012
S- 6	-0.13	HT ELEC	KWH	0.97	0.61	0.57	0.37	0.17	0.08	0.00	0.02	0.12	0.36	0.66	0.84	4.8
		HT DEM	KW	.00608	.00608	.00602	.00517	.00395	.00340	.00188	.00298	.00359	.00462	.00608	.00608	0.00608
		REM ELEC	KWH	0.80	0.71	0.80	0.78	0.85	0.89	1.04	1.01	0.88	0.81	0.76	0.79	10.1
		REM DEM	KW	.00342	.00342	.00458	.00505	.00487	.00580	.00563	.00568	.00516	.00464	.00383	.00342	0.00580
		TOT ELEC	KWH	1.77	1.32	1.37	1.15	1.02	0.97	1.04	1.03	1.00	1.17	1.42	1.63	14.9
		TOT DEM	KW	.00937	.00843	.00825	.00712	.00637	.00740	.00675	.00675	.00618	.00768	.00909	.00862	0.00937
S- 15	1.07	HT ELEC	KWH	1.03	0.65	0.61	0.40	0.18	0.09	0.00	0.03	0.13	0.39	0.70	0.90	5.1
		HT DEM	KW	.00608	.00608	.00602	.00517	.00395	.00340	.00188	.00298	.00359	.00462	.00608	.00608	0.00608
		REM ELEC	KWH	0.69	0.62	0.69	0.68	0.74	0.77	0.91	0.88	0.76	0.70	0.66	0.69	8.8
		REM DEM	KW	.00303	.00303	.00406	.00447	.00432	.00514	.00499	.00504	.00457	.00411	.00339	.00303	0.00514
		TOT ELEC	KWH	1.72	1.27	1.30	1.07	0.92	0.86	0.91	0.90	0.89	1.09	1.36	1.58	13.9
		TOT DEM	KW	.00885	.00823	.00805	.00646	.00566	.00655	.00602	.00611	.00575	.00717	.00867	.00858	0.00885
S- 31	0.23	HT ELEC	KWH	0.58	0.46	0.49	0.37	0.26	0.24	0.16	0.19	0.24	0.36	0.48	0.52	4.3
		HT DEM	KW	.00608	.00608	.00602	.00517	.00395	.00340	.00188	.00298	.00359	.00462	.00608	.00608	0.00608
		REM ELEC	KWH	0.81	0.73	0.85	0.80	0.77	0.92	1.03	1.01	0.90	0.78	0.79	0.84	10.2
		REM DEM	KW	.00346	.00346	.00421	.00458	.00437	.00533	.00512	.00517	.00464	.00453	.00346	.00346	0.00533
		TOT ELEC	KWH	1.38	1.19	1.33	1.17	1.03	1.16	1.19	1.20	1.14	1.14	1.26	1.36	14.5
		TOT DEM	KW	.00973	.00905	.00885	.00710	.00623	.00720	.00662	.00671	.00632	.00788	.00954	.00944	0.00973
S- 60	-0.38	HT ELEC	KWH	0.97	0.61	0.57	0.37	0.17	0.08	0.00	0.02	0.12	0.36	0.66	0.84	4.8
		HT DEM	KW	.00608	.00608	.00602	.00517	.00395	.00340	.00188	.00298	.00359	.00462	.00608	.00608	0.00608
		REM ELEC	KWH	0.80	0.71	0.80	0.78	0.85	0.89	1.04	1.01	0.88	0.81	0.76	0.79	10.1
		REM DEM	KW	.00342	.00342	.00458	.00505	.00487	.00580	.00563	.00568	.00516	.00464	.00383	.00342	0.00580
		TOT ELEC	KWH	1.77	1.32	1.37	1.15	1.02	0.97	1.04	1.03	1.00	1.17	1.42	1.63	14.9
		TOT DEM	KW	.00937	.00843	.00825	.00712	.00637	.00740	.00675	.00675	.00618	.00768	.00909	.00862	0.00937
BASE CASE : 225																
BASE	0.00	TOT GAS	THM	0.0442	0.0280	0.0261	0.0171	0.0079	0.0037	0.0000	0.0011	0.0055	0.0166	0.0302	0.0385	0.219
		TOT ELEC	KWH	1.01	0.90	1.01	0.99	1.08	1.13	1.32	1.28	1.11	1.02	0.96	1.00	12.8
		TOT DEM	KW	.00412	.00412	.00551	.00607	.00586	.00698	.00677	.00684	.00621	.00558	.00461	.00412	0.00698
S- 3	-0.19	TOT GAS	THM	0.0467	0.0296	0.0275	0.0180	0.0083	0.0039	0.0000	0.0012	0.0058	0.0176	0.0319	0.0407	0.231
		TOT ELEC	KWH	0.86	0.76	0.86	0.84	0.92	0.96	1.12	1.09	0.95	0.87	0.82	0.85	10.9
		TOT DEM	KW	.00379	.00379	.00508	.00559	.00540	.00643	.00624	.00630	.00572	.00514	.00424	.00379	0.00643
S- 6	-0.14	TOT GAS	THM	0.0509	0.0323	0.0300	0.0197	0.0091	0.0043	0.0000	0.0013	0.0063	0.0192	0.0348	0.0444	0.252
		TOT ELEC	KWH	0.80	0.71	0.80	0.78	0.85	0.89	1.04	1.01	0.88	0.81	0.76	0.79	10.1
		TOT DEM	KW	.00342	.00342	.00458	.00505	.00487	.00580	.00563	.00568	.00516	.00464	.00383	.00342	0.00580
S- 15	1.06	TOT GAS	THM	0.0539	0.0342	0.0318	0.0208	0.0096	0.0045	0.0000	0.0013	0.0067	0.0203	0.0368	0.0470	0.267
		TOT ELEC	KWH	0.69	0.62	0.69	0.68	0.74	0.77	0.91	0.88	0.76	0.70	0.66	0.69	8.8
		TOT DEM	KW	.00303	.00303	.00406	.00447	.00432	.00514	.00499	.00504	.00457	.00411	.00339	.00303	0.00514

STRAT SET	CONST COST $/SF	ENERGY FORM	UNITS (PER SF)	JAN	FEB	MAR	APR	MAY	JUN	JUL	AUG	SEP	OCT	NOV	DEC	ANNUAL
BASE CASE : 225 (CONTINUED)																
S- 31	0.22	TOT GAS	THM	0.0303	0.0239	0.0255	0.0196	0.0137	0.0128	0.0082	0.0100	0.0128	0.0189	0.0251	0.0274	0.228
		TOT ELEC	KWH	0.81	0.73	0.85	0.80	0.77	0.92	1.03	1.01	0.90	0.78	0.79	0.84	10.2
		TOT DEM	KW	.00346	.00346	.00421	.00458	.00437	.00533	.00512	.00517	.00464	.00453	.00346	.00346	0.00533
S- 60	-0.39	TOT GAS	THM	0.0509	0.0323	0.0300	0.0197	0.0091	0.0043	0.0000	0.0013	0.0063	0.0192	0.0348	0.0444	0.252
		TOT ELEC	KWH	0.80	0.71	0.80	0.78	0.85	0.89	1.04	1.01	0.88	0.81	0.76	0.79	10.1
		TOT DEM	KW	.00342	.00342	.00458	.00505	.00487	.00580	.00563	.00568	.00516	.00464	.00383	.00342	0.00580
BASE CASE : 226																
BASE	0.00	HT ELEC	KWH	0.62	0.39	0.36	0.24	0.11	0.05	0.00	0.02	0.08	0.23	0.42	0.54	3.0
		HT DEM	KW	.00435	.00435	.00435	.00435	.00313	.00218	.00070	.00165	.00252	.00435	.00435	.00435	0.00435
		REM ELEC	KWH	0.94	0.83	0.94	0.92	1.00	1.05	1.23	1.19	1.04	0.95	0.89	0.93	11.9
		REM DEM	KW	.00375	.00375	.00472	.00526	.00502	.00599	.00599	.00605	.00569	.00466	.00442	.00369	0.00605
		TOT ELEC	KWH	1.56	1.22	1.30	1.15	1.11	1.10	1.23	1.21	1.11	1.18	1.31	1.47	15.0
		TOT DEM	KW	.00753	.00745	.00745	.00648	.00489	.00580	.00580	.00587	.00550	.00708	.00745	.00745	0.00753
S- 3	-0.10	HT ELEC	KWH	0.66	0.42	0.39	0.26	0.12	0.06	0.00	0.02	0.08	0.25	0.45	0.58	3.3
		HT DEM	KW	.00435	.00435	.00435	.00435	.00313	.00218	.00070	.00165	.00252	.00435	.00435	.00435	0.00435
		REM ELEC	KWH	0.80	0.71	0.80	0.78	0.85	0.89	1.04	1.01	0.88	0.81	0.76	0.79	10.1
		REM DEM	KW	.00345	.00345	.00434	.00485	.00462	.00551	.00551	.00557	.00524	.00429	.00407	.00340	0.00557
		TOT ELEC	KWH	1.46	1.13	1.19	1.03	0.97	0.94	1.04	1.03	0.96	1.06	1.21	1.37	13.4
		TOT DEM	KW	.00748	.00741	.00741	.00643	.00486	.00576	.00576	.00583	.00546	.00703	.00741	.00741	0.00748
S- 6	-0.05	HT ELEC	KWH	0.71	0.45	0.42	0.27	0.13	0.06	0.00	0.02	0.09	0.27	0.48	0.62	3.5
		HT DEM	KW	.00435	.00435	.00435	.00435	.00313	.00218	.00070	.00165	.00252	.00435	.00435	.00435	0.00435
		REM ELEC	KWH	0.72	0.64	0.72	0.71	0.77	0.81	0.94	0.92	0.80	0.73	0.69	0.71	9.2
		REM DEM	KW	.00311	.00311	.00392	.00437	.00417	.00497	.00497	.00502	.00472	.00387	.00366	.00306	0.00502
		TOT ELEC	KWH	1.43	1.09	1.14	0.98	0.90	0.87	0.94	0.93	0.88	1.00	1.17	1.33	12.7
		TOT DEM	KW	.00673	.00666	.00666	.00579	.00437	.00518	.00518	.00525	.00491	.00633	.00666	.00666	0.00673
S- 15	0.60	HT ELEC	KWH	0.75	0.48	0.44	0.29	0.13	0.06	0.00	0.02	0.09	0.28	0.51	0.66	3.7
		HT DEM	KW	.00435	.00435	.00435	.00435	.00313	.00218	.00070	.00165	.00252	.00435	.00435	.00435	0.00435
		REM ELEC	KWH	0.63	0.56	0.63	0.61	0.67	0.70	0.82	0.80	0.69	0.64	0.60	0.62	8.0
		REM DEM	KW	.00277	.00277	.00349	.00389	.00371	.00443	.00443	.00447	.00420	.00344	.00326	.00273	0.00447
		TOT ELEC	KWH	1.38	1.03	1.07	0.90	0.80	0.76	0.82	0.81	0.79	0.92	1.11	1.28	11.7
		TOT DEM	KW	.00653	.00646	.00646	.00562	.00424	.00503	.00503	.00509	.00477	.00614	.00646	.00646	0.00653
S- 31	0.32	HT ELEC	KWH	0.57	0.38	0.38	0.22	0.11	0.08	0.04	0.06	0.09	0.19	0.40	0.42	2.9
		HT DEM	KW	.00435	.00435	.00435	.00435	.00313	.00218	.00070	.00165	.00252	.00435	.00435	.00435	0.00435
		REM ELEC	KWH	0.69	0.61	0.71	0.68	0.70	0.82	0.91	0.91	0.80	0.69	0.67	0.71	8.9
		REM DEM	KW	.00304	.00300	.00348	.00361	.00361	.00435	.00426	.00431	.00383	.00383	.00326	.00313	0.00435
		TOT ELEC	KWH	1.25	1.00	1.09	0.90	0.81	0.90	0.95	0.97	0.89	0.88	1.06	1.14	11.8
		TOT DEM	KW	.00715	.00708	.00708	.00601	.00372	.00436	.00429	.00436	.00386	.00658	.00708	.00708	0.00715
S- 60	-0.29	HT ELEC	KWH	0.71	0.45	0.42	0.27	0.13	0.06	0.00	0.02	0.09	0.27	0.48	0.62	3.5
		HT DEM	KW	.00435	.00435	.00435	.00435	.00313	.00218	.00070	.00165	.00252	.00435	.00435	.00435	0.00435
		REM ELEC	KWH	0.72	0.64	0.72	0.71	0.77	0.81	0.94	0.92	0.80	0.73	0.69	0.71	9.2
		REM DEM	KW	.00311	.00311	.00392	.00437	.00417	.00497	.00497	.00502	.00472	.00387	.00366	.00306	0.00502
		TOT ELEC	KWH	1.43	1.09	1.14	0.98	0.90	0.87	0.94	0.93	0.88	1.00	1.17	1.33	12.7
		TOT DEM	KW	.00673	.00666	.00666	.00579	.00437	.00518	.00518	.00525	.00491	.00633	.00666	.00666	0.00673
BASE CASE : 227																
BASE	0.00	TOT GAS	THM	0.0323	0.0205	0.0190	0.0125	0.0058	0.0027	0.0000	0.0008	0.0040	0.0122	0.0221	0.0282	0.160
		TOT ELEC	KWH	0.94	0.83	0.94	0.92	1.00	1.05	1.23	1.19	1.04	0.95	0.89	0.93	11.9
		TOT DEM	KW	.00375	.00375	.00472	.00526	.00502	.00599	.00599	.00605	.00569	.00466	.00442	.00369	0.00605
S- 3	-0.10	TOT GAS	THM	0.0347	0.0220	0.0205	0.0134	0.0062	0.0029	0.0000	0.0009	0.0043	0.0131	0.0237	0.0303	0.172
		TOT ELEC	KWH	0.80	0.71	0.80	0.78	0.85	0.89	1.04	1.01	0.88	0.81	0.76	0.79	10.1
		TOT DEM	KW	.00345	.00345	.00434	.00485	.00462	.00551	.00551	.00557	.00524	.00429	.00407	.00340	0.00557
S- 6	-0.05	TOT GAS	THM	0.0372	0.0236	0.0219	0.0144	0.0066	0.0031	0.0000	0.0009	0.0046	0.0140	0.0254	0.0324	0.184
		TOT ELEC	KWH	0.72	0.64	0.72	0.71	0.77	0.81	0.94	0.92	0.80	0.73	0.69	0.71	9.2
		TOT DEM	KW	.00311	.00311	.00392	.00437	.00417	.00497	.00497	.00502	.00472	.00387	.00366	.00306	0.00502
S- 15	0.57	TOT GAS	THM	0.0396	0.0251	0.0233	0.0153	0.0071	0.0033	0.0000	0.0010	0.0049	0.0149	0.0270	0.0345	0.196
		TOT ELEC	KWH	0.63	0.56	0.63	0.61	0.67	0.70	0.82	0.80	0.69	0.64	0.60	0.62	8.0
		TOT DEM	KW	.00277	.00277	.00349	.00389	.00371	.00443	.00443	.00447	.00420	.00344	.00326	.00273	0.00447
S- 31	0.31	TOT GAS	THM	0.0299	0.0202	0.0197	0.0117	0.0057	0.0042	0.0020	0.0031	0.0048	0.0100	0.0208	0.0223	0.154
		TOT ELEC	KWH	0.69	0.61	0.71	0.68	0.70	0.82	0.91	0.91	0.80	0.69	0.67	0.71	8.9
		TOT DEM	KW	.00304	.00304	.00348	.00361	.00361	.00435	.00426	.00431	.00383	.00383	.00326	.00313	0.00435
S- 60	-0.30	TOT GAS	THM	0.0372	0.0236	0.0219	0.0144	0.0066	0.0031	0.0000	0.0009	0.0046	0.0140	0.0254	0.0324	0.184
		TOT ELEC	KWH	0.72	0.64	0.72	0.71	0.77	0.81	0.94	0.92	0.80	0.73	0.69	0.71	9.2
		TOT DEM	KW	.00311	.00311	.00392	.00437	.00417	.00497	.00497	.00502	.00472	.00387	.00366	.00306	0.00502
BASE CASE : 228																
BASE	0.00	TOT ELEC	KWH	1.31	1.02	1.14	1.06	1.05	1.16	1.22	1.22	1.13	1.06	1.09	1.16	13.6
		TOT DEM	KW	.00789	.00726	.00655	.00466	.00473	.00537	.00537	.00529	.00497	.00489	.00718	.00734	0.00789
S- 3	-0.13	TOT ELEC	KWH	1.20	0.94	1.05	0.98	0.96	1.06	1.13	1.13	1.04	0.98	1.00	1.06	12.5
		TOT DEM	KW	.00738	.00679	.00613	.00435	.00443	.00502	.00502	.00494	.00465	.00458	.00672	.00686	0.00738
S- 6	-0.08	TOT ELEC	KWH	1.12	0.88	0.98	0.91	0.90	0.99	1.05	1.05	0.97	0.91	0.94	0.99	11.7
		TOT DEM	KW	.00688	.00633	.00571	.00406	.00413	.00468	.00468	.00461	.00433	.00427	.00626	.00640	0.00688
S- 15	1.00	TOT ELEC	KWH	1.02	0.80	0.89	0.83	0.82	0.90	0.95	0.95	0.88	0.83	0.85	0.90	10.6
		TOT DEM	KW	.00650	.00598	.00539	.00383	.00390	.00442	.00442	.00436	.00409	.00403	.00592	.00605	0.00650

(HT ELEC=ELECTRIC CONSUMPTION FOR SPACE AND WATER HEATING. REM ELEC=REMAINING ELECTRIC CONSUMPTION (NOT USED FOR HEATING). TOT ELEC=TOTAL ELECTRIC CONSUMPTION. HT DEM=PEAK ELECTRIC DEMAND FOR SPACE AND WATER HEATING. REM DEM=PEAK ELECTRIC DEMAND FOR REMAINING LOAD (NOT HEATING). TOT DEM=TOTAL PEAK ELECTRIC DEMAND. GAS=TOTAL GAS CONSUMPTION. ANNUAL FIGURES ARE SUMS OF MONTHLY CONSUMPTION VALUES OR ARE YEARLY MAXIMUMS OF MONTHLY PEAK DEMAND VALUES.)

STRAT SET	CONST COST $/SF	ENERGY FORM	UNITS (PER SF)	JAN	FEB	MAR	APR	MAY	JUN	JUL	AUG	SEP	OCT	NOV	DEC	ANNUAL
BASE CASE : 228 (CONTINUED)																
S- 31	0.28	TOT ELEC	KWH	1.04	0.81	0.91	0.84	0.83	0.92	0.97	0.97	0.90	0.84	0.86	0.92	10.8
		TOT DEM	KW	.00749	.00689	.00622	.00442	.00449	.00509	.00509	.00502	.00472	.00464	.00682	.00697	0.00749
S- 60	-0.33	TOT ELEC	KWH	1.12	0.88	0.98	0.91	0.90	0.99	1.05	1.05	0.97	0.91	0.94	0.99	11.7
		TOT DEM	KW	.00688	.00633	.00571	.00406	.00413	.00468	.00468	.00461	.00433	.00427	.00626	.00640	0.00688
BASE CASE : 229																
BASE	0.00	HT ELEC	KWH	0.41	0.26	0.24	0.16	0.07	0.03	0.00	0.01	0.05	0.15	0.28	0.36	2.0
		HT DEM	KW	.00608	.00578	.00547	.00468	.00359	.00298	.00146	.00249	.00292	.00474	.00590	.00553	0.00608
		REM ELEC	KWH	1.03	0.92	1.03	1.01	1.10	1.15	1.35	1.31	1.14	1.05	0.98	1.02	13.1
		REM DEM	KW	.00354	.00354	.00474	.00522	.00504	.00600	.00582	.00588	.00534	.00480	.00396	.00354	0.00600
		TOT ELEC	KWH	1.44	1.18	1.28	1.17	1.17	1.19	1.35	1.32	1.19	1.20	1.26	1.38	15.1
		TOT DEM	KW	.00988	.00988	.00988	.00909	.00741	.00840	.00781	.00781	.00721	.00958	.00988	.00988	0.00988
S- 3	-0.14	HT ELEC	KWH	0.43	0.27	0.25	0.17	0.08	0.04	0.00	0.01	0.05	0.16	0.29	0.37	2.1
		HT DEM	KW	.00608	.00578	.00547	.00468	.00359	.00298	.00146	.00249	.00292	.00474	.00590	.00553	0.00608
		REM ELEC	KWH	0.88	0.78	0.88	0.86	0.94	0.99	1.15	1.12	0.97	0.90	0.84	0.87	11.2
		REM DEM	KW	.00326	.00326	.00437	.00481	.00465	.00553	.00536	.00542	.00492	.00442	.00365	.00326	0.00553
		TOT ELEC	KWH	1.32	1.06	1.14	1.03	1.02	1.02	1.15	1.13	1.03	1.06	1.13	1.25	13.3
		TOT DEM	KW	.00950	.00950	.00950	.00874	.00713	.00807	.00751	.00751	.00694	.00922	.00950	.00950	0.00950
S- 6	-0.09	HT ELEC	KWH	0.47	0.30	0.28	0.18	0.08	0.04	0.00	0.01	0.06	0.18	0.32	0.41	2.3
		HT DEM	KW	.00608	.00578	.00547	.00468	.00359	.00298	.00146	.00249	.00292	.00474	.00590	.00553	0.00608
		REM ELEC	KWH	0.81	0.72	0.81	0.79	0.87	0.91	1.06	1.03	0.90	0.82	0.77	0.80	10.3
		REM DEM	KW	.00294	.00294	.00393	.00433	.00418	.00498	.00483	.00488	.00443	.00398	.00329	.00294	0.00498
		TOT ELEC	KWH	1.28	1.02	1.09	0.97	0.95	0.95	1.06	1.04	0.95	1.00	1.09	1.21	12.6
		TOT DEM	KW	.00885	.00885	.00885	.00814	.00664	.00752	.00699	.00699	.00646	.00858	.00885	.00885	0.00885
S- 15	1.18	HT ELEC	KWH	0.50	0.31	0.29	0.19	0.09	0.04	0.00	0.01	0.06	0.19	0.34	0.43	2.5
		HT DEM	KW	.00608	.00578	.00547	.00468	.00359	.00298	.00146	.00249	.00292	.00474	.00590	.00553	0.00608
		REM ELEC	KWH	0.71	0.63	0.71	0.69	0.76	0.79	0.93	0.90	0.78	0.72	0.68	0.70	9.0
		REM DEM	KW	.00261	.00261	.00350	.00385	.00372	.00443	.00430	.00434	.00394	.00354	.00292	.00261	0.00443
		TOT ELEC	KWH	1.21	0.94	1.00	0.88	0.84	0.83	0.93	0.91	0.84	0.91	1.01	1.13	11.5
		TOT DEM	KW	.00850	.00689	.00680	.00587	.00502	.00553	.00519	.00519	.00493	.00629	.00706	.00689	0.00850
S- 31	0.14	HT ELEC	KWH	0.34	0.26	0.28	0.22	0.15	0.14	0.09	0.11	0.14	0.21	0.28	0.30	2.5
		HT DEM	KW	.00608	.00578	.00547	.00468	.00359	.00298	.00146	.00249	.00292	.00474	.00590	.00553	0.00608
		REM ELEC	KWH	0.85	0.78	0.90	0.84	0.81	0.97	1.09	1.07	0.95	0.82	0.83	0.89	10.8
		REM DEM	KW	.00325	.00325	.00395	.00430	.00410	.00500	.00480	.00485	.00435	.00425	.00325	.00325	0.00500
		TOT ELEC	KWH	1.19	1.04	1.18	1.06	0.96	1.11	1.18	1.18	1.09	1.03	1.11	1.19	13.3
		TOT DEM	KW	.00948	.00768	.00758	.00654	.00559	.00616	.00578	.00578	.00550	.00702	.00787	.00768	0.00948
S- 39	0.97	HT ELEC	KWH	0.37	0.29	0.31	0.24	0.17	0.16	0.10	0.12	0.16	0.23	0.31	0.34	2.8
		HT DEM	KW	.00608	.00578	.00547	.00468	.00359	.00298	.00146	.00249	.00292	.00474	.00590	.00553	0.00608
		REM ELEC	KWH	0.77	0.70	0.81	0.76	0.73	0.88	0.99	0.97	0.86	0.74	0.75	0.80	9.8
		REM DEM	KW	.00304	.00304	.00370	.00402	.00384	.00468	.00449	.00454	.00407	.00398	.00304	.00304	0.00468
		TOT ELEC	KWH	1.15	1.00	1.13	1.00	0.90	1.04	1.09	1.09	1.02	0.98	1.06	1.14	12.6
		TOT DEM	KW	.00910	.00737	.00728	.00628	.00537	.00591	.00555	.00555	.00528	.00673	.00755	.00737	0.00910
S- 44	0.17	HT ELEC	KWH	0.35	0.28	0.30	0.23	0.16	0.15	0.10	0.12	0.15	0.22	0.29	0.32	2.7
		HT DEM	KW	.00608	.00578	.00547	.00468	.00359	.00298	.00146	.00249	.00292	.00474	.00590	.00553	0.00608
		REM ELEC	KWH	0.81	0.74	0.85	0.80	0.77	0.93	1.04	1.02	0.91	0.78	0.79	0.84	10.3
		REM DEM	KW	.00317	.00317	.00386	.00420	.00400	.00488	.00468	.00473	.00425	.00415	.00317	.00317	0.00488
		TOT ELEC	KWH	1.17	1.02	1.15	1.03	0.93	1.08	1.14	1.14	1.06	1.00	1.09	1.16	13.0
		TOT DEM	KW	.00930	.00753	.00744	.00642	.00549	.00604	.00567	.00567	.00539	.00688	.00772	.00753	0.00930
S- 60	-0.34	HT ELEC	KWH	0.47	0.30	0.28	0.18	0.08	0.04	0.00	0.01	0.06	0.18	0.32	0.41	2.3
		HT DEM	KW	.00608	.00578	.00547	.00468	.00359	.00298	.00146	.00249	.00292	.00474	.00590	.00553	0.00608
		REM ELEC	KWH	0.81	0.72	0.81	0.79	0.87	0.91	1.06	1.03	0.90	0.82	0.77	0.80	10.3
		REM DEM	KW	.00294	.00294	.00393	.00433	.00418	.00498	.00483	.00488	.00443	.00398	.00329	.00294	0.00498
		TOT ELEC	KWH	1.28	1.02	1.09	0.97	0.95	0.95	1.06	1.04	0.95	1.00	1.09	1.21	12.6
		TOT DEM	KW	.00885	.00885	.00885	.00814	.00664	.00752	.00699	.00699	.00646	.00858	.00885	.00885	0.00885
BASE CASE : 230																
BASE	0.00	TOT GAS	THM	0.0214	0.0136	0.0126	0.0083	0.0038	0.0018	0.0000	0.0005	0.0027	0.0081	0.0146	0.0187	0.106
		TOT ELEC	KWH	1.03	0.92	1.03	1.01	1.10	1.15	1.35	1.31	1.14	1.05	0.98	1.02	13.1
		TOT DEM	KW	.00354	.00354	.00474	.00522	.00504	.00600	.00582	.00588	.00534	.00480	.00396	.00354	0.00600
S- 3	-0.15	TOT GAS	THM	0.0226	0.0143	0.0133	0.0087	0.0040	0.0019	0.0000	0.0006	0.0028	0.0085	0.0155	0.0197	0.112
		TOT ELEC	KWH	0.88	0.78	0.88	0.86	0.94	0.99	1.15	1.12	0.97	0.90	0.84	0.87	11.2
		TOT DEM	KW	.00326	.00326	.00437	.00481	.00465	.00553	.00536	.00542	.00492	.00442	.00365	.00326	0.00553
S- 6	-0.10	TOT GAS	THM	0.0246	0.0156	0.0145	0.0095	0.0044	0.0021	0.0000	0.0006	0.0030	0.0093	0.0168	0.0215	0.122
		TOT ELEC	KWH	0.81	0.72	0.81	0.79	0.87	0.91	1.06	1.03	0.90	0.82	0.77	0.80	10.3
		TOT DEM	KW	.00294	.00294	.00393	.00433	.00418	.00498	.00483	.00488	.00443	.00398	.00329	.00294	0.00498
S- 15	1.17	TOT GAS	THM	0.0261	0.0165	0.0154	0.0101	0.0046	0.0022	0.0000	0.0006	0.0032	0.0098	0.0178	0.0227	0.129
		TOT ELEC	KWH	0.71	0.63	0.71	0.69	0.76	0.79	0.93	0.90	0.78	0.72	0.68	0.70	9.0
		TOT DEM	KW	.00261	.00261	.00350	.00385	.00372	.00443	.00430	.00434	.00394	.00354	.00292	.00261	0.00443
S- 31	0.13	TOT GAS	THM	0.0176	0.0139	0.0148	0.0114	0.0079	0.0074	0.0048	0.0058	0.0074	0.0110	0.0145	0.0158	0.132
		TOT ELEC	KWH	0.85	0.78	0.90	0.84	0.81	0.97	1.09	1.07	0.95	0.82	0.83	0.89	10.8
		TOT DEM	KW	.00325	.00325	.00395	.00430	.00410	.00500	.00480	.00485	.00435	.00425	.00325	.00325	0.00500
S- 39	0.96	TOT GAS	THM	0.0196	0.0154	0.0165	0.0126	0.0088	0.0082	0.0053	0.0065	0.0082	0.0122	0.0162	0.0176	0.147
		TOT ELEC	KWH	0.77	0.70	0.81	0.76	0.73	0.88	0.99	0.97	0.86	0.74	0.75	0.80	9.8
		TOT DEM	KW	.00304	.00304	.00370	.00402	.00384	.00468	.00449	.00454	.00407	.00398	.00304	.00304	0.00468
S- 44	0.15	TOT GAS	THM	0.0185	0.0146	0.0156	0.0120	0.0083	0.0078	0.0050	0.0061	0.0078	0.0115	0.0153	0.0167	0.139
		TOT ELEC	KWH	0.81	0.74	0.85	0.80	0.77	0.93	1.04	1.02	0.91	0.78	0.79	0.84	10.3
		TOT DEM	KW	.00317	.00317	.00386	.00420	.00400	.00488	.00468	.00473	.00425	.00415	.00317	.00317	0.00488
S- 60	-0.35	TOT GAS	THM	0.0246	0.0156	0.0145	0.0095	0.0044	0.0021	0.0000	0.0006	0.0030	0.0093	0.0168	0.0215	0.122
		TOT ELEC	KWH	0.81	0.72	0.81	0.79	0.87	0.91	1.06	1.03	0.90	0.82	0.77	0.80	10.3
		TOT DEM	KW	.00294	.00294	.00393	.00433	.00418	.00498	.00483	.00488	.00443	.00398	.00329	.00294	0.00498

STRAT SET	CONST COST $/SF	ENERGY FORM	UNITS (PER SF)	JAN	FEB	MAR	APR	MAY	JUN	JUL	AUG	SEP	OCT	NOV	DEC	ANNUAL
BASE CASE : 231																
BASE	0.00	HT ELEC	KWH	0.34	0.22	0.20	0.13	0.06	0.03	0.00	0.01	0.04	0.13	0.23	0.30	1.7
		HT DEM	KW	.00395	.00371	.00371	.00340	.00265	.00194	.00071	.00142	.00197	.00332	.00379	.00379	0.00395
		REM ELEC	KWH	1.00	0.89	1.00	0.98	1.07	1.12	1.31	1.27	1.10	1.02	0.95	0.99	12.7
		REM DEM	KW	.00330	.00330	.00416	.00464	.00442	.00528	.00528	.00533	.00501	.00410	.00389	.00325	0.00533
		TOT ELEC	KWH	1.34	1.10	1.20	1.11	1.13	1.15	1.31	1.28	1.15	1.14	1.18	1.29	14.4
		TOT DEM	KW	.00675	.00607	.00601	.00506	.00439	.00520	.00513	.00520	.00472	.00526	.00621	.00614	0.00675
S- 3	-0.09	HT ELEC	KWH	0.36	0.23	0.21	0.14	0.06	0.03	0.00	0.01	0.05	0.14	0.25	0.32	1.8
		HT DEM	KW	.00395	.00371	.00371	.00340	.00265	.00194	.00071	.00142	.00197	.00332	.00379	.00379	0.00395
		REM ELEC	KWH	0.85	0.76	0.85	0.83	0.91	0.95	1.11	1.08	0.94	0.86	0.81	0.84	10.8
		REM DEM	KW	.00304	.00304	.00382	.00426	.00407	.00485	.00485	.00490	.00461	.00377	.00358	.00299	0.00490
		TOT ELEC	KWH	1.22	0.99	1.07	0.97	0.97	0.98	1.11	1.09	0.98	1.00	1.06	1.16	12.6
		TOT DEM	KW	.00653	.00588	.00581	.00490	.00424	.00503	.00496	.00503	.00457	.00509	.00601	.00594	0.00653
S- 6	-0.04	HT ELEC	KWH	0.39	0.25	0.23	0.15	0.07	0.03	0.00	0.01	0.05	0.15	0.27	0.34	1.9
		HT DEM	KW	.00395	.00371	.00371	.00340	.00265	.00194	.00071	.00142	.00197	.00332	.00379	.00379	0.00395
		REM ELEC	KWH	0.77	0.68	0.77	0.75	0.82	0.86	1.01	0.98	0.85	0.78	0.73	0.76	9.8
		REM DEM	KW	.00275	.00275	.00346	.00385	.00368	.00439	.00439	.00443	.00416	.00341	.00323	.00270	0.00443
		TOT ELEC	KWH	1.16	0.93	1.00	0.90	0.89	0.89	1.01	0.99	0.90	0.93	1.00	1.10	11.7
		TOT DEM	KW	.00608	.00547	.00541	.00456	.00395	.00468	.00462	.00468	.00426	.00474	.00559	.00553	0.00608
S- 15	0.65	HT ELEC	KWH	0.42	0.26	0.25	0.16	0.07	0.04	0.00	0.01	0.05	0.16	0.28	0.36	2.1
		HT DEM	KW	.00395	.00371	.00371	.00340	.00265	.00194	.00071	.00142	.00197	.00332	.00379	.00379	0.00395
		REM ELEC	KWH	0.67	0.59	0.67	0.65	0.71	0.75	0.87	0.85	0.74	0.68	0.64	0.66	8.5
		REM DEM	KW	.00244	.00244	.00307	.00342	.00326	.00389	.00389	.00393	.00369	.00303	.00287	.00240	0.00393
		TOT ELEC	KWH	1.09	0.86	0.92	0.81	0.79	0.78	0.87	0.86	0.79	0.84	0.92	1.02	10.6
		TOT DEM	KW	.00590	.00531	.00525	.00442	.00383	.00454	.00448	.00454	.00413	.00460	.00543	.00537	0.00590
S- 31	0.19	HT ELEC	KWH	0.44	0.30	0.29	0.17	0.08	0.06	0.03	0.05	0.07	0.15	0.31	0.33	2.3
		HT DEM	KW	.00395	.00371	.00371	.00340	.00265	.00194	.00071	.00142	.00197	.00332	.00379	.00379	0.00395
		REM ELEC	KWH	0.77	0.69	0.80	0.76	0.79	0.92	1.02	1.02	0.90	0.78	0.75	0.80	10.0
		REM DEM	KW	.00293	.00288	.00334	.00347	.00347	.00418	.00410	.00414	.00368	.00368	.00313	.00301	0.00418
		TOT ELEC	KWH	1.21	0.99	1.09	0.93	0.87	0.98	1.05	1.07	0.97	0.93	1.06	1.13	12.3
		TOT DEM	KW	.00635	.00578	.00565	.00476	.00381	.00432	.00425	.00419	.00387	.00508	.00584	.00584	0.00635
S- 39	0.72	HT ELEC	KWH	0.54	0.36	0.36	0.21	0.10	0.08	0.04	0.06	0.09	0.18	0.38	0.40	2.8
		HT DEM	KW	.00395	.00371	.00371	.00340	.00265	.00194	.00071	.00142	.00197	.00332	.00379	.00379	0.00395
		REM ELEC	KWH	0.69	0.61	0.71	0.68	0.70	0.82	0.91	0.91	0.80	0.69	0.67	0.71	8.9
		REM DEM	KW	.00265	.00261	.00302	.00314	.00314	.00378	.00370	.00374	.00333	.00333	.00283	.00272	0.00378
		TOT ELEC	KWH	1.23	0.98	1.07	0.89	0.81	0.89	0.94	0.96	0.89	0.88	1.04	1.12	11.7
		TOT DEM	KW	.00610	.00555	.00543	.00458	.00366	.00415	.00409	.00403	.00372	.00488	.00561	.00561	0.00610
S- 44	0.21	HT ELEC	KWH	0.49	0.33	0.32	0.19	0.09	0.07	0.03	0.05	0.08	0.16	0.34	0.36	2.5
		HT DEM	KW	.00395	.00371	.00371	.00340	.00265	.00194	.00071	.00142	.00197	.00332	.00379	.00379	0.00395
		REM ELEC	KWH	0.73	0.65	0.76	0.72	0.75	0.87	0.96	0.96	0.85	0.74	0.71	0.76	9.5
		REM DEM	KW	.00280	.00276	.00320	.00332	.00332	.00400	.00392	.00396	.00352	.00352	.00300	.00288	0.00400
		TOT ELEC	KWH	1.21	0.98	1.08	0.91	0.84	0.94	1.00	1.01	0.93	0.90	1.05	1.12	12.0
		TOT DEM	KW	.00625	.00569	.00556	.00469	.00375	.00425	.00419	.00413	.00381	.00500	.00575	.00575	0.00625
S- 60	-0.29	HT ELEC	KWH	0.39	0.25	0.23	0.15	0.07	0.03	0.00	0.01	0.05	0.15	0.27	0.34	1.9
		HT DEM	KW	.00395	.00371	.00371	.00340	.00265	.00194	.00071	.00142	.00197	.00332	.00379	.00379	0.00395
		REM ELEC	KWH	0.77	0.68	0.77	0.75	0.82	0.86	1.01	0.98	0.85	0.78	0.73	0.76	9.8
		REM DEM	KW	.00275	.00275	.00346	.00385	.00368	.00439	.00439	.00443	.00416	.00341	.00323	.00270	0.00443
		TOT ELEC	KWH	1.16	0.93	1.00	0.90	0.89	0.89	1.01	0.99	0.90	0.93	1.00	1.10	11.7
		TOT DEM	KW	.00608	.00547	.00541	.00456	.00395	.00468	.00462	.00468	.00426	.00474	.00559	.00553	0.00608
BASE CASE : 232																
BASE	0.00	TOT GAS	THM	0.0178	0.0113	0.0105	0.0069	0.0032	0.0015	0.0000	0.0004	0.0022	0.0067	0.0121	0.0155	0.088
		TOT ELEC	KWH	1.00	0.89	1.00	0.98	1.07	1.12	1.31	1.27	1.10	1.02	0.95	0.99	12.7
		TOT DEM	KW	.00330	.00330	.00416	.00464	.00442	.00528	.00528	.00533	.00501	.00410	.00389	.00325	0.00533
S- 3	-0.10	TOT GAS	THM	0.0191	0.0121	0.0113	0.0074	0.0034	0.0016	0.0000	0.0005	0.0024	0.0072	0.0131	0.0166	0.095
		TOT ELEC	KWH	0.85	0.76	0.85	0.83	0.91	0.95	1.11	1.08	0.94	0.86	0.81	0.84	10.8
		TOT DEM	KW	.00304	.00304	.00382	.00426	.00407	.00485	.00485	.00490	.00461	.00377	.00358	.00299	0.00490
S- 6	-0.05	TOT GAS	THM	0.0204	0.0129	0.0120	0.0079	0.0036	0.0017	0.0000	0.0005	0.0025	0.0077	0.0139	0.0178	0.101
		TOT ELEC	KWH	0.77	0.68	0.77	0.75	0.82	0.86	1.01	0.98	0.85	0.78	0.73	0.76	9.8
		TOT DEM	KW	.00275	.00275	.00346	.00385	.00368	.00439	.00439	.00443	.00416	.00341	.00323	.00270	0.00443
S- 15	0.60	TOT GAS	THM	0.0218	0.0138	0.0129	0.0084	0.0039	0.0018	0.0000	0.0005	0.0027	0.0082	0.0149	0.0190	0.108
		TOT ELEC	KWH	0.67	0.59	0.67	0.65	0.71	0.75	0.87	0.85	0.74	0.68	0.64	0.66	8.5
		TOT DEM	KW	.00244	.00244	.00307	.00342	.00326	.00389	.00389	.00393	.00369	.00303	.00287	.00240	0.00393
S- 31	0.18	TOT GAS	THM	0.0233	0.0157	0.0154	0.0091	0.0044	0.0032	0.0016	0.0024	0.0037	0.0078	0.0162	0.0174	0.120
		TOT ELEC	KWH	0.77	0.69	0.80	0.76	0.79	0.92	1.02	1.02	0.90	0.78	0.75	0.80	10.0
		TOT DEM	KW	.00293	.00293	.00334	.00347	.00347	.00418	.00410	.00414	.00368	.00368	.00313	.00301	0.00418
S- 39	0.69	TOT GAS	THM	0.0283	0.0191	0.0187	0.0111	0.0054	0.0039	0.0019	0.0029	0.0045	0.0095	0.0197	0.0212	0.146
		TOT ELEC	KWH	0.69	0.61	0.71	0.68	0.70	0.82	0.91	0.91	0.80	0.69	0.67	0.71	8.9
		TOT DEM	KW	.00265	.00265	.00302	.00314	.00314	.00378	.00370	.00374	.00333	.00333	.00283	.00272	0.00378
S- 44	0.20	TOT GAS	THM	0.0256	0.0173	0.0169	0.0100	0.0049	0.0036	0.0017	0.0026	0.0041	0.0086	0.0178	0.0191	0.132
		TOT ELEC	KWH	0.73	0.65	0.76	0.72	0.75	0.87	0.96	0.96	0.85	0.74	0.71	0.76	9.5
		TOT DEM	KW	.00280	.00280	.00320	.00332	.00332	.00400	.00392	.00396	.00352	.00352	.00300	.00288	0.00400
S- 60	-0.30	TOT GAS	THM	0.0204	0.0129	0.0120	0.0079	0.0036	0.0017	0.0000	0.0005	0.0025	0.0077	0.0139	0.0178	0.101
		TOT ELEC	KWH	0.77	0.68	0.77	0.75	0.82	0.86	1.01	0.98	0.85	0.78	0.73	0.76	9.8
		TOT DEM	KW	.00275	.00275	.00346	.00385	.00368	.00439	.00439	.00443	.00416	.00341	.00323	.00270	0.00443

(HT ELEC=ELECTRIC CONSUMPTION FOR SPACE AND WATER HEATING. REM ELEC=REMAINING ELECTRIC CONSUMPTION (NOT USED FOR HEATING). TOT ELEC=TOTAL ELECTRIC CONSUMPTION. HT DEM=PEAK ELECTRIC DEMAND FOR SPACE AND WATER HEATING. REM DEM=PEAK ELECTRIC DEMAND FOR REMAINING LOAD (NOT HEATING). TOT DEM=TOTAL PEAK ELECTRIC DEMAND. GAS=TOTAL GAS CONSUMPTION. ANNUAL FIGURES ARE SUMS OF MONTHLY CONSUMPTION VALUES OR ARE YEARLY MAXIMUMS OF MONTHLY PEAK DEMAND VALUES.)

STRAT SET	CONST COST $/SF	ENERGY FORM	UNITS (PER SF)	JAN	FEB	MAR	APR	MAY	JUN	JUL	AUG	SEP	OCT	NOV	DEC	ANNUAL
BASE CASE : 233																
BASE	0.00	TOT ELEC	KWH	1.39	1.09	1.22	1.13	1.12	1.23	1.31	1.31	1.20	1.13	1.16	1.23	14.5
		TOT DEM	KW	.00573	.00527	.00476	.00338	.00344	.00390	.00390	.00384	.00361	.00355	.00521	.00533	0.00573
S- 1	0.09	TOT ELEC	KWH	1.28	1.00	1.12	1.04	1.02	1.13	1.20	1.20	1.10	1.04	1.06	1.13	13.3
		TOT DEM	KW	.00556	.00512	.00461	.00328	.00334	.00378	.00378	.00373	.00350	.00345	.00506	.00517	0.00556
S- 5	0.14	TOT ELEC	KWH	1.15	0.90	1.01	0.94	0.92	1.02	1.08	1.08	1.00	0.94	0.96	1.02	12.0
		TOT DEM	KW	.00527	.00485	.00437	.00311	.00316	.00358	.00358	.00353	.00332	.00327	.00480	.00490	0.00527
S- 14	1.45	TOT ELEC	KWH	1.06	0.83	0.92	0.86	0.85	0.94	0.99	0.99	0.91	0.86	0.88	0.94	11.0
		TOT DEM	KW	.00510	.00469	.00423	.00301	.00306	.00347	.00347	.00342	.00321	.00316	.00464	.00474	0.00510
S- 24	-0.11	TOT ELEC	KWH	1.15	0.90	1.01	0.94	0.92	1.02	1.08	1.08	1.00	0.94	0.96	1.02	12.0
		TOT DEM	KW	.00527	.00485	.00437	.00311	.00316	.00358	.00358	.00353	.00332	.00327	.00480	.00490	0.00527
S- 30	0.37	TOT ELEC	KWH	1.14	0.89	1.00	0.93	0.92	1.01	1.07	1.07	0.99	0.93	0.95	1.01	11.9
		TOT DEM	KW	.00625	.00575	.00519	.00369	.00375	.00425	.00425	.00419	.00394	.00387	.00569	.00581	0.00625
S- 37	0.39	TOT ELEC	KWH	1.10	0.86	0.97	0.90	0.89	0.98	1.04	1.04	0.95	0.90	0.92	0.98	11.5
		TOT DEM	KW	.00613	.00564	.00509	.00362	.00368	.00417	.00417	.00411	.00386	.00380	.00558	.00570	0.00613
S- 38	1.20	TOT ELEC	KWH	1.06	0.83	0.92	0.86	0.85	0.94	0.99	0.99	0.91	0.86	0.88	0.94	11.0
		TOT DEM	KW	.00603	.00555	.00500	.00356	.00362	.00410	.00410	.00404	.00380	.00374	.00549	.00561	0.00603
BASE CASE : 234																
BASE	0.00	HT ELEC	KWH	0.62	0.39	0.37	0.24	0.11	0.05	0.00	0.02	0.08	0.23	0.42	0.54	3.1
		HT DEM	KW	.00608	.00608	.00602	.00517	.00395	.00340	.00188	.00298	.00359	.00462	.00608	.00608	0.00608
		REM ELEC	KWH	1.08	0.96	1.08	1.05	1.15	1.21	1.41	1.37	1.19	1.10	1.03	1.07	13.7
		REM DEM	KW	.00392	.00392	.00525	.00579	.00559	.00665	.00645	.00652	.00592	.00532	.00439	.00392	0.00665
		TOT ELEC	KWH	1.70	1.35	1.45	1.29	1.26	1.26	1.41	1.39	1.27	1.33	1.45	1.61	16.8
		TOT DEM	KW	.01008	.00907	.00887	.00766	.00685	.00796	.00726	.00726	.00665	.00827	.00978	.00927	0.01008
S- 3	-0.14	HT ELEC	KWH	0.65	0.41	0.39	0.25	0.12	0.06	0.00	0.02	0.08	0.25	0.45	0.57	3.2
		HT DEM	KW	.00608	.00608	.00602	.00517	.00395	.00340	.00188	.00298	.00359	.00462	.00608	.00608	0.00608
		REM ELEC	KWH	0.92	0.82	0.92	0.90	0.98	1.03	1.21	1.17	1.02	0.94	0.88	0.91	11.7
		REM DEM	KW	.00362	.00362	.00484	.00533	.00515	.00613	.00595	.00601	.00546	.00490	.00405	.00362	0.00613
		TOT ELEC	KWH	1.58	1.23	1.31	1.15	1.10	1.08	1.21	1.19	1.10	1.18	1.32	1.48	14.9
		TOT DEM	KW	.01005	.00905	.00884	.00764	.00683	.00794	.00724	.00724	.00663	.00824	.00975	.00925	0.01005
S- 6	-0.09	HT ELEC	KWH	0.71	0.45	0.42	0.27	0.13	0.06	0.00	0.02	0.09	0.27	0.49	0.62	3.5
		HT DEM	KW	.00608	.00608	.00602	.00517	.00395	.00340	.00188	.00298	.00359	.00462	.00608	.00608	0.00608
		REM ELEC	KWH	0.85	0.76	0.85	0.83	0.91	0.95	1.11	1.08	0.94	0.86	0.81	0.84	10.8
		REM DEM	KW	.00326	.00326	.00437	.00481	.00465	.00553	.00536	.00542	.00492	.00442	.00365	.00326	0.00553
		TOT ELEC	KWH	1.56	1.21	1.27	1.11	1.03	1.01	1.11	1.10	1.03	1.13	1.30	1.46	14.3
		TOT DEM	KW	.00930	.00837	.00818	.00707	.00632	.00735	.00670	.00670	.00614	.00763	.00902	.00856	0.00930
S- 15	1.13	HT ELEC	KWH	0.75	0.48	0.44	0.29	0.13	0.06	0.00	0.02	0.09	0.28	0.51	0.66	3.7
		HT DEM	KW	.00608	.00608	.00602	.00517	.00395	.00340	.00188	.00298	.00359	.00462	.00608	.00608	0.00608
		REM ELEC	KWH	0.74	0.66	0.74	0.72	0.79	0.83	0.97	0.94	0.82	0.75	0.71	0.73	9.4
		REM DEM	KW	.00289	.00289	.00387	.00426	.00412	.00490	.00475	.00480	.00436	.00392	.00323	.00289	0.00490
		TOT ELEC	KWH	1.50	1.14	1.19	1.02	0.92	0.89	0.97	0.96	0.91	1.04	1.22	1.39	13.1
		TOT DEM	KW	.00878	.00817	.00799	.00641	.00562	.00650	.00597	.00606	.00571	.00711	.00860	.00852	0.00878
S- 31	0.14	HT ELEC	KWH	0.49	0.39	0.41	0.32	0.22	0.21	0.13	0.16	0.21	0.30	0.40	0.44	3.7
		HT DEM	KW	.00608	.00608	.00602	.00517	.00395	.00340	.00188	.00298	.00359	.00462	.00608	.00608	0.00608
		REM ELEC	KWH	0.87	0.79	0.91	0.86	0.83	0.99	1.11	1.09	0.97	0.84	0.85	0.90	11.0
		REM DEM	KW	.00359	.00359	.00437	.00476	.00453	.00553	.00531	.00536	.00481	.00470	.00359	.00359	0.00553
		TOT ELEC	KWH	1.36	1.18	1.32	1.17	1.05	1.20	1.24	1.25	1.17	1.14	1.25	1.34	14.7
		TOT DEM	KW	.00970	.00902	.00883	.00708	.00621	.00718	.00660	.00669	.00630	.00786	.00951	.00941	0.00970
S- 39	0.93	HT ELEC	KWH	0.52	0.41	0.44	0.34	0.23	0.22	0.14	0.17	0.22	0.32	0.43	0.47	3.9
		HT DEM	KW	.00608	.00608	.00602	.00517	.00395	.00340	.00188	.00298	.00359	.00462	.00608	.00608	0.00608
		REM ELEC	KWH	0.80	0.73	0.84	0.79	0.76	0.91	1.02	1.00	0.89	0.77	0.78	0.83	10.1
		REM DEM	KW	.00340	.00340	.00413	.00450	.00429	.00523	.00502	.00507	.00455	.00445	.00340	.00340	0.00523
		TOT ELEC	KWH	1.32	1.14	1.28	1.12	0.99	1.13	1.16	1.17	1.11	1.09	1.21	1.30	14.0
		TOT DEM	KW	.00930	.00865	.00846	.00679	.00595	.00688	.00632	.00642	.00604	.00753	.00911	.00902	0.00930
S- 44	0.16	HT ELEC	KWH	0.50	0.40	0.42	0.32	0.23	0.21	0.14	0.17	0.21	0.31	0.41	0.45	3.8
		HT DEM	KW	.00608	.00608	.00602	.00517	.00395	.00340	.00188	.00298	.00359	.00462	.00608	.00608	0.00608
		REM ELEC	KWH	0.85	0.77	0.89	0.83	0.80	0.96	1.08	1.06	0.94	0.81	0.82	0.88	10.7
		REM DEM	KW	.00350	.00350	.00425	.00463	.00441	.00538	.00516	.00522	.00468	.00457	.00350	.00350	0.00538
		TOT ELEC	KWH	1.35	1.17	1.31	1.16	1.03	1.17	1.22	1.23	1.15	1.13	1.24	1.33	14.5
		TOT DEM	KW	.00953	.00886	.00867	.00696	.00610	.00705	.00648	.00658	.00619	.00772	.00934	.00924	0.00953
S- 60	-0.34	HT ELEC	KWH	0.71	0.45	0.42	0.27	0.13	0.06	0.00	0.02	0.09	0.27	0.49	0.62	3.5
		HT DEM	KW	.00608	.00608	.00602	.00517	.00395	.00340	.00188	.00298	.00359	.00462	.00608	.00608	0.00608
		REM ELEC	KWH	0.85	0.76	0.85	0.83	0.91	0.95	1.11	1.08	0.94	0.86	0.81	0.84	10.8
		REM DEM	KW	.00326	.00326	.00437	.00481	.00465	.00553	.00536	.00542	.00492	.00442	.00365	.00326	0.00553
		TOT ELEC	KWH	1.56	1.21	1.27	1.11	1.03	1.01	1.11	1.10	1.03	1.13	1.30	1.46	14.3
		TOT DEM	KW	.00930	.00837	.00818	.00707	.00632	.00735	.00670	.00670	.00614	.00763	.00902	.00856	0.00930
BASE CASE : 235																
BASE	0.00	TOT GAS	THM	0.0325	0.0206	0.0192	0.0126	0.0058	0.0027	0.0000	0.0008	0.0040	0.0122	0.0222	0.0283	0.161
		TOT ELEC	KWH	1.08	0.96	1.08	1.05	1.15	1.21	1.41	1.37	1.19	1.10	1.03	1.07	13.7
		TOT DEM	KW	.00392	.00392	.00525	.00579	.00559	.00665	.00645	.00652	.00592	.00532	.00439	.00392	0.00665
S- 3	-0.15	TOT GAS	THM	0.0343	0.0218	0.0202	0.0133	0.0061	0.0029	0.0000	0.0008	0.0043	0.0129	0.0235	0.0299	0.170
		TOT ELEC	KWH	0.92	0.82	0.92	0.90	0.98	1.03	1.21	1.17	1.02	0.94	0.88	0.91	11.7
		TOT DEM	KW	.00362	.00362	.00484	.00533	.00515	.00613	.00595	.00601	.00546	.00490	.00405	.00362	0.00613
S- 6	-0.10	TOT GAS	THM	0.0374	0.0237	0.0220	0.0144	0.0067	0.0031	0.0000	0.0009	0.0046	0.0141	0.0255	0.0326	0.185
		TOT ELEC	KWH	0.85	0.76	0.85	0.83	0.91	0.95	1.11	1.08	0.94	0.86	0.81	0.84	10.8
		TOT DEM	KW	.00326	.00326	.00437	.00481	.00465	.00553	.00536	.00542	.00492	.00442	.00365	.00326	0.00553

STRAT SET	CONST COST $/SF	ENERGY FORM	UNITS (PER SF)	JAN	FEB	MAR	APR	MAY	JUN	JUL	AUG	SEP	OCT	NOV	DEC	ANNUAL
BASE CASE : 235 (CONTINUED)																
S- 15	1.12	TOT GAS	THM	0.0396	0.0251	0.0233	0.0153	0.0071	0.0033	0.0000	0.0010	0.0049	0.0149	0.0270	0.0345	0.196
		TOT ELEC	KWH	0.74	0.66	0.74	0.72	0.79	0.83	0.97	0.94	0.82	0.75	0.71	0.73	9.4
		TOT DEM	KW	.00289	.00289	.00387	.00426	.00412	.00490	.00475	.00480	.00436	.00392	.00323	.00289	0.00490
S- 31	0.13	TOT GAS	THM	0.0257	0.0203	0.0216	0.0166	0.0116	0.0108	0.0069	0.0085	0.0108	0.0160	0.0212	0.0232	0.193
		TOT ELEC	KWH	0.87	0.79	0.91	0.86	0.83	0.99	1.11	1.09	0.97	0.84	0.85	0.90	11.0
		TOT DEM	KW	.00359	.00359	.00437	.00476	.00453	.00553	.00531	.00536	.00481	.00470	.00359	.00359	0.00553
S- 39	0.92	TOT GAS	THM	0.0273	0.0215	0.0230	0.0176	0.0123	0.0115	0.0074	0.0090	0.0115	0.0170	0.0225	0.0246	0.205
		TOT ELEC	KWH	0.80	0.73	0.84	0.79	0.76	0.91	1.02	1.00	0.89	0.77	0.78	0.83	10.1
		TOT DEM	KW	.00340	.00340	.00413	.00450	.00429	.00523	.00502	.00507	.00455	.00445	.00340	.00340	0.00523
S- 44	0.16	TOT GAS	THM	0.0263	0.0208	0.0222	0.0170	0.0119	0.0111	0.0071	0.0087	0.0111	0.0164	0.0218	0.0238	0.198
		TOT ELEC	KWH	0.85	0.77	0.89	0.83	0.80	0.96	1.08	1.06	0.94	0.81	0.82	0.88	10.7
		TOT DEM	KW	.00350	.00350	.00425	.00463	.00441	.00538	.00516	.00522	.00468	.00457	.00350	.00350	0.00538
S- 60	-0.35	TOT GAS	THM	0.0374	0.0237	0.0220	0.0144	0.0067	0.0031	0.0000	0.0009	0.0046	0.0141	0.0255	0.0326	0.185
		TOT ELEC	KWH	0.85	0.76	0.85	0.83	0.91	0.95	1.11	1.08	0.94	0.86	0.81	0.84	10.8
		TOT DEM	KW	.00326	.00326	.00437	.00481	.00465	.00553	.00536	.00542	.00492	.00442	.00365	.00326	0.00553
BASE CASE : 236																
BASE	0.00	HT ELEC	KWH	0.44	0.28	0.26	0.17	0.08	0.04	0.00	0.01	0.05	0.17	0.30	0.39	2.2
		HT DEM	KW	.00435	.00435	.00435	.00435	.00313	.00218	.00070	.00165	.00252	.00435	.00435	.00435	0.00435
		REM ELEC	KWH	1.02	0.90	1.02	0.99	1.08	1.14	1.33	1.29	1.12	1.03	0.97	1.01	12.9
		REM DEM	KW	.00355	.00355	.00447	.00499	.00476	.00567	.00567	.00573	.00539	.00441	.00418	.00350	0.00573
		TOT ELEC	KWH	1.46	1.18	1.28	1.16	1.16	1.17	1.33	1.30	1.18	1.20	1.27	1.39	15.1
		TOT DEM	KW	.00750	.00743	.00743	.00645	.00487	.00578	.00578	.00585	.00548	.00705	.00743	.00743	0.00750
S- 3	-0.11	HT ELEC	KWH	0.48	0.30	0.28	0.18	0.08	0.04	0.00	0.01	0.06	0.18	0.33	0.42	2.4
		HT DEM	KW	.00435	.00435	.00435	.00435	.00313	.00218	.00070	.00165	.00252	.00435	.00435	.00435	0.00435
		REM ELEC	KWH	0.87	0.77	0.87	0.85	0.92	0.97	1.13	1.10	0.96	0.88	0.83	0.86	11.0
		REM DEM	KW	.00327	.00327	.00412	.00459	.00438	.00523	.00523	.00528	.00496	.00407	.00385	.00322	0.00528
		TOT ELEC	KWH	1.35	1.07	1.15	1.03	1.01	1.01	1.13	1.11	1.02	1.06	1.15	1.27	13.4
		TOT DEM	KW	.00746	.00739	.00739	.00642	.00485	.00574	.00574	.00582	.00545	.00701	.00739	.00739	0.00746
S- 6	-0.06	HT ELEC	KWH	0.51	0.32	0.30	0.20	0.09	0.04	0.00	0.01	0.06	0.19	0.35	0.44	2.5
		HT DEM	KW	.00435	.00435	.00435	.00435	.00313	.00218	.00070	.00165	.00252	.00435	.00435	.00435	0.00435
		REM ELEC	KWH	0.78	0.69	0.78	0.76	0.83	0.87	1.02	0.99	0.86	0.79	0.74	0.77	9.9
		REM DEM	KW	.00295	.00295	.00370	.00413	.00394	.00470	.00470	.00475	.00446	.00366	.00347	.00290	0.00475
		TOT ELEC	KWH	1.29	1.02	1.08	0.96	0.92	0.92	1.02	1.00	0.93	0.98	1.09	1.22	12.4
		TOT DEM	KW	.00673	.00666	.00666	.00579	.00437	.00518	.00518	.00525	.00491	.00633	.00666	.00666	0.00673
S- 15	0.49	HT ELEC	KWH	0.54	0.34	0.32	0.21	0.10	0.05	0.00	0.01	0.07	0.20	0.37	0.47	2.7
		HT DEM	KW	.00435	.00435	.00435	.00435	.00313	.00218	.00070	.00165	.00252	.00435	.00435	.00435	0.00435
		REM ELEC	KWH	0.68	0.60	0.68	0.66	0.72	0.76	0.89	0.86	0.75	0.69	0.65	0.67	8.6
		REM DEM	KW	.00262	.00262	.00330	.00368	.00351	.00419	.00419	.00423	.00398	.00326	.00309	.00258	0.00423
		TOT ELEC	KWH	1.22	0.95	1.00	0.87	0.82	0.80	0.89	0.88	0.82	0.89	1.02	1.15	11.3
		TOT DEM	KW	.00653	.00646	.00646	.00562	.00424	.00503	.00503	.00509	.00477	.00614	.00646	.00646	0.00653
S- 31	0.17	HT ELEC	KWH	0.42	0.29	0.28	0.17	0.08	0.06	0.03	0.04	0.07	0.14	0.30	0.32	2.2
		HT DEM	KW	.00435	.00435	.00435	.00435	.00313	.00218	.00070	.00165	.00252	.00435	.00435	.00435	0.00435
		REM ELEC	KWH	0.75	0.67	0.78	0.74	0.77	0.90	1.00	1.00	0.88	0.76	0.73	0.78	9.8
		REM DEM	KW	.00296	.00292	.00338	.00351	.00351	.00423	.00415	.00419	.00372	.00372	.00317	.00305	0.00423
		TOT ELEC	KWH	1.18	0.96	1.06	0.91	0.85	0.96	1.03	1.04	0.95	0.91	1.03	1.10	12.0
		TOT DEM	KW	.00710	.00703	.00703	.00596	.00369	.00433	.00426	.00433	.00383	.00653	.00703	.00703	0.00710
S- 39	0.63	HT ELEC	KWH	0.49	0.33	0.33	0.19	0.09	0.07	0.03	0.05	0.08	0.17	0.34	0.37	2.5
		HT DEM	KW	.00395	.00395	.00395	.00395	.00284	.00197	.00063	.00150	.00229	.00395	.00395	.00395	0.00395
		REM ELEC	KWH	0.67	0.60	0.70	0.66	0.69	0.81	0.89	0.89	0.79	0.68	0.66	0.70	8.8
		REM DEM	KW	.00270	.00266	.00308	.00320	.00320	.00385	.00377	.00381	.00339	.00339	.00289	.00277	0.00385
		TOT ELEC	KWH	1.17	0.94	1.03	0.86	0.79	0.87	0.93	0.94	0.87	0.85	1.00	1.07	11.3
		TOT DEM	KW	.00683	.00676	.00676	.00574	.00355	.00417	.00410	.00417	.00369	.00628	.00676	.00676	0.00683
S- 44	0.19	HT ELEC	KWH	0.46	0.31	0.30	0.18	0.09	0.06	0.03	0.05	0.07	0.15	0.32	0.34	2.4
		HT DEM	KW	.00395	.00395	.00395	.00395	.00284	.00197	.00063	.00150	.00229	.00395	.00395	.00395	0.00395
		REM ELEC	KWH	0.72	0.64	0.74	0.71	0.74	0.86	0.95	0.95	0.84	0.73	0.70	0.74	9.3
		REM DEM	KW	.00283	.00279	.00324	.00336	.00336	.00405	.00397	.00401	.00356	.00356	.00304	.00292	0.00405
		TOT ELEC	KWH	1.17	0.95	1.05	0.89	0.82	0.92	0.98	1.00	0.91	0.88	1.02	1.09	11.7
		TOT DEM	KW	.00698	.00691	.00691	.00586	.00363	.00426	.00419	.00426	.00377	.00642	.00691	.00691	0.00698
S- 60	-0.31	HT ELEC	KWH	0.51	0.32	0.30	0.20	0.09	0.04	0.00	0.01	0.06	0.19	0.35	0.44	2.5
		HT DEM	KW	.00435	.00435	.00435	.00435	.00313	.00218	.00070	.00165	.00252	.00435	.00435	.00435	0.00435
		REM ELEC	KWH	0.78	0.69	0.78	0.76	0.83	0.87	1.02	0.99	0.86	0.79	0.74	0.77	9.9
		REM DEM	KW	.00295	.00295	.00370	.00413	.00394	.00470	.00470	.00475	.00446	.00366	.00347	.00290	0.00475
		TOT ELEC	KWH	1.29	1.02	1.08	0.96	0.92	0.92	1.02	1.00	0.93	0.98	1.09	1.22	12.4
		TOT DEM	KW	.00673	.00666	.00666	.00579	.00437	.00518	.00518	.00525	.00491	.00633	.00666	.00666	0.00673
BASE CASE : 237																
BASE	0.00	TOT GAS	THM	0.0232	0.0147	0.0137	0.0090	0.0041	0.0020	0.0000	0.0006	0.0029	0.0087	0.0159	0.0202	0.115
		TOT ELEC	KWH	1.02	0.90	1.02	0.99	1.08	1.14	1.33	1.29	1.12	1.03	0.97	1.01	12.9
		TOT DEM	KW	.00355	.00355	.00447	.00499	.00476	.00567	.00567	.00573	.00539	.00441	.00418	.00350	0.00573
S- 3	-0.13	TOT GAS	THM	0.0250	0.0159	0.0148	0.0097	0.0045	0.0021	0.0000	0.0006	0.0031	0.0094	0.0171	0.0218	0.124
		TOT ELEC	KWH	0.87	0.77	0.87	0.85	0.92	0.97	1.13	1.10	0.96	0.88	0.83	0.86	11.0
		TOT DEM	KW	.00327	.00327	.00412	.00459	.00438	.00523	.00523	.00528	.00496	.00407	.00385	.00322	0.00528
S- 6	-0.08	TOT GAS	THM	0.0267	0.0169	0.0157	0.0103	0.0048	0.0022	0.0000	0.0007	0.0033	0.0100	0.0182	0.0232	0.132
		TOT ELEC	KWH	0.78	0.69	0.78	0.76	0.83	0.87	1.02	0.99	0.86	0.79	0.74	0.77	9.9
		TOT DEM	KW	.00295	.00295	.00370	.00413	.00394	.00470	.00470	.00475	.00446	.00366	.00347	.00290	0.00475

(HT ELEC=ELECTRIC CONSUMPTION FOR SPACE AND WATER HEATING. REM ELEC=REMAINING ELECTRIC CONSUMPTION (NOT USED FOR HEATING). TOT ELEC=TOTAL ELECTRIC CONSUMPTION. HT DEM=PEAK ELECTRIC DEMAND FOR SPACE AND WATER HEATING. REM DEM=PEAK ELECTRIC DEMAND FOR REMAINING LOAD (NOT HEATING). TOT DEM=TOTAL PEAK ELECTRIC DEMAND. GAS=TOTAL GAS CONSUMPTION. ANNUAL FIGURES ARE SUMS OF MONTHLY CONSUMPTION VALUES OR ARE YEARLY MAXIMUMS OF MONTHLY PEAK DEMAND VALUES.)

BASE CASE : 237 (CONTINUED)

STRAT SET	CONST COST $/SF	ENERGY FORM	UNITS (PER SF)	JAN	FEB	MAR	APR	MAY	JUN	JUL	AUG	SEP	OCT	NOV	DEC	ANNUAL
S- 15	0.43	TOT GAS	THM	0.0285	0.0180	0.0168	0.0110	0.0051	0.0024	0.0000	0.0007	0.0035	0.0107	0.0195	0.0248	0.141
		TOT ELEC	KWH	0.68	0.60	0.68	0.66	0.72	0.76	0.89	0.86	0.75	0.69	0.65	0.67	8.6
		TOT DEM	KW	.00262	.00262	.00330	.00368	.00351	.00419	.00419	.00423	.00398	.00326	.00309	.00258	0.00423
S- 31	0.15	TOT GAS	THM	0.0223	0.0151	0.0147	0.0087	0.0043	0.0031	0.0015	0.0023	0.0036	0.0075	0.0155	0.0167	0.115
		TOT ELEC	KWH	0.75	0.67	0.78	0.74	0.77	0.90	1.00	1.00	0.88	0.76	0.73	0.78	9.8
		TOT DEM	KW	.00296	.00296	.00338	.00351	.00351	.00423	.00415	.00415	.00372	.00372	.00317	.00305	0.00423
S- 39	0.59	TOT GAS	THM	0.0258	0.0174	0.0170	0.0101	0.0049	0.0036	0.0017	0.0027	0.0041	0.0086	0.0180	0.0193	0.133
		TOT ELEC	KWH	0.67	0.60	0.70	0.66	0.69	0.81	0.89	0.89	0.79	0.68	0.66	0.70	8.8
		TOT DEM	KW	.00270	.00270	.00308	.00320	.00320	.00385	.00377	.00381	.00339	.00339	.00289	.00277	0.00385
S- 44	0.18	TOT GAS	THM	0.0241	0.0162	0.0159	0.0094	0.0046	0.0033	0.0016	0.0025	0.0038	0.0081	0.0167	0.0180	0.124
		TOT ELEC	KWH	0.72	0.64	0.74	0.71	0.74	0.86	0.95	0.95	0.84	0.73	0.70	0.74	9.3
		TOT DEM	KW	.00283	.00283	.00324	.00336	.00336	.00405	.00397	.00401	.00356	.00356	.00304	.00292	0.00405
S- 60	-0.32	TOT GAS	THM	0.0267	0.0169	0.0157	0.0103	0.0048	0.0022	0.0000	0.0007	0.0033	0.0100	0.0182	0.0232	0.132
		TOT ELEC	KWH	0.78	0.69	0.78	0.76	0.83	0.87	1.02	0.99	0.86	0.79	0.74	0.77	9.9
		TOT DEM	KW	.00295	.00295	.00370	.00413	.00394	.00470	.00470	.00475	.00446	.00366	.00347	.00290	0.00475

BASE CASE : 238

STRAT SET	CONST COST $/SF	ENERGY FORM	UNITS (PER SF)	JAN	FEB	MAR	APR	MAY	JUN	JUL	AUG	SEP	OCT	NOV	DEC	ANNUAL
BASE	0.00	TOT ELEC	KWH	1.65	1.29	1.44	1.34	1.32	1.46	1.55	1.55	1.43	1.34	1.38	1.46	17.2
		TOT DEM	KW	.00708	.00651	.00588	.00418	.00425	.00481	.00481	.00474	.00446	.00439	.00644	.00658	0.00708
S- 1	0.09	TOT ELEC	KWH	1.53	1.19	1.34	1.24	1.22	1.35	1.43	1.43	1.32	1.24	1.27	1.35	15.9
		TOT DEM	KW	.00588	.00541	.00488	.00347	.00353	.00400	.00400	.00394	.00370	.00365	.00535	.00547	0.00588
S- 5	0.14	TOT ELEC	KWH	1.37	1.07	1.20	1.12	1.10	1.22	1.29	1.29	1.19	1.12	1.14	1.22	14.3
		TOT DEM	KW	.00552	.00508	.00458	.00326	.00331	.00375	.00375	.00370	.00348	.00342	.00502	.00513	0.00552
S- 14	1.45	TOT ELEC	KWH	1.26	0.98	1.10	1.02	1.01	1.11	1.18	1.18	1.09	1.02	1.05	1.11	13.1
		TOT DEM	KW	.00527	.00485	.00437	.00311	.00316	.00358	.00358	.00353	.00332	.00327	.00480	.00490	0.00527
S- 24	-0.11	TOT ELEC	KWH	1.37	1.07	1.20	1.12	1.10	1.22	1.29	1.29	1.19	1.12	1.14	1.22	14.3
		TOT DEM	KW	.00552	.00508	.00458	.00326	.00331	.00375	.00375	.00370	.00348	.00342	.00502	.00513	0.00552
S- 30	0.37	TOT ELEC	KWH	1.18	0.92	1.03	0.96	0.95	1.05	1.11	1.11	1.02	0.96	0.98	1.05	12.3
		TOT DEM	KW	.00650	.00598	.00539	.00383	.00390	.00442	.00442	.00436	.00409	.00403	.00592	.00605	0.00650
S- 37	0.39	TOT ELEC	KWH	1.14	0.89	1.00	0.93	0.92	1.01	1.07	1.07	0.99	0.93	0.95	1.01	11.9
		TOT DEM	KW	.00640	.00589	.00531	.00378	.00384	.00435	.00435	.00429	.00403	.00397	.00582	.00595	0.00640
S- 38	1.20	TOT ELEC	KWH	1.11	0.87	0.97	0.90	0.89	0.99	1.04	1.04	0.96	0.90	0.93	0.99	11.6
		TOT DEM	KW	.00630	.00580	.00523	.00372	.00378	.00428	.00428	.00422	.00397	.00391	.00573	.00586	0.00630

BASE CASE : 239

STRAT SET	CONST COST $/SF	ENERGY FORM	UNITS (PER SF)	JAN	FEB	MAR	APR	MAY	JUN	JUL	AUG	SEP	OCT	NOV	DEC	ANNUAL
BASE	0.00	HT ELEC	KWH	0.42	0.27	0.25	0.16	0.08	0.04	0.00	0.01	0.05	0.16	0.29	0.37	2.1
		HT DEM	KW	.00608	.00578	.00547	.00468	.00359	.00298	.00146	.00249	.00292	.00474	.00590	.00553	0.00608
		REM ELEC	KWH	1.03	0.92	1.03	1.01	1.10	1.15	1.35	1.31	1.14	1.05	0.98	1.02	13.1
		REM DEM	KW	.00354	.00354	.00474	.00522	.00504	.00600	.00582	.00588	.00534	.00480	.00396	.00354	0.00600
		TOT ELEC	KWH	1.46	1.19	1.28	1.17	1.18	1.19	1.35	1.32	1.19	1.21	1.27	1.39	15.2
		TOT DEM	KW	.00988	.00988	.00988	.00909	.00741	.00840	.00781	.00781	.00721	.00958	.00988	.00988	0.00988
S- 3	-0.11	HT ELEC	KWH	0.45	0.28	0.26	0.17	0.08	0.04	0.00	0.01	0.06	0.17	0.30	0.39	2.2
		HT DEM	KW	.00608	.00578	.00547	.00468	.00359	.00298	.00146	.00249	.00292	.00474	.00590	.00553	0.00608
		REM ELEC	KWH	0.88	0.78	0.88	0.86	0.94	0.99	1.15	1.12	0.97	0.90	0.84	0.87	11.2
		REM DEM	KW	.00326	.00326	.00437	.00481	.00465	.00553	.00536	.00542	.00492	.00442	.00365	.00326	0.00553
		TOT ELEC	KWH	1.33	1.07	1.15	1.03	1.02	1.02	1.15	1.13	1.03	1.06	1.14	1.26	13.4
		TOT DEM	KW	.00950	.00950	.00950	.00874	.00713	.00807	.00751	.00751	.00694	.00922	.00950	.00950	0.00950
S- 6	-0.07	HT ELEC	KWH	0.49	0.31	0.29	0.19	0.09	0.04	0.00	0.01	0.06	0.18	0.33	0.43	2.4
		HT DEM	KW	.00608	.00578	.00547	.00468	.00359	.00298	.00146	.00249	.00292	.00474	.00590	.00553	0.00608
		REM ELEC	KWH	0.81	0.72	0.81	0.79	0.87	0.91	1.06	1.03	0.90	0.82	0.77	0.80	10.3
		REM DEM	KW	.00294	.00294	.00393	.00433	.00418	.00498	.00483	.00488	.00443	.00398	.00329	.00294	0.00498
		TOT ELEC	KWH	1.30	1.03	1.10	0.98	0.95	0.95	1.06	1.04	0.96	1.01	1.11	1.23	12.7
		TOT DEM	KW	.00885	.00885	.00885	.00814	.00664	.00752	.00699	.00699	.00646	.00858	.00885	.00885	0.00885
S- 15	1.16	HT ELEC	KWH	0.52	0.33	0.30	0.20	0.09	0.04	0.00	0.01	0.06	0.19	0.35	0.45	2.5
		HT DEM	KW	.00608	.00578	.00547	.00468	.00359	.00298	.00146	.00249	.00292	.00474	.00590	.00553	0.00608
		REM ELEC	KWH	0.71	0.63	0.71	0.69	0.76	0.79	0.93	0.90	0.78	0.72	0.68	0.70	9.0
		REM DEM	KW	.00261	.00261	.00350	.00385	.00372	.00443	.00430	.00434	.00394	.00354	.00292	.00261	0.00443
		TOT ELEC	KWH	1.23	0.96	1.01	0.89	0.85	0.84	0.93	0.91	0.85	0.91	1.03	1.15	11.6
		TOT DEM	KW	.00850	.00689	.00680	.00587	.00502	.00553	.00519	.00519	.00493	.00629	.00706	.00689	0.00850
S- 31	0.25	HT ELEC	KWH	0.35	0.27	0.29	0.22	0.16	0.15	0.09	0.11	0.15	0.22	0.29	0.31	2.6
		HT DEM	KW	.00608	.00578	.00547	.00468	.00359	.00298	.00146	.00249	.00292	.00474	.00590	.00553	0.00608
		REM ELEC	KWH	0.81	0.74	0.85	0.80	0.77	0.93	1.04	1.02	0.91	0.78	0.79	0.84	10.3
		REM DEM	KW	.00325	.00325	.00395	.00430	.00410	.00500	.00480	.00485	.00435	.00425	.00325	.00325	0.00500
		TOT ELEC	KWH	1.16	1.01	1.15	1.03	0.93	1.07	1.13	1.13	1.05	1.00	1.08	1.16	12.9
		TOT DEM	KW	.00955	.00774	.00764	.00659	.00563	.00621	.00583	.00583	.00554	.00707	.00793	.00774	0.00955
S- 60	-0.31	HT ELEC	KWH	0.49	0.31	0.29	0.19	0.09	0.04	0.00	0.01	0.06	0.18	0.33	0.43	2.4
		HT DEM	KW	.00608	.00578	.00547	.00468	.00359	.00298	.00146	.00249	.00292	.00474	.00590	.00553	0.00608
		REM ELEC	KWH	0.81	0.72	0.81	0.79	0.87	0.91	1.06	1.03	0.90	0.82	0.77	0.80	10.3
		REM DEM	KW	.00294	.00294	.00393	.00433	.00418	.00498	.00483	.00488	.00443	.00398	.00329	.00294	0.00498
		TOT ELEC	KWH	1.30	1.03	1.10	0.98	0.95	0.95	1.06	1.04	0.96	1.01	1.11	1.23	12.7
		TOT DEM	KW	.00885	.00885	.00885	.00814	.00664	.00752	.00699	.00699	.00646	.00858	.00885	.00885	0.00885

BASE CASE : 240

STRAT SET	CONST COST $/SF	ENERGY FORM	UNITS (PER SF)	JAN	FEB	MAR	APR	MAY	JUN	JUL	AUG	SEP	OCT	NOV	DEC	ANNUAL
BASE	0.00	TOT GAS	THM	0.0222	0.0141	0.0131	0.0086	0.0040	0.0019	0.0000	0.0006	0.0028	0.0084	0.0152	0.0194	0.110
		TOT ELEC	KWH	1.03	0.92	1.03	1.01	1.10	1.15	1.35	1.31	1.14	1.05	0.98	1.02	13.1
		TOT DEM	KW	.00354	.00354	.00474	.00522	.00504	.00600	.00582	.00588	.00534	.00480	.00396	.00354	0.00600

BASE CASE : 240 (CONTINUED)

STRAT SET	CONST COST $/SF	ENERGY FORM	UNITS (PER SF)	JAN	FEB	MAR	APR	MAY	JUN	JUL	AUG	SEP	OCT	NOV	DEC	ANNUAL
S- 3	-0.11	TOT GAS	THM	0.0234	0.0148	0.0138	0.0090	0.0042	0.0020	0.0000	0.0006	0.0029	0.0088	0.0160	0.0204	0.116
		TOT ELEC	KWH	0.88	0.78	0.88	0.86	0.94	0.99	1.15	1.12	0.97	0.90	0.84	0.87	11.2
		TOT DEM	KW	.00326	.00326	.00437	.00481	.00465	.00553	.00536	.00542	.00492	.00442	.00365	.00326	0.00553
S- 6	-0.07	TOT GAS	THM	0.0257	0.0163	0.0151	0.0099	0.0046	0.0022	0.0000	0.0006	0.0032	0.0097	0.0175	0.0224	0.127
		TOT ELEC	KWH	0.81	0.72	0.81	0.79	0.87	0.91	1.06	1.03	0.90	0.82	0.77	0.80	10.3
		TOT DEM	KW	.00294	.00294	.00393	.00433	.00418	.00498	.00483	.00488	.00443	.00398	.00329	.00294	0.00498
S- 15	1.16	TOT GAS	THM	0.0271	0.0172	0.0159	0.0105	0.0048	0.0023	0.0000	0.0007	0.0034	0.0102	0.0185	0.0236	0.134
		TOT ELEC	KWH	0.71	0.63	0.71	0.69	0.76	0.79	0.93	0.90	0.78	0.72	0.68	0.70	9.0
		TOT DEM	KW	.00261	.00261	.00350	.00385	.00372	.00443	.00430	.00434	.00394	.00354	.00292	.00261	0.00443
S- 31	0.25	TOT GAS	THM	0.0182	0.0144	0.0153	0.0118	0.0082	0.0077	0.0049	0.0060	0.0077	0.0114	0.0151	0.0164	0.137
		TOT ELEC	KWH	0.81	0.74	0.85	0.80	0.77	0.93	1.04	1.02	0.91	0.78	0.79	0.84	10.3
		TOT DEM	KW	.00325	.00325	.00395	.00430	.00410	.00500	.00480	.00485	.00435	.00425	.00325	.00325	0.00500
S- 60	-0.31	TOT GAS	THM	0.0257	0.0163	0.0151	0.0099	0.0046	0.0022	0.0000	0.0006	0.0032	0.0097	0.0175	0.0224	0.127
		TOT ELEC	KWH	0.81	0.72	0.81	0.79	0.87	0.91	1.06	1.03	0.90	0.82	0.77	0.80	10.3
		TOT DEM	KW	.00294	.00294	.00393	.00433	.00418	.00498	.00483	.00488	.00443	.00398	.00329	.00294	0.00498

BASE CASE : 241

STRAT SET	CONST COST $/SF	ENERGY FORM	UNITS (PER SF)	JAN	FEB	MAR	APR	MAY	JUN	JUL	AUG	SEP	OCT	NOV	DEC	ANNUAL
BASE	0.00	HT ELEC	KWH	0.30	0.19	0.18	0.12	0.05	0.03	0.00	0.01	0.04	0.11	0.20	0.26	1.5
		HT DEM	KW	.00435	.00409	.00409	.00374	.00291	.00213	.00078	.00157	.00218	.00365	.00418	.00418	0.00435
		REM ELEC	KWH	1.00	0.88	1.00	0.97	1.06	1.11	1.30	1.26	1.10	1.01	0.95	0.98	12.6
		REM DEM	KW	.00343	.00343	.00431	.00481	.00459	.00547	.00547	.00553	.00520	.00426	.00404	.00337	0.00553
		TOT ELEC	KWH	1.29	1.07	1.17	1.09	1.11	1.13	1.30	1.27	1.13	1.12	1.15	1.24	14.1
		TOT DEM	KW	.00748	.00673	.00666	.00561	.00486	.00576	.00568	.00576	.00524	.00583	.00688	.00681	0.00748
S- 3	-0.08	HT ELEC	KWH	0.32	0.20	0.19	0.12	0.06	0.03	0.00	0.01	0.04	0.12	0.22	0.28	1.6
		HT DEM	KW	.00435	.00409	.00409	.00374	.00291	.00213	.00078	.00157	.00218	.00365	.00418	.00418	0.00435
		REM ELEC	KWH	0.85	0.75	0.85	0.82	0.90	0.94	1.10	1.07	0.93	0.86	0.80	0.83	10.7
		REM DEM	KW	.00315	.00315	.00396	.00442	.00422	.00503	.00503	.00508	.00478	.00391	.00371	.00310	0.00508
		TOT ELEC	KWH	1.17	0.95	1.03	0.95	0.96	0.97	1.10	1.08	0.97	0.98	1.02	1.11	12.3
		TOT DEM	KW	.00724	.00652	.00644	.00543	.00471	.00557	.00550	.00557	.00507	.00565	.00666	.00659	0.00724
S- 6	-0.03	HT ELEC	KWH	0.34	0.22	0.20	0.13	0.06	0.03	0.00	0.01	0.04	0.13	0.23	0.30	1.7
		HT DEM	KW	.00435	.00409	.00409	.00374	.00291	.00213	.00078	.00157	.00218	.00365	.00418	.00418	0.00435
		REM ELEC	KWH	0.77	0.68	0.77	0.75	0.81	0.85	1.00	0.97	0.84	0.78	0.73	0.76	9.7
		REM DEM	KW	.00285	.00285	.00359	.00400	.00382	.00455	.00455	.00460	.00432	.00354	.00336	.00281	0.00460
		TOT ELEC	KWH	1.11	0.90	0.97	0.88	0.88	0.88	1.00	0.98	0.89	0.90	0.96	1.06	11.4
		TOT DEM	KW	.00674	.00607	.00600	.00506	.00438	.00519	.00512	.00519	.00472	.00526	.00620	.00613	0.00674
S- 15	0.66	HT ELEC	KWH	0.37	0.23	0.22	0.14	0.07	0.03	0.00	0.01	0.05	0.14	0.25	0.32	1.8
		HT DEM	KW	.00435	.00409	.00409	.00374	.00291	.00213	.00078	.00157	.00218	.00365	.00418	.00418	0.00435
		REM ELEC	KWH	0.67	0.59	0.67	0.65	0.71	0.74	0.87	0.84	0.73	0.67	0.63	0.66	8.4
		REM DEM	KW	.00253	.00253	.00318	.00355	.00339	.00404	.00404	.00408	.00384	.00314	.00298	.00249	0.00408
		TOT ELEC	KWH	1.03	0.82	0.88	0.79	0.77	0.77	0.87	0.85	0.78	0.81	0.88	0.98	10.2
		TOT DEM	KW	.00654	.00589	.00582	.00490	.00425	.00504	.00497	.00504	.00458	.00510	.00602	.00595	0.00654
S- 31	0.29	HT ELEC	KWH	0.47	0.31	0.31	0.18	0.09	0.06	0.03	0.05	0.07	0.16	0.32	0.35	2.4
		HT DEM	KW	.00435	.00409	.00409	.00374	.00291	.00213	.00078	.00157	.00218	.00365	.00418	.00418	0.00435
		REM ELEC	KWH	0.75	0.67	0.78	0.74	0.77	0.89	0.99	0.99	0.87	0.76	0.73	0.78	9.7
		REM DEM	KW	.00300	.00295	.00342	.00355	.00355	.00428	.00419	.00424	.00377	.00377	.00321	.00308	0.00428
		TOT ELEC	KWH	1.21	0.98	1.08	0.92	0.85	0.96	1.02	1.04	0.95	0.91	1.05	1.12	12.1
		TOT DEM	KW	.00715	.00651	.00636	.00536	.00429	.00486	.00479	.00472	.00436	.00572	.00658	.00658	0.00715
S- 60	-0.28	HT ELEC	KWH	0.34	0.22	0.20	0.13	0.06	0.03	0.00	0.01	0.04	0.13	0.23	0.30	1.7
		HT DEM	KW	.00435	.00409	.00409	.00374	.00291	.00213	.00078	.00157	.00218	.00365	.00418	.00418	0.00435
		REM ELEC	KWH	0.77	0.68	0.77	0.75	0.81	0.85	1.00	0.97	0.84	0.78	0.73	0.76	9.7
		REM DEM	KW	.00285	.00285	.00359	.00400	.00382	.00455	.00455	.00460	.00432	.00354	.00336	.00281	0.00460
		TOT ELEC	KWH	1.11	0.90	0.97	0.88	0.88	0.88	1.00	0.98	0.89	0.90	0.96	1.06	11.4
		TOT DEM	KW	.00674	.00607	.00600	.00506	.00438	.00519	.00512	.00519	.00472	.00526	.00620	.00613	0.00674

BASE CASE : 242

STRAT SET	CONST COST $/SF	ENERGY FORM	UNITS (PER SF)	JAN	FEB	MAR	APR	MAY	JUN	JUL	AUG	SEP	OCT	NOV	DEC	ANNUAL
BASE	0.00	TOT GAS	THM	0.0157	0.0100	0.0093	0.0061	0.0028	0.0013	0.0000	0.0004	0.0019	0.0059	0.0108	0.0137	0.078
		TOT ELEC	KWH	1.00	0.88	1.00	0.97	1.06	1.11	1.30	1.26	1.10	1.01	0.95	0.98	12.6
		TOT DEM	KW	.00343	.00343	.00431	.00481	.00459	.00547	.00547	.00553	.00520	.00426	.00404	.00337	0.00553
S- 3	-0.09	TOT GAS	THM	0.0169	0.0107	0.0100	0.0065	0.0030	0.0014	0.0000	0.0004	0.0021	0.0064	0.0116	0.0147	0.084
		TOT ELEC	KWH	0.85	0.75	0.85	0.82	0.90	0.94	1.10	1.07	0.93	0.86	0.80	0.83	10.7
		TOT DEM	KW	.00315	.00315	.00396	.00442	.00422	.00503	.00503	.00508	.00478	.00391	.00371	.00310	0.00508
S- 6	-0.04	TOT GAS	THM	0.0181	0.0114	0.0106	0.0070	0.0032	0.0015	0.0000	0.0004	0.0022	0.0068	0.0123	0.0157	0.089
		TOT ELEC	KWH	0.77	0.68	0.77	0.75	0.81	0.85	1.00	0.97	0.84	0.78	0.73	0.76	9.7
		TOT DEM	KW	.00285	.00285	.00359	.00400	.00382	.00455	.00455	.00460	.00432	.00354	.00336	.00281	0.00460
S- 15	0.61	TOT GAS	THM	0.0193	0.0122	0.0114	0.0075	0.0034	0.0016	0.0000	0.0005	0.0024	0.0073	0.0132	0.0168	0.096
		TOT ELEC	KWH	0.67	0.59	0.67	0.65	0.71	0.74	0.87	0.84	0.73	0.67	0.63	0.66	8.4
		TOT DEM	KW	.00253	.00253	.00318	.00355	.00339	.00404	.00404	.00408	.00384	.00314	.00298	.00249	0.00408
S- 31	0.28	TOT GAS	THM	0.0244	0.0165	0.0161	0.0096	0.0047	0.0034	0.0016	0.0025	0.0039	0.0082	0.0170	0.0183	0.126
		TOT ELEC	KWH	0.75	0.67	0.78	0.74	0.77	0.89	0.99	0.99	0.87	0.76	0.73	0.78	9.7
		TOT DEM	KW	.00300	.00300	.00342	.00355	.00355	.00428	.00419	.00424	.00377	.00377	.00321	.00308	0.00428
S- 60	-0.29	TOT GAS	THM	0.0181	0.0114	0.0106	0.0070	0.0032	0.0015	0.0000	0.0004	0.0022	0.0068	0.0123	0.0157	0.089
		TOT ELEC	KWH	0.77	0.68	0.77	0.75	0.81	0.85	1.00	0.97	0.84	0.78	0.73	0.76	9.7
		TOT DEM	KW	.00285	.00285	.00359	.00400	.00382	.00455	.00455	.00460	.00432	.00354	.00336	.00281	0.00460

BASE CASE : 243

STRAT SET	CONST COST $/SF	ENERGY FORM	UNITS (PER SF)	JAN	FEB	MAR	APR	MAY	JUN	JUL	AUG	SEP	OCT	NOV	DEC	ANNUAL
BASE	0.00	TOT ELEC	KWH	1.43	1.12	1.25	1.16	1.15	1.27	1.34	1.34	1.24	1.16	1.19	1.27	14.9
		TOT DEM	KW	.00578	.00532	.00480	.00341	.00347	.00393	.00393	.00387	.00364	.00358	.00526	.00538	0.00578

(HT ELEC=ELECTRIC CONSUMPTION FOR SPACE AND WATER HEATING. REM ELEC=REMAINING ELECTRIC CONSUMPTION (NOT USED FOR HEATING). TOT ELEC=TOTAL ELECTRIC CONSUMPTION. HT DEM=PEAK ELECTRIC DEMAND FOR SPACE AND WATER HEATING. REM DEM=PEAK ELECTRIC DEMAND FOR REMAINING LOAD (NOT HEATING). TOT DEM=TOTAL PEAK ELECTRIC DEMAND. GAS=TOTAL GAS CONSUMPTION. ANNUAL FIGURES ARE SUMS OF MONTHLY CONSUMPTION VALUES OR ARE YEARLY MAXIMUMS OF MONTHLY PEAK DEMAND VALUES.)

STRAT SET	CONST COST $/SF	ENERGY FORM	UNITS (PER SF)	JAN	FEB	MAR	APR	MAY	JUN	JUL	AUG	SEP	OCT	NOV	DEC	ANNUAL
BASE CASE : 243 (CONTINUED)																
S- 1	0.09	TOT ELEC	KWH	1.32	1.03	1.15	1.07	1.05	1.16	1.23	1.23	1.14	1.07	1.10	1.16	13.7
		TOT DEM	KW	.00514	.00473	.00427	.00303	.00308	.00350	.00350	.00344	.00324	.00319	.00468	.00478	0.00514
S- 5	0.14	TOT ELEC	KWH	1.17	0.92	1.02	0.95	0.94	1.04	1.10	1.10	1.01	0.95	0.98	1.04	12.2
		TOT DEM	KW	.00491	.00452	.00408	.00290	.00295	.00334	.00334	.00329	.00309	.00304	.00447	.00457	0.00491
S- 14	1.45	TOT ELEC	KWH	1.06	0.83	0.92	0.86	0.85	0.94	0.99	0.99	0.91	0.86	0.88	0.94	11.0
		TOT DEM	KW	.00468	.00431	.00388	.00276	.00281	.00318	.00318	.00314	.00295	.00290	.00426	.00435	0.00468
S- 24	-0.11	TOT ELEC	KWH	1.17	0.92	1.02	0.95	0.94	1.04	1.10	1.10	1.01	0.95	0.98	1.04	12.2
		TOT DEM	KW	.00491	.00452	.00408	.00290	.00295	.00334	.00334	.00329	.00309	.00304	.00447	.00457	0.00491
S- 30	0.46	TOT ELEC	KWH	1.10	0.86	0.97	0.90	0.89	0.98	1.04	1.04	0.95	0.90	0.92	0.98	11.5
		TOT DEM	KW	.00640	.00589	.00531	.00378	.00384	.00435	.00435	.00429	.00403	.00397	.00582	.00595	0.00640
BASE CASE : 244																
BASE	0.00	HT ELEC	KWH	0.73	0.46	0.43	0.28	0.13	0.06	0.00	0.02	0.09	0.28	0.50	0.64	3.6
		HT DEM	KW	.00608	.00608	.00602	.00517	.00395	.00340	.00188	.00298	.00359	.00462	.00608	.00608	0.00608
		REM ELEC	KWH	1.09	0.97	1.09	1.06	1.16	1.21	1.42	1.38	1.20	1.10	1.04	1.08	13.8
		REM DEM	KW	.00412	.00412	.00551	.00607	.00586	.00698	.00677	.00684	.00621	.00558	.00461	.00412	0.00698
		TOT ELEC	KWH	1.82	1.43	1.52	1.34	1.29	1.28	1.42	1.40	1.29	1.38	1.53	1.71	17.4
		TOT DEM	KW	.01015	.00913	.00893	.00771	.00690	.00802	.00731	.00731	.00670	.00832	.00985	.00934	0.01015
S- 3	-0.16	HT ELEC	KWH	0.77	0.49	0.46	0.30	0.14	0.07	0.00	0.02	0.10	0.29	0.53	0.67	3.8
		HT DEM	KW	.00608	.00608	.00602	.00517	.00395	.00340	.00188	.00298	.00359	.00462	.00608	.00608	0.00608
		REM ELEC	KWH	0.93	0.83	0.93	0.91	0.99	1.04	1.22	1.18	1.03	0.94	0.89	0.92	11.8
		REM DEM	KW	.00379	.00379	.00508	.00559	.00540	.00643	.00624	.00630	.00572	.00514	.00424	.00379	0.00643
		TOT ELEC	KWH	1.71	1.32	1.39	1.21	1.13	1.10	1.22	1.20	1.12	1.24	1.41	1.59	15.6
		TOT DEM	KW	.01012	.00911	.00891	.00769	.00688	.00799	.00729	.00729	.00668	.00830	.00982	.00931	0.01012
S- 6	-0.12	HT ELEC	KWH	0.84	0.53	0.50	0.33	0.15	0.07	0.00	0.02	0.10	0.32	0.58	0.73	4.2
		HT DEM	KW	.00608	.00608	.00602	.00517	.00395	.00340	.00188	.00298	.00359	.00462	.00608	.00608	0.00608
		REM ELEC	KWH	0.86	0.76	0.86	0.84	0.92	0.96	1.12	1.09	0.95	0.87	0.82	0.85	10.9
		REM DEM	KW	.00342	.00342	.00458	.00505	.00487	.00580	.00563	.00568	.00516	.00464	.00383	.00342	0.00580
		TOT ELEC	KWH	1.70	1.30	1.36	1.16	1.07	1.03	1.12	1.11	1.05	1.19	1.39	1.58	15.1
		TOT DEM	KW	.00937	.00843	.00825	.00712	.00637	.00740	.00675	.00675	.00618	.00768	.00909	.00862	0.00937
S- 15	1.09	HT ELEC	KWH	0.89	0.56	0.52	0.34	0.16	0.07	0.00	0.02	0.11	0.33	0.61	0.77	4.4
		HT DEM	KW	.00608	.00608	.00602	.00517	.00395	.00340	.00188	.00298	.00359	.00462	.00608	.00608	0.00608
		REM ELEC	KWH	0.75	0.66	0.75	0.73	0.80	0.83	0.98	0.95	0.82	0.76	0.71	0.74	9.5
		REM DEM	KW	.00303	.00303	.00406	.00447	.00432	.00514	.00499	.00504	.00457	.00411	.00339	.00303	0.00514
		TOT ELEC	KWH	1.64	1.23	1.27	1.07	0.95	0.91	0.98	0.97	0.93	1.09	1.32	1.51	13.9
		TOT DEM	KW	.00885	.00823	.00805	.00646	.00566	.00655	.00602	.00611	.00575	.00717	.00867	.00858	0.00885
S- 31	0.20	HT ELEC	KWH	0.50	0.40	0.42	0.32	0.23	0.21	0.14	0.17	0.21	0.31	0.41	0.45	3.8
		HT DEM	KW	.00608	.00608	.00602	.00517	.00395	.00340	.00188	.00298	.00359	.00462	.00608	.00608	0.00608
		REM ELEC	KWH	0.87	0.79	0.91	0.86	0.83	0.99	1.11	1.09	0.97	0.84	0.85	0.90	11.0
		REM DEM	KW	.00346	.00346	.00421	.00458	.00437	.00533	.00512	.00517	.00464	.00453	.00346	.00346	0.00533
		TOT ELEC	KWH	1.37	1.19	1.34	1.18	1.05	1.20	1.25	1.25	1.18	1.15	1.26	1.35	14.8
		TOT DEM	KW	.00973	.00905	.00885	.00710	.00623	.00720	.00662	.00671	.00632	.00788	.00954	.00944	0.00973
S- 60	-0.36	HT ELEC	KWH	0.84	0.53	0.50	0.33	0.15	0.07	0.00	0.02	0.10	0.32	0.58	0.73	4.2
		HT DEM	KW	.00608	.00608	.00602	.00517	.00395	.00340	.00188	.00298	.00359	.00462	.00608	.00608	0.00608
		REM ELEC	KWH	0.86	0.76	0.86	0.84	0.92	0.96	1.12	1.09	0.95	0.87	0.82	0.85	10.9
		REM DEM	KW	.00342	.00342	.00458	.00505	.00487	.00580	.00563	.00568	.00516	.00464	.00383	.00342	0.00580
		TOT ELEC	KWH	1.70	1.30	1.36	1.16	1.07	1.03	1.12	1.11	1.05	1.19	1.39	1.58	15.1
		TOT DEM	KW	.00937	.00843	.00825	.00712	.00637	.00740	.00675	.00675	.00618	.00768	.00909	.00862	0.00937
BASE CASE : 245																
BASE	0.00	TOT GAS	THM	0.0384	0.0243	0.0226	0.0148	0.0068	0.0032	0.0000	0.0009	0.0047	0.0144	0.0262	0.0334	0.190
		TOT ELEC	KWH	1.09	0.97	1.09	1.06	1.16	1.21	1.42	1.38	1.20	1.10	1.04	1.08	13.8
		TOT DEM	KW	.00412	.00412	.00551	.00607	.00586	.00698	.00677	.00684	.00621	.00558	.00461	.00412	0.00698
S- 3	-0.17	TOT GAS	THM	0.0406	0.0257	0.0239	0.0157	0.0072	0.0034	0.0000	0.0010	0.0050	0.0153	0.0277	0.0354	0.201
		TOT ELEC	KWH	0.93	0.83	0.93	0.91	0.99	1.04	1.22	1.18	1.03	0.94	0.89	0.92	11.8
		TOT DEM	KW	.00379	.00379	.00508	.00559	.00540	.00643	.00624	.00630	.00572	.00514	.00424	.00379	0.00643
S- 6	-0.13	TOT GAS	THM	0.0442	0.0280	0.0261	0.0171	0.0079	0.0037	0.0000	0.0011	0.0055	0.0166	0.0302	0.0385	0.219
		TOT ELEC	KWH	0.86	0.76	0.86	0.84	0.92	0.96	1.12	1.09	0.95	0.87	0.82	0.85	10.9
		TOT DEM	KW	.00342	.00342	.00458	.00505	.00487	.00580	.00563	.00568	.00516	.00464	.00383	.00342	0.00580
S- 15	1.08	TOT GAS	THM	0.0467	0.0296	0.0275	0.0180	0.0083	0.0039	0.0000	0.0012	0.0058	0.0176	0.0319	0.0407	0.231
		TOT ELEC	KWH	0.75	0.66	0.75	0.73	0.79	0.83	0.97	0.95	0.82	0.76	0.71	0.74	9.5
		TOT DEM	KW	.00303	.00303	.00406	.00447	.00432	.00514	.00499	.00504	.00457	.00411	.00339	.00303	0.00514
S- 31	0.19	TOT GAS	THM	0.0263	0.0208	0.0222	0.0170	0.0119	0.0111	0.0071	0.0087	0.0111	0.0164	0.0218	0.0238	0.198
		TOT ELEC	KWH	0.87	0.79	0.91	0.86	0.83	0.99	1.11	1.09	0.97	0.84	0.85	0.90	11.0
		TOT DEM	KW	.00346	.00346	.00421	.00458	.00437	.00533	.00512	.00517	.00464	.00453	.00346	.00346	0.00533
S- 60	-0.37	TOT GAS	THM	0.0442	0.0280	0.0261	0.0171	0.0079	0.0037	0.0000	0.0011	0.0055	0.0166	0.0302	0.0385	0.219
		TOT ELEC	KWH	0.86	0.76	0.86	0.84	0.92	0.96	1.12	1.09	0.95	0.87	0.82	0.85	10.9
		TOT DEM	KW	.00342	.00342	.00458	.00505	.00487	.00580	.00563	.00568	.00516	.00464	.00383	.00342	0.00580
BASE CASE : 246																
BASE	0.00	HT ELEC	KWH	0.45	0.29	0.27	0.18	0.08	0.04	0.00	0.01	0.06	0.17	0.31	0.40	2.3
		HT DEM	KW	.00435	.00435	.00435	.00435	.00313	.00218	.00070	.00165	.00252	.00435	.00435	.00435	0.00435
		REM ELEC	KWH	1.03	0.91	1.03	1.00	1.09	1.14	1.34	1.30	1.13	1.04	0.98	1.01	13.0
		REM DEM	KW	.00375	.00375	.00472	.00526	.00502	.00599	.00599	.00605	.00569	.00466	.00442	.00369	0.00605
		TOT ELEC	KWH	1.48	1.20	1.29	1.18	1.17	1.18	1.34	1.31	1.19	1.21	1.29	1.41	15.3
		TOT DEM	KW	.00753	.00745	.00745	.00648	.00489	.00580	.00580	.00587	.00550	.00708	.00745	.00745	0.00753

STRAT SET	CONST COST $/SF	ENERGY FORM	UNITS (PER SF)	JAN	FEB	MAR	APR	MAY	JUN	JUL	AUG	SEP	OCT	NOV	DEC	ANNUAL
BASE CASE : 246 (CONTINUED)																
S- 3	-0.10	HT ELEC	KWH	0.49	0.31	0.29	0.19	0.09	0.04	0.00	0.01	0.06	0.18	0.33	0.43	2.4
		HT DEM	KW	.00435	.00435	.00435	.00435	.00313	.00218	.00070	.00165	.00252	.00435	.00435	.00435	0.00435
		REM ELEC	KWH	0.88	0.78	0.88	0.85	0.93	0.98	1.14	1.11	0.97	0.89	0.83	0.87	11.1
		REM DEM	KW	.00345	.00345	.00434	.00485	.00462	.00551	.00551	.00557	.00524	.00429	.00407	.00340	0.00557
		TOT ELEC	KWH	1.37	1.09	1.16	1.04	1.02	1.02	1.14	1.12	1.03	1.07	1.17	1.29	13.5
		TOT DEM	KW	.00748	.00741	.00741	.00643	.00486	.00576	.00576	.00583	.00546	.00703	.00741	.00741	0.00748
S- 6	-0.05	HT ELEC	KWH	0.52	0.33	0.31	0.20	0.09	0.04	0.00	0.01	0.06	0.20	0.35	0.45	2.6
		HT DEM	KW	.00435	.00435	.00435	.00435	.00313	.00218	.00070	.00165	.00252	.00435	.00435	.00435	0.00435
		REM ELEC	KWH	0.79	0.70	0.79	0.77	0.84	0.88	1.03	1.00	0.87	0.80	0.75	0.78	10.0
		REM DEM	KW	.00311	.00311	.00392	.00437	.00417	.00497	.00497	.00502	.00472	.00387	.00366	.00306	0.00502
		TOT ELEC	KWH	1.31	1.03	1.10	0.97	0.93	0.92	1.03	1.01	0.93	1.00	1.10	1.23	12.6
		TOT DEM	KW	.00675	.00668	.00668	.00581	.00439	.00520	.00520	.00526	.00493	.00635	.00668	.00668	0.00675
S- 15	0.55	HT ELEC	KWH	0.56	0.35	0.33	0.22	0.10	0.05	0.00	0.01	0.07	0.21	0.38	0.49	2.8
		HT DEM	KW	.00435	.00435	.00435	.00435	.00313	.00218	.00070	.00165	.00252	.00435	.00435	.00435	0.00435
		REM ELEC	KWH	0.69	0.61	0.69	0.67	0.73	0.76	0.90	0.87	0.76	0.70	0.65	0.68	8.7
		REM DEM	KW	.00277	.00277	.00349	.00389	.00371	.00443	.00443	.00447	.00420	.00344	.00326	.00273	0.00447
		TOT ELEC	KWH	1.24	0.96	1.01	0.88	0.83	0.81	0.90	0.88	0.83	0.90	1.03	1.16	11.5
		TOT DEM	KW	.00653	.00646	.00646	.00562	.00424	.00503	.00503	.00509	.00477	.00614	.00646	.00646	0.00653
S- 31	0.27	HT ELEC	KWH	0.42	0.28	0.28	0.16	0.08	0.06	0.03	0.04	0.07	0.14	0.29	0.31	2.2
		HT DEM	KW	.00435	.00435	.00435	.00435	.00313	.00218	.00070	.00165	.00252	.00435	.00435	.00435	0.00435
		REM ELEC	KWH	0.75	0.67	0.78	0.74	0.77	0.89	0.99	0.99	0.87	0.76	0.73	0.78	9.7
		REM DEM	KW	.00304	.00300	.00348	.00361	.00361	.00435	.00426	.00431	.00383	.00383	.00326	.00313	0.00435
		TOT ELEC	KWH	1.17	0.95	1.05	0.90	0.85	0.95	1.02	1.03	0.94	0.90	1.02	1.09	11.9
		TOT DEM	KW	.00715	.00708	.00708	.00601	.00372	.00436	.00429	.00436	.00386	.00658	.00708	.00708	0.00715
S- 60	-0.30	HT ELEC	KWH	0.52	0.33	0.31	0.20	0.09	0.04	0.00	0.01	0.06	0.20	0.35	0.45	2.6
		HT DEM	KW	.00435	.00435	.00435	.00435	.00313	.00218	.00070	.00165	.00252	.00435	.00435	.00435	0.00435
		REM ELEC	KWH	0.79	0.70	0.79	0.77	0.84	0.88	1.03	1.00	0.87	0.80	0.75	0.78	10.0
		REM DEM	KW	.00311	.00311	.00392	.00437	.00417	.00497	.00497	.00502	.00472	.00387	.00366	.00306	0.00502
		TOT ELEC	KWH	1.31	1.03	1.10	0.97	0.93	0.92	1.03	1.01	0.93	1.00	1.10	1.23	12.6
		TOT DEM	KW	.00675	.00668	.00668	.00581	.00439	.00520	.00520	.00526	.00493	.00635	.00668	.00668	0.00675
BASE CASE : 247																
BASE	0.00	TOT GAS	THM	0.0238	0.0151	0.0140	0.0092	0.0042	0.0020	0.0000	0.0006	0.0030	0.0090	0.0163	0.0208	0.118
		TOT ELEC	KWH	1.03	0.91	1.03	1.00	1.09	1.14	1.34	1.30	1.13	1.04	0.98	1.01	13.0
		TOT DEM	KW	.00375	.00375	.00472	.00526	.00502	.00599	.00599	.00605	.00569	.00466	.00442	.00369	0.00605
S- 3	-0.11	TOT GAS	THM	0.0257	0.0163	0.0151	0.0099	0.0046	0.0022	0.0000	0.0006	0.0032	0.0097	0.0175	0.0224	0.127
		TOT ELEC	KWH	0.88	0.78	0.88	0.85	0.93	0.98	1.14	1.11	0.97	0.89	0.83	0.87	11.1
		TOT DEM	KW	.00345	.00345	.00434	.00485	.00462	.00551	.00551	.00557	.00524	.00429	.00407	.00340	0.00557
S- 6	-0.06	TOT GAS	THM	0.0273	0.0173	0.0161	0.0105	0.0049	0.0023	0.0000	0.0007	0.0034	0.0103	0.0186	0.0238	0.135
		TOT ELEC	KWH	0.79	0.70	0.79	0.77	0.84	0.88	1.03	1.00	0.87	0.80	0.75	0.78	10.0
		TOT DEM	KW	.00311	.00311	.00392	.00437	.00417	.00497	.00497	.00502	.00472	.00387	.00366	.00306	0.00502
S- 15	0.49	TOT GAS	THM	0.0293	0.0186	0.0173	0.0113	0.0052	0.0025	0.0000	0.0007	0.0036	0.0110	0.0200	0.0255	0.145
		TOT ELEC	KWH	0.69	0.61	0.69	0.67	0.73	0.76	0.90	0.87	0.76	0.70	0.65	0.68	8.7
		TOT DEM	KW	.00277	.00277	.00349	.00389	.00371	.00443	.00443	.00447	.00420	.00344	.00326	.00273	0.00447
S- 31	0.26	TOT GAS	THM	0.0219	0.0148	0.0145	0.0086	0.0042	0.0031	0.0015	0.0023	0.0035	0.0073	0.0153	0.0164	0.113
		TOT ELEC	KWH	0.75	0.67	0.78	0.74	0.77	0.89	0.99	0.99	0.87	0.76	0.73	0.78	9.7
		TOT DEM	KW	.00304	.00304	.00348	.00361	.00361	.00435	.00426	.00431	.00383	.00383	.00326	.00313	0.00435
S- 60	-0.31	TOT GAS	THM	0.0273	0.0173	0.0161	0.0105	0.0049	0.0023	0.0000	0.0007	0.0034	0.0103	0.0186	0.0238	0.135
		TOT ELEC	KWH	0.79	0.70	0.79	0.77	0.84	0.88	1.03	1.00	0.87	0.80	0.75	0.78	10.0
		TOT DEM	KW	.00311	.00311	.00392	.00437	.00417	.00497	.00497	.00502	.00472	.00387	.00366	.00306	0.00502
BASE CASE : 248																
BASE	0.00	TOT ELEC	KWH	1.56	1.22	1.36	1.26	1.25	1.38	1.46	1.46	1.34	1.26	1.30	1.38	16.2
		TOT DEM	KW	.00620	.00570	.00515	.00366	.00372	.00422	.00422	.00415	.00391	.00384	.00564	.00577	0.00620
S- 1	0.09	TOT ELEC	KWH	1.45	1.13	1.27	1.18	1.16	1.28	1.36	1.36	1.25	1.18	1.21	1.28	15.1
		TOT DEM	KW	.00589	.00542	.00489	.00348	.00353	.00401	.00401	.00395	.00371	.00365	.00536	.00548	0.00589
S- 5	0.14	TOT ELEC	KWH	1.30	1.01	1.13	1.05	1.04	1.15	1.22	1.22	1.12	1.05	1.08	1.15	13.5
		TOT DEM	KW	.00521	.00479	.00432	.00307	.00313	.00354	.00354	.00349	.00328	.00323	.00474	.00485	0.00521
S- 14	1.45	TOT ELEC	KWH	1.18	0.92	1.03	0.96	0.95	1.05	1.11	1.11	1.02	0.96	0.98	1.05	12.3
		TOT DEM	KW	.00496	.00456	.00412	.00293	.00298	.00337	.00337	.00332	.00312	.00308	.00451	.00461	0.00496
S- 24	-0.11	TOT ELEC	KWH	1.30	1.01	1.13	1.05	1.04	1.15	1.22	1.22	1.12	1.05	1.08	1.15	13.5
		TOT DEM	KW	.00521	.00479	.00432	.00307	.00313	.00354	.00354	.00349	.00328	.00323	.00474	.00485	0.00521
S- 30	0.46	TOT ELEC	KWH	1.17	0.92	1.02	0.95	0.94	1.04	1.10	1.10	1.01	0.95	0.98	1.04	12.2
		TOT DEM	KW	.00670	.00616	.00556	.00395	.00402	.00456	.00456	.00449	.00422	.00415	.00610	.00623	0.00670

(HT ELEC=ELECTRIC CONSUMPTION FOR SPACE AND WATER HEATING. REM ELEC=REMAINING ELECTRIC CONSUMPTION (NOT USED FOR HEATING). TOT ELEC=TOTAL ELECTRIC CONSUMPTION. HT DEM=PEAK ELECTRIC DEMAND FOR SPACE AND WATER HEATING. REM DEM=PEAK ELECTRIC DEMAND FOR REMAINING LOAD (NOT HEATING). TOT DEM=TOTAL PEAK ELECTRIC DEMAND. GAS=TOTAL GAS CONSUMPTION. ANNUAL FIGURES ARE SUMS OF MONTHLY CONSUMPTION VALUES OR ARE YEARLY MAXIMUMS OF MONTHLY PEAK DEMAND VALUES.)

Hot/Cold Climate

Identify Region

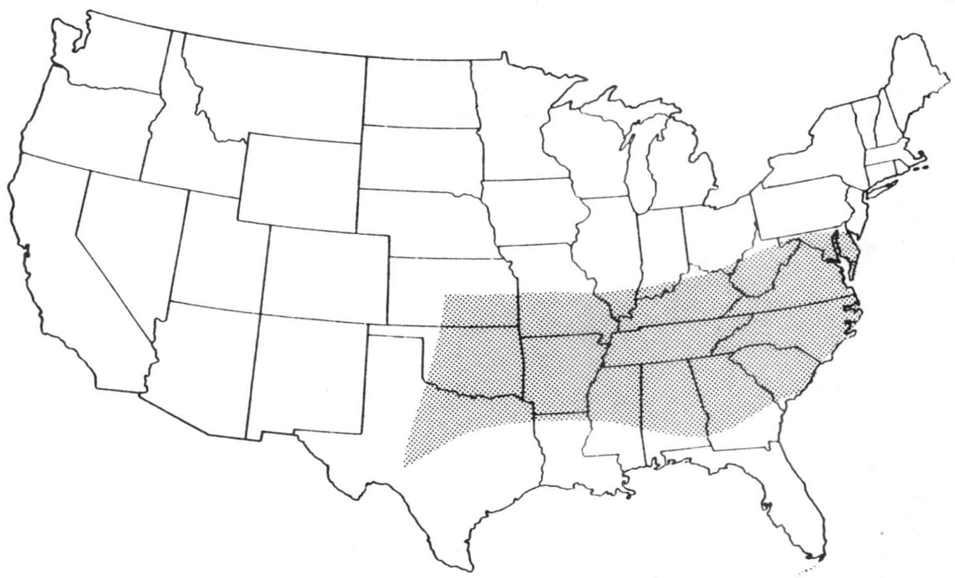

CITIES IN REGION

AL	Birmingham Huntsville Montgomery
AR	Fort Smith Little Rock
DC	Washington
DE	Dover
GA	Atlanta Augusta Columbus Macon
IL	Cairo East St. Louis
IN	Evansville
KA	Topeka Wichita
KY	Lexington Louisville
MD	Baltimore Hagerstown
MO	Kansas City Springfield St. Louis
MS	Meridian
NC	Asheville Charlotte Durham Greensboro Wilmington Winston-Salem
OK	Oklahoma City Tulsa
SC	Columbia Greenville
TN	Chattanooga Knoxville Memphis Nashville
TX	Abilene Dallas San Angelo Texarkana Wichita Falls
VA	Norfolk Richmond Roanoke
WV	Charleston Huntington

BEGIN YOUR 5 LEVELS OF STRATEGY SELECTION.

Note: See Section II for a full discussion of the five climate regions used in this manual.

Make sure your design comes up to current good practice, which includes meeting ASHRAE 90-75 Standards. The table below identifies good practice that meets 90-75 Standards as well as common design assumptions applicable to all building sizes.

ALL BUILDINGS		
	GOOD PRACTICE	**DESIGN ASSUMPTIONS**
SHELL	o Perimeter Insulation (R-5) o Infiltration: 0 when HVAC is operating (pressurized building)	o Ventilation: 14 cfm/person during occupied hours
HVAC	o Air conditioning performs to ASHRAE 90-75 (year 1980) minimum efficiencies o Boilers perform with 65% efficiency (no waste heat to space) o Heat pumps perform to ASHRAE 90-75 (year 1980) minimum efficiencies o Controls: · 72 degrees heating, 74 degrees cooling occupied periods · 55 degrees heating, 95 degrees cooling unoccupied periods · 2 degrees temperature float · outside air temperature reset for hot water boilers · low leakage dampers (less than 3%) · zone control for all heat pump and multi-zone systems	
INTERNAL LOAD		o Occupant density: 200 square feet per person o Lights: 2.25 watts/sq. ft. o Office equipment: .75 watts/sq. ft.
SCHEDULE	o HVAC comes on one hour before occupied period, turned to setback during unoccupied period	o Inside lighting: 100% during operating hours, 5% at other times o Outside lighting: 100% 6pm til midnight

Note: See Section II, page 179, for detailed information on these Level 1 strategies and assumptions.

Some aspects of ASHRAE 90-75 and current **good practice** vary according to building size and HVAC system type. Make sure your design comes up to these criteria. You may want to come back to this table as you explore HVAC options in Level 2.

		BUILDING SIZE		
		SMALL 2000-10000 sq ft	MEDIUM 10000-25000 sq ft	LARGE 25000-50000 sq ft
SHELL	ROOF INSULATION	R-8.3	R-8.3	R-8.3
	WALL INSULATION	R-7.5	R-7.5	R-7.5
	GLAZING	DOUBLE GLAZED WITH CLEAR GLASS	DOUBLE GLAZED WITH TINTED GLASS	DOUBLE GLAZED WITH TINTED GLASS
	AIR CHANGES	1/2 AC/HR UNOCCUPIED	1/3 AC/HR UNOCCUPIED	1/3 AC/HR UNOCCUPIED

HVAC	SMALL			MEDIUM			LARGE		
	PACKAGED TERMINAL AIR CONDITIONERS	ROOFTOP DX	HEAT PUMP	CONSTANT VOLUME	VARIABLE AIR VOLUME	SPLIT HEAT PUMP	CONSTANT VOLUME	VARIABLE AIR VOLUME	HYDRONIC HEAT PUMP
TEMPERATURE CONTROL	ZONE	OA RESET	ZONE	OA RESET	FIXED	ZONE	OA RESET	FIXED	ZONE
AIR CONTROL	FIXED VOLUME	ENTHALPY ECONOMIZER	FIXED VOLUME	ENTHALPY ECONOMIZER	ENTHALPY ECONOMIZER	ENTHALPY ECONOMIZER	ENTHALPY ECONOMIZER	ENTHALPY ECONOMIZER	FIXED VOLUME

Some common **design assumptions** also vary according to building size and no. of stories.

		SMALL	MEDIUM		LARGE	
		1 STORY	1 STORY	2 STORY	2 STORY	4 STORY ①
INT. LOAD	ELEVATOR & OUTDOOR LIGHTS	1345 WATTS	4035 WATTS	4535 WATTS	7860 WATTS	9360 WATTS
SCHEDULE	OCCUPANCY	80% 8am - 6pm Monday - Friday	80% 8am - 6pm Monday - Friday		80% 8am - 6pm: M-F 100% 9am - 1pm: SAT	

GO TO THE LEVEL 2 PAGE SPECIFIC TO YOUR BUILDING SIZE & NO. OF STORIES.

Note ① For 3 or 5 story structures, use "4-story" classification.

LEVEL 2 — MAKE INITIAL CHOICES

SMALL/1-STORY

Project ______________________
date __________ bldg. sq. ft. __________

The building characteristic choices shown here make a **big** difference in energy cost savings. Investigate these choices within the requirements of your design.

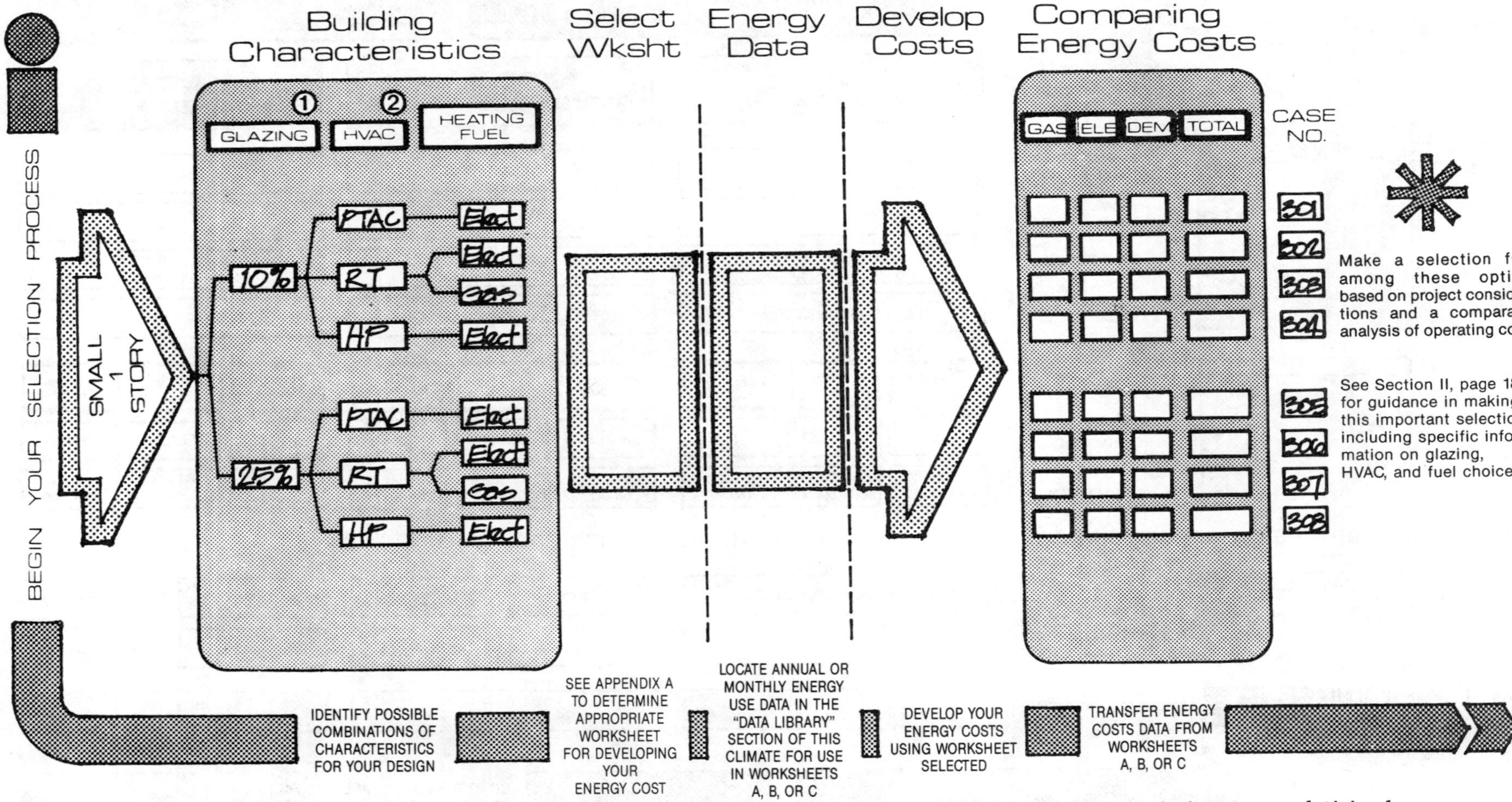

Make a selection from among these options based on project considerations and a comparative analysis of operating costs.

See Section II, page 189, for guidance in making this important selection, including specific information on glazing, HVAC, and fuel choices.

Note: See Section II, page 189, for detailed information on these Level 2 strategies and the use of this form.

① For glazing <17% use 10%, for >17% use 25%.

② PTAC = Packaged Terminal A/C, RT = Rooftop DX, HP = Packaged Terminal Heat Pump

Project ______________________________

date ____________ bldg. sq. ft. ____________

Locate the row of strategy set options that apply to the building case you chose in Level 2. Select as many of these Level 3 strategy set options as you wish for valuation.

STRATEGY SET OPTIONS

SMALL 1 STORY

SELECTED LEVEL 2 CASE NUMBER	SIMPLEST LIGHTING IMPROVEMTS	MODEST LIGHTING & MECHANICAL IMPROVEMTS	MAJOR LIGHTING & MECHANICAL IMPROVEMTS	DAYLIGHT DIMMING CONTROLS AND LIGHTING IMPROVEMTS
301	1	10 or 25	20	
302	3	7 or 61	16	
303	3	7 or 61	16	
304	1	12 or 28	22	
305	2	11 or 26	21	35
306	4	8 or 27	17	32
307	4	8 or 27	17	32
308	2	13 or 29	23	36

DESCRIPTOR KEY

1. High eff. ballasts over all bldg.
2. High eff. ballasts over all bldg. + tinted glazing.
3. High eff. ballasts over all bldg. + HVAC downsized for ballasts.
4. High eff. ballasts over all bldg. + HVAC downsized for ballasts + tinted glazing + HVAC downsized for tinted glass.
7. High eff. ballasts and lamps over all bldg. + HVAC downsized for ballasts + cooling COP improvement.
8. High eff. ballasts and lamps over all bldg. + HVAC downsized for ballasts + cooling COP improvement + tinted glazing + HVAC downsized for tinted glass.
10. High eff. ballasts and lamps over all bldg. + cooling COP improvement.
11. High eff. ballasts and lamps over all bldg. + cooling COP improvement + tinted glazing.
12. High eff. ballasts and lamps over all bldg. + heat pump COP improvement.
13. High eff. ballasts and lamps over all bldg. + heat pump COP improvement + tinted glazing.
16. High eff. ballasts, lamps and fixtures over all bldg. + HVAC downsized for ballasts and fixtures + cooling COP improvement.
17. High eff. ballasts, lamps and fixtures over all bldg. + HVAC downsized for ballasts and fixtures + cooling COP improvement + tinted glazing + HVAC downsized for tinted glass.
20. High eff. ballasts, lamps and fixtures over all bldg. + cooling COP improvement.
21. High eff. ballasts, lamps and fixtures over all bldg. + cooling COP improvement + tinted glazing.
22. High eff. ballasts, lamps and fixtures over all bldg. + heat pump COP improvement.
23. High eff. ballasts, lamps and fixtures over all bldg. + heat pump COP improvement + tinted glazing.
25. Non-uniform lighting + high eff. ballasts over all bldg. + cooling COP improvement.
61. Non-uniform lighting + high eff. ballasts over all bldg. + HVAC downsized for ballasts + cooling COP improvement.
26. Non-uniform lighting + high eff. ballasts over all bldg. + cooling COP improvement + tinted glazing.
27. Non-uniform lighting + high eff. ballasts over all bldg. + HVAC downsized for ballasts + cooling COP improvement + tinted glazing + HVAC downsized for tinted glass.
28. Non-uniform lighting + high eff. ballasts over all bldg. + heat pump COP improvement.
29. Non-uniform lighting + high eff. ballasts over all bldg. + heat pump COP improvement + tinted glazing.
32. Perimeter daylighting dimmers + high eff. ballasts over all bldg. + HVAC downsized for ballasts + cooling COP improvement.
35. Perimeter daylighting dimmers + high eff. ballasts over all bldg. + cooling COP improvement.
36. Perimeter daylighting dimmers + high eff. ballasts over all bldg. + heat pump COP improvement.

GO TO LEVEL 3 ANALYSIS SHEET.

Note: See Section II, page 209, for detailed descriptions and discussions of these and other Level 3 strategy sets.

LEVEL 2–

MAKE INITIAL CHOICES

MEDIUM/1-STORY

Project______________________________

date____________ bldg. sq. ft. ____________

The building characteristic choices shown here make a **big** difference in energy cost savings. Investigate these choices within the requirements of your design.

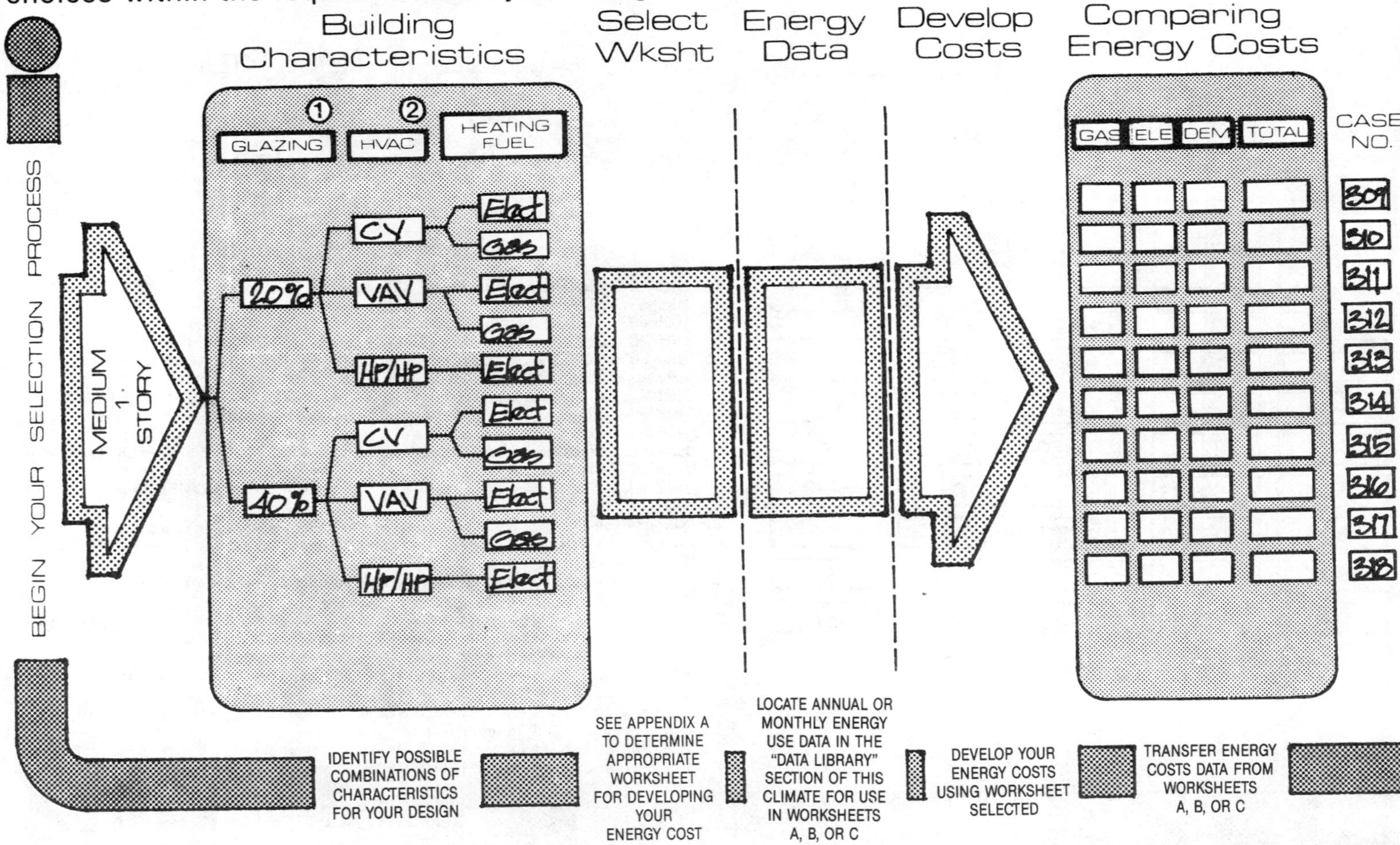

Make a selection from among these options based on project considerations and a comparative analysis of operating costs.

See Section II, page 189, for guidance in making this important selection, including specific information on glazing, HVAC, and fuel choices.

Note: See Section II, page 189, for detailed information on these Level 2 strategies and the use of this form.

① For glazing <30% use 20%, for >30% use 40%.

② CV=Constant Volume, VAV=Variable Air Volume, HHP=Split System Heat Pump.

MEDIUM/1-STORY

Project ______________________________
date ____________ bldg. sq. ft. ____________

Locate the row of strategy set options that apply to the building case you chose in Level 2. Select as many of these Level 3 strategy set options as you wish for valuation.

STRATEGY SET OPTIONS

MEDIUM 1 STORY

SELECTED LEVEL 2 CASE NUMBER	SIMPLEST LIGHTING IMPROVEMTS	MODEST LIGHTING & MECHANICAL IMPROVEMTS	MAJOR LIGHTING & MECHANICAL IMPROVEMTS	DAYLIGHT DIMMING CONTROLS AND LIGHTING IMPROVEMTS	SKYLIGHT'G, DAYLIGHT DIMMING CONTROLS & LIGHTING IMPROVEMTS
309	3	7 or 61	16	32 or 45 or 40	50
310	3	7 or 61	16	32 or 45 or 40	50
311	3	7 or 61	16	32 or 45 or 40	50
312	3	7 or 61	16	32 or 45 or 40	50
313	3	9 or 62	18	33 or 46 or 41	51
314	3	7 or 61	16	32 or 45 or 40	50
315	3	7 or 61	16	32 or 45 or 40	50
316	3	7 or 61	16	32 or 45 or 40	50
317	3	7 or 61	16	32 or 45 or 40	50
318	3	9 or 62	18	33 or 46 or 41	51

DESCRIPTOR KEY

3. High eff. ballasts over all bldg. + HVAC downsized for ballasts.
7. High eff. ballasts and lamps over all bldg. + HVAC downsized for ballasts + cooling COP improvement.
9. High eff. ballasts and lamps over all bldg. + HVAC downsized for ballasts + heat pump COP improvement.
16. High eff. ballasts, lamps and fixtures over all bldg. + HVAC downsized for ballasts and fixtures + cooling COP improvement.
18. High eff. ballasts, lamps and fixtures over all bldg. + HVAC downsized for ballasts and fixtures + heat pump COP improvement.
61. Non-uniform lighting + high eff. ballasts over all bldg. + HVAC downsized for ballasts + cooling COP improvement.
62. Non-uniform lighting + high eff. ballasts over all bldg. + HVAC downsized for ballasts + heat pump COP improvement.
32. Perimeter daylighting dimmers + high eff. ballasts over all bldg. + HVAC downsized for ballasts + cooling COP improvement.
33. Perimeter daylighting dimmers + high eff. ballasts over all bldg. + HVAC downsized for ballasts + heat pump COP improvement.
40. Perimeter daylighting dimmers + high eff. ballasts over all bldg. + high eff. lamps and fixtures in core + HVAC downsized for ballasts and fixtures + cooling COP improvement.
41. Perimeter daylighting dimmers + high eff. ballasts over all bldg. + high eff. lamps and fixtures in core + HVAC downsized for ballasts and fixtures + heat pump COP improvement.
45. Perimeter daylighting dimmers + high eff. ballasts over all bldg. + high eff. lamps in core + HVAC downsized for ballasts + cooling COP improvement.
46. Perimeter daylighting dimmers + high eff. ballasts over all bldg. + high eff. lamps in core + HVAC downsized for ballasts + heat pump COP improvement.
50. Perimeter daylighting dimmers + skylights with dimmers + high eff. ballasts over all bldg. + HVAC downsized for ballasts + cooling COP improvement.
51. Perimeter daylighting dimmers + skylights with dimmers + high eff. ballasts over all bldg. + HVAC downsized for ballasts + heat pump COP improvement.

GO TO LEVEL 3 ANALYSIS SHEET.

Note: See Section II, page 209, for detailed descriptions and discussions of these and other Level 3 strategy sets.

MEDIUM/2-STORY

Project______________________________

date____________ bldg. sq. ft. ____________

The building characteristic choices shown here make a **big** difference in energy cost savings. Investigate these choices within the requirements of your design.

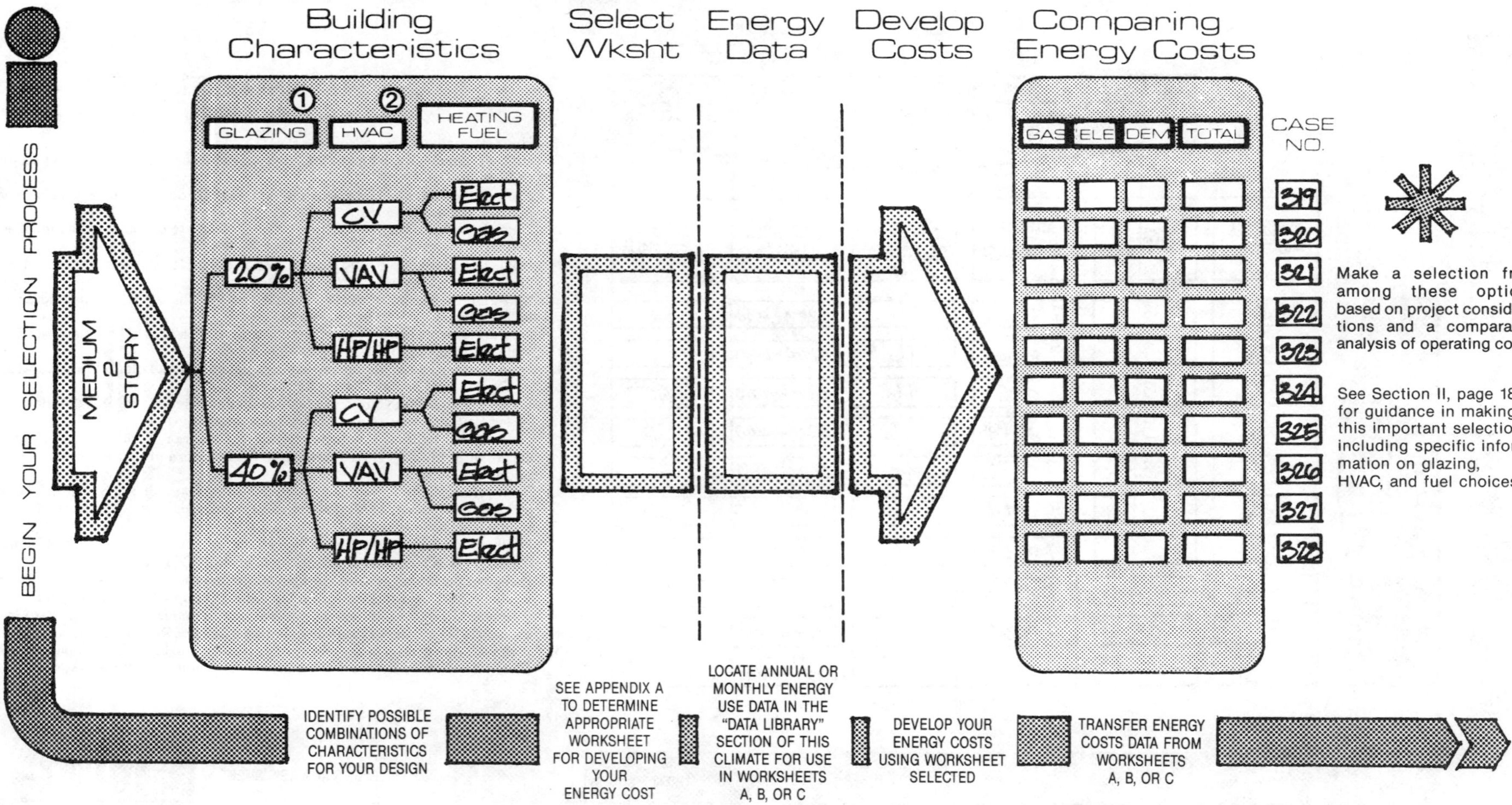

Make a selection from among these options based on project considerations and a comparative analysis of operating costs.

See Section II, page 189, for guidance in making this important selection, including specific information on glazing, HVAC, and fuel choices.

Note: See Section II, page 189, for detailed information on these Level 2 strategies and the use of this form.

① For glazing <30% use 20%, for >30% use 40%.

② CV=Constant Volume, VAV=Variable Air Volume, HHP=Split System Heat Pump.

Project ______________________________
date ____________ bldg. sq. ft. ____________

Locate the row of strategy set options that apply to the building case you chose in Level 2. Select as many of these Level 3 strategy set options as you wish for valuation.

STRATEGY SET OPTIONS

MEDIUM 2 STORY

SELECTED LEVEL 2 CASE NUMBER	SIMPLEST LIGHTING IMPROVEMTS	MODEST LIGHTING & MECHANICAL IMPROVEMTS	MAJOR LIGHTING & MECHANICAL IMPROVEMTS	DAYLIGHT DIMMING CONTROLS AND LIGHTING IMPROVEMTS
319	3	7 or 61	16	32
320	3	7 or 61	16	32
321	3	7 or 61	16	32
322	3	7 or 61	16	32
323	3	9 or 62	18	33
324	3	7 or 61	16	32
325	3	7 or 61	16	32
326	3	7 or 61	16	32
327	3	7 or 61	16	32
328	3	9 or 62	18	33

DESCRIPTOR KEY

3. High eff. ballasts over all bldg. + HVAC downsized for ballasts.
7. High eff. ballasts and lamps over all bldg. + HVAC downsized for ballasts + cooling COP improvement.
9. High eff. ballasts and lamps over all bldg. + HVAC downsized for ballasts + heat pump COP improvement.
16. High eff. ballasts, lamps and fixtures over all bldg. + HVAC downsized for ballasts and fixtures + cooling COP improvement.
18. High eff. ballasts, lamps and fixtures over all bldg. + HVAC downsized for ballasts and fixtures + heat pump COP improvement.
61. Non-uniform lighting + high eff. ballasts over all bldg. + HVAC downsized for ballasts + cooling COP improvement.
62. Non-uniform lighting + high eff. ballasts over all bldg. + HVAC downsized for ballasts + heat pump COP improvement.
32. Perimeter daylighting dimmers + high eff. ballasts over all bldg. + HVAC downsized for ballasts + cooling COP improvement.
33. Perimeter daylighting dimmers + high eff. ballasts over all bldg. + HVAC downsized for ballasts + heat pump COP improvement.

GO TO LEVEL 3 ANALYSIS SHEET.

Note: See Section II, page 209, for detailed descriptions and discussions of these and other Level 3 strategy sets.

LARGE/2-STORY

Project ______________________________
date __________ bldg. sq. ft. __________

The building characteristic choices shown here make a **big** difference in energy cost savings. Investigate these choices within the requirements of your design.

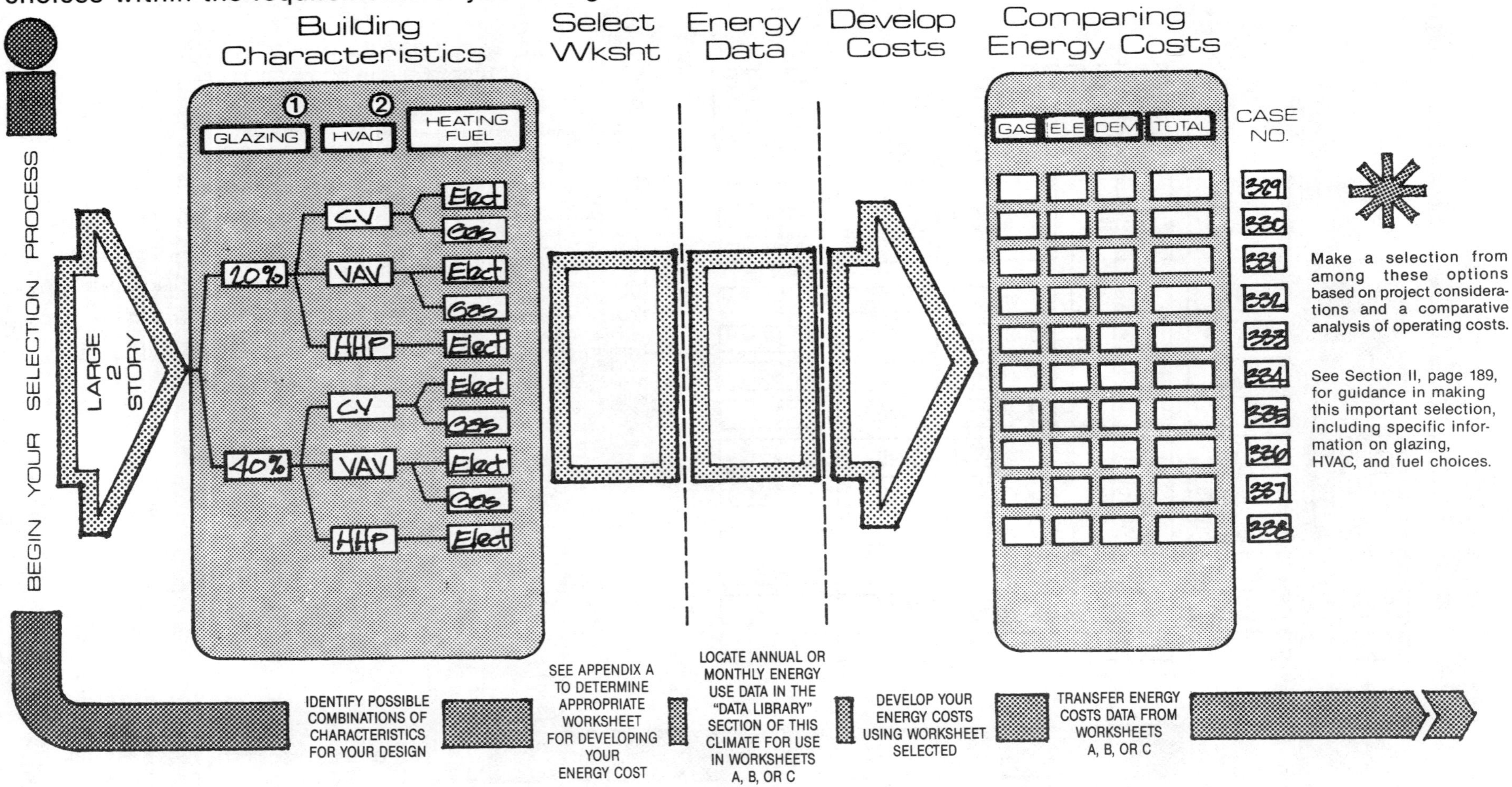

Note: See Section II, page 189, for detailed information on these Level 2 strategies and the use of this form.

① For glazing <30% use 20%, for >30% use 40%.

② CV=Constant Volume, VAV=Variable Air Volume, HHP=Hydronic Heat Pump.

Project ________________________
date ____________ bldg. sq. ft. ____________

Locate the row of strategy set options that apply to the building case you chose in Level 2. Select as many of these Level 3 strategy set options as you wish for valuation.

LARGE 2 STORY

STRATEGY SET OPTIONS

SELECTED LEVEL 2 CASE NUMBER	SIMPLEST LIGHTING IMPROVEMTS	MODEST LIGHTING & MECHANICAL IMPROVEMTS	MAJOR LIGHTING & MECHANICAL IMPROVEMTS	DAYLIGHT DIMMING CONTROLS AND LIGHTING IMPROVEMTS
329	3	7 or 61	16	32 or 45 or 40
330	3	7 or 61	16	32 or 45 or 40
331	3	7 or 61	16	32 or 45 or 40
332	3	7 or 61	16	32 or 45 or 40
333	1	12 or 28	22	36 or 48 or 43
334	3	7 or 61	16	32 or 45 or 40
335	3	7 or 61	16	32 or 45 or 40
336	3	7 or 61	16	32 or 45 or 40
337	3	7 or 61	16	32 or 45 or 40
338	1	12 or 28	22	36 or 48 or 43

DESCRIPTOR KEY

1. High eff. ballasts over all bldg.
3. High eff. ballasts over all bldg. + HVAC downsized for ballasts.
7. High eff. ballasts and lamps over all bldg. + HVAC downsized for ballasts + cooling COP improvement.
12. High eff. ballasts and lamps over all bldg. + heat pump COP improvement.
16. High eff. ballasts, lamps and fixtures over all bldg. + HVAC downsized for ballasts and fixtures + cooling COP improvement.
61. Non-uniform lighting + high eff. ballasts over all bldg. + HVAC downsized for ballasts + cooling COP improvement.
28. Non-uniform lighting + high eff. ballasts over all bldg. + heat pump COP improvement.
32. Perimeter daylighting dimmers + high eff. ballasts over all bldg. + HVAC downsized for ballasts + cooling COP improvement.
36. Perimeter daylighting dimmers + high eff. ballasts over all bldg. + heat pump COP improvement.
40. Perimeter daylighting dimmers + high eff. ballasts over all bldg. + high eff. lamps and fixtures in core + HVAC downsized for ballasts and fixtures + cooling COP improvement.
43. Perimeter daylighting dimmers + high eff. ballasts over all bldg. + high eff. lamps and fixtures in core + heat pump COP improvement.
45. Perimeter daylighting dimmers + high eff. ballasts over all bldg. + high eff. lamps in core + HVAC downsized for ballasts + cooling COP improvement.
48. Perimeter daylighting dimmers + high eff. ballasts over all bldg. + high eff. lamps in core + heat pump COP improvement.

GO TO LEVEL 3 ANALYSIS SHEET.

Note: See Section II, page 209, for detailed descriptions and discussions of these and other Level 3 strategy sets.

LARGE/4-STORY

Project__________________________________

date______________ bldg. sq. ft.______________

The building characteristic choices shown here make a **big** difference in energy cost savings. Investigate these choices within the requirements of your design.

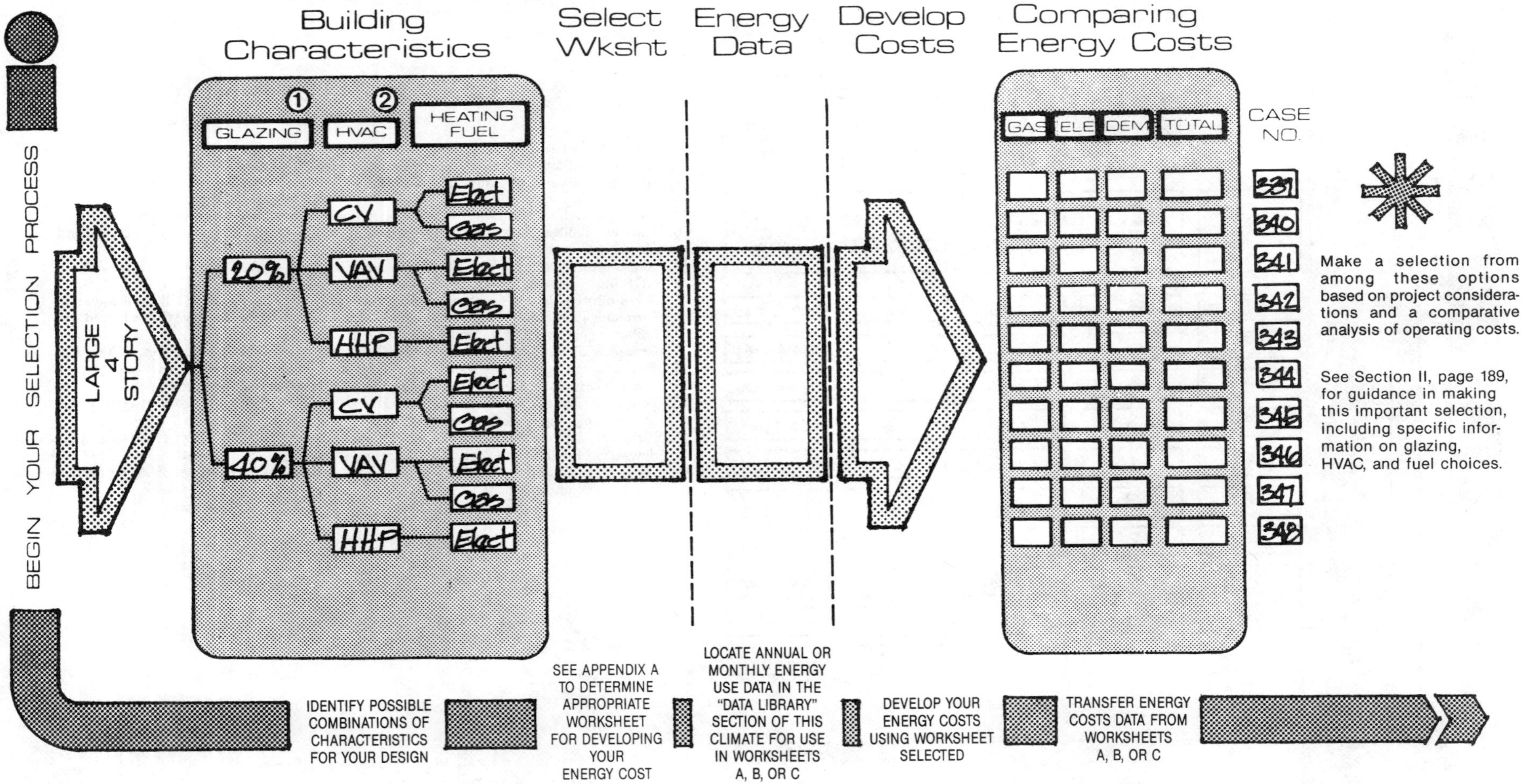

Note: See Section II, page 189, for detailed information on these Level 2 strategies and the use of this form.

① For glazing <30% use 20%, for >30% use 40%.

② CV=Constant Volume, VAV=Variable Air Volume, HHP=Hydronic Heat Pump.

Project ______________________

date __________ bldg. sq. ft. __________

Locate the row of strategy set options that apply to the building case you chose in Level 2. Select as many of these Level 3 strategy set options as you wish for valuation.

STRATEGY SET OPTIONS

LARGE 4 STORY

SELECTED LEVEL 2 CASE NUMBER	SIMPLEST LIGHTING IMPROVEMTS	MODEST LIGHTING & MECHANICAL IMPROVEMTS	MAJOR LIGHTING & MECHANICAL IMPROVEMTS	DAYLIGHT DIMMING CONTROLS AND LIGHTING IMPROVEMTS
339	3	7 or 61	16	32
340	3	7 or 61	16	32
341	3	7 or 61	16	32
342	3	7 or 61	16	32
343	1	12 or 28	22	36
344	3	7 or 61	16	32
345	3	7 or 61	16	32
346	3	7 or 61	16	32
347	3	7 or 61	16	32
348	1	12 or 28	22	36

DESCRIPTOR KEY

1. High eff. ballasts over all bldg.
3. High eff. ballasts over all bldg.+HVAC downsized for ballasts.
7. High eff. ballasts and lamps over all bldg.+HVAC downsized for ballasts+cooling COP improvement.
12. High eff. ballasts and lamps over all bldg.+heat pump COP improvement.
16. High eff. ballasts, lamps and fixtures over all bldg.+HVAC downsized for ballasts and fixtures+cooling COP improvement.
22. High eff. ballasts, lamps and fixtures over all bldg.+heat pump COP improvement.
61. Non-uniform lighting+high eff. ballasts over all bldg.+HVAC downsized for ballasts+cooling COP improvement.
28. Non-uniform lighting+high eff. ballasts over all bldg.+heat pump COP improvement.
32. Perimeter daylighting dimmers+high eff. ballasts over all bldg.+HVAC downsized for ballasts+cooling COP improvement.
36. Perimeter daylighting dimmers+high eff. ballasts over all bldg.+heat pump COP improvement.

GO TO LEVEL 3 ANALYSIS SHEET.

Note: See Section II, page 209, for detailed descriptions and discussions of these and other Level 3 strategy sets.

ANALYSIS/INSTRUCTIONS

For each strategy set you want to investigate, develop its construction costs using the upper route and its energy costs using the lower route. The combined information will allow you to select a strategy set for optimum economic benefit.

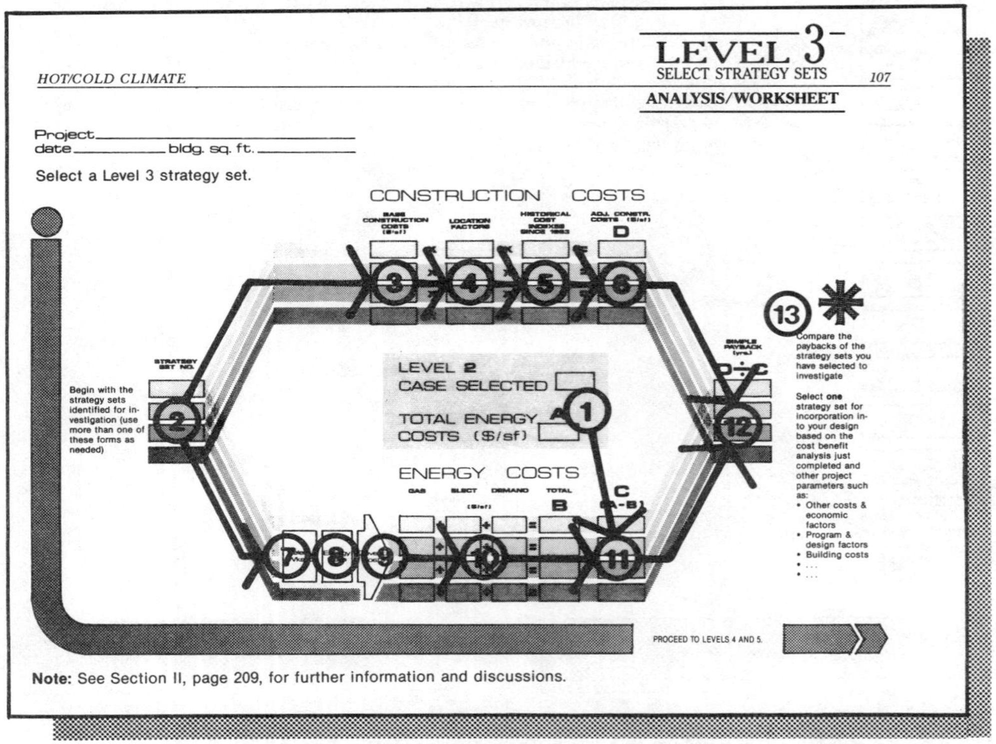

(1) Enter the Level 2 case selected and its total energy costs as developed in Level 2.

(2) As many as four strategy sets for any *one* Level 2 case can be analyzed on each sheet. You will need to use more sheets to analyze more than four strategy sets or to analyze strategy sets for another Level 2 case.

(3) Obtain *base construction costs* ($/SF) for each strategy set from the Data Library.

(4) Obtain *location factors* from commonly used cost estimating handbooks. Additional multipliers may be entered to account for above or below average construction costs due to specific project conditions.

(5) Obtain *historical cost indexes* (factors since 1983) from commonly used cost estimating handbooks.

(6) Multiply 3, 4, and 5 to obtain *adjusted construction costs* ($/SF). User may wish to use more refined construction cost data as supplied by a cost estimator for the specific project.

(7) Use the same worksheet you selected for use in Level 2.

(8) Locate energy use data for the selected strategy sets in the Data Library.

(9) Develop energy costs for each strategy set using the worksheet selected.

(10) Transfer energy cost information from worksheets A, B, or C.

(11) Find the energy savings from each strategy set by subtracting each "B" from "A."

(12) Calculate *simple payback* by dividing "D" (initial costs) by "C" (yearly savings). Alternative economic analysis techniques other than simple payback may be employed using information from columns "D" and "C."

(13) Cost factors to consider in making a decision on these strategy sets include:
- Negative construction costs (first cost savings)
- Positive construction costs
- Annual energy cost savings

Project______________________________
date____________ bldg. sq. ft. ____________

Select a Level 3 strategy set.

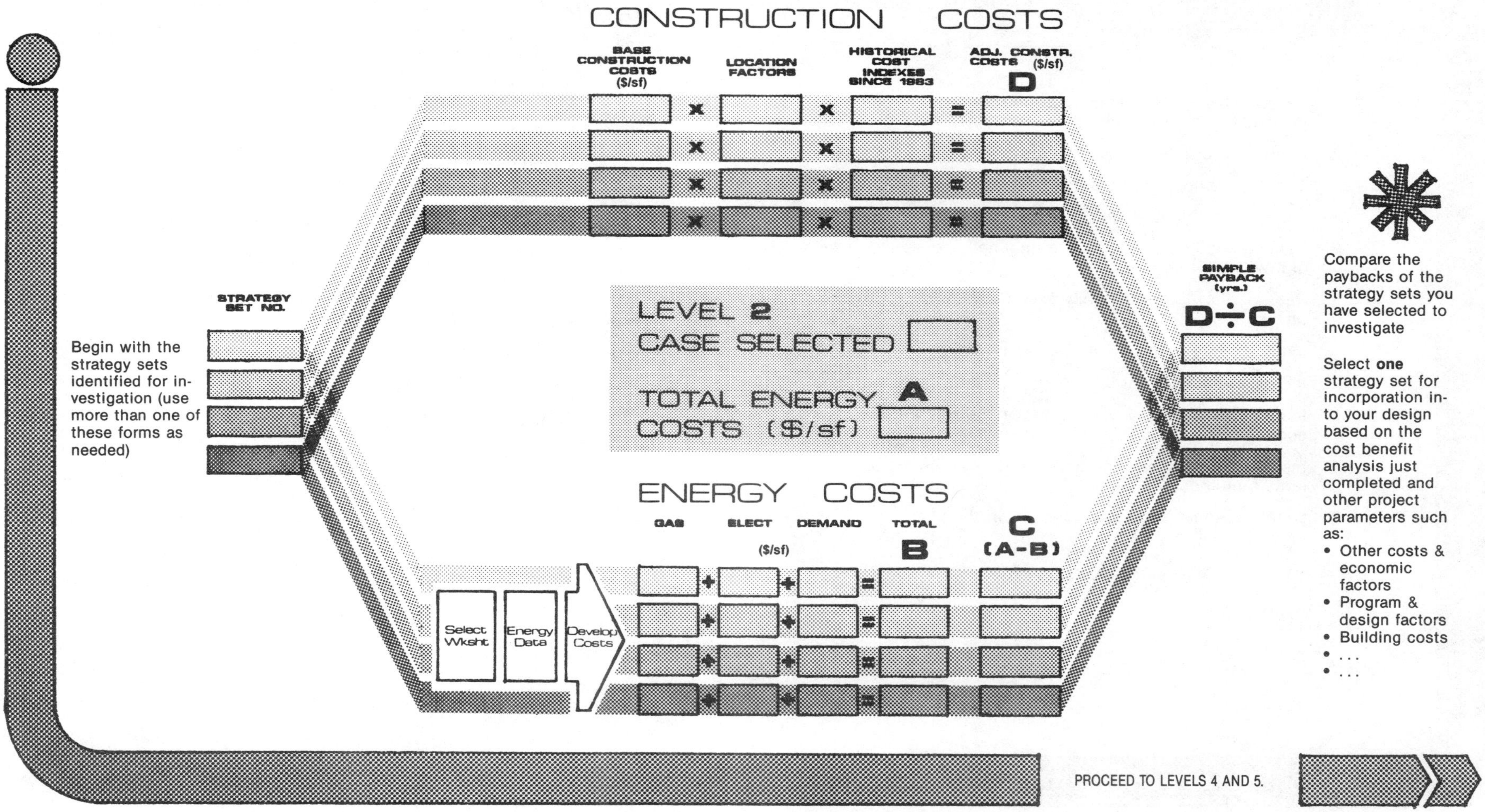

Begin with the strategy sets identified for investigation (use more than one of these forms as needed)

Compare the paybacks of the strategy sets you have selected to investigate

Select **one** strategy set for incorporation into your design based on the cost benefit analysis just completed and other project parameters such as:

- Other costs & economic factors
- Program & design factors
- Building costs
- ...
- ...

Note: See Section II, page 209, for further information and discussions.

Review these additional strategies for the possibility of further savings. Check the description in Section II, page 235, carefully for guidance on adopting any of the strategies.

A SELECT ON THE BASIS OF TECHNICAL JUDGEMENT

1. Increase thermostat deadband for occupied periods from 2°F to 8°F (70°F heating–78°F cooling)
2. Design for efficient space planning
3. Optimize building orientation
4. Minimize building height to reduce impact of glazed area
5. Reduce shading coefficient to minimally acceptable levels (not with daylighting)
6. Reduce glazing to minimally acceptable levels (not with daylighting)
7. Landscape to minimize wind induced infiltration effect
12. Increase manual light switching
13. Install operable windows (best for buildings without economizer cycle)
15. Install energy monitoring and control system (medium and large buildings only)
16. Locate outdoor cooling equipment in the shade

B VALUE MAY BE DETERMINED BY SIMPLE CALCULATION

19. High efficiency motors and pumps
20. Multiple speed compressors
21. Variable or two speed circulating fan
22. Electric resistance heaters
23. Desuperheater or stand-alone heat pump for DHW
24. Point-of-use electric domestic hot water heaters
25. External shading
26. Solar domestic hot water heater
29. Demand limiter
30. Absorption chiller-heater
31. Cogeneration

Note: The number preceding each strategy refers to the corresponding paragraph in Section II, page 235. See Section II, page 235, for further information and discussion of each strategy.

Review this checklist and adopt the procedures that are appropriate to your project. These actions are **important** if you are to achieve and maintain the savings from the selected Level 1-4 strategies.

A CHECKLIST FOR SUCCESSFUL TURNOVER AND BUILDING OPERATION

	DESIGN	CONSTRUCTION	TURNOVER	BUILDING OPERATION
DESIGN TEAM (ARCHITECTS/ ENGINEERS)	• Document all design assumptions (1) • Estimate energy use by sub-system (2) **& For Large Buildings:** • Document all systems and control strategies (3) • Specify contents of O&M Manuals (4) • Specify HVAC balancing requirements (5) • Specify equipment protection and cleaning (6) • Specify on-site instruction (7) • Specify access doors, valves, test ports (8) • Specify piping system ID (9) • Specify price quotations on all long-term maintenance contracts (10)	• Prepare "as-built" drawings (11) • Observe construction process for compliance (12)	• Conduct "punch list" inspection (15) • Visually check for air leakage (15) • If possible, participate in on-site training session(s) (7)	
CONTRACTOR & SUB-CONTRACTOR	**For Small Buildings Only:** • Document all systems and control strategies (3) • Specify contents of O&M Manuals (4) • Specify HVAC balancing requirements (5) • Specify equipment protection and cleaning (6) • Specify on-site instruction (7) • Specify access doors, valves, test ports (8) • Specify piping system ID (9) • Specify price quotations on all long-term maintenance contracts (10)	• Prepare O&M Manual (4) • Perform final calibration of all controls (5) • Hire independent balancing agent or require of subs (5) • Protect and clean equipment (6) • Provide all access doors, valves, and test ports (8) • Place piping system ID tags (9) • Prepare "as-built" drawings (11) • Use HVAC system to "flush" construction odors (13)	• Deliver O&M Manual (4) • Conduct on-site instruction and training (7)	• Honor warranty and call back as required (16)
EQUIPMENT MANUFACT.		• Support calibration of controls (5) • Provide all special tools required (14)	• Support O&M Manual preparation (4) • Support on-site instruction and training (7)	
OWNER & BLDG. STAFF	• Check all key design assumptions (1)	• Periodically check compliance on-site (12)	• Attend on-site instruction and training sessions (7)	• Monitor energy consumption against estimates (2) • Periodically test equipment efficiency (17) • Periodically sample and test water (18) • Inspect and calibrate controls (19) • Monitor occupant complains (20)

Note: See Section II, page 255, for more information and discussion. Numbers shown in ( ) refer to the numbered paragraphs in Section II, page 255.

1. Locate data by Base Case number.
2. Locate monthly or annual data for Base Case or Strategy Set.

Use monthly data for <u>Worksheets B or C</u>

Use annual data for <u>Worksheet A</u>

BASE CASE 326

STRAT SET	CONST COST $/SF	ENERGY FORM	UNITS (PER SF)	JAN	FEB	MAR	APR	MAY	JUN	JUL	AUG	SEP	OCT	NOV	DEC	ANNUAL
BASE	0.00	HT ELEC	KWH	0.76	0.53	0.31	0.09	0.03	0.03	0.03	0.03	0.03	0.05	0.23	0.65	2.8
		HT DEM	KW	.00605	.00587	.00575	.00405	.00369	.00006	.00006	.00006	.00224	.00363	.00520	.00587	0.00605
		REM ELEC	KWH	0.97	0.87	0.97	1.06	1.23	1.47	1.70	1.63	1.34	1.10	0.93	0.97	14.3
		REM DEM	KW	.00504	.00457	.00543	.00651	.00713	.00698	.00775	.00752	.00721	.00635	.00519	.00473	0.00775
		TOT ELEC	KWH	1.74	1.40	1.28	1.15	1.26	1.50	1.73	1.66	1.37	1.15	1.15	1.63	17.1
		TOT DEM	KW	.00930	.00828	.00828	.00698	.00716	.00763	.00781	.00763	.00744	.00670	.00679	.00837	0.00930
S- 3	-0.24	HT ELEC	KWH	0.81	0.56	0.32	0.10	0.03	0.03	0.03	0.03	0.03	0.05	0.24	0.69	3.0
		HT DEM	KW	.00605	.00587	.00575	.00405	.00369	.00006	.00006	.00006	.00224	.00363	.00520	.00587	0.00605
		REM ELEC	KWH	0.88	0.79	0.88	0.96	1.12	1.34	1.55	1.48	1.22	1.00	0.85	0.88	13.0
		REM DEM	KW	.00464	.00421	.00500	.00600	.00657	.00643	.00714	.00693	.00664	.00585	.00478	.00436	0.00714

BASE CASE 327

STRAT SET	CONST COST $/SF	ENERGY FORM	UNITS (PER SF)	JAN	FEB	MAR	APR	MAY	JUN	JUL	AUG	SEP	OCT	NOV	DEC	ANNUAL
BASE	0.00	TOT GAS	THM	0.0402	0.0277	0.0161	0.0048	0.0015	0.0015	0.0015	0.0015	0.0015	0.0026	0.0118	0.0343	0.146
		TOT ELEC	KWH	0.97	0.87	0.97	1.06	1.23	1.47	1.70	1.63	1.34	1.10	0.93	0.97	14.3
		TOT DEM	KW	.00504	.00457	.00543	.00651	.00713	.00698	.00775	.00752	.00721	.00635	.00519	.00473	0.00775
S- 3	-0.24	TOT GAS	THM	0.0426	0.0295	0.0171	0.0051	0.0015	0.0015	0.0015	0.0015	0.0015	0.0028	0.0126	0.0364	0.155
		TOT ELEC	KWH	0.88	0.79	0.88	0.96	1.12	1.34	1.55	1.48	1.22	1.00	0.85	0.88	13.0
		TOT DEM	KW	.00464	.00421	.00500	.00600	.00657	.00643	.00714	.00693	.00664	.00585	.00478	.00436	0.00714
S- 7	-0.17	TOT GAS	THM	0.0451	0.0312	0.0180	0.0054	0.0016	0.0016	0.0016	0.0016	0.0016	0.0030	0.0133	0.0385	0.164
		TOT ELEC	KWH	0.79	0.71	0.79	0.86	1.00	1.19	1.38	1.32	1.09	0.89	0.75	0.79	11.6
		TOT DEM	KW	.00406	.00369	.00438	.00525	.00575	.00563	.00625	.00606	.00581	.00513	.00419	.00381	0.00625
S- 16	0.40	TOT GAS	THM	0.0476	0.0329	0.0190	0.0057	0.0017	0.0017	0.0017	0.0017	0.0017	0.0031	0.0140	0.0407	0.173
		TOT ELEC	KWH	0.69	0.62	0.69	0.75	0.87	1.04	1.20	1.15	0.95	0.78	0.66	0.69	10.1
		TOT DEM	KW	.00370	.00336	.00398	.00478	.00523	.00512	.00569	.00552	.00529	.00467	.00381	.00347	0.00569
S- 25	0.38	TOT GAS	THM	0.0451	0.0312	0.0180	0.0054	0.0016	0.0016	0.0016	0.0016	0.0016	0.0030	0.0133	0.0385	0.164
		TOT ELEC	KWH	0.79	0.71	0.79	0.86	1.00	1.19	1.38	1.32	1.09	0.89	0.75	0.79	11.6
		TOT DEM	KW	.00406	.00369	.00438	.00525	.00575	.00563	.00625	.00606	.00581	.00513	.00419	.00381	0.00625
S- 32	0.09	TOT GAS	THM	0.0433	0.0258	0.0211	0.0087	0.0041	0.0018	0.0018	0.0018	0.0032	0.0070	0.0141	0.0298	0.162
		TOT ELEC	KWH	0.68	0.62	0.75	0.81	0.87	1.07	1.14	1.13	0.96	0.81	0.67	0.68	10.2
		TOT DEM	KW	.00414	.00403	.00443	.00512	.00535	.00552	.00575	.00564	.00558	.00500	.00426	.00380	0.00575

HT ELEC = Electric consumption for space and water heating.

REM ELEC = Remaining electric consumption (not used for heating).

TOT ELEC = Total electric consumption.

HT DEM = Peak electrical demand for space and water heating.

REM DEM = Peak electrical demand for remaining load (not heating).

TOT DEM = Total peak electrical demand.

GAS = Total gas consumption.

(Annual figures are sums of monthly consumption values and are yearly maximums of monthly peak demand values.)

STRAT SET	CONST COST $/SF	ENERGY FORM	UNITS (PER SF)	JAN	FEB	MAR	APR	MAY	JUN	JUL	AUG	SEP	OCT	NOV	DEC	ANNUAL
								BASE CASE : 301								
BASE	0.00	TOT ELEC	KWH	1.74	1.37	1.21	1.16	1.26	1.34	1.50	1.43	1.27	1.19	1.17	1.58	16.3
		TOT DEM	KW	.00485	.00435	.00471	.00549	.00606	.00578	.00713	.00663	.00620	.00513	.00442	.00442	0.00713
S- 1	0.09	TOT ELEC	KWH	1.63	1.28	1.12	1.08	1.17	1.25	1.40	1.34	1.19	1.11	1.09	1.47	15.2
		TOT DEM	KW	.00443	.00398	.00430	.00502	.00554	.00528	.00652	.00606	.00567	.00469	.00404	.00404	0.00652
S- 10	0.21	TOT ELEC	KWH	1.47	1.15	1.01	0.97	1.05	1.12	1.26	1.21	1.07	1.00	0.99	1.33	13.7
		TOT DEM	KW	.00561	.00488	.00376	.00438	.00482	.00460	.00561	.00522	.00488	.00404	.00387	.00511	0.00561
S- 20	1.52	TOT ELEC	KWH	1.35	1.06	0.93	0.89	0.97	1.03	1.16	1.11	0.98	0.92	0.91	1.22	12.6
		TOT DEM	KW	.00510	.00444	.00342	.00398	.00439	.00418	.00510	.00474	.00444	.00367	.00352	.00464	0.00510
S- 25	-0.04	TOT ELEC	KWH	1.47	1.15	1.01	0.97	1.05	1.12	1.26	1.21	1.07	1.00	0.99	1.33	13.7
		TOT DEM	KW	.00561	.00488	.00376	.00438	.00482	.00460	.00561	.00522	.00488	.00404	.00387	.00511	0.00561
								BASE CASE : 302								
BASE	0.00	TOT ELEC	KWH	1.44	1.17	1.11	1.03	1.14	1.37	1.57	1.52	1.25	1.03	1.02	1.35	15.0
		TOT DEM	KW	.00829	.00738	.00738	.00622	.00638	.00680	.00696	.00680	.00663	.00597	.00605	.00746	0.00829
S- 3	-0.12	TOT ELEC	KWH	1.34	1.09	1.04	0.97	1.06	1.27	1.47	1.41	1.16	0.97	0.95	1.26	14.0
		TOT DEM	KW	.00760	.00676	.00676	.00570	.00585	.00623	.00638	.00623	.00608	.00547	.00555	.00684	0.00760
S- 7	0.00	TOT ELEC	KWH	1.24	1.01	0.95	0.89	0.98	1.17	1.35	1.30	1.07	0.89	0.88	1.16	12.9
		TOT DEM	KW	.00673	.00599	.00599	.00505	.00518	.00552	.00565	.00552	.00538	.00485	.00491	.00606	0.00673
S- 16	1.21	TOT ELEC	KWH	1.15	0.94	0.89	0.83	0.91	1.09	1.26	1.21	1.00	0.83	0.82	1.08	12.0
		TOT DEM	KW	.00622	.00554	.00554	.00466	.00479	.00510	.00522	.00510	.00498	.00448	.00454	.00560	0.00622
S- 61	-0.25	TOT ELEC	KWH	1.24	1.01	0.95	0.89	0.98	1.17	1.35	1.30	1.07	0.89	0.88	1.16	12.9
		TOT DEM	KW	.00673	.00599	.00599	.00505	.00518	.00552	.00565	.00552	.00538	.00485	.00491	.00606	0.00673
								BASE CASE : 303								
BASE	0.00	TOT GAS	THM	0.0286	0.0198	0.0114	0.0034	0.0010	0.0010	0.0010	0.0010	0.0010	0.0019	0.0084	0.0244	0.104
		TOT ELEC	KWH	0.88	0.79	0.88	0.96	1.12	1.34	1.55	1.48	1.22	1.00	0.84	0.88	13.0
		TOT DEM	KW	.00283	.00283	.00498	.00567	.00615	.00594	.00691	.00663	.00622	.00532	.00442	.00283	0.00691
S- 3	-0.13	TOT GAS	THM	0.0300	0.0207	0.0120	0.0036	0.0011	0.0011	0.0011	0.0011	0.0011	0.0020	0.0088	0.0256	0.109
		TOT ELEC	KWH	0.81	0.73	0.81	0.88	1.02	1.23	1.42	1.36	1.12	0.92	0.77	0.81	11.9
		TOT DEM	KW	.00260	.00260	.00456	.00519	.00563	.00544	.00633	.00608	.00570	.00487	.00405	.00260	0.00633
S- 7	-0.01	TOT GAS	THM	0.0316	0.0218	0.0126	0.0038	0.0012	0.0012	0.0012	0.0012	0.0012	0.0021	0.0093	0.0270	0.115
		TOT ELEC	KWH	0.73	0.65	0.73	0.79	0.92	1.10	1.27	1.22	1.01	0.82	0.70	0.73	10.7
		TOT DEM	KW	.00230	.00230	.00404	.00460	.00499	.00482	.00561	.00539	.00505	.00432	.00359	.00230	0.00561
S- 16	1.20	TOT GAS	THM	0.0355	0.0245	0.0142	0.0043	0.0013	0.0013	0.0013	0.0013	0.0013	0.0023	0.0104	0.0303	0.129
		TOT ELEC	KWH	0.65	0.58	0.65	0.71	0.82	0.98	1.14	1.09	0.90	0.74	0.62	0.65	9.6
		TOT DEM	KW	.00212	.00212	.00373	.00425	.00461	.00445	.00518	.00497	.00466	.00399	.00332	.00212	0.00518
S- 61	-0.26	TOT GAS	THM	0.0316	0.0218	0.0126	0.0038	0.0012	0.0012	0.0012	0.0012	0.0012	0.0021	0.0093	0.0270	0.115
		TOT ELEC	KWH	0.73	0.65	0.73	0.79	0.92	1.10	1.27	1.22	1.01	0.82	0.70	0.73	10.7
		TOT DEM	KW	.00230	.00230	.00404	.00460	.00499	.00482	.00561	.00539	.00505	.00432	.00359	.00230	0.00561
								BASE CASE : 304								
BASE	0.00	TOT ELEC	KWH	1.46	1.19	1.21	1.16	1.29	1.38	1.55	1.49	1.30	1.21	1.13	1.33	15.7
		TOT DEM	KW	.00485	.00435	.00471	.00549	.00606	.00578	.00713	.00663	.00620	.00513	.00442	.00442	0.00713
S- 1	0.09	TOT ELEC	KWH	1.34	1.09	1.11	1.07	1.18	1.27	1.43	1.37	1.20	1.11	1.04	1.22	14.4
		TOT DEM	KW	.00443	.00398	.00430	.00502	.00554	.00528	.00652	.00606	.00567	.00469	.00404	.00404	0.00652
S- 12	0.23	TOT ELEC	KWH	1.17	0.96	0.97	0.93	1.03	1.11	1.25	1.20	1.05	0.97	0.91	1.07	12.6
		TOT DEM	KW	.00381	.00342	.00370	.00432	.00477	.00454	.00561	.00522	.00488	.00404	.00348	.00348	0.00561
S- 22	1.54	TOT ELEC	KWH	1.05	0.85	0.87	0.83	0.92	0.99	1.11	1.07	0.93	0.87	0.81	0.96	11.3
		TOT DEM	KW	.00347	.00311	.00337	.00393	.00434	.00413	.00510	.00474	.00444	.00367	.00316	.00316	0.00510
S- 28	-0.02	TOT ELEC	KWH	1.17	0.96	0.97	0.93	1.03	1.11	1.25	1.20	1.05	0.97	0.91	1.07	12.6
		TOT DEM	KW	.00381	.00342	.00370	.00432	.00477	.00454	.00561	.00522	.00488	.00404	.00348	.00348	0.00561
								BASE CASE : 305								
BASE	0.00	TOT ELEC	KWH	1.84	1.44	1.27	1.22	1.32	1.41	1.58	1.51	1.34	1.26	1.24	1.67	17.2
		TOT DEM	KW	.00534	.00479	.00519	.00605	.00668	.00637	.00786	.00731	.00684	.00566	.00487	.00487	0.00786
S- 2	0.39	TOT ELEC	KWH	1.6[illegible]	1.33	1.17	1.12	1.22	1.30	1.45	1.39	1.23	1.15	1.14	1.53	15.8
		TOT DEM	KW	.00479	.00430	.00465	.00543	.00599	.00571	.00705	.00656	.00613	.00508	.00437	.00437	0.00705
S- 11	0.53	TOT ELEC	KWH	1.54	1.21	1.07	1.02	1.11	1.18	1.32	1.27	1.12	1.05	1.04	1.40	14.4
		TOT DEM	KW	.00606	.00527	.00406	.00473	.00521	.00497	.00606	.00564	.00527	.00436	.00418	.00551	0.00606
S- 21	1.84	TOT ELEC	KWH	1.41	1.11	0.98	0.94	1.02	1.08	1.21	1.16	1.03	0.96	0.95	1.28	13.2
		TOT DEM	KW	.00554	.00482	.00371	.00432	.00476	.00454	.00554	.00515	.00482	.00399	.00382	.00504	0.00554
S- 26	0.28	TOT ELEC	KWH	1.54	1.21	1.07	1.02	1.11	1.18	1.32	1.27	1.12	1.05	1.04	1.40	14.4
		TOT DEM	KW	.00606	.00527	.00406	.00473	.00521	.00497	.00606	.00564	.00527	.00436	.00418	.00551	0.00606
S- 35	1.00	TOT ELEC	KWH	1.01	0.79	0.88	0.87	0.89	1.10	1.15	1.14	0.98	0.84	0.77	0.87	11.3
		TOT DEM	KW	.00725	.00645	.00645	.00544	.00558	.00594	.00609	.00594	.00580	.00522	.00529	.00652	0.00725
								BASE CASE : 306								
BASE	0.00	TOT ELEC	KWH	1.50	1.22	1.15	1.08	1.19	1.42	1.64	1.58	1.29	1.08	1.06	1.40	15.6
		TOT DEM	KW	.01250	.01113	.01113	.00938	.00962	.01025	.01050	.01025	.01000	.00900	.00912	.01125	0.01250
S- 4	-0.09	TOT ELEC	KWH	1.41	1.15	1.09	1.01	1.12	1.34	1.54	1.48	1.22	1.01	1.00	1.32	14.7
		TOT DEM	KW	.01196	.01064	.01064	.00897	.00921	.00981	.01005	.00981	.00957	.00861	.00873	.01076	0.01196

STRAT SET	CONST COST $/SF	ENERGY FORM	UNITS (PER SF)	JAN	FEB	MAR	APR	MAY	JUN	JUL	AUG	SEP	OCT	NOV	DEC	ANNUAL
								BASE CASE : 306 (CONTINUED)								
S- 8	0.03	TOT ELEC	KWH	1.30	1.05	1.00	0.93	1.03	1.23	1.42	1.36	1.12	0.93	0.92	1.22	13.5
		TOT DEM	KW	.01142	.01016	.01016	.00857	.00879	.00936	.00959	.00936	.00914	.00822	.00834	.01028	0.01142
S- 17	1.20	TOT ELEC	KWH	1.22	0.99	0.94	0.88	0.97	1.16	1.33	1.28	1.05	0.88	0.86	1.14	12.7
		TOT DEM	KW	.01089	.00969	.00969	.00817	.00839	.00893	.00915	.00893	.00871	.00784	.00795	.00980	0.01089
S- 27	-0.21	TOT ELEC	KWH	1.30	1.05	1.00	0.93	1.03	1.23	1.42	1.36	1.12	0.93	0.92	1.22	13.5
		TOT DEM	KW	.01142	.01016	.01016	.00857	.00879	.00936	.00959	.00936	.00914	.00822	.00834	.01028	0.01142
S- 32	0.51	TOT ELEC	KWH	1.34	1.04	1.09	0.96	0.93	1.09	1.18	1.16	1.01	0.91	0.93	1.15	12.8
		TOT DEM	KW	.00650	.00579	.00579	.00488	.00500	.00533	.00546	.00533	.00520	.00468	.00475	.00585	0.00650
								BASE CASE : 307								
BASE	0.00	TOT GAS	THM	0.0246	0.0170	0.0098	0.0029	0.0009	0.0009	0.0009	0.0009	0.0009	0.0016	0.0072	0.0210	0.089
		TOT ELEC	KWH	0.95	0.85	0.95	1.03	1.20	1.43	1.65	1.58	1.31	1.07	0.90	0.95	13.9
		TOT DEM	KW	.00315	.00315	.00553	.00630	.00684	.00660	.00768	.00737	.00691	.00591	.00492	.00315	0.00768
S- 4	-0.10	TOT GAS	THM	0.0341	0.0236	0.0136	0.0041	0.0012	0.0012	0.0012	0.0012	0.0012	0.0022	0.0100	0.0291	0.124
		TOT ELEC	KWH	0.84	0.75	0.84	0.91	1.06	1.27	1.46	1.40	1.16	0.95	0.80	0.84	12.3
		TOT DEM	KW	.00277	.00277	.00486	.00554	.00601	.00581	.00675	.00648	.00607	.00520	.00432	.00277	0.00675
S- 8	0.02	TOT GAS	THM	0.0363	0.0251	0.0145	0.0044	0.0013	0.0013	0.0013	0.0013	0.0013	0.0024	0.0107	0.0310	0.132
		TOT ELEC	KWH	0.75	0.67	0.75	0.81	0.95	1.13	1.31	1.25	1.03	0.85	0.71	0.75	11.0
		TOT DEM	KW	.00246	.00246	.00431	.00491	.00533	.00515	.00599	.00575	.00539	.00461	.00383	.00246	0.00599
S- 17	1.19	TOT GAS	THM	0.0407	0.0281	0.0163	0.0049	0.0015	0.0015	0.0015	0.0015	0.0015	0.0027	0.0120	0.0348	0.148
		TOT ELEC	KWH	0.67	0.60	0.67	0.73	0.85	1.02	1.17	1.13	0.93	0.76	0.64	0.67	9.9
		TOT DEM	KW	.00228	.00228	.00401	.00457	.00496	.00479	.00557	.00535	.00501	.00429	.00356	.00228	0.00557
S- 27	-0.23	TOT GAS	THM	0.0363	0.0251	0.0145	0.0044	0.0013	0.0013	0.0013	0.0013	0.0013	0.0024	0.0107	0.0310	0.132
		TOT ELEC	KWH	0.75	0.67	0.75	0.81	0.95	1.13	1.31	1.25	1.03	0.85	0.71	0.75	11.0
		TOT DEM	KW	.00246	.00246	.00431	.00491	.00533	.00515	.00599	.00575	.00539	.00461	.00383	.00246	0.00599
S- 32	0.50	TOT GAS	THM	0.0355	0.0245	0.0142	0.0043	0.0013	0.0013	0.0013	0.0013	0.0013	0.0023	0.0104	0.0303	0.129
		TOT ELEC	KWH	0.67	0.60	0.67	0.73	0.84	1.01	1.17	1.12	0.92	0.75	0.64	0.67	9.8
		TOT DEM	KW	.00221	.00221	.00387	.00441	.00479	.00463	.00538	.00516	.00484	.00414	.00344	.00221	0.00538
								BASE CASE : 308								
BASE	0.00	TOT ELEC	KWH	1.53	1.25	1.27	1.22	1.35	1.45	1.63	1.57	1.37	1.27	1.19	1.40	16.5
		TOT DEM	KW	.00575	.00516	.00558	.00651	.00719	.00685	.00846	.00787	.00736	.00609	.00525	.00525	0.00846
S- 2	0.39	TOT ELEC	KWH	1.39	1.14	1.15	1.11	1.23	1.32	1.49	1.42	1.25	1.15	1.08	1.27	15.0
		TOT DEM	KW	.00479	.00430	.00465	.00543	.00599	.00571	.00705	.00656	.00613	.00508	.00437	.00437	0.00705
S- 13	0.55	TOT ELEC	KWH	1.22	1.00	1.01	0.97	1.07	1.15	1.30	1.24	1.09	1.01	0.94	1.11	13.1
		TOT DEM	KW	.00412	.00370	.00400	.00467	.00515	.00491	.00606	.00564	.00527	.00436	.00376	.00376	0.00606
S- 23	1.86	TOT ELEC	KWH	1.11	0.90	0.92	0.88	0.98	1.05	1.18	1.13	0.99	0.92	0.86	1.01	11.9
		TOT DEM	KW	.00377	.00338	.00366	.00427	.00471	.00449	.00554	.00515	.00482	.00399	.00343	.00343	0.00554
S- 29	0.30	TOT ELEC	KWH	1.22	1.00	1.01	0.97	1.07	1.15	1.30	1.24	1.09	1.01	0.94	1.11	13.1
		TOT DEM	KW	.00412	.00370	.00400	.00467	.00515	.00491	.00606	.00564	.00527	.00436	.00376	.00376	0.00606
S- 36	1.02	TOT ELEC	KWH	1.11	0.88	0.98	0.96	0.99	1.21	1.27	1.26	1.09	0.93	0.85	0.96	12.5
		TOT DEM	KW	.00575	.00512	.00512	.00431	.00443	.00471	.00483	.00471	.00460	.00414	.00420	.00518	0.00575
								BASE CASE : 309								
BASE	0.00	HT ELEC	KWH	0.56	0.41	0.32	0.20	0.13	0.12	0.15	0.13	0.10	0.16	0.29	0.48	3.0
		HT DEM	KW	.00680	.00537	.00503	.00306	.00286	.00061	.00082	.00054	.00211	.00279	.00394	.00510	0.00680
		REM ELEC	KWH	0.97	0.87	0.97	1.05	1.22	1.46	1.69	1.62	1.33	1.09	0.92	0.97	14.2
		REM DEM	KW	.00393	.00312	.00461	.00556	.00610	.00590	.00678	.00651	.00624	.00508	.00420	.00393	0.00678
		TOT ELEC	KWH	1.52	1.27	1.28	1.26	1.35	1.58	1.84	1.74	1.44	1.26	1.21	1.45	17.3
		TOT DEM	KW	.01310	.01048	.01061	.00668	.00707	.00707	.00760	.00721	.00694	.00707	.00838	.01022	0.01310
S- 3	-0.09	HT ELEC	KWH	0.59	0.43	0.33	0.21	0.14	0.13	0.16	0.13	0.11	0.17	0.30	0.51	3.2
		HT DEM	KW	.00548	.00477	.00477	.00263	.00241	.00060	.00088	.00093	.00192	.00241	.00345	.00466	0.00548
		REM ELEC	KWH	0.88	0.79	0.88	0.95	1.11	1.33	1.54	1.47	1.21	0.99	0.84	0.88	12.9
		REM DEM	KW	.00309	.00245	.00362	.00437	.00480	.00464	.00533	.00512	.00490	.00400	.00330	.00309	0.00533
		TOT ELEC	KWH	1.46	1.22	1.21	1.17	1.25	1.46	1.69	1.60	1.32	1.17	1.14	1.38	16.1
		TOT DEM	KW	.00926	.00741	.00750	.00472	.00500	.00500	.00537	.00509	.00491	.00500	.00593	.00722	0.00926
S- 7	0.00	HT ELEC	KWH	0.64	0.47	0.37	0.24	0.15	0.14	0.17	0.14	0.12	0.19	0.33	0.56	3.5
		HT DEM	KW	.00548	.00477	.00477	.00263	.00241	.00060	.00088	.00093	.00192	.00241	.00345	.00466	0.00548
		REM ELEC	KWH	0.80	0.71	0.80	0.87	1.01	1.21	1.39	1.33	1.10	0.90	0.76	0.80	11.7
		REM DEM	KW	.00293	.00232	.00343	.00414	.00454	.00439	.00505	.00485	.00465	.00379	.00313	.00293	0.00505
		TOT ELEC	KWH	1.44	1.19	1.16	1.10	1.16	1.35	1.56	1.48	1.22	1.09	1.09	1.35	15.2
		TOT DEM	KW	.00896	.00717	.00726	.00457	.00484	.00484	.00520	.00493	.00475	.00484	.00573	.00699	0.00896
S- 16	1.21	HT ELEC	KWH	0.67	0.49	0.38	0.25	0.16	0.15	0.18	0.15	0.12	0.20	0.34	0.58	3.7
		HT DEM	KW	.00548	.00477	.00477	.00263	.00241	.00060	.00088	.00093	.00192	.00241	.00345	.00466	0.00548
		REM ELEC	KWH	0.71	0.63	0.71	0.77	0.89	1.07	1.24	1.19	0.98	0.80	0.68	0.71	10.4
		REM DEM	KW	.00274	.00218	.00322	.00388	.00426	.00412	.00473	.00454	.00435	.00355	.00293	.00274	0.00473
		TOT ELEC	KWH	1.38	1.12	1.09	1.01	1.05	1.22	1.42	1.34	1.10	1.00	1.02	1.29	14.1
		TOT DEM	KW	.00861	.00689	.00697	.00439	.00465	.00465	.00499	.00474	.00456	.00465	.00551	.00672	0.00861

(HT ELEC=ELECTRIC CONSUMPTION FOR SPACE AND WATER HEATING. REM ELEC=REMAINING ELECTRIC CONSUMPTION (NOT USED FOR HEATING). TOT ELEC=TOTAL ELECTRIC CONSUMPTION. HT DEM=PEAK ELECTRIC DEMAND FOR SPACE AND WATER HEATING. REM DEM=PEAK ELECTRIC DEMAND FOR REMAINING LOAD (NOT HEATING). TOT DEM=TOTAL PEAK ELECTRIC DEMAND. GAS=TOTAL GAS CONSUMPTION. ANNUAL FIGURES ARE SUMS OF MONTHLY CONSUMPTION VALUES OR ARE YEARLY MAXIMUMS OF MONTHLY PEAK DEMAND VALUES.)

STRAT SET	CONST COST $/SF	ENERGY FORM	UNITS (PER SF)	JAN	FEB	MAR	APR	MAY	JUN	JUL	AUG	SEP	OCT	NOV	DEC	ANNUAL
BASE CASE : 309 (CONTINUED)																
S- 32	0.27	HT ELEC	KWH	0.71	0.51	0.46	0.27	0.13	0.06	0.08	0.08	0.08	0.21	0.38	0.61	3.6
		HT DEM	KW	.00548	.00477	.00477	.00263	.00241	.00060	.00088	.00093	.00192	.00241	.00345	.00466	0.00548
		REM ELEC	KWH	0.78	0.71	0.87	0.92	0.99	1.23	1.31	1.30	1.10	0.92	0.77	0.78	11.7
		REM DEM	KW	.00427	.00405	.00444	.00505	.00533	.00544	.00555	.00544	.00538	.00488	.00427	.00366	0.00555
		TOT ELEC	KWH	1.50	1.22	1.33	1.19	1.13	1.29	1.39	1.37	1.18	1.14	1.15	1.39	15.3
		TOT DEM	KW	.00921	.00829	.00847	.00580	.00589	.00635	.00663	.00645	.00617	.00553	.00691	.00847	0.00921
S- 40	0.32	HT ELEC	KWH	0.73	0.52	0.47	0.27	0.14	0.07	0.08	0.08	0.08	0.22	0.39	0.62	3.7
		HT DEM	KW	.00548	.00477	.00477	.00263	.00241	.00060	.00088	.00093	.00192	.00241	.00345	.00466	0.00548
		REM ELEC	KWH	0.71	0.65	0.78	0.84	0.90	1.11	1.19	1.18	1.00	0.84	0.70	0.71	10.6
		REM DEM	KW	.00387	.00367	.00402	.00458	.00483	.00493	.00503	.00493	.00488	.00443	.00387	.00332	0.00503
		TOT ELEC	KWH	1.44	1.17	1.26	1.11	1.04	1.18	1.27	1.25	1.07	1.05	1.09	1.33	14.3
		TOT DEM	KW	.00843	.00759	.00776	.00531	.00540	.00582	.00607	.00590	.00565	.00506	.00632	.00776	0.00843
S- 45	0.97	HT ELEC	KWH	0.66	0.47	0.43	0.25	0.12	0.06	0.07	0.07	0.07	0.20	0.35	0.56	3.3
		HT DEM	KW	.00548	.00477	.00477	.00263	.00241	.00060	.00088	.00093	.00192	.00241	.00345	.00466	0.00548
		REM ELEC	KWH	0.75	0.68	0.83	0.88	0.95	1.18	1.25	1.24	1.05	0.88	0.74	0.75	11.2
		REM DEM	KW	.00423	.00402	.00440	.00501	.00528	.00539	.00550	.00539	.00534	.00484	.00423	.00363	0.00550
		TOT ELEC	KWH	1.41	1.16	1.26	1.13	1.08	1.24	1.33	1.31	1.12	1.08	1.09	1.31	14.5
		TOT DEM	KW	.00913	.00822	.00840	.00575	.00584	.00630	.00657	.00639	.00612	.00548	.00685	.00840	0.00913
S- 50	1.58	HT ELEC	KWH	0.74	0.53	0.48	0.28	0.14	0.07	0.08	0.08	0.08	0.22	0.39	0.63	3.7
		HT DEM	KW	.00548	.00477	.00477	.00263	.00241	.00060	.00088	.00093	.00192	.00241	.00345	.00466	0.00548
		REM ELEC	KWH	0.76	0.69	0.84	0.89	0.96	1.19	1.27	1.25	1.06	0.89	0.75	0.76	11.3
		REM DEM	KW	.00380	.00360	.00394	.00449	.00473	.00483	.00493	.00483	.00478	.00434	.00380	.00325	0.00493
		TOT ELEC	KWH	1.49	1.22	1.32	1.17	1.10	1.25	1.35	1.33	1.14	1.11	1.14	1.39	15.0
		TOT DEM	KW	.00840	.00756	.00773	.00529	.00538	.00580	.00605	.00588	.00563	.00504	.00630	.00773	0.00840
S- 61	-0.25	HT ELEC	KWH	0.64	0.47	0.37	0.24	0.15	0.14	0.17	0.14	0.12	0.19	0.33	0.56	3.5
		HT DEM	KW	.00548	.00477	.00477	.00263	.00241	.00060	.00088	.00093	.00192	.00241	.00345	.00466	0.00548
		REM ELEC	KWH	0.80	0.71	0.80	0.87	1.01	1.21	1.39	1.33	1.10	0.90	0.76	0.80	11.7
		REM DEM	KW	.00293	.00232	.00343	.00414	.00454	.00439	.00505	.00485	.00465	.00379	.00313	.00293	0.00505
		TOT ELEC	KWH	1.44	1.19	1.16	1.10	1.16	1.35	1.56	1.48	1.22	1.09	1.09	1.35	15.2
		TOT DEM	KW	.00896	.00717	.00726	.00457	.00484	.00484	.00520	.00493	.00475	.00484	.00573	.00699	0.00896
BASE CASE : 310																
BASE	0.00	TOT GAS	THM	0.0293	0.0214	0.0166	0.0107	0.0069	0.0064	0.0078	0.0066	0.0053	0.0086	0.0150	0.0253	0.160
		TOT ELEC	KWH	0.97	0.87	0.97	1.05	1.22	1.46	1.69	1.62	1.33	1.09	0.92	0.97	14.2
		TOT DEM	KW	.00393	.00312	.00461	.00556	.00610	.00590	.00678	.00651	.00624	.00508	.00420	.00393	0.00678
S- 3	-0.09	TOT GAS	THM	0.0307	0.0225	0.0175	0.0113	0.0072	0.0067	0.0082	0.0069	0.0055	0.0091	0.0158	0.0265	0.168
		TOT ELEC	KWH	0.88	0.79	0.88	0.95	1.11	1.33	1.54	1.47	1.21	0.99	0.84	0.88	12.9
		TOT DEM	KW	.00309	.00245	.00362	.00437	.00480	.00464	.00533	.00512	.00490	.00400	.00330	.00309	0.00533
S- 7	0.00	TOT GAS	THM	0.0339	0.0248	0.0192	0.0124	0.0080	0.0074	0.0091	0.0076	0.0061	0.0100	0.0174	0.0292	0.185
		TOT ELEC	KWH	0.80	0.71	0.80	0.87	1.01	1.21	1.39	1.33	1.10	0.90	0.76	0.80	11.7
		TOT DEM	KW	.00293	.00232	.00343	.00414	.00454	.00439	.00505	.00485	.00465	.00379	.00313	.00293	0.00505
S- 16	1.20	TOT GAS	THM	0.0351	0.0257	0.0200	0.0129	0.0083	0.0077	0.0094	0.0079	0.0063	0.0104	0.0180	0.0303	0.192
		TOT ELEC	KWH	0.71	0.63	0.71	0.77	0.89	1.07	1.24	1.19	0.98	0.80	0.68	0.71	10.4
		TOT DEM	KW	.00274	.00218	.00322	.00388	.00426	.00412	.00473	.00454	.00435	.00355	.00293	.00274	0.00473
S- 32	0.27	TOT GAS	THM	0.0374	0.0268	0.0244	0.0140	0.0070	0.0034	0.0042	0.0040	0.0040	0.0112	0.0200	0.0319	0.189
		TOT ELEC	KWH	0.78	0.71	0.87	0.92	0.99	1.23	1.31	1.30	1.10	0.92	0.77	0.78	11.7
		TOT DEM	KW	.00505	.00433	.00422	.00477	.00544	.00533	.00555	.00549	.00511	.00500	.00505	.00411	0.00555
S- 40	0.31	TOT GAS	THM	0.0382	0.0274	0.0249	0.0143	0.0071	0.0035	0.0042	0.0041	0.0041	0.0114	0.0205	0.0326	0.193
		TOT ELEC	KWH	0.71	0.65	0.78	0.84	0.90	1.11	1.19	1.18	1.00	0.84	0.70	0.71	10.6
		TOT DEM	KW	.00362	.00352	.00387	.00448	.00468	.00483	.00503	.00493	.00488	.00438	.00372	.00332	0.00503
S- 45	0.96	TOT GAS	THM	0.0348	0.0250	0.0227	0.0130	0.0065	0.0032	0.0039	0.0037	0.0037	0.0104	0.0187	0.0297	0.176
		TOT ELEC	KWH	0.75	0.68	0.83	0.88	0.95	1.18	1.25	1.24	1.05	0.88	0.74	0.75	11.2
		TOT DEM	KW	.00396	.00385	.00423	.00489	.00512	.00528	.00550	.00539	.00534	.00478	.00407	.00363	0.00550
S- 50	1.58	TOT GAS	THM	0.0388	0.0278	0.0253	0.0145	0.0073	0.0035	0.0043	0.0041	0.0041	0.0116	0.0208	0.0331	0.196
		TOT ELEC	KWH	0.76	0.69	0.84	0.89	0.96	1.19	1.27	1.25	1.06	0.89	0.75	0.76	11.3
		TOT DEM	KW	.00355	.00345	.00380	.00439	.00458	.00473	.00493	.00483	.00478	.00429	.00365	.00325	0.00493
S- 61	-0.25	TOT GAS	THM	0.0339	0.0248	0.0192	0.0124	0.0080	0.0074	0.0091	0.0076	0.0061	0.0100	0.0174	0.0292	0.185
		TOT ELEC	KWH	0.80	0.71	0.80	0.87	1.01	1.21	1.39	1.33	1.10	0.90	0.76	0.80	11.7
		TOT DEM	KW	.00293	.00232	.00343	.00414	.00454	.00439	.00505	.00485	.00465	.00379	.00313	.00293	0.00505
BASE CASE : 311																
BASE	0.00	HT ELEC	KWH	0.62	0.43	0.25	0.07	0.02	0.02	0.02	0.02	0.02	0.04	0.18	0.53	2.3
		HT DEM	KW	.00483	.00469	.00459	.00324	.00295	.00005	.00005	.00005	.00179	.00290	.00415	.00469	0.00483
		REM ELEC	KWH	0.91	0.82	0.91	0.99	1.15	1.38	1.59	1.53	1.26	1.03	0.87	0.91	13.4
		REM DEM	KW	.00416	.00378	.00448	.00538	.00589	.00576	.00640	.00621	.00595	.00525	.00429	.00390	0.00640
		TOT ELEC	KWH	1.54	1.25	1.16	1.07	1.18	1.40	1.62	1.55	1.28	1.07	1.05	1.44	15.7
		TOT DEM	KW	.00786	.00700	.00700	.00590	.00605	.00645	.00660	.00645	.00629	.00566	.00574	.00707	0.00786
S- 3	-0.11	HT ELEC	KWH	0.66	0.46	0.26	0.08	0.02	0.02	0.02	0.02	0.02	0.04	0.19	0.56	2.4
		HT DEM	KW	.00483	.00469	.00459	.00324	.00295	.00005	.00005	.00005	.00179	.00290	.00415	.00469	0.00483
		REM ELEC	KWH	0.82	0.74	0.82	0.90	1.04	1.25	1.44	1.38	1.14	0.93	0.79	0.82	12.1
		REM DEM	KW	.00378	.00343	.00407	.00489	.00535	.00524	.00582	.00565	.00541	.00477	.00390	.00355	0.00582
		TOT ELEC	KWH	1.48	1.19	1.09	0.97	1.06	1.27	1.46	1.40	1.16	0.97	0.98	1.39	14.5
		TOT DEM	KW	.00746	.00664	.00664	.00560	.00574	.00612	.00627	.00612	.00597	.00537	.00545	.00671	0.00746
S- 7	-0.03	HT ELEC	KWH	0.70	0.48	0.28	0.08	0.03	0.03	0.03	0.03	0.03	0.05	0.21	0.60	2.5
		HT DEM	KW	.00483	.00469	.00459	.00324	.00295	.00005	.00005	.00005	.00179	.00290	.00415	.00469	0.00483
		REM ELEC	KWH	0.74	0.66	0.74	0.81	0.94	1.12	1.30	1.24	1.02	0.84	0.71	0.74	10.9
		REM DEM	KW	.00337	.00306	.00363	.00436	.00477	.00467	.00519	.00503	.00483	.00426	.00348	.00317	0.00519
		TOT ELEC	KWH	1.44	1.15	1.02	0.89	0.96	1.15	1.32	1.27	1.05	0.89	0.92	1.34	13.5
		TOT DEM	KW	.00721	.00649	.00663	.00454	.00461	.00497	.00519	.00505	.00483	.00433	.00541	.00663	0.00721

STRAT SET	CONST COST $/SF	ENERGY FORM	UNITS (PER SF)	JAN	FEB	MAR	APR	MAY	JUN	JUL	AUG	SEP	OCT	NOV	DEC	ANNUAL
BASE CASE : 311 (CONTINUED)																
S- 16	1.08	HT ELEC	KWH	0.74	0.51	0.30	0.09	0.03	0.03	0.03	0.03	0.03	0.05	0.22	0.63	2.7
		HT DEM	KW	.00483	.00469	.00459	.00324	.00295	.00005	.00005	.00005	.00179	.00290	.00415	.00469	0.00483
		REM ELEC	KWH	0.65	0.58	0.65	0.70	0.82	0.98	1.13	1.08	0.89	0.73	0.62	0.65	9.5
		REM DEM	KW	.00311	.00282	.00335	.00402	.00440	.00430	.00478	.00464	.00445	.00392	.00320	.00292	0.00478
		TOT ELEC	KWH	1.39	1.09	0.94	0.79	0.84	1.01	1.16	1.11	0.92	0.78	0.84	1.28	12.2
		TOT DEM	KW	.00691	.00622	.00636	.00435	.00442	.00477	.00498	.00484	.00463	.00415	.00518	.00636	0.00691
S- 32	0.24	HT ELEC	KWH	0.89	0.53	0.43	0.18	0.08	0.04	0.04	0.04	0.07	0.14	0.29	0.61	3.3
		HT DEM	KW	.00430	.00417	.00409	.00288	.00262	.00004	.00004	.00004	.00159	.00258	.00370	.00417	0.00430
		REM ELEC	KWH	0.68	0.62	0.75	0.80	0.86	1.06	1.13	1.12	0.95	0.80	0.67	0.68	10.1
		REM DEM	KW	.00349	.00340	.00373	.00432	.00451	.00466	.00485	.00475	.00470	.00422	.00359	.00320	0.00485
		TOT ELEC	KWH	1.56	1.14	1.18	0.98	0.94	1.10	1.17	1.16	1.02	0.94	0.96	1.29	13.4
		TOT DEM	KW	.00680	.00612	.00626	.00428	.00435	.00469	.00490	.00476	.00456	.00408	.00510	.00626	0.00680
S- 40	0.29	HT ELEC	KWH	0.87	0.52	0.42	0.18	0.08	0.04	0.04	0.04	0.06	0.14	0.28	0.60	3.3
		HT DEM	KW	.00430	.00417	.00409	.00288	.00262	.00004	.00004	.00004	.00159	.00258	.00370	.00417	0.00430
		REM ELEC	KWH	0.62	0.56	0.68	0.73	0.78	0.97	1.03	1.02	0.86	0.73	0.61	0.62	9.2
		REM DEM	KW	.00320	.00312	.00343	.00396	.00414	.00427	.00445	.00436	.00432	.00387	.00329	.00294	0.00445
		TOT ELEC	KWH	1.48	1.08	1.10	0.90	0.86	1.00	1.07	1.06	0.93	0.87	0.89	1.21	12.5
		TOT DEM	KW	.00666	.00599	.00613	.00420	.00426	.00460	.00480	.00466	.00446	.00400	.00500	.00613	0.00666
S- 45	0.89	HT ELEC	KWH	0.83	0.49	0.40	0.17	0.08	0.03	0.03	0.03	0.06	0.13	0.27	0.57	3.1
		HT DEM	KW	.00430	.00417	.00409	.00288	.00262	.00004	.00004	.00004	.00159	.00258	.00370	.00417	0.00430
		REM ELEC	KWH	0.65	0.59	0.72	0.77	0.82	1.02	1.09	1.08	0.91	0.77	0.64	0.65	9.7
		REM DEM	KW	.00335	.00325	.00358	.00414	.00432	.00446	.00465	.00456	.00451	.00405	.00344	.00307	0.00465
		TOT ELEC	KWH	1.48	1.08	1.12	0.93	0.90	1.05	1.12	1.11	0.97	0.90	0.91	1.22	12.8
		TOT DEM	KW	.00683	.00615	.00628	.00430	.00437	.00471	.00492	.00478	.00458	.00410	.00512	.00628	0.00683
S- 50	1.56	HT ELEC	KWH	1.02	0.61	0.50	0.21	0.10	0.04	0.04	0.04	0.08	0.16	0.33	0.70	3.8
		HT DEM	KW	.00460	.00446	.00437	.00308	.00281	.00005	.00005	.00005	.00170	.00276	.00396	.00446	0.00460
		REM ELEC	KWH	0.66	0.60	0.73	0.78	0.84	1.04	1.11	1.10	0.93	0.78	0.65	0.66	9.9
		REM DEM	KW	.00351	.00341	.00375	.00433	.00453	.00468	.00487	.00477	.00472	.00424	.00360	.00321	0.00487
		TOT ELEC	KWH	1.68	1.21	1.23	0.99	0.94	1.08	1.15	1.14	1.01	0.95	0.99	1.37	13.7
		TOT DEM	KW	.00637	.00573	.00586	.00401	.00408	.00440	.00459	.00446	.00427	.00382	.00478	.00586	0.00637
S- 61	-0.27	HT ELEC	KWH	0.70	0.48	0.28	0.08	0.03	0.03	0.03	0.03	0.03	0.05	0.21	0.60	2.5
		HT DEM	KW	.00483	.00469	.00459	.00324	.00295	.00005	.00005	.00005	.00179	.00290	.00415	.00469	0.00483
		REM ELEC	KWH	0.74	0.66	0.74	0.81	0.94	1.12	1.30	1.24	1.02	0.84	0.71	0.74	10.9
		REM DEM	KW	.00337	.00306	.00363	.00436	.00477	.00467	.00519	.00503	.00483	.00426	.00348	.00317	0.00519
		TOT ELEC	KWH	1.44	1.15	1.02	0.89	0.96	1.15	1.32	1.27	1.05	0.89	0.92	1.34	13.5
		TOT DEM	KW	.00721	.00649	.00663	.00454	.00461	.00497	.00519	.00505	.00483	.00433	.00541	.00663	0.00721
BASE CASE : 312																
BASE	0.00	TOT GAS	THM	0.0327	0.0226	0.0131	0.0039	0.0012	0.0012	0.0012	0.0012	0.0012	0.0021	0.0096	0.0280	0.119
		TOT ELEC	KWH	0.91	0.82	0.91	0.99	1.15	1.38	1.59	1.53	1.26	1.03	0.87	0.91	13.4
		TOT DEM	KW	.00416	.00378	.00448	.00538	.00589	.00576	.00640	.00621	.00595	.00525	.00429	.00390	0.00640
S- 3	-0.11	TOT GAS	THM	0.0347	0.0239	0.0139	0.0042	0.0013	0.0013	0.0013	0.0013	0.0013	0.0023	0.0102	0.0296	0.126
		TOT ELEC	KWH	0.82	0.74	0.82	0.90	1.04	1.25	1.44	1.38	1.14	0.93	0.79	0.82	12.1
		TOT DEM	KW	.00378	.00343	.00407	.00489	.00535	.00524	.00582	.00565	.00541	.00477	.00390	.00355	0.00582
S- 7	-0.03	TOT GAS	THM	0.0369	0.0255	0.0147	0.0044	0.0013	0.0013	0.0013	0.0013	0.0013	0.0024	0.0109	0.0315	0.134
		TOT ELEC	KWH	0.74	0.66	0.74	0.81	0.94	1.12	1.30	1.24	1.02	0.84	0.71	0.74	10.9
		TOT DEM	KW	.00337	.00306	.00363	.00436	.00477	.00467	.00519	.00503	.00483	.00426	.00348	.00317	0.00519
S- 16	1.07	TOT GAS	THM	0.0388	0.0268	0.0155	0.0047	0.0014	0.0014	0.0014	0.0014	0.0014	0.0025	0.0114	0.0331	0.141
		TOT ELEC	KWH	0.65	0.58	0.65	0.70	0.82	0.98	1.13	1.08	0.89	0.73	0.62	0.65	9.5
		TOT DEM	KW	.00311	.00282	.00335	.00402	.00440	.00430	.00478	.00464	.00445	.00392	.00320	.00292	0.00478
S- 32	0.24	TOT GAS	THM	0.0465	0.0277	0.0226	0.0094	0.0044	0.0019	0.0019	0.0019	0.0035	0.0075	0.0151	0.0320	0.174
		TOT ELEC	KWH	0.68	0.62	0.75	0.80	0.86	1.06	1.13	1.12	0.95	0.80	0.67	0.68	10.1
		TOT DEM	KW	.00349	.00340	.00373	.00432	.00451	.00466	.00485	.00475	.00470	.00422	.00359	.00320	0.00485
S- 40	0.28	TOT GAS	THM	0.0457	0.0272	0.0222	0.0092	0.0043	0.0019	0.0019	0.0019	0.0034	0.0074	0.0149	0.0315	0.171
		TOT ELEC	KWH	0.62	0.56	0.68	0.73	0.78	0.97	1.03	1.02	0.86	0.73	0.61	0.62	9.2
		TOT DEM	KW	.00320	.00312	.00343	.00396	.00414	.00427	.00445	.00436	.00432	.00387	.00329	.00294	0.00445
S- 45	0.88	TOT GAS	THM	0.0435	0.0259	0.0212	0.0088	0.0041	0.0018	0.0018	0.0018	0.0033	0.0070	0.0142	0.0300	0.163
		TOT ELEC	KWH	0.65	0.59	0.72	0.77	0.82	1.02	1.09	1.08	0.91	0.77	0.64	0.65	9.7
		TOT DEM	KW	.00335	.00325	.00358	.00414	.00432	.00446	.00465	.00456	.00451	.00405	.00344	.00307	0.00465
S- 50	1.55	TOT GAS	THM	0.0537	0.0320	0.0261	0.0109	0.0050	0.0022	0.0022	0.0022	0.0040	0.0086	0.0175	0.0370	0.201
		TOT ELEC	KWH	0.66	0.60	0.73	0.78	0.84	1.04	1.11	1.10	0.93	0.78	0.65	0.66	9.9
		TOT DEM	KW	.00351	.00341	.00375	.00433	.00453	.00468	.00487	.00477	.00472	.00424	.00360	.00321	0.00487
S- 61	-0.28	TOT GAS	THM	0.0369	0.0255	0.0147	0.0044	0.0013	0.0013	0.0013	0.0013	0.0013	0.0024	0.0109	0.0315	0.134
		TOT ELEC	KWH	0.74	0.66	0.74	0.81	0.94	1.12	1.30	1.24	1.02	0.84	0.71	0.74	10.9
		TOT DEM	KW	.00337	.00306	.00363	.00436	.00477	.00467	.00519	.00503	.00483	.00426	.00348	.00317	0.00519
BASE CASE : 313																
BASE	0.00	TOT ELEC	KWH	1.28	1.01	1.00	0.99	1.17	1.48	1.65	1.54	1.19	1.04	0.96	1.17	14.5
		TOT DEM	KW	.00532	.00477	.00516	.00602	.00665	.00633	.00782	.00727	.00680	.00563	.00485	.00485	0.00782
S- 3	-0.08	TOT ELEC	KWH	1.18	0.94	0.92	0.91	1.09	1.37	1.53	1.42	1.10	0.96	0.88	1.09	13.4
		TOT DEM	KW	.00498	.00447	.00484	.00564	.00623	.00594	.00733	.00682	.00638	.00528	.00454	.00454	0.00733
S- 9	0.01	TOT ELEC	KWH	1.09	0.87	0.86	0.84	1.00	1.26	1.41	1.31	1.02	0.89	0.82	1.00	12.4
		TOT DEM	KW	.00448	.00402	.00435	.00507	.00560	.00534	.00659	.00613	.00573	.00474	.00409	.00409	0.00659
S- 18	1.14	TOT ELEC	KWH	1.00	0.80	0.79	0.78	0.92	1.16	1.30	1.21	0.93	0.82	0.75	0.92	11.4
		TOT DEM	KW	.00425	.00381	.00413	.00481	.00531	.00506	.00625	.00581	.00544	.00450	.00387	.00387	0.00625

(HT ELEC=ELECTRIC CONSUMPTION FOR SPACE AND WATER HEATING. REM ELEC=REMAINING ELECTRIC CONSUMPTION (NOT USED FOR HEATING). TOT ELEC=TOTAL ELECTRIC CONSUMPTION. HT DEM=PEAK ELECTRIC DEMAND FOR SPACE AND WATER HEATING. REM DEM=PEAK ELECTRIC DEMAND FOR REMAINING LOAD (NOT HEATING). TOT DEM=TOTAL PEAK ELECTRIC DEMAND. GAS=TOTAL GAS CONSUMPTION. ANNUAL FIGURES ARE SUMS OF MONTHLY CONSUMPTION VALUES OR ARE YEARLY MAXIMUMS OF MONTHLY PEAK DEMAND VALUES.)

BASE CASE : 313 (CONTINUED)

STRAT SET	CONST COST $/SF	ENERGY FORM	UNITS (PER SF)	JAN	FEB	MAR	APR	MAY	JUN	JUL	AUG	SEP	OCT	NOV	DEC	ANNUAL
S- 33	0.28	TOT ELEC	KWH	1.28	0.99	1.04	0.92	0.89	1.04	1.12	1.11	0.96	0.87	0.89	1.10	12.2
		TOT DEM	KW	.00627	.00564	.00577	.00395	.00401	.00433	.00451	.00439	.00420	.00376	.00470	.00577	0.00627
S- 41	0.33	TOT ELEC	KWH	1.18	0.91	0.95	0.84	0.82	0.95	1.03	1.02	0.88	0.80	0.82	1.01	11.2
		TOT DEM	KW	.00602	.00542	.00554	.00379	.00385	.00415	.00433	.00421	.00403	.00361	.00452	.00554	0.00602
S- 46	0.94	TOT ELEC	KWH	1.22	0.94	0.99	0.87	0.85	0.99	1.07	1.06	0.92	0.82	0.85	1.04	11.6
		TOT DEM	KW	.00617	.00555	.00568	.00389	.00395	.00426	.00444	.00432	.00413	.00370	.00463	.00568	0.00617
S- 51	1.59	TOT ELEC	KWH	1.10	0.85	0.89	0.79	0.77	0.89	0.97	0.96	0.83	0.75	0.77	0.95	10.5
		TOT DEM	KW	.00600	.00540	.00552	.00378	.00384	.00414	.00432	.00420	.00402	.00360	.00450	.00552	0.00600
S- 62	-0.24	TOT ELEC	KWH	1.09	0.87	0.86	0.84	1.00	1.26	1.41	1.31	1.02	0.89	0.82	1.00	12.4
		TOT DEM	KW	.00448	.00402	.00435	.00507	.00560	.00534	.00659	.00613	.00573	.00474	.00409	.00409	0.00659

BASE CASE : 314

STRAT SET	CONST COST $/SF	ENERGY FORM	UNITS (PER SF)	JAN	FEB	MAR	APR	MAY	JUN	JUL	AUG	SEP	OCT	NOV	DEC	ANNUAL
BASE	0.00	HT ELEC	KWH	0.91	0.67	0.52	0.33	0.21	0.20	0.24	0.20	0.16	0.27	0.47	0.79	5.0
		HT DEM	KW	.00763	.00603	.00565	.00343	.00320	.00069	.00092	.00061	.00237	.00313	.00443	.00572	0.00763
		REM ELEC	KWH	1.05	0.95	1.05	1.15	1.33	1.60	1.84	1.77	1.46	1.19	1.01	1.05	15.5
		REM DEM	KW	.00408	.00323	.00478	.00576	.00633	.00612	.00703	.00675	.00647	.00527	.00436	.00408	0.00703
		TOT ELEC	KWH	1.96	1.61	1.57	1.48	1.55	1.80	2.09	1.97	1.62	1.46	1.47	1.84	20.5
		TOT DEM	KW	.01350	.01080	.01094	.00769	.00810	.00810	.00864	.00824	.00796	.00729	.00864	.01053	0.01350
S- 3	-0.12	HT ELEC	KWH	0.96	0.70	0.54	0.35	0.23	0.21	0.26	0.21	0.17	0.28	0.49	0.83	5.2
		HT DEM	KW	.00763	.00664	.00664	.00366	.00336	.00084	.00122	.00130	.00267	.00336	.00481	.00649	0.00763
		REM ELEC	KWH	0.96	0.86	0.96	1.04	1.21	1.45	1.68	1.61	1.33	1.09	0.92	0.96	14.1
		REM DEM	KW	.00321	.00254	.00376	.00453	.00498	.00481	.00553	.00531	.00509	.00415	.00343	.00321	0.00553
		TOT ELEC	KWH	1.92	1.56	1.50	1.39	1.44	1.66	1.93	1.82	1.50	1.37	1.41	1.79	19.3
		TOT DEM	KW	.00938	.00750	.00760	.00535	.00563	.00563	.00600	.00572	.00553	.00507	.00600	.00732	0.00938
S- 7	-0.03	HT ELEC	KWH	1.05	0.77	0.60	0.39	0.25	0.23	0.28	0.24	0.19	0.31	0.54	0.91	5.8
		HT DEM	KW	.00763	.00664	.00664	.00366	.00336	.00084	.00122	.00130	.00267	.00336	.00481	.00649	0.00763
		REM ELEC	KWH	0.87	0.78	0.87	0.95	1.10	1.32	1.52	1.46	1.20	0.99	0.83	0.87	12.8
		REM DEM	KW	.00304	.00242	.00357	.00431	.00472	.00457	.00525	.00504	.00483	.00394	.00326	.00304	0.00525
		TOT ELEC	KWH	1.92	1.55	1.47	1.33	1.35	1.55	1.80	1.69	1.39	1.30	1.37	1.78	18.5
		TOT DEM	KW	.00905	.00815	.00833	.00570	.00579	.00624	.00652	.00634	.00606	.00543	.00679	.00833	0.00905
S- 16	1.18	HT ELEC	KWH	1.09	0.80	0.62	0.40	0.26	0.24	0.29	0.25	0.20	0.32	0.56	0.94	6.0
		HT DEM	KW	.00763	.00664	.00664	.00366	.00336	.00084	.00122	.00130	.00267	.00336	.00481	.00649	0.00763
		REM ELEC	KWH	0.78	0.70	0.78	0.84	0.98	1.17	1.36	1.30	1.07	0.88	0.74	0.78	11.4
		REM DEM	KW	.00286	.00227	.00335	.00404	.00444	.00429	.00493	.00473	.00454	.00370	.00306	.00286	0.00493
		TOT ELEC	KWH	1.87	1.50	1.40	1.24	1.24	1.41	1.65	1.54	1.27	1.20	1.30	1.72	17.4
		TOT DEM	KW	.00871	.00784	.00801	.00549	.00557	.00601	.00627	.00610	.00584	.00523	.00653	.00801	0.00871
S- 32	0.24	HT ELEC	KWH	1.14	0.82	0.74	0.43	0.21	0.10	0.13	0.12	0.12	0.34	0.61	0.97	5.8
		HT DEM	KW	.00763	.00664	.00664	.00366	.00336	.00084	.00122	.00130	.00267	.00336	.00481	.00649	0.00763
		REM ELEC	KWH	0.83	0.74	0.83	0.90	1.05	1.26	1.45	1.39	1.15	0.94	0.79	0.83	12.2
		REM DEM	KW	.00508	.00435	.00424	.00480	.00547	.00536	.00558	.00552	.00513	.00502	.00508	.00413	0.00558
		TOT ELEC	KWH	1.97	1.56	1.57	1.33	1.26	1.36	1.58	1.51	1.27	1.28	1.40	1.80	18.0
		TOT DEM	KW	.00959	.00863	.00882	.00604	.00614	.00662	.00690	.00671	.00643	.00575	.00719	.00882	0.00959
S- 40	0.29	HT ELEC	KWH	1.16	0.83	0.76	0.43	0.22	0.11	0.13	0.12	0.12	0.35	0.62	0.99	5.9
		HT DEM	KW	.00763	.00664	.00664	.00366	.00336	.00084	.00122	.00130	.00267	.00336	.00481	.00649	0.00763
		REM ELEC	KWH	0.72	0.65	0.79	0.85	0.91	1.12	1.20	1.19	1.01	0.85	0.71	0.72	10.7
		REM DEM	KW	.00366	.00356	.00392	.00453	.00473	.00489	.00509	.00499	.00494	.00443	.00377	.00336	0.00509
		TOT ELEC	KWH	1.88	1.48	1.55	1.28	1.13	1.23	1.33	1.31	1.13	1.19	1.33	1.71	16.6
		TOT DEM	KW	.00875	.00788	.00805	.00551	.00560	.00604	.00630	.00612	.00586	.00525	.00656	.00805	0.00875
S- 45	0.94	HT ELEC	KWH	1.06	0.76	0.69	0.40	0.20	0.10	0.12	0.11	0.11	0.32	0.57	0.90	5.3
		HT DEM	KW	.00763	.00664	.00664	.00366	.00336	.00084	.00122	.00130	.00267	.00336	.00481	.00649	0.00763
		REM ELEC	KWH	0.78	0.71	0.86	0.92	0.99	1.22	1.30	1.29	1.09	0.92	0.77	0.78	11.6
		REM DEM	KW	.00400	.00389	.00428	.00495	.00517	.00534	.00556	.00545	.00539	.00484	.00411	.00367	0.00556
		TOT ELEC	KWH	1.84	1.47	1.55	1.31	1.18	1.31	1.42	1.40	1.20	1.23	1.33	1.68	17.0
		TOT DEM	KW	.00957	.00861	.00880	.00603	.00612	.00660	.00689	.00670	.00641	.00574	.00718	.00880	0.00957
S- 50	1.55	HT ELEC	KWH	1.08	0.77	0.70	0.40	0.20	0.10	0.12	0.11	0.11	0.32	0.58	0.92	5.4
		HT DEM	KW	.00763	.00664	.00664	.00366	.00336	.00084	.00122	.00130	.00267	.00336	.00481	.00649	0.00763
		REM ELEC	KWH	0.80	0.73	0.88	0.94	1.01	1.25	1.33	1.32	1.12	0.94	0.79	0.80	11.9
		REM DEM	KW	.00396	.00385	.00423	.00489	.00512	.00528	.00550	.00539	.00534	.00478	.00407	.00363	0.00550
		TOT ELEC	KWH	1.87	1.50	1.58	1.34	1.21	1.35	1.45	1.44	1.23	1.26	1.36	1.72	17.3
		TOT DEM	KW	.01027	.00924	.00945	.00647	.00657	.00709	.00739	.00719	.00688	.00616	.00770	.00945	0.01027
S- 61	-0.28	HT ELEC	KWH	1.05	0.77	0.60	0.39	0.25	0.23	0.28	0.24	0.19	0.31	0.54	0.91	5.8
		HT DEM	KW	.00763	.00664	.00664	.00366	.00336	.00084	.00122	.00130	.00267	.00336	.00481	.00649	0.00763
		REM ELEC	KWH	0.87	0.78	0.87	0.95	1.10	1.32	1.52	1.46	1.20	0.99	0.83	0.87	12.8
		REM DEM	KW	.00304	.00242	.00357	.00431	.00472	.00457	.00525	.00504	.00483	.00394	.00326	.00304	0.00525
		TOT ELEC	KWH	1.92	1.55	1.47	1.33	1.35	1.55	1.80	1.69	1.39	1.30	1.37	1.78	18.5
		TOT DEM	KW	.00905	.00815	.00833	.00570	.00579	.00624	.00652	.00634	.00606	.00543	.00679	.00833	0.00905

BASE CASE : 315

STRAT SET	CONST COST $/SF	ENERGY FORM	UNITS (PER SF)	JAN	FEB	MAR	APR	MAY	JUN	JUL	AUG	SEP	OCT	NOV	DEC	ANNUAL
BASE	0.00	TOT GAS	THM	0.0478	0.0350	0.0271	0.0175	0.0112	0.0104	0.0128	0.0107	0.0086	0.0141	0.0245	0.0412	0.261
		TOT ELEC	KWH	1.05	0.95	1.05	1.15	1.33	1.60	1.84	1.77	1.46	1.19	1.01	1.05	15.5
		TOT DEM	KW	.00408	.00323	.00478	.00576	.00633	.00612	.00703	.00675	.00647	.00527	.00436	.00408	0.00703
S- 3	-0.13	TOT GAS	THM	0.0503	0.0369	0.0286	0.0184	0.0118	0.0110	0.0135	0.0113	0.0091	0.0148	0.0258	0.0435	0.275
		TOT ELEC	KWH	0.96	0.86	0.96	1.04	1.21	1.45	1.68	1.61	1.33	1.09	0.92	0.96	14.1
		TOT DEM	KW	.00321	.00254	.00376	.00453	.00498	.00481	.00553	.00531	.00509	.00415	.00343	.00321	0.00553
S- 7	-0.03	TOT GAS	THM	0.0553	0.0405	0.0314	0.0202	0.0130	0.0121	0.0148	0.0124	0.0100	0.0163	0.0284	0.0477	0.302
		TOT ELEC	KWH	0.87	0.78	0.87	0.95	1.10	1.32	1.52	1.46	1.20	0.99	0.83	0.87	12.8
		TOT DEM	KW	.00304	.00242	.00357	.00431	.00472	.00457	.00525	.00504	.00483	.00394	.00326	.00304	0.00525
S- 16	1.17	TOT GAS	THM	0.0575	0.0421	0.0327	0.0210	0.0135	0.0126	0.0154	0.0129	0.0104	0.0170	0.0295	0.0496	0.314
		TOT ELEC	KWH	0.78	0.70	0.78	0.84	0.98	1.17	1.36	1.30	1.07	0.88	0.74	0.78	11.4
		TOT DEM	KW	.00286	.00227	.00335	.00404	.00444	.00429	.00493	.00473	.00454	.00370	.00306	.00286	0.00493

BASE CASE : 315 (CONTINUED)

STRAT SET	CONST COST $/SF	ENERGY FORM	UNITS (PER SF)	JAN	FEB	MAR	APR	MAY	JUN	JUL	AUG	SEP	OCT	NOV	DEC	ANNUAL
S- 32	0.24	TOT GAS	THM	0.0598	0.0429	0.0390	0.0223	0.0112	0.0054	0.0066	0.0063	0.0063	0.0178	0.0320	0.0510	0.302
		TOT ELEC	KWH	0.83	0.74	0.83	0.90	1.05	1.26	1.45	1.39	1.15	0.94	0.79	0.83	12.2
		TOT DEM	KW	.00508	.00435	.00424	.00480	.00547	.00536	.00558	.00552	.00513	.00502	.00508	.00413	0.00558
S- 40	0.28	TOT GAS	THM	0.0610	0.0437	0.0397	0.0228	0.0114	0.0055	0.0068	0.0065	0.0065	0.0182	0.0326	0.0521	0.308
		TOT ELEC	KWH	0.72	0.65	0.79	0.85	0.91	1.12	1.20	1.19	1.01	0.85	0.71	0.72	10.7
		TOT DEM	KW	.00366	.00356	.00392	.00453	.00473	.00489	.00509	.00499	.00494	.00443	.00377	.00336	0.00509
S- 45	0.93	TOT GAS	THM	0.0556	0.0399	0.0362	0.0208	0.0104	0.0051	0.0062	0.0059	0.0059	0.0166	0.0298	0.0475	0.281
		TOT ELEC	KWH	0.78	0.71	0.86	0.92	0.99	1.22	1.30	1.29	1.09	0.92	0.77	0.78	11.6
		TOT DEM	KW	.00400	.00389	.00428	.00495	.00517	.00534	.00556	.00545	.00539	.00484	.00411	.00367	0.00556
S- 50	1.55	TOT GAS	THM	0.0566	0.0406	0.0369	0.0212	0.0106	0.0051	0.0063	0.0060	0.0060	0.0169	0.0303	0.0483	0.286
		TOT ELEC	KWH	0.80	0.73	0.88	0.94	1.01	1.25	1.33	1.32	1.12	0.94	0.79	0.80	11.9
		TOT DEM	KW	.00396	.00385	.00423	.00489	.00512	.00528	.00550	.00539	.00534	.00478	.00407	.00363	0.00550
S- 61	-0.28	TOT GAS	THM	0.0553	0.0405	0.0314	0.0202	0.0130	0.0121	0.0148	0.0124	0.0100	0.0163	0.0284	0.0477	0.302
		TOT ELEC	KWH	0.87	0.78	0.87	0.95	1.10	1.32	1.52	1.46	1.20	0.99	0.83	0.87	12.8
		TOT DEM	KW	.00304	.00242	.00357	.00431	.00472	.00457	.00525	.00504	.00483	.00394	.00326	.00304	0.00525

BASE CASE : 316

STRAT SET	CONST COST $/SF	ENERGY FORM	UNITS (PER SF)	JAN	FEB	MAR	APR	MAY	JUN	JUL	AUG	SEP	OCT	NOV	DEC	ANNUAL
BASE	0.00	HT ELEC	KWH	0.89	0.62	0.36	0.11	0.03	0.03	0.03	0.03	0.03	0.06	0.26	0.76	3.2
		HT DEM	KW	.00585	.00567	.00556	.00392	.00357	.00006	.00006	.00006	.00216	.00351	.00503	.00567	0.00585
		REM ELEC	KWH	0.95	0.85	0.95	1.04	1.20	1.44	1.67	1.60	1.32	1.08	0.91	C.95	14.0
		REM DEM	KW	.00434	.00394	.00468	.00561	.00615	.00601	.00668	.00648	.00621	.00548	.00448	.00407	0.00668
		TOT ELEC	KWH	1.84	1.47	1.31	1.14	1.24	1.47	1.70	1.63	1.35	1.14	1.17	1.71	17.2
		TOT DEM	KW	.00800	.00712	.00712	.00600	.00616	.00656	.00672	.00656	.00640	.00576	.00584	.00720	0.00800
S- 3	-0.14	HT ELEC	KWH	0.95	0.66	0.38	0.11	0.03	0.03	0.03	0.03	0.03	0.06	0.28	0.81	3.5
		HT DEM	KW	.00585	.00567	.00556	.00392	.00357	.00006	.00006	.00006	.00216	.00351	.00503	.00567	0.00585
		REM ELEC	KWH	0.86	0.77	0.86	0.94	1.09	1.31	1.51	1.45	1.19	0.98	0.83	0.86	12.7
		REM DEM	KW	.00395	.00358	.00425	.00510	.00558	.00546	.00607	.00589	.00565	.00498	.00407	.00370	0.00607
		TOT ELEC	KWH	1.81	1.43	1.24	1.05	1.13	1.34	1.55	1.48	1.23	1.04	1.10	1.67	16.1
		TOT DEM	KW	.00760	.00676	.00676	.00570	.00585	.00623	.00638	.00623	.00608	.00547	.00555	.00684	0.00760
S- 7	-0.06	HT ELEC	KWH	1.00	0.69	0.40	0.12	0.04	0.04	0.04	0.04	0.04	0.07	0.29	0.86	3.6
		HT DEM	KW	.00585	.00567	.00556	.00392	.00357	.00006	.00006	.00006	.00216	.00351	.00503	.00567	0.00585
		REM ELEC	KWH	0.78	0.70	0.78	0.84	0.98	1.17	1.36	1.30	1.07	0.88	0.74	0.78	11.4
		REM DEM	KW	.00352	.00320	.00379	.00455	.00499	.00488	.00542	.00526	.00504	.00444	.00363	.00331	0.00542
		TOT ELEC	KWH	1.78	1.39	1.18	0.96	1.02	1.21	1.39	1.34	1.11	0.94	1.04	1.63	15.0
		TOT DEM	KW	.00730	.00657	.00672	.00460	.00467	.00504	.00526	.00511	.00489	.00438	.00548	.00672	0.00730
S- 16	1.01	HT ELEC	KWH	1.05	0.72	0.42	0.13	0.04	0.04	0.04	0.04	0.04	0.07	0.31	0.90	3.8
		HT DEM	KW	.00585	.00567	.00556	.00392	.00357	.00006	.00006	.00006	.00216	.00351	.00503	.00567	0.00585
		REM ELEC	KWH	0.67	0.61	0.67	0.73	0.85	1.02	1.18	1.13	0.93	0.76	0.64	0.67	9.9
		REM DEM	KW	.00324	.00294	.00349	.00419	.00459	.00449	.00499	.00484	.00464	.00409	.00334	.00304	0.00499
		TOT ELEC	KWH	1.72	1.33	1.09	0.86	0.89	1.06	1.22	1.17	0.97	0.83	0.95	1.57	13.7
		TOT DEM	KW	.00722	.00650	.00664	.00455	.00462	.00498	.00520	.00505	.00484	.00433	.00542	.00664	0.00722
S- 32	0.21	HT ELEC	KWH	1.25	0.74	0.61	0.25	0.12	0.05	0.05	0.05	0.09	0.20	0.41	0.86	4.7
		HT DEM	KW	.00535	.00519	.00508	.00358	.00326	.00005	.00005	.00005	.00198	.00321	.00460	.00519	0.00535
		REM ELEC	KWH	0.71	0.63	0.71	0.77	0.89	1.07	1.24	1.19	0.98	0.80	0.68	0.71	10.4
		REM DEM	KW	.00340	.00330	.00381	.00452	.00462	.00493	.00508	.00493	.00483	.00432	.00361	.00330	0.00508
		TOT ELEC	KWH	1.95	1.38	1.31	1.02	1.01	1.12	1.29	1.24	1.07	1.00	1.08	1.57	15.1
		TOT DEM	KW	.00680	.00612	.00626	.00428	.00435	.00469	.00490	.00476	.00456	.00408	.00510	.00626	0.00680
S- 40	0.26	HT ELEC	KWH	1.31	0.78	0.64	0.27	0.12	0.05	0.05	0.05	0.10	0.21	0.43	0.90	4.9
		HT DEM	KW	.00535	.00519	.00508	.00358	.00326	.00005	.00005	.00005	.00198	.00321	.00460	.00519	0.00535
		REM ELEC	KWH	0.64	0.58	0.70	0.75	0.81	1.00	1.06	1.05	0.89	0.75	0.63	0.64	9.5
		REM DEM	KW	.00318	.00309	.00356	.00423	.00432	.00461	.00475	.00461	.00451	.00404	.00337	.00309	0.00475
		TOT ELEC	KWH	1.95	1.36	1.34	1.02	0.93	1.05	1.12	1.11	0.99	0.96	1.05	1.54	14.4
		TOT DEM	KW	.00655	.00590	.00603	.00413	.00419	.00452	.00472	.00458	.00439	.00393	.00491	.00603	0.00655
S- 45	0.84	HT ELEC	KWH	1.20	0.72	0.58	0.24	0.11	0.05	0.05	0.05	0.09	0.19	0.39	0.83	4.5
		HT DEM	KW	.00535	.00519	.00508	.00358	.00326	.00005	.00005	.00005	.00198	.00321	.00460	.00519	0.00535
		REM ELEC	KWH	0.67	0.61	0.74	0.79	0.85	1.05	1.12	1.11	0.94	0.79	0.66	0.67	10.0
		REM DEM	KW	.00328	.00318	.00368	.00436	.00446	.00475	.00490	.00475	.00465	.00417	.00348	.00318	0.00490
		TOT ELEC	KWH	1.87	1.33	1.33	1.03	0.96	1.10	1.17	1.16	1.03	0.98	1.05	1.50	14.5
		TOT DEM	KW	.00670	.00603	.00616	.00422	.00429	.00462	.00482	.00469	.00449	.00402	.00502	.00616	0.00670
S- 50	1.53	HT ELEC	KWH	1.09	0.65	0.53	0.22	0.10	0.04	0.04	0.04	0.08	0.18	0.35	0.75	4.1
		HT DEM	KW	.00561	.00544	.00533	.00376	.00342	.00006	.00006	.00006	.00208	.00337	.00482	.00544	0.00561
		REM ELEC	KWH	0.68	0.62	0.75	0.80	0.86	1.06	1.13	1.12	0.95	0.80	0.67	0.68	10.1
		REM DEM	KW	.00325	.00315	.00364	.00432	.00441	.00470	.00485	.00470	.00461	.00412	.00344	.00315	0.00485
		TOT ELEC	KWH	1.76	1.26	1.28	1.02	0.96	1.11	1.18	1.17	1.03	0.97	1.02	1.43	14.2
		TOT DEM	KW	.00661	.00595	.00608	.00416	.00423	.00456	.00476	.00463	.00443	.00397	.00496	.00608	0.00661
S- 61	-0.30	HT ELEC	KWH	1.00	0.69	0.40	0.12	0.04	0.04	0.04	0.04	0.04	0.07	0.29	0.86	3.6
		HT DEM	KW	.00585	.00567	.00556	.00392	.00357	.00006	.00006	.00006	.00216	.00351	.00503	.00567	0.00585
		REM ELEC	KWH	0.78	0.70	0.78	0.84	0.98	1.17	1.36	1.30	1.07	0.88	0.74	0.78	11.4
		REM DEM	KW	.00352	.00320	.00379	.00455	.00499	.00488	.00542	.00526	.00504	.00444	.00363	.00331	0.00542
		TOT ELEC	KWH	1.78	1.39	1.18	0.96	1.02	1.21	1.39	1.34	1.11	0.94	1.04	1.63	15.0
		TOT DEM	KW	.00730	.00657	.00672	.00460	.00467	.00504	.00526	.00511	.00489	.00438	.00548	.00672	0.00730

BASE CASE : 317

STRAT SET	CONST COST $/SF	ENERGY FORM	UNITS (PER SF)	JAN	FEB	MAR	APR	MAY	JUN	JUL	AUG	SEP	OCT	NOV	DEC	ANNUAL
BASE	0.00	TOT GAS	THM	0.0468	0.0323	0.0187	0.0056	0.0017	0.0017	0.0017	0.0017	0.0017	0.0031	0.0138	0.0399	0.170
		TOT ELEC	KWH	0.95	0.85	0.95	1.04	1.20	1.44	1.67	1.60	1.32	1.08	0.91	0.95	14.0
		TOT DEM	KW	.00434	.00394	.00468	.00561	.00615	.00601	.00668	.00648	.00621	.00548	.00448	.00407	0.00668

(HT ELEC=ELECTRIC CONSUMPTION FOR SPACE AND WATER HEATING. REM ELEC=REMAINING ELECTRIC CONSUMPTION (NOT USED FOR HEATING). TOT ELEC=TOTAL ELECTRIC CONSUMPTION. HT DEM=PEAK ELECTRIC DEMAND FOR SPACE AND WATER HEATING. REM DEM=PEAK ELECTRIC DEMAND FOR REMAINING LOAD (NOT HEATING). TOT DEM=TOTAL PEAK ELECTRIC DEMAND. GAS=TOTAL GAS CONSUMPTION. ANNUAL FIGURES ARE SUMS OF MONTHLY CONSUMPTION VALUES OR ARE YEARLY MAXIMUMS OF MONTHLY PEAK DEMAND VALUES.)

STRAT SET	CONST COST $/SF	ENERGY FORM	UNITS (PER SF)	JAN	FEB	MAR	APR	MAY	JUN	JUL	AUG	SEP	OCT	NOV	DEC	ANNUAL
									BASE CASE : 317 (CONTINUED)							
S- 3	-0.15	TOT GAS	THM	0.0498	0.0344	0.0199	0.0060	0.0018	0.0018	0.0018	0.0018	0.0018	0.0033	0.0147	0.0425	0.181
		TOT ELEC	KWH	0.86	0.77	0.86	0.94	1.09	1.31	1.51	1.45	1.19	0.98	0.83	0.86	12.7
		TOT DEM	KW	.00395	.00358	.00425	.00510	.00558	.00546	.00607	.00589	.00565	.00498	.00407	.00370	0.00607
S- 7	-0.06	TOT GAS	THM	0.0525	0.0363	0.0210	0.0063	0.0019	0.0019	0.0019	0.0019	0.0019	0.0034	0.0155	0.0449	0.191
		TOT ELEC	KWH	0.78	0.70	0.78	0.84	0.98	1.17	1.36	1.30	1.07	0.88	0.74	0.78	11.4
		TOT DEM	KW	.00352	.00320	.00379	.00455	.00499	.00488	.00542	.00526	.00504	.00444	.00363	.00331	0.00542
S- 16	1.00	TOT GAS	THM	0.0556	0.0384	0.0222	0.0067	0.0020	0.0020	0.0020	0.0020	0.0020	0.0036	0.0164	0.0475	0.202
		TOT ELEC	KWH	0.67	0.61	0.67	0.73	0.85	1.02	1.18	1.13	0.93	0.76	0.64	0.67	9.9
		TOT DEM	KW	.00324	.00294	.00349	.00419	.00459	.00449	.00499	.00484	.00464	.00409	.00334	.00304	0.00499
S- 32	0.21	TOT GAS	THM	0.0654	0.0390	0.0318	0.0132	0.0061	0.0027	0.0027	0.0027	0.0049	0.0105	0.0213	0.0451	0.245
		TOT ELEC	KWH	0.70	0.63	0.77	0.82	0.88	1.09	1.16	1.15	0.98	0.82	0.69	0.70	10.4
		TOT DEM	KW	.00366	.00356	.00391	.00452	.00472	.00488	.00508	.00498	.00493	.00442	.00376	.00335	0.00508
S- 40	0.25	TOT GAS	THM	0.0689	0.0410	0.0335	0.0139	0.0065	0.0028	0.0028	0.0028	0.0052	0.0111	0.0224	0.0475	0.258
		TOT ELEC	KWH	0.64	0.58	0.70	0.75	0.81	1.00	1.06	1.05	0.89	0.75	0.63	0.64	9.5
		TOT DEM	KW	.00342	.00332	.00366	.00423	.00442	.00456	.00475	.00466	.00461	.00413	.00352	.00314	0.00475
S- 45	0.82	TOT GAS	THM	0.0630	0.0375	0.0307	0.0127	0.0059	0.0026	0.0026	0.0026	0.0047	0.0101	0.0205	0.0434	0.236
		TOT ELEC	KWH	0.67	0.61	0.74	0.79	0.85	1.05	1.12	1.11	0.94	0.79	0.66	0.67	10.0
		TOT DEM	KW	.00353	.00343	.00377	.00436	.00456	.00470	.00490	.00480	.00475	.00426	.00363	.00323	0.00490
S- 50	1.52	TOT GAS	THM	0.0611	0.0364	0.0298	0.0124	0.0057	0.0025	0.0025	0.0025	0.0046	0.0098	0.0199	0.0421	0.229
		TOT ELEC	KWH	0.68	0.62	0.75	0.80	0.86	1.06	1.13	1.12	0.95	0.80	0.67	0.68	10.1
		TOT DEM	KW	.00349	.00340	.00373	.00432	.00451	.00466	.00485	.00475	.00470	.00422	.00359	.00320	0.00485
S- 61	-0.31	TOT GAS	THM	0.0525	0.0363	0.0210	0.0063	0.0019	0.0019	0.0019	0.0019	0.0019	0.0034	0.0155	0.0449	0.191
		TOT ELEC	KWH	0.78	0.70	0.78	0.84	0.98	1.17	1.36	1.30	1.07	0.88	0.74	0.78	11.4
		TOT DEM	KW	.00352	.00320	.00379	.00455	.00499	.00488	.00542	.00526	.00504	.00444	.00363	.00331	0.00542
									BASE CASE : 318							
BASE	0.00	TOT ELEC	KWH	1.40	1.11	1.10	1.08	1.29	1.62	1.81	1.69	1.30	1.14	1.05	1.29	15.9
		TOT DEM	KW	.00567	.00509	.00550	.00642	.00709	.00676	.00834	.00776	.00726	.00600	.00517	.00517	0.00834
S- 3	-0.12	TOT ELEC	KWH	1.29	1.03	1.01	1.00	1.19	1.50	1.68	1.56	1.21	1.06	0.97	1.19	14.7
		TOT DEM	KW	.00532	.00477	.00516	.00602	.00665	.00633	.00782	.00727	.00680	.00563	.00485	.00485	0.00782
S- 9	-0.02	TOT ELEC	KWH	1.19	0.94	0.93	0.92	1.09	1.38	1.54	1.43	1.11	0.97	0.89	1.09	13.5
		TOT DEM	KW	.00478	.00429	.00464	.00541	.00598	.00569	.00703	.00654	.00612	.00506	.00436	.00436	0.00703
S- 18	1.07	TOT ELEC	KWH	1.10	0.88	0.86	0.85	1.01	1.27	1.42	1.32	1.02	0.90	0.82	1.01	12.5
		TOT DEM	KW	.00453	.00406	.00440	.00513	.00566	.00539	.00666	.00619	.00579	.00480	.00413	.00413	0.00666
S- 33	0.25	TOT ELEC	KWH	1.37	1.05	1.11	0.98	0.95	1.11	1.20	1.18	1.03	0.92	0.95	1.17	13.0
		TOT DEM	KW	.00621	.00559	.00571	.00391	.00397	.00428	.00447	.00435	.00416	.00373	.00466	.00571	0.00621
S- 41	0.29	TOT ELEC	KWH	1.30	1.00	1.05	0.93	0.91	1.05	1.14	1.13	0.98	0.88	0.91	1.12	12.4
		TOT DEM	KW	.00596	.00536	.00548	.00375	.00381	.00411	.00429	.00417	.00399	.00358	.00447	.00548	0.00596
S- 46	0.88	TOT ELEC	KWH	1.33	1.03	1.08	0.95	0.93	1.08	1.17	1.16	1.00	0.90	0.93	1.14	12.7
		TOT DEM	KW	.00611	.00550	.00562	.00385	.00391	.00422	.00440	.00428	.00409	.00367	.00458	.00562	0.00611
S- 51	1.56	TOT ELEC	KWH	1.23	0.95	0.99	0.88	0.85	0.99	1.08	1.06	0.92	0.83	0.85	1.05	11.7
		TOT DEM	KW	.00594	.00535	.00546	.00374	.00380	.00410	.00428	.00416	.00398	.00356	.00446	.00546	0.00594
S- 62	-0.27	TOT ELEC	KWH	1.19	0.94	0.93	0.92	1.09	1.38	1.54	1.43	1.11	0.97	0.89	1.09	13.5
		TOT DEM	KW	.00478	.00429	.00464	.00541	.00598	.00569	.00703	.00654	.00612	.00506	.00436	.00436	0.00703
									BASE CASE : 319							
BASE	0.00	HT ELEC	KWH	0.59	0.43	0.34	0.22	0.14	0.13	0.16	0.13	0.11	0.17	0.30	0.51	3.2
		HT DEM	KW	.00705	.00613	.00613	.00338	.00310	.00078	.00113	.00120	.00247	.00310	.00444	.00599	0.00705
		REM ELEC	KWH	0.99	0.89	0.99	1.08	1.26	1.50	1.74	1.66	1.37	1.12	0.95	0.99	14.6
		REM DEM	KW	.00387	.00307	.00454	.00548	.00601	.00581	.00668	.00641	.00615	.00501	.00414	.00387	0.00668
		TOT ELEC	KWH	1.59	1.32	1.33	1.30	1.39	1.63	1.90	1.80	1.48	1.30	1.25	1.50	17.8
		TOT DEM	KW	.01275	.01020	.01033	.00650	.00689	.00689	.00739	.00701	.00676	.00689	.00816	.00994	0.01275
S- 3	-0.09	HT ELEC	KWH	0.62	0.46	0.35	0.23	0.15	0.14	0.17	0.14	0.11	0.18	0.32	0.54	3.4
		HT DEM	KW	.00705	.00613	.00613	.00338	.00310	.00078	.00113	.00120	.00247	.00310	.00444	.00599	0.00705
		REM ELEC	KWH	0.90	0.81	0.90	0.98	1.14	1.37	1.58	1.52	1.25	1.02	0.86	0.90	13.3
		REM DEM	KW	.00305	.00242	.00358	.00431	.00473	.00458	.00526	.00505	.00484	.00395	.00326	.00305	0.00526
		TOT ELEC	KWH	1.53	1.27	1.26	1.21	1.29	1.51	1.75	1.66	1.36	1.21	1.19	1.44	16.7
		TOT DEM	KW	.00897	.00718	.00727	.00457	.00484	.00484	.00520	.00493	.00475	.00484	.00574	.00700	0.00897
S- 7	0.00	HT ELEC	KWH	0.68	0.50	0.39	0.25	0.16	0.15	0.18	0.15	0.12	0.20	0.35	0.59	3.7
		HT DEM	KW	.00705	.00613	.00613	.00338	.00310	.00078	.00113	.00120	.00247	.00310	.00444	.00599	0.00705
		REM ELEC	KWH	0.82	0.73	0.82	0.89	1.03	1.24	1.43	1.37	1.13	0.92	0.78	0.82	12.0
		REM DEM	KW	.00291	.00230	.00341	.00411	.00451	.00436	.00501	.00481	.00461	.00376	.00311	.00291	0.00501
		TOT ELEC	KWH	1.50	1.23	1.20	1.14	1.19	1.39	1.61	1.52	1.25	1.13	1.13	1.41	15.7
		TOT DEM	KW	.00867	.00694	.00702	.00442	.00468	.00468	.00503	.00477	.00460	.00468	.00555	.00676	0.00867
S- 16	1.22	HT ELEC	KWH	0.71	0.52	0.40	0.26	0.17	0.16	0.19	0.16	0.13	0.21	0.37	0.61	3.9
		HT DEM	KW	.00705	.00613	.00613	.00338	.00310	.00078	.00113	.00120	.00247	.00310	.00444	.00599	0.00705
		REM ELEC	KWH	0.73	0.65	0.73	0.79	0.92	1.10	1.27	1.22	1.01	0.82	0.70	0.73	10.7
		REM DEM	KW	.00272	.00216	.00319	.00385	.00422	.00408	.00469	.00450	.00431	.00352	.00291	.00272	0.00469
		TOT ELEC	KWH	1.44	1.17	1.13	1.05	1.09	1.26	1.46	1.38	1.13	1.03	1.06	1.34	14.6
		TOT DEM	KW	.00832	.00666	.00674	.00424	.00449	.00449	.00483	.00458	.00441	.00449	.00532	.00649	0.00832
S- 32	0.36	HT ELEC	KWH	0.75	0.54	0.49	0.28	0.14	0.07	0.08	0.08	0.08	0.22	0.40	0.64	3.8
		HT DEM	KW	.00705	.00613	.00613	.00338	.00310	.00078	.00113	.00120	.00247	.00310	.00444	.00599	0.00705
		REM ELEC	KWH	0.76	0.70	0.84	0.90	0.97	1.20	1.28	1.27	1.07	0.90	0.75	0.76	11.4
		REM DEM	KW	.00387	.00377	.00414	.00479	.00500	.00516	.00538	.00527	.00522	.00468	.00398	.00355	0.00538
		TOT ELEC	KWH	1.51	1.23	1.33	1.18	1.11	1.27	1.36	1.34	1.15	1.12	1.15	1.40	15.2
		TOT DEM	KW	.00892	.00803	.00821	.00562	.00571	.00615	.00642	.00624	.00598	.00535	.00669	.00821	0.00892

STRAT SET	CONST COST $/SF	ENERGY FORM	UNITS (PER SF)	JAN	FEB	MAR	APR	MAY	JUN	JUL	AUG	SEP	OCT	NOV	DEC	ANNUAL
									BASE CASE : 319 (CONTINUED)							
S- 61	-0.25	HT ELEC	KWH	0.68	0.50	0.39	0.25	0.16	0.15	0.18	0.15	0.12	0.20	0.35	0.59	3.7
		HT DEM	KW	.00705	.00613	.00613	.00338	.00310	.00078	.00113	.00120	.00247	.00310	.00444	.00599	0.00705
		REM ELEC	KWH	0.82	0.73	0.82	0.89	1.03	1.24	1.43	1.37	1.13	0.92	0.78	0.82	12.0
		REM DEM	KW	.00291	.00230	.00341	.00411	.00451	.00436	.00501	.00481	.00461	.00376	.00311	.00291	0.00501
		TOT ELEC	KWH	1.50	1.23	1.20	1.14	1.19	1.39	1.61	1.52	1.25	1.13	1.13	1.41	15.7
		TOT DEM	KW	.00867	.00694	.00702	.00442	.00468	.00468	.00503	.00477	.00460	.00468	.00555	.00676	0.00867
									BASE CASE : 320							
BASE	0.00	TOT GAS	THM	0.0311	0.0228	0.0177	0.0114	0.0073	0.0068	0.0083	0.0070	0.0056	0.0092	0.0160	0.0269	0.170
		TOT ELEC	KWH	0.99	0.89	0.99	1.08	1.26	1.50	1.74	1.66	1.37	1.12	0.95	0.99	14.6
		TOT DEM	KW	.00387	.00307	.00454	.00548	.00601	.00581	.00668	.00641	.00615	.00501	.00414	.00387	0.00668
S- 3	-0.09	TOT GAS	THM	0.0328	0.0240	0.0186	0.0120	0.0077	0.0072	0.0088	0.0073	0.0059	0.0097	0.0168	0.0283	0.179
		TOT ELEC	KWH	0.90	0.81	0.90	0.98	1.14	1.37	1.58	1.52	1.25	1.02	0.86	0.90	13.3
		TOT DEM	KW	.00305	.00242	.00358	.00431	.00473	.00458	.00526	.00505	.00484	.00395	.00326	.00305	0.00526
S- 7	-0.01	TOT GAS	THM	0.0359	0.0263	0.0204	0.0131	0.0084	0.0078	0.0096	0.0080	0.0065	0.0106	0.0184	0.0310	0.196
		TOT ELEC	KWH	0.82	0.73	0.82	0.89	1.03	1.24	1.43	1.37	1.13	0.92	0.78	0.82	12.0
		TOT DEM	KW	.00291	.00230	.00341	.00411	.00451	.00436	.00501	.00481	.00461	.00376	.00311	.00291	0.00501
S- 16	1.21	TOT GAS	THM	0.0373	0.0273	0.0212	0.0137	0.0088	0.0082	0.0100	0.0084	0.0067	0.0110	0.0192	0.0322	0.204
		TOT ELEC	KWH	0.73	0.65	0.73	0.79	0.92	1.10	1.27	1.22	1.01	0.82	0.70	0.73	10.7
		TOT DEM	KW	.00272	.00216	.00319	.00385	.00422	.00408	.00469	.00450	.00431	.00352	.00291	.00272	0.00469
S- 32	0.36	TOT GAS	THM	0.0394	0.0283	0.0257	0.0147	0.0074	0.0036	0.0044	0.0042	0.0042	0.0117	0.0211	0.0336	0.199
		TOT ELEC	KWH	0.78	0.70	0.78	0.84	0.98	1.17	1.36	1.30	1.07	0.88	0.74	0.78	11.4
		TOT DEM	KW	.00387	.00377	.00414	.00479	.00500	.00516	.00538	.00527	.00522	.00468	.00398	.00355	0.00538
S- 61	-0.25	TOT GAS	THM	0.0359	0.0263	0.0204	0.0131	0.0084	0.0078	0.0096	0.0080	0.0065	0.0106	0.0184	0.0310	0.196
		TOT ELEC	KWH	0.82	0.73	0.82	0.89	1.03	1.24	1.43	1.37	1.13	0.92	0.78	0.82	12.0
		TOT DEM	KW	.00291	.00230	.00341	.00411	.00451	.00436	.00501	.00481	.00461	.00376	.00311	.00291	0.00501
									BASE CASE : 321							
BASE	0.00	HT ELEC	KWH	0.62	0.43	0.25	0.07	0.02	0.02	0.02	0.02	0.02	0.04	0.18	0.53	2.3
		HT DEM	KW	.00607	.00589	.00577	.00407	.00370	.00006	.00006	.00006	.00225	.00364	.00522	.00589	0.00607
		REM ELEC	KWH	0.92	0.83	0.92	1.01	1.17	1.40	1.62	1.55	1.28	1.05	0.88	0.92	13.6
		REM DEM	KW	.00368	.00334	.00396	.00475	.00521	.00509	.00566	.00549	.00526	.00464	.00379	.00345	0.00566
		TOT ELEC	KWH	1.54	1.26	1.17	1.08	1.19	1.42	1.64	1.57	1.30	1.09	1.07	1.45	15.9
		TOT DEM	KW	.00795	.00708	.00700	.00533	.00549	.00588	.00604	.00588	.00580	.00525	.00564	.00723	0.00795
S- 3	-0.12	HT ELEC	KWH	0.65	0.45	0.26	0.08	0.02	0.02	0.02	0.02	0.02	0.04	0.19	0.56	2.4
		HT DEM	KW	.00607	.00589	.00577	.00407	.00370	.00006	.00006	.00006	.00225	.00364	.00522	.00589	0.00607
		REM ELEC	KWH	0.84	0.75	0.84	0.91	1.06	1.27	1.46	1.40	1.16	0.95	0.80	0.84	12.3
		REM DEM	KW	.00335	.00304	.00361	.00433	.00474	.00463	.00515	.00500	.00479	.00422	.00345	.00314	0.00515
		TOT ELEC	KWH	1.49	1.20	1.10	0.99	1.08	1.29	1.49	1.43	1.18	0.99	0.99	1.40	14.7
		TOT DEM	KW	.00755	.00672	.00664	.00506	.00521	.00559	.00574	.00559	.00551	.00498	.00536	.00687	0.00755
S- 7	-0.04	HT ELEC	KWH	0.70	0.48	0.28	0.08	0.03	0.03	0.03	0.03	0.03	0.05	0.20	0.59	2.5
		HT DEM	KW	.00607	.00589	.00577	.00407	.00370	.00006	.00006	.00006	.00225	.00364	.00522	.00589	0.00607
		REM ELEC	KWH	0.75	0.67	0.75	0.81	0.95	1.13	1.31	1.25	1.03	0.85	0.71	0.75	11.0
		REM DEM	KW	.00298	.00271	.00321	.00386	.00422	.00413	.00459	.00445	.00427	.00376	.00308	.00280	0.00459
		TOT ELEC	KWH	1.44	1.15	1.03	0.90	0.97	1.16	1.33	1.28	1.06	0.89	0.92	1.34	13.5
		TOT DEM	KW	.00733	.00660	.00674	.00462	.00469	.00506	.00528	.00513	.00491	.00440	.00550	.00674	0.00733
S- 16	1.05	HT ELEC	KWH	0.73	0.51	0.29	0.09	0.03	0.03	0.03	0.03	0.03	0.05	0.22	0.63	2.7
		HT DEM	KW	.00607	.00589	.00577	.00407	.00370	.00006	.00006	.00006	.00225	.00364	.00522	.00589	0.00607
		REM ELEC	KWH	0.66	0.59	0.66	0.71	0.83	0.99	1.15	1.10	0.91	0.74	0.63	0.66	9.6
		REM DEM	KW	.00275	.00250	.00296	.00355	.00389	.00381	.00423	.00410	.00393	.00347	.00283	.00258	0.00423
		TOT ELEC	KWH	1.39	1.10	0.95	0.80	0.86	1.02	1.17	1.13	0.93	0.79	0.84	1.28	12.3
		TOT DEM	KW	.00700	.00630	.00644	.00441	.00448	.00483	.00504	.00490	.00469	.00420	.00525	.00644	0.00700
S- 32	0.33	HT ELEC	KWH	0.74	0.44	0.36	0.15	0.07	0.03	0.03	0.03	0.06	0.12	0.24	0.51	2.8
		HT DEM	KW	.00607	.00589	.00577	.00407	.00370	.00006	.00006	.00006	.00225	.00364	.00522	.00589	0.00607
		REM ELEC	KWH	0.68	0.61	0.68	0.74	0.86	1.03	1.19	1.14	0.94	0.77	0.65	0.68	10.0
		REM DEM	KW	.00328	.00318	.00368	.00436	.00446	.00475	.00490	.00475	.00465	.00417	.00348	.00318	0.00490
		TOT ELEC	KWH	1.42	1.05	1.04	0.89	0.93	1.06	1.23	1.18	1.00	0.89	0.89	1.19	12.8
		TOT DEM	KW	.00630	.00567	.00580	.00397	.00403	.00435	.00454	.00441	.00422	.00378	.00472	.00580	0.00630
S- 61	-0.29	HT ELEC	KWH	0.70	0.48	0.28	0.08	0.03	0.03	0.03	0.03	0.03	0.05	0.20	0.59	2.5
		HT DEM	KW	.00607	.00589	.00577	.00407	.00370	.00006	.00006	.00006	.00225	.00364	.00522	.00589	0.00607
		REM ELEC	KWH	0.75	0.67	0.75	0.81	0.95	1.13	1.31	1.25	1.03	0.85	0.71	0.75	11.0
		REM DEM	KW	.00298	.00271	.00321	.00386	.00422	.00413	.00459	.00445	.00427	.00376	.00308	.00280	0.00459
		TOT ELEC	KWH	1.44	1.15	1.03	0.90	0.97	1.16	1.33	1.28	1.06	0.89	0.92	1.34	13.5
		TOT DEM	KW	.00730	.00657	.00672	.00460	.00467	.00504	.00526	.00511	.00489	.00438	.00548	.00672	0.00730

(HT ELEC=ELECTRIC CONSUMPTION FOR SPACE AND WATER HEATING. REM ELEC=REMAINING ELECTRIC CONSUMPTION (NOT USED FOR HEATING). TOT ELEC=TOTAL ELECTRIC CONSUMPTION. HT DEM=PEAK ELECTRIC DEMAND FOR SPACE AND WATER HEATING. REM DEM=PEAK ELECTRIC DEMAND FOR REMAINING LOAD (NOT HEATING). TOT DEM=TOTAL PEAK ELECTRIC DEMAND. GAS=TOTAL GAS CONSUMPTION. ANNUAL FIGURES ARE SUMS OF MONTHLY CONSUMPTION VALUES OR ARE YEARLY MAXIMUMS OF MONTHLY PEAK DEMAND VALUES.)

STRAT SET	CONST COST $/SF	ENERGY FORM	UNITS (PER SF)	JAN	FEB	MAR	APR	MAY	JUN	JUL	AUG	SEP	OCT	NOV	DEC	ANNUAL
									BASE CASE : 322							
BASE	0.00	TOT GAS	THM	0.0325	0.0224	0.0130	0.0039	0.0012	0.0012	0.0012	0.0012	0.0012	0.0021	0.0096	0.0277	0.118
		TOT ELEC	KWH	0.92	0.83	0.92	1.01	1.17	1.40	1.62	1.55	1.28	1.05	0.88	0.92	13.6
		TOT DEM	KW	.00368	.00334	.00396	.00475	.00521	.00509	.00566	.00549	.00526	.00464	.00379	.00345	0.00566
S- 3	-0.13	TOT GAS	THM	0.0344	0.0237	0.0137	0.0041	0.0012	0.0012	0.0012	0.0012	0.0012	0.0022	0.0101	0.0294	0.125
		TOT ELEC	KWH	0.84	0.75	0.84	0.91	1.06	1.27	1.46	1.40	1.16	0.95	0.80	0.84	12.3
		TOT DEM	KW	.00335	.00304	.00361	.00433	.00474	.00463	.00515	.00500	.00479	.00422	.00345	.00314	0.00515
S- 7	-0.04	TOT GAS	THM	0.0366	0.0253	0.0146	0.0044	0.0013	0.0013	0.0013	0.0013	0.0013	0.0024	0.0108	0.0313	0.133
		TOT ELEC	KWH	0.75	0.67	0.75	0.81	0.95	1.13	1.31	1.25	1.03	0.85	0.71	0.75	11.0
		TOT DEM	KW	.00298	.00271	.00321	.00386	.00422	.00413	.00459	.00445	.00427	.00376	.00308	.00280	0.00459
S- 16	1.04	TOT GAS	THM	0.0385	0.0266	0.0154	0.0046	0.0014	0.0014	0.0014	0.0014	0.0014	0.0025	0.0113	0.0329	0.140
		TOT ELEC	KWH	0.66	0.59	0.66	0.71	0.83	0.99	1.15	1.10	0.91	0.74	0.63	0.66	9.6
		TOT DEM	KW	.00275	.00250	.00296	.00355	.00389	.00381	.00423	.00410	.00393	.00347	.00283	.00258	0.00423
S- 32	0.32	TOT GAS	THM	0.0376	0.0224	0.0183	0.0076	0.0035	0.0016	0.0016	0.0016	0.0028	0.0061	0.0123	0.0259	0.141
		TOT ELEC	KWH	0.67	0.61	0.74	0.79	0.85	1.05	1.12	1.11	0.94	0.79	0.66	0.67	10.0
		TOT DEM	KW	.00353	.00343	.00377	.00436	.00456	.00470	.00490	.00480	.00475	.00426	.00363	.00323	0.00490
S- 61	-0.29	TOT GAS	THM	0.0366	0.0253	0.0146	0.0044	0.0013	0.0013	0.0013	0.0013	0.0013	0.0024	0.0108	0.0313	0.133
		TOT ELEC	KWH	0.75	0.67	0.75	0.81	0.95	1.13	1.31	1.25	1.03	0.85	0.71	0.75	11.0
		TOT DEM	KW	.00298	.00271	.00321	.00386	.00422	.00413	.00459	.00445	.00427	.00376	.00308	.00280	0.00459
									BASE CASE : 323							
BASE	0.00	TOT ELEC	KWH	1.30	1.04	1.02	1.01	1.20	1.51	1.69	1.57	1.21	1.07	0.98	1.20	14.8
		TOT DEM	KW	.00510	.00458	.00495	.00578	.00638	.00607	.00750	.00698	.00653	.00540	.00465	.00465	0.00750
S- 3	-0.09	TOT ELEC	KWH	1.21	0.96	0.95	0.93	1.11	1.40	1.56	1.45	1.12	0.99	0.90	1.11	13.7
		TOT DEM	KW	.00478	.00429	.00464	.00541	.00598	.00569	.00703	.00654	.00612	.00506	.00436	.00436	0.00703
S- 9	0.00	TOT ELEC	KWH	1.11	0.88	0.87	0.86	1.02	1.29	1.44	1.34	1.03	0.91	0.83	1.02	12.6
		TOT DEM	KW	.00430	.00386	.00417	.00487	.00537	.00512	.00632	.00588	.00550	.00455	.00392	.00392	0.00632
S- 18	1.10	TOT ELEC	KWH	1.02	0.81	0.80	0.79	0.94	1.18	1.32	1.23	0.95	0.84	0.77	0.94	11.6
		TOT DEM	KW	.00407	.00365	.00395	.00461	.00509	.00485	.00599	.00557	.00521	.00431	.00371	.00371	0.00599
S- 33	0.36	TOT ELEC	KWH	1.26	0.97	1.02	0.90	0.88	1.02	1.10	1.09	0.95	0.85	0.88	1.08	12.0
		TOT DEM	KW	.00529	.00476	.00487	.00333	.00339	.00365	.00381	.00370	.00354	.00317	.00397	.00487	0.00529
S- 62	-0.25	TOT ELEC	KWH	1.11	0.88	0.87	0.86	1.02	1.29	1.44	1.34	1.03	0.91	0.83	1.02	12.6
		TOT DEM	KW	.00430	.00386	.00417	.00487	.00537	.00512	.00632	.00588	.00550	.00455	.00392	.00392	0.00632
									BASE CASE : 324							
BASE	0.00	HT ELEC	KWH	0.86	0.63	0.49	0.31	0.20	0.19	0.23	0.19	0.16	0.25	0.44	0.74	4.7
		HT DEM	KW	.00639	.00505	.00473	.00288	.00268	.00058	.00077	.00051	.00198	.00262	.00371	.00479	0.00639
		REM ELEC	KWH	1.08	0.97	1.08	1.18	1.37	1.64	1.89	1.81	1.49	1.22	1.03	1.08	15.9
		REM DEM	KW	.00428	.00339	.00502	.00605	.00664	.00642	.00738	.00708	.00679	.00554	.00458	.00428	0.00738
		TOT ELEC	KWH	1.94	1.60	1.57	1.49	1.57	1.83	2.12	2.01	1.65	1.48	1.48	1.82	20.6
		TOT DEM	KW	.01360	.01088	.01102	.00775	.00816	.00816	.00870	.00830	.00802	.00734	.00870	.01061	0.01360
S- 3	-0.12	HT ELEC	KWH	0.91	0.66	0.51	0.33	0.21	0.20	0.24	0.20	0.16	0.27	0.47	0.78	4.9
		HT DEM	KW	.00639	.00556	.00556	.00307	.00281	.00070	.00102	.00109	.00224	.00281	.00403	.00543	0.00639
		REM ELEC	KWH	0.99	0.88	0.99	1.07	1.25	1.49	1.73	1.65	1.36	1.12	0.94	0.99	14.5
		REM DEM	KW	.00336	.00267	.00394	.00476	.00522	.00505	.00580	.00557	.00534	.00435	.00360	.00336	0.00580
		TOT ELEC	KWH	1.89	1.55	1.50	1.40	1.46	1.69	1.97	1.86	1.53	1.38	1.41	1.77	19.5
		TOT DEM	KW	.00980	.00784	.00794	.00559	.00588	.00588	.00627	.00598	.00578	.00529	.00627	.00764	0.00980
S- 7	-0.03	HT ELEC	KWH	0.99	0.73	0.56	0.36	0.23	0.22	0.27	0.22	0.18	0.29	0.51	0.86	5.4
		HT DEM	KW	.00639	.00556	.00556	.00307	.00281	.00070	.00102	.00109	.00224	.00281	.00403	.00543	0.00639
		REM ELEC	KWH	0.89	0.80	0.89	0.97	1.13	1.35	1.56	1.49	1.23	1.01	0.85	0.89	13.1
		REM DEM	KW	.00299	.00237	.00350	.00422	.00463	.00448	.00515	.00494	.00474	.00386	.00319	.00299	0.00515
		TOT ELEC	KWH	1.88	1.53	1.46	1.33	1.36	1.57	1.82	1.72	1.41	1.30	1.36	1.75	18.5
		TOT DEM	KW	.00950	.00855	.00874	.00599	.00608	.00655	.00684	.00665	.00637	.00570	.00713	.00874	0.00950
S- 16	1.17	HT ELEC	KWH	1.04	0.76	0.59	0.38	0.24	0.23	0.28	0.23	0.19	0.31	0.53	0.90	5.7
		HT DEM	KW	.00639	.00556	.00556	.00307	.00281	.00070	.00102	.00109	.00224	.00281	.00403	.00543	0.00639
		REM ELEC	KWH	0.80	0.71	0.80	0.87	1.01	1.21	1.39	1.33	1.10	0.90	0.76	0.80	11.7
		REM DEM	KW	.00255	.00202	.00299	.00361	.00396	.00383	.00440	.00422	.00405	.00330	.00273	.00255	0.00440
		TOT ELEC	KWH	1.83	1.47	1.39	1.25	1.25	1.43	1.67	1.57	1.29	1.21	1.29	1.69	17.4
		TOT DEM	KW	.00910	.00819	.00837	.00573	.00582	.00628	.00655	.00637	.00610	.00546	.00683	.00837	0.00910
S- 32	0.34	HT ELEC	KWH	1.12	0.81	0.73	0.42	0.21	0.10	0.12	0.12	0.12	0.33	0.60	0.96	5.7
		HT DEM	KW	.00639	.00556	.00556	.00307	.00281	.00070	.00102	.00109	.00224	.00281	.00403	.00543	0.00639
		REM ELEC	KWH	0.79	0.72	0.87	0.93	1.00	1.24	1.32	1.31	1.11	0.93	0.78	0.79	11.8
		REM DEM	KW	.00416	.00405	.00445	.00514	.00538	.00555	.00578	.00566	.00561	.00503	.00428	.00381	0.00578
		TOT ELEC	KWH	1.91	1.52	1.60	1.35	1.21	1.34	1.45	1.43	1.23	1.27	1.38	1.75	17.5
		TOT DEM	KW	.00970	.00873	.00892	.00611	.00621	.00669	.00698	.00679	.00650	.00582	.00728	.00892	0.00970
S- 61	-0.28	HT ELEC	KWH	0.99	0.73	0.56	0.36	0.23	0.22	0.27	0.22	0.18	0.29	0.51	0.86	5.4
		HT DEM	KW	.00639	.00556	.00556	.00307	.00281	.00070	.00102	.00109	.00224	.00281	.00403	.00543	0.00639
		REM ELEC	KWH	0.89	0.80	0.89	0.97	1.13	1.35	1.56	1.49	1.23	1.01	0.85	0.89	13.1
		REM DEM	KW	.00299	.00237	.00350	.00422	.00463	.00448	.00515	.00494	.00474	.00386	.00319	.00299	0.00515
		TOT ELEC	KWH	1.88	1.53	1.46	1.33	1.36	1.57	1.82	1.72	1.41	1.30	1.36	1.75	18.5
		TOT DEM	KW	.00950	.00855	.00874	.00599	.00608	.00655	.00684	.00665	.00637	.00570	.00713	.00874	0.00950
									BASE CASE : 325							
BASE	0.00	TOT GAS	THM	0.0452	0.0331	0.0257	0.0165	0.0106	0.0099	0.0121	0.0101	0.0082	0.0133	0.0232	0.0390	0.247
		TOT ELEC	KWH	1.08	0.97	1.08	1.18	1.37	1.64	1.89	1.81	1.49	1.22	1.03	1.08	15.9
		TOT DEM	KW	.00428	.00339	.00502	.00605	.00664	.00642	.00738	.00708	.00679	.00554	.00458	.00428	0.00738
S- 3	-0.13	TOT GAS	THM	0.0476	0.0348	0.0270	0.0174	0.0112	0.0104	0.0127	0.0107	0.0086	0.0140	0.0244	0.0411	0.260
		TOT ELEC	KWH	0.99	0.88	0.99	1.07	1.25	1.49	1.73	1.65	1.36	1.12	0.94	0.99	14.5
		TOT DEM	KW	.00336	.00267	.00394	.00476	.00522	.00505	.00580	.00557	.00534	.00435	.00360	.00336	0.00580

STRAT SET	CONST COST $/SF	ENERGY FORM	UNITS (PER SF)	JAN	FEB	MAR	APR	MAY	JUN	JUL	AUG	SEP	OCT	NOV	DEC	ANNUAL
									BASE CASE : 325 (CONTINUED)							
S- 7	-0.03	TOT GAS	THM	0.0522	0.0382	0.0296	0.0191	0.0123	0.0114	0.0140	0.0117	0.0094	0.0154	0.0268	0.0450	0.285
		TOT ELEC	KWH	0.89	0.80	0.89	0.97	1.13	1.35	1.56	1.49	1.23	1.01	0.85	0.89	13.1
		TOT DEM	KW	.00299	.00237	.00350	.00422	.00463	.00448	.00515	.00494	.00474	.00386	.00319	.00299	0.00515
S- 16	1.16	TOT GAS	THM	0.0544	0.0398	0.0309	0.0199	0.0128	0.0119	0.0146	0.0122	0.0098	0.0160	0.0279	0.0469	0.297
		TOT ELEC	KWH	0.80	0.71	0.80	0.87	1.01	1.21	1.39	1.33	1.10	0.90	0.76	0.80	11.7
		TOT DEM	KW	.00255	.00202	.00299	.00361	.00396	.00383	.00440	.00422	.00405	.00330	.00273	.00255	0.00440
S- 32	0.33	TOT GAS	THM	0.0564	0.0405	0.0368	0.0211	0.0105	0.0051	0.0063	0.0060	0.0060	0.0168	0.0302	0.0482	0.285
		TOT ELEC	KWH	0.79	0.72	0.87	0.93	1.00	1.24	1.32	1.31	1.11	0.93	0.78	0.79	11.8
		TOT DEM	KW	.00416	.00405	.00445	.00514	.00538	.00555	.00578	.00566	.00561	.00503	.00428	.00381	0.00578
S- 61	-0.28	TOT GAS	THM	0.0522	0.0382	0.0296	0.0191	0.0123	0.0114	0.0140	0.0117	0.0094	0.0154	0.0268	0.0450	0.285
		TOT ELEC	KWH	0.89	0.80	0.89	0.97	1.13	1.35	1.56	1.49	1.23	1.01	0.85	0.89	13.1
		TOT DEM	KW	.00299	.00237	.00350	.00422	.00463	.00448	.00515	.00494	.00474	.00386	.00319	.00299	0.00515
									BASE CASE : 326							
BASE	0.00	HT ELEC	KWH	0.76	0.53	0.31	0.09	0.03	0.03	0.03	0.03	0.03	0.05	0.23	0.65	2.8
		HT DEM	KW	.00605	.00587	.00575	.00405	.00369	.00006	.00006	.00006	.00224	.00363	.00520	.00587	0.00605
		REM ELEC	KWH	0.97	0.87	0.97	1.06	1.23	1.47	1.70	1.63	1.34	1.10	0.93	0.97	14.3
		REM DEM	KW	.00452	.00410	.00487	.00584	.00639	.00625	.00695	.00674	.00646	.00570	.00466	.00424	0.00695
		TOT ELEC	KWH	1.74	1.40	1.28	1.15	1.26	1.50	1.73	1.66	1.37	1.15	1.15	1.63	17.1
		TOT DEM	KW	.00930	.00828	.00818	.00623	.00642	.00688	.00707	.00688	.00679	.00614	.00660	.00846	0.00930
S- 3	-0.16	HT ELEC	KWH	0.81	0.56	0.32	0.10	0.03	0.03	0.03	0.03	0.03	0.05	0.24	0.69	3.0
		HT DEM	KW	.00605	.00587	.00575	.00405	.00369	.00006	.00006	.00006	.00224	.00363	.00520	.00587	0.00605
		REM ELEC	KWH	0.88	0.79	0.88	0.96	1.12	1.34	1.55	1.48	1.22	1.00	0.84	0.88	13.0
		REM DEM	KW	.00411	.00373	.00442	.00531	.00581	.00569	.00632	.00613	.00588	.00518	.00423	.00386	0.00632
		TOT ELEC	KWH	1.70	1.35	1.21	1.06	1.15	1.37	1.58	1.51	1.25	1.05	1.08	1.58	16.0
		TOT DEM	KW	.00890	.00792	.00783	.00596	.00614	.00659	.00676	.00659	.00650	.00587	.00632	.00810	0.00890
S- 7	-0.07	HT ELEC	KWH	0.86	0.59	0.34	0.10	0.03	0.03	0.03	0.03	0.03	0.06	0.25	0.73	3.1
		HT DEM	KW	.00605	.00587	.00575	.00405	.00369	.00006	.00006	.00006	.00224	.00363	.00520	.00587	0.00605
		REM ELEC	KWH	0.79	0.71	0.79	0.86	1.00	1.19	1.38	1.32	1.09	0.89	0.75	0.79	11.6
		REM DEM	KW	.00367	.00333	.00395	.00474	.00519	.00508	.00564	.00547	.00525	.00462	.00378	.00344	0.00564
		TOT ELEC	KWH	1.65	1.30	1.13	0.96	1.03	1.23	1.41	1.35	1.12	0.95	1.01	1.52	14.7
		TOT DEM	KW	.00865	.00778	.00796	.00545	.00554	.00597	.00623	.00605	.00580	.00519	.00649	.00796	0.00865
S- 16	1.00	HT ELEC	KWH	0.90	0.63	0.36	0.11	0.03	0.03	0.03	0.03	0.03	0.06	0.27	0.77	3.3
		HT DEM	KW	.00605	.00587	.00575	.00405	.00369	.00006	.00006	.00006	.00224	.00363	.00520	.00587	0.00605
		REM ELEC	KWH	0.69	0.62	0.69	0.75	0.87	1.04	1.20	1.15	0.95	0.78	0.66	0.69	10.1
		REM DEM	KW	.00337	.00306	.00363	.00436	.00477	.00467	.00519	.00503	.00483	.00426	.00348	.00317	0.00519
		TOT ELEC	KWH	1.59	1.24	1.05	0.86	0.90	1.07	1.23	1.18	0.98	0.84	0.92	1.46	13.4
		TOT DEM	KW	.00835	.00751	.00768	.00526	.00534	.00576	.00601	.00585	.00559	.00501	.00626	.00768	0.00835
S- 32	0.30	HT ELEC	KWH	1.05	0.62	0.51	0.21	0.10	0.04	0.04	0.04	0.08	0.17	0.34	0.72	3.9
		HT DEM	KW	.00555	.00538	.00527	.00372	.00339	.00006	.00006	.00006	.00205	.00333	.00477	.00538	0.00555
		REM ELEC	KWH	0.68	0.62	0.75	0.81	0.87	1.07	1.14	1.13	0.96	0.81	0.67	0.68	10.2
		REM DEM	KW	.00367	.00357	.00393	.00454	.00474	.00490	.00510	.00500	.00495	.00444	.00377	.00337	0.00510
		TOT ELEC	KWH	1.73	1.25	1.27	1.02	0.97	1.11	1.19	1.18	1.04	0.97	1.02	1.41	14.1
		TOT DEM	KW	.00828	.00745	.00762	.00522	.00530	.00571	.00596	.00580	.00555	.00497	.00621	.00762	0.00828
S- 61	-0.32	HT ELEC	KWH	0.86	0.59	0.34	0.10	0.03	0.03	0.03	0.03	0.03	0.06	0.25	0.73	3.1
		HT DEM	KW	.00605	.00587	.00575	.00405	.00369	.00006	.00006	.00006	.00224	.00363	.00520	.00587	0.00605
		REM ELEC	KWH	0.79	0.71	0.79	0.86	1.00	1.19	1.38	1.32	1.09	0.89	0.75	0.79	11.6
		REM DEM	KW	.00367	.00333	.00395	.00474	.00519	.00508	.00564	.00547	.00525	.00462	.00378	.00344	0.00564
		TOT ELEC	KWH	1.65	1.30	1.13	0.96	1.03	1.23	1.41	1.35	1.12	0.95	1.01	1.52	14.7
		TOT DEM	KW	.00865	.00778	.00796	.00545	.00554	.00597	.00623	.00605	.00580	.00519	.00649	.00796	0.00865
									BASE CASE : 327							
BASE	0.00	TOT GAS	THM	0.0402	0.0277	0.0161	0.0048	0.0015	0.0015	0.0015	0.0015	0.0015	0.0026	0.0118	0.0343	0.146
		TOT ELEC	KWH	0.97	0.87	0.97	1.06	1.23	1.47	1.70	1.63	1.34	1.10	0.93	0.97	14.3
		TOT DEM	KW	.00452	.00410	.00487	.00584	.00639	.00625	.00695	.00674	.00646	.00570	.00466	.00424	0.00695
S- 3	-0.16	TOT GAS	THM	0.0426	0.0294	0.0170	0.0051	0.0015	0.0015	0.0015	0.0015	0.0015	0.0028	0.0126	0.0364	0.155
		TOT ELEC	KWH	0.88	0.79	0.88	0.96	1.12	1.34	1.55	1.48	1.22	1.00	0.84	0.88	13.0
		TOT DEM	KW	.00411	.00373	.00442	.00531	.00581	.00569	.00632	.00613	.00588	.00518	.00423	.00386	0.00632
S- 7	-0.08	TOT GAS	THM	0.0451	0.0312	0.0180	0.0054	0.0016	0.0016	0.0016	0.0016	0.0016	0.0030	0.0133	0.0385	0.164
		TOT ELEC	KWH	0.79	0.71	0.79	0.86	1.00	1.19	1.38	1.32	1.09	0.89	0.75	0.79	11.6
		TOT DEM	KW	.00367	.00333	.00395	.00474	.00519	.00508	.00564	.00547	.00525	.00462	.00378	.00344	0.00564
S- 16	0.98	TOT GAS	THM	0.0476	0.0329	0.0190	0.0057	0.0017	0.0017	0.0017	0.0017	0.0017	0.0031	0.0140	0.0407	0.173
		TOT ELEC	KWH	0.69	0.62	0.69	0.75	0.87	1.04	1.20	1.15	0.95	0.78	0.66	0.69	10.1
		TOT DEM	KW	.00337	.00306	.00363	.00436	.00477	.00467	.00519	.00503	.00483	.00426	.00348	.00317	0.00519
S- 32	0.29	TOT GAS	THM	0.0553	0.0329	0.0269	0.0112	0.0052	0.0023	0.0023	0.0023	0.0041	0.0089	0.0180	0.0381	0.207
		TOT ELEC	KWH	0.68	0.62	0.75	0.81	0.87	1.07	1.14	1.13	0.96	0.81	0.67	0.68	10.2
		TOT DEM	KW	.00367	.00357	.00393	.00454	.00474	.00490	.00510	.00500	.00495	.00444	.00377	.00337	0.00510
S- 61	-0.33	TOT GAS	THM	0.0451	0.0312	0.0180	0.0054	0.0016	0.0016	0.0016	0.0016	0.0016	0.0030	0.0133	0.0385	0.164
		TOT ELEC	KWH	0.79	0.71	0.79	0.86	1.00	1.19	1.38	1.32	1.09	0.89	0.75	0.79	11.6
		TOT DEM	KW	.00367	.00333	.00395	.00474	.00519	.00508	.00564	.00547	.00525	.00462	.00378	.00344	0.00564
									BASE CASE : 328							
BASE	0.00	TOT ELEC	KWH	1.43	1.13	1.12	1.10	1.31	1.65	1.85	1.72	1.33	1.17	1.07	1.31	16.2
		TOT DEM	KW	.00581	.00521	.00564	.00658	.00726	.00692	.00854	.00794	.00743	.00615	.00529	.00529	0.00854
S- 3	-0.13	TOT ELEC	KWH	1.32	1.05	1.03	1.02	1.22	1.53	1.71	1.59	1.23	1.08	0.99	1.22	15.0
		TOT DEM	KW	.00545	.00489	.00529	.00617	.00681	.00649	.00801	.00745	.00697	.00577	.00497	.00497	0.00801

(HT ELEC=ELECTRIC CONSUMPTION FOR SPACE AND WATER HEATING. REM ELEC=REMAINING ELECTRIC CONSUMPTION (NOT USED FOR HEATING). TOT ELEC=TOTAL ELECTRIC CONSUMPTION. HT DEM=PEAK ELECTRIC DEMAND FOR SPACE AND WATER HEATING. REM DEM=PEAK ELECTRIC DEMAND FOR REMAINING LOAD (NOT HEATING). TOT DEM=TOTAL PEAK ELECTRIC DEMAND. GAS=TOTAL GAS CONSUMPTION. ANNUAL FIGURES ARE SUMS OF MONTHLY CONSUMPTION VALUES OR ARE YEARLY MAXIMUMS OF MONTHLY PEAK DEMAND VALUES.)

STRAT SET	CONST COST $/SF	ENERGY FORM	UNITS (PER SF)	JAN	FEB	MAR	APR	MAY	JUN	JUL	AUG	SEP	OCT	NOV	DEC	ANNUAL
BASE CASE : 328 (CONTINUED)																
S- 9	-0.03	TOT ELEC	KWH	1.21	0.97	0.95	0.94	1.12	1.41	1.57	1.46	1.13	0.99	0.91	1.12	13.8
		TOT DEM	KW	.00490	.00439	.00475	.00554	.00612	.00583	.00720	.00670	.00626	.00518	.00446	.00446	0.00720
S- 18	1.04	TOT ELEC	KWH	1.12	0.89	0.88	0.86	1.03	1.30	1.45	1.35	1.04	0.91	0.84	1.03	12.7
		TOT DEM	KW	.00464	.00416	.00450	.00525	.00580	.00552	.00682	.00634	.00593	.00491	.00423	.00423	0.00682
S- 33	0.33	TOT ELEC	KWH	1.32	1.02	1.07	0.95	0.92	1.07	1.16	1.15	1.00	0.89	0.92	1.13	12.6
		TOT DEM	KW	.00687	.00618	.00632	.00433	.00440	.00474	.00495	.00481	.00460	.00412	.00515	.00632	0.00687
S- 62	-0.28	TOT ELEC	KWH	1.21	0.97	0.95	0.94	1.12	1.41	1.57	1.46	1.13	0.99	0.91	1.12	13.8
		TOT DEM	KW	.00490	.00439	.00475	.00554	.00612	.00583	.00720	.00670	.00626	.00518	.00446	.00446	0.00720
BASE CASE : 329																
BASE	0.00	HT ELEC	KWH	0.46	0.34	0.26	0.17	0.11	0.10	0.12	0.10	0.08	0.14	0.23	0.40	2.5
		HT DEM	KW	.00668	.00528	.00494	.00301	.00281	.00060	.00080	.00053	.00207	.00274	.00387	.00501	0.00668
		REM ELEC	KWH	1.00	0.90	1.00	1.09	1.26	1.51	1.75	1.68	1.38	1.13	0.96	1.00	14.7
		REM DEM	KW	.00393	.00312	.00461	.00556	.00610	.00590	.00678	.00651	.00624	.00508	.00420	.00393	0.00678
		TOT ELEC	KWH	1.46	1.23	1.26	1.26	1.37	1.61	1.87	1.78	1.46	1.27	1.19	1.39	17.2
		TOT DEM	KW	.01240	.00992	.01004	.00632	.00670	.00670	.00719	.00682	.00657	.00670	.00794	.00967	0.01240
S- 3	-0.10	HT ELEC	KWH	0.48	0.35	0.27	0.18	0.11	0.11	0.13	0.11	0.09	0.14	0.25	0.42	2.6
		HT DEM	KW	.00548	.00477	.00477	.00263	.00241	.00060	.00088	.00093	.00192	.00241	.00345	.00466	0.00548
		REM ELEC	KWH	0.91	0.82	0.91	0.99	1.15	1.38	1.59	1.53	1.26	1.03	0.87	0.91	13.4
		REM DEM	KW	.00309	.00245	.00362	.00437	.00480	.00464	.00533	.00512	.00490	.00400	.00330	.00309	0.00533
		TOT ELEC	KWH	1.39	1.17	1.18	1.17	1.27	1.49	1.72	1.64	1.35	1.17	1.12	1.33	16.0
		TOT DEM	KW	.00848	.00678	.00687	.00432	.00458	.00458	.00492	.00466	.00449	.00458	.00543	.00661	0.00848
S- 7	-0.01	HT ELEC	KWH	0.53	0.39	0.30	0.19	0.12	0.12	0.14	0.12	0.10	0.16	0.27	0.46	2.9
		HT DEM	KW	.00548	.00477	.00477	.00263	.00241	.00060	.00088	.00093	.00192	.00241	.00345	.00466	0.00548
		REM ELEC	KWH	0.82	0.74	0.82	0.90	1.04	1.25	1.44	1.38	1.14	0.93	0.79	0.82	12.1
		REM DEM	KW	.00293	.00232	.00343	.00414	.00454	.00439	.00505	.00485	.00465	.00379	.00313	.00293	0.00505
		TOT ELEC	KWH	1.35	1.13	1.12	1.09	1.16	1.36	1.58	1.50	1.23	1.09	1.06	1.28	15.0
		TOT DEM	KW	.00818	.00654	.00663	.00417	.00442	.00442	.00474	.00450	.00434	.00442	.00524	.00638	0.00818
S- 16	1.20	HT ELEC	KWH	0.55	0.40	0.31	0.20	0.13	0.12	0.15	0.12	0.10	0.16	0.28	0.47	3.0
		HT DEM	KW	.00548	.00477	.00477	.00263	.00241	.00060	.00088	.00093	.00192	.00241	.00345	.00466	0.00548
		REM ELEC	KWH	0.73	0.66	0.73	0.80	0.93	1.11	1.29	1.23	1.02	0.83	0.70	0.73	10.8
		REM DEM	KW	.00274	.00218	.00322	.00388	.00426	.00412	.00473	.00454	.00435	.00355	.00293	.00274	0.00473
		TOT ELEC	KWH	1.28	1.06	1.05	1.00	1.06	1.23	1.43	1.35	1.11	0.99	0.98	1.21	13.8
		TOT DEM	KW	.00780	.00624	.00632	.00398	.00421	.00421	.00452	.00429	.00413	.00421	.00499	.00608	0.00780
S- 32	0.22	HT ELEC	KWH	0.58	0.42	0.38	0.22	0.11	0.05	0.06	0.06	0.06	0.17	0.31	0.50	3.0
		HT DEM	KW	.00548	.00477	.00477	.00263	.00241	.00060	.00088	.00093	.00192	.00241	.00345	.00466	0.00548
		REM ELEC	KWH	0.80	0.73	0.88	0.94	1.01	1.25	1.33	1.32	1.12	0.94	0.79	0.80	11.9
		REM DEM	KW	.00427	.00405	.00444	.00505	.00533	.00544	.00555	.00544	.00538	.00488	.00427	.00366	0.00555
		TOT ELEC	KWH	1.38	1.14	1.26	1.16	1.12	1.30	1.40	1.38	1.18	1.11	1.10	1.30	14.9
		TOT DEM	KW	.00865	.00778	.00796	.00545	.00554	.00597	.00623	.00605	.00580	.00519	.00649	.00796	0.00865
S- 40	0.24	HT ELEC	KWH	0.60	0.43	0.39	0.22	0.11	0.05	0.07	0.06	0.06	0.18	0.32	0.51	3.0
		HT DEM	KW	.00548	.00477	.00477	.00263	.00241	.00060	.00088	.00093	.00192	.00241	.00345	.00466	0.00548
		REM ELEC	KWH	0.70	0.64	0.78	0.83	0.89	1.10	1.18	1.17	0.99	0.83	0.69	0.70	10.5
		REM DEM	KW	.00387	.00367	.00402	.00458	.00483	.00493	.00503	.00493	.00488	.00443	.00387	.00332	0.00503
		TOT ELEC	KWH	1.30	1.07	1.17	1.05	1.00	1.16	1.24	1.23	1.05	1.01	1.01	1.21	13.5
		TOT DEM	KW	.00759	.00683	.00698	.00478	.00486	.00524	.00546	.00531	.00509	.00455	.00569	.00698	0.00759
S- 45	1.00	HT ELEC	KWH	0.54	0.39	0.35	0.20	0.10	0.05	0.06	0.06	0.06	0.16	0.29	0.46	2.7
		HT DEM	KW	.00548	.00477	.00477	.00263	.00241	.00060	.00088	.00093	.00192	.00241	.00345	.00466	0.00548
		REM ELEC	KWH	0.76	0.70	0.84	0.90	0.97	1.20	1.28	1.27	1.07	0.90	0.75	0.76	11.4
		REM DEM	KW	.00423	.00402	.00440	.00501	.00528	.00539	.00550	.00539	.00534	.00484	.00423	.00363	0.00550
		TOT ELEC	KWH	1.31	1.08	1.20	1.10	1.07	1.25	1.34	1.32	1.13	1.06	1.04	1.23	14.1
		TOT DEM	KW	.00830	.00747	.00764	.00523	.00531	.00573	.00598	.00581	.00556	.00498	.00623	.00764	0.00830
S- 61	-0.26	HT ELEC	KWH	0.53	0.39	0.30	0.19	0.12	0.12	0.14	0.12	0.10	0.16	0.27	0.46	2.9
		HT DEM	KW	.00548	.00477	.00477	.00263	.00241	.00060	.00088	.00093	.00192	.00241	.00345	.00466	0.00548
		REM ELEC	KWH	0.82	0.74	0.82	0.90	1.04	1.25	1.44	1.38	1.14	0.93	0.79	0.82	12.1
		REM DEM	KW	.00293	.00232	.00343	.00414	.00454	.00439	.00505	.00485	.00465	.00379	.00313	.00293	0.00505
		TOT ELEC	KWH	1.35	1.13	1.12	1.09	1.16	1.36	1.58	1.50	1.23	1.09	1.06	1.28	15.0
		TOT DEM	KW	.00818	.00654	.00663	.00417	.00442	.00442	.00474	.00450	.00434	.00442	.00524	.00638	0.00818
BASE CASE : 330																
BASE	0.00	TOT GAS	THM	0.0240	0.0176	0.0137	0.0088	0.0056	0.0053	0.0064	0.0054	0.0043	0.0071	0.0123	0.0207	0.131
		TOT ELEC	KWH	1.00	0.90	1.00	1.09	1.26	1.51	1.75	1.68	1.38	1.13	0.96	1.00	14.7
		TOT DEM	KW	.00393	.00312	.00461	.00556	.00610	.00590	.00678	.00651	.00624	.00508	.00420	.00393	0.00678
S- 3	-0.11	TOT GAS	THM	0.0253	0.0185	0.0144	0.0093	0.0059	0.0055	0.0068	0.0057	0.0046	0.0075	0.0130	0.0218	0.138
		TOT ELEC	KWH	0.91	0.82	0.91	0.99	1.15	1.38	1.59	1.53	1.26	1.03	0.87	0.91	13.4
		TOT DEM	KW	.00309	.00245	.00362	.00437	.00480	.00464	.00533	.00512	.00490	.00400	.00330	.00309	0.00533
S- 7	-0.02	TOT GAS	THM	0.0278	0.0203	0.0158	0.0102	0.0065	0.0061	0.0074	0.0062	0.0050	0.0082	0.0143	0.0240	0.152
		TOT ELEC	KWH	0.82	0.74	0.82	0.90	1.04	1.25	1.44	1.38	1.14	0.93	0.79	0.82	12.1
		TOT DEM	KW	.00293	.00232	.00343	.00414	.00454	.00439	.00505	.00485	.00465	.00379	.00313	.00293	0.00505
S- 16	1.18	TOT GAS	THM	0.0289	0.0212	0.0164	0.0106	0.0068	0.0063	0.0077	0.0065	0.0052	0.0085	0.0148	0.0249	0.158
		TOT ELEC	KWH	0.73	0.66	0.73	0.80	0.93	1.11	1.29	1.23	1.02	0.83	0.70	0.73	10.8
		TOT DEM	KW	.00274	.00218	.00322	.00388	.00426	.00412	.00473	.00454	.00435	.00355	.00293	.00274	0.00473
S- 32	0.21	TOT GAS	THM	0.0307	0.0220	0.0200	0.0115	0.0057	0.0028	0.0034	0.0033	0.0033	0.0091	0.0164	0.0262	0.155
		TOT ELEC	KWH	0.78	0.71	0.87	0.92	0.99	1.23	1.31	1.30	1.10	0.92	0.77	0.78	11.7
		TOT DEM	KW	.00427	.00405	.00444	.00505	.00533	.00544	.00555	.00544	.00538	.00488	.00427	.00366	0.00555
S- 40	0.23	TOT GAS	THM	0.0313	0.0224	0.0204	0.0117	0.0058	0.0028	0.0035	0.0033	0.0033	0.0093	0.0167	0.0267	0.158
		TOT ELEC	KWH	0.74	0.67	0.81	0.87	0.94	1.15	1.23	1.22	1.03	0.87	0.73	0.74	11.0
		TOT DEM	KW	.00387	.00367	.00402	.00458	.00483	.00493	.00503	.00493	.00488	.00443	.00387	.00332	0.00503

STRAT SET	CONST COST $/SF	ENERGY FORM	UNITS (PER SF)	JAN	FEB	MAR	APR	MAY	JUN	JUL	AUG	SEP	OCT	NOV	DEC	ANNUAL
BASE CASE : 330 (CONTINUED)																
S- 45	0.98	TOT GAS	THM	0.0285	0.0204	0.0186	0.0107	0.0053	0.0026	0.0032	0.0030	0.0030	0.0085	0.0153	0.0243	0.144
		TOT ELEC	KWH	0.78	0.71	0.86	0.92	0.99	1.22	1.30	1.29	1.09	0.92	0.77	0.78	11.6
		TOT DEM	KW	.00423	.00402	.00440	.00501	.00528	.00539	.00550	.00539	.00534	.00484	.00423	.00363	0.00550
S- 61	-0.27	TOT GAS	THM	0.0278	0.0203	0.0158	0.0102	0.0065	0.0061	0.0074	0.0062	0.0050	0.0082	0.0143	0.0240	0.152
		TOT ELEC	KWH	0.82	0.74	0.82	0.90	1.04	1.25	1.44	1.38	1.14	0.93	0.79	0.82	12.1
		TOT DEM	KW	.00293	.00232	.00343	.00414	.00454	.00439	.00505	.00485	.00465	.00379	.00313	.00293	0.00505
BASE CASE : 331																
BASE	0.00	HT ELEC	KWH	0.46	0.32	0.18	0.06	0.02	0.02	0.02	0.02	0.02	0.03	0.14	0.39	1.7
		HT DEM	KW	.00358	.00347	.00340	.00240	.00218	.00004	.00004	.00004	.00132	.00215	.00308	.00347	0.00358
		REM ELEC	KWH	0.94	0.84	0.94	1.02	1.19	1.42	1.64	1.57	1.30	1.06	0.90	0.94	13.8
		REM DEM	KW	.00416	.00378	.00448	.00538	.00589	.00576	.00640	.00621	.00595	.00525	.00429	.00390	0.00640
		TOT ELEC	KWH	1.40	1.16	1.12	1.08	1.20	1.44	1.66	1.59	1.31	1.09	1.03	1.33	15.5
		TOT DEM	KW	.00778	.00692	.00692	.00584	.00599	.00638	.00654	.00638	.00622	.00560	.00568	.00700	0.00778
S- 3	-0.13	HT ELEC	KWH	0.49	0.34	0.20	0.06	0.02	0.02	0.02	0.02	0.02	0.03	0.14	0.42	1.8
		HT DEM	KW	.00358	.00347	.00340	.00240	.00218	.00004	.00004	.00004	.00132	.00215	.00308	.00347	0.00358
		REM ELEC	KWH	0.85	0.76	0.85	0.93	1.08	1.29	1.49	1.42	1.17	0.96	0.81	0.85	12.5
		REM DEM	KW	.00378	.00343	.00407	.00489	.00535	.00524	.00582	.00565	.00541	.00477	.00390	.00355	0.00582
		TOT ELEC	KWH	1.34	1.10	1.05	0.98	1.09	1.31	1.51	1.44	1.19	0.99	0.96	1.27	14.3
		TOT DEM	KW	.00738	.00657	.00657	.00554	.00568	.00605	.00620	.00605	.00590	.00531	.00539	.00664	0.00738
S- 7	-0.05	HT ELEC	KWH	0.52	0.36	0.21	0.06	0.02	0.02	0.02	0.02	0.02	0.03	0.15	0.44	1.9
		HT DEM	KW	.00358	.00347	.00340	.00240	.00218	.00004	.00004	.00004	.00132	.00215	.00308	.00347	0.00358
		REM ELEC	KWH	0.76	0.68	0.76	0.83	0.96	1.15	1.33	1.28	1.05	0.86	0.73	0.76	11.2
		REM DEM	KW	.00337	.00306	.00363	.00436	.00477	.00467	.00519	.00503	.00483	.00426	.00348	.00317	0.00519
		TOT ELEC	KWH	1.28	1.04	0.97	0.89	0.98	1.17	1.35	1.30	1.07	0.90	0.88	1.20	13.1
		TOT DEM	KW	.00713	.00642	.00656	.00449	.00456	.00492	.00513	.00499	.00478	.00428	.00535	.00656	0.00713
S- 16	1.04	HT ELEC	KWH	0.55	0.38	0.22	0.07	0.02	0.02	0.02	0.02	0.02	0.04	0.16	0.47	2.0
		HT DEM	KW	.00358	.00347	.00340	.00240	.00218	.00004	.00004	.00004	.00132	.00215	.00308	.00347	0.00358
		REM ELEC	KWH	0.67	0.60	0.67	0.72	0.84	1.01	1.16	1.11	0.92	0.75	0.64	0.67	9.8
		REM DEM	KW	.00311	.00282	.00335	.00402	.00440	.00430	.00478	.00464	.00445	.00392	.00320	.00292	0.00478
		TOT ELEC	KWH	1.21	0.97	0.88	0.79	0.86	1.03	1.18	1.13	0.94	0.79	0.80	1.13	11.8
		TOT DEM	KW	.00683	.00615	.00628	.00430	.00437	.00471	.00492	.00478	.00458	.00410	.00512	.00628	0.00683
S- 32	0.18	HT ELEC	KWH	0.65	0.39	0.32	0.13	0.06	0.03	0.03	0.03	0.05	0.10	0.21	0.45	2.4
		HT DEM	KW	.00310	.00301	.00295	.00208	.00189	.00003	.00003	.00003	.00115	.00186	.00267	.00301	0.00310
		REM ELEC	KWH	0.70	0.64	0.78	0.83	0.89	1.10	1.18	1.17	0.99	0.83	0.69	0.70	10.5
		REM DEM	KW	.00325	.00315	.00364	.00432	.00441	.00470	.00485	.00470	.00461	.00412	.00344	.00315	0.00485
		TOT ELEC	KWH	1.36	1.03	1.09	0.96	0.95	1.13	1.20	1.19	1.04	0.94	0.91	1.15	13.0
		TOT DEM	KW	.00685	.00616	.00630	.00432	.00438	.00473	.00493	.00480	.00459	.00411	.00514	.00630	0.00685
S- 40	0.21	HT ELEC	KWH	0.64	0.38	0.31	0.13	0.06	0.03	0.03	0.03	0.05	0.10	0.21	0.44	2.4
		HT DEM	KW	.00310	.00301	.00295	.00208	.00189	.00003	.00003	.00003	.00115	.00186	.00267	.00301	0.00310
		REM ELEC	KWH	0.64	0.58	0.71	0.76	0.82	1.01	1.07	1.06	0.90	0.76	0.63	0.64	9.6
		REM DEM	KW	.00297	.00288	.00332	.00394	.00403	.00430	.00443	.00430	.00421	.00377	.00315	.00288	0.00443
		TOT ELEC	KWH	1.28	0.97	1.02	0.89	0.87	1.03	1.10	1.09	0.95	0.86	0.84	1.08	12.0
		TOT DEM	KW	.00660	.00594	.00607	.00416	.00422	.00455	.00475	.00462	.00442	.00396	.00495	.00607	0.00660
S- 45	0.88	HT ELEC	KWH	0.61	0.36	0.30	0.12	0.06	0.03	0.03	0.03	0.05	0.10	0.20	0.42	2.3
		HT DEM	KW	.00310	.00301	.00295	.00208	.00189	.00003	.00003	.00003	.00115	.00186	.00267	.00301	0.00310
		REM ELEC	KWH	0.68	0.62	0.75	0.80	0.86	1.06	1.13	1.12	0.95	0.80	0.67	0.68	10.1
		REM DEM	KW	.00312	.00302	.00349	.00414	.00423	.00451	.00465	.00451	.00442	.00395	.00330	.00302	0.00465
		TOT ELEC	KWH	1.29	0.98	1.04	0.92	0.92	1.09	1.16	1.15	0.99	0.90	0.86	1.10	12.4
		TOT DEM	KW	.00675	.00607	.00621	.00425	.00432	.00466	.00486	.00472	.00452	.00405	.00506	.00621	0.00675
S- 61	-0.30	HT ELEC	KWH	0.52	0.36	0.21	0.06	0.02	0.02	0.02	0.02	0.02	0.03	0.15	0.44	1.9
		HT DEM	KW	.00358	.00347	.00340	.00240	.00218	.00004	.00004	.00004	.00132	.00215	.00308	.00347	0.00358
		REM ELEC	KWH	0.76	0.68	0.76	0.83	0.96	1.15	1.33	1.28	1.05	0.86	0.73	0.76	11.2
		REM DEM	KW	.00337	.00306	.00363	.00436	.00477	.00467	.00519	.00503	.00483	.00426	.00348	.00317	0.00519
		TOT ELEC	KWH	1.28	1.04	0.97	0.89	0.98	1.17	1.35	1.30	1.07	0.90	0.88	1.20	13.1
		TOT DEM	KW	.00713	.00642	.00656	.00449	.00456	.00492	.00513	.00499	.00478	.00428	.00535	.00656	0.00713
BASE CASE : 332																
BASE	0.00	TOT GAS	THM	0.0241	0.0167	0.0097	0.0029	0.0009	0.0009	0.0009	0.0009	0.0009	0.0016	0.0071	0.0206	0.088
		TOT ELEC	KWH	0.94	0.84	0.94	1.02	1.19	1.42	1.64	1.57	1.30	1.06	0.90	0.94	13.8
		TOT DEM	KW	.00416	.00378	.00448	.00538	.00589	.00576	.00640	.00621	.00595	.00525	.00429	.00390	0.00640
S- 3	-0.14	TOT GAS	THM	0.0257	0.0177	0.0103	0.0031	0.0009	0.0009	0.0009	0.0009	0.0009	0.0017	0.0076	0.0219	0.093
		TOT ELEC	KWH	0.85	0.76	0.85	0.93	1.08	1.29	1.49	1.42	1.17	0.96	D.81	0.85	12.5
		TOT DEM	KW	.00378	.00343	.00407	.00489	.00535	.00524	.00582	.00565	.00541	.00477	.00390	.00355	0.00582
S- 7	-0.07	TOT GAS	THM	0.0271	0.0187	0.0108	0.0033	0.0010	0.0010	0.0010	0.0010	0.0010	0.0018	0.0080	0.0232	0.099
		TOT ELEC	KWH	0.76	0.68	0.76	0.83	0.96	1.15	1.33	1.28	1.05	0.86	0.73	0.76	11.2
		TOT DEM	KW	.00337	.00306	.00363	.00436	.00477	.00467	.00519	.00503	.00483	.00426	.00348	.00317	0.00519
S- 16	1.01	TOT GAS	THM	0.0287	0.0198	0.0115	0.0034	0.0010	0.0010	0.0010	0.0010	0.0010	0.0019	0.0084	0.0245	0.104
		TOT ELEC	KWH	0.67	0.60	0.67	0.72	0.84	1.01	1.16	1.11	0.92	0.75	0.64	0.67	9.8
		TOT DEM	KW	.00311	.00282	.00335	.00402	.00440	.00430	.00478	.00464	.00445	.00392	.00320	.00292	0.00478
S- 32	0.16	TOT GAS	THM	0.0360	0.0215	0.0175	0.0073	0.0034	0.0015	0.0015	0.0015	0.0027	0.0058	0.0117	0.0248	0.135
		TOT ELEC	KWH	0.70	0.64	0.78	0.83	0.89	1.10	1.18	1.17	0.99	0.83	0.69	0.70	10.5
		TOT DEM	KW	.00349	.00340	.00373	.00432	.00451	.00466	.00485	.00475	.00470	.00422	.00359	.00320	0.00485
S- 40	0.19	TOT GAS	THM	0.0352	0.0210	0.0172	0.0071	0.0033	0.0015	0.0015	0.0015	0.0026	0.0057	0.0115	0.0243	0.132
		TOT ELEC	KWH	0.64	0.58	0.71	0.76	0.82	1.01	1.07	1.06	0.90	0.76	0.63	0.64	9.6
		TOT DEM	KW	.00319	.00310	.00341	.00394	.00412	.00425	.00443	.00434	.00430	.00385	.00328	.00292	0.00443

(HT ELEC=ELECTRIC CONSUMPTION FOR SPACE AND WATER HEATING. REM ELEC=REMAINING ELECTRIC CONSUMPTION (NOT USED FOR HEATING). TOT ELEC=TOTAL ELECTRIC CONSUMPTION. HT DEM=PEAK ELECTRIC DEMAND FOR SPACE AND WATER HEATING. REM DEM=PEAK ELECTRIC DEMAND FOR REMAINING LOAD (NOT HEATING). TOT DEM=TOTAL PEAK ELECTRIC DEMAND. GAS=TOTAL GAS CONSUMPTION. ANNUAL FIGURES ARE SUMS OF MONTHLY CONSUMPTION VALUES OR ARE YEARLY MAXIMUMS OF MONTHLY PEAK DEMAND VALUES.)

BASE CASE : 332 (CONTINUED)

STRAT SET	CONST COST $/SF	ENERGY FORM	UNITS (PER SF)	JAN	FEB	MAR	APR	MAY	JUN	JUL	AUG	SEP	OCT	NOV	DEC	ANNUAL
S- 45	0.86	TOT GAS	THM	0.0336	0.0200	0.0164	0.0068	0.0032	0.0014	0.0014	0.0014	0.0025	0.0054	0.0110	0.0232	0.126
		TOT ELEC	KWH	0.68	0.62	0.75	0.80	0.86	1.06	1.13	1.12	0.95	0.80	0.67	0.68	10.1
		TOT DEM	KW	.00335	.00325	.00358	.00414	.00432	.00446	.00465	.00456	.00451	.00405	.00344	.00307	0.00465
S- 61	-0.31	TOT GAS	THM	0.0272	0.0188	0.0109	0.0033	0.0010	0.0010	0.0010	0.0010	0.0010	0.0018	0.0080	0.0233	0.099
		TOT ELEC	KWH	0.76	0.68	0.76	0.83	0.96	1.15	1.33	1.28	1.05	0.86	0.73	0.76	11.2
		TOT DEM	KW	.00337	.00306	.00363	.00436	.00477	.00467	.00519	.00503	.00483	.00426	.00348	.00317	0.00519

BASE CASE : 333

STRAT SET	CONST COST $/SF	ENERGY FORM	UNITS (PER SF)	JAN	FEB	MAR	APR	MAY	JUN	JUL	AUG	SEP	OCT	NOV	DEC	ANNUAL
BASE	0.00	TOT ELEC	KWH	1.34	1.12	1.30	1.33	1.34	1.59	1.66	1.66	1.46	1.28	1.15	1.20	16.4
		TOT DEM	KW	.00698	.00517	.00558	.00635	.00642	.00677	.00691	.00684	.00677	.00628	.00537	.00649	0.00698
S- 1	0.09	TOT ELEC	KWH	1.24	1.03	1.19	1.22	1.24	1.46	1.53	1.53	1.34	1.18	1.06	1.10	15.1
		TOT DEM	KW	.00678	.00461	.00488	.00563	.00576	.00610	.00617	.00617	.00603	.00556	.00468	.00631	0.00678
S- 12	0.20	TOT ELEC	KWH	1.12	0.92	1.07	1.10	1.12	1.32	1.37	1.37	1.21	1.06	0.95	0.99	13.6
		TOT DEM	KW	.00643	.00457	.00527	.00482	.00489	.00514	.00527	.00527	.00514	.00476	.00392	.00579	0.00643
S- 22	1.51	TOT ELEC	KWH	1.01	0.84	0.97	1.00	1.01	1.19	1.24	1.24	1.09	0.96	0.86	0.90	12.3
		TOT DEM	KW	.00618	.00531	.00531	.00420	.00426	.00457	.00464	.00464	.00451	.00414	.00340	.00531	0.00618
S- 28	-0.05	TOT ELEC	KWH	1.12	0.92	1.07	1.10	1.12	1.32	1.37	1.37	1.21	1.06	0.95	0.99	13.6
		TOT DEM	KW	.00643	.00457	.00527	.00482	.00489	.00514	.00527	.00527	.00514	.00476	.00392	.00579	0.00643
S- 36	0.42	TOT ELEC	KWH	1.05	0.87	1.01	1.04	1.05	1.24	1.29	1.29	1.14	1.00	0.90	0.93	12.8
		TOT DEM	KW	.00635	.00616	.00616	.00457	.00495	.00502	.00521	.00533	.00489	.00451	.00432	.00635	0.00635
S- 43	1.26	TOT ELEC	KWH	0.97	0.80	0.93	0.96	0.97	1.14	1.19	1.19	1.05	0.92	0.83	0.86	11.8
		TOT DEM	KW	.00613	.00595	.00595	.00441	.00478	.00484	.00503	.00515	.00472	.00435	.00417	.00613	0.00613
S- 48	0.45	TOT ELEC	KWH	1.02	0.84	0.98	1.00	1.02	1.20	1.25	1.25	1.10	0.97	0.87	0.91	12.4
		TOT DEM	KW	.00625	.00606	.00606	.00450	.00487	.00494	.00513	.00525	.00481	.00444	.00425	.00625	0.00625

BASE CASE : 334

STRAT SET	CONST COST $/SF	ENERGY FORM	UNITS (PER SF)	JAN	FEB	MAR	APR	MAY	JUN	JUL	AUG	SEP	OCT	NOV	DEC	ANNUAL
BASE	0.00	HT ELEC	KWH	0.65	0.48	0.37	0.24	0.15	0.14	0.17	0.15	0.12	0.19	0.34	0.56	3.6
		HT DEM	KW	.00685	.00541	.00507	.00308	.00288	.00062	.00082	.00055	.00212	.00281	.00397	.00514	0.00685
		REM ELEC	KWH	1.05	0.95	1.05	1.15	1.33	1.60	1.84	1.77	1.46	1.19	1.01	1.05	15.5
		REM DEM	KW	.00408	.00323	.00478	.00576	.00633	.00612	.00703	.00675	.00647	.00527	.00436	.00408	0.00703
		TOT ELEC	KWH	1.71	1.42	1.43	1.39	1.49	1.74	2.02	1.91	1.57	1.39	1.34	1.62	19.1
		TOT DEM	KW	.01260	.01008	.01021	.00718	.00756	.00756	.00806	.00769	.00743	.00680	.00806	.00983	0.01260
S- 3	-0.13	HT ELEC	KWH	0.69	0.50	0.39	0.25	0.16	0.15	0.18	0.15	0.12	0.20	0.35	0.59	3.8
		HT DEM	KW	.00548	.00477	.00477	.00263	.00241	.00060	.00088	.00093	.00192	.00241	.00345	.00466	0.00548
		REM ELEC	KWH	0.96	0.86	0.96	1.04	1.21	1.45	1.68	1.61	1.33	1.09	0.92	0.96	14.1
		REM DEM	KW	.00321	.00254	.00376	.00453	.00498	.00481	.00553	.00531	.00509	.00415	.00343	.00321	0.00553
		TOT ELEC	KWH	1.65	1.36	1.35	1.30	1.37	1.60	1.86	1.76	1.45	1.29	1.27	1.55	17.9
		TOT DEM	KW	.00873	.00698	.00707	.00498	.00524	.00524	.00559	.00533	.00515	.00471	.00559	.00681	0.00873
S- 7	-0.04	HT ELEC	KWH	0.76	0.55	0.43	0.28	0.18	0.17	0.20	0.17	0.14	0.22	0.39	0.65	4.1
		HT DEM	KW	.00548	.00477	.00477	.00263	.00241	.00060	.00088	.00093	.00192	.00241	.00345	.00466	0.00548
		REM ELEC	KWH	0.87	0.78	0.87	0.95	1.10	1.32	1.52	1.46	1.20	0.99	0.83	0.87	12.8
		REM DEM	KW	.00304	.00242	.00357	.00431	.00472	.00457	.00525	.00504	.00483	.00394	.00326	.00304	0.00525
		TOT ELEC	KWH	1.63	1.33	1.30	1.22	1.28	1.48	1.73	1.63	1.34	1.21	1.22	1.52	16.9
		TOT DEM	KW	.00842	.00758	.00775	.00530	.00539	.00581	.00606	.00589	.00564	.00505	.00631	.00775	0.00842
S- 16	1.16	HT ELEC	KWH	0.79	0.57	0.45	0.29	0.18	0.17	0.21	0.18	0.14	0.23	0.40	0.68	4.3
		HT DEM	KW	.00548	.00477	.00477	.00263	.00241	.00060	.00088	.00093	.00192	.00241	.00345	.00466	0.00548
		REM ELEC	KWH	0.78	0.70	0.78	0.84	0.98	1.17	1.36	1.30	1.07	0.88	0.74	0.78	11.4
		REM DEM	KW	.00286	.00227	.00335	.00404	.00444	.00429	.00493	.00473	.00454	.00370	.00306	.00286	0.00493
		TOT ELEC	KWH	1.56	1.27	1.22	1.13	1.16	1.35	1.57	1.48	1.21	1.11	1.14	1.45	15.7
		TOT DEM	KW	.00810	.00729	.00745	.00510	.00518	.00559	.00583	.00567	.00543	.00486	.00607	.00745	0.00810
S- 32	0.19	HT ELEC	KWH	0.82	0.59	0.53	0.31	0.15	0.07	0.09	0.09	0.09	0.24	0.44	0.70	4.1
		HT DEM	KW	.00548	.00477	.00477	.00263	.00241	.00060	.00088	.00093	.00192	.00241	.00345	.00466	0.00548
		REM ELEC	KWH	0.80	0.73	0.89	0.95	1.02	1.26	1.34	1.33	1.13	0.95	0.79	0.80	12.0
		REM DEM	KW	.00430	.00407	.00446	.00508	.00536	.00547	.00558	.00547	.00541	.00491	.00430	.00368	0.00558
		TOT ELEC	KWH	1.62	1.32	1.42	1.25	1.17	1.33	1.43	1.42	1.21	1.19	1.23	1.50	16.1
		TOT DEM	KW	.00892	.00803	.00821	.00562	.00571	.00615	.00642	.00624	.00598	.00535	.00669	.00821	0.00892
S- 40	0.22	HT ELEC	KWH	0.83	0.60	0.54	0.31	0.16	0.08	0.09	0.09	0.09	0.25	0.45	0.71	4.2
		HT DEM	KW	.00548	.00477	.00477	.00263	.00241	.00060	.00088	.00093	.00192	.00241	.00345	.00466	0.00548
		REM ELEC	KWH	0.71	0.65	0.78	0.84	0.90	1.11	1.19	1.18	1.00	0.84	0.70	0.71	10.6
		REM DEM	KW	.00392	.00372	.00407	.00463	.00489	.00499	.00509	.00499	.00494	.00448	.00392	.00336	0.00509
		TOT ELEC	KWH	1.54	1.24	1.33	1.15	1.06	1.19	1.28	1.27	1.08	1.09	1.15	1.42	14.8
		TOT DEM	KW	.00814	.00733	.00749	.00513	.00521	.00562	.00586	.00570	.00545	.00488	.00611	.00749	0.00814
S- 45	0.96	HT ELEC	KWH	0.76	0.55	0.50	0.28	0.14	0.07	0.08	0.08	0.08	0.23	0.41	0.65	3.8
		HT DEM	KW	.00548	.00477	.00477	.00263	.00241	.00060	.00088	.00093	.00192	.00241	.00345	.00466	0.00548
		REM ELEC	KWH	0.77	0.70	0.85	0.91	0.98	1.21	1.29	1.28	1.08	0.91	0.76	0.77	11.5
		REM DEM	KW	.00428	.00406	.00445	.00506	.00534	.00545	.00556	.00545	.00539	.00489	.00428	.00367	0.00556
		TOT ELEC	KWH	1.53	1.25	1.35	1.19	1.12	1.28	1.37	1.36	1.16	1.14	1.17	1.42	15.3
		TOT DEM	KW	.00890	.00801	.00819	.00561	.00570	.00614	.00641	.00623	.00596	.00534	.00667	.00819	0.00890
S- 61	-0.29	HT ELEC	KWH	0.76	0.55	0.43	0.28	0.18	0.17	0.20	0.17	0.14	0.22	0.39	0.65	4.1
		HT DEM	KW	.00548	.00477	.00477	.00263	.00241	.00060	.00088	.00093	.00192	.00241	.00345	.00466	0.00548
		REM ELEC	KWH	0.87	0.78	0.87	0.95	1.10	1.32	1.52	1.46	1.20	0.99	0.83	0.87	12.8
		REM DEM	KW	.00304	.00242	.00357	.00431	.00472	.00457	.00525	.00504	.00483	.00394	.00326	.00304	0.00525
		TOT ELEC	KWH	1.63	1.33	1.30	1.22	1.28	1.48	1.73	1.63	1.34	1.21	1.22	1.52	16.9
		TOT DEM	KW	.00842	.00758	.00775	.00530	.00539	.00581	.00606	.00589	.00564	.00505	.00631	.00775	0.00842

BASE CASE : 335

STRAT SET	CONST COST $/SF	ENERGY FORM	UNITS (PER SF)	JAN	FEB	MAR	APR	MAY	JUN	JUL	AUG	SEP	OCT	NOV	DEC	ANNUAL
BASE	0.00	TOT GAS	THM	0.0343	0.0252	0.0195	0.0126	0.0081	0.0075	0.0092	0.0077	0.0062	0.0101	0.0176	0.0297	0.188
		TOT ELEC	KWH	1.05	0.95	1.05	1.15	1.33	1.60	1.84	1.77	1.46	1.19	1.01	1.05	15.5
		TOT DEM	KW	.00408	.00323	.00478	.00576	.00633	.00612	.00703	.00675	.00647	.00527	.00436	.00408	0.00703

BASE CASE : 335 (CONTINUED)

STRAT SET	CONST COST $/SF	ENERGY FORM	UNITS (PER SF)	JAN	FEB	MAR	APR	MAY	JUN	JUL	AUG	SEP	OCT	NOV	DEC	ANNUAL
S- 3	-0.14	TOT GAS	THM	0.0361	0.0265	0.0205	0.0132	0.0085	0.0079	0.0097	0.0081	0.0065	0.0107	0.0186	0.0312	0.197
		TOT ELEC	KWH	0.96	0.86	0.96	1.04	1.21	1.45	1.68	1.61	1.33	1.09	0.92	0.96	14.1
		TOT DEM	KW	.00321	.00254	.00376	.00453	.00498	.00481	.00553	.00531	.00509	.00415	.00343	.00321	0.00553
S- 7	-0.05	TOT GAS	THM	0.0397	0.0291	0.0226	0.0145	0.0093	0.0087	0.0106	0.0089	0.0072	0.0117	0.0204	0.0343	0.217
		TOT ELEC	KWH	0.87	0.78	0.87	0.95	1.10	1.32	1.52	1.46	1.20	0.99	0.83	0.87	12.8
		TOT DEM	KW	.00304	.00242	.00357	.00431	.00472	.00457	.00525	.00504	.00483	.00394	.00326	.00304	0.00525
S- 16	1.14	TOT GAS	THM	0.0413	0.0302	0.0235	0.0151	0.0097	0.0090	0.0111	0.0093	0.0074	0.0122	0.0212	0.0357	0.226
		TOT ELEC	KWH	0.78	0.70	0.78	0.84	0.98	1.17	1.36	1.30	1.07	0.88	0.74	0.78	11.4
		TOT DEM	KW	.00286	.00227	.00335	.00404	.00444	.00429	.00493	.00473	.00454	.00370	.00306	.00286	0.00493
S- 32	0.18	TOT GAS	THM	0.0430	0.0308	0.0280	0.0161	0.0080	0.0039	0.0048	0.0046	0.0046	0.0128	0.0230	0.0367	0.217
		TOT ELEC	KWH	0.79	0.72	0.87	0.93	1.00	1.24	1.32	1.31	1.11	0.93	0.78	0.79	11.8
		TOT DEM	KW	.00430	.00407	.00446	.00508	.00536	.00547	.00558	.00547	.00541	.00491	.00430	.00368	0.00558
S- 40	0.21	TOT GAS	THM	0.0438	0.0314	0.0285	0.0164	0.0082	0.0040	0.0049	0.0046	0.0046	0.0130	0.0234	0.0373	0.221
		TOT ELEC	KWH	0.72	0.65	0.79	0.85	0.91	1.12	1.20	1.19	1.01	0.85	0.71	0.72	10.7
		TOT DEM	KW	.00392	.00372	.00407	.00463	.00489	.00499	.00509	.00499	.00494	.00448	.00392	.00336	0.00509
S- 45	0.94	TOT GAS	THM	0.0420	0.0301	0.0273	0.0157	0.0078	0.0038	0.0047	0.0045	0.0045	0.0125	0.0225	0.0358	0.212
		TOT ELEC	KWH	0.78	0.71	0.86	0.92	0.99	1.22	1.30	1.29	1.09	0.92	0.77	0.78	11.6
		TOT DEM	KW	.00428	.00406	.00445	.00506	.00534	.00545	.00556	.00545	.00539	.00489	.00428	.00367	0.00556
S- 61	-0.30	TOT GAS	THM	0.0397	0.0291	0.0226	0.0145	0.0093	0.0087	0.0106	0.0089	0.0072	0.0117	0.0204	0.0343	0.217
		TOT ELEC	KWH	0.87	0.78	0.87	0.95	1.10	1.32	1.52	1.46	1.20	0.99	0.83	0.87	12.8
		TOT DEM	KW	.00304	.00242	.00357	.00431	.00472	.00457	.00525	.00504	.00483	.00394	.00326	.00304	0.00525

BASE CASE : 336

STRAT SET	CONST COST $/SF	ENERGY FORM	UNITS (PER SF)	JAN	FEB	MAR	APR	MAY	JUN	JUL	AUG	SEP	OCT	NOV	DEC	ANNUAL
BASE	0.00	HT ELEC	KWH	0.57	0.39	0.23	0.07	0.02	0.02	0.02	0.02	0.02	0.04	0.17	0.48	2.1
		HT DEM	KW	.00378	.00367	.00359	.00253	.00231	.00004	.00004	.00004	.00140	.00227	.00325	.00367	0.00378
		REM ELEC	KWH	0.97	0.87	0.97	1.06	1.23	1.47	1.70	1.63	1.34	1.10	0.93	0.97	14.3
		REM DEM	KW	.00434	.00394	.00468	.00561	.00615	.00601	.00668	.00648	.00621	.00548	.00448	.00407	0.00668
		TOT ELEC	KWH	1.54	1.26	1.20	1.13	1.25	1.49	1.72	1.65	1.36	1.14	1.10	1.46	16.4
		TOT DEM	KW	.00840	.00748	.00748	.00630	.00647	.00689	.00706	.00689	.00672	.00605	.00613	.00756	0.00840
S- 3	-0.16	HT ELEC	KWH	0.60	0.41	0.24	0.07	0.02	0.02	0.02	0.02	0.02	0.04	0.18	0.51	2.2
		HT DEM	KW	.00378	.00367	.00359	.00253	.00231	.00004	.00004	.00004	.00140	.00227	.00325	.00367	0.00378
		REM ELEC	KWH	0.88	0.79	0.88	0.96	1.12	1.34	1.55	1.48	1.22	1.00	0.84	0.88	13.0
		REM DEM	KW	.00395	.00358	.00425	.00510	.00558	.00546	.00607	.00589	.00565	.00498	.00407	.00370	0.00607
		TOT ELEC	KWH	1.48	1.21	1.12	1.03	1.14	1.36	1.57	1.50	1.24	1.04	1.02	1.39	15.2
		TOT DEM	KW	.00800	.00712	.00712	.00600	.00616	.00656	.00672	.00656	.00640	.00576	.00584	.00720	0.00800
S- 7	-0.07	HT ELEC	KWH	0.63	0.44	0.25	0.08	0.02	0.02	0.02	0.02	0.02	0.04	0.19	0.54	2.3
		HT DEM	KW	.00378	.00367	.00359	.00253	.00231	.00004	.00004	.00004	.00140	.00227	.00325	.00367	0.00378
		REM ELEC	KWH	0.79	0.71	0.79	0.86	1.00	1.19	1.38	1.32	1.09	0.89	0.75	0.79	11.6
		REM DEM	KW	.00352	.00320	.00379	.00455	.00499	.00488	.00542	.00526	.00504	.00444	.00363	.00331	0.00542
		TOT ELEC	KWH	1.42	1.14	1.04	0.93	1.02	1.22	1.40	1.35	1.11	0.93	0.94	1.33	13.9
		TOT DEM	KW	.00775	.00697	.00713	.00488	.00496	.00535	.00558	.00543	.00519	.00465	.00581	.00713	0.00775
S- 16	0.99	HT ELEC	KWH	0.67	0.46	0.27	0.08	0.02	0.02	0.02	0.02	0.02	0.04	0.20	0.57	2.4
		HT DEM	KW	.00378	.00367	.00359	.00253	.00231	.00004	.00004	.00004	.00140	.00227	.00325	.00367	0.00378
		REM ELEC	KWH	0.69	0.62	0.69	0.75	0.87	1.04	1.20	1.15	0.95	0.78	0.66	0.69	10.1
		REM DEM	KW	.00324	.00294	.00349	.00419	.00459	.00449	.00499	.00484	.00464	.00409	.00334	.00304	0.00499
		TOT ELEC	KWH	1.36	1.08	0.95	0.83	0.89	1.06	1.23	1.18	0.97	0.82	0.85	1.26	12.5
		TOT DEM	KW	.00745	.00670	.00685	.00469	.00477	.00514	.00536	.00521	.00499	.00447	.00559	.00685	0.00745
S- 32	0.15	HT ELEC	KWH	0.79	0.47	0.39	0.16	0.07	0.03	0.03	0.03	0.06	0.13	0.26	0.55	3.0
		HT DEM	KW	.00328	.00318	.00312	.00220	.00200	.00003	.00003	.00003	.00121	.00197	.00282	.00318	0.00328
		REM ELEC	KWH	0.72	0.65	0.79	0.85	0.91	1.12	1.20	1.19	1.01	0.85	0.71	0.72	10.7
		REM DEM	KW	.00340	.00330	.00381	.00452	.00462	.00493	.00508	.00493	.00483	.00432	.00361	.00330	0.00508
		TOT ELEC	KWH	1.51	1.12	1.18	1.01	0.98	1.16	1.23	1.22	1.07	0.97	0.96	1.26	13.7
		TOT DEM	KW	.00700	.00630	.00644	.00441	.00448	.00483	.00504	.00490	.00469	.00420	.00525	.00644	0.00700
S- 40	0.18	HT ELEC	KWH	0.83	0.50	0.41	0.17	0.08	0.03	0.03	0.03	0.06	0.13	0.27	0.57	3.1
		HT DEM	KW	.00328	.00318	.00312	.00220	.00200	.00003	.00003	.00003	.00121	.00197	.00282	.00318	0.00328
		REM ELEC	KWH	0.66	0.60	0.73	0.78	0.84	1.04	1.10	1.09	0.93	0.78	0.65	0.66	9.9
		REM DEM	KW	.00318	.00309	.00356	.00423	.00432	.00461	.00475	.00461	.00451	.00404	.00337	.00309	0.00475
		TOT ELEC	KWH	1.49	1.10	1.14	0.95	0.92	1.07	1.14	1.13	0.99	0.91	0.92	1.23	13.0
		TOT DEM	KW	.00675	.00607	.00621	.00425	.00432	.00466	.00486	.00472	.00452	.00405	.00506	.00621	0.00675
S- 45	0.84	HT ELEC	KWH	0.76	0.45	0.37	0.15	0.07	0.03	0.03	0.03	0.06	0.12	0.25	0.53	2.9
		HT DEM	KW	.00328	.00318	.00312	.00220	.00200	.00003	.00003	.00003	.00121	.00197	.00282	.00318	0.00328
		REM ELEC	KWH	0.70	0.64	0.78	0.83	0.89	1.10	1.18	1.17	0.99	0.83	0.69	0.70	10.5
		REM DEM	KW	.00328	.00318	.00368	.00436	.00446	.00475	.00490	.00475	.00465	.00417	.00348	.00318	0.00490
		TOT ELEC	KWH	1.47	1.10	1.15	0.98	0.96	1.13	1.21	1.20	1.04	0.95	0.94	1.23	13.4
		TOT DEM	KW	.00690	.00621	.00635	.00435	.00442	.00476	.00497	.00483	.00462	.00414	.00518	.00635	0.00690
S- 61	-0.32	HT ELEC	KWH	0.63	0.44	0.25	0.08	0.02	0.02	0.02	0.02	0.02	0.04	0.19	0.54	2.3
		HT DEM	KW	.00378	.00367	.00359	.00253	.00231	.00004	.00004	.00004	.00140	.00227	.00325	.00367	0.00378
		REM ELEC	KWH	0.79	0.71	0.79	0.86	1.00	1.19	1.38	1.32	1.09	0.89	0.75	0.79	11.6
		REM DEM	KW	.00352	.00320	.00379	.00455	.00499	.00488	.00542	.00526	.00504	.00444	.00363	.00331	0.00542
		TOT ELEC	KWH	1.42	1.14	1.04	0.93	1.02	1.22	1.40	1.35	1.11	0.93	0.94	1.33	13.9
		TOT DEM	KW	.00775	.00697	.00713	.00488	.00496	.00535	.00558	.00543	.00519	.00465	.00581	.00713	0.00775

BASE CASE : 337

STRAT SET	CONST COST $/SF	ENERGY FORM	UNITS (PER SF)	JAN	FEB	MAR	APR	MAY	JUN	JUL	AUG	SEP	OCT	NOV	DEC	ANNUAL
BASE	0.00	TOT GAS	THM	0.0295	0.0204	0.0118	0.0035	0.0011	0.0011	0.0011	0.0011	0.0011	0.0019	0.0087	0.0252	0.107
		TOT ELEC	KWH	0.97	0.87	0.97	1.06	1.23	1.47	1.70	1.63	1.34	1.10	0.93	0.97	14.3
		TOT DEM	KW	.00434	.00394	.00468	.00561	.00615	.00601	.00668	.00648	.00621	.00548	.00448	.00407	0.00668

(HT ELEC=ELECTRIC CONSUMPTION FOR SPACE AND WATER HEATING. REM ELEC=REMAINING ELECTRIC CONSUMPTION (NOT USED FOR HEATING). TOT ELEC=TOTAL ELECTRIC CONSUMPTION. HT DEM=PEAK ELECTRIC DEMAND FOR SPACE AND WATER HEATING. REM DEM=PEAK ELECTRIC DEMAND FOR REMAINING LOAD (NOT HEATING). TOT DEM=TOTAL PEAK ELECTRIC DEMAND. GAS=TOTAL GAS CONSUMPTION. ANNUAL FIGURES ARE SUMS OF MONTHLY CONSUMPTION VALUES OR ARE YEARLY MAXIMUMS OF MONTHLY PEAK DEMAND VALUES.)

STRAT SET	CONST COST $/SF	ENERGY FORM	UNITS (PER SF)	JAN	FEB	MAR	APR	MAY	JUN	JUL	AUG	SEP	OCT	NOV	DEC	ANNUAL
BASE CASE : 337 (CONTINUED)																
S- 3	-0.17	TOT GAS	THM	0.0314	0.0217	0.0126	0.0038	0.0011	0.0011	0.0011	0.0011	0.0011	0.0021	0.0092	0.0268	0.114
		TOT ELEC	KWH	0.88	0.79	0.88	0.96	1.12	1.34	1.55	1.48	1.22	1.00	0.84	0.88	13.0
		TOT DEM	KW	.00395	.00358	.00425	.00510	.00558	.00546	.00607	.00589	.00565	.00498	.00407	.00370	0.00607
S- 7	-0.09	TOT GAS	THM	0.0332	0.0229	0.0133	0.0040	0.0012	0.0012	0.0012	0.0012	0.0012	0.0022	0.0098	0.0283	0.121
		TOT ELEC	KWH	0.79	0.71	0.79	0.86	1.00	1.19	1.38	1.32	1.09	0.89	0.75	0.79	11.6
		TOT DEM	KW	.00352	.00320	.00379	.00455	.00499	.00488	.00542	.00526	.00504	.00444	.00363	.00331	0.00542
S- 16	0.95	TOT GAS	THM	0.0351	0.0242	0.0140	0.0042	0.0013	0.0013	0.0013	0.0013	0.0013	0.0023	0.0103	0.0300	0.128
		TOT ELEC	KWH	0.69	0.62	0.69	0.75	0.87	1.04	1.20	1.15	0.95	0.78	0.66	0.69	10.1
		TOT DEM	KW	.00324	.00294	.00349	.00419	.00459	.00449	.00499	.00484	.00464	.00409	.00334	.00304	0.00499
S- 32	0.14	TOT GAS	THM	0.0366	0.0218	0.0178	0.0074	0.0034	0.0015	0.0015	0.0015	0.0027	0.0059	0.0119	0.0252	0.137
		TOT ELEC	KWH	0.72	0.65	0.79	0.85	0.91	1.12	1.20	1.19	1.01	0.85	0.71	0.72	10.7
		TOT DEM	KW	.00366	.00356	.00391	.00452	.00472	.00488	.00508	.00498	.00493	.00442	.00376	.00335	0.00508
S- 40	0.17	TOT GAS	THM	0.0411	0.0245	0.0200	0.0083	0.0039	0.0017	0.0017	0.0017	0.0031	0.0066	0.0134	0.0283	0.154
		TOT ELEC	KWH	0.66	0.60	0.73	0.78	0.84	1.04	1.10	1.09	0.93	0.78	0.65	0.66	9.9
		TOT DEM	KW	.00342	.00332	.00366	.00423	.00442	.00456	.00475	.00466	.00461	.00413	.00352	.00314	0.00475
S- 45	0.81	TOT GAS	THM	0.0376	0.0224	0.0183	0.0076	0.0035	0.0016	0.0016	0.0016	0.0028	0.0061	0.0123	0.0259	0.141
		TOT ELEC	KWH	0.70	0.64	0.78	0.83	0.89	1.10	1.18	1.17	0.99	0.83	0.69	0.70	10.5
		TOT DEM	KW	.00353	.00343	.00377	.00436	.00456	.00470	.00490	.00480	.00475	.00426	.00363	.00323	0.00490
S- 61	-0.34	TOT GAS	THM	0.0332	0.0229	0.0133	0.0040	0.0012	0.0012	0.0012	0.0012	0.0012	0.0022	0.0098	0.0283	0.121
		TOT ELEC	KWH	0.79	0.71	0.79	0.86	1.00	1.19	1.38	1.32	1.09	0.89	0.75	0.79	11.6
		TOT DEM	KW	.00352	.00320	.00379	.00455	.00499	.00488	.00542	.00526	.00504	.00444	.00363	.00331	0.00542
BASE CASE : 338																
BASE	0.00	TOT ELEC	KWH	1.39	1.15	1.34	1.37	1.39	1.64	1.71	1.71	1.50	1.32	1.18	1.23	16.9
		TOT DEM	KW	.00763	.00565	.00610	.00694	.00702	.00740	.00755	.00748	.00740	.00687	.00588	.00710	0.00763
S- 1	0.09	TOT ELEC	KWH	1.28	1.06	1.23	1.26	1.28	1.51	1.58	1.58	1.39	1.22	1.09	1.14	15.6
		TOT DEM	KW	.00633	.00430	.00456	.00525	.00538	.00570	.00576	.00576	.00563	.00519	.00437	.00589	0.00633
S- 12	0.20	TOT ELEC	KWH	1.15	0.95	1.11	1.13	1.15	1.36	1.41	1.41	1.25	1.09	0.98	1.02	14.0
		TOT DEM	KW	.00598	.00425	.00490	.00448	.00454	.00478	.00490	.00490	.00478	.00443	.00365	.00538	0.00598
S- 22	1.51	TOT ELEC	KWH	1.05	0.87	1.01	1.04	1.05	1.24	1.29	1.29	1.14	1.00	0.90	0.93	12.8
		TOT DEM	KW	.00568	.00488	.00488	.00386	.00392	.00420	.00426	.00426	.00415	.00381	.00312	.00488	0.00568
S- 28	-0.05	TOT ELEC	KWH	1.15	0.95	1.11	1.13	1.15	1.36	1.41	1.41	1.25	1.09	0.98	1.02	14.0
		TOT DEM	KW	.00598	.00425	.00490	.00448	.00454	.00478	.00490	.00490	.00478	.00443	.00365	.00538	0.00598
S- 36	0.43	TOT ELEC	KWH	1.10	0.91	1.06	1.09	1.10	1.30	1.35	1.35	1.19	1.05	0.94	0.98	13.4
		TOT DEM	KW	.00653	.00633	.00633	.00470	.00509	.00516	.00535	.00549	.00503	.00464	.00444	.00653	0.00653
S- 43	1.26	TOT ELEC	KWH	1.03	0.86	1.00	1.02	1.03	1.22	1.27	1.27	1.12	0.98	0.88	0.92	12.6
		TOT DEM	KW	.00633	.00614	.00614	.00456	.00494	.00500	.00519	.00532	.00487	.00449	.00430	.00633	0.00633
S- 48	0.45	TOT ELEC	KWH	1.07	0.88	1.03	1.05	1.07	1.26	1.31	1.31	1.16	1.01	0.91	0.95	13.0
		TOT DEM	KW	.00643	.00624	.00624	.00463	.00502	.00508	.00527	.00540	.00495	.00457	.00437	.00643	0.00643
BASE CASE : 339																
BASE	0.00	HT ELEC	KWH	0.46	0.34	0.26	0.17	0.11	0.10	0.12	0.10	0.08	0.14	0.24	0.40	2.5
		HT DEM	KW	.00548	.00433	.00406	.00247	.00230	.00049	.00066	.00044	.00170	.00225	.00318	.00411	0.00548
		REM ELEC	KWH	1.05	0.94	1.05	1.14	1.32	1.59	1.83	1.76	1.45	1.19	1.00	1.05	15.4
		REM DEM	KW	.00409	.00324	.00479	.00578	.00634	.00613	.00705	.00677	.00649	.00529	.00437	.00409	0.00705
		TOT ELEC	KWH	1.51	1.28	1.31	1.31	1.43	1.69	1.96	1.86	1.53	1.32	1.24	1.45	17.9
		TOT DEM	KW	.01275	.01020	.01033	.00650	.00689	.00689	.00739	.00701	.00676	.00689	.00816	.00994	0.01275
S- 3	-0.10	HT ELEC	KWH	0.48	0.36	0.28	0.18	0.11	0.11	0.13	0.11	0.09	0.14	0.25	0.42	2.7
		HT DEM	KW	.00548	.00477	.00477	.00263	.00241	.00060	.00088	.00093	.00192	.00241	.00345	.00466	0.00548
		REM ELEC	KWH	0.95	0.85	0.95	1.04	1.20	1.44	1.67	1.60	1.32	1.08	0.91	0.95	14.0
		REM DEM	KW	.00322	.00255	.00377	.00455	.00500	.00483	.00555	.00533	.00511	.00416	.00344	.00322	0.00555
		TOT ELEC	KWH	1.44	1.21	1.23	1.21	1.32	1.55	1.80	1.70	1.40	1.22	1.16	1.37	16.6
		TOT DEM	KW	.00897	.00718	.00727	.00457	.00484	.00484	.00520	.00493	.00475	.00484	.00574	.00700	0.00897
S- 7	-0.01	HT ELEC	KWH	0.53	0.39	0.30	0.19	0.13	0.12	0.14	0.12	0.10	0.16	0.27	0.46	2.9
		HT DEM	KW	.00548	.00477	.00477	.00263	.00241	.00060	.00088	.00093	.00192	.00241	.00345	.00466	0.00548
		REM ELEC	KWH	0.86	0.77	0.86	0.94	1.09	1.31	1.51	1.45	1.19	0.98	0.83	0.86	12.7
		REM DEM	KW	.00306	.00243	.00359	.00433	.00475	.00459	.00528	.00507	.00486	.00396	.00327	.00306	0.00528
		TOT ELEC	KWH	1.40	1.16	1.17	1.13	1.22	1.42	1.65	1.57	1.29	1.14	1.10	1.32	15.6
		TOT DEM	KW	.00867	.00694	.00702	.00442	.00468	.00468	.00503	.00477	.00460	.00468	.00555	.00676	0.00867
S- 16	1.20	HT ELEC	KWH	0.55	0.41	0.32	0.20	0.13	0.12	0.15	0.12	0.10	0.16	0.28	0.48	3.0
		HT DEM	KW	.00548	.00477	.00477	.00263	.00241	.00060	.00088	.00093	.00192	.00241	.00345	.00466	0.00548
		REM ELEC	KWH	0.77	0.69	0.77	0.84	0.97	1.16	1.34	1.29	1.06	0.87	0.73	0.77	11.3
		REM DEM	KW	.00287	.00228	.00337	.00406	.00445	.00431	.00495	.00475	.00455	.00371	.00307	.00287	0.00495
		TOT ELEC	KWH	1.32	1.10	1.08	1.04	1.10	1.29	1.49	1.41	1.16	1.03	1.02	1.25	14.3
		TOT DEM	KW	.00832	.00666	.00674	.00424	.00449	.00449	.00483	.00458	.00441	.00449	.00532	.00649	0.00832
S- 32	0.30	HT ELEC	KWH	0.58	0.42	0.38	0.22	0.11	0.05	0.06	0.06	0.06	0.17	0.31	0.50	2.9
		HT DEM	KW	.00548	.00477	.00477	.00263	.00241	.00060	.00088	.00093	.00192	.00241	.00345	.00466	0.00548
		REM ELEC	KWH	0.80	0.73	0.89	0.95	1.02	1.26	1.34	1.33	1.13	0.95	0.79	0.80	12.0
		REM DEM	KW	.00437	.00415	.00454	.00517	.00545	.00557	.00568	.00557	.00551	.00500	.00437	.00375	0.00568
		TOT ELEC	KWH	1.39	1.15	1.27	1.17	1.13	1.31	1.41	1.39	1.19	1.12	1.10	1.30	14.9
		TOT DEM	KW	.00892	.00803	.00821	.00562	.00571	.00615	.00642	.00624	.00598	.00535	.00669	.00821	0.00892
S- 61	-0.26	HT ELEC	KWH	0.53	0.39	0.30	0.19	0.13	0.12	0.14	0.12	0.10	0.16	0.27	0.46	2.9
		HT DEM	KW	.00548	.00477	.00477	.00263	.00241	.00060	.00088	.00093	.00192	.00241	.00345	.00466	0.00548
		REM ELEC	KWH	0.86	0.77	0.86	0.94	1.09	1.31	1.51	1.45	1.19	0.98	0.83	0.86	12.7
		REM DEM	KW	.00306	.00243	.00359	.00433	.00475	.00459	.00528	.00507	.00486	.00396	.00327	.00306	0.00528
		TOT ELEC	KWH	1.40	1.16	1.17	1.13	1.22	1.42	1.65	1.57	1.29	1.14	1.10	1.32	15.6
		TOT DEM	KW	.00867	.00694	.00702	.00442	.00468	.00468	.00503	.00477	.00460	.00468	.00555	.00676	0.00867
BASE CASE : 340																
BASE	0.00	TOT GAS	THM	0.0242	0.0177	0.0137	0.0089	0.0057	0.0053	0.0065	0.0054	0.0044	0.0071	0.0124	0.0209	0.132
		TOT ELEC	KWH	1.05	0.94	1.05	1.14	1.32	1.59	1.83	1.76	1.45	1.19	1.00	1.05	15.4
		TOT DEM	KW	.00409	.00324	.00479	.00578	.00634	.00613	.00705	.00677	.00649	.00529	.00437	.00409	0.00705
S- 3	-0.11	TOT GAS	THM	0.0254	0.0186	0.0144	0.0093	0.0060	0.0056	0.0068	0.0057	0.0046	0.0075	0.0131	0.0219	0.139
		TOT ELEC	KWH	0.95	0.85	0.95	1.04	1.20	1.44	1.67	1.60	1.32	1.08	0.91	0.95	14.0
		TOT DEM	KW	.00322	.00255	.00377	.00455	.00500	.00483	.00555	.00533	.00511	.00416	.00344	.00322	0.00555
S- 7	-0.03	TOT GAS	THM	0.0279	0.0204	0.0159	0.0102	0.0066	0.0061	0.0075	0.0063	0.0050	0.0082	0.0143	0.0241	0.153
		TOT ELEC	KWH	0.86	0.77	0.86	0.94	1.09	1.31	1.51	1.45	1.19	0.98	0.83	0.86	12.7
		TOT DEM	KW	.00306	.00243	.00359	.00433	.00475	.00459	.00528	.00507	.00486	.00396	.00327	.00306	0.00528
S- 16	1.18	TOT GAS	THM	0.0291	0.0213	0.0165	0.0106	0.0068	0.0064	0.0078	0.0065	0.0052	0.0086	0.0149	0.0251	0.159
		TOT ELEC	KWH	0.77	0.69	0.77	0.84	0.97	1.16	1.34	1.29	1.06	0.87	0.73	0.77	11.3
		TOT DEM	KW	.00287	.00228	.00337	.00406	.00445	.00431	.00495	.00475	.00455	.00371	.00307	.00287	0.00495
S- 32	0.29	TOT GAS	THM	0.0307	0.0220	0.0200	0.0115	0.0057	0.0028	0.0034	0.0033	0.0033	0.0091	0.0164	0.0262	0.155
		TOT ELEC	KWH	0.80	0.73	0.89	0.95	1.02	1.26	1.34	1.33	1.13	0.95	0.79	0.80	12.0
		TOT DEM	KW	.00437	.00415	.00454	.00517	.00545	.00557	.00568	.00557	.00551	.00500	.00437	.00375	0.00568
S- 61	-0.27	TOT GAS	THM	0.0279	0.0204	0.0159	0.0102	0.0066	0.0061	0.0075	0.0063	0.0050	0.0082	0.0143	0.0241	0.153
		TOT ELEC	KWH	0.86	0.77	0.86	0.94	1.09	1.31	1.51	1.45	1.19	0.98	0.83	0.86	12.7
		TOT DEM	KW	.00306	.00243	.00359	.00433	.00475	.00459	.00528	.00507	.00486	.00396	.00327	.00306	0.00528
BASE CASE : 341																
BASE	0.00	HT ELEC	KWH	0.42	0.29	0.17	0.05	0.02	0.02	0.02	0.02	0.02	0.03	0.12	0.36	1.5
		HT DEM	KW	.00413	.00401	.00392	.00277	.00252	.00004	.00004	.00004	.00153	.00248	.00355	.00401	0.00413
		REM ELEC	KWH	0.95	0.85	0.95	1.04	1.20	1.44	1.67	1.60	1.32	1.08	0.91	0.95	14.0
		REM DEM	KW	.00368	.00334	.00396	.00475	.00521	.00509	.00566	.00549	.00526	.00464	.00379	.00345	0.00566
		TOT ELEC	KWH	1.37	1.14	1.12	1.09	1.22	1.46	1.68	1.61	1.33	1.11	1.03	1.31	15.5
		TOT DEM	KW	.00795	.00708	.00700	.00533	.00549	.00588	.00604	.00588	.00580	.00525	.00564	.00723	0.00795
S- 3	-0.14	HT ELEC	KWH	0.45	0.31	0.18	0.05	0.02	0.02	0.02	0.02	0.02	0.03	0.13	0.38	1.6
		HT DEM	KW	.00413	.00401	.00392	.00277	.00252	.00004	.00004	.00004	.00153	.00248	.00355	.00401	0.00413
		REM ELEC	KWH	0.86	0.77	0.86	0.94	1.09	1.31	1.51	1.45	1.19	0.98	0.83	0.86	12.7
		REM DEM	KW	.00335	.00304	.00361	.00433	.00474	.00463	.00515	.00500	.00479	.00422	.00345	.00314	0.00515
		TOT ELEC	KWH	1.31	1.08	1.04	0.99	1.11	1.32	1.53	1.46	1.21	1.01	0.96	1.25	14.3
		TOT DEM	KW	.00755	.00672	.00664	.00506	.00521	.00559	.00574	.00559	.00551	.00498	.00536	.00687	0.00755
S- 7	-0.06	HT ELEC	KWH	0.47	0.33	0.19	0.06	0.02	0.02	0.02	0.02	0.02	0.03	0.14	0.40	1.7
		HT DEM	KW	.00413	.00401	.00392	.00277	.00252	.00004	.00004	.00004	.00153	.00248	.00355	.00401	0.00413
		REM ELEC	KWH	0.78	0.70	0.78	0.84	0.98	1.17	1.36	1.30	1.07	0.88	0.74	0.78	11.4
		REM DEM	KW	.00298	.00271	.00321	.00386	.00422	.00413	.00459	.00445	.00427	.00376	.00308	.00280	0.00459
		TOT ELEC	KWH	1.25	1.02	0.96	0.90	1.00	1.19	1.37	1.32	1.09	0.91	0.88	1.18	13.1
		TOT DEM	KW	.00730	.00657	.00672	.00460	.00467	.00504	.00526	.00511	.00489	.00438	.00548	.00672	0.00730
S- 16	1.03	HT ELEC	KWH	0.50	0.35	0.20	0.06	0.02	0.02	0.02	0.02	0.02	0.03	0.15	0.43	1.8
		HT DEM	KW	.00413	.00401	.00392	.00277	.00252	.00004	.00004	.00004	.00153	.00248	.00355	.00401	0.00413
		REM ELEC	KWH	0.67	0.60	0.67	0.73	0.85	1.02	1.18	1.13	0.93	0.76	0.64	0.67	9.9
		REM DEM	KW	.00275	.00250	.00296	.00355	.00389	.00381	.00423	.00410	.00393	.00347	.00283	.00258	0.00423
		TOT ELEC	KWH	1.17	0.95	0.87	0.79	0.87	1.04	1.20	1.15	0.95	0.80	0.79	1.10	11.7
		TOT DEM	KW	.00700	.00630	.00644	.00441	.00448	.00483	.00504	.00490	.00469	.00420	.00525	.00644	0.00700
S- 32	0.26	HT ELEC	KWH	0.50	0.30	0.25	0.10	0.05	0.02	0.02	0.02	0.04	0.08	0.16	0.35	1.9
		HT DEM	KW	.00363	.00352	.00345	.00243	.00221	.00004	.00004	.00004	.00134	.00218	.00312	.00352	0.00363
		REM ELEC	KWH	0.69	0.63	0.76	0.81	0.87	1.08	1.15	1.14	0.97	0.81	0.68	0.69	10.3
		REM DEM	KW	.00328	.00318	.00368	.00436	.00446	.00475	.00490	.00475	.00465	.00417	.00348	.00318	0.00490
		TOT ELEC	KWH	1.19	0.93	1.01	0.91	0.92	1.10	1.17	1.16	1.00	0.89	0.84	1.04	12.2
		TOT DEM	KW	.00630	.00567	.00580	.00357	.00403	.00435	.00454	.00441	.00422	.00378	.00472	.00580	0.00630
S- 61	-0.31	HT ELEC	KWH	0.47	0.33	0.19	0.06	0.02	0.02	0.02	0.02	0.02	0.03	0.14	0.40	1.7
		HT DEM	KW	.00413	.00401	.00392	.00277	.00252	.00004	.00004	.00004	.00153	.00248	.00355	.00401	0.00413
		REM ELEC	KWH	0.78	0.70	0.78	0.84	0.98	1.17	1.36	1.30	1.07	0.88	0.74	0.78	11.4
		REM DEM	KW	.00298	.00271	.00321	.00386	.00422	.00413	.00459	.00445	.00427	.00376	.00308	.00280	0.00459
		TOT ELEC	KWH	1.25	1.02	0.96	0.90	1.00	1.19	1.37	1.32	1.09	0.91	0.88	1.18	13.1
		TOT DEM	KW	.00730	.00657	.00672	.00460	.00467	.00504	.00526	.00511	.00489	.00438	.00548	.00672	0.00730
BASE CASE : 342																
BASE	0.00	TOT GAS	THM	0.0221	0.0153	0.0089	0.0027	0.0008	0.0008	0.0008	0.0008	0.0008	0.0014	0.0065	0.0189	0.081
		TOT ELEC	KWH	0.95	0.85	0.95	1.04	1.20	1.44	1.67	1.60	1.32	1.08	0.91	0.95	14.0
		TOT DEM	KW	.00368	.00334	.00396	.00475	.00521	.00509	.00566	.00549	.00526	.00464	.00379	.00345	0.00566
S- 3	-0.15	TOT GAS	THM	0.0235	0.0162	0.0094	0.0028	0.0009	0.0009	0.0009	0.0009	0.0009	0.0015	0.0069	0.0201	0.085
		TOT ELEC	KWH	0.86	0.77	0.86	0.94	1.09	1.31	1.51	1.45	1.19	0.98	0.83	0.86	12.7
		TOT DEM	KW	.00335	.00304	.00361	.00433	.00474	.00463	.00515	.00500	.00479	.00422	.00345	.00314	0.00515
S- 7	-0.07	TOT GAS	THM	0.0249	0.0172	0.0099	0.0030	0.0009	0.0009	0.0009	0.0009	0.0009	0.0016	0.0073	0.0212	0.090
		TOT ELEC	KWH	0.78	0.70	0.78	0.84	0.98	1.17	1.36	1.30	1.07	0.88	0.74	0.78	11.4
		TOT DEM	KW	.00298	.00271	.00321	.00386	.00422	.00413	.00459	.00445	.00427	.00376	.00308	.00280	0.00459
S- 16	1.00	TOT GAS	THM	0.0263	0.0182	0.0105	0.0032	0.0010	0.0010	0.0010	0.0010	0.0010	0.0017	0.0077	0.0225	0.096
		TOT ELEC	KWH	0.67	0.60	0.67	0.73	0.85	1.02	1.18	1.13	0.93	0.76	0.64	0.67	9.9
		TOT DEM	KW	.00275	.00250	.00296	.00355	.00389	.00381	.00423	.00410	.00393	.00347	.00283	.00258	0.00423
S- 32	0.25	TOT GAS	THM	0.0256	0.0153	0.0125	0.0052	0.0024	0.0011	0.0011	0.0011	0.0019	0.0041	0.0084	0.0177	0.096
		TOT ELEC	KWH	0.68	0.62	0.75	0.80	0.86	1.06	1.13	1.12	0.95	0.80	0.67	0.68	10.1
		TOT DEM	KW	.00353	.00343	.00377	.00436	.00456	.00470	.00490	.00480	.00475	.00426	.00363	.00323	0.00490
S- 61	-0.32	TOT GAS	THM	0.0249	0.0172	0.0099	0.0030	0.0009	0.0009	0.0009	0.0009	0.0009	0.0016	0.0073	0.0212	0.090
		TOT ELEC	KWH	0.78	0.70	0.78	0.84	0.98	1.17	1.36	1.30	1.07	0.88	0.74	0.78	11.4
		TOT DEM	KW	.00298	.00271	.00321	.00386	.00422	.00413	.00459	.00445	.00427	.00376	.00308	.00280	0.00459

(HT ELEC=ELECTRIC CONSUMPTION FOR SPACE AND WATER HEATING. REM ELEC=REMAINING ELECTRIC CONSUMPTION (NOT USED FOR HEATING). TOT ELEC=TOTAL ELECTRIC CONSUMPTION. HT DEM=PEAK ELECTRIC DEMAND FOR SPACE AND WATER HEATING. REM DEM=PEAK ELECTRIC DEMAND FOR REMAINING LOAD (NOT HEATING). TOT DEM=TOTAL PEAK ELECTRIC DEMAND. GAS=TOTAL GAS CONSUMPTION. ANNUAL FIGURES ARE SUMS OF MONTHLY CONSUMPTION VALUES OR ARE YEARLY MAXIMUMS OF MONTHLY PEAK DEMAND VALUES.)

STRAT SET	CONST COST $/SF	ENERGY FORM	UNITS (PER SF)	JAN	FEB	MAR	APR	MAY	JUN	JUL	AUG	SEP	OCT	NOV	DEC	ANNUAL
BASE CASE : 343																
BASE	0.00	TOT ELEC	KWH	1.38	1.14	1.33	1.36	1.38	1.63	1.70	1.70	1.50	1.31	1.18	1.23	16.8
		TOT DEM	KW	.00698	.00517	.00558	.00635	.00642	.00677	.00691	.00684	.00677	.00628	.00537	.00649	0.00698
S- 1	0.09	TOT ELEC	KWH	1.26	1.05	1.22	1.25	1.26	1.49	1.56	1.56	1.37	1.20	1.08	1.12	15.4
		TOT DEM	KW	.00620	.00422	.00446	.00515	.00527	.00558	.00564	.00564	.00552	.00508	.00428	.00577	0.00620
S- 12	0.20	TOT ELEC	KWH	1.12	0.93	1.08	1.11	1.12	1.33	1.38	1.38	1.22	1.07	0.96	1.00	13.7
		TOT DEM	KW	.00593	.00421	.00486	.00445	.00451	.00474	.00486	.00486	.00474	.00439	.00362	.00534	0.00593
S- 22	1.51	TOT ELEC	KWH	1.04	0.86	1.00	1.03	1.04	1.23	1.28	1.28	1.13	0.99	0.89	0.93	12.7
		TOT DEM	KW	.00563	.00484	.00484	.00383	.00388	.00417	.00422	.00422	.00411	.00377	.00310	.00484	0.00563
S- 28	-0.05	TOT ELEC	KWH	1.12	0.93	1.08	1.11	1.12	1.33	1.38	1.38	1.22	1.07	0.96	1.00	13.7
		TOT DEM	KW	.00593	.00421	.00486	.00445	.00451	.00474	.00486	.00486	.00474	.00439	.00362	.00534	0.00593
S- 36	0.51	TOT ELEC	KWH	1.02	0.84	0.98	1.00	1.02	1.20	1.25	1.25	1.10	0.97	0.87	0.91	12.4
		TOT DEM	KW	.00630	.00611	.00611	.00454	.00491	.00498	.00517	.00529	.00485	.00447	.00428	.00630	0.00630
BASE CASE : 344																
BASE	0.00	HT ELEC	KWH	0.74	0.54	0.42	0.27	0.17	0.16	0.20	0.17	0.13	0.22	0.38	0.64	4.0
		HT DEM	KW	.00900	.00711	.00666	.00405	.00378	.00081	.00108	.00072	.00279	.00369	.00522	.00675	0.00900
		REM ELEC	KWH	1.09	0.98	1.09	1.19	1.38	1.66	1.92	1.84	1.51	1.24	1.05	1.09	16.1
		REM DEM	KW	.00428	.00339	.00502	.00605	.00664	.00642	.00738	.00708	.00679	.00554	.00458	.00428	0.00738
		TOT ELEC	KWH	1.83	1.52	1.51	1.46	1.56	1.82	2.11	2.00	1.65	1.46	1.43	1.73	20.1
		TOT DEM	KW	.01300	.01040	.01053	.00741	.00780	.00780	.00832	.00793	.00767	.00702	.00832	.01014	0.01300
S- 3	-0.15	HT ELEC	KWH	0.74	0.54	0.42	0.27	0.17	0.16	0.20	0.17	0.13	0.22	0.38	0.64	4.1
		HT DEM	KW	.00548	.00477	.00477	.00263	.00241	.00060	.00088	.00093	.00192	.00241	.00345	.00466	0.00548
		REM ELEC	KWH	1.00	0.90	1.00	1.09	1.26	1.51	1.75	1.68	1.38	1.13	0.96	1.00	14.7
		REM DEM	KW	.00336	.00267	.00394	.00476	.00522	.00505	.00580	.00557	.00534	.00435	.00360	.00336	0.00580
		TOT ELEC	KWH	1.74	1.44	1.42	1.36	1.44	1.68	1.95	1.84	1.52	1.35	1.34	1.64	18.8
		TOT DEM	KW	.00905	.00724	.00733	.00516	.00543	.00543	.00579	.00552	.00534	.00489	.00579	.00706	0.00905
S- 7	-0.05	HT ELEC	KWH	0.85	0.62	0.48	0.31	0.20	0.19	0.23	0.19	0.15	0.25	0.44	0.74	4.7
		HT DEM	KW	.00548	.00477	.00477	.00263	.00241	.00060	.00088	.00093	.00192	.00241	.00345	.00466	0.00548
		REM ELEC	KWH	0.90	0.81	0.90	0.98	1.14	1.37	1.58	1.52	1.25	1.02	0.86	0.90	13.3
		REM DEM	KW	.00299	.00237	.00350	.00422	.00463	.00448	.00515	.00494	.00474	.00386	.00319	.00299	0.00515
		TOT ELEC	KWH	1.76	1.44	1.39	1.30	1.34	1.56	1.81	1.71	1.40	1.28	1.30	1.64	18.0
		TOT DEM	KW	.00875	.00788	.00805	.00551	.00560	.00604	.00630	.00612	.00586	.00525	.00656	.00805	0.00875
S- 16	1.13	HT ELEC	KWH	0.89	0.65	0.50	0.32	0.21	0.19	0.24	0.20	0.16	0.26	0.46	0.77	4.8
		HT DEM	KW	.00548	.00477	.00477	.00263	.00241	.00060	.00088	.00093	.00192	.00241	.00345	.00466	0.00548
		REM ELEC	KWH	0.80	0.72	0.80	0.87	1.01	1.22	1.40	1.35	1.11	0.91	0.77	0.80	11.8
		REM DEM	KW	.00255	.00202	.00299	.00361	.00396	.00383	.00440	.00422	.00405	.00330	.00273	.00255	0.00440
		TOT ELEC	KWH	1.69	1.37	1.31	1.20	1.22	1.41	1.64	1.54	1.27	1.17	1.22	1.57	16.6
		TOT DEM	KW	.00840	.00756	.00773	.00529	.00538	.00580	.00605	.00588	.00563	.00504	.00630	.00773	0.00840
S- 32	0.27	HT ELEC	KWH	0.96	0.69	0.63	0.36	0.18	0.09	0.11	0.10	0.10	0.29	0.52	0.82	4.9
		HT DEM	KW	.00548	.00477	.00477	.00263	.00241	.00060	.00088	.00093	.00192	.00241	.00345	.00466	0.00548
		REM ELEC	KWH	0.80	0.73	0.88	0.94	1.01	1.25	1.33	1.32	1.12	0.94	0.79	0.80	11.9
		REM DEM	KW	.00526	.00451	.00439	.00497	.00566	.00555	.00578	.00572	.00532	.00520	.00526	.00428	0.00578
		TOT ELEC	KWH	1.76	1.42	1.51	1.30	1.19	1.34	1.44	1.42	1.22	1.23	1.30	1.62	16.8
		TOT DEM	KW	.00900	.00810	.00828	.00567	.00576	.00621	.00648	.00630	.00603	.00540	.00675	.00828	0.00900
S- 61	-0.30	HT ELEC	KWH	0.85	0.62	0.48	0.31	0.20	0.19	0.23	0.19	0.15	0.25	0.44	0.74	4.7
		HT DEM	KW	.00548	.00477	.00477	.00263	.00241	.00060	.00088	.00093	.00192	.00241	.00345	.00466	0.00548
		REM ELEC	KWH	0.90	0.81	0.90	0.98	1.14	1.37	1.58	1.52	1.25	1.02	0.86	0.90	13.3
		REM DEM	KW	.00299	.00237	.00350	.00422	.00463	.00448	.00515	.00494	.00474	.00386	.00319	.00299	0.00515
		TOT ELEC	KWH	1.76	1.44	1.39	1.30	1.34	1.56	1.81	1.71	1.40	1.28	1.30	1.64	18.0
		TOT DEM	KW	.00875	.00788	.00805	.00551	.00560	.00604	.00630	.00612	.00586	.00525	.00656	.00805	0.00875
BASE CASE : 345																
BASE	0.00	TOT GAS	THM	0.0388	0.0284	0.0220	0.0142	0.0091	0.0085	0.0104	0.0087	0.0070	0.0114	0.0199	0.0335	0.212
		TOT ELEC	KWH	1.09	0.98	1.09	1.19	1.38	1.66	1.92	1.84	1.51	1.24	1.05	1.09	16.1
		TOT DEM	KW	.00428	.00339	.00502	.00605	.00664	.00642	.00738	.00708	.00679	.00554	.00458	.00428	0.00738
S- 3	-0.16	TOT GAS	THM	0.0389	0.0285	0.0221	0.0142	0.0091	0.0085	0.0104	0.0087	0.0070	0.0115	0.0200	0.0336	0.213
		TOT ELEC	KWH	1.00	0.90	1.00	1.09	1.26	1.51	1.75	1.68	1.38	1.13	0.96	1.00	14.7
		TOT DEM	KW	.00336	.00267	.00394	.00476	.00522	.00505	.00580	.00557	.00534	.00435	.00360	.00336	0.00580
S- 7	-0.06	TOT GAS	THM	0.0448	0.0328	0.0254	0.0164	0.0105	0.0098	0.0120	0.0100	0.0081	0.0132	0.0230	0.0386	0.245
		TOT ELEC	KWH	0.90	0.81	0.90	0.98	1.14	1.37	1.58	1.52	1.25	1.02	0.86	0.90	13.3
		TOT DEM	KW	.00299	.00237	.00350	.00422	.00463	.00448	.00515	.00494	.00474	.00386	.00319	.00299	0.00515
S- 16	1.11	TOT GAS	THM	0.0466	0.0341	0.0265	0.0171	0.0109	0.0102	0.0125	0.0104	0.0084	0.0137	0.0239	0.0402	0.255
		TOT ELEC	KWH	0.80	0.72	0.80	0.87	1.01	1.22	1.40	1.35	1.11	0.91	0.77	0.80	11.8
		TOT DEM	KW	.00255	.00202	.00299	.00361	.00396	.00383	.00440	.00422	.00405	.00330	.00273	.00255	0.00440
S- 32	0.25	TOT GAS	THM	0.0527	0.0378	0.0343	0.0197	0.0098	0.0048	0.0059	0.0056	0.0056	0.0157	0.0282	0.0450	0.266
		TOT ELEC	KWH	0.81	0.73	0.81	0.88	1.02	1.23	1.42	1.36	1.12	0.92	0.77	0.81	11.9
		TOT DEM	KW	.00445	.00422	.00462	.00526	.00555	.00566	.00578	.00566	.00561	.00509	.00445	.00381	0.00578
S- 61	-0.31	TOT GAS	THM	0.0448	0.0328	0.0255	0.0164	0.0105	0.0098	0.0120	0.0100	0.0081	0.0132	0.0230	0.0387	0.245
		TOT ELEC	KWH	0.90	0.81	0.90	0.98	1.14	1.37	1.58	1.52	1.25	1.02	0.86	0.90	13.3
		TOT DEM	KW	.00299	.00237	.00350	.00422	.00463	.00448	.00515	.00494	.00474	.00386	.00319	.00299	0.00515
BASE CASE : 346																
BASE	0.00	HT ELEC	KWH	0.56	0.39	0.23	0.07	0.02	0.02	0.02	0.02	0.02	0.04	0.17	0.48	2.0
		HT DEM	KW	.00485	.00470	.00461	.00325	.00296	.00005	.00005	.00005	.00179	.00291	.00417	.00470	0.00485
		REM ELEC	KWH	0.99	0.89	0.99	1.08	1.26	1.50	1.74	1.66	1.37	1.12	0.95	0.99	14.6
		REM DEM	KW	.00452	.00410	.00487	.00584	.00639	.00625	.00695	.00674	.00646	.00570	.00466	.00424	0.00695
		TOT ELEC	KWH	1.56	1.28	1.22	1.15	1.28	1.52	1.76	1.68	1.39	1.16	1.12	1.47	16.6
		TOT DEM	KW	.00832	.00740	.00732	.00557	.00574	.00616	.00632	.00616	.00607	.00549	.00591	.00757	0.00832

STRAT SET	CONST COST $/SF	ENERGY FORM	UNITS (PER SF)	JAN	FEB	MAR	APR	MAY	JUN	JUL	AUG	SEP	OCT	NOV	DEC	ANNUAL
BASE CASE : 346 (CONTINUED)																
S- 3	-0.17	HT ELEC	KWH	0.60	0.41	0.24	0.07	0.02	0.02	0.02	0.02	0.02	0.04	0.18	0.51	2.2
		HT DEM	KW	.00465	.00451	.00442	.00312	.00284	.00005	.00005	.00005	.00172	.00279	.00400	.00451	0.00465
		REM ELEC	KWH	0.90	0.81	0.90	0.98	1.14	1.36	1.57	1.50	1.24	1.02	0.86	0.90	13.2
		REM DEM	KW	.00411	.00373	.00442	.00531	.00581	.00569	.00632	.00613	.00588	.00518	.00423	.00386	0.00632
		TOT ELEC	KWH	1.50	1.22	1.14	1.05	1.16	1.38	1.59	1.53	1.26	1.06	1.03	1.41	15.4
		TOT DEM	KW	.00785	.00699	.00691	.00526	.00542	.00581	.00597	.00581	.00573	.00518	.00557	.00714	0.00785
S- 7	-0.08	HT ELEC	KWH	0.64	0.44	0.25	0.08	0.02	0.02	0.02	0.02	0.02	0.04	0.19	0.54	2.3
		HT DEM	KW	.00485	.00470	.00461	.00325	.00296	.00005	.00005	.00005	.00179	.00291	.00417	.00470	0.00485
		REM ELEC	KWH	0.80	0.72	0.80	0.87	1.01	1.22	1.40	1.35	1.11	0.91	0.77	0.80	11.8
		REM DEM	KW	.00367	.00333	.00395	.00474	.00519	.00508	.00564	.00547	.00525	.00462	.00378	.00344	0.00564
		TOT ELEC	KWH	1.44	1.16	1.06	0.95	1.04	1.24	1.43	1.37	1.13	0.95	0.95	1.35	14.1
		TOT DEM	KW	.00760	.00684	.00699	.00479	.00486	.00524	.00547	.00532	.00509	.00456	.00570	.00699	0.00760
S- 16	0.96	HT ELEC	KWH	0.67	0.46	0.27	0.08	0.02	0.02	0.02	0.02	0.02	0.04	0.20	0.57	2.4
		HT DEM	KW	.00485	.00470	.00461	.00325	.00296	.00005	.00005	.00005	.00179	.00291	.00417	.00470	0.00485
		REM ELEC	KWH	0.70	0.63	0.70	0.76	0.89	1.06	1.23	1.17	0.97	0.79	0.67	0.70	10.3
		REM DEM	KW	.00337	.00306	.00363	.00436	.00477	.00467	.00519	.00503	.00483	.00426	.00348	.00317	0.00519
		TOT ELEC	KWH	1.37	1.09	0.97	0.84	0.91	1.09	1.25	1.20	0.99	0.84	0.87	1.27	12.7
		TOT DEM	KW	.00730	.00657	.00672	.00460	.00467	.00504	.00526	.00511	.00489	.00438	.00548	.00672	0.00730
S- 32	0.23	HT ELEC	KWH	0.77	0.46	0.38	0.16	0.07	0.03	0.03	0.03	0.06	0.12	0.25	0.53	2.9
		HT DEM	KW	.00435	.00422	.00413	.00291	.00265	.00004	.00004	.00004	.00161	.00261	.00374	.00422	0.00435
		REM ELEC	KWH	0.70	0.63	0.77	0.82	0.88	1.09	1.16	1.15	0.98	0.82	0.69	0.70	10.4
		REM DEM	KW	.00342	.00331	.00383	.00454	.00464	.00495	.00510	.00495	.00484	.00434	.00362	.00331	0.00510
		TOT ELEC	KWH	1.47	1.10	1.15	0.98	0.96	1.12	1.20	1.19	1.04	0.95	0.94	1.23	13.3
		TOT DEM	KW	.00735	.00661	.00676	.00463	.00470	.00507	.00529	.00514	.00492	.00441	.00551	.00676	0.00735
S- 61	-0.33	HT ELEC	KWH	0.64	0.44	0.25	0.08	0.02	0.02	0.02	0.02	0.02	0.04	0.19	0.54	2.3
		HT DEM	KW	.00485	.00470	.00461	.00325	.00296	.00005	.00005	.00005	.00179	.00291	.00417	.00470	0.00485
		REM ELEC	KWH	0.80	0.72	0.80	0.87	1.01	1.22	1.40	1.35	1.11	0.91	0.77	0.80	11.8
		REM DEM	KW	.00367	.00333	.00395	.00474	.00519	.00508	.00564	.00547	.00525	.00462	.00378	.00344	0.00564
		TOT ELEC	KWH	1.44	1.16	1.06	0.95	1.04	1.24	1.43	1.37	1.13	0.95	0.95	1.35	14.1
		TOT DEM	KW	.00760	.00684	.00699	.00479	.00486	.00524	.00547	.00532	.00509	.00456	.00570	.00699	0.00760
BASE CASE : 347																
BASE	0.00	TOT GAS	THM	0.0296	0.0205	0.0119	0.0036	0.0011	0.0011	0.0011	0.0011	0.0011	0.0019	0.0087	0.0253	0.108
		TOT ELEC	KWH	0.99	0.89	0.99	1.08	1.26	1.50	1.74	1.66	1.37	1.12	0.95	0.99	14.6
		TOT DEM	KW	.00452	.00410	.00487	.00584	.00639	.00625	.00695	.00674	.00646	.00570	.00466	.00424	0.00695
S- 3	-0.19	TOT GAS	THM	0.0315	0.0218	0.0126	0.0038	0.0011	0.0011	0.0011	0.0011	0.0011	0.0021	0.0093	0.0269	0.115
		TOT ELEC	KWH	0.90	0.81	0.90	0.98	1.14	1.36	1.57	1.50	1.24	1.02	0.86	0.90	13.2
		TOT DEM	KW	.00411	.00373	.00442	.00531	.00581	.00569	.00632	.00613	.00588	.00518	.00423	.00386	0.00632
S- 7	-0.10	TOT GAS	THM	0.0333	0.0230	0.0133	0.0040	0.0012	0.0012	0.0012	0.0012	0.0012	0.0022	0.0098	0.0285	0.121
		TOT ELEC	KWH	0.80	0.72	0.80	0.87	1.01	1.22	1.40	1.35	1.11	0.91	0.77	0.80	11.8
		TOT DEM	KW	.00367	.00333	.00395	.00474	.00519	.00508	.00564	.00547	.00525	.00462	.00378	.00344	0.00564
S- 16	0.93	TOT GAS	THM	0.0352	0.0243	0.0141	0.0042	0.0013	0.0013	0.0013	0.0013	0.0013	0.0023	0.0104	0.0301	0.128
		TOT ELEC	KWH	0.70	0.63	0.70	0.76	0.89	1.06	1.23	1.17	0.97	0.79	0.67	0.70	10.3
		TOT DEM	KW	.00337	.00306	.00363	.00436	.00477	.00467	.00519	.00503	.00483	.00426	.00348	.00317	0.00519
S- 32	0.22	TOT GAS	THM	0.0406	0.0242	0.0198	0.0082	0.0038	0.0017	0.0017	0.0017	0.0030	0.0065	0.0132	0.0280	0.152
		TOT ELEC	KWH	0.70	0.63	0.77	0.82	0.88	1.09	1.16	1.15	0.98	0.82	0.69	0.70	10.4
		TOT DEM	KW	.00367	.00357	.00393	.00454	.00474	.00490	.00510	.00500	.00495	.00444	.00377	.00337	0.00510
S- 61	-0.35	TOT GAS	THM	0.0333	0.0230	0.0133	0.0040	0.0012	0.0012	0.0012	0.0012	0.0012	0.0022	0.0098	0.0285	0.121
		TOT ELEC	KWH	0.80	0.72	0.80	0.87	1.01	1.22	1.40	1.35	1.11	0.91	0.77	0.80	11.8
		TOT DEM	KW	.00367	.00333	.00395	.00474	.00519	.00508	.00564	.00547	.00525	.00462	.00378	.00344	0.00564
BASE CASE : 348																
BASE	0.00	TOT ELEC	KWH	1.50	1.24	1.45	1.48	1.50	1.78	1.85	1.85	1.63	1.43	1.28	1.34	18.3
		TOT DEM	KW	.00740	.00548	.00592	.00673	.00681	.00718	.00733	.00725	.00718	.00666	.00570	.00688	0.00740
S- 1	0.09	TOT ELEC	KWH	1.40	1.16	1.35	1.39	1.40	1.66	1.73	1.73	1.52	1.33	1.20	1.25	17.1
		TOT DEM	KW	.00670	.00456	.00482	.00556	.00570	.00603	.00610	.00610	.00596	.00549	.00462	.00623	0.00670
S- 12	0.21	TOT ELEC	KWH	1.25	1.04	1.21	1.24	1.25	1.48	1.55	1.55	1.36	1.19	1.07	1.12	15.3
		TOT DEM	KW	.00620	.00440	.00508	.00465	.00471	.00496	.00508	.00508	.00496	.00459	.00378	.00558	0.00620
S- 22	1.52	TOT ELEC	KWH	1.14	0.95	1.10	1.13	1.14	1.35	1.40	1.40	1.24	1.08	0.97	1.01	13.9
		TOT DEM	KW	.00590	.00507	.00507	.00401	.00407	.00437	.00442	.00442	.00431	.00395	.00325	.00507	0.00590
S- 28	-0.04	TOT ELEC	KWH	1.25	1.04	1.21	1.24	1.25	1.48	1.55	1.55	1.36	1.19	1.07	1.12	15.3
		TOT DEM	KW	.00620	.00440	.00508	.00465	.00471	.00496	.00508	.00508	.00496	.00459	.00378	.00558	0.00620
S- 36	0.52	TOT ELEC	KWH	1.08	0.90	1.04	1.07	1.08	1.28	1.33	1.33	1.17	1.03	0.92	0.96	13.2
		TOT DEM	KW	.00653	.00633	.00633	.00470	.00509	.00516	.00535	.00549	.00503	.00464	.00444	.00653	0.00653

(HT ELEC=ELECTRIC CONSUMPTION FOR SPACE AND WATER HEATING. REM ELEC=REMAINING ELECTRIC CONSUMPTION (NOT USED FOR HEATING). TOT ELEC=TOTAL ELECTRIC CONSUMPTION. HT DEM=PEAK ELECTRIC DEMAND FOR SPACE AND WATER HEATING. REM DEM=PEAK ELECTRIC DEMAND FOR REMAINING LOAD (NOT HEATING). TOT DEM=TOTAL PEAK ELECTRIC DEMAND. GAS=TOTAL GAS CONSUMPTION. ANNUAL FIGURES ARE SUMS OF MONTHLY CONSUMPTION VALUES OR ARE YEARLY MAXIMUMS OF MONTHLY PEAK DEMAND VALUES.)

Hot/Humid Climate

Identify Region

CITIES IN REGION

AL	Mobile
FL	Jacksonville Miami Orlando Pensacola Tampa
GA	Albany Savannah
LA	Baton Rouge New Orleans Shreveport
MS	Biloxi Jackson
SC	Charleston
TX	Austin Beaumont Brownsville Corpus Christi Houston Laredo San Antonio Waco

BEGIN YOUR 5 LEVELS OF STRATEGY SELECTION.

Note: See Section II for a full discussion of the five climate regions used in this manual.

Make sure your design comes up to current good practice, which includes meeting ASHRAE 90–75 Standards. The table below identifies good practice that meets 90–75 Standards as well as common design assumptions applicable to all building sizes.

	ALL BUILDINGS	
	GOOD PRACTICE	**DESIGN ASSUMPTIONS**
SHELL	o Perimeter Insulation (R-5) o Infiltration: 0 when HVAC is operating (pressurized building)	o Ventilation: 14 cfm/person during occupied hours
HVAC	o Air conditioning performs to ASHRAE 90-75 (year 1980) minimum efficiencies o Boilers perform with 65% efficiency (no waste heat to space) o Heat pumps perform to ASHRAE 90-75 (year 1980) minimum efficiencies o Controls: · 72 degrees heating, 74 degrees cooling occupied periods · 55 degrees heating, 95 degrees cooling unoccupied periods · 2 degrees temperature float · outside air temperature reset for hot water boilers · low leakage dampers (less than 3%) · zone control for all heat pump and multi-zone systems	
INTERNAL LOAD		o Occupant density: 200 square feet per person o Lights: 2.25 watts/sq. ft. o Office equipment: .75 watts/sq. ft.
SCHEDULE	o HVAC comes on one hour before occupied period, turned to setback during unoccupied period	o Inside lighting: 100% during operating hours, 5% at other times o Outside lighting: 100% 6pm til midnight

Note: See Section II, page 179, for detailed information on these Level 1 strategies and assumptions.

Some aspects of ASHRAE 90-75 and current **good practice** vary according to building size and HVAC system type. Make sure your design comes up to these criteria. You may want to come back to this table as you explore HVAC options in Level 2.

BUILDING SIZE

		SMALL 2000-10000 sq ft	MEDIUM 10000-25000 sq ft	LARGE 25000-50000 sq ft
SHELL	ROOF INSULATION	R-8.3	R-8.3	R-8.3
	WALL INSULATION	R-3.8: 10% GLAZED NONE : 25% GLAZED	R-3.8	R-3.8
	GLAZING	SINGLE GLAZED TINTED OR SINGLE GLAZED REFLECTIVE	DOUBLE GLAZED TINTED OR SINGLE GLAZED REFLECTIVE	DOUBLE GLAZED TINTED OR SINGLE GLAZED REFLECTIVE
	AIR CHANGES	1/2 AC/HR UNOCCUPIED	1/3 AC/HR UNOCCUPIED	1/3 AC/HR UNOCCUPIED

HVAC		SMALL			MEDIUM			LARGE	
		PACKAGED TERMINAL AIR CONDITIONERS	ROOFTOP DX	HEAT PUMP	CONSTANT VOLUME	VARIABLE AIR VOLUME	HYDRONIC HEAT PUMP	VARIABLE AIR VOLUME	HYDRONIC HEAT PUMP
	TEMPERATURE CONTROL	ZONE	OA RESET	ZONE	FIXED	ZONE	ZONE	FIXED	ZONE
	AIR CONTROL	FIXED VOLUME	FIXED VOLUME	FIXED VOLUME	FIXED VOLUME	FIXED VOLUME	FIXED VOLUME	FIXED VOLUME	FIXED VOLUME

Some common **design assumptions** also vary according to building size and no. of stories.

		SMALL	MEDIUM		LARGE	
INT. LOAD		1 STORY	1 STORY	2 STORY	2 STORY	4 STORY ①
	ELEVATOR & OUTDOOR LIGHTS	1345 WATTS	4035 WATTS	4535 WATTS	7860 WATTS	9360 WATTS
SCHEDULE	OCCUPANCY	80% 8am - 6pm Monday - Friday	80% 8am - 6pm Monday - Friday		80% 8am - 6pm: M-F 100% 9am - 1pm: SAT	

GO TO THE LEVEL 2 PAGE SPECIFIC TO YOUR BUILDING SIZE & NO. OF STORIES.

Note ① For 3 or 5 story structures, use "4-story" classification.

LEVEL 2
MAKE INITIAL CHOICES
SMALL/1-STORY

Project ________________________________
date ____________ bldg. sq. ft. __________

The building characteristic choices shown here make a **big** difference in energy cost savings. Investigate these choices within the requirements of your design.

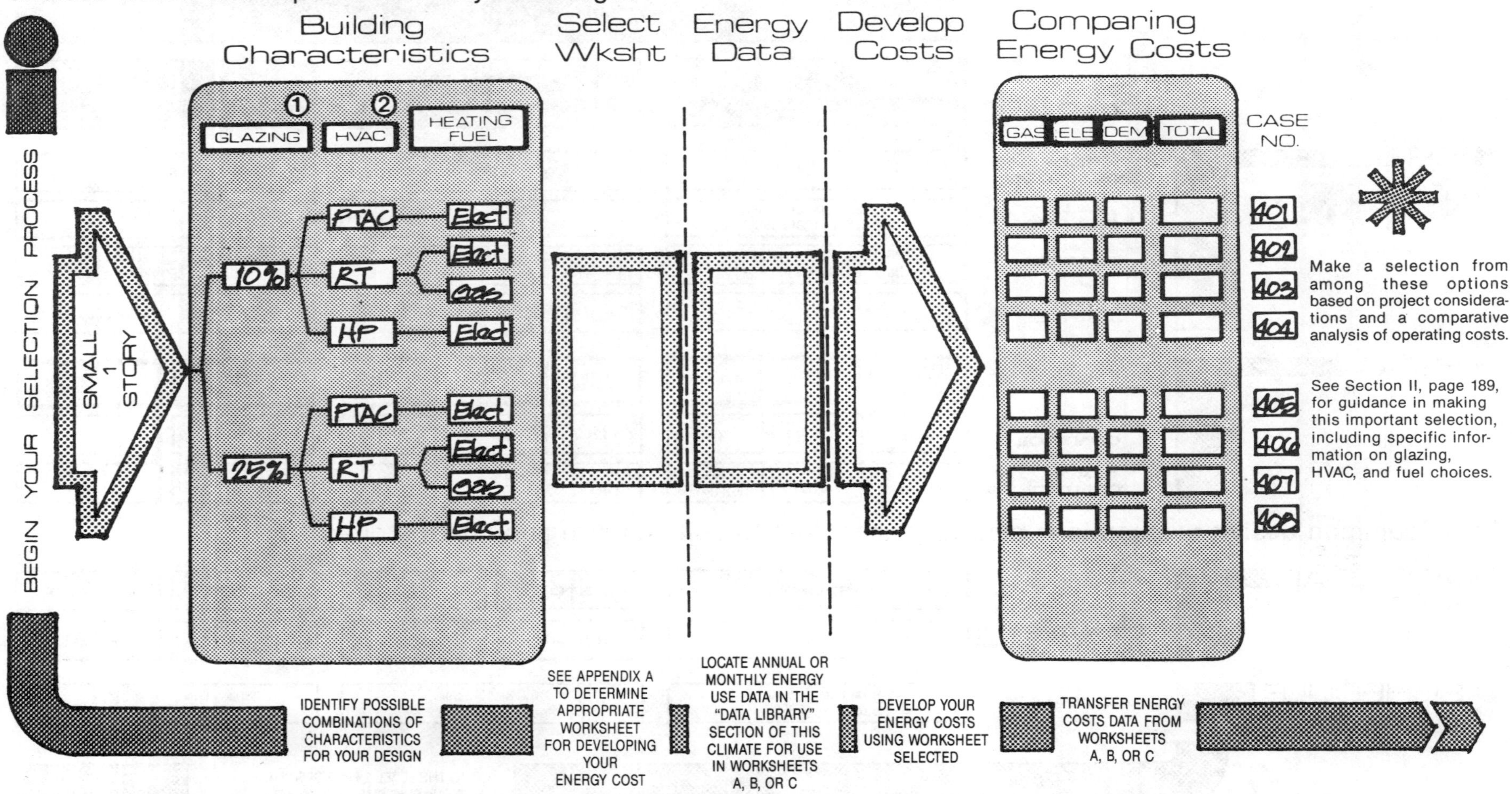

Note: See Section II, page 189, for detailed information on these Level 2 strategies and the use of this form.

① For glazing <17% use 10%, for >17% use 25%.

② PTAC = Packaged Terminal A/C, RT = Rooftop DX, HP = Packaged Terminal Heat Pump

Project ______________________
date __________ bldg. sq. ft. __________

Locate the row of strategy set options that apply to the building case you chose in Level 2. Select as many of these Level 3 strategy set options as you wish for valuation.

STRATEGY SET OPTIONS

SMALL 1 STORY

SELECTED LEVEL 2 CASE NUMBER	SIMPLEST LIGHTING IMPROVEMTS	MODEST LIGHTING & MECHANICAL IMPROVEMTS	MAJOR LIGHTING & MECHANICAL IMPROVEMTS	DAYLIGHT DIMMING CONTROLS AND LIGHTING IMPROVEMTS
401	1	10 or 25	20	
402	3	7 or 61	16	
403	3	7 or 61	16	
404	1	12 or 28	22	
405	1	10 or 25	20	35
406	3	7 or 61	16	32
407	3	7 or 61	16	32
408	1	12 or 28	22	36

DESCRIPTOR KEY

1. High eff. ballasts over all bldg.
3. High eff. ballasts over all bldg. + HVAC downsized for ballasts.
7. High eff. ballasts and lamps over all bldg. + HVAC downsized for ballasts + cooling COP improvement.
10. High eff. ballasts and lamps over all bldg. + cooling COP improvement.
12. High eff. ballasts and lamps over all bldg. + heat pump COP improvement.
16. High eff. ballasts, lamps and fixtures over all bldg. + HVAC downsized for ballasts and fixtures + cooling COP improvement.
20. High eff. ballasts, lamps and fixtures over all bldg. + cooling COP improvement.
22. High eff. ballasts, lamps and fixtures over all bldg. + heat pump COP improvement.
25. Non-uniform lighting + high eff. ballasts over all bldg. + cooling COP improvement.
61. Non-uniform lighting + high eff. ballasts over all bldg. + HVAC downsized for ballasts + cooling COP improvement.
28. Non-uniform lighting + high eff. ballasts over all bldg. + heat pump COP improvement.
32. Perimeter daylighting dimmers + high eff. ballasts over all bldg. + HVAC downsized for ballasts + cooling COP improvement.
35. Perimeter daylighting dimmers + high eff. ballasts over all bldg. + cooling COP improvement.
36. Perimeter daylighting dimmers + high eff. ballasts over all bldg. + heat pump COP improvement.

GO TO LEVEL 3 ANALYSIS SHEET.

Note: See Section II, page 209, for detailed descriptions and discussions of these and other Level 3 strategy sets.

LEVEL 2
MAKE INITIAL CHOICES
MEDIUM/1-STORY

Project ______________________________
date ____________ bldg. sq. ft. ____________

The building characteristic choices shown here make a **big** difference in energy cost savings. Investigate these choices within the requirements of your design.

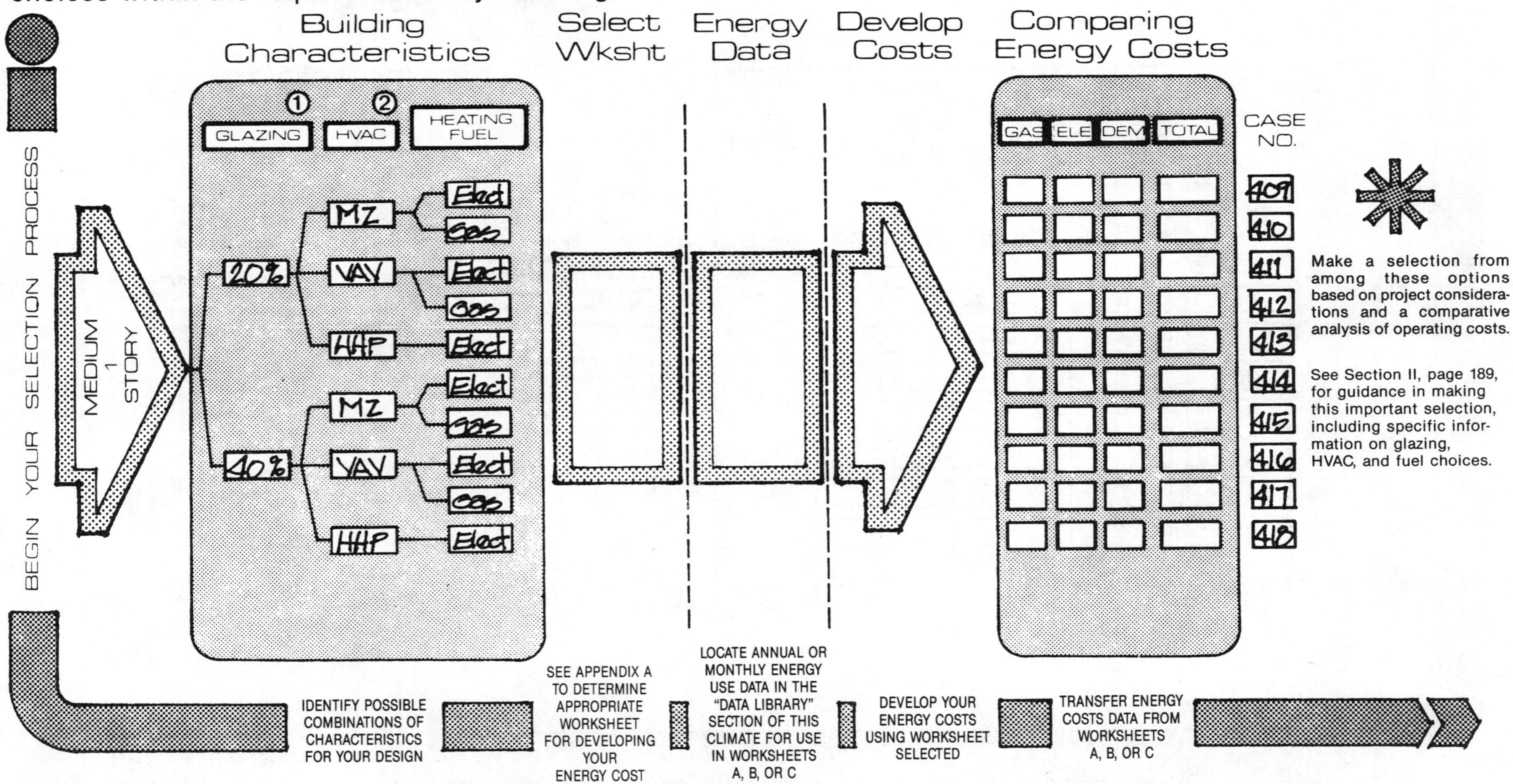

Note: See Section II, page 189, for detailed information on these Level 2 strategies and the use of this form.

① For glazing <30% use 20%, for >30% use 40%.

② MZ=Multi-zone, VAV=Variable Air Volume, HHP=Hydronic Heat Pump.

Project________________________________
date________________ bldg. sq. ft. ________________

Locate the row of strategy set options that apply to the building case you chose in Level 2. Select as many of these Level 3 strategy set options as you wish for valuation.

STRATEGY SET OPTIONS

MEDIUM 1 STORY

SELECTED LEVEL 2 CASE NUMBER	SIMPLEST LIGHTING IMPROVEMTS	MODEST LIGHTING & MECHANICAL IMPROVEMTS	MAJOR LIGHTING & MECHANICAL IMPROVEMTS	DAYLIGHT DIMMING CONTROLS AND LIGHTING IMPROVEMTS	SKYLIGHT'G, DAYLIGHT DIMMING CONTROLS & LIGHTING IMPROVEMTS
409	3	7 or 61	16	32 or 45 or 40	50
410	3	7 or 61	16	32 or 45 or 40	50
411	3	7 or 61	16	32 or 45 or 40	50
412	3	7 or 61	16	32 or 45 or 40	50
413	1	12 or 28	18	33 or 46 or 41	51
414	3	7 or 61	16	32 or 45 or 40	50
415	3	7 or 61	16	32 or 45 or 40	50
416	3	7 or 61	16	32 or 45 or 40	50
417	3	7 or 61	16	32 or 45 or 40	50
418	1	12 or 28	18	33 or 46 or 41	51

DESCRIPTOR KEY

1. High eff. ballasts over all bldg.
3. High eff. ballasts over all bldg. + HVAC downsized for ballasts.
7. High eff. ballasts and lamps over all bldg. + HVAC downsized for ballasts + cooling COP improvement.
12. High eff. ballasts and lamps over all bldg. + heat pump COP improvement.
16. High eff. ballasts, lamps and fixtures over all bldg. + HVAC downsized for ballasts and fixtures + cooling COP improvement.
18. High eff. ballasts, lamps and fixtures over all bldg. + HVAC downsized for ballasts and fixtures + heat pump COP improvement.
61. Non-uniform lighting + high eff. ballasts over all bldg. + HVAC downsized for ballasts + cooling COP improvement.
28. Non-uniform lighting + high eff. ballasts over all bldg. + heat pump COP improvement.
32. Perimeter daylighting dimmers + high eff. ballasts over all bldg. + HVAC downsized for ballasts + cooling COP improvement.
33. Perimeter daylighting dimmers + high eff. ballasts over all bldg. + HVAC downsized for ballasts + heat pump COP improvement.
40. Perimeter daylighting dimmers + high eff. ballasts over all bldg. + high eff. lamps and fixtures in core + HVAC downsized for ballasts and fixtures + cooling COP improvement.
41. Perimeter daylighting dimmers + high eff. ballasts over all bldg. + high eff. lamps and fixtures in core + HVAC downsized for ballasts and fixtures + heat pump COP improvement.
45. Perimeter daylighting dimmers + high eff. ballasts over all bldg. + high eff. lamps in core + HVAC downsized for ballasts + cooling COP improvement.
46. Perimeter daylighting dimmers + high eff. ballasts over all bldg. + high eff. lamps in core + HVAC downsized for ballasts + heat pump COP improvement.
50. Perimeter daylighting dimmers + skylights with dimmers + high eff. ballasts over all bldg. + HVAC downsized for ballasts + cooling COP improvement.
51. Perimeter daylighting dimmers + skylights with dimmers + high eff. ballasts over all bldg. + HVAC downsized for ballasts + heat pump COP improvement.

GO TO LEVEL 3 ANALYSIS SHEET.

Note: See Section II, page 209, for detailed descriptions and discussions of these and other Level 3 strategy sets.

LEVEL 2

MAKE INITIAL CHOICES

MEDIUM/2-STORY

Project ______________________
date ____________ bldg. sq. ft. ____________

The building characteristic choices shown here make a **big** difference in energy cost savings. Investigate these choices within the requirements of your design.

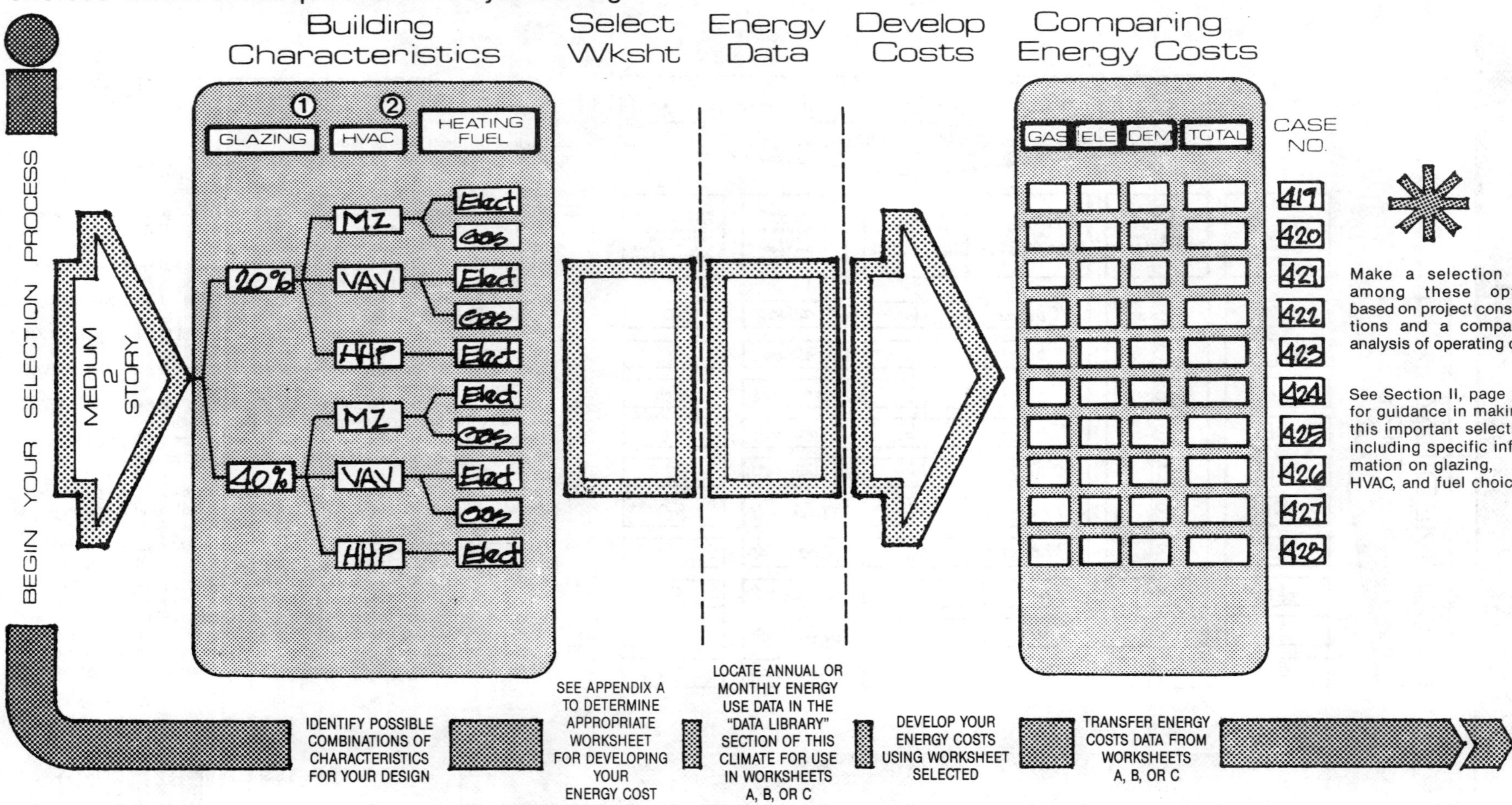

Make a selection from among these options based on project considerations and a comparative analysis of operating costs.

See Section II, page 189, for guidance in making this important selection, including specific information on glazing, HVAC, and fuel choices.

Note: See Section II, page 189, for detailed information on these Level 2 strategies and the use of this form.

① For glazing <30% use 20%, for >30% use 40%.

② MZ=Multi-zone, VAV=Variable Air Volume, HHP=Hydronic Heat Pump.

Project ____________________

date ____________ bldg. sq. ft. ____________

Locate the row of strategy set options that apply to the building case you chose in Level 2. Select as many of these Level 3 strategy set options as you wish for valuation.

STRATEGY SET OPTIONS

MEDIUM 2 STORY

SELECTED LEVEL 2 CASE NUMBER	SIMPLEST LIGHTING IMPROVEMTS	MODEST LIGHTING & MECHANICAL IMPROVEMTS	MAJOR LIGHTING & MECHANICAL IMPROVEMTS	DAYLIGHT DIMMING CONTROLS AND LIGHTING IMPROVEMTS
419	3	7 or 61	16	32
420	3	7 or 61	16	32
421	3	7 or 61	16	32
422	3	7 or 61	16	32
423	1	12 or 28	18	33
424	3	7 or 61	16	32
425	3	7 or 61	16	32
426	3	7 or 61	16	32
427	3	7 or 61	16	32
428	1	12 or 28	18	33

DESCRIPTOR KEY

1. High eff. ballasts over all bldg.
3. High eff. ballasts over all bldg. + HVAC downsized for ballasts.
7. High eff. ballasts and lamps over all bldg. + HVAC downsized for ballasts + cooling COP improvement.
12. High eff. ballasts and lamps over all bldg. + heat pump COP improvement.
16. High eff. ballasts, lamps and fixtures over all bldg. + HVAC downsized for ballasts and fixtures + cooling COP improvement.
18. High eff. ballasts, lamps and fixtures over all bldg. + HVAC downsized for ballasts and fixtures + heat pump COP improvement.
61. Non-uniform lighting + high eff. ballasts over all bldg. + HVAC downsized for ballasts + cooling COP improvement.
28. Non-uniform lighting + high eff. ballasts over all bldg. + heat pump COP improvement.
32. Perimeter daylighting dimmers + high eff. ballasts over all bldg. + HVAC downsized for ballasts + cooling COP improvement.
33. Perimeter daylighting dimmers + high eff. ballasts over all bldg. + HVAC downsized for ballasts + heat pump COP improvement.

GO TO LEVEL 3 ANALYSIS SHEET.

Note: See Section II, page 209, for detailed descriptions and discussions of these and other Level 3 strategy sets.

LEVEL 2

MAKE INITIAL CHOICES

LARGE/2-STORY

Project ______________________________

date ____________ bldg. sq. ft. ____________

The building characteristic choices shown here make a **big** difference in energy cost savings. Investigate these choices within the requirements of your design.

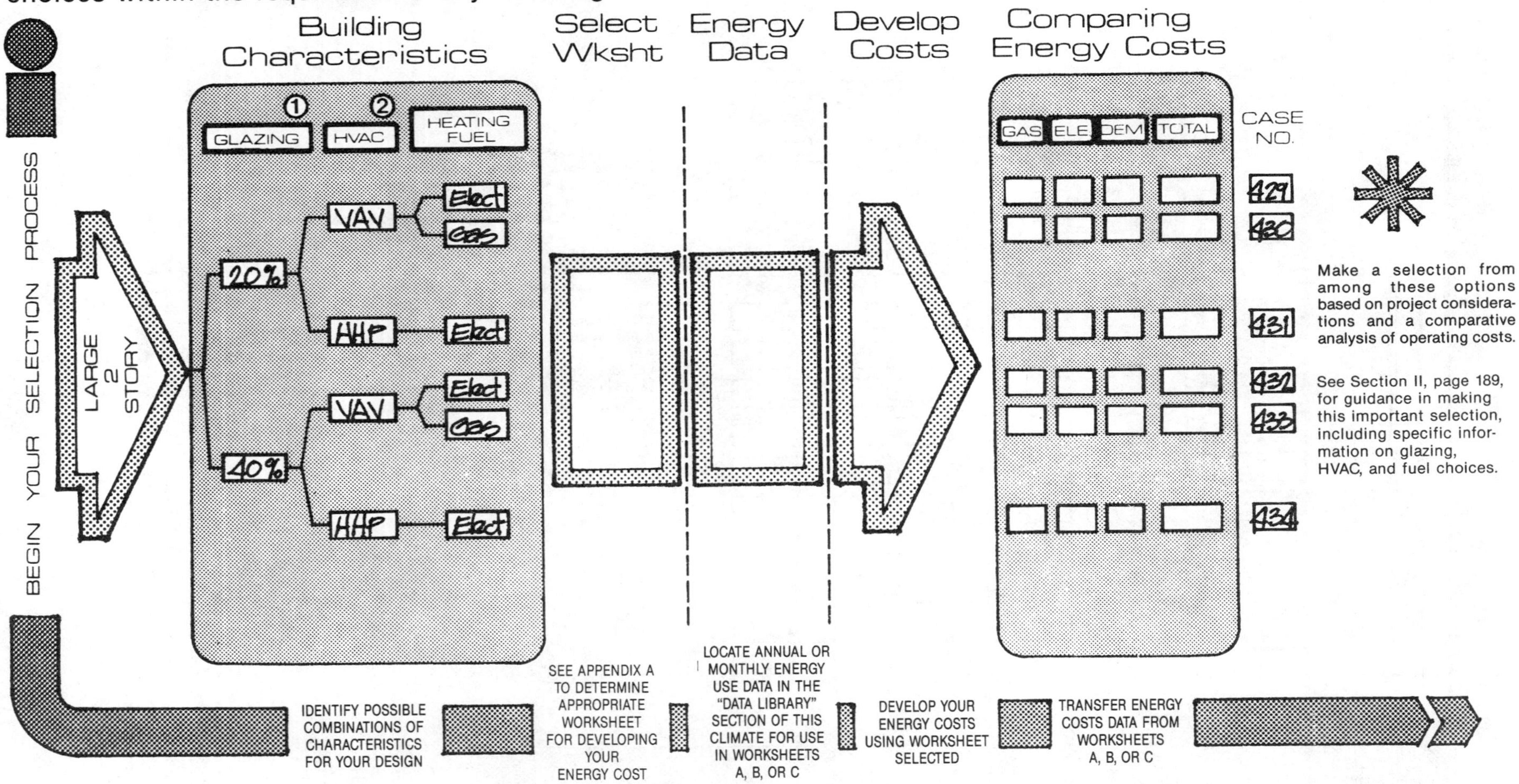

Note: See Section II, page 189, for detailed information on these Level 2 strategies and the use of this form.

① For glazing <30% use 20%, for >30% use 40%.

② MZ=Multi-zone, VAV=Variable Air Volume, HHP=Hydronic Heat Pump.

LARGE/2-STORY

Project ____________________

date __________ bldg. sq. ft. __________

Locate the row of strategy set options that apply to the building case you chose in Level 2. Select as many of these Level 3 strategy set options as you wish for valuation.

STRATEGY SET OPTIONS

LARGE 2 STORY

SELECTED LEVEL 2 CASE NUMBER	SIMPLEST LIGHTING IMPROVEMTS	MODEST LIGHTING & MECHANICAL IMPROVEMTS	MAJOR LIGHTING & MECHANICAL IMPROVEMTS	DAYLIGHT DIMMING CONTROLS AND LIGHTING IMPROVEMTS
429	3	7 or 61	16	32 or 45 or 40
430	3	7 or 61	16	32 or 45 or 40
431	1	12 or 28	22	36 or 48 or 43
432	3	7 or 61	16	32 or 45 or 40
433	3	7 or 61	16	32 or 45 or 40
434	1	12 or 28	22	36 or 48 or 43

DESCRIPTOR KEY

1. High eff. ballasts over all bldg.
3. High eff. ballasts over all bldg. + HVAC downsized for ballasts.
7. High eff. ballasts and lamps over all bldg. + HVAC downsized for ballasts + cooling COP improvement.
12. High eff. ballasts and lamps over all bldg. + heat pump COP improvement.
16. High eff. ballasts, lamps and fixtures over all bldg. + HVAC downsized for ballasts and fixtures + cooling COP improvement.
22. High eff. ballasts, lamps and fixtures over all bldg. + heat pump COP improvement.
61. Non-uniform lighting + high eff. ballasts over all bldg. + HVAC downsized for ballasts + cooling COP improvement.
28. Non-uniform lighting + high eff. ballasts over all bldg. + heat pump COP improvement.
32. Perimeter daylighting dimmers + high eff. ballasts over all bldg. + HVAC downsized for ballasts + cooling COP improvement.
36. Perimeter daylighting dimmers + high eff. ballasts over all bldg. + heat pump COP improvement.
40. Perimeter daylighting dimmers + high eff. ballasts over all bldg. + high eff. lamps and fixtures in core + HVAC downsized for ballasts and fixtures + cooling COP improvement.
43. Perimeter daylighting dimmers + high eff. ballasts over all bldg. + high eff. lamps and fixtures in core + heat pump COP improvement.
45. Perimeter daylighting dimmers + high eff. ballasts over all bldg. + high eff. lamps in core + HVAC downsized for ballasts + cooling COP improvement.
48. Perimeter daylighting dimmers + high eff. ballasts over all bldg. + high eff. lamps in core + heat pump COP improvement.

GO TO LEVEL 3 ANALYSIS SHEET.

Note: See Section II, page 209, for detailed descriptions and discussions of these and other Level 3 strategy sets.

LARGE/4-STORY

Project________________________
date__________ bldg. sq. ft. __________

The building characteristic choices shown here make a **big** difference in energy cost savings. Investigate these choices within the requirements of your design.

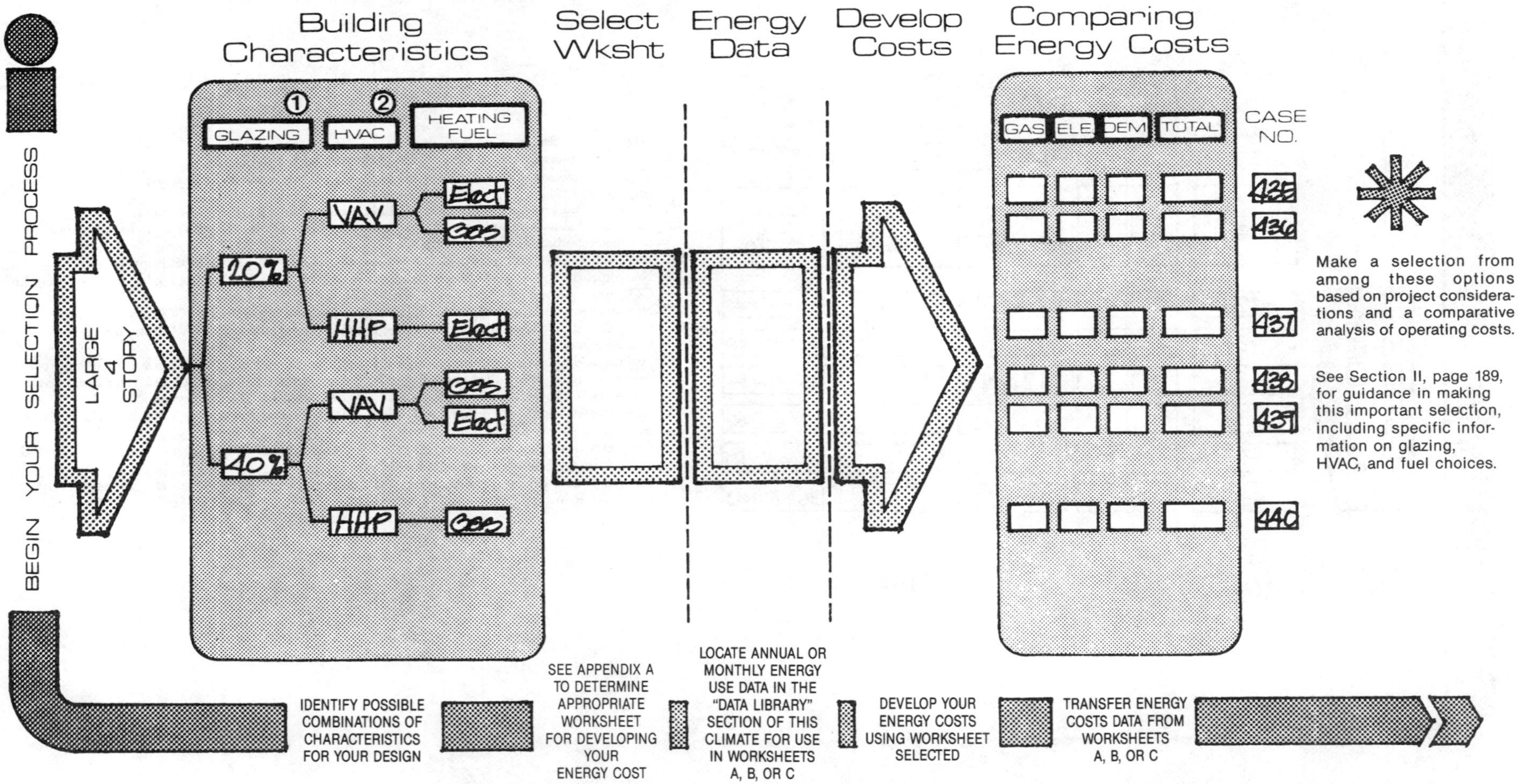

Note: See Section II, page 189, for detailed information on these Level 2 strategies and the use of this form.

① For glazing <30% use 20%, for >30% use 40%.
② VAV=Variable Air Volume, HHP=Hydronic Heat Pump.

Project ____________________
date __________ bldg. sq. ft. __________

Locate the row of strategy set options that apply to the building case you chose in Level 2. Select as many of these Level 3 strategy set options as you wish for valuation.

STRATEGY SET OPTIONS

LARGE 4 STORY

SELECTED LEVEL 2 CASE NUMBER	SIMPLEST LIGHTING IMPROVEMTS	MODEST LIGHTING & MECHANICAL IMPROVEMTS	MAJOR LIGHTING & MECHANICAL IMPROVEMTS	DAYLIGHT DIMMING CONTROLS AND LIGHTING IMPROVEMTS
435	3	7 or 61	16	32
436	3	7 or 61	16	32
437	1	12 or 28	22	36
438	3	7 or 61	16	32
439	3	7 or 61	16	32
440	1	12 or 28	22	36

DESCRIPTOR KEY

1. High eff. ballasts over all bldg.
3. High eff. ballasts over all bldg. + HVAC downsized for ballasts.
7. High eff. ballasts and lamps over all bldg. + HVAC downsized for ballasts + cooling COP improvement.
12. High eff. ballasts and lamps over all bldg. + heat pump COP improvement.
16. High eff. ballasts, lamps and fixtures over all bldg. + HVAC downsized for ballasts and fixtures + cooling COP improvement.
22. High eff. ballasts, lamps and fixtures over all bldg. + heat pump COP improvement.
61. Non-uniform lighting + high eff. ballasts over all bldg. + HVAC downsized for ballasts + cooling COP improvement.
28. Non-uniform lighting + high eff. ballasts over all bldg. + heat pump COP improvement.
32. Perimeter daylighting dimmers + high eff. ballasts over all bldg. + HVAC downsized for ballasts + cooling COP improvement.
36. Perimeter daylighting dimmers + high eff. ballasts over all bldg. + heat pump COP improvement.

GO TO LEVEL 3 ANALYSIS SHEET.

Note: See Section II, page 209, for detailed descriptions and discussions of these and other Level 3 strategy sets.

ANALYSIS/INSTRUCTIONS

For each strategy set you want to investigate, develop its construction costs using the upper route and its energy costs using the lower route. The combined information will allow you to select a strategy set for optimum economic benefit.

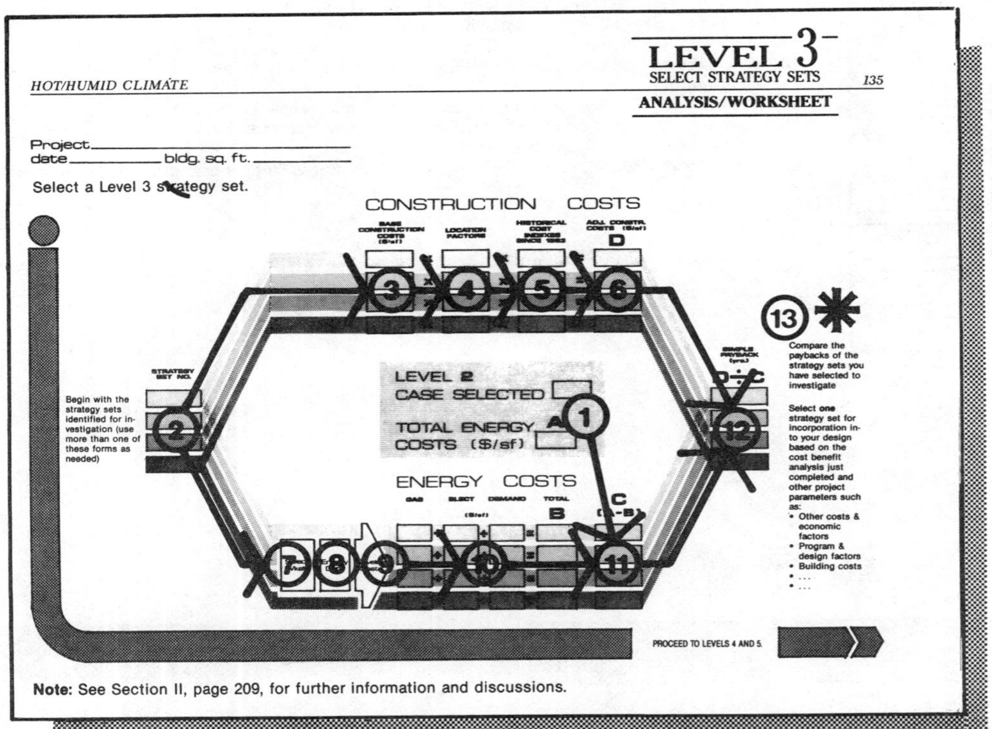

(1) Enter the Level 2 case selected and its total energy costs as developed in Level 2.

(2) As many as four strategy sets for any *one* Level 2 case can be analyzed on each sheet. You will need to use more sheets to analyze more than four strategy sets or to analyze strategy sets for another Level 2 case.

(3) Obtain *base construction costs* ($/SF) for each strategy set from the Data Library.

(4) Obtain *location factors* from commonly used cost estimating handbooks. Additional multipliers may be entered to account for above or below average construction costs due to specific project conditions.

(5) Obtain *historical cost indexes* (factors since 1983) from commonly used cost estimating handbooks.

(6) Multiply 3, 4, and 5 to obtain *adjusted construction costs* ($/SF). User may wish to use more refined construction cost data as supplied by a cost estimator for the specific project.

(7) Use the same worksheet you selected for use in Level 2.

(8) Locate energy use data for the selected strategy sets in the Data Library.

(9) Develop energy costs for each strategy set using the worksheet selected.

(10) Transfer energy cost information from worksheets A, B, or C.

(11) Find the energy savings from each strategy set by subtracting each "B" from "A."

(12) Calculate *simple payback* by dividing "D" (initial costs) by "C" (yearly savings). Alternative economic analysis techniques other than simple payback may be employed using information from columns "D" and "C."

(13) Cost factors to consider in making a decision on these strategy sets include:
- Negative construction costs (first cost savings)
- Positive construction costs
- Annual energy cost savings

Project ______________________________

date ____________ bldg. sq. ft. ____________

Select a Level 3 strategy set.

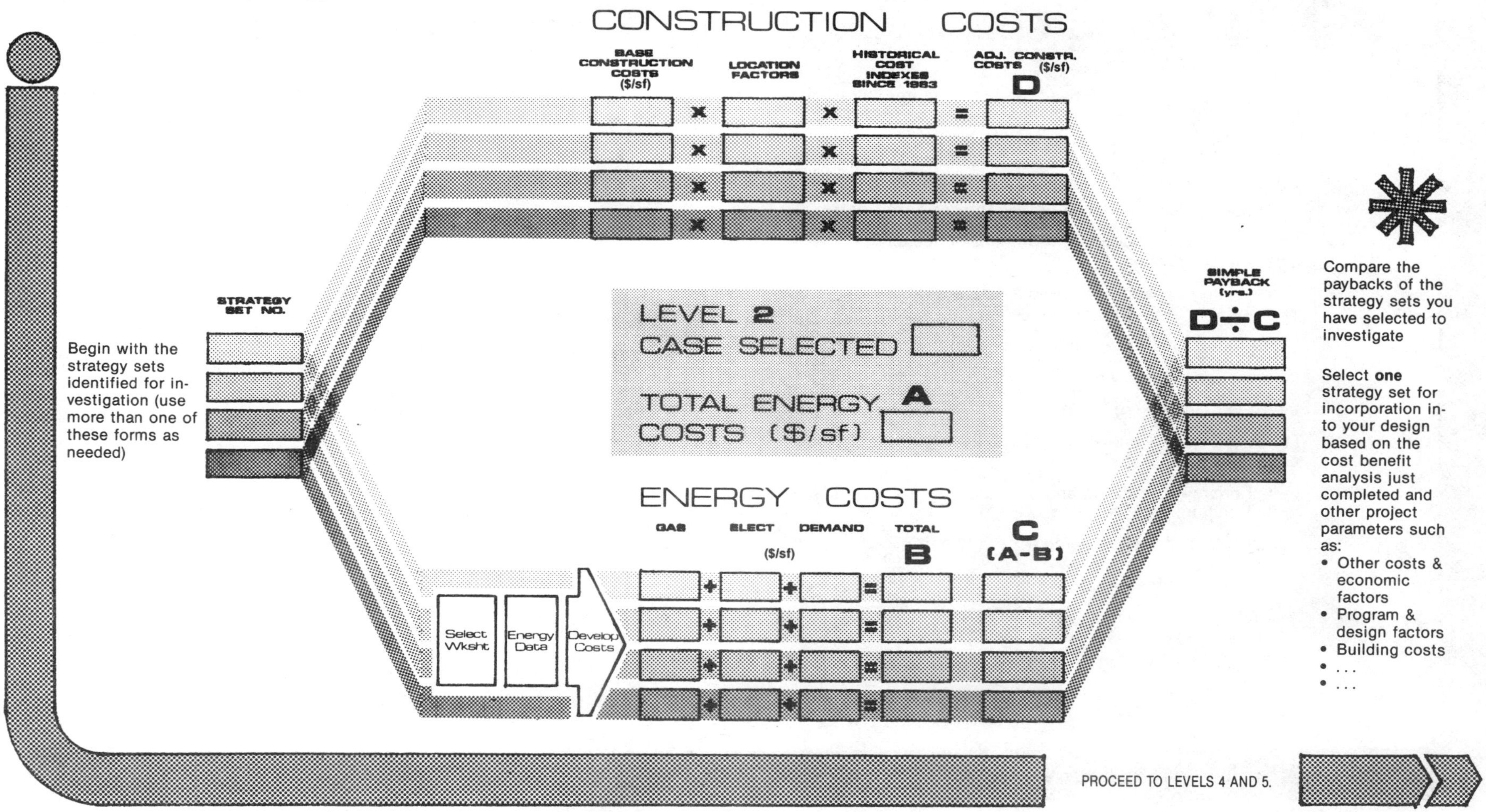

Note: See Section II, page 209, for further information and discussions.

Review these additional strategies for the possibility of further savings. Check the description in Section II, page 235, carefully for guidance on adopting any of the strategies.

A SELECT ON THE BASIS OF TECHNICAL JUDGEMENT

1. Increase thermostat deadband for occupied periods from 2°F to 8°F (70°F heating–78°F cooling)
2. Design for efficient space planning
3. Optimize building orientation
4. Minimize building height to reduce impact of glazed area
5. Reduce shading coefficient to minimally acceptable levels (not with daylighting)
6. Reduce glazing to minimally acceptable levels (not with daylighting)
7. Landscape to minimize wind induced infiltration effect
11. Increase roof reflectivity
12. Increase manual light switching
13. Install operable windows (best for buildings without economizer cycle)
15. Install energy monitoring and control system (medium and large buildings only)
16. Locate outdoor cooling equipment in shade

B VALUE MAY BE DETERMINED BY SIMPLE CALCULATION

19. High efficiency motors and pumps
20. Multiple speed compressors
21. Variable or two speed circulating fan
22. Electric resistance heateers
23. Desuperheater or stand-alone heat pump for DHW
24. Point-of-use electric domestic hot water heaters
25. External shading
26. Solar domestic hot water heater
27. Chilled water/ice storage
29. Demand limiter
30. Absorption chiller–heaters
31. Cogeneration

Note: The number preceding each strategy refers to the corresponding paragraph in Section II, page 235. See Section II, page 235, for further information and discussion of each strategy.

Review this checklist and adopt the procedures that are appropriate to your project. These actions are **important** if you are to achieve and maintain the savings from the selected Level 1–4 strategies.

A CHECKLIST FOR SUCCESSFUL TURNOVER AND BUILDING OPERATION

	DESIGN	CONSTRUCTION	TURNOVER	BUILDING OPERATION
DESIGN TEAM (ARCHITECTS/ ENGINEERS)	• Document all design assumptions (1) • Estimate energy use by sub-system (2) **& For Large Buildings:** • Document all systems and control strategies (3) • Specify contents of O&M Manuals (4) • Specify HVAC balancing requirements (5) • Specify equipment protection and cleaning (6) • Specify on-site instruction (7) • Specify access doors, valves, test ports (8) • Specify piping system ID (9) • Specify price quotations on all long-term maintenance contracts (10)	• Prepare "as-built" drawings (11) • Observe construction process for compliance (12)	• Conduct "punch list" inspection (15) • Visually check for air leakage (15) • If possible, participate in on-site training session(s) (7)	
CONTRACTOR & SUB-CONTRACTOR	**For Small Buildings Only:** • Document all systems and control strategies (3) • Specify contents of O&M Manuals (4) • Specify HVAC balancing requirements (5) • Specify equipment protection and cleaning (6) • Specify on-site instruction (7) • Specify access doors, valves, test ports (8) • Specify piping system ID (9) • Specify price quotations on all long-term maintenance contracts (10)	• Prepare O&M Manual (4) • Perform final calibration of all controls (5) • Hire independent balancing agent or require of subs (5) • Protect and clean equipment (6) • Provide all access doors, valves, and test ports (8) • Place piping system ID tags (9) • Prepare "as-built" drawings (11) • Use HVAC system to "flush" construction odors (13)	• Deliver O&M Manual (4) • Conduct on-site instruction and training (7)	• Honor warranty and call back as required (16)
EQUIPMENT MANUFACT.		• Support calibration of controls (5) • Provide all special tools required (14)	• Support O&M Manual preparation (4) • Support on-site instruction and training (7)	
OWNER & BLDG. STAFF	• Check all key design assumptions (1)	• Periodically check compliance on-site (12)	• Attend on-site instruction and training sessions (7)	• Monitor energy consumption against estimates (2) • Periodically test equipment efficiency (17) • Periodically sample and test water (18) • Inspect and calibrate controls (19) • Monitor occupant complains (20)

Note: See Section II, page 255, for more information and discussion. Numbers shown in ( ) refer to the numbered paragraphs in Section II, page 255.

1. Locate data by Base Case number.
2. Locate monthly or annual data for Base Case or Strategy Set.

Use monthly data for <u>Worksheets B or C</u>

Use annual data for <u>Worksheet A</u>

BASE CASE 326

STRAT SET	CONST COST $/SF	ENERGY FORM	UNITS (PER SF)	JAN	FEB	MAR	APR	MAY	JUN	JUL	AUG	SEP	OCT	NOV	DEC	ANNUAL
BASE	0.00	HT ELEC	KWH	0.76	0.53	0.31	0.09	0.03	0.03	0.03	0.03	0.03	0.05	0.23	0.65	2.8
		HT DEM	KW	.00605	.00587	.00575	.00405	.00369	.00006	.00006	.00006	.00224	.00363	.00520	.00587	0.00605
		REM ELEC	KWH	0.97	0.87	0.97	1.06	1.23	1.47	1.70	1.63	1.34	1.10	0.93	0.97	14.3
		REM DEM	KW	.00504	.00457	.00543	.00651	.00713	.00698	.00775	.00752	.00721	.00635	.00519	.00473	0.00775
		TOT ELEC	KWH	1.74	1.40	1.28	1.15	1.26	1.50	1.73	1.66	1.37	1.15	1.15	1.63	17.1
		TOT DEM	KW	.00930	.00828	.00828	.00698	.00716	.00763	.00781	.00763	.00744	.00670	.00679	.00837	0.00930
S- 3	-0.24	HT ELEC	KWH	0.81	0.56	0.32	0.10	0.03	0.03	0.03	0.03	0.03	0.05	0.24	0.69	3.0
		HT DEM	KW	.00605	.00587	.00575	.00405	.00369	.00006	.00006	.00006	.00224	.00363	.00520	.00587	0.00605
		REM ELEC	KWH	0.88	0.79	0.88	0.96	1.12	1.34	1.55	1.48	1.22	1.00	0.85	0.88	13.0
		REM DEM	KW	.00464	.00421	.00500	.00600	.00657	.00643	.00714	.00693	.00664	.00585	.00478	.00436	0.00714
S-																

BASE CASE 327

STRAT SET	CONST COST $/SF	ENERGY FORM	UNITS (PER SF)	JAN	FEB	MAR	APR	MAY	JUN	JUL	AUG	SEP	OCT	NOV	DEC	ANNUAL
BASE	0.00	TOT GAS	THM	0.0402	0.0277	0.0161	0.0048	0.0015	0.0015	0.0015	0.0015	0.0015	0.0026	0.0118	0.0343	0.146
		TOT ELEC	KWH	0.97	0.87	0.97	1.06	1.23	1.47	1.70	1.63	1.34	1.10	0.93	0.97	14.3
		TOT DEM	KW	.00504	.00457	.00543	.00651	.00713	.00698	.00775	.00752	.00721	.00635	.00519	.00473	0.00775
S- 3	-0.24	TOT GAS	THM	0.0426	0.0295	0.0171	0.0051	0.0015	0.0015	0.0015	0.0015	0.0015	0.0028	0.0126	0.0364	0.155
		TOT ELEC	KWH	0.88	0.79	0.88	0.96	1.12	1.34	1.55	1.48	1.22	1.00	0.85	0.88	13.0
		TOT DEM	KW	.00464	.00421	.00500	.00600	.00657	.00643	.00714	.00693	.00664	.00585	.00478	.00436	0.00714
S- 7	-0.17	TOT GAS	THM	0.0451	0.0312	0.0180	0.0054	0.0016	0.0016	0.0016	0.0016	0.0016	0.0030	0.0133	0.0385	0.164
		TOT ELEC	KWH	0.79	0.71	0.79	0.86	1.00	1.19	1.38	1.32	1.09	0.89	0.75	0.79	11.6
		TOT DEM	KW	.00406	.00369	.00438	.00525	.00575	.00563	.00625	.00606	.00581	.00513	.00419	.00381	0.00625
S- 16	0.40	TOT GAS	THM	0.0476	0.0329	0.0190	0.0057	0.0017	0.0017	0.0017	0.0017	0.0017	0.0031	0.0140	0.0407	0.173
		TOT ELEC	KWH	0.69	0.62	0.69	0.75	0.87	1.04	1.20	1.15	0.95	0.78	0.66	0.69	10.1
		TOT DEM	KW	.00370	.00336	.00398	.00478	.00523	.00512	.00569	.00552	.00529	.00467	.00381	.00347	0.00569
S- 25	0.38	TOT GAS	THM	0.0451	0.0312	0.0180	0.0054	0.0016	0.0016	0.0016	0.0016	0.0016	0.0030	0.0133	0.0385	0.164
		TOT ELEC	KWH	0.79	0.71	0.79	0.86	1.00	1.19	1.38	1.32	1.09	0.89	0.75	0.79	11.6
		TOT DEM	KW	.00406	.00369	.00438	.00525	.00575	.00563	.00625	.00606	.00581	.00513	.00419	.00381	0.00625
S- 32	0.09	TOT GAS	THM	0.0433	0.0258	0.0211	0.0087	0.0041	0.0018	0.0018	0.0018	0.0032	0.0070	0.0141	0.0298	0.162
		TOT ELEC	KWH	0.68	0.62	0.75	0.81	0.87	1.07	1.14	1.13	0.96	0.81	0.67	0.68	10.2
		TOT DEM	KW	.00414	.00403	.00443	.00512	.00535	.00552	.00575	.00564	.00558	.00500	.00426	.00380	0.00575

HT ELEC = Electric consumption for space and water heating.

REM ELEC = Remaining electric consumption (not used for heating).

TOT ELEC = Total electric consumption.

HT DEM = Peak electrical demand for space and water heating.

REM DEM = Peak electrical demand for remaining load (not heating).

TOT DEM = Total peak electrical demand.

GAS = Total gas consumption.

(Annual figures are sums of monthly consumption values and are yearly maximums of monthly peak demand values.)

STRAT SET	CONST COST $/SF	ENERGY FORM	UNITS (PER SF)	JAN	FEB	MAR	APR	MAY	JUN	JUL	AUG	SEP	OCT	NOV	DEC	ANNUAL
BASE CASE : 401																
BASE	0.00	TOT ELEC	KWH	1.41	1.19	1.28	1.29	1.46	1.56	1.70	1.67	1.48	1.34	1.28	1.34	17.0
		TOT DEM	KW	.00611	.00581	.00581	.00633	.00671	.00731	.00724	.00754	.00731	.00694	.00603	.00588	0.00754
S- 1	0.09	TOT ELEC	KWH	1.29	1.09	1.17	1.19	1.34	1.44	1.56	1.53	1.36	1.23	1.17	1.23	15.6
		TOT DEM	KW	.00571	.00543	.00543	.00592	.00627	.00684	.00677	.00705	.00684	.00649	.00564	.00550	0.00705
S- 10	0.22	TOT ELEC	KWH	1.15	0.97	1.04	1.06	1.20	1.28	1.39	1.36	1.21	1.10	1.04	1.10	13.9
		TOT DEM	KW	.00492	.00468	.00468	.00511	.00541	.00590	.00584	.00608	.00590	.00559	.00486	.00474	0.00608
S- 20	1.53	TOT ELEC	KWH	1.05	0.89	0.95	0.97	1.09	1.17	1.27	1.24	1.10	1.00	0.95	1.00	12.7
		TOT DEM	KW	.00454	.00431	.00431	.00470	.00498	.00543	.00538	.00560	.00543	.00515	.00448	.00437	0.00560
S- 25	-0.03	TOT ELEC	KWH	1.15	0.97	1.04	1.06	1.20	1.28	1.39	1.36	1.21	1.10	1.04	1.10	13.9
		TOT DEM	KW	.00492	.00468	.00468	.00511	.00541	.00590	.00584	.00608	.00590	.00559	.00486	.00474	0.00608
BASE CASE : 402																
BASE	0.00	TOT ELEC	KWH	1.52	1.34	1.41	1.36	1.52	1.58	1.66	1.63	1.54	1.47	1.38	1.49	17.9
		TOT DEM	KW	.00857	.00761	.00551	.00507	.00516	.00569	.00551	.00578	.00543	.00534	.00534	.00875	0.00875
S- 3	-0.08	TOT ELEC	KWH	1.39	1.18	1.26	1.28	1.44	1.55	1.68	1.65	1.46	1.33	1.26	1.33	16.8
		TOT DEM	KW	.00802	.00712	.00515	.00474	.00483	.00532	.00515	.00540	.00507	.00499	.00499	.00818	0.00818
S- 7	0.05	TOT ELEC	KWH	1.28	1.13	1.19	1.15	1.28	1.33	1.40	1.37	1.30	1.24	1.16	1.25	15.1
		TOT DEM	KW	.00691	.00613	.00444	.00409	.00416	.00458	.00444	.00465	.00437	.00430	.00430	.00705	0.00705
S- 16	1.29	TOT ELEC	KWH	1.15	0.97	1.04	1.06	1.20	1.28	1.39	1.36	1.21	1.10	1.04	1.10	13.9
		TOT DEM	KW	.00637	.00566	.00409	.00377	.00383	.00422	.00409	.00429	.00403	.00397	.00397	.00650	0.00650
S- 61	-0.20	TOT ELEC	KWH	1.28	1.13	1.19	1.15	1.28	1.33	1.40	1.37	1.30	1.24	1.16	1.25	15.1
		TOT DEM	KW	.00691	.00613	.00444	.00409	.00416	.00458	.00444	.00465	.00437	.00430	.00430	.00705	0.00705
BASE CASE : 403																
BASE	0.00	TOT GAS	THM	0.0136	0.0084	0.0041	0.0013	0.0013	0.0013	0.0013	0.0013	0.0013	0.0015	0.0046	0.0110	0.051
		TOT ELEC	KWH	1.20	1.15	1.34	1.35	1.50	1.57	1.66	1.62	1.54	1.45	1.28	1.23	16.9
		TOT DEM	KW	.00716	.00635	.00465	.00399	.00421	.00465	.00465	.00487	.00443	.00443	.00443	.00738	0.00738
S- 3	-0.09	TOT GAS	THM	0.0147	0.0091	0.0044	0.0014	0.0014	0.0014	0.0014	0.0014	0.0014	0.0016	0.0049	0.0119	0.055
		TOT ELEC	KWH	1.11	1.07	1.24	1.26	1.40	1.46	1.54	1.51	1.43	1.35	1.19	1.15	15.7
		TOT DEM	KW	.00671	.00595	.00436	.00374	.00394	.00436	.00436	.00457	.00415	.00415	.00415	.00692	0.00692
S- 7	0.04	TOT GAS	THM	0.0156	0.0097	0.0047	0.0015	0.0015	0.0015	0.0015	0.0015	0.0015	0.0017	0.0052	0.0127	0.058
		TOT ELEC	KWH	0.99	0.95	1.11	1.12	1.25	1.30	1.37	1.34	1.27	1.20	1.06	1.02	14.0
		TOT DEM	KW	.00598	.00531	.00389	.00333	.00352	.00389	.00389	.00407	.00370	.00370	.00370	.00617	0.00617
S- 16	1.29	TOT GAS	THM	0.0171	0.0106	0.0051	0.0016	0.0016	0.0016	0.0016	0.0016	0.0016	0.0019	0.0058	0.0139	0.064
		TOT ELEC	KWH	0.90	0.86	1.00	1.02	1.13	1.18	1.24	1.22	1.16	1.09	0.97	0.93	12.7
		TOT DEM	KW	.00555	.00492	.00360	.00309	.00326	.00360	.00360	.00378	.00343	.00343	.00343	.00572	0.00572
S- 61	-0.21	TOT GAS	THM	0.0156	0.0097	0.0047	0.0015	0.0015	0.0015	0.0015	0.0015	0.0015	0.0017	0.0052	0.0127	0.058
		TOT ELEC	KWH	0.99	0.95	1.11	1.12	1.25	1.30	1.37	1.34	1.27	1.20	1.06	1.02	14.0
		TOT DEM	KW	.00598	.00531	.00389	.00333	.00352	.00389	.00389	.00407	.00370	.00370	.00370	.00617	0.00617
BASE CASE : 404																
BASE	0.00	TOT ELEC	KWH	1.19	1.14	1.32	1.34	1.49	1.55	1.64	1.60	1.52	1.44	1.27	1.22	16.7
		TOT DEM	KW	.00596	.00596	.00596	.00664	.00664	.00731	.00724	.00754	.00709	.00694	.00633	.00596	0.00754
S- 1	0.09	TOT ELEC	KWH	1.10	1.05	1.22	1.24	1.38	1.44	1.52	1.49	1.41	1.33	1.18	1.13	15.5
		TOT DEM	KW	.00557	.00557	.00557	.00620	.00620	.00684	.00677	.00705	.00663	.00649	.00592	.00557	0.00705
S- 12	0.25	TOT ELEC	KWH	0.96	0.92	1.07	1.08	1.20	1.26	1.32	1.30	1.23	1.16	1.03	0.99	13.5
		TOT DEM	KW	.00480	.00480	.00480	.00535	.00535	.00590	.00584	.00608	.00572	.00559	.00511	.00480	0.00608
S- 22	1.56	TOT ELEC	KWH	0.87	0.83	0.96	0.98	1.09	1.13	1.20	1.17	1.11	1.05	0.93	0.89	12.2
		TOT DEM	KW	.00442	.00442	.00442	.00493	.00493	.00543	.00538	.00560	.00526	.00515	.00470	.00442	0.00560
S- 28	0.00	TOT ELEC	KWH	0.96	0.92	1.07	1.08	1.20	1.26	1.32	1.30	1.23	1.16	1.03	0.99	13.5
		TOT DEM	KW	.00480	.00480	.00480	.00535	.00535	.00590	.00584	.00608	.00572	.00559	.00511	.00480	0.00608
BASE CASE : 405																
BASE	0.00	TOT ELEC	KWH	1.46	1.23	1.32	1.34	1.51	1.62	1.76	1.72	1.53	1.39	1.32	1.39	17.6
		TOT DEM	KW	.00607	.00578	.00578	.00630	.00667	.00728	.00720	.00750	.00728	.00690	.00600	.00585	0.00750
S- 1	0.09	TOT ELEC	KWH	1.34	1.13	1.22	1.23	1.39	1.49	1.62	1.59	1.41	1.28	1.22	1.28	16.2
		TOT DEM	KW	.00568	.00540	.00540	.00589	.00624	.00680	.00673	.00701	.00680	.00645	.00561	.00547	0.00701
S- 10	0.23	TOT ELEC	KWH	1.20	1.01	1.08	1.09	1.24	1.32	1.44	1.41	1.25	1.14	1.08	1.14	14.4
		TOT DEM	KW	.00489	.00465	.00465	.00507	.00538	.00586	.00580	.00604	.00586	.00556	.00483	.00471	0.00604
S- 20	1.54	TOT ELEC	KWH	1.09	0.92	0.98	1.00	1.13	1.21	1.31	1.28	1.14	1.03	0.98	1.03	13.1
		TOT DEM	KW	.00451	.00429	.00429	.00468	.00496	.00540	.00535	.00557	.00540	.00512	.00446	.00434	0.00557
S- 25	-0.02	TOT ELEC	KWH	1.20	1.01	1.08	1.09	1.24	1.32	1.44	1.41	1.25	1.14	1.08	1.14	14.4
		TOT DEM	KW	.00489	.00465	.00465	.00507	.00538	.00586	.00580	.00604	.00586	.00556	.00483	.00471	0.00604
S- 35	0.70	TOT ELEC	KWH	0.98	0.85	0.98	1.06	1.07	1.26	1.26	1.29	1.19	1.02	0.90	0.99	12.9
		TOT DEM	KW	.00446	.00423	.00423	.00462	.00489	.00534	.00528	.00550	.00534	.00506	.00440	.00429	0.00550
BASE CASE : 406																
BASE	0.00	TOT ELEC	KWH	1.56	1.38	1.45	1.40	1.56	1.62	1.71	1.67	1.58	1.51	1.42	1.53	18.4
		TOT DEM	KW	.00857	.00761	.00551	.00507	.00516	.00569	.00551	.00578	.00543	.00534	.00534	.00875	0.00875
S- 3	-0.08	TOT ELEC	KWH	1.45	1.28	1.35	1.30	1.45	1.50	1.59	1.56	1.47	1.40	1.32	1.42	17.1
		TOT DEM	KW	.00802	.00712	.00515	.00474	.00483	.00532	.00515	.00540	.00507	.00499	.00499	.00818	0.00818
BASE CASE : 406 (CONTINUED)																
S- 7	0.06	TOT ELEC	KWH	1.32	1.16	1.22	1.18	1.32	1.36	1.44	1.41	1.33	1.27	1.19	1.29	15.5
		TOT DEM	KW	.00691	.00613	.00444	.00409	.00416	.00458	.00444	.00465	.00437	.00430	.00430	.00705	0.00705
S- 16	1.27	TOT ELEC	KWH	1.21	1.07	1.12	1.08	1.21	1.25	1.32	1.29	1.22	1.16	1.09	1.18	14.2
		TOT DEM	KW	.00637	.00566	.00409	.00377	.00383	.00422	.00409	.00429	.00403	.00397	.00397	.00650	0.00650
S- 32	0.53	TOT ELEC	KWH	1.04	0.90	1.04	1.12	1.14	1.34	1.34	1.37	1.26	1.08	0.96	1.05	13.7
		TOT DEM	KW	.00833	.00740	.00536	.00493	.00502	.00553	.00536	.00561	.00527	.00519	.00519	.00850	0.00850
S- 61	-0.19	TOT ELEC	KWH	1.32	1.16	1.22	1.18	1.32	1.36	1.44	1.41	1.33	1.27	1.19	1.29	15.5
		TOT DEM	KW	.00691	.00613	.00444	.00409	.00416	.00458	.00444	.00465	.00437	.00430	.00430	.00705	0.00705
BASE CASE : 407																
BASE	0.00	TOT GAS	THM	0.0120	0.0075	0.0036	0.0011	0.0011	0.0011	0.0011	0.0011	0.0011	0.0013	0.0041	0.0098	0.045
		TOT ELEC	KWH	1.24	1.19	1.38	1.40	1.56	1.63	1.71	1.68	1.59	1.51	1.33	1.28	17.5
		TOT DEM	KW	.00847	.00751	.00550	.00471	.00498	.00550	.00550	.00576	.00524	.00524	.00524	.00873	0.00873
S- 3	-0.08	TOT GAS	THM	0.0130	0.0081	0.0039	0.0012	0.0012	0.0012	0.0012	0.0012	0.0012	0.0015	0.0044	0.0105	0.049
		TOT ELEC	KWH	1.15	1.10	1.28	1.30	1.44	1.51	1.59	1.56	1.47	1.39	1.23	1.18	16.2
		TOT DEM	KW	.00791	.00701	.00513	.00440	.00465	.00513	.00513	.00538	.00489	.00489	.00489	.00815	0.00815
S- 7	0.05	TOT GAS	THM	0.0138	0.0086	0.0041	0.0013	0.0013	0.0013	0.0013	0.0013	0.0013	0.0015	0.0046	0.0112	0.052
		TOT ELEC	KWH	1.03	0.99	1.15	1.16	1.29	1.35	1.42	1.39	1.32	1.25	1.10	1.06	14.5
		TOT DEM	KW	.00681	.00604	.00442	.00379	.00400	.00442	.00442	.00463	.00421	.00421	.00421	.00702	0.00702
S- 16	1.26	TOT GAS	THM	0.0151	0.0094	0.0045	0.0014	0.0014	0.0014	0.0014	0.0014	0.0014	0.0017	0.0051	0.0123	0.057
		TOT ELEC	KWH	0.93	0.89	1.03	1.05	1.17	1.22	1.28	1.26	1.19	1.13	1.00	0.96	13.1
		TOT DEM	KW	.00628	.00556	.00408	.00349	.00369	.00408	.00408	.00427	.00388	.00388	.00388	.00647	0.00647
S- 32	0.53	TOT GAS	THM	0.0038	0.0038	0.0056	0.0056	0.0038	0.0056	0.0038	0.0056	0.0038	0.0038	0.0038	0.0038	0.054
		TOT ELEC	KWH	0.97	0.84	0.97	1.04	1.05	1.24	1.24	1.27	1.17	1.00	0.89	0.98	12.7
		TOT DEM	KW	.00823	.00729	.00534	.00458	.00483	.00534	.00534	.00560	.00509	.00509	.00509	.00848	0.00848
S- 61	-0.20	TOT GAS	THM	0.0138	0.0086	0.0041	0.0013	0.0013	0.0013	0.0013	0.0013	0.0013	0.0015	0.0046	0.0112	0.052
		TOT ELEC	KWH	1.03	0.99	1.15	1.16	1.29	1.35	1.42	1.39	1.32	1.25	1.10	1.06	14.5
		TOT DEM	KW	.00681	.00604	.00442	.00379	.00400	.00442	.00442	.00463	.00421	.00421	.00421	.00702	0.00702
BASE CASE : 408																
BASE	0.00	TOT ELEC	KWH	1.23	1.18	1.37	1.38	1.54	1.61	1.70	1.66	1.57	1.49	1.31	1.26	17.3
		TOT DEM	KW	.00643	.00643	.00643	.00716	.00716	.00790	.00781	.00814	.00765	.00749	.00684	.00643	0.00814
S- 1	0.09	TOT ELEC	KWH	1.14	1.09	1.26	1.28	1.42	1.49	1.57	1.54	1.46	1.38	1.22	1.17	16.0
		TOT DEM	KW	.00601	.00601	.00601	.00670	.00670	.00738	.00731	.00761	.00715	.00700	.00639	.00601	0.00761
S- 12	0.25	TOT ELEC	KWH	0.99	0.95	1.11	1.12	1.25	1.30	1.37	1.34	1.27	1.20	1.06	1.02	14.0
		TOT DEM	KW	.00518	.00518	.00518	.00577	.00577	.00636	.00630	.00656	.00617	.00604	.00551	.00518	0.00656
S- 22	1.56	TOT ELEC	KWH	0.89	0.86	1.00	1.01	1.12	1.17	1.23	1.21	1.15	1.08	0.96	0.92	12.6
		TOT DEM	KW	.00477	.00477	.00477	.00532	.00532	.00586	.00580	.00604	.00568	.00556	.00507	.00477	0.00604
S- 28	0.00	TOT ELEC	KWH	0.99	0.95	1.11	1.12	1.25	1.30	1.37	1.34	1.27	1.20	1.06	1.02	14.0
		TOT DEM	KW	.00518	.00518	.00518	.00577	.00577	.00636	.00630	.00656	.00617	.00604	.00551	.00518	0.00656
S- 36	0.73	TOT ELEC	KWH	1.04	0.90	1.04	1.12	1.14	1.34	1.34	1.37	1.26	1.08	0.96	1.05	13.7
		TOT DEM	KW	.00415	.00415	.00415	.00462	.00462	.00509	.00504	.00525	.00494	.00483	.00441	.00415	0.00525
BASE CASE : 409																
BASE	0.00	HT ELEC	KWH	0.23	0.12	0.07	0.06	0.06	0.06	0.06	0.06	0.06	0.06	0.08	0.21	1.1
		HT DEM	KW	.00415	.00315	.00212	.00034	.00002	.00002	.00001	.00002	.00003	.00071	.00228	.00344	0.00415
		REM ELEC	KWH	1.29	1.24	1.44	1.46	1.62	1.69	1.78	1.75	1.66	1.57	1.38	1.33	18.2
		REM DEM	KW	.00548	.00555	.00576	.00611	.00659	.00694	.00687	.00694	.00687	.00666	.00583	.00562	0.00694
		TOT ELEC	KWH	1.52	1.36	1.50	1.51	1.68	1.75	1.84	1.80	1.71	1.62	1.46	1.54	19.3
		TOT DEM	KW	.00810	.00778	.00632	.00672	.00705	.00737	.00729	.00737	.00737	.00721	.00640	.00786	0.00810
S- 3	-0.07	HT ELEC	KWH	0.25	0.14	0.07	0.06	0.06	0.06	0.06	0.06	0.06	0.06	0.08	0.23	1.2
		HT DEM	KW	.00415	.00315	.00212	.00034	.00002	.00002	.00001	.00002	.00003	.00071	.00228	.00344	0.00415
		REM ELEC	KWH	1.18	1.13	1.31	1.33	1.48	1.54	1.63	1.59	1.51	1.43	1.26	1.21	16.6
		REM DEM	KW	.00514	.00521	.00540	.00573	.00618	.00651	.00644	.00651	.00644	.00625	.00547	.00527	0.00651
		TOT ELEC	KWH	1.43	1.26	1.39	1.39	1.54	1.61	1.69	1.66	1.57	1.49	1.35	1.44	17.8
		TOT DEM	KW	.00761	.00731	.00594	.00632	.00662	.00693	.00685	.00693	.00693	.00677	.00601	.00738	0.00761
S- 7	0.02	HT ELEC	KWH	0.26	0.14	0.08	0.06	0.06	0.06	0.06	0.06	0.06	0.06	0.09	0.24	1.3
		HT DEM	KW	.00415	.00315	.00212	.00034	.00002	.00002	.00001	.00002	.00003	.00071	.00228	.00344	0.00415
		REM ELEC	KWH	1.07	1.03	1.19	1.21	1.34	1.40	1.48	1.45	1.37	1.30	1.15	1.10	15.1
		REM DEM	KW	.00461	.00467	.00485	.00514	.00555	.00584	.00578	.00584	.00578	.00561	.00491	.00473	0.00584
		TOT ELEC	KWH	1.33	1.17	1.27	1.27	1.41	1.47	1.54	1.51	1.44	1.36	1.24	1.34	16.4
		TOT DEM	KW	.00669	.00642	.00522	.00555	.00582	.00609	.00602	.00609	.00609	.00595	.00529	.00649	0.00669
S- 16	1.18	HT ELEC	KWH	0.28	0.15	0.08	0.07	0.07	0.07	0.07	0.07	0.07	0.07	0.09	0.25	1.3
		HT DEM	KW	.00415	.00315	.00212	.00034	.00002	.00002	.00001	.00002	.00003	.00071	.00228	.00344	0.00415
		REM ELEC	KWH	0.98	0.94	1.09	1.10	1.23	1.28	1.35	1.32	1.26	1.19	1.05	1.01	13.8
		REM DEM	KW	.00428	.00434	.00450	.00477	.00515	.00542	.00537	.00542	.00537	.00520	.00455	.00439	0.00542
		TOT ELEC	KWH	1.25	1.09	1.17	1.17	1.30	1.35	1.42	1.39	1.32	1.25	1.14	1.26	15.1
		TOT DEM	KW	.00648	.00622	.00505	.00538	.00564	.00590	.00583	.00590	.00590	.00577	.00512	.00629	0.00648

(HT ELEC=ELECTRIC CONSUMPTION FOR SPACE AND WATER HEATING. REM ELEC=REMAINING ELECTRIC CONSUMPTION (NOT USED FOR HEATING). TOT ELEC=TOTAL ELECTRIC CONSUMPTION. HT DEM=PEAK ELECTRIC DEMAND FOR SPACE AND WATER HEATING. REM DEM=PEAK ELECTRIC DEMAND FOR REMAINING LOAD (NOT HEATING). TOT DEM=TOTAL PEAK ELECTRIC DEMAND. GAS=TOTAL GAS CONSUMPTION. ANNUAL FIGURES ARE SUMS OF MONTHLY CONSUMPTION VALUES OR ARE YEARLY MAXIMUMS OF MONTHLY PEAK DEMAND VALUES.)

STRAT SET	CONST COST $/SF	ENERGY FORM	UNITS (PER SF)	JAN	FEB	MAR	APR	MAY	JUN	JUL	AUG	SEP	OCT	NOV	DEC	ANNUAL
BASE CASE : 409 (CONTINUED)																
S- 32	0.28	HT ELEC	KWH	0.29	0.19	0.11	0.04	0.04	0.04	0.04	0.04	0.04	0.04	0.11	0.22	1.2
		HT DEM	KW	.00415	.00315	.00212	.00034	.00002	.00002	.00001	.00002	.00003	.00071	.00228	.00344	0.00415
		REM ELEC	KWH	0.76	0.73	0.95	1.07	1.09	1.28	1.28	1.25	1.20	1.02	0.87	0.82	12.4
		REM DEM	KW	.00451	.00457	.00474	.00502	.00542	.00571	.00565	.00571	.00565	.00548	.00480	.00463	0.00571
		TOT ELEC	KWH	1.05	0.92	1.06	1.11	1.13	1.32	1.32	1.29	1.24	1.06	0.98	1.04	13.6
		TOT DEM	KW	.00762	.00732	.00594	.00632	.00663	.00693	.00686	.00693	.00693	.00678	.00602	.00739	0.00762
S- 40	0.96	HT ELEC	KWH	0.35	0.22	0.13	0.05	0.04	0.05	0.05	0.05	0.05	0.05	0.13	0.26	1.4
		HT DEM	KW	.00415	.00315	.00212	.00034	.00002	.00002	.00001	.00002	.00003	.00071	.00228	.00344	0.00415
		REM ELEC	KWH	0.68	0.65	0.85	0.95	0.98	1.14	1.14	1.12	1.08	0.91	0.78	0.73	11.1
		REM DEM	KW	.00419	.00425	.00441	.00467	.00504	.00531	.00526	.00531	.00526	.00510	.00446	.00430	0.00531
		TOT ELEC	KWH	1.02	0.88	0.98	1.00	1.02	1.19	1.19	1.17	1.12	0.96	0.90	1.00	12.5
		TOT DEM	KW	.00714	.00685	.00557	.00593	.00621	.00650	.00643	.00650	.00650	.00635	.00564	.00693	0.00714
S- 45	0.33	HT ELEC	KWH	0.33	0.21	0.12	0.05	0.04	0.04	0.04	0.04	0.04	0.05	0.12	0.25	1.4
		HT DEM	KW	.00415	.00315	.00212	.00034	.00002	.00002	.00001	.00002	.00003	.00071	.00228	.00344	0.00415
		REM ELEC	KWH	0.73	0.70	0.92	1.02	1.05	1.23	1.23	1.20	1.15	0.98	0.83	0.79	11.9
		REM DEM	KW	.00435	.00441	.00457	.00485	.00523	.00551	.00545	.00551	.00545	.00529	.00463	.00446	0.00551
		TOT ELEC	KWH	1.06	0.91	1.04	1.07	1.09	1.27	1.27	1.25	1.20	1.03	0.95	1.04	13.3
		TOT DEM	KW	.00742	.00712	.00579	.00616	.00646	.00675	.00668	.00675	.00675	.00660	.00586	.00720	0.00742
S- 50	1.60	HT ELEC	KWH	0.32	0.20	0.12	0.05	0.04	0.04	0.04	0.04	0.04	0.05	0.12	0.24	1.3
		HT DEM	KW	.00415	.00315	.00212	.00034	.00002	.00002	.00001	.00002	.00003	.00071	.00228	.00344	0.00415
		REM ELEC	KWH	0.71	0.68	0.89	1.00	1.02	1.19	1.19	1.17	1.13	0.95	0.81	0.77	11.6
		REM DEM	KW	.00446	.00452	.00469	.00497	.00537	.00565	.00559	.00565	.00559	.00542	.00475	.00458	0.00565
		TOT ELEC	KWH	1.03	0.89	1.01	1.04	1.06	1.24	1.24	1.21	1.17	1.00	0.93	1.01	12.9
		TOT DEM	KW	.00755	.00725	.00589	.00627	.00657	.00687	.00679	.00687	.00687	.00672	.00596	.00732	0.00755
S- 61	-0.23	HT ELEC	KWH	0.26	0.14	0.08	0.06	0.06	0.06	0.06	0.06	0.06	0.06	0.09	0.24	1.3
		HT DEM	KW	.00415	.00315	.00212	.00034	.00002	.00002	.00001	.00002	.00003	.00071	.00228	.00344	0.00415
		REM ELEC	KWH	1.07	1.03	1.19	1.21	1.34	1.40	1.48	1.45	1.37	1.30	1.15	1.10	15.1
		REM DEM	KW	.00461	.00467	.00485	.00514	.00555	.00584	.00578	.00584	.00578	.00561	.00491	.00473	0.00584
		TOT ELEC	KWH	1.33	1.17	1.27	1.27	1.41	1.47	1.54	1.51	1.44	1.36	1.24	1.34	16.4
		TOT DEM	KW	.00669	.00642	.00522	.00555	.00582	.00609	.00602	.00609	.00609	.00595	.00529	.00649	0.00669
BASE CASE : 410																
BASE	0.00	TOT GAS	THM	0.0117	0.0062	0.0034	0.0028	0.0028	0.0028	0.0028	0.0028	0.0028	0.0028	0.0039	0.0106	0.056
		TOT ELEC	KWH	1.29	1.24	1.44	1.46	1.62	1.69	1.76	1.75	1.66	1.57	1.38	1.33	18.2
		TOT DEM	KW	.00548	.00555	.00576	.00611	.00659	.00694	.00687	.00694	.00687	.00666	.00583	.00562	0.00694
S- 3	-0.07	TOT GAS	THM	0.0130	0.0069	0.0038	0.0032	0.0032	0.0032	0.0032	0.0032	0.0032	0.0032	0.0043	0.0117	0.062
		TOT ELEC	KWH	1.18	1.13	1.31	1.33	1.48	1.54	1.63	1.59	1.51	1.43	1.26	1.21	16.6
		TOT DEM	KW	.00514	.00521	.00540	.00573	.00618	.00651	.00644	.00651	.00644	.00625	.00547	.00527	0.00651
S- 7	0.01	TOT GAS	THM	0.0134	0.0071	0.0039	0.0033	0.0033	0.0033	0.0033	0.0033	0.0033	0.0033	0.0045	0.0121	0.064
		TOT ELEC	KWH	1.07	1.03	1.19	1.21	1.34	1.40	1.48	1.45	1.37	1.30	1.15	1.10	15.1
		TOT DEM	KW	.00461	.00467	.00485	.00514	.00555	.00584	.00578	.00584	.00578	.00561	.00491	.00473	0.00584
S- 16	1.18	TOT GAS	THM	0.0140	0.0075	0.0041	0.0034	0.0034	0.0034	0.0034	0.0034	0.0034	0.0034	0.0047	0.0127	0.067
		TOT ELEC	KWH	0.98	0.94	1.09	1.10	1.23	1.26	1.35	1.32	1.26	1.19	1.05	1.01	13.8
		TOT DEM	KW	.00428	.00434	.00450	.00477	.00515	.00542	.00537	.00542	.00537	.00520	.00455	.00439	0.00542
S- 32	0.28	TOT GAS	THM	0.0148	0.0095	0.0055	0.0021	0.0018	0.0019	0.0019	0.0019	0.0019	0.0022	0.0055	0.0113	0.061
		TOT ELEC	KWH	0.76	0.73	0.95	1.07	1.09	1.28	1.28	1.25	1.20	1.02	0.87	0.82	12.4
		TOT DEM	KW	.00451	.00457	.00474	.00502	.00542	.00571	.00565	.00571	.00565	.00548	.00480	.00463	0.00571
S- 40	0.96	TOT GAS	THM	0.0177	0.0113	0.0065	0.0025	0.0022	0.0023	0.0023	0.0023	0.0023	0.0027	0.0065	0.0135	0.073
		TOT ELEC	KWH	0.68	0.65	0.85	0.95	0.98	1.14	1.14	1.12	1.08	0.91	0.78	0.73	11.1
		TOT DEM	KW	.00419	.00425	.00441	.00467	.00504	.00531	.00526	.00531	.00526	.00510	.00446	.00430	0.00531
S- 45	0.33	TOT GAS	THM	0.0168	0.0107	0.0062	0.0024	0.0021	0.0022	0.0022	0.0022	0.0022	0.0025	0.0062	0.0128	0.069
		TOT ELEC	KWH	0.73	0.70	0.92	1.02	1.05	1.23	1.23	1.20	1.15	0.98	0.83	0.79	11.9
		TOT DEM	KW	.00435	.00441	.00457	.00485	.00523	.00551	.00545	.00551	.00545	.00529	.00463	.00446	0.00551
S- 50	1.60	TOT GAS	THM	0.0163	0.0104	0.0060	0.0023	0.0020	0.0021	0.0021	0.0021	0.0021	0.0025	0.0060	0.0124	0.067
		TOT ELEC	KWH	0.71	0.68	0.89	1.00	1.02	1.19	1.19	1.17	1.13	0.95	0.81	0.77	11.6
		TOT DEM	KW	.00446	.00452	.00469	.00497	.00537	.00565	.00559	.00565	.00559	.00542	.00475	.00458	0.00565
S- 61	-0.23	TOT GAS	THM	0.0134	0.0071	0.0039	0.0033	0.0033	0.0033	0.0033	0.0033	0.0033	0.0033	0.0045	0.0121	0.064
		TOT ELEC	KWH	1.07	1.03	1.19	1.21	1.34	1.40	1.48	1.45	1.37	1.30	1.15	1.10	15.1
		TOT DEM	KW	.00461	.00467	.00485	.00514	.00555	.00584	.00578	.00584	.00578	.00561	.00491	.00473	0.00584
BASE CASE : 411																
BASE	0.00	HT ELEC	KWH	0.31	0.19	0.09	0.03	0.03	0.03	0.03	0.03	0.03	0.04	0.11	0.25	1.2
		HT DEM	KW	.00358	.00311	.00272	.00107	.00029	.00007	.00007	.00007	.00007	.00158	.00265	.00354	0.00358
		REM ELEC	KWH	1.20	1.15	1.34	1.35	1.50	1.57	1.66	1.62	1.54	1.45	1.28	1.23	16.9
		REM DEM	KW	.00514	.00514	.00520	.00566	.00598	.00637	.00637	.00650	.00637	.00611	.00539	.00520	0.00650
		TOT ELEC	KWH	1.51	1.34	1.43	1.38	1.53	1.60	1.69	1.65	1.57	1.49	1.39	1.49	18.1
		TOT DEM	KW	.00634	.00528	.00528	.00568	.00614	.00647	.00647	.00660	.00647	.00627	.00541	.00640	0.00660
S- 3	-0.12	HT ELEC	KWH	0.33	0.21	0.10	0.03	0.03	0.03	0.03	0.03	0.03	0.04	0.11	0.27	1.2
		HT DEM	KW	.00358	.00311	.00272	.00107	.00029	.00007	.00007	.00007	.00007	.00158	.00265	.00354	0.00358
		REM ELEC	KWH	1.09	1.05	1.22	1.23	1.37	1.43	1.51	1.48	1.40	1.32	1.17	1.12	15.4
		REM DEM	KW	.00472	.00472	.00478	.00520	.00550	.00586	.00586	.00598	.00586	.00562	.00496	.00478	0.00598
		TOT ELEC	KWH	1.42	1.25	1.32	1.26	1.40	1.46	1.54	1.51	1.43	1.36	1.28	1.39	16.6
		TOT DEM	KW	.00595	.00496	.00496	.00533	.00577	.00608	.00608	.00620	.00608	.00589	.00508	.00601	0.00620
S- 7	-0.04	HT ELEC	KWH	0.35	0.22	0.11	0.03	0.03	0.03	0.03	0.03	0.03	0.04	0.12	0.29	1.3
		HT DEM	KW	.00358	.00311	.00272	.00107	.00029	.00007	.00007	.00007	.00007	.00158	.00265	.00354	0.00358
		REM ELEC	KWH	0.97	0.92	1.07	1.09	1.21	1.26	1.33	1.31	1.24	1.17	1.03	0.99	13.6
		REM DEM	KW	.00419	.00419	.00424	.00461	.00488	.00519	.00519	.00530	.00519	.00498	.00440	.00424	0.00530
		TOT ELEC	KWH	1.32	1.14	1.18	1.12	1.24	1.30	1.37	1.34	1.27	1.21	1.15	1.28	14.9
		TOT DEM	KW	.00523	.00436	.00436	.00469	.00507	.00534	.00534	.00545	.00534	.00518	.00447	.00529	0.00545

STRAT SET	CONST COST $/SF	ENERGY FORM	UNITS (PER SF)	JAN	FEB	MAR	APR	MAY	JUN	JUL	AUG	SEP	OCT	NOV	DEC	ANNUAL
BASE CASE : 411 (CONTINUED)																
S- 16	1.08	HT ELEC	KWH	0.37	0.23	0.11	0.03	0.03	0.03	0.03	0.03	0.03	0.04	0.13	0.30	1.4
		HT DEM	KW	.00358	.00311	.00272	.00107	.00029	.00007	.00007	.00007	.00007	.00158	.00265	.00354	0.00358
		REM ELEC	KWH	0.85	0.82	0.95	0.96	1.07	1.12	1.18	1.15	1.09	1.03	0.91	0.88	12.0
		REM DEM	KW	.00379	.00379	.00384	.00418	.00442	.00470	.00470	.00480	.00470	.00451	.00398	.00384	0.00480
		TOT ELEC	KWH	1.22	1.05	1.06	0.99	1.10	1.15	1.21	1.19	1.13	1.07	1.04	1.18	13.4
		TOT DEM	KW	.00507	.00422	.00422	.00454	.00491	.00517	.00517	.00528	.00517	.00502	.00433	.00512	0.00528
S- 32	0.23	HT ELEC	KWH	0.43	0.26	0.17	0.05	0.04	0.04	0.04	0.04	0.04	0.06	0.16	0.35	1.7
		HT DEM	KW	.00358	.00311	.00272	.00107	.00029	.00007	.00007	.00007	.00007	.00158	.00265	.00354	0.00358
		REM ELEC	KWH	0.86	0.83	1.03	1.09	1.14	1.30	1.31	1.31	1.21	1.08	0.91	0.90	13.0
		REM DEM	KW	.00407	.00407	.00412	.00448	.00474	.00505	.00505	.00515	.00505	.00484	.00427	.00412	0.00515
		TOT ELEC	KWH	1.28	1.10	1.20	1.14	1.18	1.34	1.35	1.35	1.24	1.14	1.07	1.25	14.7
		TOT DEM	KW	.00611	.00514	.00429	.00468	.00494	.00520	.00520	.00533	.00520	.00500	.00442	.00650	0.00650
S- 40	0.89	HT ELEC	KWH	0.51	0.32	0.21	0.06	0.04	0.04	0.04	0.04	0.04	0.07	0.19	0.42	2.0
		HT DEM	KW	.00358	.00311	.00272	.00107	.00029	.00007	.00007	.00007	.00007	.00158	.00265	.00354	0.00358
		REM ELEC	KWH	0.78	0.76	0.93	0.99	1.04	1.18	1.19	1.19	1.10	0.98	0.83	0.81	11.8
		REM DEM	KW	.00370	.00370	.00374	.00407	.00431	.00459	.00459	.00468	.00459	.00440	.00388	.00374	0.00468
		TOT ELEC	KWH	1.29	1.07	1.14	1.05	1.08	1.22	1.24	1.24	1.14	1.05	1.01	1.24	13.8
		TOT DEM	KW	.00600	.00504	.00421	.00459	.00485	.00510	.00510	.00523	.00510	.00491	.00434	.00638	0.00638
S- 45	0.27	HT ELEC	KWH	0.47	0.29	0.19	0.06	0.04	0.04	0.04	0.04	0.04	0.07	0.17	0.39	1.8
		HT DEM	KW	.00358	.00311	.00272	.00107	.00029	.00007	.00007	.00007	.00007	.00158	.00265	.00354	0.00358
		REM ELEC	KWH	0.82	0.79	0.98	1.04	1.09	1.24	1.25	1.25	1.15	1.03	0.87	0.86	12.4
		REM DEM	KW	.00389	.00389	.00394	.00429	.00454	.00483	.00483	.00493	.00483	.00463	.00409	.00394	0.00493
		TOT ELEC	KWH	1.29	1.08	1.17	1.10	1.13	1.28	1.29	1.29	1.19	1.10	1.04	1.24	14.2
		TOT DEM	KW	.00606	.00510	.00426	.00464	.00490	.00516	.00516	.00529	.00516	.00497	.00439	.00645	0.00645
S- 50	1.54	HT ELEC	KWH	0.47	0.29	0.19	0.06	0.04	0.04	0.04	0.04	0.04	0.07	0.17	0.39	1.8
		HT DEM	KW	.00358	.00311	.00272	.00107	.00029	.00007	.00007	.00007	.00007	.00158	.00265	.00354	0.00358
		REM ELEC	KWH	0.81	0.78	0.96	1.02	1.07	1.22	1.23	1.23	1.13	1.01	0.85	0.84	12.2
		REM DEM	KW	.00407	.00407	.00412	.00448	.00474	.00505	.00505	.00515	.00505	.00484	.00427	.00412	0.00515
		TOT ELEC	KWH	1.27	1.07	1.15	1.08	1.11	1.26	1.27	1.27	1.17	1.08	1.03	1.23	14.0
		TOT DEM	KW	.00624	.00520	.00520	.00559	.00605	.00637	.00637	.00650	.00637	.00617	.00533	.00631	0.00650
S- 61	-0.29	HT ELEC	KWH	0.35	0.22	0.11	0.03	0.03	0.03	0.03	0.03	0.03	0.04	0.12	0.29	1.3
		HT DEM	KW	.00358	.00311	.00272	.00107	.00029	.00007	.00007	.00007	.00007	.00158	.00265	.00354	0.00358
		REM ELEC	KWH	0.97	0.92	1.07	1.09	1.21	1.26	1.33	1.31	1.24	1.17	1.03	0.99	13.6
		REM DEM	KW	.00419	.00419	.00424	.00461	.00488	.00519	.00519	.00530	.00519	.00498	.00440	.00424	0.00530
		TOT ELEC	KWH	1.32	1.14	1.18	1.12	1.24	1.30	1.37	1.34	1.27	1.21	1.15	1.28	14.9
		TOT DEM	KW	.00523	.00436	.00436	.00469	.00507	.00534	.00534	.00545	.00534	.00518	.00447	.00529	0.00545
BASE CASE : 412																
BASE	0.00	TOT GAS	THM	0.0163	0.0102	0.0049	0.0015	0.0015	0.0015	0.0015	0.0015	0.0015	0.0018	0.0055	0.0133	0.061
		TOT ELEC	KWH	1.20	1.15	1.34	1.35	1.50	1.57	1.66	1.62	1.54	1.45	1.28	1.23	16.9
		TOT DEM	KW	.00514	.00514	.00520	.00566	.00598	.00637	.00637	.00650	.00637	.00611	.00539	.00520	0.00650
S- 3	-0.13	TOT GAS	THM	0.0174	0.0108	0.0052	0.0016	0.0016	0.0016	0.0016	0.0016	0.0016	0.0020	0.0059	0.0142	0.065
		TOT ELEC	KWH	1.09	1.05	1.22	1.23	1.37	1.43	1.51	1.48	1.40	1.32	1.17	1.12	15.4
		TOT DEM	KW	.00472	.00472	.00478	.00520	.00550	.00586	.00586	.00598	.00586	.00562	.00496	.00478	0.00598
S- 7	-0.05	TOT GAS	THM	0.0185	0.0115	0.0055	0.0017	0.0017	0.0017	0.0017	0.0017	0.0017	0.0021	0.0062	0.0150	0.069
		TOT ELEC	KWH	0.97	0.92	1.07	1.09	1.21	1.26	1.33	1.31	1.24	1.17	1.03	0.99	13.6
		TOT DEM	KW	.00419	.00419	.00424	.00461	.00488	.00519	.00519	.00530	.00519	.00498	.00440	.00424	0.00530
S- 16	1.07	TOT GAS	THM	0.0194	0.0121	0.0058	0.0018	0.0018	0.0018	0.0018	0.0018	0.0018	0.0022	0.0066	0.0158	0.073
		TOT ELEC	KWH	0.85	0.82	0.95	0.96	1.07	1.12	1.18	1.15	1.09	1.03	0.91	0.88	12.0
		TOT DEM	KW	.00379	.00379	.00384	.00418	.00442	.00470	.00470	.00480	.00470	.00451	.00398	.00384	0.00480
S- 32	0.22	TOT GAS	THM	0.0223	0.0138	0.0089	0.0027	0.0018	0.0019	0.0019	0.0019	0.0018	0.0032	0.0081	0.0185	0.087
		TOT ELEC	KWH	0.86	0.83	1.03	1.09	1.14	1.30	1.31	1.31	1.21	1.08	0.91	0.90	13.0
		TOT DEM	KW	.00407	.00407	.00412	.00448	.00474	.00505	.00505	.00515	.00505	.00484	.00427	.00412	0.00515
S- 40	0.88	TOT GAS	THM	0.0269	0.0166	0.0107	0.0033	0.0022	0.0023	0.0023	0.0023	0.0022	0.0039	0.0098	0.0222	0.105
		TOT ELEC	KWH	0.78	0.76	0.93	0.99	1.04	1.18	1.19	1.19	1.10	0.98	0.83	0.81	11.8
		TOT DEM	KW	.00370	.00370	.00374	.00407	.00431	.00459	.00459	.00468	.00459	.00440	.00388	.00374	0.00468
S- 45	0.27	TOT GAS	THM	0.0246	0.0152	0.0098	0.0030	0.0020	0.0021	0.0021	0.0021	0.0020	0.0036	0.0090	0.0203	0.096
		TOT ELEC	KWH	0.82	0.79	0.98	1.04	1.09	1.24	1.25	1.25	1.15	1.03	0.87	0.86	12.4
		TOT DEM	KW	.00389	.00389	.00394	.00429	.00454	.00483	.00483	.00493	.00483	.00463	.00409	.00394	0.00493
S- 50	1.53	TOT GAS	THM	0.0246	0.0152	0.0098	0.0030	0.0020	0.0021	0.0021	0.0021	0.0020	0.0036	0.0090	0.0203	0.096
		TOT ELEC	KWH	0.81	0.78	0.96	1.02	1.07	1.22	1.23	1.23	1.13	1.01	0.85	0.84	12.2
		TOT DEM	KW	.00407	.00407	.00412	.00448	.00474	.00505	.00505	.00515	.00505	.00484	.00427	.00412	0.00515
S- 61	-0.30	TOT GAS	THM	0.0185	0.0115	0.0055	0.0017	0.0017	0.0017	0.0017	0.0017	0.0017	0.0021	0.0062	0.0150	0.069
		TOT ELEC	KWH	0.97	0.92	1.07	1.09	1.21	1.26	1.33	1.31	1.24	1.17	1.03	0.99	13.6
		TOT DEM	KW	.00419	.00419	.00424	.00461	.00488	.00519	.00519	.00530	.00519	.00498	.00440	.00424	0.00530
BASE CASE : 413																
BASE	0.00	TOT ELEC	KWH	1.29	1.16	1.41	1.52	1.56	1.75	1.77	1.77	1.65	1.47	1.29	1.29	17.9
		TOT DEM	KW	.00575	.00561	.00568	.00631	.00631	.00665	.00658	.00693	.00665	.00644	.00603	.00575	0.00693
S- 1	0.09	TOT ELEC	KWH	1.19	1.07	1.30	1.40	1.44	1.62	1.63	1.63	1.52	1.35	1.19	1.19	16.5
		TOT DEM	KW	.00558	.00544	.00551	.00612	.00612	.00645	.00638	.00672	.00645	.00625	.00585	.00558	0.00672
S- 12	0.16	TOT ELEC	KWH	1.07	0.96	1.17	1.26	1.29	1.45	1.47	1.47	1.36	1.21	1.07	1.07	14.8
		TOT DEM	KW	.00530	.00517	.00523	.00581	.00581	.00612	.00606	.00638	.00612	.00593	.00555	.00530	0.00638
S- 18	1.47	TOT ELEC	KWH	0.98	0.88	1.07	1.16	1.18	1.33	1.35	1.35	1.25	1.12	0.98	0.98	13.6
		TOT DEM	KW	.00512	.00500	.00506	.00561	.00561	.00592	.00586	.00617	.00592	.00574	.00537	.00512	0.00617

(HT ELEC=ELECTRIC CONSUMPTION FOR SPACE AND WATER HEATING. REM ELEC=REMAINING ELECTRIC CONSUMPTION (NOT USED FOR HEATING). TOT ELEC=TOTAL ELECTRIC CONSUMPTION. HT DEM=PEAK ELECTRIC DEMAND FOR SPACE AND WATER HEATING. REM DEM=PEAK ELECTRIC DEMAND FOR REMAINING LOAD (NOT HEATING). TOT DEM=TOTAL PEAK ELECTRIC DEMAND. GAS=TOTAL GAS CONSUMPTION. ANNUAL FIGURES ARE SUMS OF MONTHLY CONSUMPTION VALUES OR ARE YEARLY MAXIMUMS OF MONTHLY PEAK DEMAND VALUES.)

STRAT SET	CONST COST $/SF	ENERGY FORM	UNITS (PER SF)	JAN	FEB	MAR	APR	MAY	JUN	JUL	AUG	SEP	OCT	NOV	DEC	ANNUAL
BASE CASE : 413 (CONTINUED)																
S- 28	-0.09	TOT ELEC	KWH	1.07	0.96	1.17	1.26	1.29	1.45	1.47	1.47	1.36	1.21	1.07	1.07	14.8
		TOT DEM	KW	.00530	.00517	.00523	.00581	.00581	.00612	.00606	.00638	.00612	.00593	.00555	.00530	0.00638
S- 33	0.43	TOT ELEC	KWH	0.98	0.88	1.07	1.16	1.18	1.33	1.35	1.35	1.25	1.12	0.98	0.98	13.6
		TOT DEM	KW	.00438	.00428	.00433	.00480	.00480	.00507	.00502	.00528	.00507	.00491	.00459	.00438	0.00528
S- 41	1.19	TOT ELEC	KWH	0.90	0.81	0.99	1.06	1.09	1.23	1.24	1.24	1.15	1.02	0.90	0.90	12.5
		TOT DEM	KW	.00423	.00413	.00418	.00464	.00464	.00490	.00484	.00510	.00490	.00474	.00444	.00423	0.00510
S- 46	0.48	TOT ELEC	KWH	0.94	0.85	1.03	1.11	1.14	1.28	1.30	1.30	1.21	1.07	0.94	0.94	13.1
		TOT DEM	KW	.00430	.00420	.00425	.00471	.00471	.00497	.00492	.00518	.00497	.00482	.00451	.00430	0.00518
S- 51	1.74	TOT ELEC	KWH	0.95	0.86	1.04	1.12	1.15	1.29	1.31	1.31	1.21	1.08	0.95	0.95	13.2
		TOT DEM	KW	.00438	.00428	.00433	.00480	.00480	.00507	.00502	.00528	.00507	.00491	.00459	.00438	0.00528
BASE CASE : 414																
BASE	0.00	HT ELEC	KWH	0.22	0.12	0.06	0.05	0.05	0.05	0.05	0.05	0.05	0.05	0.07	0.20	1.0
		HT DEM	KW	.00415	.00315	.00212	.00034	.00002	.00002	.00001	.00002	.00003	.00071	.00228	.00344	0.00415
		REM ELEC	KWH	1.37	1.31	1.52	1.54	1.72	1.79	1.89	1.85	1.76	1.66	1.47	1.41	19.3
		REM DEM	KW	.00602	.00610	.00632	.00671	.00724	.00762	.00754	.00762	.00754	.00732	.00640	.00617	0.00762
		TOT ELEC	KWH	1.59	1.43	1.59	1.60	1.77	1.85	1.94	1.91	1.81	1.71	1.54	1.61	20.4
		TOT DEM	KW	.00850	.00816	.00663	.00706	.00740	.00774	.00765	.00774	.00774	.00757	.00672	.00825	0.00850
S- 3	-0.08	HT ELEC	KWH	0.24	0.13	0.07	0.06	0.06	0.06	0.06	0.06	0.06	0.06	0.08	0.22	1.2
		HT DEM	KW	.00415	.00315	.00212	.00034	.00002	.00002	.00001	.00002	.00003	.00071	.00228	.00344	0.00415
		REM ELEC	KWH	1.25	1.20	1.39	1.41	1.57	1.64	1.72	1.69	1.60	1.51	1.34	1.28	17.6
		REM DEM	KW	.00554	.00561	.00582	.00617	.00666	.00701	.00694	.00701	.00694	.00673	.00589	.00568	0.00701
		TOT ELEC	KWH	1.49	1.33	1.46	1.47	1.63	1.70	1.78	1.75	1.66	1.57	1.42	1.51	18.8
		TOT DEM	KW	.00800	.00768	.00624	.00664	.00696	.00728	.00720	.00728	.00728	.00712	.00632	.00776	0.00800
S- 7	0.01	HT ELEC	KWH	0.25	0.13	0.07	0.06	0.06	0.06	0.06	0.06	0.06	0.06	0.08	0.23	1.2
		HT DEM	KW	.00415	.00315	.00212	.00034	.00002	.00002	.00001	.00002	.00003	.00071	.00228	.00344	0.00415
		REM ELEC	KWH	1.14	1.09	1.26	1.28	1.42	1.49	1.57	1.54	1.46	1.38	1.22	1.17	16.0
		REM DEM	KW	.00491	.00498	.00516	.00547	.00591	.00622	.00616	.00622	.00616	.00597	.00522	.00504	0.00622
		TOT ELEC	KWH	1.39	1.22	1.34	1.34	1.49	1.55	1.63	1.60	1.52	1.44	1.30	1.40	17.2
		TOT DEM	KW	.00702	.00674	.00548	.00583	.00611	.00639	.00632	.00639	.00639	.00625	.00555	.00681	0.00702
S- 16	1.16	HT ELEC	KWH	0.26	0.14	0.08	0.06	0.06	0.06	0.06	0.06	0.06	0.06	0.09	0.24	1.3
		HT DEM	KW	.00415	.00315	.00212	.00034	.00002	.00002	.00001	.00002	.00003	.00071	.00228	.00344	0.00415
		REM ELEC	KWH	1.04	0.99	1.15	1.17	1.30	1.36	1.43	1.40	1.33	1.26	1.11	1.07	14.6
		REM DEM	KW	.00445	.00450	.00467	.00495	.00535	.00563	.00557	.00563	.00557	.00540	.00473	.00456	0.00563
		TOT ELEC	KWH	1.30	1.13	1.23	1.23	1.36	1.42	1.50	1.47	1.39	1.32	1.20	1.31	15.9
		TOT DEM	KW	.00680	.00653	.00530	.00564	.00592	.00619	.00612	.00619	.00619	.00605	.00537	.00660	0.00680
S- 32	0.28	HT ELEC	KWH	0.28	0.18	0.10	0.04	0.03	0.04	0.04	0.04	0.04	0.04	0.10	0.21	1.1
		HT DEM	KW	.00415	.00315	.00212	.00034	.00002	.00002	.00001	.00002	.00003	.00071	.00228	.00344	0.00415
		REM ELEC	KWH	0.93	0.90	1.17	1.31	1.34	1.57	1.57	1.54	1.47	1.25	1.06	1.00	15.2
		REM DEM	KW	.00495	.00501	.00520	.00551	.00595	.00626	.00620	.00626	.00620	.00601	.00526	.00507	0.00626
		TOT ELEC	KWH	1.20	1.07	1.27	1.35	1.37	1.60	1.60	1.57	1.51	1.29	1.17	1.21	16.3
		TOT DEM	KW	.00769	.00738	.00600	.00638	.00669	.00700	.00692	.00700	.00700	.00684	.00608	.00746	0.00769
S- 40	0.96	HT ELEC	KWH	0.29	0.18	0.11	0.04	0.04	0.04	0.04	0.04	0.04	0.04	0.11	0.27	1.2
		HT DEM	KW	.00415	.00315	.00212	.00034	.00002	.00002	.00001	.00002	.00003	.00071	.00228	.00344	0.00415
		REM ELEC	KWH	0.84	0.81	1.06	1.19	1.21	1.42	1.42	1.39	1.34	1.13	0.97	0.91	13.8
		REM DEM	KW	.00468	.00474	.00491	.00521	.00562	.00592	.00586	.00592	.00586	.00568	.00497	.00480	0.00592
		TOT ELEC	KWH	1.13	1.00	1.17	1.23	1.25	1.46	1.46	1.43	1.38	1.18	1.07	1.13	15.0
		TOT DEM	KW	.00755	.00725	.00589	.00627	.00657	.00687	.00679	.00687	.00687	.00672	.00596	.00732	0.00755
S- 45	0.32	HT ELEC	KWH	0.28	0.18	0.10	0.04	0.03	0.04	0.04	0.04	0.04	0.04	0.10	0.21	1.1
		HT DEM	KW	.00415	.00315	.00212	.00034	.00002	.00002	.00001	.00002	.00003	.00071	.00228	.00344	0.00415
		REM ELEC	KWH	0.89	0.86	1.12	1.26	1.28	1.50	1.50	1.47	1.42	1.20	1.02	0.96	14.6
		REM DEM	KW	.00483	.00490	.00508	.00539	.00581	.00612	.00606	.00612	.00606	.00588	.00514	.00496	0.00612
		TOT ELEC	KWH	1.17	1.04	1.23	1.30	1.32	1.54	1.54	1.51	1.45	1.24	1.13	1.18	15.8
		TOT DEM	KW	.00748	.00718	.00583	.00621	.00651	.00681	.00673	.00681	.00681	.00666	.00591	.00726	0.00748
S- 50	1.59	HT ELEC	KWH	0.28	0.18	0.10	0.04	0.03	0.04	0.04	0.04	0.04	0.04	0.10	0.21	1.1
		HT DEM	KW	.00415	.00315	.00212	.00034	.00002	.00002	.00001	.00002	.00003	.00071	.00228	.00344	0.00415
		REM ELEC	KWH	0.87	0.84	1.10	1.23	1.26	1.47	1.47	1.44	1.39	1.17	1.00	0.94	14.3
		REM DEM	KW	.00489	.00495	.00514	.00545	.00588	.00619	.00613	.00619	.00613	.00594	.00520	.00501	0.00619
		TOT ELEC	KWH	1.15	1.02	1.20	1.27	1.29	1.51	1.51	1.48	1.42	1.21	1.10	1.15	15.4
		TOT DEM	KW	.00769	.00738	.00600	.00638	.00669	.00700	.00692	.00700	.00700	.00684	.00608	.00746	0.00769
S- 61	-0.24	HT ELEC	KWH	0.25	0.13	0.07	0.06	0.06	0.06	0.06	0.06	0.06	0.06	0.08	0.23	1.2
		HT DEM	KW	.00415	.00315	.00212	.00034	.00002	.00002	.00001	.00002	.00003	.00071	.00228	.00344	0.00415
		REM ELEC	KWH	1.14	1.09	1.26	1.28	1.42	1.49	1.57	1.54	1.46	1.38	1.22	1.17	16.0
		REM DEM	KW	.00491	.00498	.00516	.00547	.00591	.00622	.00616	.00622	.00616	.00597	.00522	.00504	0.00622
		TOT ELEC	KWH	1.39	1.22	1.34	1.34	1.49	1.55	1.63	1.60	1.52	1.44	1.30	1.40	17.2
		TOT DEM	KW	.00702	.00674	.00548	.00583	.00611	.00639	.00632	.00639	.00639	.00625	.00555	.00681	0.00702
BASE CASE : 415																
BASE	0.00	TOT GAS	THM	0.0113	0.0060	0.0033	0.0027	0.0027	0.0027	0.0027	0.0027	0.0027	0.0027	0.0038	0.0102	0.054
		TOT ELEC	KWH	1.37	1.31	1.52	1.54	1.72	1.79	1.89	1.85	1.76	1.66	1.47	1.41	19.3
		TOT DEM	KW	.00602	.00610	.00632	.00671	.00724	.00762	.00754	.00762	.00754	.00732	.00640	.00617	0.00762
S- 3	-0.08	TOT GAS	THM	0.0124	0.0066	0.0036	0.0030	0.0030	0.0030	0.0030	0.0030	0.0030	0.0030	0.0041	0.0112	0.059
		TOT ELEC	KWH	1.25	1.20	1.39	1.41	1.57	1.64	1.72	1.69	1.60	1.51	1.34	1.28	17.6
		TOT DEM	KW	.00554	.00561	.00582	.00617	.00666	.00701	.00694	.00701	.00694	.00673	.00589	.00568	0.00701
S- 7	0.01	TOT GAS	THM	0.0129	0.0069	0.0037	0.0031	0.0031	0.0031	0.0031	0.0031	0.0031	0.0031	0.0043	0.0116	0.061
		TOT ELEC	KWH	1.14	1.09	1.26	1.28	1.42	1.49	1.57	1.54	1.46	1.38	1.22	1.17	16.0
		TOT DEM	KW	.00491	.00498	.00516	.00547	.00591	.00622	.00616	.00622	.00616	.00597	.00522	.00504	0.00622
S- 16	1.16	TOT GAS	THM	0.0135	0.0072	0.0039	0.0033	0.0033	0.0033	0.0033	0.0033	0.0033	0.0033	0.0045	0.0122	0.064
		TOT ELEC	KWH	1.04	0.99	1.15	1.17	1.30	1.36	1.43	1.40	1.33	1.26	1.11	1.07	14.6
		TOT DEM	KW	.00445	.00450	.00467	.00495	.00535	.00563	.00557	.00563	.00557	.00540	.00473	.00456	0.00563

STRAT SET	CONST COST $/SF	ENERGY FORM	UNITS (PER SF)	JAN	FEB	MAR	APR	MAY	JUN	JUL	AUG	SEP	OCT	NOV	DEC	ANNUAL
BASE CASE : 415 (CONTINUED)																
S- 32	0.27	TOT GAS	THM	0.0141	0.0090	0.0052	0.0020	0.0017	0.0018	0.0018	0.0018	0.0018	0.0021	0.0052	0.0107	0.058
		TOT ELEC	KWH	0.93	0.90	1.17	1.31	1.34	1.57	1.57	1.54	1.47	1.25	1.06	1.00	15.2
		TOT DEM	KW	.00495	.00501	.00520	.00551	.00595	.00626	.00620	.00626	.00620	.00601	.00526	.00507	0.00626
S- 40	0.95	TOT GAS	THM	0.0147	0.0094	0.0054	0.0021	0.0018	0.0019	0.0019	0.0019	0.0019	0.0022	0.0054	0.0112	0.060
		TOT ELEC	KWH	0.84	0.81	1.06	1.19	1.21	1.42	1.42	1.39	1.34	1.13	0.97	0.91	13.8
		TOT DEM	KW	.00468	.00474	.00491	.00521	.00562	.00592	.00586	.00592	.00586	.00568	.00497	.00480	0.00592
S- 45	0.32	TOT GAS	THM	0.0143	0.0092	0.0053	0.0021	0.0018	0.0019	0.0019	0.0019	0.0019	0.0022	0.0053	0.0109	0.059
		TOT ELEC	KWH	0.89	0.86	1.12	1.26	1.28	1.50	1.50	1.47	1.42	1.20	1.02	0.96	14.6
		TOT DEM	KW	.00483	.00490	.00508	.00539	.00581	.00612	.00606	.00612	.00606	.00588	.00514	.00496	0.00612
S- 50	1.59	TOT GAS	THM	0.0141	0.0090	0.0052	0.0020	0.0017	0.0018	0.0018	0.0018	0.0018	0.0021	0.0052	0.0107	0.058
		TOT ELEC	KWH	0.87	0.84	1.10	1.23	1.26	1.47	1.47	1.44	1.39	1.17	1.00	0.94	14.3
		TOT DEM	KW	.00489	.00495	.00514	.00545	.00588	.00619	.00613	.00619	.00613	.00594	.00520	.00501	0.00619
S- 61	-0.24	TOT GAS	THM	0.0129	0.0069	0.0037	0.0031	0.0031	0.0031	0.0031	0.0031	0.0031	0.0031	0.0043	0.0116	0.061
		TOT ELEC	KWH	1.14	1.09	1.26	1.28	1.42	1.49	1.57	1.54	1.46	1.38	1.22	1.17	16.0
		TOT DEM	KW	.00491	.00498	.00516	.00547	.00591	.00622	.00616	.00622	.00616	.00597	.00522	.00504	0.00622
BASE CASE : 416																
BASE	0.00	HT ELEC	KWH	0.33	0.21	0.10	0.03	0.03	0.03	0.03	0.03	0.03	0.04	0.11	0.27	1.3
		HT DEM	KW	.00358	.00311	.00272	.00107	.00029	.00007	.00007	.00007	.00007	.00158	.00265	.00354	0.00358
		REM ELEC	KWH	1.25	1.20	1.39	1.41	1.57	1.64	1.72	1.69	1.60	1.51	1.34	1.28	17.6
		REM DEM	KW	.00545	.00545	.00552	.00600	.00635	.00676	.00676	.00690	.00676	.00649	.00573	.00552	0.00690
		TOT ELEC	KWH	1.58	1.40	1.49	1.44	1.60	1.67	1.76	1.72	1.63	1.55	1.45	1.56	18.9
		TOT DEM	KW	.00763	.00725	.00580	.00610	.00671	.00694	.00702	.00702	.00687	.00671	.00588	.00763	0.00763
S- 3	-0.15	HT ELEC	KWH	0.36	0.22	0.11	0.03	0.03	0.03	0.03	0.03	0.03	0.04	0.12	0.29	1.3
		HT DEM	KW	.00358	.00311	.00272	.00107	.00029	.00007	.00007	.00007	.00007	.00158	.00265	.00354	0.00358
		REM ELEC	KWH	1.14	1.09	1.26	1.28	1.42	1.49	1.57	1.54	1.46	1.38	1.22	1.17	16.0
		REM DEM	KW	.00502	.00502	.00508	.00552	.00584	.00622	.00622	.00635	.00622	.00597	.00527	.00508	0.00635
		TOT ELEC	KWH	1.49	1.31	1.37	1.31	1.46	1.52	1.60	1.57	1.49	1.42	1.34	1.46	17.3
		TOT DEM	KW	.00718	.00682	.00546	.00574	.00632	.00653	.00661	.00661	.00646	.00632	.00553	.00718	0.00718
S- 7	-0.06	HT ELEC	KWH	0.38	0.24	0.11	0.04	0.04	0.04	0.04	0.04	0.04	0.04	0.13	0.31	1.4
		HT DEM	KW	.00358	.00311	.00272	.00107	.00029	.00007	.00007	.00007	.00007	.00158	.00265	.00354	0.00358
		REM ELEC	KWH	1.01	0.97	1.12	1.14	1.26	1.32	1.39	1.36	1.29	1.22	1.08	1.04	14.2
		REM DEM	KW	.00445	.00445	.00450	.00490	.00518	.00552	.00552	.00563	.00552	.00529	.00467	.00450	0.00563
		TOT ELEC	KWH	1.39	1.20	1.24	1.17	1.30	1.36	1.43	1.40	1.33	1.26	1.21	1.34	15.6
		TOT DEM	KW	.00630	.00598	.00479	.00504	.00554	.00573	.00580	.00580	.00567	.00554	.00485	.00630	0.00630
S- 16	1.05	HT ELEC	KWH	0.40	0.25	0.12	0.04	0.04	0.04	0.04	0.04	0.04	0.04	0.13	0.32	1.5
		HT DEM	KW	.00358	.00311	.00272	.00107	.00029	.00007	.00007	.00007	.00007	.00158	.00265	.00354	0.00358
		REM ELEC	KWH	0.89	0.85	0.99	1.00	1.11	1.16	1.23	1.20	1.14	1.08	0.95	0.91	12.5
		REM DEM	KW	.00403	.00403	.00408	.00444	.00469	.00500	.00500	.00510	.00500	.00479	.00423	.00408	0.00510
		TOT ELEC	KWH	1.29	1.10	1.11	1.04	1.15	1.20	1.26	1.24	1.17	1.12	1.08	1.24	14.0
		TOT DEM	KW	.00610	.00580	.00464	.00488	.00537	.00555	.00561	.00561	.00549	.00537	.00470	.00610	0.00610
S- 32	0.21	HT ELEC	KWH	0.32	0.20	0.13	0.04	0.03	0.03	0.03	0.03	0.03	0.05	0.12	0.26	1.3
		HT DEM	KW	.00358	.00311	.00272	.00107	.00029	.00007	.00007	.00007	.00007	.00158	.00265	.00354	0.00358
		REM ELEC	KWH	0.84	0.82	1.01	1.08	1.13	1.28	1.29	1.29	1.19	1.06	0.90	0.88	12.8
		REM DEM	KW	.00403	.00403	.00408	.00444	.00469	.00500	.00500	.00510	.00500	.00479	.00423	.00408	0.00510
		TOT ELEC	KWH	1.16	1.02	1.14	1.11	1.15	1.31	1.32	1.32	1.22	1.11	1.01	1.15	14.1
		TOT DEM	KW	.00611	.00514	.00429	.00468	.00494	.00520	.00520	.00533	.00520	.00500	.00442	.00650	0.00650
S- 40	0.87	HT ELEC	KWH	0.35	0.22	0.14	0.04	0.03	0.03	0.03	0.03	0.03	0.05	0.13	0.29	1.4
		HT DEM	KW	.00358	.00311	.00272	.00107	.00029	.00007	.00007	.00007	.00007	.00158	.00265	.00354	0.00358
		REM ELEC	KWH	0.78	0.76	0.93	0.99	1.04	1.18	1.19	1.19	1.10	0.98	0.83	0.81	11.8
		REM DEM	KW	.00387	.00387	.00392	.00426	.00451	.00480	.00480	.00490	.00480	.00461	.00407	.00392	0.00490
		TOT ELEC	KWH	1.13	0.97	1.07	1.03	1.07	1.21	1.22	1.22	1.13	1.03	0.95	1.11	13.2
		TOT DEM	KW	.00590	.00496	.00414	.00452	.00477	.00502	.00502	.00515	.00502	.00484	.00427	.00628	0.00628
S- 45	0.25	HT ELEC	KWH	0.35	0.22	0.14	0.04	0.03	0.03	0.03	0.03	0.03	0.05	0.13	0.29	1.4
		HT DEM	KW	.00358	.00311	.00272	.00107	.00029	.00007	.00007	.00007	.00007	.00158	.00265	.00354	0.00358
		REM ELEC	KWH	0.81	0.79	0.97	1.03	1.08	1.23	1.24	1.24	1.14	1.02	0.86	0.85	12.3
		REM DEM	KW	.00401	.00401	.00406	.00442	.00467	.00498	.00498	.00508	.00498	.00478	.00422	.00406	0.00508
		TOT ELEC	KWH	1.16	1.01	1.11	1.08	1.11	1.26	1.27	1.27	1.17	1.07	0.99	1.14	13.7
		TOT DEM	KW	.00602	.00506	.00422	.00461	.00486	.00512	.00512	.00525	.00512	.00493	.00435	.00640	0.00640
S- 50	1.52	HT ELEC	KWH	0.31	0.19	0.12	0.04	0.03	0.03	0.03	0.03	0.03	0.04	0.11	0.26	1.2
		HT DEM	KW	.00358	.00311	.00272	.00107	.00029	.00007	.00007	.00007	.00007	.00158	.00265	.00354	0.00358
		REM ELEC	KWH	0.82	0.79	0.98	1.04	1.09	1.24	1.25	1.25	1.15	1.03	0.87	0.86	12.4
		REM DEM	KW	.00403	.00403	.00408	.00444	.00469	.00500	.00500	.00510	.00500	.00479	.00423	.00408	0.00510
		TOT ELEC	KWH	1.13	0.98	1.10	1.08	1.12	1.27	1.28	1.28	1.18	1.07	0.98	1.11	13.6
		TOT DEM	KW	.00650	.00617	.00494	.00520	.00572	.00592	.00598	.00598	.00585	.00572	.00500	.00650	0.00650
S- 61	-0.31	HT ELEC	KWH	0.38	0.24	0.11	0.04	0.04	0.04	0.04	0.04	0.04	0.04	0.13	0.31	1.4
		HT DEM	KW	.00358	.00311	.00272	.00107	.00029	.00007	.00007	.00007	.00007	.00158	.00265	.00354	0.00358
		REM ELEC	KWH	1.01	0.97	1.12	1.14	1.26	1.32	1.39	1.36	1.29	1.22	1.08	1.04	14.2
		REM DEM	KW	.00445	.00445	.00450	.00490	.00518	.00552	.00552	.00563	.00552	.00529	.00467	.00450	0.00563
		TOT ELEC	KWH	1.39	1.20	1.24	1.17	1.30	1.36	1.43	1.40	1.33	1.26	1.21	1.34	15.6
		TOT DEM	KW	.00630	.00598	.00479	.00504	.00554	.00573	.00580	.00580	.00567	.00554	.00485	.00630	0.00630

(HT ELEC=ELECTRIC CONSUMPTION FOR SPACE AND WATER HEATING. REM ELEC=REMAINING ELECTRIC CONSUMPTION (NOT USED FOR HEATING). TOT ELEC=TOTAL ELECTRIC CONSUMPTION. HT DEM=PEAK ELECTRIC DEMAND FOR SPACE AND WATER HEATING. REM DEM=PEAK ELECTRIC DEMAND FOR REMAINING LOAD (NOT HEATING). TOT DEM=TOTAL PEAK ELECTRIC DEMAND. GAS=TOTAL GAS CONSUMPTION. ANNUAL FIGURES ARE SUMS OF MONTHLY CONSUMPTION VALUES OR ARE YEARLY MAXIMUMS OF MONTHLY PEAK DEMAND VALUES.)

STRAT SET	CONST COST $/SF	ENERGY FORM	UNITS (PER SF)	JAN	FEB	MAR	APR	MAY	JUN	JUL	AUG	SEP	OCT	NOV	DEC	ANNUAL
BASE CASE : 417																
BASE	0.00	TOT GAS	THM	0.0172	0.0107	0.0052	0.0016	0.0016	0.0016	0.0016	0.0016	0.0016	0.0019	0.0058	0.0140	0.064
		TOT ELEC	KWH	1.25	1.20	1.39	1.41	1.57	1.64	1.72	1.69	1.60	1.51	1.34	1.28	17.6
		TOT DEM	KW	.00545	.00545	.00552	.00600	.00635	.00676	.00676	.00690	.00676	.00649	.00573	.00552	0.00690
S- 3	-0.16	TOT GAS	THM	0.0183	0.0114	0.0055	0.0017	0.0017	0.0017	0.0017	0.0017	0.0017	0.0021	0.0062	0.0149	0.069
		TOT ELEC	KWH	1.14	1.09	1.26	1.28	1.42	1.49	1.57	1.54	1.46	1.38	1.22	1.17	16.0
		TOT DEM	KW	.00502	.00502	.00508	.00552	.00584	.00622	.00622	.00635	.00622	.00597	.00527	.00508	0.00635
S- 7	-0.07	TOT GAS	THM	0.0195	0.0121	0.0058	0.0018	0.0018	0.0018	0.0018	0.0018	0.0018	0.0022	0.0066	0.0158	0.073
		TOT ELEC	KWH	1.01	0.97	1.12	1.14	1.26	1.32	1.39	1.36	1.29	1.22	1.08	1.04	14.2
		TOT DEM	KW	.00445	.00445	.00450	.00490	.00518	.00552	.00552	.00563	.00552	.00529	.00467	.00450	0.00563
S- 16	1.04	TOT GAS	THM	0.0203	0.0126	0.0061	0.0019	0.0019	0.0019	0.0019	0.0019	0.0019	0.0023	0.0068	0.0165	0.076
		TOT ELEC	KWH	0.89	0.85	0.99	1.00	1.11	1.16	1.23	1.20	1.14	1.08	0.95	0.91	12.5
		TOT DEM	KW	.00403	.00403	.00408	.00444	.00469	.00500	.00500	.00510	.00500	.00479	.00423	.00408	0.00510
S- 32	0.20	TOT GAS	THM	0.0163	0.0101	0.0065	0.0020	0.0013	0.0014	0.0014	0.0014	0.0013	0.0024	0.0059	0.0135	0.064
		TOT ELEC	KWH	0.84	0.82	1.01	1.08	1.13	1.28	1.29	1.29	1.19	1.06	0.90	0.88	12.8
		TOT DEM	KW	.00403	.00403	.00408	.00444	.00469	.00500	.00500	.00510	.00500	.00479	.00423	.00408	0.00510
S- 40	0.86	TOT GAS	THM	0.0180	0.0111	0.0072	0.0022	0.0015	0.0015	0.0015	0.0015	0.0015	0.0026	0.0065	0.0149	0.070
		TOT ELEC	KWH	0.78	0.76	0.93	0.99	1.04	1.18	1.19	1.19	1.10	0.98	0.83	0.81	11.8
		TOT DEM	KW	.00387	.00387	.00392	.00426	.00451	.00480	.00480	.00490	.00480	.00461	.00407	.00392	0.00490
S- 45	0.25	TOT GAS	THM	0.0180	0.0111	0.0072	0.0022	0.0015	0.0015	0.0015	0.0015	0.0015	0.0026	0.0065	0.0149	0.070
		TOT ELEC	KWH	0.81	0.79	0.97	1.03	1.08	1.23	1.24	1.24	1.14	1.02	0.86	0.85	12.3
		TOT DEM	KW	.00401	.00401	.00406	.00442	.00467	.00498	.00498	.00508	.00498	.00478	.00422	.00406	0.00508
S- 50	1.51	TOT GAS	THM	0.0157	0.0097	0.0063	0.0019	0.0013	0.0014	0.0014	0.0014	0.0013	0.0023	0.0057	0.0130	0.062
		TOT ELEC	KWH	0.82	0.79	0.98	1.04	1.09	1.24	1.25	1.25	1.15	1.03	0.87	0.86	12.4
		TOT DEM	KW	.00403	.00403	.00408	.00444	.00469	.00500	.00500	.00510	.00500	.00479	.00423	.00408	0.00510
S- 61	-0.32	TOT GAS	THM	0.0195	0.0121	0.0058	0.0018	0.0018	0.0018	0.0018	0.0018	0.0018	0.0022	0.0066	0.0158	0.073
		TOT ELEC	KWH	1.01	0.97	1.12	1.14	1.26	1.32	1.39	1.36	1.29	1.22	1.08	1.04	14.2
		TOT DEM	KW	.00445	.00445	.00450	.00490	.00518	.00552	.00552	.00563	.00552	.00529	.00467	.00450	0.00563
BASE CASE : 418																
BASE	0.00	TOT ELEC	KWH	1.32	1.20	1.45	1.56	1.60	1.80	1.82	1.82	1.69	1.51	1.32	1.32	18.4
		TOT DEM	KW	.00596	.00582	.00589	.00653	.00653	.00689	.00682	.00718	.00689	.00668	.00625	.00596	0.00718
S- 1	0.09	TOT ELEC	KWH	1.22	1.11	1.34	1.45	1.48	1.67	1.68	1.68	1.56	1.39	1.22	1.22	17.0
		TOT DEM	KW	.00495	.00483	.00489	.00542	.00542	.00572	.00566	.00596	.00572	.00554	.00519	.00495	0.00596
S- 12	0.21	TOT ELEC	KWH	1.10	0.99	1.21	1.30	1.33	1.50	1.51	1.51	1.41	1.25	1.10	1.10	15.3
		TOT DEM	KW	.00465	.00454	.00459	.00510	.00510	.00538	.00532	.00560	.00538	.00521	.00487	.00465	0.00560
S- 18	1.52	TOT ELEC	KWH	1.01	0.91	1.11	1.19	1.22	1.37	1.39	1.39	1.29	1.15	1.01	1.01	14.0
		TOT DEM	KW	.00449	.00438	.00444	.00492	.00492	.00519	.00514	.00541	.00519	.00503	.00471	.00449	0.00541
S- 28	-0.04	TOT ELEC	KWH	1.10	0.99	1.21	1.30	1.33	1.50	1.51	1.51	1.41	1.25	1.10	1.10	15.3
		TOT DEM	KW	.00465	.00454	.00459	.00510	.00510	.00538	.00532	.00560	.00538	.00521	.00487	.00465	0.00560
S- 33	0.47	TOT ELEC	KWH	1.04	0.94	1.15	1.23	1.26	1.42	1.44	1.44	1.33	1.19	1.04	1.04	14.5
		TOT DEM	KW	.00477	.00466	.00471	.00523	.00523	.00552	.00546	.00575	.00552	.00535	.00500	.00477	0.00575
S- 41	1.23	TOT ELEC	KWH	0.98	0.88	1.07	1.16	1.18	1.33	1.35	1.35	1.25	1.12	0.98	0.98	13.6
		TOT DEM	KW	.00461	.00450	.00455	.00505	.00505	.00533	.00527	.00555	.00533	.00516	.00483	.00461	0.00555
S- 46	0.52	TOT ELEC	KWH	1.01	0.91	1.11	1.19	1.22	1.37	1.39	1.39	1.29	1.15	1.01	1.01	14.0
		TOT DEM	KW	.00468	.00457	.00462	.00513	.00513	.00541	.00536	.00564	.00541	.00525	.00491	.00468	0.00564
S- 51	1.79	TOT ELEC	KWH	0.95	0.86	1.04	1.12	1.15	1.29	1.31	1.31	1.21	1.08	0.95	0.95	13.2
		TOT DEM	KW	.00477	.00466	.00471	.00523	.00523	.00552	.00546	.00575	.00552	.00535	.00500	.00477	0.00575
BASE CASE : 419																
BASE	0.00	HT ELEC	KWH	0.20	0.11	0.06	0.05	0.05	0.05	0.05	0.05	0.05	0.05	0.07	0.18	1.0
		HT DEM	KW	.00415	.00315	.00212	.00034	.00002	.00002	.00001	.00002	.00003	.00071	.00228	.00344	0.00415
		REM ELEC	KWH	1.29	1.23	1.43	1.45	1.61	1.68	1.77	1.74	1.65	1.56	1.38	1.32	18.1
		REM DEM	KW	.00570	.00577	.00598	.00634	.00685	.00721	.00714	.00721	.00714	.00692	.00606	.00584	0.00721
		TOT ELEC	KWH	1.49	1.34	1.49	1.50	1.66	1.73	1.82	1.79	1.70	1.61	1.44	1.51	19.1
		TOT DEM	KW	.00830	.00797	.00647	.00689	.00722	.00755	.00747	.00755	.00755	.00739	.00656	.00805	0.00830
S- 3	-0.07	HT ELEC	KWH	0.23	0.12	0.07	0.06	0.06	0.06	0.06	0.06	0.06	0.06	0.08	0.21	1.1
		HT DEM	KW	.00415	.00315	.00212	.00034	.00002	.00002	.00001	.00002	.00003	.00071	.00228	.00344	0.00415
		REM ELEC	KWH	1.18	1.13	1.31	1.33	1.48	1.54	1.63	1.59	1.51	1.43	1.26	1.21	16.6
		REM DEM	KW	.00534	.00541	.00561	.00595	.00642	.00676	.00669	.00676	.00669	.00649	.00568	.00548	0.00676
		TOT ELEC	KWH	1.41	1.25	1.38	1.38	1.53	1.60	1.68	1.65	1.57	1.48	1.34	1.42	17.7
		TOT DEM	KW	.00780	.00749	.00608	.00647	.00679	.00710	.00702	.00710	.00710	.00694	.00616	.00757	0.00780
S- 7	0.02	HT ELEC	KWH	0.23	0.12	0.07	0.06	0.06	0.06	0.06	0.06	0.06	0.06	0.08	0.21	1.1
		HT DEM	KW	.00415	.00315	.00212	.00034	.00002	.00002	.00001	.00002	.00003	.00071	.00228	.00344	0.00415
		REM ELEC	KWH	1.07	1.02	1.19	1.20	1.34	1.39	1.47	1.44	1.37	1.29	1.14	1.10	15.0
		REM DEM	KW	.00480	.00486	.00504	.00534	.00577	.00607	.00601	.00607	.00601	.00583	.00510	.00492	0.00607
		TOT ELEC	KWH	1.30	1.14	1.25	1.26	1.39	1.45	1.53	1.50	1.42	1.35	1.22	1.31	16.1
		TOT DEM	KW	.00686	.00659	.00535	.00569	.00597	.00624	.00617	.00624	.00624	.00611	.00542	.00665	0.00686
S- 16	1.19	HT ELEC	KWH	0.24	0.13	0.07	0.06	0.06	0.06	0.06	0.06	0.06	0.06	0.08	0.22	1.2
		HT DEM	KW	.00415	.00315	.00212	.00034	.00002	.00002	.00001	.00002	.00003	.00071	.00228	.00344	0.00415
		REM ELEC	KWH	0.97	0.93	1.08	1.10	1.22	1.27	1.34	1.32	1.25	1.18	1.04	1.00	13.7
		REM DEM	KW	.00445	.00450	.00467	.00495	.00535	.00563	.00557	.00563	.00557	.00540	.00473	.00456	0.00563
		TOT ELEC	KWH	1.22	1.06	1.15	1.16	1.28	1.33	1.40	1.37	1.31	1.24	1.12	1.22	14.9
		TOT DEM	KW	.00664	.00637	.00518	.00551	.00578	.00604	.00598	.00604	.00604	.00591	.00525	.00644	0.00664
BASE CASE : 419 (CONTINUED)																
S- 32	0.38	HT ELEC	KWH	0.28	0.18	0.10	0.04	0.03	0.04	0.04	0.04	0.04	0.04	0.10	0.21	1.1
		HT DEM	KW	.00415	.00315	.00212	.00034	.00002	.00002	.00001	.00002	.00003	.00071	.00228	.00344	0.00415
		REM ELEC	KWH	0.81	0.78	1.02	1.14	1.16	1.36	1.36	1.33	1.28	1.08	0.92	0.87	13.2
		REM DEM	KW	.00457	.00462	.00480	.00509	.00549	.00578	.00572	.00578	.00572	.00555	.00486	.00468	0.00578
		TOT ELEC	KWH	1.08	0.96	1.12	1.17	1.20	1.40	1.40	1.37	1.32	1.12	1.03	1.08	14.3
		TOT DEM	KW	.00748	.00718	.00583	.00621	.00651	.00681	.00673	.00681	.00681	.00666	.00591	.00726	0.00748
S- 61	-0.23	HT ELEC	KWH	0.23	0.12	0.07	0.06	0.06	0.06	0.06	0.06	0.06	0.06	0.08	0.21	1.1
		HT DEM	KW	.00415	.00315	.00212	.00034	.00002	.00002	.00001	.00002	.00003	.00071	.00228	.00344	0.00415
		REM ELEC	KWH	1.07	1.02	1.19	1.20	1.34	1.39	1.47	1.44	1.37	1.29	1.14	1.10	15.0
		REM DEM	KW	.00480	.00486	.00504	.00534	.00577	.00607	.00601	.00607	.00601	.00583	.00510	.00492	0.00607
		TOT ELEC	KWH	1.30	1.14	1.25	1.26	1.39	1.45	1.53	1.50	1.42	1.35	1.22	1.31	16.1
		TOT DEM	KW	.00686	.00659	.00535	.00569	.00597	.00624	.00617	.00624	.00624	.00611	.00542	.00665	0.00686
BASE CASE : 420																
BASE	0.00	TOT GAS	THM	0.0104	0.0055	0.0030	0.0025	0.0025	0.0025	0.0025	0.0025	0.0025	0.0025	0.0035	0.0094	0.049
		TOT ELEC	KWH	1.29	1.23	1.43	1.45	1.61	1.68	1.77	1.74	1.65	1.56	1.38	1.32	18.1
		TOT DEM	KW	.00570	.00577	.00598	.00634	.00685	.00721	.00714	.00721	.00714	.00692	.00606	.00584	0.00721
S- 3	-0.07	TOT GAS	THM	0.0116	0.0062	0.0034	0.0028	0.0028	0.0028	0.0028	0.0028	0.0028	0.0028	0.0039	0.0105	0.055
		TOT ELEC	KWH	1.18	1.13	1.31	1.33	1.48	1.54	1.63	1.59	1.51	1.43	1.26	1.21	16.6
		TOT DEM	KW	.00534	.00541	.00561	.00595	.00642	.00676	.00669	.00676	.00669	.00649	.00568	.00548	0.00676
S- 7	0.01	TOT GAS	THM	0.0119	0.0063	0.0035	0.0029	0.0029	0.0029	0.0029	0.0029	0.0029	0.0029	0.0040	0.0108	0.057
		TOT ELEC	KWH	1.07	1.02	1.19	1.20	1.34	1.39	1.47	1.44	1.37	1.29	1.14	1.10	15.0
		TOT DEM	KW	.00480	.00486	.00504	.00534	.00577	.00607	.00601	.00607	.00601	.00583	.00510	.00492	0.00607
S- 16	1.19	TOT GAS	THM	0.0124	0.0066	0.0036	0.0030	0.0030	0.0030	0.0030	0.0030	0.0030	0.0030	0.0041	0.0112	0.059
		TOT ELEC	KWH	0.97	0.93	1.08	1.10	1.22	1.27	1.34	1.32	1.25	1.18	1.04	1.00	13.7
		TOT DEM	KW	.00445	.00450	.00467	.00495	.00535	.00563	.00557	.00563	.00557	.00540	.00473	.00456	0.00563
S- 32	0.38	TOT GAS	THM	0.0141	0.0090	0.0052	0.0020	0.0017	0.0018	0.0018	0.0018	0.0018	0.0021	0.0052	0.0107	0.058
		TOT ELEC	KWH	0.81	0.78	1.02	1.14	1.16	1.36	1.36	1.33	1.28	1.08	0.92	0.87	13.2
		TOT DEM	KW	.00457	.00462	.00480	.00509	.00549	.00578	.00572	.00578	.00572	.00555	.00486	.00468	0.00578
S- 61	-0.24	TOT GAS	THM	0.0119	0.0063	0.0035	0.0029	0.0029	0.0029	0.0029	0.0029	0.0029	0.0029	0.0040	0.0108	0.057
		TOT ELEC	KWH	1.07	1.02	1.19	1.20	1.34	1.39	1.47	1.44	1.37	1.29	1.14	1.10	15.0
		TOT DEM	KW	.00480	.00486	.00504	.00534	.00577	.00607	.00601	.00607	.00601	.00583	.00510	.00492	0.00607
BASE CASE : 421																
BASE	0.00	HT ELEC	KWH	0.28	0.17	0.08	0.03	0.03	0.03	0.03	0.03	0.03	0.03	0.09	0.22	1.0
		HT DEM	KW	.00358	.00311	.00272	.00107	.00029	.00007	.00007	.00007	.00007	.00158	.00265	.00354	0.00358
		REM ELEC	KWH	1.21	1.16	1.35	1.37	1.52	1.59	1.68	1.64	1.56	1.47	1.30	1.25	17.1
		REM DEM	KW	.00524	.00524	.00530	.00577	.00610	.00650	.00650	.00663	.00650	.00623	.00550	.00530	0.00663
		TOT ELEC	KWH	1.49	1.33	1.43	1.39	1.55	1.62	1.70	1.67	1.58	1.50	1.39	1.47	18.1
		TOT DEM	KW	.00728	.00692	.00553	.00582	.00641	.00662	.00670	.00670	.00655	.00641	.00561	.00728	0.00728
S- 3	-0.14	HT ELEC	KWH	0.29	0.18	0.09	0.03	0.03	0.03	0.03	0.03	0.03	0.03	0.10	0.24	1.1
		HT DEM	KW	.00358	.00311	.00272	.00107	.00029	.00007	.00007	.00007	.00007	.00158	.00265	.00354	0.00358
		REM ELEC	KWH	1.11	1.06	1.23	1.25	1.39	1.45	1.53	1.50	1.42	1.34	1.19	1.14	15.6
		REM DEM	KW	.00481	.00481	.00487	.00530	.00560	.00597	.00597	.00609	.00597	.00572	.00505	.00487	0.00609
		TOT ELEC	KWH	1.40	1.24	1.32	1.28	1.42	1.48	1.56	1.52	1.45	1.37	1.28	1.38	16.7
		TOT DEM	KW	.00682	.00648	.00518	.00546	.00600	.00621	.00627	.00627	.00614	.00600	.00525	.00682	0.00682
S- 7	-0.05	HT ELEC	KWH	0.31	0.19	0.09	0.03	0.03	0.03	0.03	0.03	0.03	0.03	0.10	0.25	1.2
		HT DEM	KW	.00358	.00311	.00272	.00107	.00029	.00007	.00007	.00007	.00007	.00158	.00265	.00354	0.00358
		REM ELEC	KWH	0.98	0.94	1.09	1.10	1.23	1.28	1.35	1.32	1.26	1.19	1.05	1.01	13.8
		REM DEM	KW	.00427	.00427	.00432	.00470	.00497	.00529	.00529	.00540	.00529	.00508	.00448	.00432	0.00540
		TOT ELEC	KWH	1.29	1.13	1.18	1.13	1.26	1.31	1.38	1.35	1.28	1.22	1.15	1.26	15.0
		TOT DEM	KW	.00601	.00571	.00457	.00481	.00529	.00547	.00553	.00553	.00541	.00529	.00463	.00601	0.00601
S- 16	1.08	HT ELEC	KWH	0.33	0.20	0.10	0.03	0.03	0.03	0.03	0.03	0.03	0.04	0.11	0.26	1.2
		HT DEM	KW	.00358	.00311	.00272	.00107	.00029	.00007	.00007	.00007	.00007	.00158	.00265	.00354	0.00358
		REM ELEC	KWH	0.86	0.82	0.96	0.97	1.08	1.13	1.19	1.16	1.10	1.04	0.92	0.88	12.1
		REM DEM	KW	.00386	.00386	.00391	.00425	.00450	.00479	.00479	.00489	.00479	.00460	.00406	.00391	0.00489
		TOT ELEC	KWH	1.18	1.03	1.05	1.00	1.11	1.16	1.22	1.19	1.13	1.08	1.03	1.15	13.3
		TOT DEM	KW	.00581	.00552	.00442	.00465	.00511	.00529	.00535	.00535	.00523	.00511	.00447	.00581	0.00581
S- 32	0.31	HT ELEC	KWH	0.38	0.24	0.15	0.05	0.03	0.03	0.03	0.03	0.03	0.06	0.14	0.32	1.5
		HT DEM	KW	.00358	.00311	.00272	.00107	.00029	.00007	.00007	.00007	.00007	.00158	.00265	.00354	0.00358
		REM ELEC	KWH	0.83	0.81	1.00	1.06	1.11	1.26	1.27	1.27	1.17	1.05	0.88	0.87	12.6
		REM DEM	KW	.00411	.00411	.00416	.00452	.00478	.00510	.00510	.00520	.00510	.00489	.00432	.00416	0.00520
		TOT ELEC	KWH	1.21	1.04	1.15	1.10	1.14	1.29	1.31	1.31	1.20	1.10	1.02	1.19	14.1
		TOT DEM	KW	.00637	.00536	.00447	.00488	.00515	.00542	.00542	.00556	.00542	.00522	.00461	.00678	0.00678
S- 61	-0.30	HT ELEC	KWH	0.31	0.19	0.09	0.03	0.03	0.03	0.03	0.03	0.03	0.03	0.10	0.25	1.2
		HT DEM	KW	.00358	.00311	.00272	.00107	.00029	.00007	.00007	.00007	.00007	.00158	.00265	.00354	0.00358
		REM ELEC	KWH	0.98	0.94	1.09	1.10	1.23	1.28	1.35	1.32	1.26	1.19	1.05	1.01	13.8
		REM DEM	KW	.00427	.00427	.00432	.00470	.00497	.00529	.00529	.00540	.00529	.00508	.00448	.00432	0.00540
		TOT ELEC	KWH	1.29	1.13	1.18	1.13	1.26	1.31	1.38	1.35	1.28	1.22	1.15	1.26	15.0
		TOT DEM	KW	.00601	.00571	.00457	.00481	.00529	.00547	.00553	.00553	.00541	.00529	.00463	.00601	0.00601
BASE CASE : 422																
BASE	0.00	TOT GAS	THM	0.0144	0.0089	0.0043	0.0013	0.0013	0.0013	0.0013	0.0013	0.0013	0.0016	0.0049	0.0117	0.054
		TOT ELEC	KWH	1.21	1.16	1.35	1.37	1.52	1.59	1.68	1.64	1.56	1.47	1.30	1.25	17.1
		TOT DEM	KW	.00524	.00524	.00530	.00577	.00610	.00650	.00650	.00663	.00650	.00623	.00550	.00530	0.00663
S- 3	-0.15	TOT GAS	THM	0.0154	0.0095	0.0046	0.0014	0.0014	0.0014	0.0014	0.0014	0.0014	0.0017	0.0052	0.0125	0.057
		TOT ELEC	KWH	1.11	1.06	1.23	1.25	1.39	1.45	1.53	1.50	1.42	1.34	1.19	1.14	15.6
		TOT DEM	KW	.00481	.00481	.00487	.00530	.00560	.00597	.00597	.00609	.00597	.00572	.00505	.00487	0.00609

(HT ELEC=ELECTRIC CONSUMPTION FOR SPACE AND WATER HEATING. REM ELEC=REMAINING ELECTRIC CONSUMPTION (NOT USED FOR HEATING). TOT ELEC=TOTAL ELECTRIC CONSUMPTION. HT DEM=PEAK ELECTRIC DEMAND FOR SPACE AND WATER HEATING. REM DEM=PEAK ELECTRIC DEMAND FOR REMAINING LOAD (NOT HEATING). TOT DEM=TOTAL PEAK ELECTRIC DEMAND. GAS=TOTAL GAS CONSUMPTION. ANNUAL FIGURES ARE SUMS OF MONTHLY CONSUMPTION VALUES OR ARE YEARLY MAXIMUMS OF MONTHLY PEAK DEMAND VALUES.)

STRAT SET	CONST COST $/SF	ENERGY FORM	UNITS (PER SF)	JAN	FEB	MAR	APR	MAY	JUN	JUL	AUG	SEP	OCT	NOV	DEC	ANNUAL
BASE CASE : 422 (CONTINUED)																
S- 7	-0.06	TOT GAS	THM	0.0163	0.0101	0.0049	0.0015	0.0015	0.0015	0.0015	0.0015	0.0015	0.0018	0.0055	0.0133	0.061
		TOT ELEC	KWH	0.98	0.94	1.09	1.10	1.23	1.28	1.35	1.32	1.26	1.19	1.05	1.01	13.8
		TOT DEM	KW	.00427	.00427	.00432	.00470	.00497	.00529	.00529	.00540	.00529	.00508	.00448	.00432	0.00540
S- 16	1.07	TOT GAS	THM	0.0171	0.0106	0.0051	0.0016	0.0016	0.0016	0.0016	0.0016	0.0016	0.0019	0.0058	0.0139	0.064
		TOT ELEC	KWH	0.86	0.82	0.96	0.97	1.08	1.13	1.19	1.16	1.10	1.04	0.92	0.88	12.1
		TOT DEM	KW	.00386	.00386	.00391	.00425	.00450	.00479	.00479	.00489	.00479	.00460	.00406	.00391	0.00489
S- 32	0.30	TOT GAS	THM	0.0201	0.0125	0.0080	0.0024	0.0017	0.0017	0.0017	0.0017	0.0017	0.0029	0.0073	0.0166	0.079
		TOT ELEC	KWH	0.83	0.81	1.00	1.06	1.11	1.26	1.27	1.27	1.17	1.05	0.88	0.87	12.6
		TOT DEM	KW	.00411	.00411	.00416	.00452	.00478	.00510	.00510	.00520	.00510	.00489	.00432	.00416	0.00520
S- 61	-0.31	TOT GAS	THM	0.0163	0.0101	0.0049	0.0015	0.0015	0.0015	0.0015	0.0015	0.0015	0.0018	0.0055	0.0133	0.061
		TOT ELEC	KWH	0.98	0.94	1.09	1.10	1.23	1.28	1.35	1.32	1.26	1.19	1.05	1.01	13.8
		TOT DEM	KW	.00427	.00427	.00432	.00470	.00497	.00529	.00529	.00540	.00529	.00508	.00448	.00432	0.00540
BASE CASE : 423																
BASE	0.00	TOT ELEC	KWH	1.31	1.18	1.44	1.55	1.58	1.78	1.80	1.80	1.67	1.49	1.31	1.31	18.2
		TOT DEM	KW	.00596	.00582	.00589	.00653	.00653	.00689	.00682	.00718	.00689	.00668	.00625	.00596	0.00718
S- 1	0.09	TOT ELEC	KWH	1.20	1.09	1.32	1.42	1.45	1.64	1.65	1.65	1.54	1.37	1.20	1.20	16.7
		TOT DEM	KW	.00530	.00518	.00524	.00581	.00581	.00613	.00607	.00639	.00613	.00594	.00556	.00530	0.00639
S- 12	0.20	TOT ELEC	KWH	1.07	0.97	1.18	1.27	1.30	1.46	1.48	1.48	1.37	1.22	1.07	1.07	14.9
		TOT DEM	KW	.00506	.00494	.00500	.00555	.00555	.00586	.00580	.00610	.00586	.00567	.00531	.00506	0.00610
S- 18	1.51	TOT ELEC	KWH	0.97	0.88	1.07	1.15	1.17	1.32	1.34	1.34	1.24	1.11	0.97	0.97	13.5
		TOT DEM	KW	.00483	.00471	.00477	.00530	.00530	.00559	.00553	.00582	.00559	.00541	.00506	.00483	0.00582
S- 28	-0.05	TOT ELEC	KWH	1.07	0.97	1.18	1.27	1.30	1.46	1.48	1.48	1.37	1.22	1.07	1.07	14.9
		TOT DEM	KW	.00506	.00494	.00500	.00555	.00555	.00586	.00580	.00610	.00586	.00567	.00531	.00506	0.00610
S- 33	0.56	TOT ELEC	KWH	0.86	0.77	0.94	1.01	1.04	1.17	1.18	1.18	1.09	0.98	0.86	0.86	11.9
		TOT DEM	KW	.00480	.00468	.00474	.00526	.00526	.00555	.00549	.00578	.00555	.00538	.00503	.00480	0.00578
BASE CASE : 424																
BASE	0.00	HT ELEC	KWH	0.21	0.11	0.06	0.05	0.05	0.05	0.05	0.05	0.05	0.05	0.07	0.19	1.0
		HT DEM	KW	.00415	.00315	.00212	.00034	.00002	.00002	.00001	.00002	.00003	.00071	.00228	.00344	0.00415
		REM ELEC	KWH	1.41	1.35	1.56	1.58	1.76	1.84	1.94	1.90	1.80	1.70	1.50	1.45	19.8
		REM DEM	KW	.00650	.00658	.00683	.00724	.00782	.00823	.00815	.00823	.00815	.00790	.00691	.00667	0.00823
		TOT ELEC	KWH	1.62	1.46	1.63	1.63	1.81	1.89	1.99	1.95	1.85	1.75	1.57	1.64	20.8
		TOT DEM	KW	.00973	.00934	.00759	.00808	.00847	.00885	.00876	.00885	.00885	.00866	.00769	.00944	0.00973
S- 3	-0.09	HT ELEC	KWH	0.23	0.12	0.07	0.06	0.06	0.06	0.06	0.06	0.06	0.06	0.08	0.21	1.1
		HT DEM	KW	.00415	.00315	.00212	.00034	.00002	.00002	.00001	.00002	.00003	.00071	.00228	.00344	0.00415
		REM ELEC	KWH	1.29	1.23	1.43	1.45	1.61	1.68	1.77	1.74	1.65	1.56	1.38	1.32	18.1
		REM DEM	KW	.00598	.00606	.00628	.00666	.00719	.00757	.00749	.00757	.00749	.00727	.00636	.00613	0.00757
		TOT ELEC	KWH	1.52	1.36	1.50	1.50	1.67	1.74	1.83	1.79	1.70	1.61	1.45	1.53	19.2
		TOT DEM	KW	.00916	.00879	.00714	.00760	.00797	.00834	.00824	.00834	.00834	.00815	.00724	.00889	0.00916
S- 7	-0.00	HT ELEC	KWH	0.24	0.13	0.07	0.06	0.06	0.06	0.06	0.06	0.06	0.06	0.08	0.21	1.1
		HT DEM	KW	.00415	.00315	.00212	.00034	.00002	.00002	.00001	.00002	.00003	.00071	.00228	.00344	0.00415
		REM ELEC	KWH	1.16	1.12	1.30	1.31	1.46	1.53	1.61	1.57	1.49	1.41	1.25	1.20	16.4
		REM DEM	KW	.00531	.00538	.00558	.00591	.00638	.00672	.00665	.00672	.00665	.00645	.00564	.00544	0.00672
		TOT ELEC	KWH	1.40	1.24	1.36	1.37	1.52	1.58	1.66	1.63	1.55	1.47	1.33	1.41	17.5
		TOT DEM	KW	.00804	.00772	.00627	.00667	.00699	.00732	.00724	.00732	.00732	.00716	.00635	.00780	0.00804
S- 16	1.15	HT ELEC	KWH	0.25	0.13	0.07	0.06	0.06	0.06	0.06	0.06	0.06	0.06	0.08	0.23	1.2
		HT DEM	KW	.00415	.00315	.00212	.00034	.00002	.00002	.00001	.00002	.00003	.00071	.00228	.00344	0.00415
		REM ELEC	KWH	1.07	1.02	1.19	1.20	1.34	1.39	1.47	1.44	1.37	1.29	1.14	1.10	15.0
		REM DEM	KW	.00480	.00486	.00505	.00535	.00578	.00608	.00602	.00608	.00602	.00584	.00511	.00492	0.00608
		TOT ELEC	KWH	1.32	1.15	1.26	1.26	1.40	1.46	1.53	1.50	1.43	1.35	1.22	1.32	16.2
		TOT DEM	KW	.00778	.00747	.00607	.00646	.00677	.00708	.00700	.00708	.00708	.00692	.00615	.00755	0.00778
S- 32	0.36	HT ELEC	KWH	0.26	0.17	0.10	0.04	0.03	0.03	0.03	0.03	0.03	0.04	0.10	0.20	1.1
		HT DEM	KW	.00415	.00315	.00212	.00034	.00002	.00002	.00001	.00002	.00003	.00071	.00228	.00344	0.00415
		REM ELEC	KWH	0.91	0.88	1.15	1.28	1.31	1.53	1.53	1.50	1.45	1.22	1.04	0.98	14.9
		REM DEM	KW	.00510	.00517	.00536	.00568	.00614	.00646	.00640	.00646	.00640	.00620	.00543	.00523	0.00646
		TOT ELEC	KWH	1.17	1.05	1.24	1.32	1.34	1.57	1.57	1.54	1.48	1.26	1.14	1.18	16.0
		TOT DEM	KW	.00782	.00751	.00610	.00649	.00680	.00712	.00704	.00712	.00712	.00696	.00618	.00759	0.00782
S- 61	-0.25	HT ELEC	KWH	0.24	0.13	0.07	0.06	0.06	0.06	0.06	0.06	0.06	0.06	0.08	0.21	1.1
		HT DEM	KW	.00415	.00315	.00212	.00034	.00002	.00002	.00001	.00002	.00003	.00071	.00228	.00344	0.00415
		REM ELEC	KWH	1.16	1.12	1.30	1.31	1.46	1.53	1.61	1.57	1.49	1.41	1.25	1.20	16.4
		REM DEM	KW	.00531	.00538	.00558	.00591	.00638	.00672	.00665	.00672	.00665	.00645	.00564	.00544	0.00672
		TOT ELEC	KWH	1.40	1.24	1.36	1.37	1.52	1.58	1.66	1.63	1.55	1.47	1.33	1.41	17.5
		TOT DEM	KW	.00804	.00772	.00627	.00667	.00699	.00732	.00724	.00732	.00732	.00716	.00635	.00780	0.00804
BASE CASE : 425																
BASE	0.00	TOT GAS	THM	0.0107	0.0057	0.0031	0.0026	0.0026	0.0026	0.0026	0.0026	0.0026	0.0026	0.0036	0.0097	0.051
		TOT ELEC	KWH	1.41	1.35	1.56	1.58	1.76	1.84	1.94	1.90	1.80	1.70	1.50	1.45	19.8
		TOT DEM	KW	.00650	.00658	.00683	.00724	.00782	.00823	.00815	.00823	.00815	.00790	.00691	.00667	0.00823
S- 3	-0.10	TOT GAS	THM	0.0119	0.0063	0.0035	0.0029	0.0029	0.0029	0.0029	0.0029	0.0029	0.0029	0.0040	0.0108	0.057
		TOT ELEC	KWH	1.29	1.23	1.43	1.45	1.61	1.68	1.77	1.74	1.65	1.56	1.38	1.32	18.1
		TOT DEM	KW	.00598	.00606	.00628	.00666	.00719	.00757	.00749	.00757	.00749	.00727	.00636	.00613	0.00757
S- 7	-0.01	TOT GAS	THM	0.0121	0.0065	0.0035	0.0029	0.0029	0.0029	0.0029	0.0029	0.0029	0.0029	0.0040	0.0109	0.058
		TOT ELEC	KWH	1.16	1.12	1.30	1.31	1.46	1.53	1.61	1.57	1.49	1.41	1.25	1.20	16.4
		TOT DEM	KW	.00531	.00538	.00558	.00591	.00638	.00672	.00665	.00672	.00665	.00645	.00564	.00544	0.00672
S- 16	1.14	TOT GAS	THM	0.0129	0.0069	0.0037	0.0031	0.0031	0.0031	0.0031	0.0031	0.0031	0.0031	0.0043	0.0116	0.061
		TOT ELEC	KWH	1.07	1.02	1.19	1.20	1.34	1.39	1.47	1.44	1.37	1.29	1.14	1.10	15.0
		TOT DEM	KW	.00480	.00486	.00505	.00535	.00578	.00608	.00602	.00608	.00602	.00584	.00511	.00492	0.00608

STRAT SET	CONST COST $/SF	ENERGY FORM	UNITS (PER SF)	JAN	FEB	MAR	APR	MAY	JUN	JUL	AUG	SEP	OCT	NOV	DEC	ANNUAL
BASE CASE : 425 (CONTINUED)																
S- 32	0.36	TOT GAS	THM	0.0133	0.0085	0.0049	0.0019	0.0016	0.0017	0.0017	0.0017	0.0017	0.0020	0.0049	0.0102	0.055
		TOT ELEC	KWH	0.91	0.88	1.15	1.28	1.31	1.53	1.53	1.50	1.45	1.22	1.04	0.98	14.9
		TOT DEM	KW	.00510	.00517	.00536	.00568	.00614	.00646	.00640	.00646	.00640	.00620	.00543	.00523	0.00646
S- 61	-0.25	TOT GAS	THM	0.0121	0.0065	0.0035	0.0029	0.0029	0.0029	0.0029	0.0029	0.0029	0.0029	0.0040	0.0109	0.058
		TOT ELEC	KWH	1.16	1.12	1.30	1.31	1.46	1.53	1.61	1.57	1.49	1.41	1.25	1.20	16.4
		TOT DEM	KW	.00531	.00538	.00558	.00591	.00638	.00672	.00665	.00672	.00665	.00645	.00564	.00544	0.00672
BASE CASE : 426																
BASE	0.00	HT ELEC	KWH	0.30	0.18	0.09	0.03	0.03	0.03	0.03	0.03	0.03	0.03	0.10	0.24	1.1
		HT DEM	KW	.00492	.00428	.00374	.00148	.00039	.00010	.00010	.00010	.00010	.00216	.00364	.00487	0.00492
		REM ELEC	KWH	1.29	1.23	1.43	1.45	1.61	1.68	1.77	1.74	1.65	1.56	1.38	1.32	18.1
		REM DEM	KW	.00571	.00571	.00578	.00629	.00665	.00709	.00709	.00723	.00709	.00680	.00600	.00578	0.00723
		TOT ELEC	KWH	1.58	1.42	1.52	1.48	1.64	1.71	1.80	1.77	1.67	1.59	1.48	1.56	19.2
		TOT DEM	KW	.00770	.00762	.00622	.00661	.00708	.00747	.00739	.00755	.00708	.00700	.00638	.00778	0.00778
S- 3	-0.16	HT ELEC	KWH	0.32	0.20	0.09	0.03	0.03	0.03	0.03	0.03	0.03	0.04	0.11	0.26	1.2
		HT DEM	KW	.00492	.00428	.00374	.00148	.00039	.00010	.00010	.00010	.00010	.00216	.00364	.00487	0.00492
		REM ELEC	KWH	1.17	1.12	1.30	1.32	1.47	1.53	1.62	1.58	1.50	1.42	1.25	1.20	16.5
		REM DEM	KW	.00525	.00525	.00532	.00579	.00612	.00652	.00652	.00665	.00652	.00625	.00552	.00532	0.00665
		TOT ELEC	KWH	1.49	1.32	1.40	1.35	1.50	1.56	1.65	1.61	1.53	1.45	1.36	1.46	17.7
		TOT DEM	KW	.00725	.00717	.00586	.00622	.00666	.00703	.00695	.00710	.00666	.00659	.00600	.00732	0.00732
S- 7	-0.07	HT ELEC	KWH	0.34	0.21	0.10	0.03	0.03	0.03	0.03	0.03	0.03	0.04	0.11	0.27	1.3
		HT DEM	KW	.00492	.00428	.00374	.00148	.00039	.00010	.00010	.00010	.00010	.00216	.00364	.00487	0.00492
		REM ELEC	KWH	1.04	0.99	1.15	1.17	1.30	1.36	1.43	1.40	1.33	1.26	1.11	1.07	14.6
		REM DEM	KW	.00466	.00466	.00472	.00513	.00543	.00578	.00578	.00590	.00578	.00555	.00490	.00472	0.00590
		TOT ELEC	KWH	1.37	1.20	1.25	1.20	1.33	1.39	1.46	1.43	1.36	1.29	1.22	1.34	15.9
		TOT DEM	KW	.00637	.00630	.00514	.00547	.00585	.00617	.00611	.00624	.00585	.00579	.00527	.00643	0.00643
S- 16	1.02	HT ELEC	KWH	0.35	0.22	0.11	0.03	0.03	0.03	0.03	0.03	0.03	0.04	0.12	0.29	1.3
		HT DEM	KW	.00492	.00428	.00374	.00148	.00039	.00010	.00010	.00010	.00010	.00216	.00364	.00487	0.00492
		REM ELEC	KWH	0.92	0.88	1.02	1.03	1.15	1.20	1.26	1.24	1.17	1.11	0.98	0.94	12.9
		REM DEM	KW	.00422	.00422	.00427	.00465	.00491	.00523	.00523	.00534	.00523	.00502	.00443	.00427	0.00534
		TOT ELEC	KWH	1.27	1.10	1.12	1.06	1.18	1.23	1.30	1.27	1.21	1.15	1.10	1.23	14.2
		TOT DEM	KW	.00615	.00609	.00497	.00528	.00565	.00596	.00590	.00602	.00565	.00559	.00509	.00621	0.00621
S- 32	0.29	HT ELEC	KWH	0.33	0.20	0.13	0.04	0.03	0.03	0.03	0.03	0.03	0.05	0.12	0.27	1.3
		HT DEM	KW	.00492	.00428	.00374	.00148	.00039	.00010	.00010	.00010	.00010	.00216	.00364	.00487	0.00492
		REM ELEC	KWH	0.83	0.81	1.00	1.06	1.11	1.26	1.27	1.27	1.17	1.05	0.88	0.87	12.6
		REM DEM	KW	.00411	.00411	.00416	.00452	.00478	.00510	.00510	.00520	.00510	.00489	.00432	.00416	0.00520
		TOT ELEC	KWH	1.16	1.01	1.13	1.10	1.14	1.29	1.30	1.30	1.20	1.09	1.00	1.14	13.9
		TOT DEM	KW	.00650	.00547	.00457	.00498	.00526	.00554	.00554	.00567	.00554	.00533	.00471	.00692	0.00692
S- 61	-0.32	HT ELEC	KWH	0.34	0.21	0.10	0.03	0.03	0.03	0.03	0.03	0.03	0.04	0.11	0.27	1.3
		HT DEM	KW	.00492	.00428	.00374	.00148	.00039	.00010	.00010	.00010	.00010	.00216	.00364	.00487	0.00492
		REM ELEC	KWH	1.04	0.99	1.15	1.17	1.30	1.36	1.43	1.40	1.33	1.26	1.11	1.07	14.6
		REM DEM	KW	.00466	.00466	.00472	.00513	.00543	.00578	.00578	.00590	.00578	.00555	.00490	.00472	0.00590
		TOT ELEC	KWH	1.37	1.20	1.25	1.20	1.33	1.39	1.46	1.43	1.36	1.29	1.22	1.34	15.9
		TOT DEM	KW	.00637	.00630	.00514	.00547	.00585	.00617	.00611	.00624	.00585	.00579	.00527	.00643	0.00643
BASE CASE : 427																
BASE	0.00	TOT GAS	THM	0.0156	0.0097	0.0047	0.0015	0.0015	0.0015	0.0015	0.0015	0.0015	0.0017	0.0052	0.0127	0.058
		TOT ELEC	KWH	1.29	1.23	1.43	1.45	1.61	1.68	1.77	1.74	1.65	1.56	1.38	1.32	18.1
		TOT DEM	KW	.00571	.00571	.00578	.00629	.00665	.00709	.00709	.00723	.00709	.00680	.00600	.00578	0.00723
S- 3	-0.17	TOT GAS	THM	0.0166	0.0103	0.0050	0.0016	0.0016	0.0016	0.0016	0.0016	0.0016	0.0019	0.0056	0.0135	0.062
		TOT ELEC	KWH	1.17	1.12	1.30	1.32	1.47	1.53	1.62	1.58	1.50	1.42	1.25	1.20	16.5
		TOT DEM	KW	.00525	.00525	.00532	.00579	.00612	.00652	.00652	.00665	.00652	.00625	.00552	.00532	0.00665
S- 7	-0.08	TOT GAS	THM	0.0176	0.0110	0.0053	0.0017	0.0017	0.0017	0.0017	0.0017	0.0017	0.0020	0.0059	0.0143	0.066
		TOT ELEC	KWH	1.04	0.99	1.15	1.17	1.30	1.36	1.43	1.40	1.33	1.26	1.11	1.07	14.6
		TOT DEM	KW	.00466	.00466	.00472	.00513	.00543	.00578	.00578	.00590	.00578	.00555	.00490	.00472	0.00590
S- 16	1.01	TOT GAS	THM	0.0185	0.0115	0.0056	0.0017	0.0017	0.0017	0.0017	0.0017	0.0017	0.0021	0.0062	0.0151	0.069
		TOT ELEC	KWH	0.92	0.88	1.02	1.03	1.15	1.20	1.26	1.24	1.17	1.11	0.98	0.94	12.9
		TOT DEM	KW	.00422	.00422	.00427	.00465	.00491	.00523	.00523	.00534	.00523	.00502	.00443	.00427	0.00534
S- 32	0.28	TOT GAS	THM	0.0173	0.0107	0.0069	0.0021	0.0014	0.0015	0.0015	0.0015	0.0014	0.0025	0.0063	0.0143	0.068
		TOT ELEC	KWH	0.83	0.81	1.00	1.06	1.11	1.26	1.27	1.27	1.17	1.05	0.88	0.87	12.6
		TOT DEM	KW	.00411	.00411	.00416	.00452	.00478	.00510	.00510	.00520	.00510	.00489	.00432	.00416	0.00520
S- 61	-0.33	TOT GAS	THM	0.0176	0.0110	0.0053	0.0017	0.0017	0.0017	0.0017	0.0017	0.0017	0.0020	0.0059	0.0143	0.066
		TOT ELEC	KWH	1.04	0.99	1.15	1.17	1.30	1.36	1.43	1.40	1.33	1.26	1.11	1.07	14.6
		TOT DEM	KW	.00466	.00466	.00472	.00513	.00543	.00578	.00578	.00590	.00578	.00555	.00490	.00472	0.00590
BASE CASE : 428																
BASE	0.00	TOT ELEC	KWH	1.44	1.30	1.58	1.70	1.74	1.96	1.98	1.98	1.84	1.64	1.44	1.44	20.0
		TOT DEM	KW	.00654	.00638	.00646	.00717	.00717	.00756	.00749	.00788	.00756	.00733	.00686	.00654	0.00788
S- 1	0.09	TOT ELEC	KWH	1.34	1.21	1.47	1.58	1.62	1.82	1.84	1.84	1.71	1.53	1.34	1.34	18.6
		TOT DEM	KW	.00622	.00607	.00614	.00682	.00682	.00719	.00712	.00749	.00719	.00697	.00652	.00622	0.00749
S- 12	0.21	TOT ELEC	KWH	1.20	1.09	1.32	1.42	1.45	1.64	1.65	1.65	1.54	1.37	1.20	1.20	16.7
		TOT DEM	KW	.00549	.00536	.00543	.00602	.00602	.00636	.00629	.00662	.00636	.00616	.00576	.00549	0.00662
S- 18	1.52	TOT ELEC	KWH	1.09	0.99	1.20	1.29	1.32	1.49	1.50	1.50	1.40	1.25	1.09	1.09	15.2
		TOT DEM	KW	.00523	.00510	.00517	.00573	.00573	.00605	.00598	.00630	.00605	.00586	.00548	.00523	0.00630

(HT ELEC=ELECTRIC CONSUMPTION FOR SPACE AND WATER HEATING. REM ELEC=REMAINING ELECTRIC CONSUMPTION (NOT USED FOR HEATING). TOT ELEC=TOTAL ELECTRIC CONSUMPTION. HT DEM=PEAK ELECTRIC DEMAND FOR SPACE AND WATER HEATING. REM DEM=PEAK ELECTRIC DEMAND FOR REMAINING LOAD (NOT HEATING). TOT DEM=TOTAL PEAK ELECTRIC DEMAND. GAS=TOTAL GAS CONSUMPTION. ANNUAL FIGURES ARE SUMS OF MONTHLY CONSUMPTION VALUES OR ARE YEARLY MAXIMUMS OF MONTHLY PEAK DEMAND VALUES.)

STRAT SET	CONST COST $/SF	ENERGY FORM	UNITS (PER SF)	JAN	FEB	MAR	APR	MAY	JUN	JUL	AUG	SEP	OCT	NOV	DEC	ANNUAL
BASE CASE : 428 (CONTINUED)																
S- 28	-0.04	TOT ELEC	KWH	1.20	1.09	1.32	1.42	1.45	1.64	1.65	1.65	1.54	1.37	1.20	1.20	16.7
		TOT DEM	KW	.00549	.00536	.00543	.00602	.00602	.00636	.00629	.00662	.00636	.00616	.00576	.00549	0.00662
S- 33	0.57	TOT ELEC	KWH	1.02	0.92	1.11	1.20	1.23	1.38	1.40	1.40	1.30	1.16	1.02	1.02	14.1
		TOT DEM	KW	.00509	.00497	.00503	.00558	.00558	.00588	.00582	.00613	.00588	.00570	.00533	.00509	0.00613
BASE CASE : 429																
BASE	0.00	HT ELEC	KWH	0.21	0.13	0.06	0.02	0.02	0.02	0.02	0.02	0.02	0.02	0.07	0.17	0.8
		HT DEM	KW	.00358	.00311	.00272	.00107	.00029	.00007	.00007	.00007	.00007	.00158	.00265	.00354	0.00358
		REM ELEC	KWH	1.20	1.15	1.34	1.35	1.50	1.57	1.66	1.62	1.54	1.45	1.28	1.23	16.9
		REM DEM	KW	.00514	.00514	.00520	.00566	.00598	.00637	.00637	.00650	.00637	.00611	.00539	.00520	0.00650
		TOT ELEC	KWH	1.41	1.28	1.40	1.37	1.52	1.59	1.68	1.64	1.56	1.48	1.35	1.40	17.7
		TOT DEM	KW	.00634	.00528	.00528	.00568	.00614	.00647	.00647	.00660	.00647	.00627	.00541	.00640	0.00660
S- 3	-0.11	HT ELEC	KWH	0.22	0.14	0.07	0.02	0.02	0.02	0.02	0.02	0.02	0.02	0.07	0.18	0.8
		HT DEM	KW	.00358	.00311	.00272	.00107	.00029	.00007	.00007	.00007	.00007	.00158	.00265	.00354	0.00358
		REM ELEC	KWH	1.09	1.05	1.22	1.23	1.37	1.43	1.51	1.48	1.40	1.32	1.17	1.12	15.4
		REM DEM	KW	.00472	.00472	.00478	.00520	.00550	.00586	.00586	.00598	.00586	.00562	.00496	.00478	0.00598
		TOT ELEC	KWH	1.31	1.18	1.28	1.25	1.39	1.45	1.53	1.50	1.42	1.35	1.24	1.30	16.2
		TOT DEM	KW	.00595	.00496	.00496	.00533	.00577	.00608	.00608	.00620	.00608	.00589	.00508	.00601	0.00620
S- 7	-0.03	HT ELEC	KWH	0.23	0.15	0.07	0.02	0.02	0.02	0.02	0.02	0.02	0.03	0.08	0.19	0.9
		HT DEM	KW	.00358	.00311	.00272	.00107	.00029	.00007	.00007	.00007	.00007	.00158	.00265	.00354	0.00358
		REM ELEC	KWH	0.97	0.92	1.07	1.09	1.21	1.26	1.33	1.31	1.24	1.17	1.03	0.99	13.6
		REM DEM	KW	.00419	.00419	.00424	.00461	.00488	.00519	.00519	.00530	.00519	.00498	.00440	.00424	0.00530
		TOT ELEC	KWH	1.20	1.07	1.14	1.11	1.23	1.29	1.35	1.33	1.26	1.20	1.11	1.18	14.5
		TOT DEM	KW	.00523	.00436	.00436	.00469	.00507	.00534	.00534	.00545	.00534	.00518	.00447	.00529	0.00545
S- 16	1.12	HT ELEC	KWH	0.25	0.15	0.07	0.02	0.02	0.02	0.02	0.02	0.02	0.03	0.08	0.20	0.9
		HT DEM	KW	.00358	.00311	.00272	.00107	.00029	.00007	.00007	.00007	.00007	.00158	.00265	.00354	0.00358
		REM ELEC	KWH	0.85	0.82	0.95	0.96	1.07	1.12	1.18	1.15	1.09	1.03	0.91	0.88	12.0
		REM DEM	KW	.00379	.00379	.00384	.00418	.00442	.00470	.00470	.00480	.00470	.00451	.00398	.00384	0.00480
		TOT ELEC	KWH	1.10	0.97	1.02	0.98	1.09	1.14	1.20	1.17	1.12	1.06	0.99	1.08	12.9
		TOT DEM	KW	.00507	.00422	.00422	.00454	.00491	.00517	.00517	.00528	.00517	.00502	.00433	.00512	0.00528
S- 32	0.20	HT ELEC	KWH	0.28	0.17	0.11	0.03	0.02	0.02	0.02	0.02	0.02	0.04	0.10	0.23	1.1
		HT DEM	KW	.00358	.00311	.00272	.00107	.00029	.00007	.00007	.00007	.00007	.00158	.00265	.00354	0.00358
		REM ELEC	KWH	0.86	0.83	1.03	1.09	1.14	1.30	1.31	1.31	1.21	1.08	0.91	0.90	13.0
		REM DEM	KW	.00407	.00407	.00412	.00448	.00474	.00505	.00505	.00515	.00505	.00484	.00427	.00412	0.00515
		TOT ELEC	KWH	1.14	1.01	1.14	1.13	1.17	1.32	1.34	1.34	1.23	1.12	1.01	1.13	14.1
		TOT DEM	KW	.00611	.00514	.00429	.00468	.00494	.00520	.00520	.00533	.00520	.00500	.00442	.00650	0.00650
S- 40	0.92	HT ELEC	KWH	0.34	0.21	0.14	0.04	0.03	0.03	0.03	0.03	0.03	0.05	0.12	0.28	1.3
		HT DEM	KW	.00358	.00311	.00272	.00107	.00029	.00007	.00007	.00007	.00007	.00158	.00265	.00354	0.00358
		REM ELEC	KWH	0.78	0.76	0.93	0.99	1.04	1.18	1.19	1.19	1.10	0.98	0.83	0.81	11.8
		REM DEM	KW	.00370	.00370	.00374	.00407	.00431	.00459	.00459	.00468	.00459	.00440	.00388	.00374	0.00468
		TOT ELEC	KWH	1.12	0.96	1.07	1.03	1.07	1.21	1.22	1.22	1.13	1.03	0.95	1.09	13.1
		TOT DEM	KW	.00600	.00504	.00421	.00459	.00485	.00510	.00510	.00523	.00510	.00491	.00434	.00638	0.00638
S- 45	0.23	HT ELEC	KWH	0.31	0.19	0.12	0.04	0.03	0.03	0.03	0.03	0.03	0.04	0.11	0.26	1.2
		HT DEM	KW	.00358	.00311	.00272	.00107	.00029	.00007	.00007	.00007	.00007	.00158	.00265	.00354	0.00358
		REM ELEC	KWH	0.82	0.79	0.98	1.04	1.09	1.24	1.25	1.25	1.15	1.03	0.87	0.86	12.4
		REM DEM	KW	.00389	.00389	.00394	.00429	.00454	.00483	.00483	.00493	.00483	.00463	.00409	.00394	0.00493
		TOT ELEC	KWH	1.13	0.99	1.10	1.08	1.12	1.27	1.28	1.28	1.18	1.07	0.98	1.11	13.6
		TOT DEM	KW	.00606	.00510	.00426	.00464	.00490	.00516	.00516	.00529	.00516	.00497	.00439	.00645	0.00645
S- 61	-0.27	HT ELEC	KWH	0.23	0.15	0.07	0.02	0.02	0.02	0.02	0.02	0.02	0.03	0.08	0.19	0.9
		HT DEM	KW	.00358	.00311	.00272	.00107	.00029	.00007	.00007	.00007	.00007	.00158	.00265	.00354	0.00358
		REM ELEC	KWH	0.97	0.92	1.07	1.09	1.21	1.26	1.33	1.31	1.24	1.17	1.03	0.99	13.6
		REM DEM	KW	.00419	.00419	.00424	.00461	.00488	.00519	.00519	.00530	.00519	.00498	.00440	.00424	0.00530
		TOT ELEC	KWH	1.20	1.07	1.14	1.11	1.23	1.29	1.35	1.33	1.26	1.20	1.11	1.18	14.5
		TOT DEM	KW	.00523	.00436	.00436	.00469	.00507	.00534	.00534	.00545	.00534	.00518	.00447	.00529	0.00545
BASE CASE : 430																
BASE	0.00	TOT GAS	THM	0.0108	0.0067	0.0032	0.0010	0.0010	0.0010	0.0010	0.0010	0.0010	0.0012	0.0037	0.0088	0.041
		TOT ELEC	KWH	1.20	1.15	1.34	1.35	1.50	1.57	1.66	1.62	1.54	1.45	1.28	1.23	16.9
		TOT DEM	KW	.00514	.00514	.00520	.00566	.00598	.00637	.00637	.00650	.00637	.00611	.00539	.00520	0.00650
S- 3	-0.12	TOT GAS	THM	0.0116	0.0072	0.0035	0.0011	0.0011	0.0011	0.0011	0.0011	0.0011	0.0013	0.0039	0.0094	0.043
		TOT ELEC	KWH	1.09	1.05	1.22	1.23	1.37	1.43	1.51	1.48	1.40	1.32	1.17	1.12	15.4
		TOT DEM	KW	.00472	.00472	.00478	.00520	.00550	.00586	.00586	.00598	.00586	.00562	.00496	.00478	0.00598
S- 7	-0.03	TOT GAS	THM	0.0123	0.0076	0.0037	0.0012	0.0012	0.0012	0.0012	0.0012	0.0012	0.0014	0.0041	0.0100	0.046
		TOT ELEC	KWH	0.97	0.92	1.07	1.09	1.21	1.26	1.33	1.31	1.24	1.17	1.03	0.99	13.6
		TOT DEM	KW	.00419	.00419	.00424	.00461	.00488	.00519	.00519	.00530	.00519	.00498	.00440	.00424	0.00530
S- 16	1.11	TOT GAS	THM	0.0129	0.0080	0.0039	0.0012	0.0012	0.0012	0.0012	0.0012	0.0012	0.0014	0.0043	0.0105	0.048
		TOT ELEC	KWH	0.85	0.82	0.95	0.96	1.07	1.12	1.18	1.15	1.09	1.03	0.91	0.88	12.0
		TOT DEM	KW	.00379	.00379	.00384	.00418	.00442	.00470	.00470	.00480	.00470	.00451	.00398	.00384	0.00480
S- 32	0.20	TOT GAS	THM	0.0148	0.0092	0.0059	0.0018	0.0012	0.0013	0.0013	0.0013	0.0012	0.0021	0.0054	0.0122	0.058
		TOT ELEC	KWH	0.86	0.83	1.03	1.09	1.14	1.30	1.31	1.31	1.21	1.08	0.91	0.90	13.0
		TOT DEM	KW	.00407	.00407	.00412	.00448	.00474	.00505	.00505	.00515	.00505	.00484	.00427	.00412	0.00515
S- 40	0.91	TOT GAS	THM	0.0177	0.0110	0.0071	0.0022	0.0015	0.0015	0.0015	0.0015	0.0015	0.0026	0.0065	0.0147	0.070
		TOT ELEC	KWH	0.78	0.76	0.93	0.99	1.04	1.18	1.19	1.19	1.10	0.98	0.83	0.81	11.8
		TOT DEM	KW	.00370	.00370	.00374	.00407	.00431	.00459	.00459	.00468	.00459	.00440	.00388	.00374	0.00468
S- 45	0.22	TOT GAS	THM	0.0163	0.0101	0.0065	0.0020	0.0013	0.0014	0.0014	0.0014	0.0013	0.0024	0.0059	0.0135	0.064
		TOT ELEC	KWH	0.82	0.79	0.98	1.04	1.09	1.24	1.25	1.25	1.15	1.03	0.87	0.86	12.4
		TOT DEM	KW	.00389	.00389	.00394	.00429	.00454	.00483	.00483	.00493	.00483	.00463	.00409	.00394	0.00493
S- 61	-0.28	TOT GAS	THM	0.0123	0.0076	0.0037	0.0012	0.0012	0.0012	0.0012	0.0012	0.0012	0.0014	0.0041	0.0100	0.046
		TOT ELEC	KWH	0.97	0.92	1.07	1.09	1.21	1.26	1.33	1.31	1.24	1.17	1.03	0.99	13.6
		TOT DEM	KW	.00419	.00419	.00424	.00461	.00488	.00519	.00519	.00530	.00519	.00498	.00440	.00424	0.00530

STRAT SET	CONST COST $/SF	ENERGY FORM	UNITS (PER SF)	JAN	FEB	MAR	APR	MAY	JUN	JUL	AUG	SEP	OCT	NOV	DEC	ANNUAL
BASE CASE : 431																
BASE	0.00	TOT ELEC	KWH	1.29	1.16	1.41	1.52	1.56	1.75	1.77	1.77	1.65	1.47	1.29	1.29	17.9
		TOT DEM	KW	.00575	.00561	.00568	.00631	.00631	.00665	.00658	.00693	.00665	.00644	.00603	.00575	0.00693
S- 1	0.09	TOT ELEC	KWH	1.19	1.07	1.30	1.40	1.44	1.62	1.63	1.63	1.52	1.35	1.19	1.19	16.5
		TOT DEM	KW	.00558	.00544	.00551	.00612	.00612	.00645	.00638	.00672	.00645	.00625	.00585	.00558	0.00672
S- 12	0.29	TOT ELEC	KWH	1.07	0.96	1.17	1.26	1.29	1.45	1.47	1.47	1.36	1.21	1.07	1.07	14.8
		TOT DEM	KW	.00530	.00517	.00523	.00581	.00581	.00612	.00606	.00638	.00612	.00593	.00555	.00530	0.00638
S- 22	1.60	TOT ELEC	KWH	0.98	0.88	1.07	1.16	1.18	1.33	1.35	1.35	1.25	1.12	0.98	0.98	13.6
		TOT DEM	KW	.00512	.00500	.00506	.00561	.00561	.00592	.00586	.00617	.00592	.00574	.00537	.00512	0.00617
S- 28	-0.05	TOT ELEC	KWH	1.07	0.96	1.17	1.26	1.29	1.45	1.47	1.47	1.36	1.21	1.07	1.07	14.8
		TOT DEM	KW	.00530	.00517	.00523	.00581	.00581	.00612	.00606	.00638	.00612	.00593	.00555	.00530	0.00638
S- 36	0.52	TOT ELEC	KWH	0.98	0.88	1.07	1.16	1.18	1.33	1.35	1.35	1.25	1.12	0.98	0.98	13.6
		TOT DEM	KW	.00438	.00428	.00433	.00480	.00480	.00507	.00502	.00528	.00507	.00491	.00459	.00438	0.00528
S- 43	1.36	TOT ELEC	KWH	0.90	0.81	0.99	1.06	1.09	1.23	1.24	1.24	1.15	1.02	0.90	0.90	12.5
		TOT DEM	KW	.00423	.00413	.00418	.00464	.00464	.00490	.00484	.00510	.00490	.00474	.00444	.00423	0.00510
S- 48	0.55	TOT ELEC	KWH	0.94	0.85	1.03	1.11	1.14	1.28	1.30	1.30	1.21	1.07	0.94	0.94	13.1
		TOT DEM	KW	.00430	.00420	.00425	.00471	.00471	.00497	.00492	.00518	.00497	.00482	.00451	.00430	0.00518
BASE CASE : 432																
BASE	0.00	HT ELEC	KWH	0.22	0.14	0.07	0.02	0.02	0.02	0.02	0.02	0.02	0.02	0.07	0.18	0.8
		HT DEM	KW	.00385	.00335	.00293	.00116	.00031	.00008	.00008	.00008	.00008	.00169	.00285	.00381	0.00385
		REM ELEC	KWH	1.24	1.19	1.38	1.40	1.56	1.63	1.71	1.68	1.59	1.51	1.33	1.28	17.5
		REM DEM	KW	.00545	.00545	.00552	.00600	.00635	.00676	.00676	.00690	.00676	.00649	.00573	.00552	0.00690
		TOT ELEC	KWH	1.46	1.33	1.45	1.42	1.58	1.65	1.74	1.70	1.61	1.53	1.40	1.46	18.3
		TOT DEM	KW	.00763	.00725	.00580	.00610	.00671	.00694	.00702	.00702	.00687	.00671	.00588	.00763	0.00763
S- 3	-0.13	HT ELEC	KWH	0.24	0.15	0.07	0.02	0.02	0.02	0.02	0.02	0.02	0.03	0.08	0.19	0.9
		HT DEM	KW	.00358	.00311	.00272	.00107	.00029	.00007	.00007	.00007	.00007	.00158	.00265	.00354	0.00358
		REM ELEC	KWH	1.14	1.09	1.26	1.28	1.42	1.49	1.57	1.54	1.46	1.38	1.22	1.17	16.0
		REM DEM	KW	.00502	.00502	.00508	.00552	.00584	.00622	.00622	.00635	.00622	.00597	.00527	.00508	0.00635
		TOT ELEC	KWH	1.37	1.23	1.33	1.30	1.45	1.51	1.59	1.56	1.48	1.40	1.30	1.36	16.9
		TOT DEM	KW	.00718	.00682	.00546	.00574	.00632	.00653	.00661	.00661	.00646	.00632	.00553	.00718	0.00718
S- 7	-0.04	HT ELEC	KWH	0.25	0.16	0.07	0.02	0.02	0.02	0.02	0.02	0.02	0.03	0.08	0.20	0.9
		HT DEM	KW	.00358	.00311	.00272	.00107	.00029	.00007	.00007	.00007	.00007	.00158	.00265	.00354	0.00358
		REM ELEC	KWH	1.00	0.96	1.11	1.13	1.25	1.31	1.38	1.35	1.28	1.21	1.07	1.03	14.1
		REM DEM	KW	.00445	.00445	.00450	.00490	.00518	.00552	.00552	.00563	.00552	.00529	.00467	.00450	0.00563
		TOT ELEC	KWH	1.25	1.11	1.19	1.15	1.28	1.33	1.41	1.38	1.31	1.24	1.16	1.23	15.0
		TOT DEM	KW	.00630	.00598	.00479	.00504	.00554	.00573	.00580	.00580	.00567	.00554	.00485	.00630	0.00630
S- 16	1.10	HT ELEC	KWH	0.26	0.16	0.08	0.02	0.02	0.02	0.02	0.02	0.02	0.03	0.09	0.21	1.0
		HT DEM	KW	.00358	.00311	.00272	.00107	.00029	.00007	.00007	.00007	.00007	.00158	.00265	.00354	0.00358
		REM ELEC	KWH	0.88	0.84	0.98	0.99	1.10	1.15	1.22	1.19	1.13	1.07	0.94	0.91	12.4
		REM DEM	KW	.00403	.00403	.00408	.00444	.00469	.00500	.00500	.00510	.00500	.00479	.00423	.00408	0.00510
		TOT ELEC	KWH	1.14	1.01	1.06	1.02	1.13	1.18	1.24	1.21	1.15	1.10	1.03	1.12	13.4
		TOT DEM	KW	.00610	.00580	.00464	.00488	.00537	.00555	.00561	.00561	.00549	.00537	.00470	.00610	0.00610
S- 32	0.19	HT ELEC	KWH	0.21	0.13	0.08	0.03	0.02	0.02	0.02	0.02	0.02	0.03	0.08	0.17	0.8
		HT DEM	KW	.00358	.00311	.00272	.00107	.00029	.00007	.00007	.00007	.00007	.00158	.00265	.00354	0.00358
		REM ELEC	KWH	0.84	0.82	1.01	1.08	1.13	1.28	1.29	1.29	1.19	1.06	0.90	0.88	12.8
		REM DEM	KW	.00403	.00403	.00408	.00444	.00469	.00500	.00500	.00510	.00500	.00479	.00423	.00408	0.00510
		TOT ELEC	KWH	1.06	0.95	1.10	1.10	1.14	1.30	1.31	1.31	1.21	1.09	0.97	1.06	13.6
		TOT DEM	KW	.00611	.00514	.00429	.00468	.00494	.00520	.00520	.00533	.00520	.00500	.00442	.00650	0.00650
S- 40	0.94	HT ELEC	KWH	0.23	0.14	0.09	0.03	0.02	0.02	0.02	0.02	0.02	0.03	0.08	0.19	0.9
		HT DEM	KW	.00358	.00311	.00272	.00107	.00029	.00007	.00007	.00007	.00007	.00158	.00265	.00354	0.00358
		REM ELEC	KWH	0.78	0.76	0.93	0.99	1.04	1.18	1.19	1.19	1.10	0.98	0.83	0.81	11.8
		REM DEM	KW	.00387	.00387	.00392	.00426	.00451	.00480	.00480	.00490	.00480	.00461	.00407	.00392	0.00490
		TOT ELEC	KWH	1.01	0.90	1.03	1.02	1.06	1.20	1.21	1.21	1.12	1.01	0.91	1.01	12.7
		TOT DEM	KW	.00590	.00496	.00414	.00452	.00477	.00502	.00502	.00515	.00502	.00484	.00427	.00628	0.00628
S- 45	0.21	HT ELEC	KWH	0.23	0.14	0.09	0.03	0.02	0.02	0.02	0.02	0.02	0.03	0.08	0.19	0.9
		HT DEM	KW	.00358	.00311	.00272	.00107	.00029	.00007	.00007	.00007	.00007	.00158	.00265	.00354	0.00358
		REM ELEC	KWH	0.81	0.79	0.97	1.03	1.08	1.23	1.24	1.24	1.14	1.02	0.86	0.85	12.3
		REM DEM	KW	.00401	.00401	.00406	.00442	.00467	.00498	.00498	.00508	.00498	.00478	.00422	.00406	0.00508
		TOT ELEC	KWH	1.04	0.93	1.06	1.06	1.10	1.25	1.26	1.26	1.16	1.05	0.95	1.04	13.2
		TOT DEM	KW	.00602	.00506	.00422	.00461	.00486	.00512	.00512	.00525	.00512	.00493	.00435	.00640	0.00640
S- 61	-0.29	HT ELEC	KWH	0.25	0.16	0.07	0.02	0.02	0.02	0.02	0.02	0.02	0.03	0.08	0.20	0.9
		HT DEM	KW	.00358	.00311	.00272	.00107	.00029	.00007	.00007	.00007	.00007	.00158	.00265	.00354	0.00358
		REM ELEC	KWH	1.00	0.96	1.11	1.13	1.25	1.31	1.38	1.35	1.28	1.21	1.07	1.03	14.1
		REM DEM	KW	.00445	.00445	.00450	.00490	.00518	.00552	.00552	.00563	.00552	.00529	.00467	.00450	0.00563
		TOT ELEC	KWH	1.25	1.11	1.19	1.15	1.28	1.33	1.41	1.38	1.31	1.24	1.16	1.23	15.0
		TOT DEM	KW	.00630	.00598	.00479	.00504	.00554	.00573	.00580	.00580	.00567	.00554	.00485	.00630	0.00630
BASE CASE : 433																
BASE	0.00	TOT GAS	THM	0.0116	0.0072	0.0035	0.0011	0.0011	0.0011	0.0011	0.0011	0.0011	0.0013	0.0039	0.0094	0.043
		TOT ELEC	KWH	1.24	1.19	1.38	1.40	1.56	1.63	1.71	1.68	1.59	1.51	1.33	1.28	17.5
		TOT DEM	KW	.00545	.00545	.00552	.00600	.00635	.00676	.00676	.00690	.00676	.00649	.00573	.00552	0.00690
S- 3	-0.13	TOT GAS	THM	0.0124	0.0077	0.0037	0.0012	0.0012	0.0012	0.0012	0.0012	0.0012	0.0014	0.0042	0.0100	0.046
		TOT ELEC	KWH	1.14	1.09	1.26	1.28	1.42	1.49	1.57	1.54	1.46	1.38	1.22	1.17	16.0
		TOT DEM	KW	.00502	.00502	.00508	.00552	.00584	.00622	.00622	.00635	.00622	.00597	.00527	.00508	0.00635

(HT ELEC=ELECTRIC CONSUMPTION FOR SPACE AND WATER HEATING. REM ELEC=REMAINING ELECTRIC CONSUMPTION (NOT USED FOR HEATING). TOT ELEC=TOTAL ELECTRIC CONSUMPTION. HT DEM=PEAK ELECTRIC DEMAND FOR SPACE AND WATER HEATING. REM DEM=PEAK ELECTRIC DEMAND FOR REMAINING LOAD (NOT HEATING). TOT DEM=TOTAL PEAK ELECTRIC DEMAND. GAS=TOTAL GAS CONSUMPTION. ANNUAL FIGURES ARE SUMS OF MONTHLY CONSUMPTION VALUES OR ARE YEARLY MAXIMUMS OF MONTHLY PEAK DEMAND VALUES.)

STRAT SET	CONST COST $/SF	ENERGY FORM	UNITS (PER SF)	JAN	FEB	MAR	APR	MAY	JUN	JUL	AUG	SEP	OCT	NOV	DEC	ANNUAL
BASE CASE : 433 (CONTINUED)																
S- 7	-0.05	TOT GAS	THM	0.0131	0.0082	0.0039	0.0012	0.0012	0.0012	0.0012	0.0012	0.0012	0.0015	0.0044	0.0107	0.049
		TOT ELEC	KWH	1.00	0.96	1.11	1.13	1.25	1.31	1.38	1.35	1.28	1.21	1.07	1.03	14.1
		TOT DEM	KW	.00445	.00445	.00450	.00490	.00518	.00552	.00552	.00563	.00552	.00529	.00467	.00450	0.00563
S- 16	1.08	TOT GAS	THM	0.0138	0.0086	0.0041	0.0013	0.0013	0.0013	0.0013	0.0013	0.0013	0.0015	0.0046	0.0112	0.052
		TOT ELEC	KWH	0.88	0.84	0.98	0.99	1.10	1.15	1.22	1.19	1.13	1.07	0.94	0.91	12.4
		TOT DEM	KW	.00403	.00403	.00408	.00444	.00469	.00500	.00500	.00510	.00500	.00479	.00423	.00408	0.00510
S- 32	0.18	TOT GAS	THM	0.0111	0.0069	0.0044	0.0013	0.0009	0.0010	0.0010	0.0010	0.0009	0.0016	0.0040	0.0092	0.043
		TOT ELEC	KWH	0.84	0.82	1.01	1.08	1.13	1.28	1.29	1.29	1.19	1.06	0.90	0.88	12.8
		TOT DEM	KW	.00403	.00403	.00408	.00444	.00469	.00500	.00500	.00510	.00500	.00479	.00423	.00408	0.00510
S- 40	0.93	TOT GAS	THM	0.0122	0.0076	0.0049	0.0015	0.0010	0.0011	0.0011	0.0011	0.0010	0.0018	0.0044	0.0101	0.048
		TOT ELEC	KWH	0.78	0.76	0.93	0.99	1.04	1.18	1.19	1.19	1.10	0.98	0.83	0.81	11.8
		TOT DEM	KW	.00387	.00387	.00392	.00426	.00451	.00480	.00480	.00490	.00480	.00461	.00407	.00392	0.00490
S- 45	0.21	TOT GAS	THM	0.0122	0.0076	0.0049	0.0015	0.0010	0.0011	0.0011	0.0011	0.0010	0.0018	0.0044	0.0101	0.048
		TOT ELEC	KWH	0.81	0.79	0.97	1.03	1.08	1.23	1.24	1.24	1.14	1.02	0.86	0.85	12.3
		TOT DEM	KW	.00401	.00401	.00406	.00442	.00467	.00498	.00498	.00508	.00498	.00478	.00422	.00406	0.00508
S- 61	-0.30	TOT GAS	THM	0.0131	0.0082	0.0039	0.0012	0.0012	0.0012	0.0012	0.0012	0.0012	0.0015	0.0044	0.0107	0.049
		TOT ELEC	KWH	1.00	0.96	1.11	1.13	1.25	1.31	1.38	1.35	1.28	1.21	1.07	1.03	14.1
		TOT DEM	KW	.00445	.00445	.00450	.00490	.00518	.00552	.00552	.00563	.00552	.00529	.00467	.00450	0.00563
BASE CASE : 434																
BASE	0.00	TOT ELEC	KWH	1.32	1.20	1.45	1.56	1.60	1.80	1.82	1.82	1.69	1.51	1.32	1.32	18.4
		TOT DEM	KW	.00596	.00582	.00589	.00653	.00653	.00689	.00682	.00718	.00689	.00668	.00625	.00596	0.00718
S- 1	0.09	TOT ELEC	KWH	1.22	1.11	1.34	1.45	1.48	1.67	1.68	1.68	1.56	1.39	1.22	1.22	17.0
		TOT DEM	KW	.00495	.00483	.00489	.00542	.00542	.00572	.00566	.00596	.00572	.00554	.00519	.00495	0.00596
S- 12	0.31	TOT ELEC	KWH	1.10	0.99	1.21	1.30	1.33	1.50	1.51	1.51	1.41	1.25	1.10	1.10	15.3
		TOT DEM	KW	.00465	.00454	.00459	.00510	.00510	.00538	.00532	.00560	.00538	.00521	.00487	.00465	0.00560
S- 22	1.62	TOT ELEC	KWH	1.01	0.91	1.11	1.19	1.22	1.37	1.39	1.39	1.29	1.15	1.01	1.01	14.0
		TOT DEM	KW	.00449	.00438	.00444	.00492	.00492	.00519	.00514	.00541	.00519	.00503	.00471	.00449	0.00541
S- 28	-0.04	TOT ELEC	KWH	1.10	0.99	1.21	1.30	1.33	1.50	1.51	1.51	1.41	1.25	1.10	1.10	15.3
		TOT DEM	KW	.00465	.00454	.00459	.00510	.00510	.00538	.00532	.00560	.00538	.00521	.00487	.00465	0.00560
S- 36	0.54	TOT ELEC	KWH	1.04	0.94	1.15	1.23	1.26	1.42	1.44	1.44	1.33	1.19	1.04	1.04	14.5
		TOT DEM	KW	.00477	.00466	.00471	.00523	.00523	.00552	.00546	.00575	.00552	.00535	.00500	.00477	0.00575
S- 43	1.38	TOT ELEC	KWH	0.98	0.88	1.07	1.16	1.18	1.33	1.35	1.35	1.25	1.12	0.98	0.98	13.6
		TOT DEM	KW	.00461	.00450	.00455	.00505	.00505	.00533	.00527	.00555	.00533	.00516	.00483	.00461	0.00555
S- 48	0.57	TOT ELEC	KWH	1.01	0.91	1.11	1.19	1.22	1.37	1.39	1.39	1.29	1.15	1.01	1.01	14.0
		TOT DEM	KW	.00468	.00457	.00462	.00513	.00513	.00541	.00536	.00564	.00541	.00525	.00491	.00468	0.00564
BASE CASE : 435																
BASE	0.00	HT ELEC	KWH	0.21	0.13	0.06	0.02	0.02	0.02	0.02	0.02	0.02	0.02	0.07	0.17	0.8
		HT DEM	KW	.00358	.00311	.00272	.00107	.00029	.00007	.00007	.00007	.00007	.00158	.00265	.00354	0.00358
		REM ELEC	KWH	1.22	1.17	1.36	1.38	1.53	1.60	1.69	1.65	1.57	1.48	1.31	1.26	17.2
		REM DEM	KW	.00524	.00524	.00530	.00577	.00610	.00650	.00650	.00663	.00650	.00623	.00550	.00530	0.00663
		TOT ELEC	KWH	1.43	1.30	1.42	1.40	1.55	1.62	1.70	1.67	1.58	1.50	1.38	1.42	18.0
		TOT DEM	KW	.00721	.00553	.00553	.00582	.00626	.00670	.00662	.00677	.00662	.00641	.00561	.00728	0.00728
S- 3	-0.12	HT ELEC	KWH	0.22	0.14	0.07	0.02	0.02	0.02	0.02	0.02	0.02	0.02	0.07	0.18	0.8
		HT DEM	KW	.00358	.00311	.00272	.00107	.00029	.00007	.00007	.00007	.00007	.00158	.00265	.00354	0.00358
		REM ELEC	KWH	1.11	1.07	1.24	1.26	1.40	1.46	1.54	1.51	1.43	1.35	1.19	1.15	15.7
		REM DEM	KW	.00481	.00481	.00487	.00530	.00560	.00597	.00597	.00609	.00597	.00572	.00505	.00487	0.00609
		TOT ELEC	KWH	1.33	1.20	1.31	1.28	1.42	1.48	1.56	1.53	1.45	1.37	1.27	1.32	16.5
		TOT DEM	KW	.00675	.00518	.00518	.00546	.00587	.00627	.00621	.00634	.00621	.00600	.00525	.00682	0.00682
S- 7	-0.04	HT ELEC	KWH	0.23	0.14	0.07	0.02	0.02	0.02	0.02	0.02	0.02	0.03	0.08	0.19	0.9
		HT DEM	KW	.00358	.00311	.00272	.00107	.00029	.00007	.00007	.00007	.00007	.00158	.00265	.00354	0.00358
		REM ELEC	KWH	0.98	0.94	1.09	1.10	1.23	1.28	1.35	1.32	1.26	1.19	1.05	1.01	13.8
		REM DEM	KW	.00427	.00427	.00432	.00470	.00497	.00529	.00529	.00540	.00529	.00508	.00448	.00432	0.00540
		TOT ELEC	KWH	1.21	1.08	1.16	1.13	1.25	1.31	1.37	1.35	1.28	1.21	1.13	1.20	14.7
		TOT DEM	KW	.00595	.00457	.00457	.00481	.00517	.00553	.00547	.00559	.00547	.00529	.00463	.00601	0.00601
S- 16	1.10	HT ELEC	KWH	0.24	0.15	0.07	0.02	0.02	0.02	0.02	0.02	0.02	0.03	0.08	0.20	0.9
		HT DEM	KW	.00358	.00311	.00272	.00107	.00029	.00007	.00007	.00007	.00007	.00158	.00265	.00354	0.00358
		REM ELEC	KWH	0.87	0.83	0.96	0.98	1.09	1.13	1.20	1.17	1.11	1.05	0.93	0.89	12.2
		REM DEM	KW	.00386	.00386	.00391	.00425	.00450	.00479	.00479	.00489	.00479	.00460	.00406	.00391	0.00489
		TOT ELEC	KWH	1.11	0.98	1.04	1.00	1.11	1.16	1.22	1.19	1.13	1.08	1.01	1.09	13.1
		TOT DEM	KW	.00575	.00442	.00442	.00465	.00500	.00535	.00529	.00540	.00529	.00511	.00447	.00581	0.00581
S- 32	0.28	HT ELEC	KWH	0.28	0.18	0.11	0.03	0.02	0.02	0.02	0.02	0.02	0.04	0.10	0.24	1.1
		HT DEM	KW	.00358	.00311	.00272	.00107	.00029	.00007	.00007	.00007	.00007	.00158	.00265	.00354	0.00358
		REM ELEC	KWH	0.83	0.81	1.00	1.06	1.11	1.26	1.27	1.27	1.17	1.05	0.88	0.87	12.6
		REM DEM	KW	.00411	.00411	.00416	.00452	.00478	.00510	.00510	.00520	.00510	.00489	.00432	.00416	0.00520
		TOT ELEC	KWH	1.12	0.98	1.11	1.09	1.13	1.28	1.30	1.30	1.20	1.09	0.99	1.10	13.7
		TOT DEM	KW	.00678	.00644	.00515	.00542	.00597	.00617	.00624	.00624	.00610	.00597	.00522	.00678	0.00678
S- 61	-0.28	HT ELEC	KWH	0.23	0.14	0.07	0.02	0.02	0.02	0.02	0.02	0.02	0.03	0.08	0.19	0.9
		HT DEM	KW	.00358	.00311	.00272	.00107	.00029	.00007	.00007	.00007	.00007	.00158	.00265	.00354	0.00358
		REM ELEC	KWH	0.98	0.94	1.09	1.10	1.23	1.28	1.35	1.32	1.26	1.19	1.05	1.01	13.8
		REM DEM	KW	.00427	.00427	.00432	.00470	.00497	.00529	.00529	.00540	.00529	.00508	.00448	.00432	0.00540
		TOT ELEC	KWH	1.21	1.08	1.16	1.13	1.25	1.31	1.37	1.35	1.28	1.21	1.13	1.20	14.7
		TOT DEM	KW	.00595	.00457	.00457	.00481	.00517	.00553	.00547	.00559	.00547	.00529	.00463	.00601	0.00601
BASE CASE : 436																
BASE	0.00	TOT GAS	THM	0.0108	0.0067	0.0032	0.0010	0.0010	0.0010	0.0010	0.0010	0.0010	0.0012	0.0036	0.0087	0.040
		TOT ELEC	KWH	1.22	1.17	1.36	1.38	1.53	1.60	1.69	1.65	1.57	1.48	1.31	1.26	17.2
		TOT DEM	KW	.00524	.00524	.00530	.00577	.00610	.00650	.00650	.00663	.00650	.00623	.00550	.00530	0.00663
BASE CASE : 436 (CONTINUED)																
S- 3	-0.13	TOT GAS	THM	0.0115	0.0071	0.0034	0.0011	0.0011	0.0011	0.0011	0.0011	0.0011	0.0013	0.0039	0.0093	0.043
		TOT ELEC	KWH	1.11	1.07	1.24	1.26	1.40	1.46	1.54	1.51	1.43	1.35	1.19	1.15	15.7
		TOT DEM	KW	.00481	.00481	.00487	.00530	.00560	.00597	.00597	.00609	.00597	.00572	.00505	.00487	0.00609
S- 7	-0.05	TOT GAS	THM	0.0122	0.0076	0.0037	0.0011	0.0011	0.0011	0.0011	0.0011	0.0011	0.0014	0.0041	0.0099	0.046
		TOT ELEC	KWH	0.98	0.94	1.09	1.10	1.23	1.28	1.35	1.32	1.26	1.19	1.05	1.01	13.8
		TOT DEM	KW	.00427	.00427	.00432	.00470	.00497	.00529	.00529	.00540	.00529	.00508	.00448	.00432	0.00540
S- 16	1.08	TOT GAS	THM	0.0128	0.0080	0.0038	0.0012	0.0012	0.0012	0.0012	0.0012	0.0012	0.0014	0.0043	0.0104	0.048
		TOT ELEC	KWH	0.87	0.83	0.96	0.98	1.09	1.13	1.20	1.17	1.11	1.05	0.93	0.89	12.2
		TOT DEM	KW	.00386	.00386	.00391	.00425	.00450	.00479	.00479	.00489	.00479	.00460	.00406	.00391	0.00489
S- 32	0.28	TOT GAS	THM	0.0149	0.0093	0.0060	0.0018	0.0012	0.0013	0.0013	0.0013	0.0012	0.0022	0.0054	0.0124	0.059
		TOT ELEC	KWH	0.83	0.81	1.00	1.06	1.11	1.26	1.27	1.27	1.17	1.05	0.88	0.87	12.6
		TOT DEM	KW	.00411	.00411	.00416	.00452	.00478	.00510	.00510	.00520	.00510	.00489	.00432	.00416	0.00520
S- 61	-0.29	TOT GAS	THM	0.0122	0.0076	0.0037	0.0011	0.0011	0.0011	0.0011	0.0011	0.0011	0.0014	0.0041	0.0099	0.046
		TOT ELEC	KWH	0.98	0.94	1.09	1.10	1.23	1.28	1.35	1.32	1.26	1.19	1.05	1.01	13.8
		TOT DEM	KW	.00427	.00427	.00432	.00470	.00497	.00529	.00529	.00540	.00529	.00508	.00448	.00432	0.00540
BASE CASE : 437																
BASE	0.00	TOT ELEC	KWH	1.31	1.18	1.44	1.55	1.58	1.78	1.80	1.80	1.67	1.49	1.31	1.31	18.2
		TOT DEM	KW	.00596	.00582	.00589	.00653	.00653	.00689	.00682	.00718	.00689	.00668	.00625	.00596	0.00718
S- 1	0.09	TOT ELEC	KWH	1.20	1.09	1.32	1.42	1.45	1.64	1.65	1.65	1.54	1.37	1.20	1.20	16.7
		TOT DEM	KW	.00530	.00518	.00524	.00581	.00581	.00613	.00607	.00639	.00613	.00594	.00556	.00530	0.00639
S- 12	0.20	TOT ELEC	KWH	1.07	0.97	1.18	1.27	1.30	1.46	1.48	1.48	1.37	1.22	1.07	1.07	14.9
		TOT DEM	KW	.00506	.00494	.00500	.00555	.00555	.00586	.00580	.00610	.00586	.00567	.00531	.00506	0.00610
S- 22	1.51	TOT ELEC	KWH	0.97	0.88	1.07	1.15	1.17	1.32	1.34	1.34	1.24	1.11	0.97	0.97	13.5
		TOT DEM	KW	.00483	.00471	.00477	.00530	.00530	.00559	.00553	.00582	.00559	.00541	.00506	.00483	0.00582
S- 28	-0.05	TOT ELEC	KWH	1.07	0.97	1.18	1.27	1.30	1.46	1.48	1.48	1.37	1.22	1.07	1.07	14.9
		TOT DEM	KW	.00506	.00494	.00500	.00555	.00555	.00586	.00580	.00610	.00586	.00567	.00531	.00506	0.00610
S- 36	0.52	TOT ELEC	KWH	0.86	0.77	0.94	1.01	1.04	1.17	1.18	1.18	1.09	0.98	0.86	0.86	11.9
		TOT DEM	KW	.00480	.00468	.00474	.00526	.00526	.00555	.00549	.00578	.00555	.00538	.00503	.00480	0.00578
BASE CASE : 438																
BASE	0.00	HT ELEC	KWH	0.23	0.14	0.07	0.02	0.02	0.02	0.02	0.02	0.02	0.03	0.08	0.19	0.9
		HT DEM	KW	.00385	.00335	.00293	.00116	.00031	.00008	.00008	.00008	.00008	.00169	.00285	.00381	0.00385
		REM ELEC	KWH	1.28	1.22	1.42	1.44	1.60	1.67	1.76	1.73	1.64	1.55	1.37	1.31	18.0
		REM DEM	KW	.00571	.00571	.00578	.00629	.00665	.00709	.00709	.00723	.00709	.00680	.00600	.00578	0.00723
		TOT ELEC	KWH	1.51	1.37	1.49	1.46	1.62	1.70	1.79	1.75	1.66	1.57	1.45	1.50	18.9
		TOT DEM	KW	.00757	.00750	.00612	.00650	.00696	.00734	.00727	.00742	.00696	.00689	.00627	.00765	0.00765
S- 3	-0.15	HT ELEC	KWH	0.25	0.15	0.07	0.02	0.02	0.02	0.02	0.02	0.02	0.03	0.08	0.20	0.9
		HT DEM	KW	.00385	.00335	.00293	.00116	.00031	.00008	.00008	.00008	.00008	.00169	.00285	.00381	0.00385
		REM ELEC	KWH	1.16	1.12	1.30	1.31	1.46	1.53	1.61	1.57	1.49	1.41	1.25	1.20	16.4
		REM DEM	KW	.00525	.00525	.00532	.00579	.00612	.00652	.00652	.00665	.00652	.00625	.00552	.00532	0.00665
		TOT ELEC	KWH	1.41	1.27	1.37	1.34	1.48	1.55	1.63	1.60	1.52	1.44	1.33	1.40	17.3
		TOT DEM	KW	.00712	.00705	.00575	.00611	.00654	.00690	.00683	.00697	.00654	.00647	.00590	.00719	0.00719
S- 7	-0.06	HT ELEC	KWH	0.26	0.16	0.08	0.02	0.02	0.02	0.02	0.02	0.02	0.03	0.09	0.21	1.0
		HT DEM	KW	.00385	.00335	.00293	.00116	.00031	.00008	.00008	.00008	.00008	.00169	.00285	.00381	0.00385
		REM ELEC	KWH	1.03	0.99	1.15	1.16	1.29	1.35	1.42	1.39	1.32	1.25	1.10	1.06	14.5
		REM DEM	KW	.00466	.00466	.00472	.00513	.00543	.00578	.00578	.00590	.00578	.00555	.00490	.00472	0.00590
		TOT ELEC	KWH	1.29	1.15	1.22	1.18	1.32	1.37	1.45	1.42	1.34	1.28	1.19	1.27	15.5
		TOT DEM	KW	.00626	.00619	.00506	.00537	.00575	.00607	.00600	.00613	.00575	.00569	.00518	.00632	0.00632
S- 16	1.05	HT ELEC	KWH	0.28	0.17	0.08	0.03	0.03	0.03	0.03	0.03	0.03	0.03	0.09	0.22	1.0
		HT DEM	KW	.00385	.00335	.00293	.00116	.00031	.00008	.00008	.00008	.00008	.00169	.00285	.00381	0.00385
		REM ELEC	KWH	0.91	0.87	1.01	1.02	1.14	1.19	1.25	1.23	1.16	1.10	0.97	0.93	12.8
		REM DEM	KW	.00422	.00422	.00427	.00465	.00491	.00523	.00523	.00534	.00523	.00502	.00443	.00427	0.00534
		TOT ELEC	KWH	1.18	1.04	1.09	1.05	1.16	1.22	1.28	1.25	1.19	1.13	1.07	1.16	13.8
		TOT DEM	KW	.00604	.00598	.00488	.00519	.00555	.00586	.00580	.00592	.00555	.00549	.00500	.00610	0.00610
S- 32	0.26	HT ELEC	KWH	0.26	0.16	0.10	0.03	0.02	0.02	0.02	0.02	0.02	0.04	0.09	0.21	1.0
		HT DEM	KW	.00385	.00335	.00293	.00116	.00031	.00008	.00008	.00008	.00008	.00169	.00285	.00381	0.00385
		REM ELEC	KWH	0.82	0.80	0.99	1.05	1.10	1.25	1.26	1.26	1.16	1.04	0.88	0.86	12.5
		REM DEM	KW	.00411	.00411	.00416	.00452	.00478	.00510	.00510	.00520	.00510	.00489	.00432	.00416	0.00520
		TOT ELEC	KWH	1.08	0.96	1.09	1.08	1.12	1.27	1.28	1.28	1.18	1.07	0.97	1.08	13.5
		TOT DEM	KW	.00639	.00537	.00449	.00490	.00517	.00544	.00544	.00558	.00544	.00524	.00462	.00680	0.00680
S- 61	-0.30	HT ELEC	KWH	0.26	0.16	0.08	0.02	0.02	0.02	0.02	0.02	0.02	0.03	0.09	0.21	1.0
		HT DEM	KW	.00385	.00335	.00293	.00116	.00031	.00008	.00008	.00008	.00008	.00169	.00285	.00381	0.00385
		REM ELEC	KWH	1.03	0.99	1.15	1.16	1.29	1.35	1.42	1.39	1.32	1.25	1.10	1.06	14.5
		REM DEM	KW	.00466	.00466	.00472	.00513	.00543	.00578	.00578	.00590	.00578	.00555	.00490	.00472	0.00590
		TOT ELEC	KWH	1.29	1.15	1.22	1.18	1.32	1.37	1.45	1.42	1.34	1.28	1.19	1.27	15.5
		TOT DEM	KW	.00626	.00619	.00506	.00537	.00575	.00607	.00600	.00613	.00575	.00569	.00518	.00632	0.00632
BASE CASE : 439																
BASE	0.00	TOT GAS	THM	0.0122	0.0076	0.0036	0.0011	0.0011	0.0011	0.0011	0.0011	0.0011	0.0014	0.0041	0.0099	0.046
		TOT ELEC	KWH	1.28	1.22	1.42	1.44	1.60	1.67	1.76	1.73	1.64	1.55	1.37	1.31	18.0
		TOT DEM	KW	.00571	.00571	.00578	.00629	.00665	.00709	.00709	.00723	.00709	.00680	.00600	.00578	0.00723
S- 3	-0.16	TOT GAS	THM	0.0125	0.0078	0.0037	0.0012	0.0012	0.0012	0.0012	0.0012	0.0012	0.0014	0.0042	0.0102	0.047
		TOT ELEC	KWH	1.16	1.12	1.30	1.31	1.46	1.53	1.61	1.57	1.49	1.41	1.25	1.20	16.4
		TOT DEM	KW	.00525	.00525	.00532	.00579	.00612	.00652	.00652	.00665	.00652	.00625	.00552	.00532	0.00665

(HT ELEC=ELECTRIC CONSUMPTION FOR SPACE AND WATER HEATING. REM ELEC=REMAINING ELECTRIC CONSUMPTION (NOT USED FOR HEATING). TOT ELEC=TOTAL ELECTRIC CONSUMPTION. HT DEM=PEAK ELECTRIC DEMAND FOR SPACE AND WATER HEATING. REM DEM=PEAK ELECTRIC DEMAND FOR REMAINING LOAD (NOT HEATING). TOT DEM=TOTAL PEAK ELECTRIC DEMAND. GAS=TOTAL GAS CONSUMPTION. ANNUAL FIGURES ARE SUMS OF MONTHLY CONSUMPTION VALUES OR ARE YEARLY MAXIMUMS OF MONTHLY PEAK DEMAND VALUES.)

STRAT SET	CONST COST $/SF	ENERGY FORM	UNITS (PER SF)	JAN	FEB	MAR	APR	MAY	JUN	JUL	AUG	SEP	OCT	NOV	DEC	ANNUAL
								BASE CASE : 439 (CONTINUED)								
S- 7	-0.07	TOT GAS	THM	0.0138	0.0086	0.0041	0.0013	0.0013	0.0013	0.0013	0.0013	0.0013	0.0015	0.0046	0.0112	0.052
		TOT ELEC	KWH	1.03	0.99	1.15	1.16	1.29	1.35	1.42	1.39	1.32	1.25	1.10	1.06	14.5
		TOT DEM	KW	.00466	.00466	.00472	.00513	.00543	.00578	.00578	.00590	.00578	.00555	.00490	.00472	0.00590
S- 16	1.04	TOT GAS	THM	0.0145	0.0090	0.0043	0.0014	0.0014	0.0014	0.0014	0.0014	0.0014	0.0016	0.0049	0.0118	0.054
		TOT ELEC	KWH	0.91	0.87	1.01	1.02	1.14	1.19	1.25	1.23	1.16	1.10	0.97	0.93	12.8
		TOT DEM	KW	.00422	.00422	.00427	.00465	.00491	.00523	.00523	.00534	.00523	.00502	.00443	.00427	0.00534
S- 32	0.25	TOT GAS	THM	0.0135	0.0084	0.0054	0.0016	0.0011	0.0012	0.0012	0.0012	0.0011	0.0020	0.0049	0.0112	0.053
		TOT ELEC	KWH	0.82	0.80	0.99	1.05	1.10	1.25	1.26	1.26	1.16	1.04	0.88	0.86	12.5
		TOT DEM	KW	.00411	.00411	.00416	.00452	.00478	.00510	.00510	.00520	.00510	.00489	.00432	.00416	0.00520
S- 61	-0.31	TOT GAS	THM	0.0138	0.0086	0.0041	0.0013	0.0013	0.0013	0.0013	0.0013	0.0013	0.0015	0.0046	0.0112	0.052
		TOT ELEC	KWH	1.03	0.99	1.15	1.16	1.29	1.35	1.42	1.39	1.32	1.25	1.10	1.06	14.5
		TOT DEM	KW	.00466	.00466	.00472	.00513	.00543	.00578	.00578	.00590	.00578	.00555	.00490	.00472	0.00590
								BASE CASE : 440								
BASE	0.00	TOT ELEC	KWH	1.44	1.30	1.58	1.70	1.74	1.96	1.98	1.98	1.84	1.64	1.44	1.44	20.0
		TOT DEM	KW	.00654	.00638	.00646	.00717	.00717	.00756	.00749	.00788	.00756	.00733	.00686	.00654	0.00788
S- 1	0.09	TOT ELEC	KWH	1.34	1.21	1.47	1.58	1.62	1.82	1.84	1.84	1.71	1.53	1.34	1.34	18.6
		TOT DEM	KW	.00622	.00607	.00614	.00682	.00682	.00719	.00712	.00749	.00719	.00697	.00652	.00622	0.00749
S- 12	0.21	TOT ELEC	KWH	1.20	1.09	1.32	1.42	1.45	1.64	1.65	1.65	1.54	1.37	1.20	1.20	16.7
		TOT DEM	KW	.00549	.00536	.00543	.00602	.00602	.00636	.00629	.00662	.00636	.00616	.00576	.00549	0.00662
S- 22	1.52	TOT ELEC	KWH	1.09	0.99	1.20	1.29	1.32	1.49	1.50	1.50	1.40	1.25	1.09	1.09	15.2
		TOT DEM	KW	.00523	.00510	.00517	.00573	.00573	.00605	.00598	.00630	.00605	.00586	.00548	.00523	0.00630
S- 28	-0.04	TOT ELEC	KWH	1.20	1.09	1.32	1.42	1.45	1.64	1.65	1.65	1.54	1.37	1.20	1.20	16.7
		TOT DEM	KW	.00549	.00536	.00543	.00602	.00602	.00636	.00629	.00662	.00636	.00616	.00576	.00549	0.00662
S- 36	0.52	TOT ELEC	KWH	1.02	0.92	1.11	1.20	1.23	1.38	1.40	1.40	1.30	1.16	1.02	1.02	14.1
		TOT DEM	KW	.00509	.00497	.00503	.00558	.00558	.00588	.00582	.00613	.00588	.00570	.00533	.00509	0.00613

(HT ELEC=ELECTRIC CONSUMPTION FOR SPACE AND WATER HEATING. REM ELEC=REMAINING ELECTRIC CONSUMPTION (NOT USED FOR HEATING). TOT ELEC=TOTAL ELECTRIC CONSUMPTION. HT DEM=PEAK ELECTRIC DEMAND FOR SPACE AND WATER HEATING. REM DEM=PEAK ELECTRIC DEMAND FOR REMAINING LOAD (NOT HEATING). TOT DEM=TOTAL PEAK ELECTRIC DEMAND. GAS=TOTAL GAS CONSUMPTION. ANNUAL FIGURES ARE SUMS OF MONTHLY CONSUMPTION VALUES OR ARE YEARLY MAXIMUMS OF MONTHLY PEAK DEMAND VALUES.)

Hot/Arid Climate

Identify Region

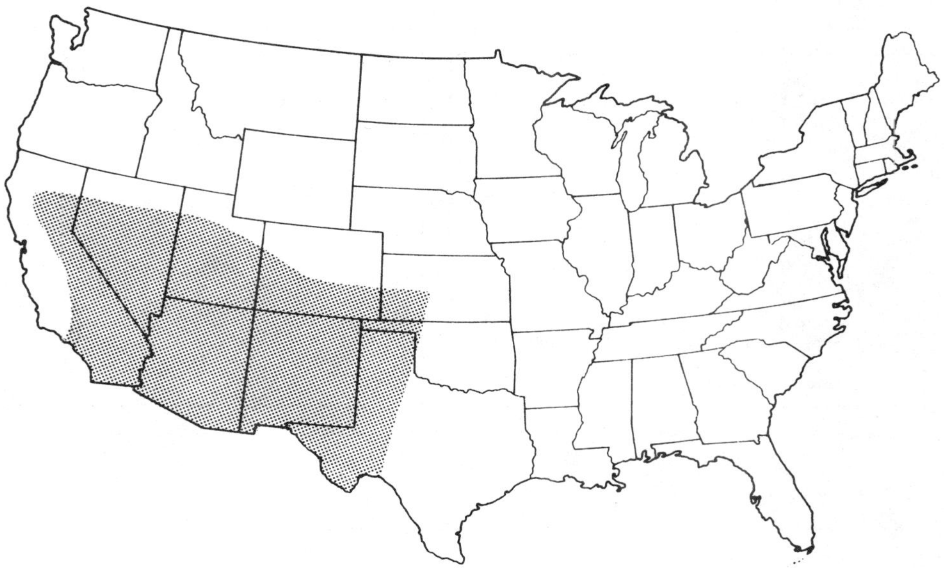

CITIES IN REGION

AZ	Flagstaff Phoenix Tucson
CA	Bakersfield Fresno Los Angeles San Bernardino San Diego
CO	Durango Grand Junction Pueblo
KA	Dodge City
NM	Albuquerque Roswell Santa Fe
NV	Las Vegas Reno
TX	Amarillo El Paso Lubbock
UT	Salt Lake City

BEGIN YOUR 5 LEVELS OF STRATEGY SELECTION.

Note: See Section II for a full discussion of the five climate regions used in this manual.

Make sure your design comes up to current good practice, which includes meeting ASHRAE 90-75 Standards. The table below identifies good practice that meets 90-75 Standards as well as common design assumptions applicable to all building sizes.

	ALL BUILDINGS	
	GOOD PRACTICE	DESIGN ASSUMPTIONS
SHELL	o Perimeter Insulation (R-5) o Infiltration: 0 when HVAC is operating (pressurized building)	o Ventilation: 14 cfm/person during occupied hours
HVAC	o Air conditioning performs to ASHRAE 90-75 (year 1980) minimum efficiencies o Boilers perform with 65% efficiency (no waste heat to space) o Heat pumps perform to ASHRAE 90-75 (year 1980) minimum efficiencies o Controls: · 72 degrees heating, 74 degrees cooling occupied periods · 55 degrees heating, 95 degrees cooling unoccupied periods · 2 degrees temperature float · outside air temperature reset for hot water boilers · low leakage dampers (less than 3%) · zone control for all heat pump and multi-zone systems	
INTERNAL LOAD		o Occupant density: 200 square feet per person o Lights: 2.25 watts/sq. ft. o Office equipment: .75 watts/sq. ft.
SCHEDULE	o HVAC comes on one hour before occupied period, turned to setback during unoccupied period	o Inside lighting: 100% during operating hours, 5% at other times o Outside lighting: 100% 6pm til midnight

Note: See Section II, page 179, for detailed information on these Level 1 strategies and assumptions.

Some aspects of ASHRAE 90-75 and current **good practice** vary according to building size and HVAC system type. Make sure your design comes up to these criteria. You may want to come back to this table as you explore HVAC options in Level 2.

BUILDING SIZE

SHELL

	SMALL 2000-10000 sq ft	MEDIUM 10000-25000 sq ft	LARGE 25000-50000 sq ft
ROOF INSULATION	R-8.3	R-8.3	R-8.3
WALL INSULATION	R-7.5	R-7.5	NONE
GLAZING	SINGLE GLAZED WITH TINTED GLASS	DOUBLE GLAZED WITH TINTED GLASS	DOUBLE GLAZED WITH TINTED GLASS
AIR CHANGES	1/2 AC/HR UNOCCUPIED	1/3 AC/HR UNOCCUPIED	1/3 AC/HR UNOCCUPIED

HVAC

	SMALL			MEDIUM			LARGE	
	PACKAGED TERMINAL AIR CONDITIONERS	ROOFTOP DX	HEAT PUMP	CONSTANT VOLUME	VARIABLE AIR VOLUME	HYDRONIC HEAT PUMP	VARIABLE AIR VOLUME	HYDRONIC HEAT PUMP
TEMPERATURE CONTROL	ZONE	OA RESET	ZONE	FIXED	ZONE	ZONE	FIXED	ZONE
AIR CONTROL	FIXED VOLUME	ENTHALPY ECONOMIZER	FIXED VOLUME	ENTHALPY ECONOMIZER	ENTHALPY ECONOMIZER	FIXED VOLUME	ENTHALPY ECONOMIZER	FIXED VOLUME

Some common **design assumptions** also vary according to building size and no. of stories.

INT. LOAD / SCHEDULE

	SMALL	MEDIUM		LARGE	
	1 STORY	1 STORY	2 STORY	2 STORY	4 STORY ①
ELEVATOR & OUTDOOR LIGHTS	1345 WATTS	4035 WATTS	4535 WATTS	7860 WATTS	9360 WATTS
OCCUPANCY	80% 8am - 6pm Monday - Friday	80% 8am - 6pm Monday - Friday		80% 8am - 6pm: M-F 100% 9am - 1pm: SAT	

GO TO THE LEVEL 2 PAGE SPECIFIC TO YOUR BUILDING SIZE & NO. OF STORIES.

Note ① For 3 or 5 story structures, use "4-story" classification.

LEVEL 2 –
MAKE INITIAL CHOICES
SMALL/1-STORY

Project ____________________
date ____________ bldg. sq. ft. ____________

The building characteristic choices shown here make a **big** difference in energy cost savings. Investigate these choices within the requirements of your design.

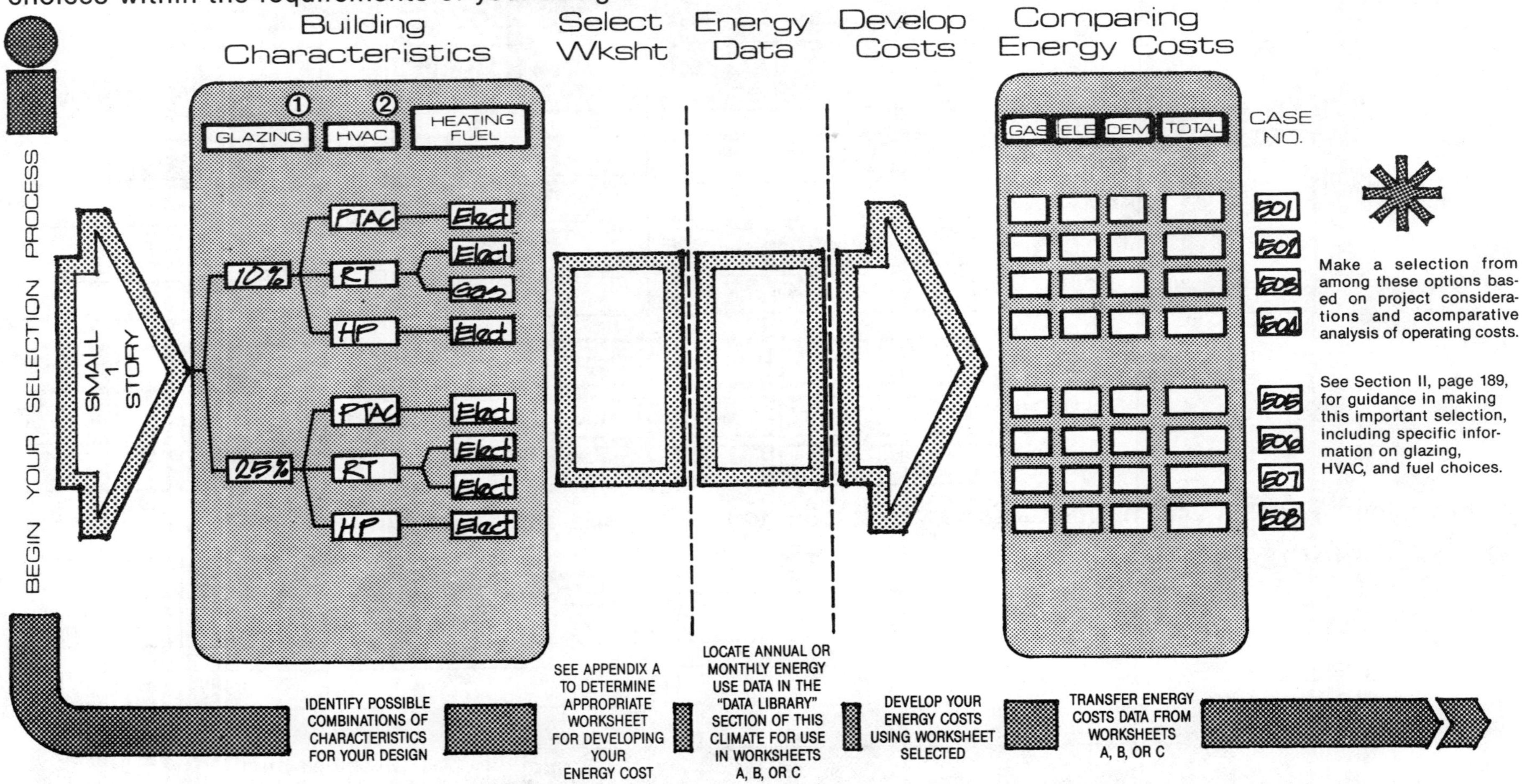

Note: See Section II, page 189, for detailed information on these Level 2 strategies and the use of this form.
① For glazing <17% use 10%, for >17% use 25%.
② PTAC=Packaged Terminal A/C, RT=Rooftop DX, HP=Packaged Terminal Heat Pump.

Project______________________________

date____________ bldg. sq. ft. ____________

Locate the row of strategy set options that apply to the building case you chose in Level 2. Select as many of these Level 3 strategy set options as you wish for valuation.

SMALL 1 STORY

STRATEGY SET OPTIONS

SELECTED LEVEL 2 CASE NUMBER	SIMPLEST LIGHTING IMPROVEMTS	MODEST LIGHTING & MECHANICAL IMPROVEMTS	MAJOR LIGHTING & MECHANICAL IMPROVEMTS	DAYLIGHT DIMMING CONTROLS AND LIGHTING IMPROVEMTS	SKYLIGHT'G, DAYLIGHT DIMMING CONTROLS & LIGHTING IMPROVEMTS
501	1	10 or 25	20		
502	3	7 or 61	16		53
503	3	7 or 61	16		53
504	1	12 or 28	22		
505	1	10 or 25	20	35	
506	3	7 or 61	16	32	53
507	3	7 or 61	16	32	53
508	1	12 or 28	22	36	

DESCRIPTOR KEY

1. High eff. ballasts over all bldg.
3. High eff. ballasts over all bldg.+HVAC downsized for ballasts.
7. High eff. ballasts and lamps over all bldg.+HVAC downsized for ballasts+cooling COP improvement.
10. High eff. ballasts and lamps over all bldg.+cooling COP improvement.
12. High eff. ballasts and lamps over all bldg.+heat pump COP improvement.
16. High eff. ballasts, lamps and fixtures over all bldg.+HVAC downsized for ballasts and fixtures+cooling COP improvement.
20. High eff. ballasts, lamps and fixtures over all bldg.+cooling COP improvement.
22. High eff. ballasts, lamps and fixtures over all bldg.+heat pump COP improvement.
25. Non-uniform lighting+high eff. ballasts over all bldg.+cooling COP improvement.
61. Non-uniform lighting+high eff. ballasts over all bldg.+HVAC downsized for ballasts+cooling COP improvement.
28. Non-uniform lighting+high eff. ballasts over all bldg.+heat pump COP improvement.
32. Perimeter daylighting dimmers+high eff. ballasts over all bldg.+HVAC downsized for ballasts+cooling COP improvement.
35. Perimeter daylighting dimmers+high eff. ballasts over all bldg.+cooling COP improvement.
36. Perimeter daylighting dimmers+high eff. ballasts over all bldg.+heat pump COP improvement.
53. High eff. ballasts and lamps over all bldg.+HVAC downsized for ballasts+evaporative cooler.

GO TO LEVEL 3 ANALYSIS SHEET.

Note: See Section II, page 209, for detailed descriptions and discussions of these and other Level 3 strategy sets.

LEVEL 2
MAKE INITIAL CHOICES
MEDIUM/1-STORY

Project ______________________________
date ____________ bldg. sq. ft. ____________

The building characteristic choices shown here make a **big** difference in energy cost savings. Investigate these choices within the requirements of your design.

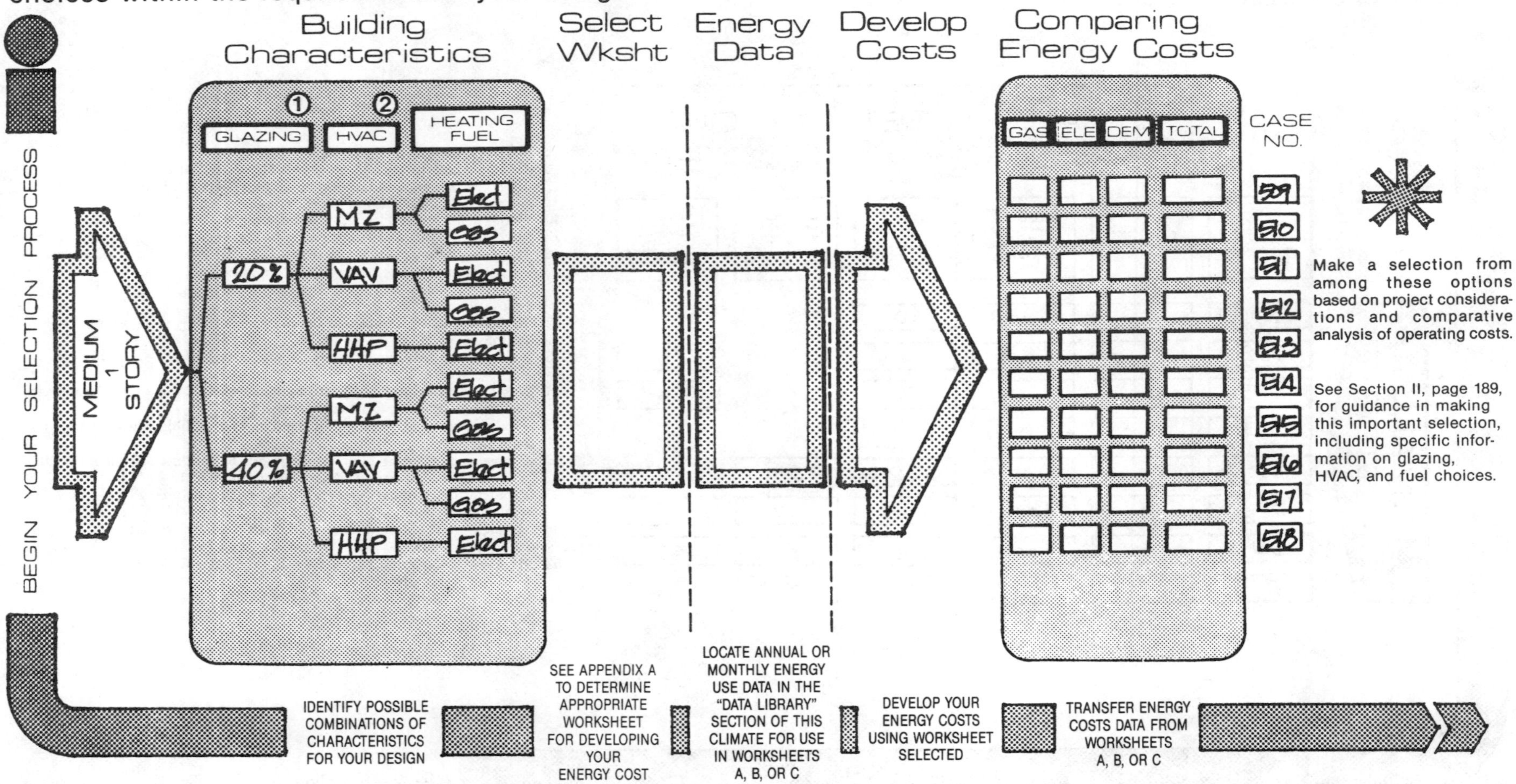

Note: See Section II, page 189, for detailed information on these Level 2 strategies and the use of this form.

① For glazing <30% use 20%, for >30% use 40%.

② MZ=Multi-zone, VAV=Variable Air Volume, HHP=Hydronic Heat Pump.

Project____________________________________
date______________ bldg. sq. ft. ______________

Locate the row of strategy set options that apply to the building case you chose in Level 2. Select as many of these Level 3 strategy set options as you wish for valuation.

STRATEGY SET OPTIONS

MEDIUM 1 STORY

SELECTED LEVEL 2 CASE NUMBER	SIMPLEST LIGHTING IMPROVEMTS	MODEST LIGHTING & MECHANICAL IMPROVEMTS	MAJOR LIGHTING & MECHANICAL IMPROVEMTS	DAYLIGHT DIMMING CONTROLS AND LIGHTING IMPROVEMTS	SKYLIGHT'G, DAYLIGHT DIMMING CONTROLS & LIGHTING IMPROVEMTS	EVAPORATIVE COOLING AND LIGHTING IMPROVEMTS
509	3	7 or 61	16	32 or 45 or 40	50	53
510	3	7 or 61	16	32 or 45 or 40	50	53
511	3	7 or 61	16	32 or 45 or 40	50	53
512	3	7 or 61	16	32 or 45 or 40	50	53
513	1	12 or 28	18	33 or 46 or 41	51	
514	3	7 or 61	16	32 or 45 or 40	50	53
515	3	7 or 61	16	32 or 45 or 40	50	53
516	3	7 or 61	16	32 or 45 or 40	50	53
517	3	7 or 61	16	32 or 45 or 40	50	53
518	1	12 or 28	18	33 or 46 or 41	51	

DESCRIPTOR KEY

1. High eff. ballasts over all bldg.
3. High eff. ballasts over all bldg. + HVAC downsized for ballasts.
7. High eff. ballasts and lamps over all bldg. + HVAC downsized for ballasts + cooling COP improvement.
12. High eff. ballasts and lamps over all bldg. + heat pump COP improvement.
16. High eff. ballasts, lamps and fixtures over all bldg. + HVAC downsized for ballasts and fixtures + cooling COP improvement.
18. High eff. ballasts, lamps and fixtures over all bldg. + HVAC downsized for ballasts and fixtures + heat pump COP improvement.
61. Non-uniform lighting + high eff. ballasts over all bldg. + HVAC downsized for ballasts + cooling COP improvement.
28. Non-uniform lighting + high eff. ballasts over all bldg. + heat pump COP improvement.
32. Perimeter daylighting dimmers + high eff. ballasts over all bldg. + HVAC downsized for ballasts + cooling COP improvement.
33. Perimeter daylighting dimmers + high eff. ballasts over all bldg. + HVAC downsized for ballasts + heat pump COP improvement.
40. Perimeter daylighting dimmers + high eff. ballasts over all bldg. + high eff. lamps and fixtures in core + HVAC downsized for ballasts and fixtures + cooling COP improvement.
41. Perimeter daylighting dimmers + high eff. ballasts over all bldg. + high eff. lamps and fixtures in core + HVAC downsized for ballasts and fixtures + heat pump COP improvement.
45. Perimeter daylighting dimmers + high eff. ballasts over all bldg. + high eff. lamps in core + HVAC downsized for ballasts + cooling COP improvement.
46. Perimeter daylighting dimmers + high eff. ballasts over all bldg. + high eff. lamps in core + HVAC downsized for ballasts + heat pump COP improvement.
50. Perimeter daylighting dimmers + skylights with dimmers + high eff. ballasts over all bldg. + HVAC downsized for ballasts + cooling COP improvement.
51. Perimeter daylighting dimmers + skylights with dimmers + high eff. ballasts over all bldg. + HVAC downsized for ballasts + heat pump COP improvement.
53. High eff. ballasts and lamps over all bldg. + HVAC downsized for ballasts + evaporative cooler.

GO TO LEVEL 3 ANALYSIS SHEET.

Note: See Section II, page 209, for detailed descriptions and discussions of these and other Level 3 strategy sets.

LEVEL 2

MAKE INITIAL CHOICES

MEDIUM/2-STORY

Project______________________________

date___________ bldg. sq. ft.___________

The building characteristic choices shown here make a **big** difference in energy cost savings. Investigate these choices within the requirements of your design.

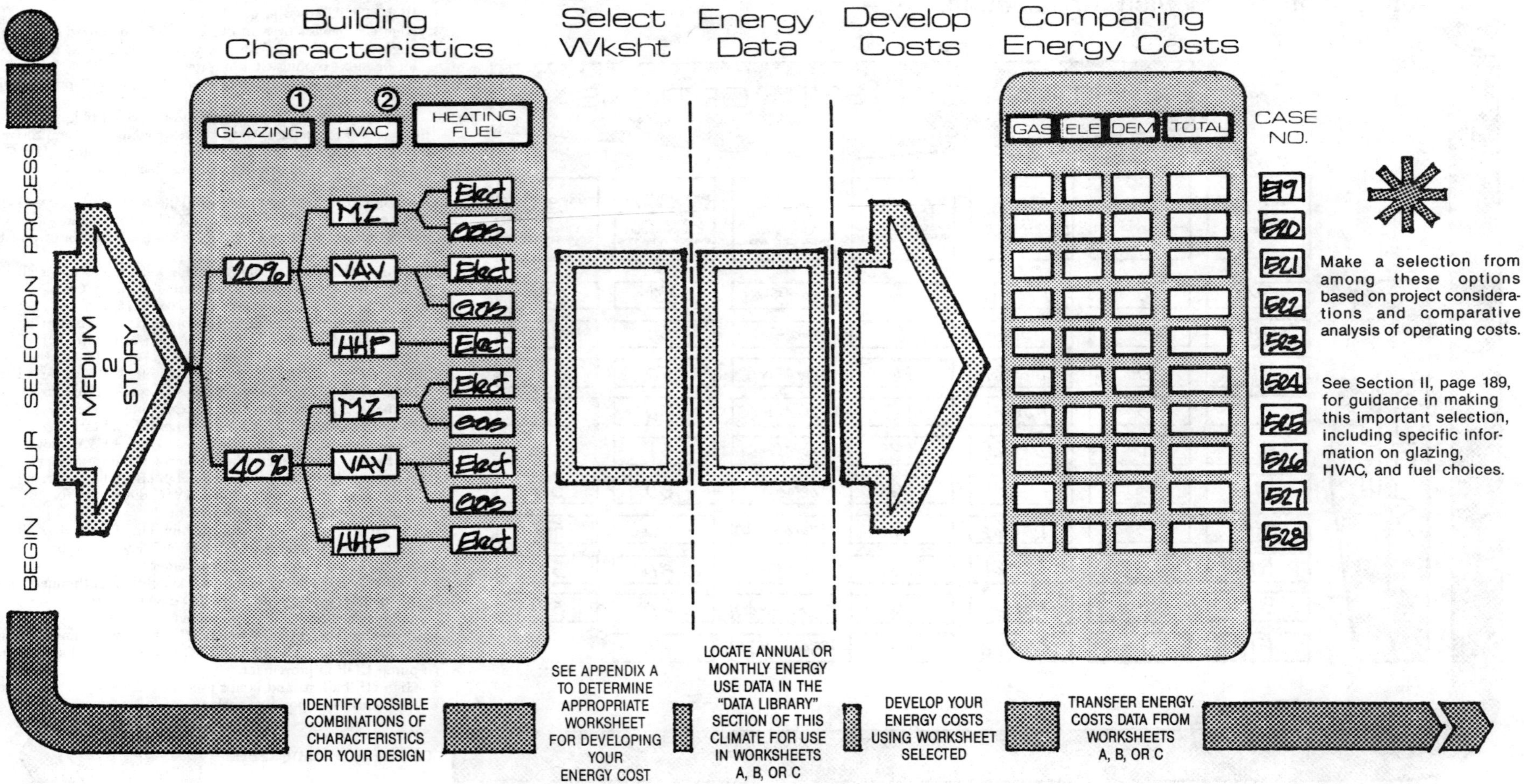

Note: See Section II, page 189, for detailed information on these Level 2 strategies and the use of this form.

① For glazing <30% use 20%, for >30% use 40%.

② MZ=Multi-zone, VAV=Variable Air Volume, HHP=Hydronic Heat Pump.

Project ____________________
date __________ bldg. sq. ft. __________

Locate the row of strategy set options that apply to the building case you chose in Level 2. Select as many of these Level 3 strategy set options as you wish for valuation.

STRATEGY SET OPTIONS

MEDIUM 2 STORY

SELECTED LEVEL 2 CASE NUMBER	SIMPLEST LIGHTING IMPROVEMTS	MODEST LIGHTING & MECHANICAL IMPROVEMTS	MAJOR LIGHTING & MECHANICAL IMPROVEMTS	DAYLIGHT DIMMING CONTROLS AND LIGHTING IMPROVEMTS
519	3	7 or 61	16	32
520	3	7 or 61	16	32
521	3	7 or 61	16	32
522	3	7 or 61	16	32
523	1	12 or 28	18	33
524	3	7 or 61	16	32
525	3	7 or 61	16	32
526	3	7 or 61	16	32
527	3	7 or 61	16	32
528	1	12 or 28	18	33

DESCRIPTOR KEY

1. High eff. ballasts over all bldg.
3. High eff. ballasts over all bldg. + HVAC downsized for ballasts.
7. High eff. ballasts and lamps over all bldg. + HVAC downsized for ballasts + cooling COP improvement.
12. High eff. ballasts and lamps over all bldg. + heat pump COP improvement.
16. High eff. ballasts, lamps and fixtures over all bldg. + HVAC downsized for ballasts and fixtures + cooling COP improvement.
18. High eff. ballasts, lamps and fixtures over all bldg. + HVAC downsized for ballasts and fixtures + heat pump COP improvement.
61. Non-uniform lighting + high eff. ballasts over all bldg. + HVAC downsized for ballasts + cooling COP improvement.
28. Non-uniform lighting + high eff. ballasts over all bldg. + heat pump COP improvement.
32. Perimeter daylighting dimmers + high eff. ballasts over all bldg. + HVAC downsized for ballasts + cooling COP improvement.
33. Perimeter daylighting dimmers + high eff. ballasts over all bldg. + HVAC downsized for ballasts + heat pump COP improvement.

GO TO LEVEL 3 ANALYSIS SHEET.

Note: See Section II, page 209, for detailed descriptions and discussions of these and other Level 3 strategy sets.

LEVEL 2
MAKE INITIAL CHOICES
LARGE/2-STORY

Project ______________________
date __________ bldg. sq. ft. __________

The building characteristic choices shown here make a **big** difference in energy cost savings. Investigate these choices within the requirements of your design.

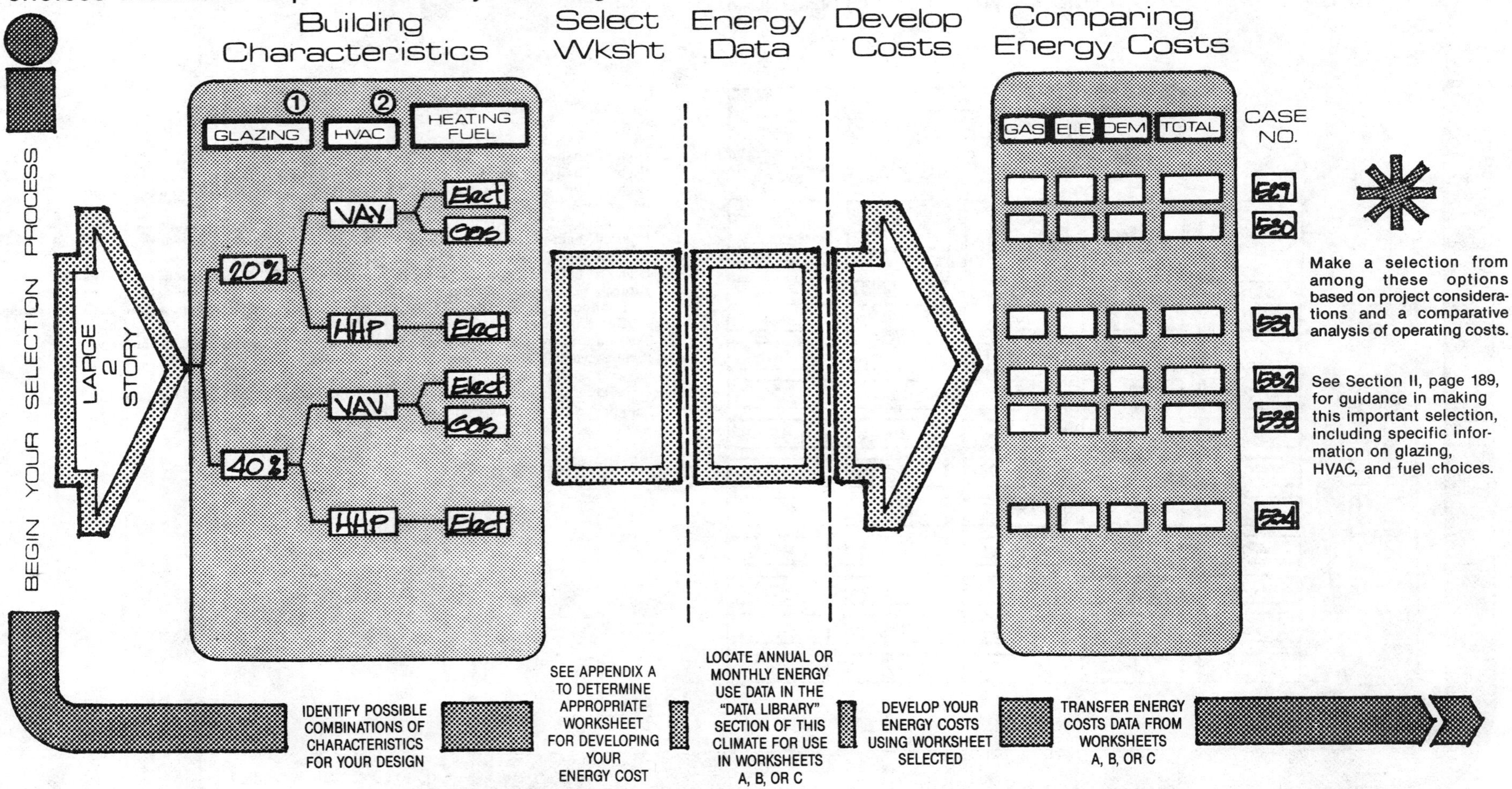

Note: See Section II, page 189, for detailed information on these Level 2 strategies and the use of this form.

① For glazing <30% use 20%, for >30% use 40%.

② VAV=Variable Air Volume, HHP=Hydronic Heat Pump.

LARGE/2-STORY

Project ______________________

date __________ bldg. sq. ft. __________

Locate the row of strategy set options that apply to the building case you chose in Level 2. Select as many of these Level 3 strategy set options as you wish for valuation.

STRATEGY SET OPTIONS

LARGE 2 STORY

SELECTED LEVEL 2 CASE NUMBER	SIMPLEST LIGHTING IMPROVEMTS	MODEST LIGHTING & MECHANICAL IMPROVEMTS	MAJOR LIGHTING & MECHANICAL IMPROVEMTS	DAYLIGHT DIMMING CONTROLS AND LIGHTING IMPROVEMTS
529	3	7 or 61	16	32 or 45 or 40
530	3	7 or 61	16	32 or 45 or 40
531	1	12 or 28	22	36 or 48 or 43
532	3	7 or 61	16	32 or 45 or 40
533	3	7 or 61	16	32 or 45 or 40
534	1	12 or 28	22	36 or 48 or 43

DESCRIPTOR KEY

1. High eff. ballasts over all bldg.
3. High eff. ballasts over all bldg. + HVAC downsized for ballasts.
7. High eff. ballasts and lamps over all bldg. + HVAC downsized for ballasts + cooling COP improvement.
12. High eff. ballasts and lamps over all bldg. + heat pump COP improvement.
16. High eff. ballasts, lamps and fixtures over all bldg. + HVAC downsized for ballasts and fixtures + cooling COP improvement.
22. High eff. ballasts, lamps and fixtures over all bldg. + heat pump COP improvement.
61. Non-uniform lighting + high eff. ballasts over all bldg. + HVAC downsized for ballasts + cooling COP improvement.
28. Non-uniform lighting + high eff. ballasts over all bldg. + heat pump COP improvement.
32. Perimeter daylighting dimmers + high eff. ballasts over all bldg. + HVAC downsized for ballasts + cooling COP improvement.
36. Perimeter daylighting dimmers + high eff. ballasts over all bldg. + heat pump COP improvement.
40. Perimeter daylighting dimmers + high eff. ballasts over all bldg. + high eff. lamps and fixtures in core + HVAC downsized for ballasts and fixtures + cooling COP improvement.
43. Perimeter daylighting dimmers + high eff. ballasts over all bldg. + high eff. lamps and fixtures in core + heat pump COP improvement.
45. Perimeter daylighting dimmers + high eff. ballasts over all bldg. + high eff. lamps in core + HVAC downsized for ballasts + cooling COP improvement.
48. Perimeter daylighting dimmers + high eff. ballasts over all bldg. + high eff. lamps in core + heat pump COP improvement.

GO TO LEVEL 3 ANALYSIS SHEET.

Note: See Section II, page 209, for detailed descriptions and discussions of these and other Level 3 strategy sets.

LARGE/4-STORY

Project______________________________

date______________ bldg. sq. ft.______________

The building characteristic choices shown here make a **big** difference in energy cost savings. Investigate these choices within the requirements of your design.

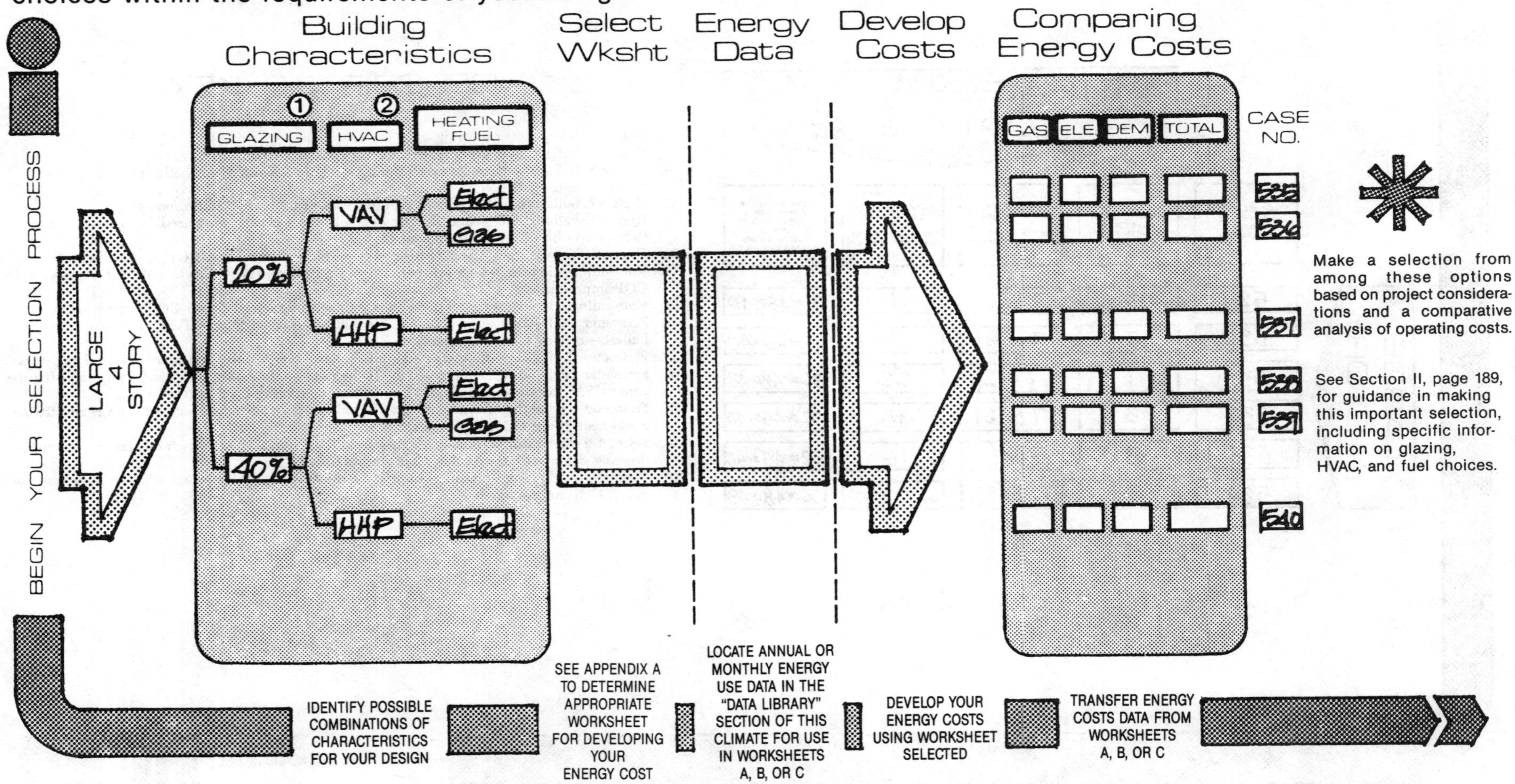

Note: See Section II, page 189, for detailed information on these Level 2 strategies and the use of this form.

① For glazing <30% use 20%, for >30% use 40%.

② VAV=Variable Air Volume, HHP=Hydronic Heat Pump.

Project ____________________
date __________ bldg. sq. ft. __________

Locate the row of strategy set options that apply to the building case you chose in Level 2. Select as many of these Level 3 strategy set options as you wish for valuation.

LARGE 4 STORY

STRATEGY SET OPTIONS

SELECTED LEVEL 2 CASE NUMBER	SIMPLEST LIGHTING IMPROVEMTS	MODEST LIGHTING & MECHANICAL IMPROVEMTS	MAJOR LIGHTING & MECHANICAL IMPROVEMTS	DAYLIGHT DIMMING CONTROLS AND LIGHTING IMPROVEMTS
535	3	7 or 61	16	32
536	3	7 or 61	16	32
537	1	12 or 28	22	36
538	3	7 or 61	16	32
539	3	7 or 61	16	32
540	1	12 or 28	22	36

DESCRIPTOR KEY

1. High eff. ballasts over all bldg.
3. High eff. ballasts over all bldg. + HVAC downsized for ballasts.
7. High eff. ballasts and lamps over all bldg. + HVAC downsized for ballasts + cooling COP improvement.
12. High eff. ballasts and lamps over all bldg. + heat pump COP improvement.
16. High eff. ballasts, lamps and fixtures over all bldg. + HVAC downsized for ballasts and fixtures + cooling COP improvement.
22. High eff. ballasts, lamps and fixtures over all bldg. + heat pump COP improvement.
61. Non-uniform lighting + high eff. ballasts over all bldg. + HVAC downsized for ballasts + cooling COP improvement.
28. Non-uniform lighting + high eff. ballasts over all bldg. + heat pump COP improvement.
32. Perimeter daylighting dimmers + high eff. ballasts over all bldg. + HVAC downsized for ballasts + cooling COP improvement.
36. Perimeter daylighting dimmers + high eff. ballasts over all bldg. + heat pump COP improvement.

GO TO LEVEL 3 ANALYSIS SHEET.

Note: See Section II, page 209, for detailed descriptions and discussions of these and other Level 3 strategy sets.

ANALYSIS/INSTRUCTIONS

For each strategy set you want to investigate, develop its construction costs using the upper route and its energy costs using the lower route. The combined information will allow you to select a strategy set for optimum economic benefit.

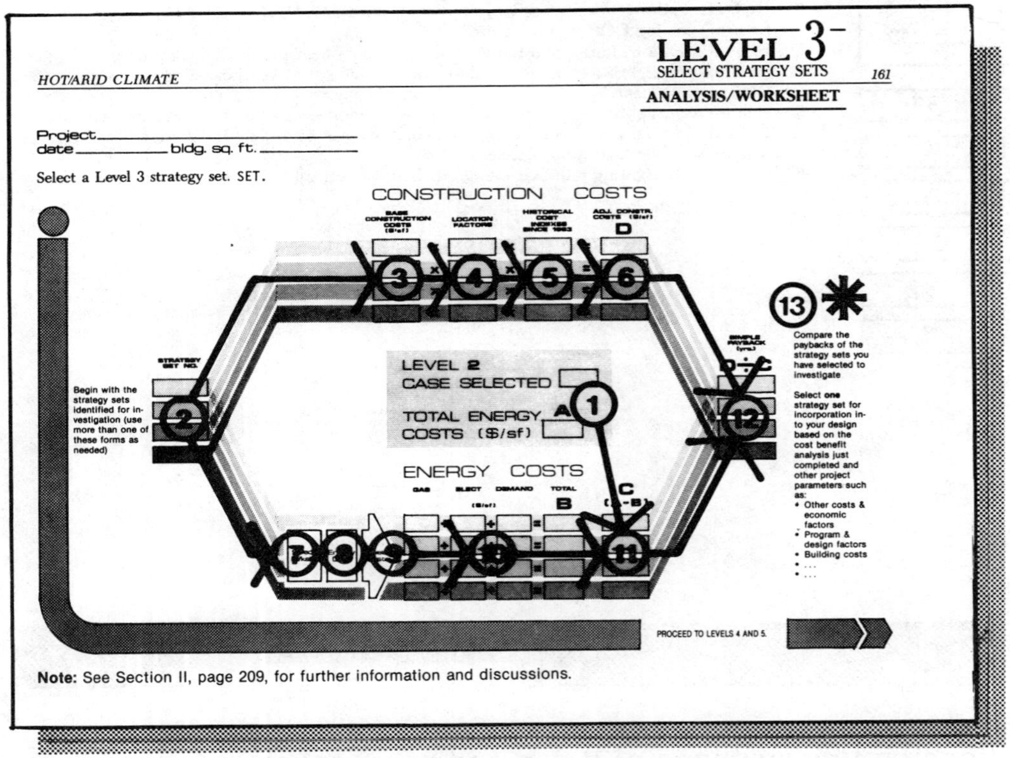

1. Enter the Level 2 case selected and its total energy costs as developed in Level 2.
2. As many as four strategy sets for any *one* Level 2 case can be analyzed on each sheet. You will need to use more sheets to analyze more than four strategy sets or to analyze strategy sets for another Level 2 case.
3. Obtain *base construction costs* ($/SF) for each strategy set from the Data Library.
4. Obtain *location factors* from commonly used cost estimating handbooks. Additional multipliers may be entered to account for above or below average construction costs due to specific project conditions.
5. Obtain *historical cost indexes* (factors since 1983) from commonly used cost estimating handbooks.
6. Multiply 3, 4, and 5 to obtain *adjusted construction costs* ($/SF). User may wish to use more refined construction cost data as supplied by a cost estimator for the specific project.
7. Use the same worksheet you selected for use in Level 2.
8. Locate energy use data for the selected strategy sets in the Data Library.
9. Develop energy costs for each strategy set using the worksheet selected.
10. Transfer energy cost information from worksheets A, B, or C.
11. Find the energy savings from each strategy set by subtracting each "B" from "A."
12. Calculate *simple payback* by dividing "D" (initial costs) by "C" (yearly savings). Alternative economic analysis techniques other than simple payback may be employed using information from columns "D" and "C."
13. Cost factors to consider in making a decision on these strategy sets include:
    - Negative construction costs (first cost savings)
    - Positive construction costs
    - Annual energy cost savings

ANALYSIS/WORKSHEET

Project ______________________
date __________ bldg. sq. ft. __________

Select a Level 3 strategy set.

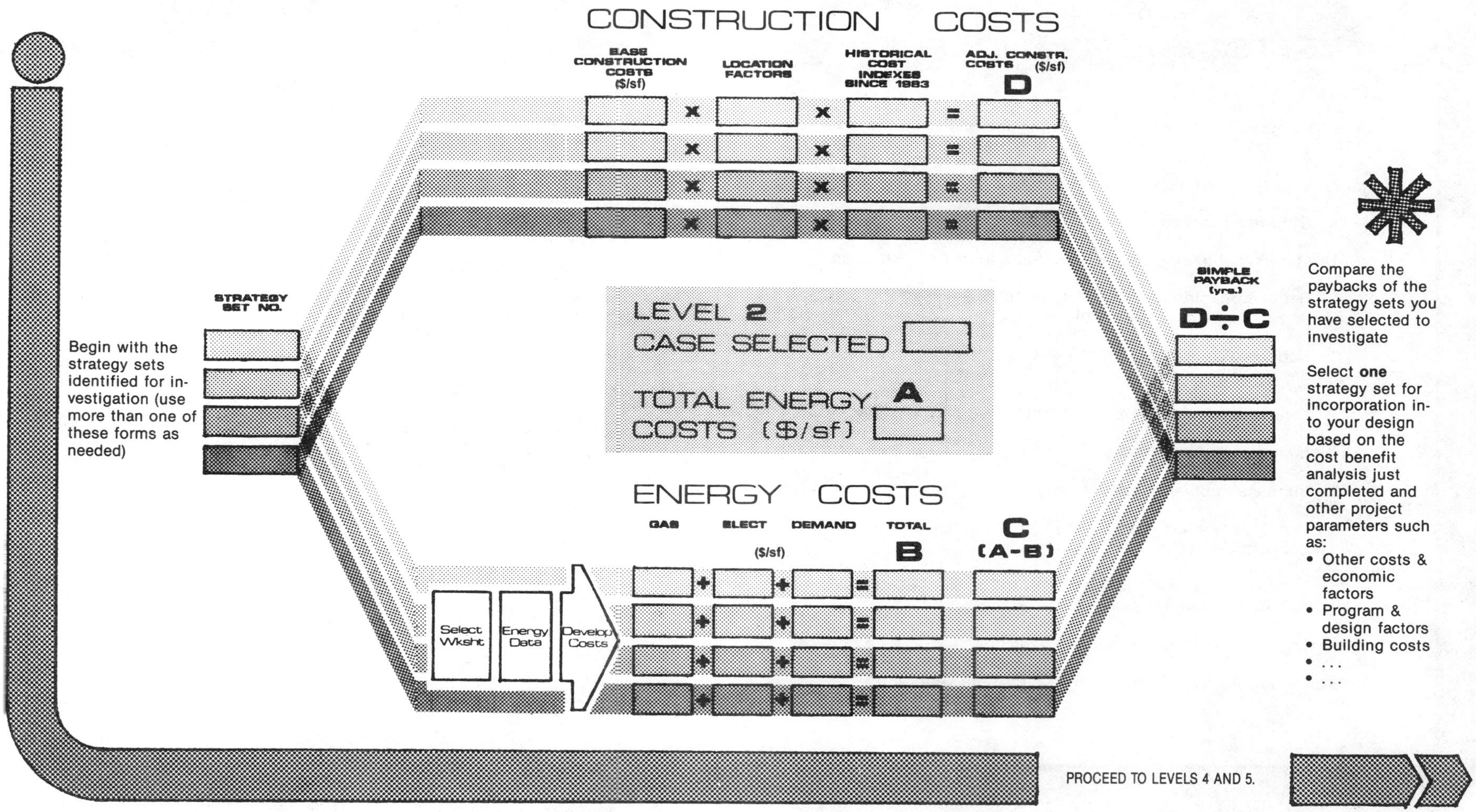

Note: See Section II, page 209, for further information and discussions.

Review these additional strategies for the possibility of further savings. Check the description in Section II, page 235, carefully for guidance on adopting any of the strategies.

A SELECT ON THE BASIS OF TECHNICAL JUDGEMENT

1. Increase thermostat deadband for occupied periods from 2°F to 8°F (70°F heating-78°F cooling)
2. Design for efficient space planning
3. Optimize building orientation
4. Mimimize building height to reduce impact of glazed area
5. Reduce shading coefficient to minimally acceptable levels (not with daylighting)
6. Reduce glazing to minimally acceptable levels (not with daylighting)
7. Landscape to minimize wind induced infiltration effect
11. Increase roof reflectivity
12. Increase manual light switching
13. Install operable windows (best for buildings without economizer cycle)
15. Install energy monitoring and control system (medium and large buildings only)
16. Locate outdoor cooling equipment in shade
18. Design for building mass with nighttime ventilation

B VALUE MAY BE DETERMINED BY SIMPLE CALCULATION

19. High efficiency motors and pumps
20. Multiple speed compressors
21. Variable or two speed circulating fan
22. Electric resistance heaters
23. Desuperheater or stand-alone heat pump for DHW
24. Point-of-use electric domestic hot water heaters
25. External shading
26. Solar domestic hot water heater
27. Chilled water/ice storage
29. Demand limiter
30. Absorption chiller—heaters
31. Cogeneration

Note: The number preceding each strategy refers to the corresponding paragraph in Section II, page 235. See Section II, page 235, for further information and discussion of each strategy.

Review this checklist and adopt the procedures that are appropriate to your project. These actions are **important** if you are to achieve and maintain the savings from the selected Level 1-4 strategies.

A CHECKLIST FOR SUCCESSFUL TURNOVER AND BUILDING OPERATION

	DESIGN	CONSTRUCTION	TURNOVER	BUILDING OPERATION
DESIGN TEAM (ARCHITECTS/ ENGINEERS)	• Document all design assumptions (1) • Estimate energy use by sub-system (2) **& For Large Buildings:** • Document all systems and control strategies (3) • Specify contents of O&M Manuals (4) • Specify HVAC balancing requirements (5) • Specify equipment protection and cleaning (6) • Specify on-site instruction (7) • Specify access doors, valves, test ports (8) • Specify piping system ID (9) • Specify price quotations on all long-term maintenance contracts (10)	• Prepare "as-built" drawings (11) • Observe construction process for compliance (12)	• Conduct "punch list" inspection (15) • Visually check for air leakage (15) • If possible, participate in on-site training session(s) (7)	
CONTRACTOR & SUB-CONTRACTOR	**For Small Buildings Only:** • Document all systems and control strategies (3) • Specify contents of O&M Manuals (4) • Specify HVAC balancing requirements (5) • Specify equipment protection and cleaning (6) • Specify on-site instruction (7) • Specify access doors, valves, test ports (8) • Specify piping system ID (9) • Specify price quotations on all long-term maintenance contracts (10)	• Prepare O&M Manual (4) • Perform final calibration of all controls (5) • Hire independent balancing agent or require of subs (5) • Protect and clean equipment (6) • Provide all access doors, valves, and test ports (8) • Place piping system ID tags (9) • Prepare "as-built" drawings (11) • Use HVAC system to "flush" construction odors (13)	• Deliver O&M Manual (4) • Conduct on-site instruction and training (7)	• Honor warranty and call back as required (16)
EQUIPMENT MANUFACT.		• Support calibration of controls (5) • Provide all special tools required (14)	• Support O&M Manual preparation (4) • Support on-site instruction and training (7)	
OWNER & BLDG. STAFF	• Check all key design assumptions (1)	• Periodically check compliance on-site (12)	• Attend on-site instruction and training sessions (7)	• Monitor energy consumption against estimates (2) • Periodically test equipment efficiency (17) • Periodically sample and test water (18) • Inspect and calibrate controls (19) • Monitor occupant complains (20)

Note: See Section II, page 255, for more information and discussion. Numbers shown in ( ) refer to the numbered paragraphs in Section II, page 255.

1. Locate data by Base Case number.
2. Locate monthly or annual data for Base Case or Strategy Set.

Use monthly data for <u>Worksheets B or C</u>

Use annual data for <u>Worksheet A</u>

BASE CASE 326

STRAT SET	CONST COST $/SF	ENERGY FORM	UNITS (PER SF)	JAN	FEB	MAR	APR	MAY	JUN	JUL	AUG	SEP	OCT	NOV	DEC	ANNUAL
BASE	0.00	HT ELEC	KWH	0.76	0.53	0.31	0.09	0.03	0.03	0.03	0.03	0.03	0.05	0.23	0.65	2.8
		HT DEM	KW	.00605	.00587	.00575	.00405	.00369	.00006	.00006	.00006	.00224	.00363	.00520	.00587	0.00605
		REM ELEC	KWH	0.97	0.87	0.97	1.06	1.23	1.47	1.70	1.63	1.34	1.10	0.93	0.97	14.3
		REM DEM	KW	.00504	.00457	.00543	.00651	.00713	.00698	.00775	.00752	.00721	.00635	.00519	.00473	0.00775
		TOT ELEC	KWH	1.74	1.40	1.28	1.15	1.26	1.50	1.73	1.66	1.37	1.15	1.15	1.63	17.1
		TOT DEM	KW	.00930	.00828	.00828	.00698	.00716	.00763	.00781	.00763	.00744	.00670	.00679	.00837	0.00930
S- 3	-0.24	HT ELEC	KWH	0.81	0.56	0.32	0.10	0.03	0.03	0.03	0.03	0.03	0.05	0.24	0.69	3.0
		HT DEM	KW	.00605	.00587	.00575	.00405	.00369	.00006	.00006	.00006	.00224	.00363	.00520	.00587	0.00605
		REM ELEC	KWH	0.88	0.79	0.88	0.96	1.12	1.34	1.55	1.48	1.22	1.00	0.85	0.88	13.0
		REM DEM	KW	.00464	.00421	.00500	.00600	.00657	.00643	.00714	.00693	.00664	.00585	.00478	.00436	0.00714
S-																

BASE CASE 327

STRAT SET	CONST COST $/SF	ENERGY FORM	UNITS (PER SF)	JAN	FEB	MAR	APR	MAY	JUN	JUL	AUG	SEP	OCT	NOV	DEC	ANNUAL
BASE	0.00	TOT GAS	THM	0.0402	0.0277	0.0161	0.0048	0.0015	0.0015	0.0015	0.0015	0.0015	0.0026	0.0118	0.0343	0.146
		TOT ELEC	KWH	0.97	0.87	0.97	1.06	1.23	1.47	1.70	1.63	1.34	1.10	0.93	0.97	14.3
		TOT DEM	KW	.00504	.00457	.00543	.00651	.00713	.00698	.00775	.00752	.00721	.00635	.00519	.00473	0.00775
S- 3	-0.24	TOT GAS	THM	0.0426	0.0295	0.0171	0.0051	0.0015	0.0015	0.0015	0.0015	0.0015	0.0028	0.0126	0.0364	0.155
		TOT ELEC	KWH	0.88	0.79	0.88	0.96	1.12	1.34	1.55	1.48	1.22	1.00	0.85	0.88	13.0
		TOT DEM	KW	.00464	.00421	.00500	.00600	.00657	.00643	.00714	.00693	.00664	.00585	.00478	.00436	0.00714
S- 7	-0.17	TOT GAS	THM	0.0451	0.0312	0.0180	0.0054	0.0016	0.0016	0.0016	0.0016	0.0016	0.0030	0.0133	0.0385	0.164
		TOT ELEC	KWH	0.79	0.71	0.79	0.86	1.00	1.19	1.38	1.32	1.09	0.89	0.75	0.79	11.6
		TOT DEM	KW	.00406	.00369	.00438	.00525	.00575	.00563	.00625	.00606	.00581	.00513	.00419	.00381	0.00625
S- 16	0.40	TOT GAS	THM	0.0476	0.0329	0.0190	0.0057	0.0017	0.0017	0.0017	0.0017	0.0017	0.0031	0.0140	0.0407	0.173
		TOT ELEC	KWH	0.69	0.62	0.69	0.75	0.87	1.04	1.20	1.15	0.95	0.78	0.66	0.69	10.1
		TOT DEM	KW	.00370	.00336	.00398	.00478	.00523	.00512	.00569	.00552	.00529	.00467	.00381	.00347	0.00569
S- 25	0.38	TOT GAS	THM	0.0451	0.0312	0.0180	0.0054	0.0016	0.0016	0.0016	0.0016	0.0016	0.0030	0.0133	0.0385	0.164
		TOT ELEC	KWH	0.79	0.71	0.79	0.86	1.00	1.19	1.38	1.32	1.09	0.89	0.75	0.79	11.6
		TOT DEM	KW	.00406	.00369	.00438	.00525	.00575	.00563	.00625	.00606	.00581	.00513	.00419	.00381	0.00625
S- 32	0.09	TOT GAS	THM	0.0433	0.0258	0.0211	0.0087	0.0041	0.0018	0.0018	0.0018	0.0032	0.0070	0.0141	0.0298	0.162
		TOT ELEC	KWH	0.68	0.62	0.75	0.81	0.87	1.07	1.14	1.13	0.96	0.81	0.67	0.68	10.2
		TOT DEM	KW	.00414	.00403	.00443	.00512	.00535	.00552	.00575	.00564	.00558	.00500	.00426	.00380	0.00575

HT ELEC = Electric consumption for space and water heating.

REM ELEC = Remaining electric consumption (not used for heating).

TOT ELEC = Total electric consumption

HT DEM = Peak electrical demand for space and water heating.

REM DEM = Peak electrical demand for remaining load (not heating).

TOT DEM = Total peak electrical demand.

GAS = Total gas consumption.

(Annual figures are sums of monthly consumption values and are yearly maximums of monthly peak demand values.)

STRAT SET	CONST COST $/SF	ENERGY FORM	UNITS (PER SF)	JAN	FEB	MAR	APR	MAY	JUN	JUL	AUG	SEP	OCT	NOV	DEC	ANNUAL
BASE CASE : 501																
BASE	0.00	TOT ELEC	KWH	1.36	1.14	1.24	1.26	1.46	1.60	1.68	1.66	1.43	1.31	1.24	1.41	16.8
		TOT DEM	KW	.00567	.00464	.00477	.00522	.00605	.00638	.00644	.00638	.00605	.00547	.00451	.00528	0.00644
S- 1	0.09	TOT ELEC	KWH	1.26	1.06	1.15	1.17	1.36	1.48	1.56	1.54	1.33	1.22	1.15	1.31	15.6
		TOT DEM	KW	.00539	.00441	.00453	.00496	.00575	.00606	.00612	.00606	.00575	.00520	.00428	.00502	0.00612
S- 10	0.22	TOT ELEC	KWH	1.13	0.95	1.03	1.04	1.21	1.32	1.39	1.38	1.18	1.08	1.03	1.17	13.9
		TOT DEM	KW	.00517	.00423	.00435	.00476	.00553	.00582	.00588	.00582	.00553	.00500	.00412	.00482	0.00588
S- 20	1.53	TOT ELEC	KWH	1.04	0.87	0.95	0.96	1.11	1.22	1.28	1.27	1.09	1.00	0.95	1.08	12.8
		TOT DEM	KW	.00486	.00397	.00408	.00447	.00519	.00546	.00552	.00546	.00519	.00469	.00386	.00453	0.00552
S- 25	-0.03	TOT ELEC	KWH	1.13	0.95	1.03	1.04	1.21	1.32	1.39	1.38	1.18	1.08	1.03	1.17	13.9
		TOT DEM	KW	.00517	.00423	.00435	.00476	.00553	.00582	.00588	.00582	.00553	.00500	.00412	.00482	0.00588
BASE CASE : 502																
BASE	0.00	TOT ELEC	KWH	1.38	1.05	1.12	1.12	1.32	1.47	1.55	1.50	1.29	1.16	1.13	1.38	15.5
		TOT DEM	KW	.00870	.00635	.00531	.00505	.00566	.00618	.00618	.00618	.00574	.00522	.00635	.00696	0.00870
S- 3	-0.02	TOT ELEC	KWH	1.29	0.99	1.04	1.04	1.23	1.38	1.45	1.41	1.20	1.09	1.06	1.29	14.5
		TOT DEM	KW	.00822	.00600	.00501	.00477	.00534	.00584	.00584	.00584	.00543	.00493	.00600	.00658	0.00822
S- 7	0.11	TOT ELEC	KWH	1.17	0.89	0.94	0.94	1.11	1.24	1.31	1.27	1.09	0.98	0.96	1.17	13.1
		TOT DEM	KW	.00793	.00579	.00484	.00460	.00515	.00563	.00563	.00563	.00523	.00476	.00579	.00634	0.00793
S- 16	1.35	TOT ELEC	KWH	1.09	0.83	0.88	0.88	1.04	1.16	1.22	1.18	1.01	0.92	0.89	1.09	12.2
		TOT DEM	KW	.00746	.00545	.00455	.00433	.00485	.00530	.00530	.00530	.00492	.00448	.00545	.00597	0.00746
S- 53	2.50	TOT ELEC	KWH	0.99	0.74	0.72	0.62	0.66	0.73	1.24	1.19	0.78	0.66	0.77	1.00	10.1
		TOT DEM	KW	.00793	.00579	.00484	.00460	.00515	.00563	.00563	.00563	.00523	.00476	.00579	.00634	0.00793
S- 61	-0.14	TOT ELEC	KWH	1.17	0.89	0.94	0.94	1.11	1.24	1.31	1.27	1.09	0.98	0.96	1.17	13.1
		TOT DEM	KW	.00793	.00579	.00484	.00460	.00515	.00563	.00563	.00563	.00523	.00476	.00579	.00634	0.00793
BASE CASE : 503																
BASE	0.00	TOT GAS	THM	0.0161	0.0083	0.0048	0.0016	0.0013	0.0012	0.0012	0.0012	0.0012	0.0019	0.0078	0.0164	0.063
		TOT ELEC	KWH	0.92	0.84	1.02	1.13	1.36	1.52	1.60	1.56	1.33	1.17	0.94	0.92	14.3
		TOT DEM	KW	.00611	.00446	.00367	.00354	.00403	.00428	.00428	.00428	.00403	.00367	.00458	.00495	0.00611
S- 3	-0.02	TOT GAS	THM	0.0173	0.0088	0.0051	0.0018	0.0014	0.0013	0.0013	0.0013	0.0013	0.0020	0.0084	0.0175	0.067
		TOT ELEC	KWH	0.84	0.78	0.94	1.04	1.25	1.40	1.48	1.44	1.23	1.08	0.87	0.84	13.2
		TOT DEM	KW	.00561	.00410	.00337	.00325	.00370	.00393	.00393	.00393	.00370	.00337	.00421	.00454	0.00561
S- 7	0.11	TOT GAS	THM	0.0188	0.0096	0.0056	0.0019	0.0015	0.0014	0.0014	0.0014	0.0014	0.0022	0.0091	0.0191	0.073
		TOT ELEC	KWH	0.75	0.69	0.83	0.92	1.11	1.24	1.31	1.28	1.09	0.96	0.77	0.75	11.7
		TOT DEM	KW	.00505	.00369	.00303	.00293	.00333	.00353	.00353	.00353	.00333	.00303	.00379	.00409	0.00505
S- 16	1.35	TOT GAS	THM	0.0199	0.0102	0.0059	0.0020	0.0016	0.0015	0.0015	0.0015	0.0015	0.0023	0.0096	0.0202	0.078
		TOT ELEC	KWH	0.68	0.63	0.76	0.85	1.02	1.13	1.20	1.17	1.00	0.88	0.71	0.68	10.7
		TOT DEM	KW	.00485	.00354	.00291	.00281	.00320	.00340	.00340	.00340	.00320	.00291	.00364	.00393	0.00485
S- 53	2.49	TOT GAS	THM	0.0204	0.0102	0.0058	0.0009	0.0009	0.0009	0.0009	0.0009	0.0009	0.0009	0.0096	0.0209	0.073
		TOT ELEC	KWH	0.61	0.55	0.61	0.61	0.64	0.71	1.22	1.15	0.77	0.64	0.59	0.62	8.7
		TOT DEM	KW	.00505	.00369	.00303	.00293	.00333	.00353	.00353	.00353	.00333	.00303	.00379	.00409	0.00505
S- 61	-0.14	TOT GAS	THM	0.0188	0.0096	0.0056	0.0019	0.0015	0.0014	0.0014	0.0014	0.0014	0.0022	0.0091	0.0191	0.073
		TOT ELEC	KWH	0.75	0.69	0.83	0.92	1.11	1.24	1.31	1.28	1.09	0.96	0.77	0.75	11.7
		TOT DEM	KW	.00505	.00369	.00303	.00293	.00333	.00353	.00353	.00353	.00333	.00303	.00379	.00409	0.00505
BASE CASE : 504																
BASE	0.00	TOT ELEC	KWH	1.34	1.12	1.22	1.24	1.44	1.57	1.65	1.63	1.40	1.29	1.22	1.39	16.5
		TOT DEM	KW	.00393	.00411	.00459	.00495	.00568	.00598	.00604	.00604	.00568	.00519	.00435	.00399	0.00604
S- 1	0.09	TOT ELEC	KWH	1.23	1.03	1.12	1.14	1.32	1.44	1.52	1.50	1.29	1.19	1.12	1.28	15.2
		TOT DEM	KW	.00373	.00390	.00436	.00471	.00540	.00568	.00574	.00574	.00540	.00494	.00413	.00379	0.00574
S- 12	0.24	TOT ELEC	KWH	1.08	0.90	0.98	1.00	1.16	1.26	1.33	1.32	1.13	1.04	0.98	1.12	13.3
		TOT DEM	KW	.00358	.00375	.00419	.00452	.00518	.00545	.00551	.00551	.00518	.00474	.00397	.00364	0.00551
S- 22	1.55	TOT ELEC	KWH	0.98	0.82	0.90	0.91	1.05	1.15	1.21	1.20	1.03	0.94	0.90	1.02	12.1
		TOT DEM	KW	.00337	.00352	.00394	.00425	.00487	.00513	.00518	.00518	.00487	.00445	.00373	.00342	0.00518
S- 28	-0.01	TOT ELEC	KWH	1.08	0.90	0.98	1.00	1.16	1.26	1.33	1.32	1.13	1.04	0.98	1.12	13.3
		TOT DEM	KW	.00358	.00375	.00419	.00452	.00518	.00545	.00551	.00551	.00518	.00474	.00397	.00364	0.00551
BASE CASE : 505																
BASE	0.00	TOT ELEC	KWH	1.49	1.25	1.36	1.38	1.60	1.75	1.84	1.82	1.56	1.44	1.36	1.55	18.4
		TOT DEM	KW	.00704	.00576	.00592	.00648	.00752	.00792	.00800	.00792	.00752	.00680	.00560	.00656	0.00800
S- 1	0.09	TOT ELEC	KWH	1.39	1.16	1.27	1.28	1.49	1.62	1.71	1.69	1.45	1.33	1.27	1.44	17.1
		TOT DEM	KW	.00669	.00547	.00562	.00616	.00714	.00752	.00760	.00752	.00714	.00646	.00532	.00623	0.00760
S- 10	0.24	TOT ELEC	KWH	1.23	1.03	1.12	1.14	1.32	1.44	1.52	1.50	1.29	1.19	1.12	1.28	15.2
		TOT DEM	KW	.00730	.00613	.00453	.00460	.00511	.00569	.00569	.00569	.00504	.00445	.00577	.00664	0.00730
S- 20	1.55	TOT ELEC	KWH	1.13	0.95	1.04	1.05	1.22	1.33	1.40	1.39	1.19	1.09	1.04	1.18	14.0
		TOT DEM	KW	.00686	.00576	.00425	.00432	.00480	.00535	.00535	.00535	.00473	.00418	.00542	.00624	0.00686
S- 25	0.01	TOT ELEC	KWH	1.23	1.03	1.12	1.14	1.32	1.44	1.52	1.50	1.29	1.19	1.12	1.28	15.2
		TOT DEM	KW	.00730	.00613	.00453	.00460	.00511	.00569	.00569	.00569	.00504	.00445	.00577	.00664	0.00730
S- 35	0.71	TOT ELEC	KWH	1.27	0.89	0.96	0.98	1.01	1.30	1.26	1.32	1.13	0.90	0.90	1.17	13.1
		TOT DEM	KW	.00700	.00588	.00434	.00441	.00490	.00546	.00546	.00546	.00483	.00427	.00553	.00637	0.00700

STRAT SET	CONST COST $/SF	ENERGY FORM	UNITS (PER SF)	JAN	FEB	MAR	APR	MAY	JUN	JUL	AUG	SEP	OCT	NOV	DEC	ANNUAL
BASE CASE : 506																
BASE	0.00	TOT ELEC	KWH	1.50	1.15	1.22	1.22	1.44	1.61	1.69	1.64	1.40	1.27	1.23	1.50	16.9
		TOT DEM	KW	.01175	.00858	.00717	.00682	.00764	.00834	.00834	.00834	.00776	.00705	.00858	.00940	0.01175
S- 3	-0.05	TOT ELEC	KWH	1.41	1.07	1.14	1.14	1.34	1.50	1.58	1.53	1.31	1.19	1.15	1.41	15.8
		TOT DEM	KW	.01110	.00810	.00677	.00644	.00721	.00788	.00788	.00788	.00733	.00666	.00810	.00888	0.01110
S- 7	0.10	TOT ELEC	KWH	1.28	0.98	1.04	1.04	1.22	1.37	1.44	1.40	1.20	1.08	1.05	1.28	14.4
		TOT DEM	KW	.01070	.00856	.00631	.00471	.00492	.00556	.00546	.00556	.00492	.00449	.00877	.00910	0.01070
S- 16	1.30	TOT ELEC	KWH	1.18	0.90	0.96	0.96	1.13	1.26	1.33	1.29	1.10	1.00	0.97	1.18	13.3
		TOT DEM	KW	.01007	.00806	.00594	.00443	.00463	.00524	.00514	.00524	.00463	.00423	.00826	.00856	0.01007
S- 32	0.57	TOT ELEC	KWH	1.20	0.84	0.91	0.93	0.95	1.23	1.19	1.25	1.07	0.86	0.86	1.10	12.4
		TOT DEM	KW	.01080	.00864	.00637	.00475	.00497	.00562	.00551	.00562	.00497	.00454	.00886	.00918	0.01080
S- 53	3.13	TOT ELEC	KWH	1.09	0.82	0.79	0.68	0.72	0.81	1.36	1.30	0.86	0.72	0.85	1.10	11.1
		TOT DEM	KW	.01020	.00816	.00602	.00449	.00469	.00530	.00520	.00530	.00469	.00428	.00836	.00867	0.01020
S- 61	-0.15	TOT ELEC	KWH	1.28	0.98	1.04	1.04	1.22	1.37	1.44	1.40	1.20	1.08	1.05	1.28	14.4
		TOT DEM	KW	.01070	.00856	.00631	.00471	.00492	.00556	.00546	.00556	.00492	.00449	.00877	.00910	0.01070
BASE CASE : 507																
BASE	0.00	TOT GAS	THM	0.0179	0.0092	0.0053	0.0018	0.0015	0.0013	0.0013	0.0013	0.0013	0.0021	0.0087	0.0182	0.070
		TOT ELEC	KWH	1.00	0.92	1.11	1.23	1.48	1.65	1.75	1.70	1.45	1.28	1.03	1.00	15.6
		TOT DEM	KW	.00825	.00602	.00495	.00478	.00545	.00578	.00578	.00578	.00545	.00495	.00619	.00668	0.00825
S- 3	-0.05	TOT GAS	THM	0.0192	0.0098	0.0057	0.0019	0.0016	0.0014	0.0014	0.0014	0.0014	0.0022	0.0093	0.0195	0.075
		TOT ELEC	KWH	0.92	0.85	1.02	1.14	1.37	1.53	1.61	1.57	1.34	1.18	0.95	0.92	14.4
		TOT DEM	KW	.00757	.00553	.00454	.00439	.00500	.00530	.00530	.00530	.00500	.00454	.00568	.00613	0.00757
S- 7	0.09	TOT GAS	THM	0.0208	0.0107	0.0062	0.0021	0.0017	0.0015	0.0015	0.0015	0.0015	0.0024	0.0101	0.0211	0.081
		TOT ELEC	KWH	0.82	0.76	0.91	1.01	1.22	1.36	1.43	1.40	1.19	1.05	0.84	0.82	12.8
		TOT DEM	KW	.00681	.00497	.00409	.00395	.00449	.00477	.00477	.00477	.00449	.00409	.00511	.00552	0.00681
S- 16	1.30	TOT GAS	THM	0.0221	0.0113	0.0066	0.0022	0.0018	0.0016	0.0016	0.0016	0.0016	0.0026	0.0107	0.0224	0.086
		TOT ELEC	KWH	0.75	0.69	0.83	0.92	1.11	1.24	1.31	1.28	1.09	0.96	0.77	0.75	11.7
		TOT DEM	KW	.00654	.00477	.00392	.00379	.00432	.00458	.00458	.00458	.00432	.00392	.00490	.00530	0.00654
S- 32	0.57	TOT GAS	THM	0.0346	0.0122	0.0103	0.0009	0.0005	0.0000	0.0000	0.0000	0.0000	0.0017	0.0094	0.0265	0.096
		TOT ELEC	KWH	1.03	0.72	0.77	0.80	0.82	1.05	1.02	1.07	0.91	0.73	0.73	0.94	10.6
		TOT DEM	KW	.00550	.00402	.00330	.00319	.00363	.00385	.00385	.00385	.00363	.00330	.00413	.00446	0.00550
S- 53	3.13	TOT GAS	THM	0.0226	0.0113	0.0064	0.0010	0.0010	0.0010	0.0010	0.0010	0.0010	0.0010	0.0107	0.0232	0.081
		TOT ELEC	KWH	0.67	0.60	0.67	0.67	0.69	0.78	1.33	1.26	0.84	0.69	0.65	0.68	9.5
		TOT DEM	KW	.00623	.00455	.00374	.00361	.00411	.00436	.00436	.00436	.00411	.00374	.00467	.00505	0.00623
S- 61	-0.16	TOT GAS	THM	0.0208	0.0107	0.0062	0.0021	0.0017	0.0015	0.0015	0.0015	0.0015	0.0024	0.0101	0.0211	0.081
		TOT ELEC	KWH	0.82	0.76	0.91	1.01	1.22	1.36	1.43	1.40	1.19	1.05	0.84	0.82	12.8
		TOT DEM	KW	.00681	.00497	.00409	.00395	.00449	.00477	.00477	.00477	.00449	.00409	.00511	.00552	0.00681
BASE CASE : 508																
BASE	0.00	TOT ELEC	KWH	1.46	1.22	1.33	1.35	1.57	1.71	1.80	1.78	1.53	1.40	1.33	1.51	18.0
		TOT DEM	KW	.00487	.00510	.00570	.00615	.00705	.00743	.00750	.00750	.00705	.00645	.00540	.00495	0.00750
S- 1	0.09	TOT ELEC	KWH	1.34	1.13	1.23	1.25	1.44	1.58	1.66	1.64	1.41	1.29	1.23	1.39	16.6
		TOT DEM	KW	.00463	.00485	.00542	.00585	.00670	.00706	.00713	.00713	.00670	.00613	.00513	.00471	0.00713
S- 12	0.26	TOT ELEC	KWH	1.17	0.99	1.07	1.09	1.26	1.38	1.45	1.44	1.23	1.13	1.07	1.22	14.5
		TOT DEM	KW	.00568	.00472	.00513	.00561	.00609	.00684	.00677	.00684	.00609	.00540	.00506	.00561	0.00684
S- 22	1.58	TOT ELEC	KWH	1.07	0.90	0.98	0.99	1.15	1.25	1.32	1.31	1.12	1.03	0.98	1.11	13.2
		TOT DEM	KW	.00534	.00444	.00482	.00527	.00572	.00643	.00637	.00643	.00572	.00508	.00476	.00527	0.00643
S- 28	0.02	TOT ELEC	KWH	1.17	0.99	1.07	1.09	1.26	1.38	1.45	1.44	1.23	1.13	1.07	1.22	14.5
		TOT DEM	KW	.00568	.00472	.00513	.00561	.00609	.00684	.00677	.00684	.00609	.00540	.00506	.00561	0.00684
S- 36	0.74	TOT ELEC	KWH	1.13	0.79	0.85	0.87	0.89	1.15	1.11	1.17	1.00	0.80	0.80	1.03	11.6
		TOT DEM	KW	.00436	.00362	.00394	.00431	.00467	.00525	.00520	.00525	.00467	.00415	.00389	.00431	0.00525
BASE CASE : 509																
BASE	0.00	HT ELEC	KWH	0.46	0.24	0.14	0.05	0.04	0.03	0.03	0.03	0.03	0.05	0.22	0.47	1.8
		HT DEM	KW	.00864	.00855	.00449	.00207	.00207	.00028	.00032	.00032	.00138	.00207	.00786	.00838	0.00864
		REM ELEC	KWH	0.98	0.90	1.09	1.21	1.45	1.62	1.71	1.67	1.42	1.25	1.01	0.98	15.3
		REM DEM	KW	.00403	.00416	.00534	.00583	.00632	.00687	.00694	.00694	.00645	.00583	.00541	.00389	0.00694
		TOT ELEC	KWH	1.44	1.14	1.22	1.26	1.49	1.66	1.75	1.70	1.46	1.31	1.23	1.45	17.1
		TOT DEM	KW	.01225	.00955	.00662	.00662	.00710	.00759	.00772	.00759	.00710	.00674	.00882	.01078	0.01225
S- 3	-0.07	HT ELEC	KWH	0.47	0.24	0.14	0.05	0.04	0.03	0.03	0.03	0.03	0.05	0.23	0.48	1.8
		HT DEM	KW	.00864	.00855	.00449	.00207	.00207	.00028	.00032	.00032	.00138	.00207	.00786	.00838	0.00864
		REM ELEC	KWH	0.91	0.84	1.01	1.12	1.35	1.51	1.59	1.55	1.32	1.16	0.94	0.91	14.2
		REM DEM	KW	.00369	.00382	.00490	.00534	.00579	.00630	.00636	.00636	.00591	.00534	.00496	.00356	0.00636
		TOT ELEC	KWH	1.38	1.08	1.15	1.17	1.39	1.54	1.63	1.58	1.36	1.22	1.16	1.38	16.0
		TOT DEM	KW	.01174	.00916	.00634	.00634	.00681	.00728	.00740	.00728	.00681	.00646	.00845	.01033	0.01174
S- 7	0.01	HT ELEC	KWH	0.49	0.25	0.14	0.05	0.04	0.04	0.04	0.04	0.04	0.06	0.24	0.49	1.9
		HT DEM	KW	.00864	.00864	.00518	.00199	.00207	.00016	.00018	.00016	.00121	.00225	.00838	.00855	0.00864
		REM ELEC	KWH	0.83	0.76	0.92	1.02	1.23	1.37	1.44	1.41	1.20	1.06	0.85	0.83	12.9
		REM DEM	KW	.00333	.00344	.00442	.00482	.00522	.00568	.00574	.00574	.00534	.00482	.00448	.00321	0.00574
		TOT ELEC	KWH	1.31	1.01	1.06	1.07	1.27	1.40	1.48	1.44	1.24	1.11	1.09	1.32	14.8
		TOT DEM	KW	.01129	.00960	.00587	.00497	.00542	.00587	.00587	.00587	.00553	.00508	.00881	.01016	0.01129

(HT ELEC=ELECTRIC CONSUMPTION FOR SPACE AND WATER HEATING. REM ELEC=REMAINING ELECTRIC CONSUMPTION (NOT USED FOR HEATING). TOT ELEC=TOTAL ELECTRIC CONSUMPTION. HT DEM=PEAK ELECTRIC DEMAND FOR SPACE AND WATER HEATING. REM DEM=PEAK ELECTRIC DEMAND FOR REMAINING LOAD (NOT HEATING). TOT DEM=TOTAL PEAK ELECTRIC DEMAND. GAS=TOTAL GAS CONSUMPTION. ANNUAL FIGURES ARE SUMS OF MONTHLY CONSUMPTION VALUES OR ARE YEARLY MAXIMUMS OF MONTHLY PEAK DEMAND VALUES.)

STRAT SET	CONST COST $/SF	ENERGY FORM	UNITS (PER SF)	JAN	FEB	MAR	APR	MAY	JUN	JUL	AUG	SEP	OCT	NOV	DEC	ANNUAL
				BASE CASE : 509 (CONTINUED)												
S- 16	1.20	HT ELEC	KWH	0.49	0.25	0.15	0.05	0.04	0.04	0.04	0.04	0.04	0.06	0.24	0.50	1.9
		HT DEM	KW	.00864	.00864	.00518	.00199	.00207	.00016	.00018	.00016	.00121	.00225	.00838	.00855	0.00864
		REM ELEC	KWH	0.76	0.70	0.84	0.93	1.12	1.25	1.32	1.29	1.10	0.97	0.78	0.76	11.8
		REM DEM	KW	.00319	.00330	.00423	.00462	.00501	.00545	.00550	.00550	.00512	.00462	.00429	.00308	0.00550
		TOT ELEC	KWH	1.25	0.95	0.98	0.98	1.16	1.29	1.36	1.32	1.13	1.03	1.02	1.25	13.7
		TOT DEM	KW	.01067	.00907	.00555	.00469	.00512	.00555	.00555	.00555	.00523	.00480	.00832	.00960	0.01067
S- 32	0.28	HT ELEC	KWH	0.41	0.22	0.19	0.08	0.06	0.04	0.04	0.04	0.04	0.08	0.19	0.33	1.7
		HT DEM	KW	.00864	.00864	.00518	.00199	.00207	.00016	.00018	.00016	.00121	.00225	.00838	.00855	0.00864
		REM ELEC	KWH	0.71	0.65	0.85	1.00	1.07	1.39	1.36	1.40	1.19	0.92	0.74	0.72	12.0
		REM DEM	KW	.00311	.00322	.00413	.00451	.00489	.00532	.00537	.00537	.00499	.00451	.00419	.00301	0.00537
		TOT ELEC	KWH	1.12	0.87	1.04	1.08	1.13	1.43	1.39	1.44	1.23	1.01	0.93	1.05	13.7
		TOT DEM	KW	.01157	.00983	.00602	.00509	.00555	.00602	.00602	.00602	.00567	.00521	.00902	.01041	0.01157
S- 40	0.98	HT ELEC	KWH	0.42	0.23	0.20	0.09	0.06	0.04	0.04	0.04	0.05	0.09	0.19	0.34	1.8
		HT DEM	KW	.00864	.00864	.00518	.00199	.00207	.00016	.00018	.00016	.00121	.00225	.00838	.00855	0.00864
		REM ELEC	KWH	0.64	0.58	0.77	0.90	0.96	1.25	1.22	1.26	1.07	0.83	0.67	0.65	10.8
		REM DEM	KW	.00288	.00298	.00383	.00417	.00452	.00492	.00497	.00497	.00462	.00417	.00388	.00278	0.00497
		TOT ELEC	KWH	1.06	0.81	0.96	0.98	1.02	1.29	1.26	1.30	1.12	0.92	0.86	0.99	12.6
		TOT DEM	KW	.01116	.00949	.00580	.00491	.00536	.00580	.00580	.00580	.00547	.00502	.00870	.01004	0.01116
S- 45	0.32	HT ELEC	KWH	0.41	0.22	0.19	0.08	0.06	0.04	0.04	0.04	0.04	0.08	0.19	0.33	1.7
		HT DEM	KW	.00864	.00864	.00518	.00199	.00207	.00016	.00018	.00016	.00121	.00225	.00838	.00855	0.00864
		REM ELEC	KWH	0.68	0.62	0.82	0.95	1.02	1.33	1.30	1.35	1.14	0.89	0.71	0.69	11.5
		REM DEM	KW	.00304	.00314	.00403	.00440	.00477	.00519	.00524	.00524	.00487	.00440	.00409	.00293	0.00524
		TOT ELEC	KWH	1.09	0.85	1.01	1.04	1.08	1.37	1.34	1.38	1.18	0.97	0.90	1.02	13.2
		TOT DEM	KW	.01136	.00966	.00591	.00500	.00545	.00591	.00591	.00591	.00557	.00511	.00886	.01022	0.01136
S- 50	1.59	HT ELEC	KWH	0.40	0.22	0.19	0.08	0.06	0.04	0.04	0.04	0.04	0.08	0.19	0.33	1.7
		HT DEM	KW	.00864	.00864	.00518	.00199	.00207	.00016	.00018	.00016	.00121	.00225	.00838	.00855	0.00864
		REM ELEC	KWH	0.65	0.59	0.78	0.91	0.98	1.28	1.24	1.29	1.09	0.85	0.68	0.66	11.0
		REM DEM	KW	.00316	.00326	.00419	.00457	.00495	.00539	.00544	.00544	.00506	.00457	.00424	.00305	0.00544
		TOT ELEC	KWH	1.05	0.82	0.97	1.00	1.04	1.32	1.28	1.32	1.13	0.93	0.87	0.99	12.7
		TOT DEM	KW	.01122	.00954	.00583	.00494	.00539	.00583	.00583	.00583	.00550	.00505	.00875	.01010	0.01122
S- 53	1.51	HT ELEC	KWH	0.56	0.28	0.16	0.02	0.02	0.02	0.02	0.02	0.02	0.02	0.26	0.57	2.0
		HT DEM	KW	.00864	.00864	.00518	.00199	.00207	.00016	.00018	.00016	.00121	.00225	.00838	.00855	0.00864
		REM ELEC	KWH	0.67	0.61	0.67	0.67	0.70	0.79	1.35	1.27	0.85	0.70	0.65	0.68	9.6
		REM DEM	KW	.00367	.00380	.00487	.00532	.00576	.00627	.00633	.00633	.00589	.00532	.00494	.00354	0.00633
		TOT ELEC	KWH	1.23	0.89	0.83	0.70	0.73	0.81	1.37	1.30	0.87	0.73	0.92	1.26	11.6
		TOT DEM	KW	.01188	.01010	.00618	.00523	.00570	.00618	.00618	.00618	.00582	.00535	.00927	.01069	0.01188
S- 61	-0.24	HT ELEC	KWH	0.49	0.25	0.14	0.05	0.04	0.04	0.04	0.04	0.04	0.06	0.24	0.49	1.9
		HT DEM	KW	.00864	.00864	.00518	.00199	.00207	.00016	.00018	.00016	.00121	.00225	.00838	.00855	0.00864
		REM ELEC	KWH	0.83	0.76	0.92	1.02	1.23	1.37	1.44	1.41	1.20	1.06	0.85	0.83	12.9
		REM DEM	KW	.00333	.00344	.00442	.00482	.00522	.00568	.00574	.00574	.00534	.00482	.00448	.00321	0.00574
		TOT ELEC	KWH	1.31	1.01	1.06	1.07	1.27	1.40	1.48	1.44	1.24	1.11	1.09	1.32	14.8
		TOT DEM	KW	.01129	.00960	.00587	.00497	.00542	.00587	.00587	.00587	.00553	.00508	.00881	.01016	0.01129
				BASE CASE : 510												
BASE	0.00	TOT GAS	THM	0.0242	0.0124	0.0072	0.0025	0.0020	0.0018	0.0018	0.0018	0.0018	0.0028	0.0117	0.0246	0.095
		TOT ELEC	KWH	0.98	0.90	1.09	1.21	1.45	1.62	1.71	1.67	1.42	1.25	1.01	0.98	15.3
		TOT DEM	KW	.00403	.00416	.00534	.00583	.00632	.00687	.00694	.00694	.00645	.00583	.00541	.00389	0.00694
S- 3	-0.08	TOT GAS	THM	0.0247	0.0126	0.0073	0.0025	0.0020	0.0018	0.0018	0.0018	0.0018	0.0029	0.0119	0.0250	0.096
		TOT ELEC	KWH	0.91	0.84	1.01	1.12	1.35	1.51	1.59	1.55	1.32	1.16	0.94	0.91	14.2
		TOT DEM	KW	.00369	.00382	.00490	.00534	.00579	.00630	.00636	.00636	.00591	.00534	.00496	.00356	0.00636
S- 7	0.01	TOT GAS	THM	0.0255	0.0131	0.0076	0.0026	0.0021	0.0019	0.0019	0.0019	0.0019	0.0030	0.0124	0.0259	0.100
		TOT ELEC	KWH	0.83	0.76	0.92	1.02	1.23	1.37	1.44	1.41	1.20	1.06	0.85	0.83	12.9
		TOT DEM	KW	.00333	.00344	.00442	.00482	.00522	.00568	.00574	.00574	.00534	.00482	.00448	.00321	0.00574
S- 16	1.19	TOT GAS	THM	0.0258	0.0132	0.0076	0.0026	0.0021	0.0019	0.0019	0.0019	0.0019	0.0030	0.0125	0.0262	0.101
		TOT ELEC	KWH	0.76	0.70	0.84	0.93	1.12	1.25	1.32	1.29	1.10	0.97	0.78	0.76	11.8
		TOT DEM	KW	.00319	.00330	.00423	.00462	.00501	.00545	.00550	.00550	.00512	.00462	.00429	.00308	0.00550
S- 32	0.28	TOT GAS	THM	0.0214	0.0118	0.0101	0.0043	0.0030	0.0021	0.0020	0.0020	0.0024	0.0043	0.0099	0.0174	0.091
		TOT ELEC	KWH	0.71	0.65	0.85	1.00	1.07	1.39	1.36	1.40	1.19	0.92	0.74	0.72	12.0
		TOT DEM	KW	.00311	.00322	.00413	.00451	.00489	.00532	.00537	.00537	.00499	.00451	.00419	.00301	0.00537
S- 40	0.98	TOT GAS	THM	0.0220	0.0121	0.0104	0.0045	0.0031	0.0021	0.0021	0.0021	0.0024	0.0045	0.0102	0.0179	0.093
		TOT ELEC	KWH	0.64	0.58	0.77	0.90	0.96	1.25	1.22	1.26	1.07	0.83	0.67	0.65	10.8
		TOT DEM	KW	.00288	.00298	.00383	.00417	.00452	.00492	.00497	.00497	.00462	.00417	.00388	.00278	0.00497
S- 45	0.32	TOT GAS	THM	0.0214	0.0118	0.0101	0.0043	0.0030	0.0021	0.0020	0.0020	0.0024	0.0043	0.0099	0.0174	0.091
		TOT ELEC	KWH	0.68	0.62	0.82	0.95	1.02	1.33	1.30	1.35	1.14	0.89	0.71	0.69	11.5
		TOT DEM	KW	.00304	.00314	.00403	.00440	.00477	.00519	.00524	.00524	.00487	.00440	.00409	.00293	0.00524
S- 50	1.59	TOT GAS	THM	0.0211	0.0116	0.0099	0.0043	0.0029	0.0021	0.0020	0.0020	0.0023	0.0043	0.0097	0.0171	0.089
		TOT ELEC	KWH	0.65	0.59	0.78	0.91	0.98	1.28	1.24	1.29	1.09	0.85	0.68	0.66	11.0
		TOT DEM	KW	.00316	.00326	.00419	.00457	.00495	.00539	.00544	.00544	.00506	.00457	.00424	.00305	0.00544
S- 53	1.51	TOT GAS	THM	0.0293	0.0147	0.0083	0.0013	0.0013	0.0013	0.0013	0.0013	0.0013	0.0013	0.0138	0.0300	0.105
		TOT ELEC	KWH	0.67	0.61	0.67	0.67	0.70	0.79	1.35	1.27	0.85	0.70	0.65	0.68	9.6
		TOT DEM	KW	.00367	.00380	.00487	.00532	.00576	.00627	.00633	.00633	.00589	.00532	.00494	.00354	0.00633
S- 61	-0.24	TOT GAS	THM	0.0255	0.0131	0.0076	0.0026	0.0021	0.0019	0.0019	0.0019	0.0019	0.0030	0.0124	0.0259	0.100
		TOT ELEC	KWH	0.83	0.76	0.92	1.02	1.23	1.37	1.44	1.41	1.20	1.06	0.85	0.83	12.9
		TOT DEM	KW	.00333	.00344	.00442	.00482	.00522	.00568	.00574	.00574	.00534	.00482	.00448	.00321	0.00574
				BASE CASE : 511												
BASE	0.00	HT ELEC	KWH	0.36	0.18	0.11	0.04	0.03	0.03	0.03	0.03	0.03	0.04	0.17	0.36	1.4
		HT DEM	KW	.00385	.00350	.00266	.00189	.00139	.00000	.00000	.00000	.00012	.00235	.00385	.00385	0.00385
		REM ELEC	KWH	0.93	0.86	1.04	1.15	1.39	1.55	1.64	1.59	1.36	1.20	0.96	0.93	14.6
		REM DEM	KW	.00411	.00430	.00492	.00542	.00573	.00617	.00623	.00611	.00586	.00542	.00480	.00436	0.00623
		TOT ELEC	KWH	1.29	1.04	1.14	1.19	1.42	1.57	1.66	1.62	1.38	1.24	1.14	1.30	16.0
		TOT DEM	KW	.00700	.00483	.00518	.00567	.00609	.00623	.00637	.00623	.00609	.00574	.00490	.00567	0.00700
				BASE CASE : 511 (CONTINUED)												
S- 3	-0.15	HT ELEC	KWH	0.38	0.20	0.11	0.04	0.03	0.03	0.03	0.03	0.03	0.04	0.18	0.39	1.5
		HT DEM	KW	.00385	.00350	.00266	.00189	.00139	.00000	.00000	.00000	.00012	.00235	.00385	.00385	0.00385
		REM ELEC	KWH	0.81	0.75	0.90	1.00	1.21	1.35	1.42	1.38	1.18	1.04	0.84	0.81	12.7
		REM DEM	KW	.00389	.00407	.00466	.00513	.00543	.00584	.00590	.00578	.00555	.00513	.00454	.00413	0.00590
		TOT ELEC	KWH	1.19	0.94	1.01	1.04	1.24	1.37	1.45	1.41	1.21	1.09	1.02	1.20	14.2
		TOT DEM	KW	.00660	.00455	.00488	.00535	.00574	.00587	.00601	.00587	.00574	.00541	.00462	.00535	0.00660
S- 7	-0.07	HT ELEC	KWH	0.41	0.21	0.12	0.04	0.03	0.03	0.03	0.03	0.03	0.05	0.20	0.42	1.6
		HT DEM	KW	.00385	.00350	.00266	.00189	.00139	.00000	.00000	.00000	.00012	.00235	.00385	.00385	0.00385
		REM ELEC	KWH	0.72	0.66	0.80	0.88	1.06	1.19	1.25	1.22	1.04	0.92	0.74	0.72	11.2
		REM DEM	KW	.00345	.00361	.00413	.00455	.00481	.00518	.00523	.00513	.00492	.00455	.00403	.00366	0.00523
		TOT ELEC	KWH	1.13	0.87	0.92	0.93	1.10	1.22	1.28	1.25	1.07	0.97	0.94	1.13	12.8
		TOT DEM	KW	.00635	.00438	.00470	.00514	.00552	.00565	.00578	.00565	.00552	.00521	.00445	.00514	0.00635
S- 16	1.04	HT ELEC	KWH	0.44	0.23	0.13	0.04	0.04	0.03	0.03	0.03	0.03	0.05	0.21	0.45	1.7
		HT DEM	KW	.00385	.00350	.00266	.00189	.00139	.00000	.00000	.00000	.00012	.00235	.00385	.00385	0.00385
		REM ELEC	KWH	0.64	0.59	0.71	0.79	0.95	1.06	1.12	1.09	0.93	0.82	0.66	0.64	10.0
		REM DEM	KW	.00310	.00324	.00371	.00409	.00432	.00465	.00470	.00461	.00442	.00409	.00362	.00329	0.00470
		TOT ELEC	KWH	1.08	0.81	0.84	0.83	0.98	1.09	1.15	1.12	0.96	0.87	0.87	1.09	11.7
		TOT DEM	KW	.00600	.00414	.00444	.00486	.00522	.00534	.00546	.00534	.00522	.00492	.00420	.00486	0.00600
S- 32	0.20	HT ELEC	KWH	0.50	0.23	0.20	0.08	0.04	0.03	0.03	0.03	0.03	0.09	0.21	0.38	1.8
		HT DEM	KW	.00385	.00350	.00266	.00189	.00139	.00000	.00000	.00000	.00012	.00235	.00385	.00385	0.00385
		REM ELEC	KWH	0.67	0.66	0.81	0.92	0.99	1.21	1.20	1.20	1.04	0.88	0.71	0.69	11.0
		REM DEM	KW	.00323	.00338	.00387	.00426	.00451	.00485	.00490	.00480	.00461	.00426	.00377	.00343	0.00490
		TOT ELEC	KWH	1.17	0.89	1.01	1.00	1.03	1.24	1.23	1.23	1.08	0.97	0.92	1.07	12.8
		TOT DEM	KW	.00655	.00485	.00406	.00439	.00452	.00498	.00498	.00498	.00465	.00426	.00472	.00531	0.00655
S- 40	0.86	HT ELEC	KWH	0.60	0.27	0.24	0.09	0.05	0.03	0.03	0.03	0.04	0.11	0.25	0.45	2.2
		HT DEM	KW	.00385	.00350	.00266	.00189	.00139	.00000	.00000	.00000	.00012	.00235	.00385	.00385	0.00385
		REM ELEC	KWH	0.61	0.60	0.74	0.84	0.90	1.10	1.09	1.09	0.95	0.80	0.65	0.63	10.0
		REM DEM	KW	.00299	.00313	.00358	.00394	.00417	.00448	.00453	.00444	.00426	.00394	.00349	.00317	0.00453
		TOT ELEC	KWH	1.21	0.87	0.98	0.93	0.95	1.13	1.12	1.12	0.99	0.91	0.90	1.08	12.2
		TOT DEM	KW	.00620	.00459	.00384	.00415	.00428	.00471	.00471	.00471	.00440	.00403	.00446	.00502	0.00620
S- 45	0.24	HT ELEC	KWH	0.54	0.24	0.22	0.08	0.05	0.03	0.03	0.03	0.03	0.10	0.23	0.41	2.0
		HT DEM	KW	.00385	.00350	.00266	.00189	.00139	.00000	.00000	.00000	.00012	.00235	.00385	.00385	0.00385
		REM ELEC	KWH	0.65	0.64	0.78	0.89	0.95	1.17	1.16	1.16	1.01	0.85	0.69	0.67	10.6
		REM DEM	KW	.00312	.00326	.00374	.00412	.00435	.00468	.00473	.00464	.00445	.00412	.00364	.00331	0.00473
		TOT ELEC	KWH	1.19	0.88	1.00	0.97	1.00	1.20	1.19	1.19	1.04	0.95	0.92	1.08	12.6
		TOT DEM	KW	.00640	.00474	.00397	.00429	.00442	.00486	.00486	.00486	.00454	.00416	.00461	.00518	0.00640
S- 50	1.51	HT ELEC	KWH	0.50	0.22	0.20	0.08	0.04	0.03	0.03	0.03	0.03	0.09	0.21	0.37	1.8
		HT DEM	KW	.00385	.00350	.00266	.00189	.00139	.00000	.00000	.00000	.00012	.00235	.00385	.00385	0.00385
		REM ELEC	KWH	0.62	0.61	0.75	0.86	0.92	1.12	1.11	1.11	0.97	0.82	0.66	0.64	10.2
		REM DEM	KW	.00327	.00342	.00392	.00432	.00456	.00491	.00496	.00486	.00466	.00432	.00382	.00347	0.00496
		TOT ELEC	KWH	1.12	0.84	0.95	0.93	0.96	1.15	1.14	1.14	1.00	0.91	0.87	1.02	12.0
		TOT DEM	KW	.00635	.00470	.00394	.00425	.00438	.00483	.00483	.00483	.00451	.00413	.00457	.00514	0.00635
S- 53	1.41	HT ELEC	KWH	0.41	0.21	0.12	0.02	0.02	0.02	0.02	0.02	0.02	0.02	0.19	0.42	1.5
		HT DEM	KW	.00385	.00350	.00266	.00189	.00139	.00000	.00000	.00000	.00012	.00235	.00385	.00385	0.00385
		REM ELEC	KWH	0.67	0.60	0.67	0.67	0.70	0.79	1.34	1.27	0.84	0.70	0.65	0.68	9.6
		REM DEM	KW	.00392	.00410	.00469	.00517	.00546	.00588	.00594	.00582	.00558	.00517	.00457	.00416	0.00594
		TOT ELEC	KWH	1.08	0.81	0.79	0.69	0.72	0.80	1.36	1.28	0.86	0.72	0.85	1.10	11.1
		TOT DEM	KW	.00706	.00487	.00522	.00572	.00614	.00628	.00642	.00628	.00614	.00579	.00494	.00572	0.00706
S- 61	-0.32	HT ELEC	KWH	0.41	0.21	0.12	0.04	0.03	0.03	0.03	0.03	0.03	0.05	0.20	0.42	1.6
		HT DEM	KW	.00385	.00350	.00266	.00189	.00139	.00000	.00000	.00000	.00012	.00235	.00385	.00385	0.00385
		REM ELEC	KWH	0.72	0.66	0.80	0.88	1.06	1.19	1.25	1.22	1.04	0.92	0.74	0.72	11.2
		REM DEM	KW	.00345	.00361	.00413	.00455	.00481	.00518	.00523	.00513	.00492	.00455	.00403	.00366	0.00523
		TOT ELEC	KWH	1.13	0.87	0.92	0.93	1.10	1.22	1.28	1.25	1.07	0.97	0.94	1.13	12.8
		TOT DEM	KW	.00635	.00438	.00470	.00514	.00552	.00565	.00578	.00565	.00552	.00521	.00445	.00514	0.00635
				BASE CASE : 512												
BASE	0.00	TOT GAS	THM	0.0187	0.0096	0.0056	0.0019	0.0015	0.0014	0.0014	0.0014	0.0014	0.0022	0.0091	0.0190	0.073
		TOT ELEC	KWH	0.93	0.86	1.04	1.15	1.39	1.55	1.64	1.59	1.36	1.20	0.96	0.93	14.6
		TOT DEM	KW	.00411	.00430	.00492	.00542	.00573	.00617	.00623	.00611	.00586	.00542	.00480	.00436	0.00623
S- 3	-0.16	TOT GAS	THM	0.0200	0.0103	0.0060	0.0020	0.0016	0.0015	0.0015	0.0015	0.0015	0.0023	0.0097	0.0204	0.078
		TOT ELEC	KWH	0.81	0.75	0.90	1.00	1.21	1.35	1.42	1.38	1.18	1.04	0.84	0.81	12.7
		TOT DEM	KW	.00389	.00407	.00466	.00513	.00543	.00584	.00590	.00578	.00555	.00513	.00454	.00413	0.00590
S- 7	-0.08	TOT GAS	THM	0.0216	0.0110	0.0064	0.0022	0.0018	0.0016	0.0016	0.0016	0.0016	0.0025	0.0104	0.0219	0.084
		TOT ELEC	KWH	0.72	0.66	0.80	0.88	1.06	1.19	1.25	1.22	1.04	0.92	0.74	0.72	11.2
		TOT DEM	KW	.00345	.00361	.00413	.00455	.00481	.00518	.00523	.00513	.00492	.00455	.00403	.00366	0.00523
S- 16	1.02	TOT GAS	THM	0.0231	0.0118	0.0068	0.0023	0.0019	0.0017	0.0017	0.0017	0.0017	0.0027	0.0112	0.0234	0.090
		TOT ELEC	KWH	0.63	0.58	0.70	0.78	0.94	1.05	1.11	1.08	0.92	0.81	0.65	0.63	9.9
		TOT DEM	KW	.00310	.00324	.00371	.00409	.00432	.00465	.00470	.00461	.00442	.00409	.00362	.00329	0.00470
S- 32	0.19	TOT GAS	THM	0.0263	0.0118	0.0105	0.0040	0.0023	0.0014	0.0014	0.0014	0.0016	0.0049	0.0110	0.0197	0.096
		TOT ELEC	KWH	0.67	0.66	0.81	0.92	0.99	1.21	1.20	1.20	1.04	0.88	0.71	0.69	11.0
		TOT DEM	KW	.00323	.00338	.00387	.00426	.00451	.00485	.00490	.00480	.00461	.00426	.00377	.00343	0.00490
S- 40	0.85	TOT GAS	THM	0.0313	0.0141	0.0125	0.0048	0.0027	0.0017	0.0017	0.0017	0.0019	0.0058	0.0131	0.0235	0.114
		TOT ELEC	KWH	0.61	0.60	0.74	0.84	0.90	1.10	1.09	1.09	0.95	0.80	0.65	0.63	10.0
		TOT DEM	KW	.00299	.00313	.00358	.00394	.00417	.00448	.00453	.00444	.00426	.00394	.00349	.00317	0.00453
S- 45	0.24	TOT GAS	THM	0.0284	0.0128	0.0113	0.0044	0.0025	0.0016	0.0016	0.0016	0.0018	0.0053	0.0119	0.0213	0.104
		TOT ELEC	KWH	0.65	0.64	0.78	0.89	0.95	1.17	1.16	1.16	1.01	0.85	0.69	0.67	10.6
		TOT DEM	KW	.00312	.00326	.00374	.00412	.00435	.00468	.00473	.00464	.00445	.00412	.00364	.00331	0.00473

(HT ELEC=ELECTRIC CONSUMPTION FOR SPACE AND WATER HEATING. REM ELEC=REMAINING ELECTRIC CONSUMPTION (NOT USED FOR HEATING). TOT ELEC=TOTAL ELECTRIC CONSUMPTION. HT DEM=PEAK ELECTRIC DEMAND FOR SPACE AND WATER HEATING. REM DEM=PEAK ELECTRIC DEMAND FOR REMAINING LOAD (NOT HEATING). TOT DEM=TOTAL PEAK ELECTRIC DEMAND. GAS=TOTAL GAS CONSUMPTION. ANNUAL FIGURES ARE SUMS OF MONTHLY CONSUMPTION VALUES OR ARE YEARLY MAXIMUMS OF MONTHLY PEAK DEMAND VALUES.)

STRAT SET	CONST COST $/SF	ENERGY FORM	UNITS (PER SF)	JAN	FEB	MAR	APR	MAY	JUN	JUL	AUG	SEP	OCT	NOV	DEC	ANNUAL
						BASE CASE : 512 (CONTINUED)										
S- 50	1.50	TOT GAS	THM	0.0260	0.0117	0.0104	0.0040	0.0023	0.0014	0.0014	0.0014	0.0016	0.0049	0.0109	0.0195	0.095
		TOT ELEC	KWH	0.62	0.61	0.75	0.86	0.92	1.12	1.11	1.11	0.97	0.82	0.66	0.64	10.2
		TOT DEM	KW	.00327	.00342	.00392	.00432	.00456	.00491	.00496	.00486	.00466	.00432	.00382	.00347	0.00496
S- 53	1.41	TOT GAS	THM	0.0216	0.0108	0.0061	0.0009	0.0009	0.0009	0.0009	0.0009	0.0009	0.0009	0.0102	0.0221	0.077
		TOT ELEC	KWH	0.67	0.60	0.67	0.67	0.70	0.79	1.34	1.27	0.84	0.70	0.65	0.68	9.6
		TOT DEM	KW	.00392	.00410	.00469	.00517	.00546	.00588	.00594	.00582	.00558	.00517	.00457	.00416	0.00594
S- 61	-0.33	TOT GAS	THM	0.0216	0.0110	0.0064	0.0022	0.0018	0.0016	0.0016	0.0016	0.0016	0.0025	0.0104	0.0219	0.084
		TOT ELEC	KWH	0.72	0.66	0.80	0.88	1.06	1.19	1.25	1.22	1.04	0.92	0.74	0.72	11.2
		TOT DEM	KW	.00345	.00361	.00413	.00455	.00481	.00518	.00523	.00513	.00492	.00455	.00403	.00366	0.00523
						BASE CASE : 513										
BASE	0.00	TOT ELEC	KWH	1.24	1.13	1.33	1.39	1.43	1.66	1.66	1.70	1.51	1.34	1.19	1.23	16.8
		TOT DEM	KW	.00646	.00522	.00555	.00588	.00588	.00640	.00653	.00640	.00640	.00581	.00555	.00575	0.00653
S- 1	0.09	TOT ELEC	KWH	1.15	1.04	1.22	1.29	1.32	1.53	1.53	1.57	1.40	1.24	1.10	1.13	15.5
		TOT DEM	KW	.00575	.00465	.00494	.00523	.00523	.00569	.00581	.00569	.00569	.00517	.00494	.00511	0.00581
S- 12	0.19	TOT ELEC	KWH	1.03	0.93	1.10	1.15	1.18	1.38	1.38	1.40	1.25	1.11	0.99	1.01	13.9
		TOT DEM	KW	.00549	.00444	.00472	.00500	.00500	.00544	.00555	.00544	.00544	.00494	.00472	.00488	0.00555
S- 18	1.50	TOT ELEC	KWH	0.95	0.86	1.01	1.06	1.09	1.27	1.27	1.29	1.15	1.02	0.91	0.93	12.8
		TOT DEM	KW	.00524	.00423	.00450	.00476	.00476	.00518	.00529	.00518	.00518	.00471	.00450	.00466	0.00529
S- 28	-0.06	TOT ELEC	KWH	1.03	0.93	1.10	1.15	1.18	1.38	1.38	1.40	1.25	1.11	0.99	1.01	13.9
		TOT DEM	KW	.00549	.00444	.00472	.00500	.00500	.00544	.00555	.00544	.00544	.00494	.00472	.00488	0.00555
S- 33	0.46	TOT ELEC	KWH	0.95	0.86	1.01	1.06	1.09	1.27	1.27	1.29	1.15	1.02	0.91	0.93	12.8
		TOT DEM	KW	.00514	.00415	.00441	.00467	.00467	.00509	.00519	.00509	.00509	.00462	.00441	.00457	0.00519
S- 41	1.22	TOT ELEC	KWH	0.87	0.79	0.93	0.98	1.00	1.17	1.17	1.19	1.06	0.94	0.84	0.86	11.8
		TOT DEM	KW	.00496	.00401	.00426	.00451	.00451	.00491	.00501	.00491	.00491	.00446	.00426	.00441	0.00501
S- 46	0.50	TOT ELEC	KWH	0.92	0.83	0.98	1.03	1.05	1.23	1.23	1.25	1.12	0.99	0.88	0.91	12.4
		TOT DEM	KW	.00506	.00409	.00434	.00460	.00460	.00501	.00511	.00501	.00501	.00455	.00434	.00450	0.00511
S- 51	1.77	TOT ELEC	KWH	0.89	0.80	0.95	1.00	1.02	1.19	1.19	1.21	1.08	0.96	0.85	0.88	12.0
		TOT DEM	KW	.00498	.00402	.00428	.00453	.00453	.00493	.00503	.00493	.00493	.00448	.00428	.00443	0.00503
						BASE CASE : 514										
BASE	0.00	HT ELEC	KWH	0.56	0.29	0.17	0.06	0.05	0.04	0.04	0.04	0.04	0.07	0.27	0.57	2.2
		HT DEM	KW	.00864	.00855	.00449	.00207	.00207	.00028	.00032	.00032	.00138	.00207	.00786	.00838	0.00864
		REM ELEC	KWH	1.04	0.96	1.16	1.29	1.55	1.73	1.83	1.78	1.52	1.34	1.08	1.04	16.3
		REM DEM	KW	.00462	.00478	.00613	.00669	.00724	.00788	.00796	.00796	.00740	.00669	.00621	.00446	0.00796
		TOT ELEC	KWH	1.60	1.25	1.32	1.34	1.59	1.77	1.87	1.82	1.56	1.40	1.35	1.61	18.5
		TOT DEM	KW	.01245	.00971	.00672	.00672	.00722	.00772	.00784	.00772	.00722	.00685	.00896	.01096	0.01245
S- 3	-0.10	HT ELEC	KWH	0.57	0.29	0.17	0.06	0.05	0.04	0.04	0.04	0.04	0.07	0.28	0.58	2.2
		HT DEM	KW	.00864	.00855	.00449	.00207	.00207	.00028	.00032	.00032	.00138	.00207	.00786	.00838	0.00864
		REM ELEC	KWH	0.97	0.89	1.07	1.19	1.43	1.60	1.69	1.65	1.40	1.24	1.00	0.97	15.1
		REM DEM	KW	.00423	.00438	.00562	.00613	.00664	.00723	.00730	.00730	.00679	.00613	.00569	.00409	0.00730
		TOT ELEC	KWH	1.53	1.18	1.24	1.25	1.48	1.64	1.73	1.69	1.45	1.30	1.27	1.54	17.3
		TOT DEM	KW	.01174	.00916	.00634	.00634	.00681	.00728	.00740	.00728	.00681	.00646	.00845	.01033	0.01174
S- 7	-0.01	HT ELEC	KWH	0.59	0.30	0.17	0.06	0.05	0.04	0.04	0.04	0.04	0.07	0.29	0.60	2.3
		HT DEM	KW	.00864	.00864	.00518	.00199	.00207	.00016	.00018	.00016	.00121	.00225	.00838	.00855	0.00864
		REM ELEC	KWH	0.88	0.81	0.98	1.09	1.31	1.46	1.55	1.50	1.28	1.13	0.91	0.88	13.8
		REM DEM	KW	.00381	.00394	.00506	.00552	.00598	.00650	.00657	.00657	.00611	.00552	.00512	.00368	0.00657
		TOT ELEC	KWH	1.47	1.12	1.15	1.15	1.36	1.51	1.59	1.55	1.33	1.20	1.20	1.48	16.1
		TOT DEM	KW	.01145	.00973	.00595	.00504	.00550	.00595	.00595	.00595	.00561	.00515	.00893	.01030	0.01145
S- 16	0.69	HT ELEC	KWH	0.59	0.30	0.18	0.06	0.05	0.04	0.04	0.04	0.04	0.07	0.29	0.60	2.3
		HT DEM	KW	.00864	.00864	.00518	.00199	.00207	.00016	.00018	.00016	.00121	.00225	.00838	.00855	0.00864
		REM ELEC	KWH	0.80	0.74	0.89	0.99	1.19	1.32	1.40	1.36	1.16	1.02	0.82	0.80	12.5
		REM DEM	KW	.00366	.00379	.00486	.00530	.00574	.00625	.00631	.00631	.00587	.00530	.00492	.00353	0.00631
		TOT ELEC	KWH	1.39	1.04	1.06	1.05	1.24	1.37	1.44	1.41	1.21	1.09	1.11	1.40	14.8
		TOT DEM	KW	.01085	.00922	.00564	.00477	.00521	.00564	.00564	.00564	.00532	.00488	.00846	.00977	0.01085
S- 32	0.26	HT ELEC	KWH	0.47	0.26	0.22	0.10	0.07	0.05	0.04	0.04	0.05	0.10	0.22	0.39	2.0
		HT DEM	KW	.00864	.00864	.00518	.00199	.00207	.00016	.00018	.00016	.00121	.00225	.00838	.00855	0.00864
		REM ELEC	KWH	0.74	0.68	0.89	1.05	1.12	1.46	1.42	1.47	1.25	0.97	0.78	0.76	12.6
		REM DEM	KW	.00359	.00371	.00477	.00520	.00563	.00613	.00619	.00619	.00576	.00520	.00483	.00347	0.00619
		TOT ELEC	KWH	1.22	0.94	1.12	1.14	1.19	1.51	1.47	1.52	1.30	1.07	1.00	1.14	14.6
		TOT DEM	KW	.01163	.00989	.00605	.00512	.00558	.00605	.00605	.00605	.00570	.00523	.00907	.01047	0.01163
S- 40	0.72	HT ELEC	KWH	0.49	0.27	0.23	0.10	0.07	0.05	0.05	0.05	0.05	0.10	0.23	0.40	2.1
		HT DEM	KW	.00864	.00864	.00518	.00199	.00207	.00016	.00018	.00016	.00121	.00225	.00838	.00855	0.00864
		REM ELEC	KWH	0.67	0.62	0.81	0.95	1.01	1.32	1.29	1.33	1.13	0.88	0.71	0.68	11.4
		REM DEM	KW	.00335	.00347	.00445	.00486	.00526	.00572	.00578	.00578	.00538	.00486	.00451	.00324	0.00578
		TOT ELEC	KWH	1.16	0.89	1.04	1.05	1.08	1.37	1.33	1.38	1.18	0.98	0.93	1.08	13.5
		TOT DEM	KW	.01130	.00961	.00588	.00497	.00542	.00588	.00588	.00588	.00554	.00508	.00881	.01017	0.01130
S- 45	0.31	HT ELEC	KWH	0.48	0.26	0.23	0.10	0.07	0.05	0.04	0.04	0.05	0.10	0.22	0.39	2.0
		HT DEM	KW	.00864	.00864	.00518	.00199	.00207	.00016	.00018	.00016	.00121	.00225	.00838	.00855	0.00864
		REM ELEC	KWH	0.71	0.65	0.86	1.00	1.08	1.40	1.37	1.42	1.20	0.93	0.75	0.73	12.1
		REM DEM	KW	.00351	.00363	.00466	.00508	.00551	.00599	.00605	.00605	.00563	.00508	.00472	.00339	0.00605
		TOT ELEC	KWH	1.19	0.92	1.08	1.10	1.14	1.45	1.41	1.46	1.25	1.03	0.97	1.12	14.1
		TOT DEM	KW	.01150	.00978	.00598	.00506	.00552	.00598	.00598	.00598	.00564	.00518	.00897	.01035	0.01150
S- 50	1.57	HT ELEC	KWH	0.48	0.26	0.22	0.10	0.07	0.05	0.04	0.04	0.05	0.10	0.22	0.39	2.0
		HT DEM	KW	.00864	.00864	.00518	.00199	.00207	.00016	.00018	.00016	.00121	.00225	.00838	.00855	0.00864
		REM ELEC	KWH	0.68	0.63	0.82	0.96	1.03	1.35	1.31	1.36	1.15	0.89	0.72	0.70	11.6
		REM DEM	KW	.00363	.00376	.00482	.00526	.00570	.00620	.00626	.00626	.00582	.00526	.00488	.00351	0.00626
		TOT ELEC	KWH	1.16	0.89	1.05	1.06	1.10	1.39	1.36	1.40	1.20	0.99	0.94	1.08	13.6
		TOT DEM	KW	.01122	.00954	.00583	.00494	.00539	.00583	.00583	.00583	.00550	.00505	.00875	.01010	0.01122
						BASE CASE : 514 (CONTINUED)										
S- 53	1.71	HT ELEC	KWH	0.68	0.34	0.19	0.03	0.03	0.03	0.03	0.03	0.03	0.03	0.32	0.69	2.4
		HT DEM	KW	.00864	.00864	.00518	.00199	.00207	.00016	.00018	.00016	.00121	.00225	.00838	.00855	0.00864
		REM ELEC	KWH	0.72	0.65	0.72	0.72	0.75	0.84	1.44	1.36	0.91	0.75	0.70	0.73	10.3
		REM DEM	KW	.00403	.00417	.00535	.00584	.00632	.00688	.00695	.00695	.00646	.00584	.00542	.00389	0.00695
		TOT ELEC	KWH	1.40	0.99	0.91	0.75	0.78	0.87	1.47	1.39	0.94	0.78	1.02	1.42	12.7
		TOT DEM	KW	.01190	.01012	.00619	.00524	.00571	.00619	.00619	.00619	.00583	.00536	.00928	.01071	0.01190
S- 61	-0.26	HT ELEC	KWH	0.59	0.30	0.17	0.06	0.05	0.04	0.04	0.04	0.04	0.07	0.29	0.60	2.3
		HT DEM	KW	.00864	.00864	.00518	.00199	.00207	.00016	.00018	.00016	.00121	.00225	.00838	.00855	0.00864
		REM ELEC	KWH	0.88	0.81	0.98	1.09	1.31	1.46	1.55	1.50	1.28	1.13	0.91	0.88	13.8
		REM DEM	KW	.00381	.00394	.00506	.00552	.00598	.00650	.00657	.00657	.00611	.00552	.00512	.00368	0.00657
		TOT ELEC	KWH	1.47	1.12	1.15	1.15	1.36	1.51	1.59	1.55	1.33	1.20	1.20	1.48	16.1
		TOT DEM	KW	.01145	.00973	.00595	.00504	.00550	.00595	.00595	.00595	.00561	.00515	.00893	.01030	0.01145
						BASE CASE : 515										
BASE	0.00	TOT GAS	THM	0.0294	0.0151	0.0087	0.0030	0.0024	0.0022	0.0022	0.0022	0.0022	0.0035	0.0143	0.0299	0.115
		TOT ELEC	KWH	1.04	0.96	1.16	1.29	1.55	1.73	1.83	1.78	1.52	1.34	1.08	1.04	16.3
		TOT DEM	KW	.00462	.00478	.00613	.00669	.00724	.00788	.00796	.00796	.00740	.00669	.00621	.00446	0.00796
S- 3	-0.10	TOT GAS	THM	0.0300	0.0153	0.0089	0.0030	0.0025	0.0022	0.0022	0.0022	0.0022	0.0035	0.0145	0.0304	0.117
		TOT ELEC	KWH	0.97	0.89	1.07	1.19	1.43	1.60	1.69	1.65	1.40	1.24	1.00	0.97	15.1
		TOT DEM	KW	.00423	.00438	.00562	.00613	.00664	.00723	.00730	.00730	.00679	.00613	.00569	.00409	0.00730
S- 7	-0.01	TOT GAS	THM	0.0310	0.0159	0.0092	0.0031	0.0025	0.0023	0.0023	0.0023	0.0023	0.0036	0.0150	0.0315	0.121
		TOT ELEC	KWH	0.88	0.81	0.98	1.09	1.31	1.46	1.55	1.50	1.28	1.13	0.91	0.88	13.8
		TOT DEM	KW	.00381	.00394	.00506	.00552	.00598	.00650	.00657	.00657	.00611	.00552	.00512	.00368	0.00657
S- 16	0.69	TOT GAS	THM	0.0312	0.0160	0.0093	0.0032	0.0026	0.0023	0.0023	0.0023	0.0023	0.0037	0.0151	0.0317	0.122
		TOT ELEC	KWH	0.80	0.74	0.89	0.99	1.19	1.32	1.40	1.36	1.16	1.02	0.82	0.80	12.5
		TOT DEM	KW	.00366	.00379	.00486	.00530	.00574	.00625	.00631	.00631	.00587	.00530	.00492	.00353	0.00631
S- 32	0.26	TOT GAS	THM	0.0250	0.0138	0.0118	0.0051	0.0035	0.0024	0.0023	0.0023	0.0028	0.0051	0.0116	0.0204	0.106
		TOT ELEC	KWH	0.74	0.68	0.89	1.05	1.12	1.46	1.42	1.47	1.25	0.97	0.78	0.76	12.6
		TOT DEM	KW	.00359	.00371	.00477	.00520	.00563	.00613	.00619	.00619	.00576	.00520	.00483	.00347	0.00619
S- 40	0.71	TOT GAS	THM	0.0257	0.0142	0.0121	0.0052	0.0036	0.0025	0.0024	0.0024	0.0028	0.0052	0.0119	0.0209	0.109
		TOT ELEC	KWH	0.67	0.62	0.81	0.95	1.01	1.32	1.29	1.33	1.13	0.88	0.71	0.68	11.4
		TOT DEM	KW	.00335	.00347	.00445	.00486	.00526	.00572	.00578	.00578	.00538	.00486	.00451	.00324	0.00578
S- 45	0.30	TOT GAS	THM	0.0253	0.0139	0.0119	0.0051	0.0035	0.0025	0.0024	0.0024	0.0028	0.0051	0.0117	0.0205	0.107
		TOT ELEC	KWH	0.71	0.65	0.86	1.00	1.08	1.40	1.37	1.42	1.20	0.93	0.75	0.73	12.1
		TOT DEM	KW	.00351	.00363	.00466	.00508	.00551	.00599	.00605	.00605	.00563	.00508	.00472	.00339	0.00605
S- 50	1.57	TOT GAS	THM	0.0250	0.0138	0.0118	0.0051	0.0035	0.0024	0.0023	0.0023	0.0028	0.0051	0.0116	0.0204	0.106
		TOT ELEC	KWH	0.68	0.63	0.82	0.96	1.03	1.35	1.31	1.36	1.15	0.89	0.72	0.70	11.6
		TOT DEM	KW	.00363	.00376	.00482	.00526	.00570	.00620	.00626	.00626	.00582	.00526	.00488	.00351	0.00626
S- 53	1.70	TOT GAS	THM	0.0354	0.0178	0.0100	0.0015	0.0015	0.0015	0.0015	0.0015	0.0015	0.0015	0.0168	0.0363	0.127
		TOT ELEC	KWH	0.72	0.65	0.72	0.72	0.75	0.84	1.44	1.36	0.91	0.75	0.70	0.73	10.3
		TOT DEM	KW	.00403	.00417	.00535	.00584	.00632	.00688	.00695	.00695	.00646	.00584	.00542	.00389	0.00695
S- 61	-0.26	TOT GAS	THM	0.0310	0.0159	0.0092	0.0031	0.0025	0.0023	0.0023	0.0023	0.0023	0.0036	0.0150	0.0315	0.121
		TOT ELEC	KWH	0.88	0.81	0.98	1.09	1.31	1.46	1.55	1.50	1.28	1.13	0.91	0.88	13.8
		TOT DEM	KW	.00381	.00394	.00506	.00552	.00598	.00650	.00657	.00657	.00611	.00552	.00512	.00368	0.00657

(HT ELEC=ELECTRIC CONSUMPTION FOR SPACE AND WATER HEATING. REM ELEC=REMAINING ELECTRIC CONSUMPTION (NOT USED FOR HEATING). TOT ELEC=TOTAL ELECTRIC CONSUMPTION. HT DEM=PEAK ELECTRIC DEMAND FOR SPACE AND WATER HEATING. REM DEM=PEAK ELECTRIC DEMAND FOR REMAINING LOAD (NOT HEATING). TOT DEM=TOTAL PEAK ELECTRIC DEMAND. GAS=TOTAL GAS CONSUMPTION. ANNUAL FIGURES ARE SUMS OF MONTHLY CONSUMPTION VALUES OR ARE YEARLY MAXIMUMS OF MONTHLY PEAK DEMAND VALUES.)

STRAT SET	CONST COST $/SF	ENERGY FORM	UNITS (PER SF)	JAN	FEB	MAR	APR	MAY	JUN	JUL	AUG	SEP	OCT	NOV	DEC	ANNUAL
BASE CASE : 516																
BASE	0.00	HT ELEC	KWH	0.38	0.20	0.11	0.04	0.03	0.03	0.03	0.03	0.03	0.04	0.19	0.39	1.5
		HT DEM	KW	.00385	.00385	.00289	.00189	.00158	.00000	.00000	.00000	.00089	.00227	.00381	.00385	0.00385
		REM ELEC	KWH	0.97	0.90	1.08	1.20	1.44	1.61	1.70	1.66	1.41	1.25	1.00	0.97	15.2
		REM DEM	KW	.00452	.00473	.00541	.00596	.00630	.00678	.00685	.00671	.00644	.00596	.00527	.00480	0.00685
		TOT ELEC	KWH	1.36	1.09	1.19	1.24	1.48	1.64	1.73	1.69	1.44	1.29	1.19	1.36	16.7
		TOT DEM	KW	.00705	.00571	.00564	.00620	.00663	.00691	.00691	.00684	.00677	.00634	.00564	.00705	0.00705
S- 3	-0.18	HT ELEC	KWH	0.41	0.21	0.12	0.04	0.03	0.03	0.03	0.03	0.03	0.05	0.20	0.42	1.6
		HT DEM	KW	.00385	.00385	.00289	.00189	.00158	.00000	.00000	.00000	.00089	.00227	.00381	.00385	0.00385
		REM ELEC	KWH	0.84	0.78	0.94	1.04	1.25	1.40	1.48	1.44	1.23	1.08	0.87	0.84	13.2
		REM DEM	KW	.00429	.00448	.00514	.00566	.00598	.00643	.00650	.00637	.00611	.00566	.00500	.00455	0.00650
		TOT ELEC	KWH	1.26	0.99	1.06	1.08	1.29	1.43	1.51	1.47	1.26	1.13	1.07	1.26	14.8
		TOT DEM	KW	.00665	.00539	.00532	.00585	.00625	.00652	.00652	.00645	.00638	.00598	.00532	.00665	0.00665
S- 7	-0.09	HT ELEC	KWH	0.44	0.23	0.13	0.04	0.04	0.03	0.03	0.03	0.03	0.05	0.21	0.45	1.7
		HT DEM	KW	.00385	.00385	.00289	.00189	.00158	.00000	.00000	.00000	.00089	.00227	.00381	.00385	0.00385
		REM ELEC	KWH	0.75	0.69	0.83	0.92	1.11	1.24	1.31	1.28	1.09	0.96	0.77	0.75	11.7
		REM DEM	KW	.00380	.00397	.00454	.00500	.00529	.00569	.00575	.00564	.00540	.00500	.00443	.00403	0.00575
		TOT ELEC	KWH	1.19	0.92	0.96	0.97	1.15	1.27	1.34	1.31	1.12	1.01	0.99	1.20	13.4
		TOT DEM	KW	.00640	.00518	.00512	.00563	.00602	.00627	.00627	.00621	.00614	.00576	.00512	.00640	0.00640
S- 16	0.98	HT ELEC	KWH	0.47	0.24	0.14	0.05	0.04	0.04	0.04	0.04	0.04	0.06	0.23	0.48	1.9
		HT DEM	KW	.00385	.00385	.00289	.00189	.00158	.00000	.00000	.00000	.00089	.00227	.00381	.00385	0.00385
		REM ELEC	KWH	0.66	0.61	0.73	0.81	0.98	1.09	1.15	1.12	0.96	0.84	0.68	0.66	10.3
		REM DEM	KW	.00342	.00357	.00409	.00451	.00477	.00513	.00518	.00508	.00487	.00451	.00399	.00363	0.00518
		TOT ELEC	KWH	1.13	0.85	0.87	0.86	1.02	1.13	1.19	1.16	0.99	0.90	0.91	1.14	12.1
		TOT DEM	KW	.00605	.00490	.00484	.00532	.00569	.00593	.00593	.00587	.00581	.00544	.00484	.00605	0.00605
S- 32	0.18	HT ELEC	KWH	0.50	0.23	0.20	0.08	0.04	0.03	0.03	0.03	0.03	0.09	0.21	0.38	1.8
		HT DEM	KW	.00385	.00385	.00289	.00189	.00158	.00000	.00000	.00000	.00089	.00227	.00381	.00385	0.00385
		REM ELEC	KWH	0.68	0.67	0.82	0.93	1.00	1.22	1.21	1.21	1.05	0.89	0.72	0.70	11.1
		REM DEM	KW	.00343	.00359	.00411	.00452	.00478	.00515	.00520	.00510	.00489	.00452	.00400	.00364	0.00520
		TOT ELEC	KWH	1.18	0.89	1.02	1.01	1.04	1.25	1.24	1.24	1.09	0.98	0.93	1.07	12.9
		TOT DEM	KW	.00655	.00550	.00419	.00452	.00478	.00524	.00531	.00524	.00485	.00445	.00537	.00616	0.00655
S- 40	0.82	HT ELEC	KWH	0.58	0.26	0.23	0.09	0.05	0.03	0.03	0.03	0.04	0.11	0.24	0.43	2.1
		HT DEM	KW	.00385	.00385	.00289	.00189	.00158	.00000	.00000	.00000	.00089	.00227	.00381	.00385	0.00385
		REM ELEC	KWH	0.62	0.61	0.75	0.85	0.91	1.11	1.10	1.10	0.96	0.81	0.66	0.64	10.1
		REM DEM	KW	.00319	.00333	.00382	.00420	.00444	.00478	.00483	.00473	.00454	.00420	.00372	.00338	0.00483
		TOT ELEC	KWH	1.19	0.87	0.98	0.94	0.96	1.14	1.13	1.13	1.00	0.92	0.90	1.07	12.2
		TOT DEM	KW	.00623	.00523	.00399	.00430	.00455	.00498	.00505	.00498	.00461	.00424	.00511	.00586	0.00623
S- 45	0.22	HT ELEC	KWH	0.53	0.24	0.21	0.08	0.05	0.03	0.03	0.03	0.03	0.10	0.22	0.40	1.9
		HT DEM	KW	.00385	.00385	.00289	.00189	.00158	.00000	.00000	.00000	.00089	.00227	.00381	.00385	0.00385
		REM ELEC	KWH	0.65	0.64	0.78	0.89	0.95	1.17	1.16	1.16	1.01	0.85	0.69	0.67	10.6
		REM DEM	KW	.00332	.00347	.00397	.00438	.00463	.00498	.00503	.00493	.00473	.00438	.00387	.00352	0.00503
		TOT ELEC	KWH	1.18	0.87	1.00	0.97	1.00	1.20	1.18	1.18	1.04	0.95	0.91	1.07	12.5
		TOT DEM	KW	.00634	.00533	.00406	.00437	.00463	.00507	.00514	.00507	.00469	.00431	.00520	.00596	0.00634
S- 50	1.49	HT ELEC	KWH	0.50	0.23	0.20	0.08	0.04	0.03	0.03	0.03	0.03	0.09	0.21	0.38	1.8
		HT DEM	KW	.00385	.00385	.00289	.00189	.00158	.00000	.00000	.00000	.00089	.00227	.00381	.00385	0.00385
		REM ELEC	KWH	0.63	0.62	0.76	0.87	0.93	1.13	1.12	1.12	0.98	0.82	0.67	0.65	10.3
		REM DEM	KW	.00345	.00361	.00413	.00455	.00481	.00518	.00523	.00513	.00492	.00455	.00403	.00366	0.00523
		TOT ELEC	KWH	1.13	0.84	0.96	0.94	0.97	1.16	1.15	1.15	1.01	0.92	0.88	1.03	12.1
		TOT DEM	KW	.00632	.00531	.00404	.00436	.00461	.00506	.00512	.00506	.00468	.00430	.00518	.00594	0.00632
S- 53	1.60	HT ELEC	KWH	0.44	0.22	0.13	0.02	0.02	0.02	0.02	0.02	0.02	0.02	0.21	0.45	1.6
		HT DEM	KW	.00385	.00385	.00289	.00189	.00158	.00000	.00000	.00000	.00089	.00227	.00381	.00385	0.00385
		REM ELEC	KWH	0.70	0.63	0.70	0.70	0.73	0.82	1.40	1.32	0.88	0.73	0.68	0.71	10.0
		REM DEM	KW	.00380	.00397	.00454	.00500	.00529	.00569	.00575	.00564	.00540	.00500	.00443	.00403	0.00575
		TOT ELEC	KWH	1.14	0.85	0.83	0.72	0.75	0.84	1.42	1.34	0.90	0.75	0.89	1.16	11.6
		TOT DEM	KW	.00778	.00630	.00622	.00685	.00731	.00762	.00762	.00755	.00747	.00700	.00622	.00778	0.00778
S- 61	-0.34	HT ELEC	KWH	0.44	0.23	0.13	0.04	0.04	0.03	0.03	0.03	0.03	0.05	0.21	0.45	1.7
		HT DEM	KW	.00385	.00385	.00289	.00189	.00158	.00000	.00000	.00000	.00089	.00227	.00381	.00385	0.00385
		REM ELEC	KWH	0.75	0.69	0.83	0.92	1.11	1.24	1.31	1.28	1.09	0.96	0.77	0.75	11.7
		REM DEM	KW	.00380	.00397	.00454	.00500	.00529	.00569	.00575	.00564	.00540	.00500	.00443	.00403	0.00575
		TOT ELEC	KWH	1.19	0.92	0.96	0.97	1.15	1.27	1.34	1.31	1.12	1.01	0.99	1.20	13.4
		TOT DEM	KW	.00640	.00518	.00512	.00563	.00602	.00627	.00627	.00621	.00614	.00576	.00512	.00640	0.00640
BASE CASE : 517																
BASE	0.00	TOT GAS	THM	0.0202	0.0103	0.0060	0.0021	0.0017	0.0015	0.0015	0.0015	0.0015	0.0024	0.0098	0.0205	0.079
		TOT ELEC	KWH	0.97	0.90	1.08	1.20	1.44	1.61	1.70	1.66	1.41	1.25	1.00	0.97	15.2
		TOT DEM	KW	.00452	.00473	.00541	.00596	.00630	.00678	.00685	.00671	.00644	.00596	.00527	.00480	0.00685
S- 3	-0.19	TOT GAS	THM	0.0216	0.0111	0.0064	0.0022	0.0018	0.0016	0.0016	0.0016	0.0016	0.0025	0.0105	0.0220	0.085
		TOT ELEC	KWH	0.84	0.78	0.94	1.04	1.25	1.40	1.48	1.44	1.23	1.08	0.87	0.84	13.2
		TOT DEM	KW	.00429	.00448	.00514	.00566	.00598	.00643	.00650	.00637	.00611	.00566	.00500	.00455	0.00650
S- 7	-0.10	TOT GAS	THM	0.0233	0.0119	0.0069	0.0024	0.0019	0.0017	0.0017	0.0017	0.0017	0.0027	0.0113	0.0236	0.091
		TOT ELEC	KWH	0.75	0.69	0.83	0.92	1.11	1.24	1.31	1.28	1.09	0.96	0.77	0.75	11.7
		TOT DEM	KW	.00380	.00397	.00454	.00500	.00529	.00569	.00575	.00564	.00540	.00500	.00443	.00403	0.00575
S- 16	0.96	TOT GAS	THM	0.0249	0.0127	0.0074	0.0025	0.0020	0.0018	0.0018	0.0018	0.0018	0.0029	0.0121	0.0253	0.097
		TOT ELEC	KWH	0.66	0.61	0.73	0.81	0.98	1.09	1.15	1.12	0.96	0.84	0.68	0.66	10.3
		TOT DEM	KW	.00342	.00357	.00409	.00451	.00477	.00513	.00518	.00508	.00487	.00451	.00399	.00363	0.00518
S- 32	0.17	TOT GAS	THM	0.0262	0.0118	0.0105	0.0040	0.0023	0.0014	0.0014	0.0014	0.0016	0.0049	0.0109	0.0197	0.096
		TOT ELEC	KWH	0.68	0.67	0.82	0.93	1.00	1.22	1.21	1.21	1.05	0.89	0.72	0.70	11.1
		TOT DEM	KW	.00343	.00359	.00411	.00452	.00478	.00515	.00520	.00510	.00489	.00452	.00400	.00364	0.00520
S- 40	0.80	TOT GAS	THM	0.0304	0.0137	0.0121	0.0047	0.0027	0.0017	0.0017	0.0017	0.0019	0.0057	0.0127	0.0228	0.111
		TOT ELEC	KWH	0.62	0.61	0.75	0.85	0.91	1.11	1.10	1.10	0.96	0.81	0.66	0.64	10.1
		TOT DEM	KW	.00319	.00333	.00382	.00420	.00444	.00478	.00483	.00473	.00454	.00420	.00372	.00338	0.00483

STRAT SET	CONST COST $/SF	ENERGY FORM	UNITS (PER SF)	JAN	FEB	MAR	APR	MAY	JUN	JUL	AUG	SEP	OCT	NOV	DEC	ANNUAL
BASE CASE : 517 (CONTINUED)																
S- 45	0.22	TOT GAS	THM	0.0278	0.0125	0.0111	0.0043	0.0024	0.0015	0.0015	0.0015	0.0017	0.0052	0.0116	0.0209	0.102
		TOT ELEC	KWH	0.65	0.64	0.78	0.89	0.95	1.17	1.16	1.16	1.01	0.85	0.69	0.67	10.6
		TOT DEM	KW	.00332	.00347	.00397	.00438	.00463	.00498	.00503	.00493	.00473	.00438	.00387	.00352	0.00503
S- 50	1.48	TOT GAS	THM	0.0259	0.0117	0.0104	0.0040	0.0023	0.0014	0.0014	0.0014	0.0016	0.0048	0.0108	0.0195	0.095
		TOT ELEC	KWH	0.63	0.62	0.76	0.87	0.93	1.13	1.12	1.12	0.98	0.82	0.67	0.65	10.3
		TOT DEM	KW	.00345	.00361	.00413	.00455	.00481	.00518	.00523	.00513	.00492	.00455	.00403	.00366	0.00523
S- 53	1.59	TOT GAS	THM	0.0233	0.0117	0.0066	0.0010	0.0010	0.0010	0.0010	0.0010	0.0010	0.0010	0.0110	0.0239	0.083
		TOT ELEC	KWH	0.70	0.63	0.70	0.70	0.73	0.82	1.40	1.32	0.88	0.73	0.68	0.71	10.0
		TOT DEM	KW	.00471	.00492	.00563	.00620	.00656	.00706	.00713	.00699	.00670	.00620	.00549	.00499	0.00713
S- 61	-0.35	TOT GAS	THM	0.0233	0.0119	0.0069	0.0024	0.0019	0.0017	0.0017	0.0017	0.0017	0.0027	0.0113	0.0236	0.091
		TOT ELEC	KWH	0.75	0.69	0.83	0.92	1.11	1.24	1.31	1.28	1.09	0.96	0.77	0.75	11.7
		TOT DEM	KW	.00380	.00397	.00454	.00500	.00529	.00569	.00575	.00564	.00540	.00500	.00443	.00403	0.00575
BASE CASE : 518																
BASE	0.00	TOT ELEC	KWH	1.35	1.22	1.44	1.51	1.55	1.80	1.80	1.84	1.64	1.46	1.29	1.33	18.2
		TOT DEM	KW	.00683	.00552	.00587	.00621	.00621	.00676	.00690	.00676	.00676	.00614	.00587	.00607	0.00690
S- 1	0.09	TOT ELEC	KWH	1.24	1.13	1.33	1.39	1.43	1.66	1.66	1.70	1.51	1.34	1.19	1.23	16.8
		TOT DEM	KW	.00567	.00458	.00487	.00516	.00516	.00562	.00573	.00562	.00562	.00510	.00487	.00504	0.00573
S- 12	0.20	TOT ELEC	KWH	1.12	1.02	1.20	1.26	1.29	1.50	1.50	1.54	1.37	1.22	1.08	1.11	15.2
		TOT DEM	KW	.00533	.00430	.00457	.00484	.00484	.00527	.00538	.00527	.00527	.00479	.00457	.00473	0.00538
S- 18	1.51	TOT ELEC	KWH	1.03	0.93	1.10	1.15	1.18	1.38	1.38	1.40	1.25	1.11	0.99	1.01	13.9
		TOT DEM	KW	.00508	.00410	.00436	.00462	.00462	.00503	.00513	.00503	.00503	.00457	.00436	.00451	0.00513
S- 28	-0.05	TOT ELEC	KWH	1.12	1.02	1.20	1.26	1.29	1.50	1.50	1.54	1.37	1.22	1.08	1.11	15.2
		TOT DEM	KW	.00533	.00430	.00457	.00484	.00484	.00527	.00538	.00527	.00527	.00479	.00457	.00473	0.00538
S- 33	0.47	TOT ELEC	KWH	0.98	0.89	1.05	1.10	1.13	1.32	1.32	1.34	1.20	1.06	0.94	0.97	13.3
		TOT DEM	KW	.00515	.00416	.00442	.00468	.00468	.00510	.00520	.00510	.00510	.00463	.00442	.00458	0.00520
S- 41	1.23	TOT ELEC	KWH	0.93	0.84	0.99	1.04	1.06	1.24	1.24	1.26	1.13	1.00	0.89	0.91	12.5
		TOT DEM	KW	.00499	.00403	.00428	.00454	.00454	.00494	.00504	.00494	.00494	.00449	.00428	.00444	0.00504
S- 46	0.51	TOT ELEC	KWH	0.95	0.86	1.02	1.07	1.10	1.28	1.28	1.30	1.16	1.03	0.92	0.94	12.9
		TOT DEM	KW	.00507	.00410	.00435	.00461	.00461	.00502	.00512	.00502	.00502	.00456	.00435	.00451	0.00512
S- 51	1.78	TOT ELEC	KWH	0.92	0.83	0.98	1.03	1.05	1.23	1.23	1.25	1.12	0.99	0.88	0.91	12.4
		TOT DEM	KW	.00500	.00404	.00429	.00454	.00454	.00495	.00505	.00495	.00495	.00449	.00429	.00444	0.00505
BASE CASE : 519																
BASE	0.00	HT ELEC	KWH	0.37	0.19	0.11	0.04	0.03	0.03	0.03	0.03	0.03	0.04	0.18	0.38	1.5
		HT DEM	KW	.00864	.00855	.00449	.00207	.00207	.00028	.00032	.00032	.00138	.00207	.00786	.00838	0.00864
		REM ELEC	KWH	0.98	0.90	1.09	1.21	1.45	1.62	1.71	1.67	1.42	1.25	1.01	0.98	15.3
		REM DEM	KW	.00414	.00428	.00550	.00600	.00650	.00707	.00714	.00714	.00664	.00600	.00557	.00400	0.00714
		TOT ELEC	KWH	1.35	1.09	1.20	1.25	1.48	1.65	1.74	1.70	1.45	1.30	1.19	1.36	16.8
		TOT DEM	KW	.01244	.00970	.00672	.00672	.00722	.00771	.00784	.00771	.00722	.00684	.00896	.01095	0.01244
S- 3	-0.06	HT ELEC	KWH	0.38	0.19	0.11	0.04	0.03	0.03	0.03	0.03	0.03	0.04	0.18	0.38	1.5
		HT DEM	KW	.00864	.00855	.00449	.00207	.00207	.00028	.00032	.00032	.00138	.00207	.00786	.00838	0.00864
		REM ELEC	KWH	0.91	0.84	1.01	1.12	1.35	1.51	1.59	1.55	1.32	1.16	0.94	0.91	14.2
		REM DEM	KW	.00380	.00393	.00504	.00550	.00596	.00648	.00655	.00655	.00609	.00550	.00511	.00367	0.00655
		TOT ELEC	KWH	1.29	1.03	1.12	1.16	1.38	1.53	1.62	1.58	1.35	1.21	1.12	1.29	15.7
		TOT DEM	KW	.01173	.00915	.00633	.00633	.00680	.00727	.00739	.00727	.00680	.00645	.00845	.01032	0.01173
S- 7	0.02	HT ELEC	KWH	0.39	0.20	0.12	0.04	0.03	0.03	0.03	0.03	0.03	0.05	0.19	0.40	1.5
		HT DEM	KW	.00864	.00864	.00518	.00199	.00207	.00016	.00018	.00016	.00121	.00225	.00838	.00855	0.00864
		REM ELEC	KWH	0.83	0.76	0.92	1.02	1.23	1.37	1.44	1.41	1.20	1.06	0.85	0.83	12.9
		REM DEM	KW	.00342	.00353	.00454	.00495	.00536	.00583	.00589	.00589	.00548	.00495	.00459	.00330	0.00589
		TOT ELEC	KWH	1.22	0.96	1.03	1.06	1.26	1.40	1.47	1.44	1.23	1.10	1.04	1.23	14.4
		TOT DEM	KW	.01128	.00959	.00587	.00496	.00541	.00587	.00587	.00587	.00553	.00508	.00880	.01015	0.01128
S- 16	1.19	HT ELEC	KWH	0.40	0.20	0.12	0.04	0.03	0.03	0.03	0.03	0.03	0.05	0.19	0.40	1.5
		HT DEM	KW	.00864	.00864	.00518	.00199	.00207	.00016	.00018	.00016	.00121	.00225	.00838	.00855	0.00864
		REM ELEC	KWH	0.76	0.70	0.84	0.93	1.12	1.25	1.32	1.29	1.10	0.97	0.78	0.76	11.8
		REM DEM	KW	.00328	.00340	.00436	.00475	.00515	.00560	.00566	.00566	.00526	.00475	.00441	.00317	0.00566
		TOT ELEC	KWH	1.15	0.90	0.96	0.97	1.15	1.28	1.35	1.32	1.13	1.01	0.97	1.16	13.4
		TOT DEM	KW	.01066	.00906	.00554	.00469	.00512	.00554	.00554	.00554	.00522	.00480	.00831	.00959	0.01066
S- 32	0.38	HT ELEC	KWH	0.31	0.17	0.15	0.06	0.04	0.03	0.03	0.03	0.03	0.06	0.14	0.25	1.3
		HT DEM	KW	.00864	.00864	.00518	.00199	.00207	.00016	.00018	.00016	.00121	.00225	.00838	.00855	0.00864
		REM ELEC	KWH	0.68	0.63	0.82	0.96	1.03	1.35	1.31	1.36	1.15	0.89	0.72	0.70	11.6
		REM DEM	KW	.00311	.00322	.00413	.00451	.00489	.00532	.00537	.00537	.00499	.00451	.00419	.00301	0.00537
		TOT ELEC	KWH	1.00	0.80	0.97	1.03	1.08	1.38	1.34	1.39	1.16	0.96	0.86	0.95	12.9
		TOT DEM	KW	.01177	.01000	.00612	.00518	.00565	.00612	.00612	.00612	.00577	.00530	.00918	.01059	0.01177
S- 61	-0.23	HT ELEC	KWH	0.39	0.20	0.12	0.04	0.03	0.03	0.03	0.03	0.03	0.05	0.19	0.40	1.5
		HT DEM	KW	.00864	.00864	.00518	.00199	.00207	.00016	.00018	.00016	.00121	.00225	.00838	.00855	0.00864
		REM ELEC	KWH	0.83	0.76	0.92	1.02	1.23	1.37	1.44	1.41	1.20	1.06	0.85	0.83	12.9
		REM DEM	KW	.00342	.00353	.00454	.00495	.00536	.00583	.00589	.00589	.00548	.00495	.00459	.00330	0.00589
		TOT ELEC	KWH	1.22	0.96	1.03	1.06	1.26	1.40	1.47	1.44	1.23	1.10	1.04	1.23	14.4
		TOT DEM	KW	.01128	.00959	.00587	.00496	.00541	.00587	.00587	.00587	.00553	.00508	.00880	.01015	0.01128
BASE CASE : 520																
BASE	0.00	TOT GAS	THM	0.0196	0.0100	0.0058	0.0020	0.0016	0.0015	0.0015	0.0015	0.0015	0.0023	0.0095	0.0199	0.077
		TOT ELEC	KWH	0.98	0.90	1.09	1.21	1.45	1.62	1.71	1.67	1.42	1.25	1.01	0.98	15.3
		TOT DEM	KW	.00414	.00428	.00550	.00600	.00650	.00707	.00714	.00714	.00664	.00600	.00557	.00400	0.00714

(HT ELEC=ELECTRIC CONSUMPTION FOR SPACE AND WATER HEATING. REM ELEC=REMAINING ELECTRIC CONSUMPTION (NOT USED FOR HEATING). TOT ELEC=TOTAL ELECTRIC CONSUMPTION. HT DEM=PEAK ELECTRIC DEMAND FOR SPACE AND WATER HEATING. REM DEM=PEAK ELECTRIC DEMAND FOR REMAINING LOAD (NOT HEATING). TOT DEM=TOTAL PEAK ELECTRIC DEMAND. GAS=TOTAL GAS CONSUMPTION. ANNUAL FIGURES ARE SUMS OF MONTHLY CONSUMPTION VALUES OR ARE YEARLY MAXIMUMS OF MONTHLY PEAK DEMAND VALUES.)

STRAT SET	CONST COST $/SF	ENERGY FORM	UNITS (PER SF)	JAN	FEB	MAR	APR	MAY	JUN	JUL	AUG	SEP	OCT	NOV	DEC	ANNUAL
BASE CASE : 520 (CONTINUED)																
S- 3	-0.07	TOT GAS	THM	0.0199	0.0102	0.0059	0.0020	0.0016	0.0015	0.0015	0.0015	0.0015	0.0023	0.0097	0.0203	0.078
		TOT ELEC	KWH	0.91	0.84	1.01	1.12	1.35	1.51	1.59	1.55	1.32	1.16	0.94	0.91	14.2
		TOT DEM	KW	.00380	.00393	.00504	.00550	.00596	.00648	.00655	.00655	.00609	.00550	.00511	.00367	0.00655
S- 7	0.02	TOT GAS	THM	0.0207	0.0106	0.0061	0.0021	0.0017	0.0015	0.0015	0.0015	0.0015	0.0024	0.0100	0.0210	0.081
		TOT ELEC	KWH	0.83	0.76	0.92	1.02	1.23	1.37	1.44	1.41	1.20	1.06	0.85	0.83	12.9
		TOT DEM	KW	.00342	.00353	.00454	.00495	.00536	.00583	.00589	.00589	.00548	.00495	.00459	.00330	0.00589
S- 16	1.19	TOT GAS	THM	0.0208	0.0107	0.0062	0.0021	0.0017	0.0015	0.0015	0.0015	0.0015	0.0024	0.0101	0.0212	0.081
		TOT ELEC	KWH	0.76	0.70	0.84	0.93	1.12	1.25	1.32	1.29	1.10	0.97	0.78	0.76	11.8
		TOT DEM	KW	.00328	.00340	.00436	.00475	.00515	.00560	.00566	.00566	.00526	.00475	.00441	.00317	0.00566
S- 32	0.38	TOT GAS	THM	0.0164	0.0090	0.0077	0.0033	0.0023	0.0016	0.0015	0.0015	0.0018	0.0033	0.0076	0.0133	0.069
		TOT ELEC	KWH	0.68	0.63	0.82	0.96	1.03	1.35	1.31	1.36	1.15	0.89	0.72	0.70	11.6
		TOT DEM	KW	.00311	.00322	.00413	.00451	.00489	.00532	.00537	.00537	.00499	.00451	.00419	.00301	0.00537
S- 61	-0.23	TOT GAS	THM	0.0207	0.0106	0.0061	0.0021	0.0017	0.0015	0.0015	0.0015	0.0015	0.0024	0.0100	0.0210	0.081
		TOT ELEC	KWH	0.83	0.76	0.92	1.02	1.23	1.37	1.44	1.41	1.20	1.06	0.85	0.83	12.9
		TOT DEM	KW	.00342	.00353	.00454	.00495	.00536	.00583	.00589	.00589	.00548	.00495	.00459	.00330	0.00589
BASE CASE : 521																
BASE	0.00	HT ELEC	KWH	0.25	0.13	0.08	0.03	0.02	0.02	0.02	0.02	0.02	0.03	0.12	0.26	1.0
		HT DEM	KW	.00385	.00350	.00266	.00189	.00139	.00000	.00000	.00000	.00012	.00235	.00385	.00385	0.00385
		REM ELEC	KWH	0.93	0.86	1.04	1.15	1.39	1.55	1.64	1.59	1.36	1.20	0.96	0.93	14.6
		REM DEM	KW	.00431	.00451	.00516	.00568	.00601	.00646	.00653	.00640	.00614	.00568	.00503	.00457	0.00653
		TOT ELEC	KWH	1.19	0.99	1.11	1.18	1.41	1.57	1.65	1.61	1.38	1.23	1.09	1.19	15.6
		TOT DEM	KW	.00705	.00486	.00522	.00571	.00613	.00627	.00642	.00627	.00613	.00578	.00494	.00571	0.00705
S- 3	-0.14	HT ELEC	KWH	0.27	0.14	0.08	0.03	0.02	0.02	0.02	0.02	0.02	0.03	0.13	0.28	1.1
		HT DEM	KW	.00385	.00350	.00266	.00189	.00139	.00000	.00000	.00000	.00012	.00235	.00385	.00385	0.00385
		REM ELEC	KWH	0.81	0.75	0.90	1.00	1.21	1.35	1.42	1.38	1.18	1.04	0.84	0.81	12.7
		REM DEM	KW	.00408	.00426	.00488	.00538	.00569	.00612	.00618	.00606	.00581	.00538	.00476	.00433	0.00618
		TOT ELEC	KWH	1.09	0.89	0.98	1.03	1.23	1.37	1.44	1.40	1.20	1.07	0.97	1.09	13.8
		TOT DEM	KW	.00660	.00455	.00488	.00535	.00574	.00587	.00601	.00587	.00574	.00541	.00462	.00535	0.00660
S- 7	-0.06	HT ELEC	KWH	0.29	0.15	0.09	0.03	0.02	0.02	0.02	0.02	0.02	0.03	0.14	0.30	1.1
		HT DEM	KW	.00385	.00350	.00266	.00189	.00139	.00000	.00000	.00000	.00012	.00235	.00385	.00385	0.00385
		REM ELEC	KWH	0.72	0.66	0.80	0.88	1.06	1.19	1.25	1.22	1.04	0.92	0.74	0.72	11.2
		REM DEM	KW	.00362	.00378	.00433	.00477	.00504	.00543	.00548	.00537	.00515	.00477	.00422	.00384	0.00548
		TOT ELEC	KWH	1.01	0.81	0.88	0.91	1.09	1.21	1.28	1.24	1.06	0.95	0.88	1.02	12.4
		TOT DEM	KW	.00635	.00438	.00470	.00514	.00552	.00565	.00578	.00565	.00552	.00521	.00445	.00514	0.00635
S- 16	1.04	HT ELEC	KWH	0.31	0.16	0.09	0.03	0.03	0.02	0.02	0.02	0.02	0.04	0.15	0.32	1.2
		HT DEM	KW	.00385	.00350	.00266	.00189	.00139	.00000	.00000	.00000	.00012	.00235	.00385	.00385	0.00385
		REM ELEC	KWH	0.63	0.58	0.70	0.78	0.94	1.05	1.11	1.08	0.92	0.81	0.65	0.63	9.9
		REM DEM	KW	.00325	.00340	.00389	.00429	.00454	.00488	.00493	.00483	.00463	.00429	.00380	.00345	0.00493
		TOT ELEC	KWH	0.95	0.74	0.80	0.81	0.97	1.07	1.13	1.10	0.94	0.85	0.81	0.95	11.1
		TOT DEM	KW	.00600	.00414	.00444	.00486	.00522	.00534	.00546	.00534	.00522	.00492	.00420	.00486	0.00600
S- 32	0.30	HT ELEC	KWH	0.38	0.17	0.15	0.06	0.03	0.02	0.02	0.02	0.02	0.07	0.16	0.28	1.4
		HT DEM	KW	.00385	.00350	.00266	.00189	.00139	.00000	.00000	.00000	.00012	.00235	.00385	.00385	0.00385
		REM ELEC	KWH	0.64	0.63	0.78	0.88	0.95	1.15	1.14	1.14	1.00	0.84	0.68	0.66	10.5
		REM DEM	KW	.00312	.00326	.00374	.00412	.00435	.00468	.00473	.00464	.00445	.00412	.00364	.00331	0.00473
		TOT ELEC	KWH	1.02	0.80	0.93	0.94	0.98	1.18	1.17	1.17	1.02	0.91	0.84	0.94	11.9
		TOT DEM	KW	.00655	.00485	.00406	.00439	.00452	.00498	.00498	.00498	.00465	.00426	.00472	.00531	0.00655
S- 61	-0.31	HT ELEC	KWH	0.29	0.15	0.09	0.03	0.02	0.02	0.02	0.02	0.02	0.03	0.14	0.30	1.1
		HT DEM	KW	.00385	.00350	.00266	.00189	.00139	.00000	.00000	.00000	.00012	.00235	.00385	.00385	0.00385
		REM ELEC	KWH	0.72	0.66	0.80	0.88	1.06	1.19	1.25	1.22	1.04	0.92	0.74	0.72	11.2
		REM DEM	KW	.00362	.00378	.00433	.00477	.00504	.00543	.00548	.00537	.00515	.00477	.00422	.00384	0.00548
		TOT ELEC	KWH	1.01	0.81	0.88	0.91	1.09	1.21	1.28	1.24	1.06	0.95	0.88	1.02	12.4
		TOT DEM	KW	.00635	.00438	.00470	.00514	.00552	.00565	.00578	.00565	.00552	.00521	.00445	.00514	0.00635
BASE CASE : 522																
BASE	0.00	TOT GAS	THM	0.0134	0.0068	0.0040	0.0014	0.0011	0.0010	0.0010	0.0010	0.0010	0.0016	0.0065	0.0136	0.052
		TOT ELEC	KWH	0.93	0.86	1.04	1.15	1.39	1.55	1.64	1.59	1.36	1.20	0.96	0.93	14.6
		TOT DEM	KW	.00431	.00451	.00516	.00568	.00601	.00646	.00653	.00640	.00614	.00568	.00503	.00457	0.00653
S- 3	-0.15	TOT GAS	THM	0.0143	0.0073	0.0042	0.0015	0.0012	0.0011	0.0011	0.0011	0.0011	0.0017	0.0069	0.0145	0.056
		TOT ELEC	KWH	0.81	0.75	0.90	1.00	1.21	1.35	1.42	1.38	1.18	1.04	0.84	0.81	12.7
		TOT DEM	KW	.00408	.00426	.00488	.00538	.00569	.00612	.00618	.00606	.00581	.00538	.00476	.00433	0.00618
S- 7	-0.07	TOT GAS	THM	0.0154	0.0079	0.0046	0.0016	0.0013	0.0011	0.0011	0.0011	0.0011	0.0018	0.0075	0.0156	0.060
		TOT ELEC	KWH	0.72	0.66	0.80	0.88	1.06	1.19	1.25	1.22	1.04	0.92	0.74	0.72	11.2
		TOT DEM	KW	.00362	.00378	.00433	.00477	.00504	.00543	.00548	.00537	.00515	.00477	.00422	.00384	0.00548
S- 16	1.03	TOT GAS	THM	0.0165	0.0084	0.0049	0.0017	0.0014	0.0012	0.0012	0.0012	0.0012	0.0019	0.0080	0.0167	0.064
		TOT ELEC	KWH	0.63	0.58	0.70	0.78	0.94	1.05	1.11	1.08	0.92	0.81	0.65	0.63	9.9
		TOT DEM	KW	.00325	.00340	.00389	.00429	.00454	.00488	.00493	.00483	.00463	.00429	.00380	.00345	0.00493
S- 32	0.30	TOT GAS	THM	0.0197	0.0089	0.0079	0.0030	0.0017	0.0011	0.0011	0.0011	0.0012	0.0037	0.0082	0.0148	0.072
		TOT ELEC	KWH	0.64	0.63	0.78	0.88	0.95	1.15	1.14	1.14	1.00	0.84	0.68	0.66	10.5
		TOT DEM	KW	.00312	.00326	.00374	.00412	.00435	.00468	.00473	.00464	.00445	.00412	.00364	.00331	0.00473
S- 61	-0.32	TOT GAS	THM	0.0154	0.0079	0.0046	0.0016	0.0013	0.0011	0.0011	0.0011	0.0011	0.0018	0.0075	0.0156	0.060
		TOT ELEC	KWH	0.72	0.66	0.80	0.88	1.06	1.19	1.25	1.22	1.04	0.92	0.74	0.72	11.2
		TOT DEM	KW	.00362	.00378	.00433	.00477	.00504	.00543	.00548	.00537	.00515	.00477	.00422	.00384	0.00548
BASE CASE : 523																
BASE	0.00	TOT ELEC	KWH	1.01	0.91	1.07	1.13	1.16	1.35	1.35	1.37	1.22	1.09	0.97	0.99	13.6
		TOT DEM	KW	.00646	.00522	.00555	.00588	.00588	.00640	.00653	.00640	.00640	.00581	.00555	.00575	0.00653
S- 1	0.09	TOT ELEC	KWH	0.93	0.84	0.99	1.04	1.06	1.24	1.24	1.26	1.13	1.00	0.89	0.91	12.5
		TOT DEM	KW	.00576	.00465	.00494	.00523	.00523	.00569	.00581	.00569	.00569	.00517	.00494	.00511	0.00581
BASE CASE : 523 (CONTINUED)																
S- 12	0.19	TOT ELEC	KWH	0.82	0.74	0.88	0.92	0.94	1.10	1.10	1.12	1.00	0.89	0.79	0.81	11.1
		TOT DEM	KW	.00549	.00444	.00472	.00500	.00500	.00544	.00555	.00544	.00544	.00494	.00472	.00488	0.00555
S- 18	1.50	TOT ELEC	KWH	0.75	0.68	0.80	0.84	0.86	1.00	1.00	1.02	0.91	0.81	0.72	0.74	10.1
		TOT DEM	KW	.00524	.00423	.00450	.00476	.00476	.00518	.00529	.00518	.00518	.00471	.00450	.00466	0.00529
S- 28	-0.06	TOT ELEC	KWH	0.82	0.74	0.88	0.92	0.94	1.10	1.10	1.12	1.00	0.89	0.79	0.81	11.1
		TOT DEM	KW	.00549	.00444	.00472	.00500	.00500	.00544	.00555	.00544	.00544	.00494	.00472	.00488	0.00555
S- 33	0.55	TOT ELEC	KWH	0.81	0.73	0.86	0.90	0.93	1.08	1.08	1.10	0.98	0.87	0.77	0.80	10.9
		TOT DEM	KW	.00513	.00414	.00440	.00466	.00466	.00508	.00518	.00508	.00508	.00461	.00440	.00456	0.00518
BASE CASE : 524																
BASE	0.00	HT ELEC	KWH	0.54	0.28	0.16	0.05	0.04	0.04	0.04	0.04	0.04	0.06	0.26	0.55	2.1
		HT DEM	KW	.00864	.00855	.00449	.00207	.00207	.00028	.00032	.00032	.00138	.00207	.00786	.00838	0.00864
		REM ELEC	KWH	1.08	0.99	1.19	1.33	1.60	1.78	1.86	1.83	1.56	1.38	1.11	1.08	16.8
		REM DEM	KW	.00490	.00506	.00650	.00709	.00768	.00836	.00844	.00844	.00785	.00709	.00658	.00473	0.00844
		TOT ELEC	KWH	1.61	1.27	1.35	1.38	1.64	1.82	1.92	1.87	1.60	1.44	1.37	1.62	18.9
		TOT DEM	KW	.01265	.00987	.00683	.00683	.00734	.00784	.00797	.00784	.00734	.00696	.00911	.01113	0.01265
S- 3	-0.10	HT ELEC	KWH	0.29	0.15	0.09	0.03	0.02	0.02	0.02	0.02	0.02	0.03	0.14	0.29	1.1
		HT DEM	KW	.00864	.00855	.00449	.00207	.00207	.00028	.00032	.00032	.00138	.00207	.00786	.00838	0.00864
		REM ELEC	KWH	1.00	0.92	1.11	1.23	1.48	1.65	1.75	1.70	1.45	1.28	1.03	1.00	15.6
		REM DEM	KW	.00450	.00465	.00597	.00651	.00705	.00767	.00775	.00775	.00721	.00651	.00604	.00434	0.00775
		TOT ELEC	KWH	1.29	1.07	1.19	1.26	1.51	1.68	1.77	1.72	1.47	1.31	1.17	1.29	16.7
		TOT DEM	KW	.01193	.00931	.00644	.00644	.00692	.00740	.00752	.00740	.00692	.00656	.00859	.01050	0.01193
S- 7	-0.01	HT ELEC	KWH	0.57	0.29	0.17	0.06	0.05	0.04	0.04	0.04	0.04	0.07	0.27	0.57	2.2
		HT DEM	KW	.00864	.00864	.00518	.00199	.00207	.00016	.00018	.00016	.00121	.00225	.00838	.00855	0.00864
		REM ELEC	KWH	0.91	0.84	1.01	1.12	1.35	1.51	1.59	1.55	1.32	1.16	0.94	0.91	14.2
		REM DEM	KW	.00405	.00419	.00537	.00586	.00635	.00691	.00698	.00698	.00649	.00586	.00544	.00391	0.00698
		TOT ELEC	KWH	1.47	1.13	1.18	1.18	1.40	1.55	1.63	1.59	1.36	1.23	1.21	1.48	16.4
		TOT DEM	KW	.01148	.00976	.00597	.00505	.00551	.00597	.00597	.00597	.00563	.00517	.00895	.01033	0.01148
S- 16	1.14	HT ELEC	KWH	0.57	0.29	0.17	0.06	0.05	0.04	0.04	0.04	0.04	0.07	0.28	0.58	2.2
		HT DEM	KW	.00864	.00864	.00518	.00199	.00207	.00016	.00018	.00016	.00121	.00225	.00838	.00855	0.00864
		REM ELEC	KWH	0.83	0.76	0.92	1.02	1.23	1.37	1.44	1.41	1.20	1.06	0.85	0.83	12.9
		REM DEM	KW	.00388	.00401	.00515	.00562	.00609	.00662	.00669	.00669	.00622	.00562	.00522	.00375	0.00669
		TOT ELEC	KWH	1.40	1.05	1.09	1.08	1.27	1.41	1.49	1.45	1.24	1.12	1.13	1.41	15.1
		TOT DEM	KW	.01084	.00921	.00564	.00477	.00520	.00564	.00564	.00564	.00531	.00488	.00846	.00976	0.01084
S- 32	0.36	HT ELEC	KWH	0.45	0.25	0.21	0.09	0.06	0.04	0.04	0.04	0.05	0.09	0.21	0.37	1.9
		HT DEM	KW	.00864	.00864	.00518	.00199	.00207	.00016	.00018	.00016	.00121	.00225	.00838	.00855	0.00864
		REM ELEC	KWH	0.74	0.68	0.89	1.05	1.12	1.46	1.42	1.47	1.25	0.97	0.78	0.76	12.6
		REM DEM	KW	.00379	.00392	.00503	.00549	.00594	.00646	.00653	.00653	.00607	.00549	.00509	.00366	0.00653
		TOT ELEC	KWH	1.20	0.93	1.11	1.14	1.18	1.51	1.47	1.52	1.30	1.06	0.99	1.12	14.5
		TOT DEM	KW	.01184	.01006	.00616	.00521	.00568	.00616	.00616	.00616	.00580	.00533	.00924	.01066	0.01184
S- 61	-0.26	HT ELEC	KWH	0.57	0.29	0.17	0.06	0.05	0.04	0.04	0.04	0.04	0.07	0.27	0.57	2.2
		HT DEM	KW	.00864	.00864	.00518	.00199	.00207	.00016	.00018	.00016	.00121	.00225	.00838	.00855	0.00864
		REM ELEC	KWH	0.91	0.84	1.01	1.12	1.35	1.51	1.59	1.55	1.32	1.16	0.94	0.91	14.2
		REM DEM	KW	.00405	.00419	.00537	.00586	.00635	.00691	.00698	.00698	.00649	.00586	.00544	.00391	0.00698
		TOT ELEC	KWH	1.47	1.13	1.18	1.18	1.40	1.55	1.63	1.59	1.36	1.23	1.21	1.48	16.4
		TOT DEM	KW	.01148	.00976	.00597	.00505	.00551	.00597	.00597	.00597	.00563	.00517	.00895	.01033	0.01148
BASE CASE : 525																
BASE	0.00	TOT GAS	THM	0.0282	0.0144	0.0084	0.0029	0.0023	0.0021	0.0021	0.0021	0.0021	0.0033	0.0136	0.0286	0.110
		TOT ELEC	KWH	1.08	0.99	1.19	1.33	1.60	1.78	1.88	1.83	1.56	1.38	1.11	1.08	16.8
		TOT DEM	KW	.00490	.00506	.00650	.00709	.00768	.00836	.00844	.00844	.00785	.00709	.00658	.00473	0.00844
S- 3	-0.10	TOT GAS	THM	0.0287	0.0147	0.0085	0.0029	0.0024	0.0021	0.0021	0.0021	0.0021	0.0034	0.0139	0.0291	0.112
		TOT ELEC	KWH	1.00	0.92	1.11	1.23	1.48	1.65	1.75	1.70	1.45	1.28	1.03	1.00	15.6
		TOT DEM	KW	.00450	.00465	.00597	.00651	.00705	.00767	.00775	.00775	.00721	.00651	.00604	.00434	0.00775
S- 7	-0.01	TOT GAS	THM	0.0297	0.0152	0.0088	0.0030	0.0024	0.0022	0.0022	0.0022	0.0022	0.0035	0.0144	0.0302	0.116
		TOT ELEC	KWH	0.91	0.84	1.01	1.12	1.35	1.51	1.59	1.55	1.32	1.16	0.94	0.91	14.2
		TOT DEM	KW	.00405	.00419	.00537	.00586	.00635	.00691	.00698	.00698	.00649	.00586	.00544	.00391	0.00698
S- 16	1.14	TOT GAS	THM	0.0300	0.0153	0.0089	0.0030	0.0025	0.0022	0.0022	0.0022	0.0022	0.0035	0.0145	0.0304	0.117
		TOT ELEC	KWH	0.83	0.76	0.92	1.02	1.23	1.37	1.44	1.41	1.20	1.06	0.85	0.83	12.9
		TOT DEM	KW	.00388	.00401	.00515	.00562	.00609	.00662	.00669	.00669	.00622	.00562	.00522	.00375	0.00669
S- 32	0.35	TOT GAS	THM	0.0238	0.0131	0.0112	0.0048	0.0033	0.0023	0.0022	0.0022	0.0026	0.0048	0.0110	0.0193	0.101
		TOT ELEC	KWH	0.74	0.68	0.89	1.05	1.12	1.46	1.42	1.47	1.25	0.97	0.78	0.76	12.6
		TOT DEM	KW	.00379	.00392	.00503	.00549	.00594	.00646	.00653	.00653	.00607	.00549	.00509	.00366	0.00653
S- 61	-0.26	TOT GAS	THM	0.0297	0.0152	0.0088	0.0030	0.0024	0.0022	0.0022	0.0022	0.0022	0.0035	0.0144	0.0302	0.116
		TOT ELEC	KWH	0.91	0.84	1.01	1.12	1.35	1.51	1.59	1.55	1.32	1.16	0.94	0.91	14.2
		TOT DEM	KW	.00405	.00419	.00537	.00586	.00635	.00691	.00698	.00698	.00649	.00586	.00544	.00391	0.00698
BASE CASE : 526																
BASE	0.00	HT ELEC	KWH	0.32	0.16	0.09	0.03	0.03	0.02	0.02	0.02	0.02	0.04	0.16	0.32	1.3
		HT DEM	KW	.00385	.00385	.00289	.00189	.00158	.00000	.00000	.00000	.00089	.00227	.00381	.00385	0.00385
		REM ELEC	KWH	1.00	0.93	1.11	1.24	1.49	1.66	1.76	1.71	1.46	1.29	1.04	1.00	15.7
		REM DEM	KW	.00446	.00466	.00533	.00587	.00621	.00668	.00675	.00662	.00635	.00587	.00520	.00472	0.00675
		TOT ELEC	KWH	1.32	1.09	1.21	1.27	1.52	1.69	1.78	1.74	1.48	1.32	1.19	1.33	17.0
		TOT DEM	KW	.00708	.00573	.00566	.00623	.00666	.00694	.00694	.00687	.00680	.00637	.00566	.00708	0.00708

(HT ELEC=ELECTRIC CONSUMPTION FOR SPACE AND WATER HEATING. REM ELEC=REMAINING ELECTRIC CONSUMPTION (NOT USED FOR HEATING). TOT ELEC=TOTAL ELECTRIC CONSUMPTION. HT DEM=PEAK ELECTRIC DEMAND FOR SPACE AND WATER HEATING. REM DEM=PEAK ELECTRIC DEMAND FOR REMAINING LOAD (NOT HEATING). TOT DEM=TOTAL PEAK ELECTRIC DEMAND. GAS=TOTAL GAS CONSUMPTION. ANNUAL FIGURES ARE SUMS OF MONTHLY CONSUMPTION VALUES OR ARE YEARLY MAXIMUMS OF MONTHLY PEAK DEMAND VALUES.)

BASE CASE : 526 (CONTINUED)

STRAT SET	CONST COST $/SF	ENERGY FORM	UNITS (PER SF)	JAN	FEB	MAR	APR	MAY	JUN	JUL	AUG	SEP	OCT	NOV	DEC	ANNUAL
S- 3	-0.19	HT ELEC	KWH	0.34	0.18	0.10	0.03	0.03	0.03	0.03	0.03	0.03	0.04	0.17	0.35	1.3
		HT DEM	KW	.00385	.00385	.00289	.00189	.00158	.00000	.00000	.00000	.00089	.00227	.00381	.00385	0.00385
		REM ELEC	KWH	0.87	0.80	0.97	1.07	1.29	1.44	1.52	1.48	1.26	1.12	0.90	0.87	13.6
		REM DEM	KW	.00423	.00442	.00506	.00558	.00590	.00635	.00641	.00628	.00603	.00558	.00494	.00449	0.00641
		TOT ELEC	KWH	1.21	0.98	1.07	1.11	1.32	1.47	1.55	1.51	1.29	1.16	1.06	1.22	14.9
		TOT DEM	KW	.00665	.00539	.00532	.00585	.00625	.00652	.00652	.00645	.00638	.00598	.00532	.00665	0.00665
S- 7	-0.10	HT ELEC	KWH	0.37	0.19	0.11	0.04	0.03	0.03	0.03	0.03	0.03	0.04	0.18	0.37	1.4
		HT DEM	KW	.00385	.00385	.00289	.00189	.00158	.00000	.00000	.00000	.00089	.00227	.00381	.00385	0.00385
		REM ELEC	KWH	0.77	0.71	0.86	0.96	1.15	1.28	1.36	1.32	1.13	0.99	0.80	0.77	12.1
		REM DEM	KW	.00374	.00391	.00448	.00493	.00522	.00561	.00567	.00556	.00533	.00493	.00437	.00397	0.00567
		TOT ELEC	KWH	1.14	0.90	0.97	0.99	1.18	1.31	1.38	1.35	1.15	1.04	0.98	1.15	13.5
		TOT DEM	KW	.00640	.00518	.00512	.00563	.00602	.00627	.00627	.00621	.00614	.00576	.00512	.00640	0.00640
S- 16	0.96	HT ELEC	KWH	0.39	0.20	0.12	0.04	0.03	0.03	0.03	0.03	0.03	0.05	0.19	0.40	1.5
		HT DEM	KW	.00385	.00385	.00289	.00189	.00158	.00000	.00000	.00000	.00089	.00227	.00381	.00385	0.00385
		REM ELEC	KWH	0.68	0.63	0.75	0.84	1.01	1.12	1.19	1.16	0.99	0.87	0.70	0.68	10.6
		REM DEM	KW	.00337	.00352	.00403	.00444	.00469	.00505	.00510	.00500	.00479	.00444	.00393	.00357	0.00510
		TOT ELEC	KWH	1.07	0.83	0.87	0.88	1.04	1.15	1.22	1.18	1.02	0.92	0.89	1.08	12.1
		TOT DEM	KW	.00605	.00490	.00484	.00532	.00569	.00593	.00593	.00587	.00581	.00544	.00484	.00605	0.00605
S- 32	0.26	HT ELEC	KWH	0.44	0.20	0.18	0.07	0.04	0.02	0.02	0.02	0.03	0.08	0.18	0.33	1.6
		HT DEM	KW	.00385	.00385	.00289	.00189	.00158	.00000	.00000	.00000	.00089	.00227	.00381	.00385	0.00385
		REM ELEC	KWH	0.68	0.67	0.82	0.93	1.00	1.22	1.21	1.21	1.05	0.89	0.72	0.70	11.1
		REM DEM	KW	.00333	.00348	.00399	.00439	.00465	.00500	.00505	.00495	.00475	.00439	.00389	.00353	0.00505
		TOT ELEC	KWH	1.12	0.86	1.00	1.00	1.04	1.25	1.23	1.23	1.08	0.97	0.91	1.03	12.7
		TOT DEM	KW	.00658	.00553	.00421	.00454	.00480	.00526	.00533	.00526	.00487	.00447	.00540	.00619	0.00658
S- 61	-0.35	HT ELEC	KWH	0.37	0.19	0.11	0.04	0.03	0.03	0.03	0.03	0.03	0.04	0.18	0.37	1.4
		HT DEM	KW	.00385	.00385	.00289	.00189	.00158	.00000	.00000	.00000	.00089	.00227	.00381	.00385	0.00385
		REM ELEC	KWH	0.77	0.71	0.86	0.96	1.15	1.28	1.36	1.32	1.13	0.99	0.80	0.77	12.1
		REM DEM	KW	.00374	.00391	.00448	.00493	.00522	.00561	.00567	.00556	.00533	.00493	.00437	.00397	0.00567
		TOT ELEC	KWH	1.14	0.90	0.97	0.99	1.18	1.31	1.38	1.35	1.15	1.04	0.98	1.15	13.5
		TOT DEM	KW	.00640	.00518	.00512	.00563	.00602	.00627	.00627	.00621	.00614	.00576	.00512	.00640	0.00640

BASE CASE : 527

STRAT SET	CONST COST $/SF	ENERGY FORM	UNITS (PER SF)	JAN	FEB	MAR	APR	MAY	JUN	JUL	AUG	SEP	OCT	NOV	DEC	ANNUAL
BASE	0.00	TOT GAS	THM	0.0168	0.0086	0.0050	0.0017	0.0014	0.0012	0.0012	0.0012	0.0012	0.0020	0.0081	0.0171	0.066
		TOT ELEC	KWH	1.00	0.93	1.11	1.24	1.49	1.66	1.76	1.71	1.46	1.29	1.04	1.00	15.7
		TOT DEM	KW	.00446	.00466	.00533	.00587	.00621	.00668	.00675	.00662	.00635	.00587	.00520	.00472	0.00675
S- 3	-0.20	TOT GAS	THM	0.0180	0.0092	0.0053	0.0018	0.0015	0.0013	0.0013	0.0013	0.0013	0.0021	0.0087	0.0183	0.070
		TOT ELEC	KWH	0.87	0.80	0.97	1.07	1.29	1.44	1.52	1.48	1.26	1.12	0.90	0.87	13.6
		TOT DEM	KW	.00423	.00442	.00506	.00558	.00590	.00635	.00641	.00628	.00603	.00558	.00494	.00449	0.00641
S- 7	-0.11	TOT GAS	THM	0.0193	0.0099	0.0057	0.0020	0.0016	0.0014	0.0014	0.0014	0.0014	0.0023	0.0094	0.0196	0.075
		TOT ELEC	KWH	0.77	0.71	0.86	0.96	1.15	1.28	1.36	1.32	1.13	0.99	0.80	0.77	12.1
		TOT DEM	KW	.00374	.00391	.00448	.00493	.00522	.00561	.00567	.00556	.00533	.00493	.00437	.00397	0.00567
S- 16	0.95	TOT GAS	THM	0.0207	0.0106	0.0061	0.0021	0.0017	0.0015	0.0015	0.0015	0.0015	0.0024	0.0100	0.0210	0.081
		TOT ELEC	KWH	0.68	0.63	0.75	0.84	1.01	1.12	1.19	1.16	0.99	0.87	0.70	0.68	10.6
		TOT DEM	KW	.00337	.00352	.00403	.00444	.00469	.00505	.00510	.00500	.00479	.00444	.00393	.00357	0.00510
S- 32	0.25	TOT GAS	THM	0.0231	0.0104	0.0092	0.0035	0.0020	0.0013	0.0013	0.0013	0.0014	0.0043	0.0096	0.0173	0.085
		TOT ELEC	KWH	0.68	0.67	0.82	0.93	1.00	1.22	1.21	1.21	1.05	0.89	0.72	0.70	11.1
		TOT DEM	KW	.00333	.00348	.00399	.00439	.00465	.00500	.00505	.00495	.00475	.00439	.00389	.00353	0.00505
S- 61	-0.24	TOT GAS	THM	0.0193	0.0099	0.0057	0.0020	0.0016	0.0014	0.0014	0.0014	0.0014	0.0023	0.0094	0.0196	0.075
		TOT ELEC	KWH	0.77	0.71	0.86	0.96	1.15	1.28	1.36	1.32	1.13	0.99	0.80	0.77	12.1
		TOT DEM	KW	.00374	.00391	.00448	.00493	.00522	.00561	.00567	.00556	.00533	.00493	.00437	.00397	0.00567

BASE CASE : 528

STRAT SET	CONST COST $/SF	ENERGY FORM	UNITS (PER SF)	JAN	FEB	MAR	APR	MAY	JUN	JUL	AUG	SEP	OCT	NOV	DEC	ANNUAL
BASE	0.00	TOT ELEC	KWH	1.32	1.20	1.41	1.49	1.52	1.77	1.77	1.81	1.61	1.43	1.27	1.31	17.9
		TOT DEM	KW	.00726	.00586	.00623	.00660	.00660	.00718	.00733	.00718	.00718	.00652	.00623	.00645	0.00733
S- 3	0.09	TOT ELEC	KWH	1.24	1.12	1.92	1.39	1.42	1.65	1.65	1.69	1.50	1.34	1.19	1.22	16.7
		TOT DEM	KW	.00689	.00557	.00592	.00626	.00626	.00682	.00696	.00682	.00682	.00619	.00592	.00612	0.00696
S- 12	0.20	TOT ELEC	KWH	1.10	1.00	1.18	1.24	1.27	1.48	1.48	1.50	1.34	1.19	1.06	1.09	14.9
		TOT DEM	KW	.00610	.00493	.00524	.00554	.00554	.00604	.00616	.00604	.00604	.00548	.00524	.00542	0.00616
S- 18	1.51	TOT ELEC	KWH	1.01	0.92	1.08	1.14	1.16	1.36	1.36	1.38	1.23	1.10	0.97	1.00	13.7
		TOT DEM	KW	.00556	.00450	.00478	.00506	.00506	.00551	.00562	.00551	.00551	.00500	.00478	.00495	0.00562
S- 28	-0.05	TOT ELEC	KWH	1.10	1.00	1.18	1.24	1.27	1.48	1.48	1.50	1.34	1.19	1.06	1.09	14.9
		TOT DEM	KW	.00610	.00493	.00524	.00554	.00554	.00604	.00616	.00604	.00604	.00548	.00524	.00542	0.00616
S- 33	0.56	TOT ELEC	KWH	0.90	0.82	0.96	1.01	1.04	1.21	1.21	1.23	1.10	0.98	0.87	0.89	12.2
		TOT DEM	KW	.00528	.00426	.00453	.00480	.00480	.00522	.00533	.00522	.00522	.00474	.00453	.00469	0.00533

BASE CASE : 529

STRAT SET	CONST COST $/SF	ENERGY FORM	UNITS (PER SF)	JAN	FEB	MAR	APR	MAY	JUN	JUL	AUG	SEP	OCT	NOV	DEC	ANNUAL
BASE	0.00	HT ELEC	KWH	0.21	0.11	0.06	0.02	0.02	0.02	0.02	0.02	0.02	0.02	0.10	0.21	0.8
		HT DEM	KW	.00385	.00350	.00266	.00189	.00139	.00000	.00000	.00000	.00012	.00235	.00385	.00385	0.00385
		REM ELEC	KWH	0.97	0.90	1.08	1.20	1.44	1.61	1.70	1.66	1.41	1.25	1.00	0.97	15.2
		REM DEM	KW	.00411	.00430	.00492	.00542	.00573	.00617	.00623	.00611	.00586	.00542	.00480	.00436	0.00623
		TOT ELEC	KWH	1.18	1.00	1.14	1.22	1.46	1.63	1.72	1.67	1.43	1.27	1.10	1.18	16.0
		TOT DEM	KW	.00700	.00483	.00518	.00567	.00609	.00623	.00637	.00623	.00609	.00574	.00490	.00567	0.00700
S- 3	-0.15	HT ELEC	KWH	0.22	0.11	0.07	0.02	0.02	0.02	0.02	0.02	0.02	0.03	0.11	0.22	0.9
		HT DEM	KW	.00385	.00350	.00266	.00189	.00139	.00000	.00000	.00000	.00012	.00235	.00385	.00385	0.00385
		REM ELEC	KWH	0.84	0.78	0.94	1.04	1.25	1.40	1.48	1.44	1.23	1.08	0.87	0.84	13.2
		REM DEM	KW	.00389	.00407	.00466	.00513	.00543	.00584	.00590	.00578	.00555	.00513	.00454	.00413	0.00590
		TOT ELEC	KWH	1.07	0.89	1.00	1.07	1.27	1.42	1.49	1.46	1.24	1.11	0.98	1.07	14.1
		TOT DEM	KW	.00660	.00455	.00488	.00535	.00574	.00587	.00601	.00587	.00574	.00541	.00462	.00535	0.00660

BASE CASE : 529 (CONTINUED)

STRAT SET	CONST COST $/SF	ENERGY FORM	UNITS (PER SF)	JAN	FEB	MAR	APR	MAY	JUN	JUL	AUG	SEP	OCT	NOV	DEC	ANNUAL
S- 7	-0.07	HT ELEC	KWH	0.24	0.12	0.07	0.02	0.02	0.02	0.02	0.02	0.02	0.03	0.11	0.24	0.9
		HT DEM	KW	.00385	.00350	.00266	.00189	.00139	.00000	.00000	.00000	.00012	.00235	.00385	.00385	0.00385
		REM ELEC	KWH	0.75	0.69	0.83	0.92	1.11	1.24	1.31	1.28	1.09	0.96	0.77	0.75	11.7
		REM DEM	KW	.00345	.00361	.00413	.00455	.00481	.00518	.00523	.00513	.00492	.00455	.00403	.00366	0.00523
		TOT ELEC	KWH	0.99	0.81	0.90	0.95	1.13	1.26	1.33	1.29	1.11	0.99	0.89	0.99	12.6
		TOT DEM	KW	.00635	.00438	.00470	.00514	.00552	.00565	.00578	.00565	.00552	.00521	.00445	.00514	0.00635
S- 16	1.04	HT ELEC	KWH	0.25	0.13	0.08	0.03	0.02	0.02	0.02	0.02	0.02	0.03	0.12	0.26	1.0
		HT DEM	KW	.00385	.00350	.00266	.00189	.00139	.00000	.00000	.00000	.00012	.00235	.00385	.00385	0.00385
		REM ELEC	KWH	0.66	0.61	0.73	0.81	0.98	1.09	1.15	1.12	0.96	0.84	0.68	0.66	10.3
		REM DEM	KW	.00310	.00324	.00371	.00409	.00432	.00465	.00470	.00461	.00442	.00409	.00362	.00329	0.00470
		TOT ELEC	KWH	0.91	0.74	0.81	0.84	1.00	1.11	1.17	1.14	0.98	0.87	0.80	0.92	11.3
		TOT DEM	KW	.00600	.00414	.00444	.00486	.00522	.00534	.00546	.00534	.00522	.00492	.00420	.00486	0.00600
S- 32	0.16	HT ELEC	KWH	0.29	0.13	0.12	0.04	0.03	0.02	0.02	0.02	0.02	0.05	0.12	0.22	1.1
		HT DEM	KW	.00385	.00350	.00266	.00189	.00139	.00000	.00000	.00000	.00012	.00235	.00385	.00385	0.00385
		REM ELEC	KWH	0.70	0.69	0.85	0.97	1.04	1.26	1.25	1.25	1.09	0.92	0.75	0.72	11.5
		REM DEM	KW	.00323	.00338	.00387	.00426	.00451	.00485	.00490	.00480	.00461	.00426	.00377	.00343	0.00490
		TOT ELEC	KWH	0.99	0.82	0.97	1.01	1.06	1.28	1.27	1.27	1.11	0.97	0.87	0.94	12.6
		TOT DEM	KW	.00655	.00485	.00406	.00439	.00452	.00498	.00498	.00498	.00465	.00426	.00472	.00531	0.00655
S- 40	0.90	HT ELEC	KWH	0.35	0.16	0.14	0.05	0.03	0.02	0.02	0.02	0.02	0.06	0.14	0.26	1.3
		HT DEM	KW	.00385	.00350	.00266	.00189	.00139	.00000	.00000	.00000	.00012	.00235	.00385	.00385	0.00385
		REM ELEC	KWH	0.63	0.62	0.77	0.87	0.94	1.14	1.13	1.13	0.99	0.83	0.68	0.66	10.4
		REM DEM	KW	.00299	.00313	.00358	.00394	.00417	.00448	.00453	.00444	.00426	.00394	.00349	.00317	0.00453
		TOT ELEC	KWH	0.98	0.78	0.91	0.93	0.97	1.16	1.15	1.15	1.01	0.90	0.82	0.91	11.7
		TOT DEM	KW	.00620	.00459	.00384	.00415	.00428	.00471	.00471	.00471	.00440	.00403	.00446	.00502	0.00620
S- 45	0.19	HT ELEC	KWH	0.31	0.14	0.13	0.05	0.03	0.02	0.02	0.02	0.02	0.06	0.13	0.24	1.1
		HT DEM	KW	.00385	.00350	.00266	.00189	.00139	.00000	.00000	.00000	.00012	.00235	.00385	.00385	0.00385
		REM ELEC	KWH	0.67	0.66	0.81	0.92	0.99	1.21	1.20	1.20	1.04	0.88	0.71	0.69	11.0
		REM DEM	KW	.00312	.00326	.00374	.00412	.00435	.00468	.00473	.00464	.00445	.00412	.00364	.00331	0.00473
		TOT ELEC	KWH	0.98	0.80	0.94	0.97	1.02	1.23	1.22	1.22	1.06	0.94	0.85	0.93	12.1
		TOT DEM	KW	.00640	.00474	.00397	.00429	.00442	.00486	.00486	.00486	.00454	.00416	.00461	.00518	0.00640
S- 61	-0.31	HT ELEC	KWH	0.24	0.12	0.07	0.02	0.02	0.02	0.02	0.02	0.02	0.03	0.11	0.24	0.9
		HT DEM	KW	.00385	.00350	.00266	.00189	.00139	.00000	.00000	.00000	.00012	.00235	.00385	.00385	0.00385
		REM ELEC	KWH	0.75	0.69	0.83	0.92	1.11	1.24	1.31	1.28	1.09	0.96	0.77	0.75	11.7
		REM DEM	KW	.00345	.00361	.00413	.00455	.00481	.00518	.00523	.00513	.00492	.00455	.00403	.00366	0.00523
		TOT ELEC	KWH	0.99	0.81	0.90	0.95	1.13	1.26	1.33	1.29	1.11	0.99	0.89	0.99	12.6
		TOT DEM	KW	.00635	.00438	.00470	.00514	.00552	.00565	.00578	.00565	.00552	.00521	.00445	.00514	0.00635

BASE CASE : 530

STRAT SET	CONST COST $/SF	ENERGY FORM	UNITS (PER SF)	JAN	FEB	MAR	APR	MAY	JUN	JUL	AUG	SEP	OCT	NOV	DEC	ANNUAL
BASE	0.00	TOT GAS	THM	0.0108	0.0055	0.0032	0.0011	0.0009	0.0008	0.0008	0.0008	0.0008	0.0013	0.0052	0.0110	0.042
		TOT ELEC	KWH	0.97	0.90	1.08	1.20	1.44	1.61	1.70	1.66	1.41	1.25	1.00	0.97	15.2
		TOT DEM	KW	.00411	.00430	.00492	.00542	.00573	.00617	.00623	.00611	.00586	.00542	.00480	.00436	0.00623
S- 3	-0.16	TOT GAS	THM	0.0116	0.0059	0.0034	0.0012	0.0009	0.0009	0.0009	0.0009	0.0009	0.0014	0.0056	0.0118	0.045
		TOT ELEC	KWH	0.84	0.78	0.94	1.04	1.25	1.40	1.48	1.44	1.23	1.08	0.87	0.84	13.2
		TOT DEM	KW	.00389	.00407	.00466	.00513	.00543	.00584	.00590	.00578	.00555	.00513	.00454	.00413	0.00590
S- 7	-0.08	TOT GAS	THM	0.0124	0.0064	0.0037	0.0013	0.0010	0.0009	0.0009	0.0009	0.0009	0.0015	0.0060	0.0126	0.049
		TOT ELEC	KWH	0.75	0.69	0.83	0.92	1.11	1.24	1.31	1.28	1.09	0.96	0.77	0.75	11.7
		TOT DEM	KW	.00345	.00361	.00413	.00455	.00481	.00518	.00523	.00513	.00492	.00455	.00403	.00366	0.00523
S- 16	1.01	TOT GAS	THM	0.0133	0.0068	0.0040	0.0014	0.0011	0.0010	0.0010	0.0010	0.0010	0.0016	0.0064	0.0135	0.052
		TOT ELEC	KWH	0.66	0.61	0.73	0.81	0.98	1.09	1.15	1.12	0.96	0.84	0.68	0.66	10.3
		TOT DEM	KW	.00310	.00324	.00371	.00409	.00432	.00465	.00470	.00461	.00442	.00409	.00362	.00329	0.00470
S- 32	0.15	TOT GAS	THM	0.0152	0.0069	0.0061	0.0023	0.0013	0.0008	0.0008	0.0008	0.0009	0.0028	0.0064	0.0114	0.056
		TOT ELEC	KWH	0.70	0.69	0.85	0.97	1.04	1.26	1.25	1.25	1.09	0.92	0.75	0.72	11.5
		TOT DEM	KW	.00323	.00338	.00387	.00426	.00451	.00485	.00490	.00480	.00461	.00426	.00377	.00343	0.00490
S- 40	0.87	TOT GAS	THM	0.0181	0.0082	0.0072	0.0028	0.0016	0.0010	0.0010	0.0010	0.0011	0.0034	0.0076	0.0136	0.066
		TOT ELEC	KWH	0.64	0.63	0.77	0.88	0.94	1.15	1.14	1.14	0.99	0.83	0.68	0.66	10.4
		TOT DEM	KW	.00299	.00313	.00358	.00394	.00417	.00448	.00453	.00444	.00426	.00394	.00349	.00317	0.00453
S- 45	0.17	TOT GAS	THM	0.0165	0.0074	0.0066	0.0025	0.0014	0.0009	0.0009	0.0009	0.0010	0.0031	0.0069	0.0124	0.060
		TOT ELEC	KWH	0.67	0.66	0.81	0.92	0.99	1.21	1.20	1.20	1.04	0.88	0.71	0.69	11.0
		TOT DEM	KW	.00312	.00326	.00374	.00412	.00435	.00468	.00473	.00464	.00445	.00412	.00364	.00331	0.00473
S- 61	-0.08	TOT GAS	THM	0.0124	0.0064	0.0037	0.0013	0.0010	0.0009	0.0009	0.0009	0.0009	0.0015	0.0060	0.0126	0.049
		TOT ELEC	KWH	0.75	0.69	0.83	0.92	1.11	1.24	1.31	1.28	1.09	0.96	0.77	0.75	11.7
		TOT DEM	KW	.00345	.00361	.00413	.00455	.00481	.00518	.00523	.00513	.00492	.00455	.00403	.00366	0.00523

BASE CASE : 531

STRAT SET	CONST COST $/SF	ENERGY FORM	UNITS (PER SF)	JAN	FEB	MAR	APR	MAY	JUN	JUL	AUG	SEP	OCT	NOV	DEC	ANNUAL
BASE	0.00	TOT ELEC	KWH	1.24	1.13	1.33	1.39	1.43	1.66	1.66	1.70	1.51	1.34	1.19	1.23	16.8
		TOT DEM	KW	.00622	.00502	.00534	.00565	.00565	.00615	.00628	.00615	.00615	.00559	.00534	.00553	0.00628
S- 1	0.09	TOT ELEC	KWH	1.15	1.04	1.22	1.29	1.32	1.53	1.53	1.57	1.40	1.24	1.10	1.13	15.5
		TOT DEM	KW	.00603	.00487	.00518	.00548	.00548	.00597	.00609	.00597	.00597	.00542	.00518	.00536	0.00609
S- 12	0.19	TOT ELEC	KWH	1.03	0.93	1.10	1.15	1.18	1.38	1.38	1.40	1.25	1.11	0.99	1.01	13.9
		TOT DEM	KW	.00572	.00462	.00491	.00520	.00520	.00566	.00578	.00566	.00566	.00514	.00491	.00509	0.00578
S- 22	1.50	TOT ELEC	KWH	0.95	0.86	1.01	1.06	1.09	1.27	1.27	1.29	1.15	1.02	0.91	0.93	12.8
		TOT DEM	KW	.00553	.00447	.00475	.00503	.00503	.00548	.00559	.00548	.00548	.00498	.00475	.00492	0.00559
S- 28	-0.06	TOT ELEC	KWH	1.03	0.93	1.10	1.15	1.18	1.38	1.38	1.40	1.25	1.11	0.99	1.01	13.9
		TOT DEM	KW	.00572	.00462	.00491	.00520	.00520	.00566	.00578	.00566	.00566	.00514	.00491	.00509	0.00578

(HT ELEC=ELECTRIC CONSUMPTION FOR SPACE AND WATER HEATING. REM ELEC=REMAINING ELECTRIC CONSUMPTION (NOT USED FOR HEATING). TOT ELEC=TOTAL ELECTRIC CONSUMPTION. HT DEM=PEAK ELECTRIC DEMAND FOR SPACE AND WATER HEATING. REM DEM=PEAK ELECTRIC DEMAND FOR REMAINING LOAD (NOT HEATING). TOT DEM=TOTAL PEAK ELECTRIC DEMAND. GAS=TOTAL GAS CONSUMPTION. ANNUAL FIGURES ARE SUMS OF MONTHLY CONSUMPTION VALUES OR ARE YEARLY MAXIMUMS OF MONTHLY PEAK DEMAND VALUES.)

STRAT SET	CONST COST $/SF	ENERGY FORM	UNITS (PER SF)	JAN	FEB	MAR	APR	MAY	JUN	JUL	AUG	SEP	OCT	NOV	DEC	ANNUAL
BASE CASE : 531 (CONTINUED)																
S- 36	0.42	TOT ELEC	KWH	0.95	0.86	1.01	1.06	1.09	1.27	1.27	1.29	1.15	1.02	0.91	0.93	12.8
		TOT DEM	KW	.00500	.00404	.00429	.00454	.00454	.00495	.00505	.00495	.00495	.00449	.00429	.00444	0.00505
S- 43	1.26	TOT ELEC	KWH	0.87	0.79	0.93	0.98	1.00	1.17	1.17	1.19	1.06	0.94	0.84	0.86	11.8
		TOT DEM	KW	.00482	.00390	.00414	.00438	.00438	.00477	.00487	.00477	.00477	.00433	.00414	.00429	0.00487
S- 48	0.45	TOT ELEC	KWH	0.92	0.83	0.98	1.03	1.05	1.23	1.23	1.25	1.12	0.99	0.88	0.91	12.4
		TOT DEM	KW	.00492	.00398	.00422	.00447	.00447	.00487	.00497	.00487	.00487	.00442	.00422	.00437	0.00497
BASE CASE : 532																
BASE	0.00	HT ELEC	KWH	0.24	0.12	0.07	0.02	0.02	0.02	0.02	0.02	0.02	0.03	0.12	0.25	0.9
		HT DEM	KW	.00385	.00385	.00289	.00189	.00158	.00000	.00000	.00000	.00089	.00227	.00381	.00385	0.00385
		REM ELEC	KWH	1.02	0.94	1.13	1.26	1.51	1.69	1.78	1.73	1.48	1.30	1.05	1.02	15.9
		REM DEM	KW	.00452	.00473	.00541	.00596	.00630	.00678	.00685	.00671	.00644	.00596	.00527	.00480	0.00685
		TOT ELEC	KWH	1.26	1.06	1.20	1.28	1.53	1.70	1.80	1.75	1.50	1.33	1.17	1.26	16.9
		TOT DEM	KW	.00705	.00571	.00564	.00620	.00663	.00691	.00691	.00684	.00677	.00634	.00564	.00705	0.00705
S- 3	-0.18	HT ELEC	KWH	0.26	0.13	0.08	0.03	0.02	0.02	0.02	0.02	0.02	0.03	0.13	0.27	1.0
		HT DEM	KW	.00385	.00385	.00289	.00189	.00158	.00000	.00000	.00000	.00089	.00227	.00381	.00385	0.00385
		REM ELEC	KWH	0.88	0.81	0.98	1.09	1.31	1.46	1.55	1.50	1.28	1.13	0.91	0.88	13.8
		REM DEM	KW	.00429	.00448	.00514	.00566	.00598	.00643	.00650	.00637	.00611	.00566	.00500	.00455	0.00650
		TOT ELEC	KWH	1.14	0.95	1.06	1.12	1.33	1.48	1.56	1.52	1.30	1.16	1.04	1.15	14.8
		TOT DEM	KW	.00665	.00539	.00532	.00585	.00625	.00652	.00652	.00645	.00638	.00598	.00532	.00665	0.00665
S- 7	-0.09	HT ELEC	KWH	0.28	0.14	0.08	0.03	0.02	0.02	0.02	0.02	0.02	0.03	0.14	0.29	1.1
		HT DEM	KW	.00385	.00385	.00289	.00189	.00158	.00000	.00000	.00000	.00089	.00227	.00381	.00385	0.00385
		REM ELEC	KWH	0.78	0.72	0.87	0.96	1.16	1.29	1.37	1.33	1.13	1.00	0.81	0.78	12.2
		REM DEM	KW	.00380	.00397	.00454	.00500	.00529	.00569	.00575	.00564	.00540	.00500	.00443	.00403	0.00575
		TOT ELEC	KWH	1.06	0.86	0.95	0.99	1.18	1.31	1.39	1.35	1.16	1.03	0.94	1.07	13.3
		TOT DEM	KW	.00640	.00518	.00512	.00563	.00602	.00627	.00627	.00621	.00614	.00576	.00512	.00640	0.00640
S- 16	0.98	HT ELEC	KWH	0.30	0.15	0.09	0.03	0.02	0.02	0.02	0.02	0.02	0.04	0.15	0.31	1.2
		HT DEM	KW	.00385	.00385	.00289	.00189	.00158	.00000	.00000	.00000	.00089	.00227	.00381	.00385	0.00385
		REM ELEC	KWH	0.69	0.64	0.77	0.85	1.03	1.14	1.21	1.18	1.00	0.89	0.71	0.69	10.8
		REM DEM	KW	.00342	.00357	.00409	.00451	.00477	.00513	.00518	.00508	.00487	.00451	.00399	.00363	0.00518
		TOT ELEC	KWH	0.99	0.79	0.86	0.88	1.05	1.17	1.23	1.20	1.03	0.92	0.86	1.00	12.0
		TOT DEM	KW	.00605	.00490	.00484	.00532	.00569	.00593	.00593	.00587	.00581	.00544	.00484	.00605	0.00605
S- 32	0.13	HT ELEC	KWH	0.32	0.14	0.13	0.05	0.03	0.02	0.02	0.02	0.02	0.06	0.13	0.24	1.2
		HT DEM	KW	.00385	.00385	.00289	.00189	.00158	.00000	.00000	.00000	.00089	.00227	.00381	.00385	0.00385
		REM ELEC	KWH	0.71	0.70	0.86	0.97	1.04	1.28	1.26	1.26	1.10	0.93	0.75	0.73	11.6
		REM DEM	KW	.00343	.00359	.00411	.00452	.00478	.00515	.00520	.00510	.00489	.00452	.00400	.00364	0.00520
		TOT ELEC	KWH	1.02	0.84	0.98	1.02	1.07	1.29	1.28	1.28	1.12	0.99	0.89	0.97	12.8
		TOT DEM	KW	.00655	.00550	.00419	.00452	.00478	.00524	.00531	.00524	.00485	.00445	.00537	.00616	0.00655
S- 40	0.85	HT ELEC	KWH	0.37	0.16	0.15	0.06	0.03	0.02	0.02	0.02	0.02	0.07	0.15	0.27	1.3
		HT DEM	KW	.00385	.00385	.00289	.00189	.00158	.00000	.00000	.00000	.00089	.00227	.00381	.00385	0.00385
		REM ELEC	KWH	0.65	0.64	0.78	0.89	0.95	1.17	1.16	1.16	1.01	0.85	0.69	0.67	10.6
		REM DEM	KW	.00319	.00333	.00382	.00420	.00444	.00478	.00483	.00473	.00454	.00420	.00372	.00338	0.00483
		TOT ELEC	KWH	1.01	0.80	0.93	0.95	0.99	1.19	1.18	1.18	1.03	0.92	0.84	0.94	11.9
		TOT DEM	KW	.00623	.00523	.00399	.00430	.00455	.00498	.00505	.00498	.00461	.00424	.00511	.00586	0.00623
S- 45	0.16	HT ELEC	KWH	0.34	0.15	0.13	0.05	0.03	0.02	0.02	0.02	0.02	0.06	0.14	0.25	1.2
		HT DEM	KW	.00385	.00385	.00289	.00189	.00158	.00000	.00000	.00000	.00089	.00227	.00381	.00385	0.00385
		REM ELEC	KWH	0.68	0.67	0.82	0.93	1.00	1.22	1.21	1.21	1.05	0.89	0.72	0.70	11.1
		REM DEM	KW	.00332	.00347	.00397	.00438	.00463	.00498	.00503	.00493	.00473	.00438	.00387	.00352	0.00503
		TOT ELEC	KWH	1.01	0.82	0.96	0.98	1.03	1.24	1.23	1.23	1.08	0.95	0.86	0.95	12.3
		TOT DEM	KW	.00634	.00533	.00406	.00437	.00463	.00507	.00514	.00507	.00469	.00431	.00520	.00596	0.00634
S- 61	-0.34	HT ELEC	KWH	0.28	0.14	0.08	0.03	0.02	0.02	0.02	0.02	0.02	0.03	0.14	0.29	1.1
		HT DEM	KW	.00385	.00385	.00289	.00189	.00158	.00000	.00000	.00000	.00089	.00227	.00381	.00385	0.00385
		REM ELEC	KWH	0.78	0.72	0.87	0.96	1.16	1.29	1.37	1.33	1.13	1.00	0.81	0.78	12.2
		REM DEM	KW	.00380	.00397	.00454	.00500	.00529	.00569	.00575	.00564	.00540	.00500	.00443	.00403	0.00575
		TOT ELEC	KWH	1.06	0.86	0.95	0.99	1.18	1.31	1.39	1.35	1.16	1.03	0.94	1.07	13.3
		TOT DEM	KW	.00640	.00518	.00512	.00563	.00602	.00627	.00627	.00621	.00614	.00576	.00512	.00640	0.00640
BASE CASE : 533																
BASE	0.00	TOT GAS	THM	0.0128	0.0066	0.0038	0.0013	0.0011	0.0010	0.0010	0.0010	0.0010	0.0015	0.0062	0.0130	0.050
		TOT ELEC	KWH	1.02	0.94	1.13	1.26	1.51	1.69	1.78	1.73	1.48	1.30	1.05	1.02	15.9
		TOT DEM	KW	.00452	.00473	.00541	.00596	.00630	.00678	.00685	.00671	.00644	.00596	.00527	.00480	0.00685
S- 3	-0.20	TOT GAS	THM	0.0137	0.0070	0.0041	0.0014	0.0011	0.0010	0.0010	0.0010	0.0010	0.0016	0.0067	0.0140	0.054
		TOT ELEC	KWH	0.88	0.81	0.98	1.09	1.31	1.46	1.55	1.50	1.28	1.13	0.91	0.88	13.8
		TOT DEM	KW	.00429	.00448	.00514	.00566	.00598	.00643	.00650	.00637	.00611	.00566	.00500	.00455	0.00650
S- 7	-0.11	TOT GAS	THM	0.0148	0.0076	0.0044	0.0015	0.0012	0.0011	0.0011	0.0011	0.0011	0.0017	0.0072	0.0150	0.058
		TOT ELEC	KWH	0.78	0.72	0.87	0.96	1.16	1.29	1.37	1.33	1.13	1.00	0.81	0.78	12.2
		TOT DEM	KW	.00380	.00397	.00454	.00500	.00529	.00569	.00575	.00564	.00540	.00500	.00443	.00403	0.00575
S- 16	0.94	TOT GAS	THM	0.0158	0.0081	0.0047	0.0016	0.0013	0.0012	0.0012	0.0012	0.0012	0.0019	0.0077	0.0160	0.062
		TOT ELEC	KWH	0.69	0.64	0.77	0.85	1.03	1.14	1.21	1.18	1.00	0.89	0.71	0.69	10.8
		TOT DEM	KW	.00342	.00357	.00409	.00451	.00477	.00513	.00518	.00508	.00487	.00451	.00399	.00363	0.00518
S- 32	0.12	TOT GAS	THM	0.0166	0.0075	0.0066	0.0025	0.0015	0.0009	0.0009	0.0009	0.0010	0.0031	0.0069	0.0124	0.061
		TOT ELEC	KWH	0.71	0.70	0.86	0.97	1.04	1.28	1.26	1.26	1.10	0.93	0.75	0.73	11.6
		TOT DEM	KW	.00343	.00359	.00411	.00452	.00478	.00515	.00520	.00510	.00489	.00452	.00400	.00364	0.00520
S- 40	0.83	TOT GAS	THM	0.0191	0.0086	0.0076	0.0029	0.0017	0.0011	0.0011	0.0011	0.0012	0.0036	0.0080	0.0144	0.070
		TOT ELEC	KWH	0.65	0.64	0.78	0.89	0.95	1.17	1.16	1.16	1.01	0.85	0.69	0.67	10.6
		TOT DEM	KW	.00319	.00333	.00382	.00420	.00444	.00478	.00483	.00473	.00454	.00420	.00372	.00338	0.00483
S- 45	0.14	TOT GAS	THM	0.0176	0.0075	0.0070	0.0027	0.0015	0.0010	0.0010	0.0010	0.0011	0.0033	0.0073	0.0132	0.064
		TOT ELEC	KWH	0.68	0.67	0.82	0.93	1.00	1.22	1.21	1.21	1.05	0.89	0.72	0.70	11.1
		TOT DEM	KW	.00332	.00347	.00397	.00438	.00463	.00498	.00503	.00493	.00473	.00438	.00387	.00352	0.00503

STRAT SET	CONST COST $/SF	ENERGY FORM	UNITS (PER SF)	JAN	FEB	MAR	APR	MAY	JUN	JUL	AUG	SEP	OCT	NOV	DEC	ANNUAL
BASE CASE : 533 (CONTINUED)																
S- 61	-0.36	TOT GAS	THM	0.0148	0.0076	0.0044	0.0015	0.0012	0.0011	0.0011	0.0011	0.0011	0.0017	0.0072	0.0150	0.058
		TOT ELEC	KWH	0.78	0.72	0.87	0.96	1.16	1.29	1.37	1.33	1.13	1.00	0.81	0.78	12.2
		TOT DEM	KW	.00380	.00397	.00454	.00500	.00529	.00569	.00575	.00564	.00540	.00500	.00443	.00403	0.00575
BASE CASE : 534																
BASE	0.00	TOT ELEC	KWH	1.36	1.23	1.45	1.53	1.56	1.82	1.82	1.86	1.66	1.47	1.31	1.34	18.4
		TOT DEM	KW	.00683	.00552	.00587	.00621	.00621	.00676	.00690	.00676	.00676	.00614	.00587	.00607	0.00690
S- 1	0.09	TOT ELEC	KWH	1.26	1.14	1.34	1.41	1.45	1.68	1.68	1.72	1.53	1.36	1.21	1.24	17.0
		TOT DEM	KW	.00567	.00458	.00487	.00516	.00516	.00562	.00573	.00562	.00562	.00510	.00487	.00504	0.00573
S- 12	0.20	TOT ELEC	KWH	1.13	1.03	1.21	1.27	1.30	1.51	1.51	1.55	1.38	1.22	1.09	1.12	15.3
		TOT DEM	KW	.00533	.00430	.00457	.00484	.00484	.00527	.00538	.00527	.00527	.00479	.00457	.00473	0.00538
S- 22	1.51	TOT ELEC	KWH	1.04	0.94	1.11	1.16	1.19	1.39	1.39	1.41	1.26	1.12	0.99	1.02	14.0
		TOT DEM	KW	.00508	.00410	.00436	.00462	.00462	.00503	.00513	.00503	.00503	.00457	.00436	.00451	0.00513
S- 28	-0.05	TOT ELEC	KWH	1.13	1.03	1.21	1.27	1.30	1.51	1.51	1.55	1.38	1.22	1.09	1.12	15.3
		TOT DEM	KW	.00533	.00430	.00457	.00484	.00484	.00527	.00538	.00527	.00527	.00479	.00457	.00473	0.00538
S- 36	0.43	TOT ELEC	KWH	0.99	0.90	1.06	1.11	1.14	1.33	1.33	1.35	1.21	1.07	0.95	0.98	13.4
		TOT DEM	KW	.00515	.00416	.00442	.00468	.00468	.00510	.00520	.00510	.00510	.00463	.00442	.00458	0.00520
S- 43	1.27	TOT ELEC	KWH	0.93	0.84	1.00	1.05	1.07	1.25	1.25	1.27	1.13	1.01	0.89	0.92	12.6
		TOT DEM	KW	.00499	.00403	.00428	.00454	.00454	.00494	.00504	.00494	.00494	.00449	.00428	.00444	0.00504
S- 48	0.45	TOT ELEC	KWH	0.96	0.87	1.03	1.08	1.11	1.29	1.29	1.31	1.17	1.04	0.92	0.95	13.0
		TOT DEM	KW	.00507	.00410	.00435	.00461	.00461	.00502	.00512	.00502	.00502	.00456	.00435	.00451	0.00512
BASE CASE : 535																
BASE	0.00	HT ELEC	KWH	0.43	0.22	0.13	0.04	0.04	0.03	0.03	0.03	0.03	0.05	0.21	0.44	1.7
		HT DEM	KW	.00385	.00350	.00266	.00189	.00139	.00000	.00000	.00000	.00012	.00235	.00385	.00385	0.00385
		REM ELEC	KWH	1.03	0.95	1.14	1.27	1.53	1.71	1.80	1.75	1.50	1.32	1.06	1.03	16.1
		REM DEM	KW	.00431	.00451	.00516	.00568	.00601	.00646	.00653	.00640	.00614	.00568	.00503	.00457	0.00653
		TOT ELEC	KWH	1.46	1.17	1.27	1.32	1.56	1.74	1.84	1.79	1.53	1.37	1.27	1.47	17.8
		TOT DEM	KW	.00705	.00486	.00522	.00571	.00613	.00627	.00642	.00627	.00613	.00578	.00494	.00571	0.00705
S- 3	-0.18	HT ELEC	KWH	0.46	0.24	0.14	0.05	0.04	0.03	0.03	0.03	0.03	0.05	0.22	0.47	1.8
		HT DEM	KW	.00385	.00350	.00266	.00189	.00139	.00000	.00000	.00000	.00012	.00235	.00385	.00385	0.00385
		REM ELEC	KWH	0.90	0.83	0.99	1.11	1.33	1.48	1.57	1.53	1.30	1.15	0.92	0.90	14.0
		REM DEM	KW	.00408	.00426	.00488	.00538	.00569	.00612	.00616	.00606	.00581	.00538	.00476	.00433	0.00618
		TOT ELEC	KWH	1.36	1.06	1.13	1.15	1.37	1.52	1.60	1.56	1.34	1.20	1.15	1.36	15.8
		TOT DEM	KW	.00660	.00455	.00488	.00535	.00574	.00587	.00601	.00587	.00574	.00541	.00462	.00535	0.00660
S- 7	-0.10	HT ELEC	KWH	0.50	0.25	0.15	0.05	0.04	0.04	0.04	0.04	0.04	0.06	0.24	0.50	1.9
		HT DEM	KW	.00385	.00350	.00266	.00189	.00139	.00000	.00000	.00000	.00012	.00235	.00385	.00385	0.00385
		REM ELEC	KWH	0.79	0.73	0.88	0.98	1.18	1.31	1.39	1.35	1.15	1.02	0.82	0.79	12.4
		REM DEM	KW	.00362	.00378	.00433	.00477	.00504	.00543	.00548	.00537	.00515	.00477	.00422	.00384	0.00548
		TOT ELEC	KWH	1.29	0.99	1.03	1.03	1.22	1.35	1.43	1.39	1.19	1.08	1.06	1.30	14.3
		TOT DEM	KW	.00635	.00438	.00470	.00514	.00552	.00565	.00578	.00565	.00552	.00521	.00445	.00514	0.00635
S- 16	0.97	HT ELEC	KWH	0.53	0.27	0.16	0.05	0.04	0.04	0.04	0.04	0.04	0.06	0.26	0.54	2.1
		HT DEM	KW	.00385	.00350	.00266	.00189	.00139	.00000	.00000	.00000	.00012	.00235	.00385	.00385	0.00385
		REM ELEC	KWH	0.70	0.64	0.77	0.86	1.04	1.16	1.22	1.19	1.01	0.89	0.72	0.70	10.9
		REM DEM	KW	.00325	.00340	.00389	.00429	.00454	.00488	.00493	.00483	.00463	.00429	.00380	.00345	0.00493
		TOT ELEC	KWH	1.23	0.92	0.93	0.92	1.08	1.19	1.26	1.23	1.05	0.96	0.98	1.24	13.0
		TOT DEM	KW	.00600	.00414	.00444	.00486	.00522	.00534	.00546	.00534	.00522	.00492	.00420	.00486	0.00600
S- 32	0.22	HT ELEC	KWH	0.64	0.29	0.25	0.10	0.06	0.03	0.03	0.03	0.04	0.12	0.27	0.48	2.3
		HT DEM	KW	.00385	.00350	.00266	.00189	.00139	.00000	.00000	.00000	.00012	.00235	.00385	.00385	0.00385
		REM ELEC	KWH	0.71	0.70	0.86	0.97	1.04	1.28	1.26	1.26	1.10	0.93	0.75	0.73	11.6
		REM DEM	KW	.00312	.00326	.00374	.00412	.00435	.00468	.00473	.00464	.00445	.00412	.00364	.00331	0.00473
		TOT ELEC	KWH	1.34	0.98	1.11	1.07	1.10	1.31	1.30	1.30	1.14	1.05	1.02	1.21	13.9
		TOT DEM	KW	.00655	.00485	.00406	.00439	.00452	.00498	.00498	.00498	.00465	.00426	.00472	.00531	0.00655
S- 61	-0.35	HT ELEC	KWH	0.50	0.25	0.15	0.05	0.04	0.04	0.04	0.04	0.04	0.06	0.24	0.50	1.9
		HT DEM	KW	.00385	.00350	.00266	.00189	.00139	.00000	.00000	.00000	.00012	.00235	.00385	.00385	0.00385
		REM ELEC	KWH	0.79	0.73	0.88	0.98	1.18	1.31	1.39	1.35	1.15	1.02	0.82	0.79	12.4
		REM DEM	KW	.00362	.00378	.00433	.00477	.00504	.00543	.00548	.00537	.00515	.00477	.00422	.00384	0.00548
		TOT ELEC	KWH	1.29	0.99	1.03	1.03	1.22	1.35	1.43	1.39	1.19	1.08	1.06	1.30	14.3
		TOT DEM	KW	.00635	.00438	.00470	.00514	.00552	.00565	.00578	.00565	.00552	.00521	.00445	.00514	0.00635
BASE CASE : 536																
BASE	0.00	TOT GAS	THM	0.0226	0.0116	0.0067	0.0023	0.0019	0.0017	0.0017	0.0017	0.0017	0.0026	0.0109	0.0229	0.088
		TOT ELEC	KWH	1.03	0.95	1.14	1.27	1.53	1.71	1.80	1.75	1.50	1.32	1.06	1.03	16.1
		TOT DEM	KW	.00431	.00451	.00516	.00568	.00601	.00646	.00653	.00640	.00614	.00568	.00503	.00457	0.00653
S- 3	-0.20	TOT GAS	THM	0.0242	0.0124	0.0072	0.0025	0.0020	0.0018	0.0018	0.0018	0.0018	0.0028	0.0117	0.0246	0.095
		TOT ELEC	KWH	0.90	0.83	0.99	1.11	1.33	1.48	1.57	1.53	1.30	1.15	0.92	0.90	14.0
		TOT DEM	KW	.00408	.00426	.00488	.00538	.00569	.00612	.00618	.00606	.00581	.00538	.00476	.00433	0.00618
S- 7	-0.12	TOT GAS	THM	0.0261	0.0134	0.0078	0.0027	0.0021	0.0019	0.0019	0.0019	0.0019	0.0031	0.0126	0.0265	0.102
		TOT ELEC	KWH	0.79	0.73	0.88	0.98	1.18	1.31	1.39	1.35	1.15	1.02	0.82	0.79	12.4
		TOT DEM	KW	.00362	.00378	.00433	.00477	.00504	.00543	.00548	.00537	.00515	.00477	.00422	.00384	0.00548
S- 16	0.94	TOT GAS	THM	0.0279	0.0143	0.0083	0.0028	0.0023	0.0021	0.0021	0.0021	0.0021	0.0033	0.0135	0.0283	0.109
		TOT ELEC	KWH	0.70	0.64	0.77	0.86	1.04	1.16	1.22	1.19	1.01	0.89	0.72	0.70	10.9
		TOT DEM	KW	.00325	.00340	.00389	.00429	.00454	.00488	.00493	.00483	.00463	.00429	.00380	.00345	0.00493

(HT ELEC=ELECTRIC CONSUMPTION FOR SPACE AND WATER HEATING. REM ELEC=REMAINING ELECTRIC CONSUMPTION (NOT USED FOR HEATING). TOT ELEC=TOTAL ELECTRIC CONSUMPTION. HT DEM=PEAK ELECTRIC DEMAND FOR SPACE AND WATER HEATING. REM DEM=PEAK ELECTRIC DEMAND FOR REMAINING LOAD (NOT HEATING). TOT DEM=TOTAL PEAK ELECTRIC DEMAND. GAS=TOTAL GAS CONSUMPTION. ANNUAL FIGURES ARE SUMS OF MONTHLY CONSUMPTION VALUES OR ARE YEARLY MAXIMUMS OF MONTHLY PEAK DEMAND VALUES.)

STRAT SET	CONST COST $/SF	ENERGY FORM	UNITS (PER SF)	JAN	FEB	MAR	APR	MAY	JUN	JUL	AUG	SEP	OCT	NOV	DEC	ANNUAL
BASE CASE : 536 (CONTINUED)																
S- 32	0.20	TOT GAS	THM	0.0333	0.0150	0.0133	0.0051	0.0029	0.0018	0.0018	0.0018	0.0021	0.0062	0.0139	0.0250	0.122
		TOT ELEC	KWH	0.71	0.70	0.86	0.97	1.04	1.28	1.26	1.26	1.10	0.93	0.75	0.73	11.6
		TOT DEM	KW	.00312	.00326	.00374	.00412	.00435	.00468	.00473	.00464	.00445	.00412	.00364	.00331	0.00473
S- 61	-0.37	TOT GAS	THM	0.0261	0.0134	0.0078	0.0027	0.0021	0.0019	0.0019	0.0019	0.0019	0.0031	0.0126	0.0265	0.102
		TOT ELEC	KWH	0.79	0.73	0.88	0.98	1.18	1.31	1.39	1.35	1.15	1.02	0.82	0.79	12.4
		TOT DEM	KW	.00362	.00378	.00433	.00477	.00504	.00543	.00548	.00537	.00515	.00477	.00422	.00384	0.00548
BASE CASE : 537																
BASE	0.00	TOT ELEC	KWH	1.15	1.04	1.22	1.29	1.32	1.53	1.53	1.57	1.40	1.24	1.10	1.13	15.5
		TOT DEM	KW	.00646	.00522	.00555	.00588	.00588	.00640	.00653	.00640	.00640	.00581	.00555	.00575	0.00653
S- 1	0.09	TOT ELEC	KWH	1.05	0.95	1.12	1.18	1.21	1.41	1.41	1.43	1.28	1.14	1.01	1.04	14.2
		TOT DEM	KW	.00575	.00465	.00494	.00523	.00523	.00569	.00581	.00569	.00569	.00517	.00494	.00511	0.00581
S- 12	0.19	TOT ELEC	KWH	0.94	0.85	1.00	1.05	1.08	1.26	1.26	1.28	1.14	1.02	0.90	0.93	12.7
		TOT DEM	KW	.00549	.00444	.00472	.00500	.00500	.00544	.00555	.00544	.00544	.00494	.00472	.00488	0.00555
S- 22	1.50	TOT ELEC	KWH	0.85	0.77	0.91	0.95	0.98	1.14	1.14	1.16	1.04	0.92	0.82	0.84	11.5
		TOT DEM	KW	.00524	.00423	.00450	.00476	.00476	.00518	.00529	.00518	.00518	.00471	.00450	.00466	0.00529
S- 28	-0.06	TOT ELEC	KWH	0.94	0.85	1.00	1.05	1.08	1.26	1.26	1.28	1.14	1.02	0.90	0.93	12.7
		TOT DEM	KW	.00549	.00444	.00472	.00500	.00500	.00544	.00555	.00544	.00544	.00494	.00472	.00488	0.00555
S- 36	0.51	TOT ELEC	KWH	0.92	0.83	0.98	1.03	1.05	1.23	1.23	1.25	1.12	0.99	0.88	0.91	12.4
		TOT DEM	KW	.00513	.00414	.00440	.00466	.00466	.00508	.00518	.00508	.00508	.00461	.00440	.00456	0.00518
BASE CASE : 538																
BASE	0.00	HT ELEC	KWH	0.41	0.21	0.12	0.04	0.03	0.03	0.03	0.03	0.03	0.05	0.20	0.42	1.6
		HT DEM	KW	.00385	.00385	.00289	.00189	.00158	.00000	.00000	.00000	.00089	.00227	.00381	.00385	0.00385
		REM ELEC	KWH	1.08	1.00	1.20	1.34	1.61	1.79	1.89	1.84	1.57	1.39	1.12	1.08	16.9
		REM DEM	KW	.00446	.00466	.00533	.00587	.00621	.00668	.00675	.00662	.00635	.00587	.00520	.00472	0.00675
		TOT ELEC	KWH	1.49	1.21	1.32	1.38	1.64	1.82	1.92	1.87	1.60	1.43	1.31	1.50	18.5
		TOT DEM	KW	.00708	.00573	.00566	.00623	.00666	.00694	.00694	.00687	.00680	.00637	.00566	.00708	0.00708
S- 3	-0.23	HT ELEC	KWH	0.44	0.22	0.13	0.04	0.04	0.03	0.03	0.03	0.03	0.05	0.21	0.44	1.7
		HT DEM	KW	.00385	.00385	.00289	.00189	.00158	.00000	.00000	.00000	.00089	.00227	.00381	.00385	0.00385
		REM ELEC	KWH	0.94	0.87	1.04	1.16	1.40	1.56	1.65	1.60	1.37	1.21	0.97	0.94	14.7
		REM DEM	KW	.00423	.00442	.00506	.00558	.00590	.00635	.00641	.00628	.00603	.00558	.00494	.00449	0.00641
		TOT ELEC	KWH	1.38	1.09	1.17	1.21	1.43	1.59	1.68	1.63	1.40	1.26	1.18	1.39	16.4
		TOT DEM	KW	.00668	.00541	.00534	.00588	.00628	.00655	.00655	.00648	.00641	.00601	.00534	.00668	0.00668
S- 7	-0.14	HT ELEC	KWH	0.47	0.24	0.14	0.05	0.04	0.03	0.03	0.03	0.03	0.06	0.23	0.48	1.8
		HT DEM	KW	.00385	.00385	.00289	.00189	.00158	.00000	.00000	.00000	.00089	.00227	.00381	.00385	0.00385
		REM ELEC	KWH	0.83	0.77	0.92	1.03	1.24	1.38	1.46	1.42	1.21	1.07	0.86	0.83	13.0
		REM DEM	KW	.00374	.00391	.00448	.00493	.00522	.00561	.00567	.00556	.00533	.00493	.00437	.00397	0.00567
		TOT ELEC	KWH	1.30	1.01	1.06	1.07	1.27	1.41	1.49	1.45	1.24	1.12	1.09	1.31	14.8
		TOT DEM	KW	.00643	.00521	.00514	.00566	.00604	.00630	.00630	.00624	.00617	.00579	.00514	.00643	0.00643
S- 16	0.89	HT ELEC	KWH	0.50	0.26	0.15	0.05	0.04	0.04	0.04	0.04	0.04	0.06	0.24	0.51	2.0
		HT DEM	KW	.00385	.00385	.00289	.00189	.00158	.00000	.00000	.00000	.00089	.00227	.00381	.00385	0.00385
		REM ELEC	KWH	0.74	0.68	0.82	0.91	1.09	1.22	1.29	1.25	1.07	0.94	0.76	0.74	11.5
		REM DEM	KW	.00337	.00352	.00403	.00444	.00469	.00505	.00510	.00500	.00479	.00444	.00393	.00357	0.00510
		TOT ELEC	KWH	1.24	0.94	0.97	0.96	1.13	1.26	1.33	1.29	1.11	1.00	1.00	1.25	13.5
		TOT DEM	KW	.00608	.00492	.00486	.00535	.00572	.00596	.00596	.00590	.00584	.00547	.00486	.00608	0.00608
S- 32	0.18	HT ELEC	KWH	0.54	0.24	0.21	0.08	0.05	0.03	0.03	0.03	0.03	0.10	0.22	0.40	2.0
		HT DEM	KW	.00385	.00385	.00289	.00189	.00158	.00000	.00000	.00000	.00089	.00227	.00381	.00385	0.00385
		REM ELEC	KWH	0.73	0.71	0.88	1.00	1.07	1.31	1.30	1.30	1.13	0.95	0.77	0.75	11.9
		REM DEM	KW	.00333	.00348	.00399	.00439	.00465	.00500	.00505	.00495	.00475	.00439	.00389	.00353	0.00505
		TOT ELEC	KWH	1.26	0.96	1.09	1.08	1.12	1.34	1.33	1.33	1.16	1.05	1.00	1.15	13.9
		TOT DEM	KW	.00658	.00553	.00421	.00454	.00480	.00526	.00533	.00526	.00487	.00447	.00540	.00619	0.00658
S- 61	-0.38	HT ELEC	KWH	0.47	0.24	0.14	0.05	0.04	0.03	0.03	0.03	0.03	0.06	0.23	0.48	1.8
		HT DEM	KW	.00385	.00385	.00289	.00189	.00158	.00000	.00000	.00000	.00089	.00227	.00381	.00385	0.00385
		REM ELEC	KWH	0.83	0.77	0.92	1.03	1.24	1.38	1.46	1.42	1.21	1.07	0.86	0.83	13.0
		REM DEM	KW	.00374	.00391	.00448	.00493	.00522	.00561	.00567	.00556	.00533	.00493	.00437	.00397	0.00567
		TOT ELEC	KWH	1.30	1.01	1.06	1.07	1.27	1.41	1.49	1.45	1.24	1.12	1.09	1.31	14.8
		TOT DEM	KW	.00643	.00521	.00514	.00566	.00604	.00630	.00630	.00624	.00617	.00579	.00514	.00643	0.00643
BASE CASE : 539																
BASE	0.00	TOT GAS	THM	0.0215	0.0110	0.0064	0.0022	0.0018	0.0016	0.0016	0.0016	0.0016	0.0025	0.0104	0.0218	0.084
		TOT ELEC	KWH	1.08	1.00	1.20	1.34	1.61	1.79	1.89	1.84	1.57	1.39	1.12	1.08	16.9
		TOT DEM	KW	.00446	.00466	.00533	.00587	.00621	.00668	.00675	.00662	.00635	.00587	.00520	.00472	0.00675
S- 3	-0.25	TOT GAS	THM	0.0230	0.0118	0.0068	0.0023	0.0019	0.0017	0.0017	0.0017	0.0017	0.0027	0.0111	0.0233	0.090
		TOT ELEC	KWH	0.94	0.87	1.04	1.16	1.40	1.56	1.65	1.60	1.37	1.21	0.97	0.94	14.7
		TOT DEM	KW	.00423	.00442	.00506	.00558	.00590	.00635	.00641	.00628	.00603	.00558	.00494	.00449	0.00641
S- 7	-0.16	TOT GAS	THM	0.0247	0.0126	0.0073	0.0025	0.0020	0.0018	0.0018	0.0018	0.0018	0.0029	0.0120	0.0251	0.097
		TOT ELEC	KWH	0.83	0.77	0.92	1.03	1.24	1.38	1.46	1.42	1.21	1.07	0.86	0.83	13.0
		TOT DEM	KW	.00374	.00391	.00448	.00493	.00522	.00561	.00567	.00556	.00533	.00493	.00437	.00397	0.00567
S- 16	0.85	TOT GAS	THM	0.0264	0.0135	0.0078	0.0027	0.0022	0.0020	0.0020	0.0020	0.0020	0.0031	0.0128	0.0268	0.103
		TOT ELEC	KWH	0.74	0.68	0.82	0.91	1.09	1.22	1.29	1.25	1.07	0.94	0.76	0.74	11.5
		TOT DEM	KW	.00337	.00352	.00403	.00444	.00469	.00505	.00510	.00500	.00479	.00444	.00393	.00357	0.00510
S- 32	0.16	TOT GAS	THM	0.0281	0.0127	0.0112	0.0043	0.0025	0.0015	0.0015	0.0015	0.0018	0.0053	0.0117	0.0211	0.103
		TOT ELEC	KWH	0.73	0.71	0.88	1.00	1.07	1.31	1.30	1.30	1.13	0.95	0.77	0.75	11.9
		TOT DEM	KW	.00333	.00348	.00399	.00439	.00465	.00500	.00505	.00495	.00475	.00439	.00389	.00353	0.00505
S- 61	-0.38	TOT GAS	THM	0.0247	0.0126	0.0073	0.0025	0.0020	0.0018	0.0018	0.0018	0.0018	0.0029	0.0120	0.0251	0.097
		TOT ELEC	KWH	0.83	0.77	0.92	1.03	1.24	1.38	1.46	1.42	1.21	1.07	0.86	0.83	13.0
		TOT DEM	KW	.00374	.00391	.00448	.00493	.00522	.00561	.00567	.00556	.00533	.00493	.00437	.00397	0.00567

STRAT SET	CONST COST $/SF	ENERGY FORM	UNITS (PER SF)	JAN	FEB	MAR	APR	MAY	JUN	JUL	AUG	SEP	OCT	NOV	DEC	ANNUAL
BASE CASE : 540																
BASE	0.00	TOT ELEC	KWH	1.44	1.31	1.54	1.62	1.66	1.93	1.93	1.97	1.76	1.56	1.38	1.42	19.5
		TOT DEM	KW	.00726	.00586	.00623	.00660	.00660	.00718	.00733	.00718	.00718	.00652	.00623	.00645	0.00733
S- 1	0.09	TOT ELEC	KWH	1.35	1.22	1.44	1.51	1.55	1.80	1.80	1.84	1.64	1.46	1.29	1.33	18.2
		TOT DEM	KW	.00689	.00557	.00592	.00626	.00626	.00682	.00696	.00682	.00682	.00619	.00592	.00612	0.00696
S- 12	0.20	TOT ELEC	KWH	1.21	1.09	1.29	1.35	1.39	1.61	1.61	1.65	1.47	1.30	1.16	1.19	16.3
		TOT DEM	KW	.00610	.00493	.00524	.00554	.00554	.00604	.00616	.00604	.00604	.00548	.00524	.00542	0.00616
S- 22	1.51	TOT ELEC	KWH	1.10	1.00	1.18	1.24	1.27	1.48	1.48	1.50	1.34	1.19	1.06	1.09	14.9
		TOT DEM	KW	.00556	.00450	.00478	.00506	.00506	.00551	.00562	.00551	.00551	.00500	.00478	.00495	0.00562
S- 28	-0.05	TOT ELEC	KWH	1.21	1.09	1.29	1.35	1.39	1.61	1.61	1.65	1.47	1.30	1.16	1.19	16.3
		TOT DEM	KW	.00610	.00493	.00524	.00554	.00554	.00604	.00616	.00604	.00604	.00548	.00524	.00542	0.00616
S- 36	0.52	TOT ELEC	KWH	0.98	0.89	1.05	1.10	1.13	1.32	1.32	1.34	1.20	1.06	0.94	0.97	13.3
		TOT DEM	KW	.00528	.00426	.00453	.00480	.00480	.00522	.00533	.00522	.00522	.00474	.00453	.00469	0.00533

(HT ELEC=ELECTRIC CONSUMPTION FOR SPACE AND WATER HEATING. REM ELEC=REMAINING ELECTRIC CONSUMPTION (NOT USED FOR HEATING). TOT ELEC=TOTAL ELECTRIC CONSUMPTION. HT DEM=PEAK ELECTRIC DEMAND FOR SPACE AND WATER HEATING. REM DEM=PEAK ELECTRIC DEMAND FOR REMAINING LOAD (NOT HEATING). TOT DEM=TOTAL PEAK ELECTRIC DEMAND. GAS=TOTAL GAS CONSUMPTION. ANNUAL FIGURES ARE SUMS OF MONTHLY CONSUMPTION VALUES OR ARE YEARLY MAXIMUMS OF MONTHLY PEAK DEMAND VALUES.)

SECTION **II**

DETAILS OF STRATEGY LEVELS

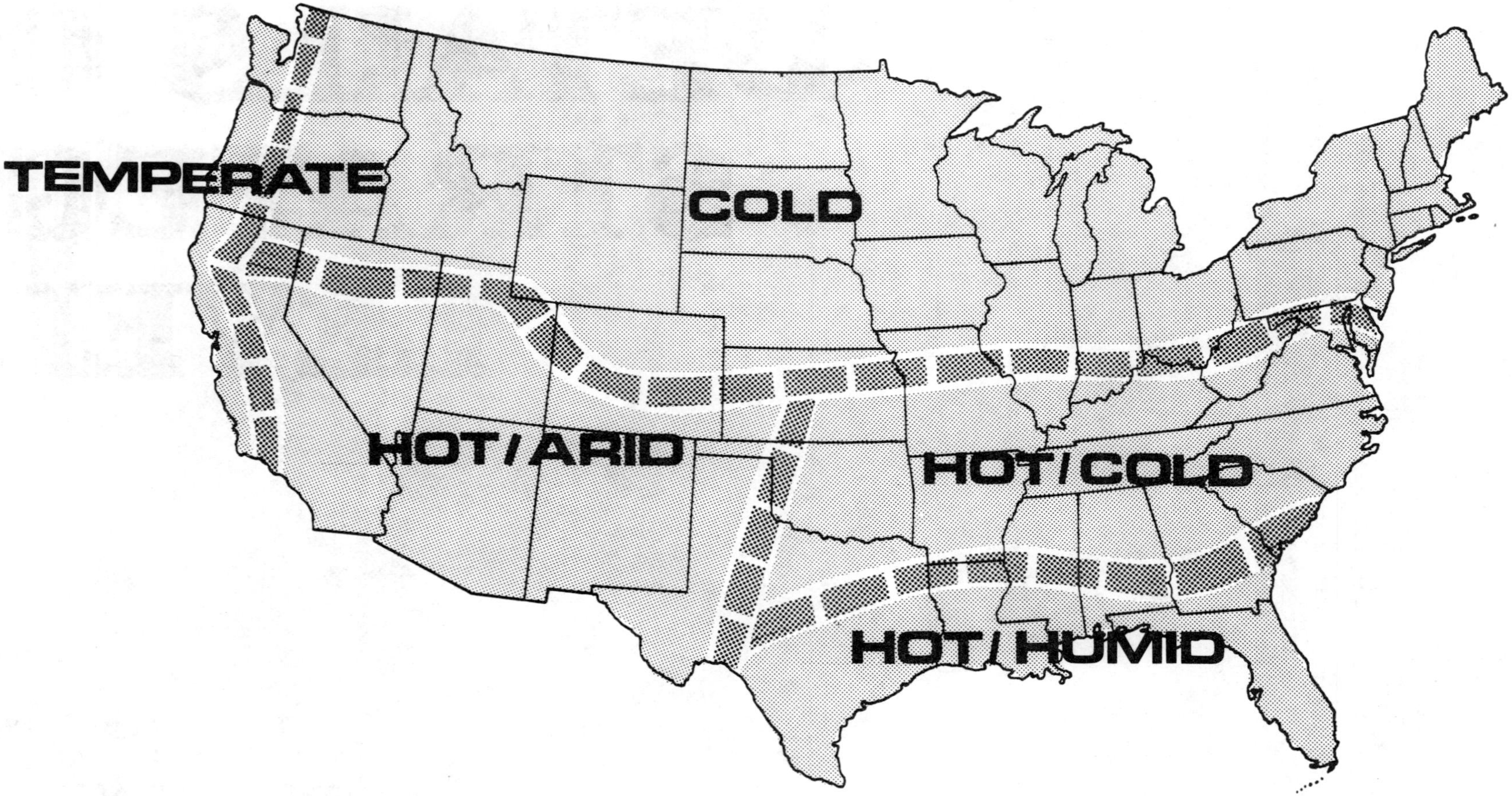
TEMPERATE
COLD
HOT / ARID
HOT / COLD
HOT / HUMID

Climate Regions

COLD

CO	Denver Colorado Springs
CT	Hartford New Haven
DE	Wilmington
IA	Cedar Rapids Des Moines Sioux City
ID	Boise Lewiston
IL	Chicago Rockford Springfield
IN	Ft. Wayne Indianapolis South Bend
KA	Concordia
MA	Boston Fall River Springfield
ME	Portland Presque Isle
MI	Detroit Flint Grand Rapids Sault Ste. Marie
MN	Duluth Minneapolis/ St. Paul
MO	St. Joseph
MT	Billings Butte Great Falls
NB	Omaha Scottsbluff
ND	Bismarck Fargo
NH	Manchester
NJ	Atlantic City Camden Newark Trenton
NV	Elko
NY	Albany Binghamton Buffalo New York City Plattsburg Rochester Syracuse
OH	Akron Cincinnati Cleveland Columbus Dayton Toledo
OR	Pendleton Klamath Falls
PA	Erie Harrisburg Pittsburg Philadelphia Scranton/ Wilkes Barre
RI	Providence
SD	Rapid City Sioux Falls
UT	Ogden
VT	Burlington Rutland
WA	Spokane Yakima
WI	Appleton Eau Claire Green Bay Madison Milwaukee
WV	Wheeling
WY	Casper Cheyenne

Energy requirements for small office building design depend on climate characteristics at the building location. Different climate characteristics may suggest different design features. A procedure is therefore needed to identify climate regions across the United States. A continuous spectrum of climate conditions occurs across the United States, limiting the ability to categorize the country into a relatively few (5–10) climate regions. For this handbook on small office building design, a "climate region map" is constructed to reflect general climate characteristics while allowing for subjective input.

Five climate regions are used to represent the range of conditions observed in the contiguous United States. The adjacent map presents the five regions covering the United States. Each climate region is represented by a city used in developing energy strategies. The representative cities are: Madison, WI (cold); Seattle, WA (temperate); Washington, D.C. (hot/cold); Lake Charles, LA (hot/humid); and El Paso, TX (hot/dry). Using a specific statistical criterion, the criteria included in each of the five regions are more similar to other cities in the same region than to cities in other regions. Cities near climate region boundaries tend to be similar to nearby cities from another region. This similarity must be recognized by users of the climate region map and factored into the energy strategy selection process. It is possible that a city near a climate region boundary could be better represented by the adjacent climate. Specific knowledge of the local climate should be used in choosing a representative climate region. Effects of elevation are particularly important to consider. The climate map is intended as a guide in the selection process and not as a definitive statement on climate regions.

The remainder of this section describes briefly the process used to construct the five regions. The data used to characterize the climate for each city are hourly climate data for a "typical" year, called Typical Meteorological Year (TMY) data. TMY data are derived using a procedure developed by Sandia Laboratories and are available for over 200 United States locations.

TMY consists of hourly weather values suitable for use in building energy-load calculations. The approach adopted for selecting a TMY for a given location is as follows: a typical month for each of the 12 calendar months from the long term period of record (23 years for most locations) is chosen, and then these 12 months are linked to form the TMY. The data set generated to form the basis for selection of a typical month consists of 13 daily indices calculated from the hourly values of dry bulb temperatures, dew point temperature, wind velocity and solar radiation. Month/year combinations having statistics that are "close" to the longer term statistics are candidates for typical months. Final selection of a typical month includes consideration of persistence of the weather patterns.

Ten climate indicators are constructed from the hourly TMY data and are the basis for the development of the climate regions. The indicators are annual summaries as follows:

Average dry bulb temperature (°F)
Average daily dry bulb maximum (°F)
Average daily dry bulb minimum (°F)
Average wet bulb temperature (°F)
Average humidity ratio (lbs. moisture/lb. dry air)
Cooling degree hours, base 57°F (°F-hr.)
Heating degree hours, base 45°F (°F-hr.)
Cooling humidity ratio hours, base 55°F, 0.010 humidity ratio
Heating humidity ratio hours, base 45°F, 0.005 humidity ratio
Average daily solar radiation on horizontal surface

The climate indicators are selected using the procedures derived by Bowen, Moreno and Olsen (1982). They developed the indicators for their relationship to building energy use. Heating and cooling degree hours measure how much a building must be heated and cooled. Heating and

TEMPERATE

CA	Sacramento San Francisco San Jose Santa Barbara
OR	Eugene Portland
WA	Seattle Tacoma

HOT/COLD

AL	Birmingham Huntsville Montgomery
AR	Fort Smith Little Rock
DC	Washington
DE	Dover
GA	Atlanta Augusta Columbus Macon
IL	Cairo East St. Louis
IN	Evansville
KA	Topeka Wichita
KY	Lexington Louisville
MD	Baltimore Hagerstown
MO	Kansas City Springfield St. Louis
MS	Meridian
NC	Asheville Charlotte Durham Greensboro Wilmington Winston-Salem
OK	Oklahoma City Tulsa
SC	Columbia Greenville
TN	Chattanooga Knoxville Memphis Nashville
TX	Abilene Dallas San Angelo Texarkana Wichita Falls
VA	Norfolk Richmond Roanoke
WV	Charleston Huntington

HOT/HUMID

AL	Mobile
FL	Jacksonville Miami Orlando Pensacola Tampa
GA	Albany Savannah
LA	Baton Rouge New Orleans Shreveport
MS	Biloxi Jackson
SC	Charleston
TX	Austin Beaumont Brownsville Corpus Christi Houston Laredo San Antonio Waco

HOT/ARID

AZ	Flagstaff Phoenix Tucson
CA	Bakersfield Fresno Los Angeles San Bernardino San Diego
CO	Durango Grand Junction Pueblo
KA	Dodge City
NM	Albuquerque Roswell Santa Fe
NV	Las Vegas Reno
TX	Amarillo El Paso Lubbock
UT	Salt Lake City

cooling humidity ratio hours measure low humidity during the periods a building must be heated and excessive humidity during the periods when the buildings must be cooled.

The ten climate indicators are constructed for the 183 cities available and are used to construct the climate regions. A statistical procedure, K-means cluster analysis, is used to develop clusters of cities similar in climate. Similarity is based on the similarity between cities on the ten climate indicators. If only two climate indicators were used, then the procedure defines clusters using cities that are "close" together on a plot of the cities using the indicators. The final five climate regions have been based on only the last six of the ten climate indicators listed. These six are more directly related to climate effects on building energy loads. Final assignment of cities to the five climate regions considered the results of K-means cluster analysis using ten and fifteen zones and other subsets of climate indicators.

LEVEL 1: Start Right

Current practice in office building design represents a marked improvement in energy efficiency over design of the recent past. Many of the sources of heavy energy use have been eliminated by the modest upgrading in building design that has occurred in the last few years. To a considerable degree, this has resulted from the guidance given by energy efficiency standards of the American Society of Heating, Refrigerating and Air-Conditioning Engineers, ASHRAE Standard 90–75, "Energy Conservation in New Building Design." Appreciable improvement has also occurred, however, because of an increased understanding of the importance of the quality of the office work place for productivity and personal satisfaction. In particular, this understanding has led to improvements in glass characteristics and in interior lighting.

Today, all 50 states have some form of code or standard for energy conservation in new buildings, and these codes all draw at least some of their criteria and/or format from ASHRAE 90–75 or ASHRAE 90A-1980 (from National Conference of States on Building Codes and Standards, Inc. *State Codes for Energy Conservation in New Buildings; January 1984*, Washington, D.C., Edison Electric Institute, 1983). Recent surveys indicate, however, that the compliance level in the states is only about 60%, with the remaining buildings being less energy efficient. After consultation with a variety of technical experts representing both designers and building owners, it was determined that the energy conservation level indicated by ASHRAE 90–75 should be selected as the baseline or starting point for using this handbook. Although this standard is not difficult to achieve by conventional design and construction practice, the fact that so many buildings that do not meet this level are still constructed indicates the importance of assuring that the design you are investigating is in compliance.

The increase in energy use that can result from substandard building and mechanical characteristics is substantial. In an analysis done on a 40,000 sq. ft. four story building with 40% glazing and a constant volume/gas HVAC system, the following differences were seen for three of the representative climates:

CLIMATE	COLD		HOT/COLD		HOT/ARID	
Construction Quality	Base Case	Poor Qual.	Base Case	Poor Qual.	Base Case	Poor Qual.
Glazing	Double Tinted	Double Clear	Double Tinted	Single Clear	Double Tinted	Single Clear
Wall-R	R-10	R-10	R-10	R-3	R-10	R-3
Roof-R	R-17.5	R-11	R-11	R-7.5	R-11	R-7.5
Night Infilt.	.33 ACH	.5 ACH	.33 ACH	.5 ACH	.33 ACH	.5 ACH
Economizer	Yes	No	Yes	No	Yes	No
Annual energy use Btu/SF/yr.	78.6	102.1	76.3	118.3	83.9	116.1
% increase	30%		55%		38%	

This chapter describes the characteristics that are included in the energy simulations for the buildings in Level 1.

INSULATION LEVELS—WALLS AND ROOF

According to ASHRAE Standard 90–75, the insulation requirements for wall and roof areas vary as a function of climate, building height, and glass area and type. In the simulations in this handbook, the resistance (R) values for the wall and roof sections were selected to comply with the Overall Thermal Transfer Value (OTTV) requirements of the Standard. Specific values for each of the Level 1 building choices are shown in Section I. The required values for wall insulation are not difficult to achieve through con-

BASE CASE BUILDINGS WALL SECTIONS

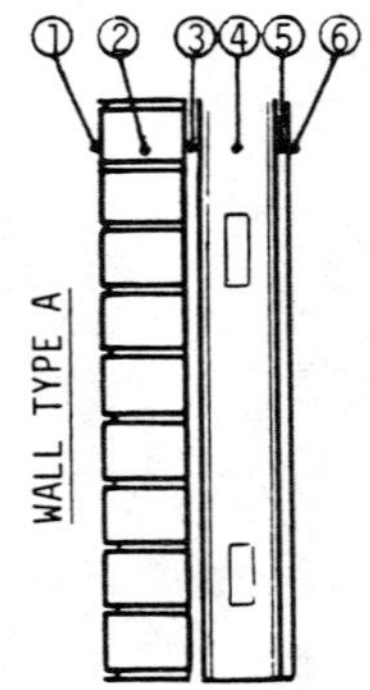

	R-VALUE W/O INSUL	R-VALUE W/INSUL
① Air Film	0.17	0.17
② Brick 4"	0.80	0.80
③ Air Space 1/2"	0.81	0.75
④ Fiberglass Batt Insulation & Steel Studs	--	7.53
⑤ Gypsum 1/2"	0.45	0.45
⑥ Air Film	0.68	0.68
	2.91	10.38

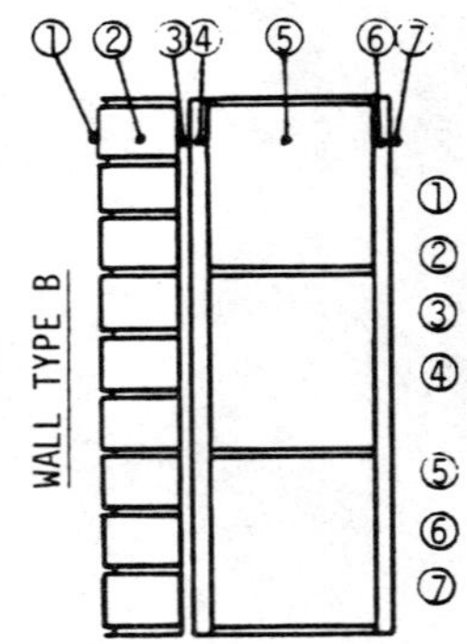

	R-VALUE W/O INSUL	R-VALUE W/INSUL
① Air Film	0.17	0.17
② Brick 4"	0.80	0.80
③ Air Space 1/2"	0.75	0.75
④ Polystyrene Insulation 3/4"	--	3.8
⑤ Block 8"	1.72	1.72
⑥ Gypsum	0.45	0.45
⑦ Air Film	0.68	0.68
	4.57	8.37

BASE CASE BUILDINGS ROOF SECTIONS

TYPE A

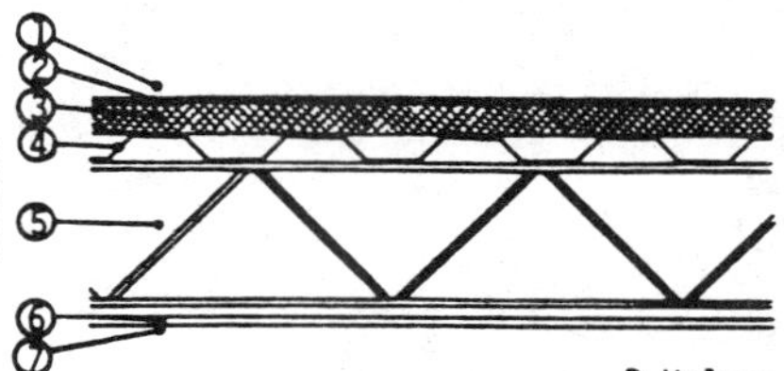

		R-Value
1	Exterior Air Film	.20
2	Built-Up Roofing	.33
3	Rigid Insulation	8.33
4	Metal Deck	.01
5	Air Space	.92
6	Acoustical Tile	1.25
7	Interior Air Film	.62
	TOTAL	11.66

TYPE B

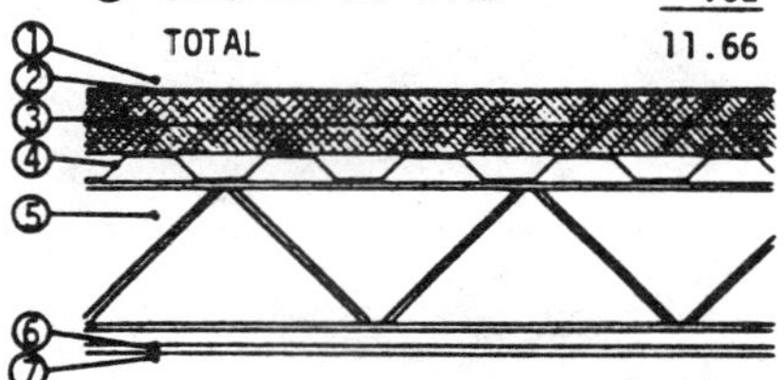

		R-Value
1	Exterior Air Film	.20
2	Built-Up Roofing	.33
3	Rigid Insulation	14.30
4	Metal Deck	.01
5	Air Space	.92
6	Acoustical Tile	1.25
7	Interior Air Film	.62
	TOTAL	17.63

ventional construction practice. The accompanying sections are examples of two wall types that can accommodate the required level of thermal resistance.

The masonry wall section was taken as representative for the hot/humid climate buildings up to two stories in height. Above two stories, it is felt that the lighter construction of the stud wall plus insulation cross section, with or without masonry cladding, would be most common.

In all other climates the stud wall plus insulation cross section is believed to be most representative of contemporary construction practices. As the 3½" batt insulation makes up most of the thermal resistance of this wall, inclusion or elimination of the masonry exterior cladding will have little effect on the calculated energy values in the handbook. Thus, these values can be used with a variety of wall sections as long as the insulation layer is included.

In the uninsulated wall, this is not true; the masonry accounts for roughly one-half of the thermal resistance of the wall. If it is desired to eliminate masonry, then insulation should be provided in the wall—an inexpensive option.

Except in the uninsulated wall, the insulation thicknesses have been determined by selecting the least costly common construction practice that will meet the ASHRAE 90–75 requirements. In most cases this results in a level of insulation (3½" batt) that is above the minimums by an appreciable amount. The addition of more insulation will generally have very little beneficial effect on energy costs, because relatively little of these costs is attributable to heat loss or gain through the opaque surfaces. At this level of wall insulation, most remaining heat gain or loss is through the glass areas.

Similarly, for the roof, once the least cost insulation levels indicated have been provided, very little utility cost improvement will be achieved by increasing to the next common level of insulation.

The use of perimeter insulation was included for locations where slab-on-grade construction was used. The resistance of the perimeter insulation is equal to R-5.

GLAZING

The type and amount of glazing chosen for a particular building is an important consideration for both architectural and energy reasons. In our simulations, the glazing type was selected based upon typical practice for buildings of different sizes in each region. These selections were made after extensive discussion with designers and building owners from each of the climatic regions. With the exceptions of both tinted single glazing in small buildings in hot climates and reflective single glazing in the hot/humid climate, all of the glazing types are double glazed. In addition to thermal insulation, which is valuable in all climate regions, double glazing controls condensation under hot/humid conditions. Depending upon the building size and location, the glazing may be clear (shading coefficient 0.81) or tinted (shading coefficient 0.55). The reflective glass has a shading coefficient of 0.36. The percentage of the exterior walls that is glazed is a major design factor selected in Level 2 by the user of this handbook. The choices vary with building size and location, based upon the typical practice in each case. A sufficient range has been provided to allow a variety of architectural expression.

INFILTRATION

The infiltration of outside air into a building as a function of pressure differential can be a significant influence on energy use, as the outside air must be conditioned to satisfy interior comfort requirements. All of the buildings included in these simulations have ventilation systems that draw in outside air under controlled conditions through operable dampers and distribute that air through the mechanical supply system. During periods when the ventilation system is operating, the building is assumed to be under positive pressure compared to exterior conditions. Normally, this results in an exfiltration of conditioned air through the building skin rather than an uncontrolled infiltration into the building. To represent this,the infiltration level has been set at zero during those occupied periods when

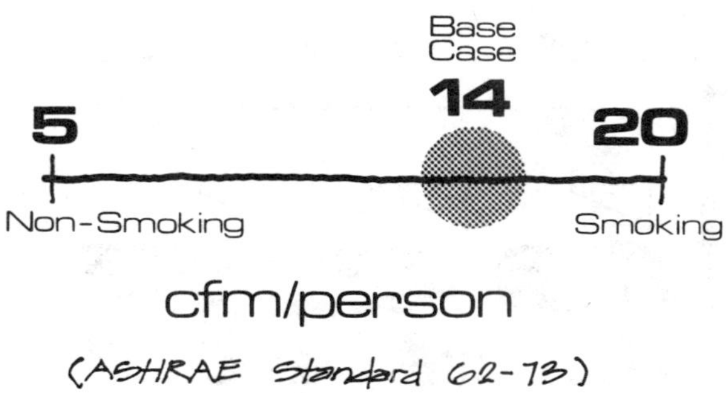

the ventilation system is operational. Because the ventilation system is not operational during periods when the building is not occupied, a positive pressure will not be maintained and infiltration will likely result. In these calculations, the infiltration rate that was simulated is equal to one-third to one-half of an air change per hour, varying with building size. As shown in the detailed descriptions in Section I, smaller buildings having a higher skin to volume ratio are assumed to have a slightly higher infiltration rate.

In order to achieve these levels of control of the infiltration of outside air, careful construction practice is required, including thorough caulking of cracks and reasonable quality gasketed assemblies (windows and curtain wall sections). It is recommended that buildings include vestibules at major entrances and that exterior glazing and wall construction meet industry standards.

VENTILATION

As described previously, all of the buildings in this handbook are equipped with a ventilation air system that draws in outside air under controlled conditions and distributes it through the mechanical air supply system during occupied periods. In our simulations, outside air is drawn into the building at the rate of 14 cubic feet per minute per person, a weighted average based upon the requirements for smokers and non-smokers. When the building is not occupied, the ventilation system is assumed to be off.

In office buildings throughout most parts of the country, an economizer cycle is employed as a part of the ventilation system to reduce energy requirements. The economizer cycle draws in outside air, when ambient conditions permit, to provide required cooling to the interior of the building without the operation of the mechanical cooling system. The fans that are included in the ventilation system operate to distribute the outside air through the supply air ductwork. In our simulations, the control of the economizer cycle is either through the use of a sensible (dry bulb) temperature indicator or through an enthalpy sensor, which includes both temperature and moisture considerations. The economizer cycle outside air dampers are assumed to be low leakage; when fully closed and at standard conditions, leakage should not exceed 3%.

An economizer cycle has not been included in the hot/humid climate region, as it is not standard practice for office buildings in those locations. Additionally, designers may wish *not* to use the economizer cycle in hot/arid climates if local climate conditions (such as dust storms) make its use inappropriate. Some mechanical system choices do not readily permit the incorporation of an economizer cycle. In this handbook, small packaged terminal air conditioning units, small heat pumps, and the hydronic loop heat pumps fall into this category. Accordingly, in our simulations, an economizer cycle was not included with these mechanical system choices.

In any economizer cycle system, it is important to make careful selection of the controls and damper equipment. Controller drift and damper leakage can seriously compromise the energy-saving potential of economizer cycles. It is estimated that 10–20% or more of the effectiveness of an economizer may be lost through lack of maintenance and system performance degradation. Economizers offer major savings, but they *must* be well maintained.

LIGHTING AND EQUIPMENT

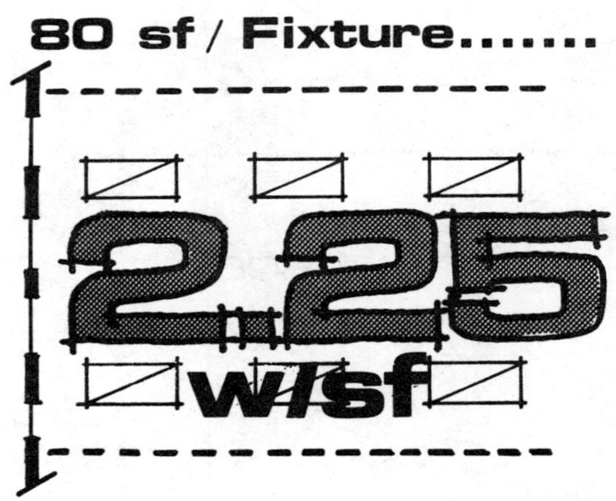

Artificial lights and office equipment are major factors in both the direct consumption of electricity and the generation of heat that must be accommodated by the heating/cooling mechanical systems within the building. In our simulations, the overall average for lighting is assumed to be 2.25 watts/sq. ft. This is equivalent to one, 2 ft. × 4 ft., four-lamp fluorescent fixture for each 80 sq. ft. of office space. The fixtures are assumed to use standard lamps and ballasts and a prismatic lens. This combination will provide in excess of 60 fc in open office configurations. In addition, the simulations include 0.75 watts/ sq. ft. for conventional office equipment such as copiers, word processors, coffee makers, and micro computers. This assumption includes diversity factors for all of this equipment, as it is unlikely to be operating on a continuous basis. Caution should be employed

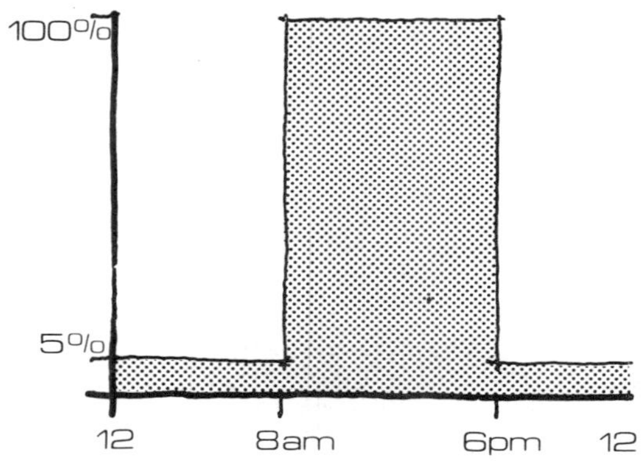

occupied schedule

if substantially more office equipment is expected to be in use in the office building under consideration. If mainframe computers or other similar energy intensive equipment will be used in the office, you should consider the use of the alternate design approach recommended in Section III of this handbook.

The interior lighting is assumed to be on continuously during the operating hours of the building. At all other times, 5% of the lights are assumed to be on for safety and security purposes, as well as to accommodate the requirements of cleaning.

The use of outside lights for illumination of building and grounds is also included in these simulations. Furthmore, in buildings that are greater than one floor in height, an elevator was assumed as part of the design, and the use of electricity for elevator operation was included in the same category as the outdoor lighting. Specific values regarding these assumptions are included in Section I of this handbook.

MECHANICAL SYSTEMS

The choice of the mechanical system to provide heating and cooling in the building strongly influences the amount of energy required. After careful investigation, typical mechanical system choices for each building size and climate location were made and simulations performed based upon their operating characteristics. Substantial detail regarding the mechanical system choices is included in Section II–2 of this handbook; however, all of the system choices are assumed to meet prevailing requirements for energy efficient operation, including ASHRAE 90–75.

AUTOMATIC CONTROLS

In the Level 1 buildings, standard electronic thermostats are used to control the mechanical heating and cooling systems. Simulations include a heating set point of 72°F and a cooling set point of 74°F. These points, based on recommendations of building owners, were selected to maintain highest productivity and employee comfort. Additional energy savings can occur if the temperature range is broader. In Level 4, a wide deadband strategy with a temperature range up to 8°F (70°F–78°F) is described. The thermostats are assumed to have a temperature float of 2° (plus or minus one degree from the set point).

Night temperature setback is an important strategy that is part of the Level 1 "good practice" building. In the analysis described at the beginning of this chapter, comparing the base case building to a non-energy efficient building, differences across climates ranged from 30% to 55% savings. These non-energy efficient buildings were modeled to include night temperature setback. If the non-energy conserving building did not have night temperature setback and outside air temperature reset, the resulting annual energy consumption could be more than 2½ times that of today's "good practice" building.

Night temperature setback to 55°F during the heating season is recommended. When indoor temperatures fall below this level, heat should be provided to maintain 55°F. In the morning, a warm-up period of one hour or more should be scheduled before the occupancy period. In the summer, it is recommended that the system be shut off during unoccupied periods, again, however, allowing a one hour or greater cool-down period before occupancy in the morning.

In many mechanical system types included in this handbook, the temperature of the hot and cold decks is adjustable as a function of outside air conditions. In our simulations, the hot deck supply air temperature ramp varies from 120°F supply air at 0°F outside air to 70°F supply air at 70°F outside air. The cold deck supply air ramp varies from 55°F supply air at 90°F outside air to 65°F supply air at 70°F outside air. These temperature ramps were selected as standard for the equipment as it is typically adjusted. Depending upon local climatic conditions and building characteristics, adjustments could be made to these control ramps to achieve greater energy savings. Consultation with equipment suppliers is recommended to select the optimum control strategy in each climate region. The simulations in this handbook include zone control on all heat pumps and multi-zone mechanical systems.

This section has described the details of the assumptions made in Level 1. These assumptions represent the minimum starting point for use of this handbook and for the selection of strategy sets that will further reduce energy requirements. The assumptions made are typical for standard practice in each area of the country and vary for building size. In the next section of this handbook, Level 2, the Level 1 assumptions are combined with choices made by you regarding specific architectural and mechanical system options.

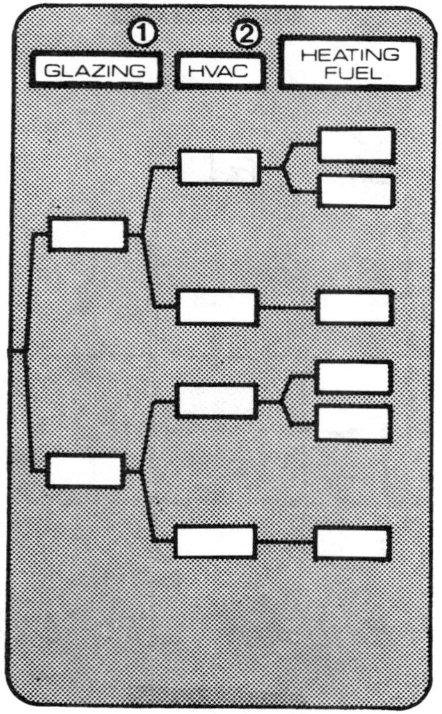

LEVEL 2: Make Initial Choices

Level 2 analysis focuses on the selection of key building characteristics in order to improve energy performance. Whereas analysis of Level 1 was principally concerned with aspects of the building design that are guided by ASHRAE Standard 90–75, Level 2 is concerned with those building characteristics that can be varied within the general constraints of 90–75 and good practice.

Levels 1 and 2 represent most of the decisions that would be made through the schematic design of a building. Level 3, which follows, deals with issues generally covered in early design development. The importance of Level 1 and Level 2 decisions cannot be overestimated. Taken together, they can account for up to 50% of the potential first cost and energy savings available in most projects.

This section is written to parallel the sequence of the Level 2 worksheets in the selector process. Major portions of this section include:

- Selection of Key Building Characteristics
- Use of the Data Library
- Discussion of Cost Development Methods
- Displaying Results and Selecting a Base Case

As can be seen on the principal worksheets for this section, all key variables are laid out in a simple decision tree. This graphic technique makes it is easy to see the consequences of selecting different sets of variables. In effect, each set of key variables represents a separate "path" through Level 2 of the selector.

The user may investigate as few or as many "paths" as time and interest permit. Once several options have been filled out, this approach makes it possible to make a reasonable connection between building design, system, or fuel choices and economic consequences. In effect, this analysis can become a justification for a design or for proposed design changes.

The values chosen to represent each key variable will be explained in greater detail in the next part of this chapter. In general, however, it is important to realize that the values are in the middle of the range of all possible values. These values, we believe, cover the majority of the situations found by designers. They are not the extremes. If insufficient range is presented in the key variables and a major departure is desired in the design, it may be appropriate to use the alternative process explained in Section III.

In Level 2, selections are made to identify basic building characteristics. Taken together, size, height, glass area, HVAC system type and heating fuel type make a large difference in the energy use of a building. It is felt that the range of choices offered will cover the basic characteristics of a vast majority of typical small office building designs. Following is a specific discussion of these building characteristics.

The handbook is directed at office buildings of less than 50,000 sq. ft. floor area. The total small office building category has been subdivided into three size ranges: 2,000–10,000 sq. ft.; 10,000–25,000 sq. ft.; and 25,000–50,000 sq. ft. These ranges are defined by break points that have been determined from analysis of building code requirements, basic structural system size ranges, useful HVAC system application ranges, and on a survey of what designers considered to be most typical buildings as a function of climate and size. This analysis showed that in four climates the most characteristic small end building had light frame construction and a mechanical system that operated as if the building were a single zone. In the hot/humid climate this small building is typically of masonry construction. At the upper end there was a clear need for a building with more fire resistive construction and an HVAC system with multiple zones. Further analysis of HVAC system characteristics (see chart) suggested a middle size category.

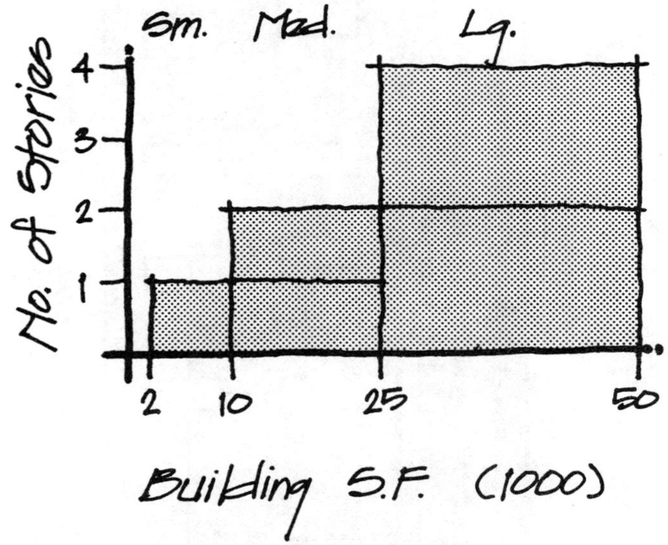

HEIGHT

An analysis of building proportions for the range of building sizes represented in the handbook concluded that roof/wall ratio and surface/volume ratio both vary with the number of floors in the building but change little for a given combination of number of floors and floor area. Thus it was determined that the number of floors was one key variable in describing the envelope-related energy characteristics of the building. Considering the practical ranges of floor layout for leasability, the combinations of building stories and size led to the establishment of the ranges that are offered:

- Small building—one story
- Medium building—one or two stories
- Large building—two or four stories

For a large building of three or five stories, the user should work with the four story path—the strategy selection will be little affected. For a small building of two stories, analysis based on the one story model will yield satisfactory results.

GLASS AREA

Windows

Windows, if properly selected and controlled, can save substantial energy and improve the quality of the office space at the same time. Reasonable amounts of window area (up to 30 to 40 percent of the wall surface) save energy when used in conjunction with daylight dimming controls. To control heat loss, double glazing is usually adequate in most northern climates. Control of cooling loads from windows is very important even in northern climates. External shading devices and internal shading devices (drapes or venetian blinds) are significantly better than either heavily tinted or reflective glass for user comfort and view. Shading devices require careful design, particularly for east and west windows, and drapes or blinds must be conscientiously operated for reliable effect. The Uniform Building Code is considering a code change to permit external shading to project over public use space.

The range of glass areas tabulated in the handbook has been developed with guidance from a review panel of building owners and managers to be representative of window sizes that are acceptable from a general leasability standpoint. The glazing percentages vary depending on building size, with the small buildings having lower percent glass areas. Slightly smaller glass percentage is also given for the medium building in the cold climate, acknowledging the generally more severe winter exposure.

The area of glass in the building is one of the important major variables in determining energy cost impact. Its influence extends to heating, cooling, and lighting. The shading coefficient (SC) of the glass has an effect as part of this influence. In the typical building treated by this handbook, we have generally considered the shading coefficient to be a fixed value representative of usual design practice. Tinted glass has been assumed at a moderate level SC (see Sec. II–1). It is further assumed that window treatment is fixed and that no blinds are used. The combined impact of glass area and shading coefficient is thus represented by glass area alone.

An analysis of the energy data shows, for example, that changing the glass area on a large four-story building from 20% to 40% can have a substantial impact. The impact varies by climate and by HVAC system type.

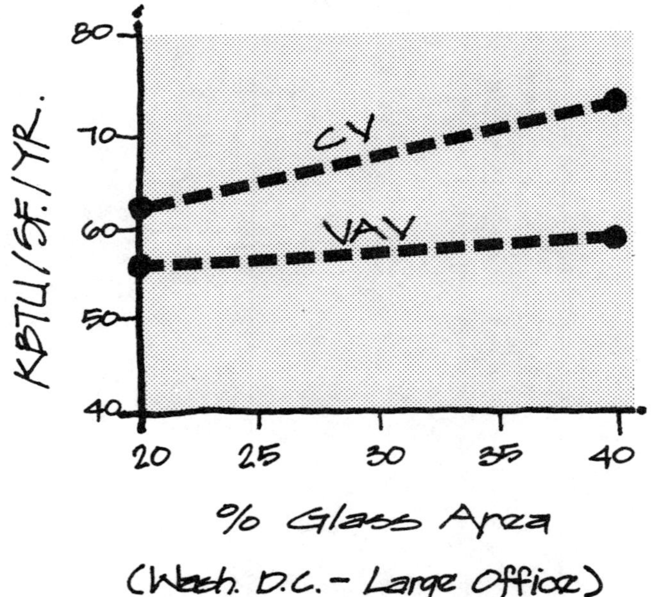

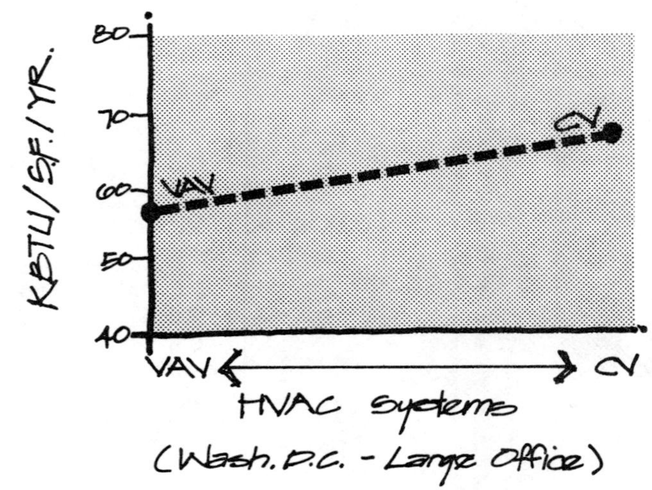

HVAC SYSTEMS

HVAC system selection has important consequences for energy efficiency in building operation. This choice has as much as or more impact on energy cost than the selection of glass area. An analysis of energy data shows, for example, that varying the choice of HVAC system among the recommended systems can have a substantial impact on a large four-story building. The most typical systems for the three building sizes have been selected after researching the practice of professionals experienced in the design of HVAC

HANDBOOK SYSTEMS

BUILDING SIZE	HVAC SYSTEM	CLIMATE				
		COLD	TEMP.	HOT/COLD	HOT/HUMID	HOT/ARID
SMALL 4,000 - 10,000 SF	PTAC	•	•	•	•	•
	HP	•	•	•	•	•
	RT	•	•	•	•	•
MEDIUM 10,000 - 25,000 SF	MZ				•	•
	CV	•	•	•		
	VAV	•	•	•	•	•
	HP/HP	•	•	•		
	hyd HP				•	•
LARGE 25,000 - 50,000 SF	CV	•	•	•		
	VAV	•	•	•	•	•
	hyd HP	•	•	•	•	•

TECH. PANEL RECOMMENDATIONS

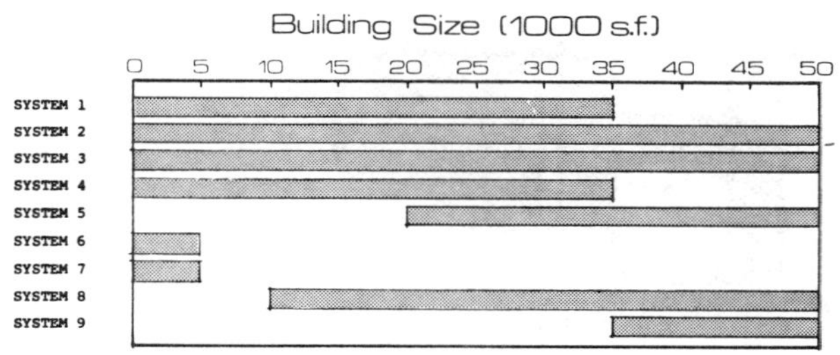

SYSTEM 1: Rooftop - Electric VC, Gas Heating - Single Unit
SYSTEM 2: Rooftop - Electric VC, Gas Heating - Multiple Units - 3 Stories Maximum
SYSTEM 3: Unitary Air To Air Heat Pump
SYSTEM 4: Split System Heat Pump
SYSTEM 5: Hydronic Distribution - Heat Pump
SYSTEM 6: Through The Wall Heat Pump
SYSTEM 7: Central Heating (Gas Or Electric) With Through The Wall A.C.
SYSTEM 8: Central Cooling Only With Reheat (Constant Vol., VAV, Multi-Zone)
SYSTEM 9: Central Chilled Water And Heating With Fan Coils (2,3,4 Pipe Systems)

PTAC

Packaged Terminal Air Conditioner

systems in each climate region. A technical advisory panel has further presented recommendations (see chart) of effective HVAC systems for various building sizes.

The medium and large building HVAC systems were assumed to be five-zone systems—a core and four perimeter exposures. This is reflective of a multi-tenant office building in which air movement is restricted by partitions. For the small building, however, it is assumed that the building effectively acts as a single zone. Thus, some load sharing (between a colder north side and a too hot south side, for instance) will occur. This single zone vs. multi-zone assumption has a noticeable impact on energy consumption. The effect is to cause the net energy consumption (on a Btu/sq. ft. basis) on the small building to be lower than that which would be predicted based on an extrapolation of the Btu/sq. ft. trends indicated for medium and small buildings. Were a small building in reality to be partitioned off into multiple HVAC zones, then the energy consumption would be greater than that indicated here. In fact, given otherwise equal conditions, the consumption on a Btu/sq. ft. basis would probably be somewhat greater than that of the medium building analyzed here.

The following detailed description covers each of the systems selected for the various size buildings.

Small Building

A. Packaged Terminal Air Conditioners

This system is comprised of a group of self-contained PTAC units. Each unit, which is mounted through a sleeve on an outside wall, serves a relatively small space. However, due to air interchange and the relatively small area involved, the entire building is treated as a single zone in this handbook. The individual units range in size (in ¼-ton increments) from ¾- to 1½-tons cooling capacity. The exact size and total number of units vary depending on building layout and load. Rated cooling COP is 2.0, the ASHRAE 90–75 (1980) minimum. Heating is provided by electric resistance coils. The units have a small circulating fan that draws in supply air and blows it over the coils and then into the space. There is no supply ductwork or plenum.

Each self-contained PTAC unit can, of course, be individually controlled. However, the average set temperature during occupied hours is assumed to be 72°F for heating and 74°F for cooling. In addition, all units should be connected to a central time controller that insures that a 55°F setback temperature for heating during unoccupied hours and turns the units off during summer unoccupied hours. (Note: this controller is normally an option that must be specified.)

Each unit is adjusted to provide the specified fixed volume of outside ventilation air but does not have an economizer capability, a significant shortcoming.

B. Packaged Terminal Heat Pumps

This system is identical to the PTAC system described above, except that each packaged terminal unit is now a heat pump instead of an air conditioner and electric heating coil. Thus, the units provide more efficient space heating as well as space cooling. Once again, the units are wall mounted through a sleeve, range in size from ¾ to 1½-ton cooling capacity, and have the ASHRAE 90–75 (1980) COPs of 1.5 at 17°F and 2.5 at 47°F for heating and 2.0 for cooling. When the heat pumps are unable to meet the space heating load, backup electric resistance coils (integral to the units) are utilized. Each unit has it own circulating fan.

As with the PTAC system, each unit can be individually controlled, but it is assumed that the average set temperature is 72°F for heating and 74°F for cooling. A central time controller provides a heating setback to 55°F and prevents cooling during unoccupied hours. It is recommended that, to avoid high power draw due to suddenly increasing the set temperature by a large amount, the thermostat have a step control that normally provides morning heatup without the use of electric resistance heat. Each unit provides the specified fixed volume of outside ventilation air but does not have an economizer cycle, a significant shortcoming.

Packaged Rooftop DX

C. Packaged Rooftop DX

This system involves a single packaged DX (direct expansion) cooler that is roof mounted on a curb. All components, options, fittings, and controls are installed at the factory, allowing simple and quick installation at the site. An air duct distribution system with open plenum return is employed to serve the single zone. The units have separate constant volume supply and return fans, low leakage dampers (less than 3%), and an integral gas or electric furnace. Reheating is not possible.

The unit has the ASHRAE 90–75 (1980) minimum cooling COP of 2.2. The gas furnace has an annual efficiency of 65%, while the electric furnace is 100% efficient.

The entire space is served by one central thermostat set to 72°F for heating and 74°F for cooling. This thermostat should have a time clock that sets it back to 55°F for heating and off for cooling during unoccupied hours. In appropriate climates, the unit has a factory installed economizer (either sensible or enthalpy, as specified elsewhere). Otherwise, a fixed outside air fraction is maintained.

Medium Building

Constant Volume

A. Constant Volume DX

This system is very similar to the packaged rooftop system specified for the small building, except that the constant volume air duct distribution/plenum return system now serves multiple zones. In addition, when gas is the heating source, it is provided by baseboard hot water heaters supplied by a boiler. When electricity is the heating source, it is provided by terminal electric resistance coils in the ducts. Thus, "reheating" is possible with either form of heating.

The DX cooler is factory assembled and ready for hookup and roof mounting on a curb. Separate supply and return air fans with low leakage (i.e., less than 3%) dampers are provided. Where specified, a factory installed economizer (sensible or enthalpy) is included. Otherwise, the specified fixed outdoor air fraction is maintained. The DX cooler has the ASHRAE 90–75 (1980) minimum cooling COP of 2.2. The gas boiler has a seasonal efficiency of 65%, while the electric resistance coils have an efficiency of 100%.

A central controller should be provided to maintain a minimum space temperature of 55°F during unoccupied hours and to prevent space cooling during such times. During occupied hours the space is maintained at 72°F for heating, 74°F for cooling. Outside air reset is employed to control the supply air delivery temperature.

B. Split Heat Pump

HP/HP

Split Heat Pump

This system is comprised of two packaged, roof-mounted heat pumps that condition the entire building. One unit serves the core and the other supplies the perimeter zones. Both utilize a constant volume air duct distribution system with open plenum return. Primary heating is provided by the heat pumps, with backup terminal electric resistance heating/reheating by electric coils in the ducts.

The heat pumps are packaged, factory assembled units requiring minimal on-site effort to install. Each has separate supply and return air fans with low leakage (i.e., less than 3%) dampers. Where specified, a factory installed economizer (sensible or enthalpy) is included. Otherwise, the specified fixed outside air fraction is maintained. The heat pumps have the ASHRAE 90–75 (1980) minimum cooling COP of 2.2 and heating COP of 1.5 at 15°F and 2.5 at 47°F.

A central controller should be provided to maintain a minimum space temperature of 55°F during unoccupied hours and to prevent space cooling during such times. It is recommended, to avoid high power draw due to suddenly increasing the set temperature by a large amount, that the thermostat be of the type that has a step control that normally provides morning heatup without the use of electric resistance heat. During occupied hours the space is maintained at 72°F heating, 74°F cooling. A zone type control system is employed to moderate supply air temperature according to the zone with the greatest demand.

VAV
Variable Air Volume

C. Variable Air Volume DX

This system is essentially the same as the medium building Constant Volume DX, except that a variable air volume duct distribution/plenum return system is provided. The system serves multiple zones, and space heating may be provided either by hot water baseboard heaters supplied by a gas boiler or by terminal electric resistance coils in the ducts. The simple air pressure activated VAV boxes have a minimum turn down ratio of 30%. Reheating is significantly minimized but not eliminated.

The factory packaged DX cooler has separate supply and return fans, with variable inlet vanes and constant speed motor on the supply fan. Low leakage (i.e., less than 3%) dampers are provided. In appropriate climates, a factory installed economizer (sensible or enthalpy) is included. Otherwise, the specified outside air fraction is maintained. The DX cooler has the ASHRAE 90–75 (1980) minimum cooling COP of 2.2. The gas boiler has a seasonal efficiency of 65%, while the electric resistance coils are 100% efficient.

A central controller should be provided to maintain a minimum space temperature of 55°F during unoccupied hours and also to prevent cooling during such times. During occupied hours the space is maintained at 72°F for heating, 74°F for cooling. Since the supply air quantity is varied, a fixed cold/hot deck temperature is employed to enhance reliability and simplicity of operation.

MZ
Three Deck Multi-zone

D. Three Deck Multi-Zone

This system employs a packaged multi-zone unit with three decks: one hot, one cold, and one a return bypass. Dampers are used at the multiple zones. A single duct air distribution/plenum return system is utilized. The cold deck provides design load cooling, while the hot deck provides design load heating. Loads less than these maximums are met by mixing either the cold or hot deck air with return air, thus eliminating reheating. Either gas or electricity is used as a heating source to coils in the hot deck.

The factory-built unit has separate supply and return fans with low leakage (i.e., less than 3%) dampers. An economizer (either sensible or enthalpy) is included, depending on climate. Otherwise, a fixed outside air fraction is maintained. The DX cooler has the ASHRAE 90–75 (1980) minimum cooling COP of 2.2. The gas boiler has a seasonal efficiency of 65%, while the electric resistance coils are 100% efficient.

A central controller should be provided to maintain 72°F for heating and 74°F for cooling during occupied hours and 55°F during unoccupied hours. No space cooling is provided during unoccupied hours. A zone type control system is employed to modulate the supply air temperature according to the zone with greatest demand.

Large Building

A. Constant Volume DX

This system is very similar to the rooftop DX specified for the medium building, except that it is a higher quality central system. It employs a single duct distribution/plenum return system serving multiple zones. When gas is the heating source, it is provided by baseboard hot water heaters supplied by a boiler. When electricity is the heating source, it is provided by terminal electric resistance coils in the ducts. Thus, reheating is possible with either heating subsystem.

The central DX cooler is partially factory assembled. The evaporator is installed inside the building (in a penthouse or mechanical space), while the condenser may be roof or ground mounted. Separate supply and return air fans with low leakage (i.e., less than 3%) dampers are provided. Where specified, an economizer (sensible or enthalpy) is included. Otherwise, the specified fixed outdoor air fraction is maintained. The DX cooler has the ASHRAE 90–75 (1980) minimum cooling COP of 2.5. The gas boiler has a seasonal efficiency of 65%, while the electric resistance coils have an efficiency of 100%.

A central controller should be provided to maintain a minimum space temperature of 55°F during unoccupied hours and also to prevent space cooling during such times. During occupied hours the space is maintained at 72°F for heating, 74°F for cooling. Outside air reset is employed to control the supply air delivery temperature.

VAV

Variable Air Volume

B. Variable Air Volume DX

This system is essentially the same as the variable air volume DX system described earlier for the medium building. Once again, the system serves multiple zones through a single duct variable volume distribution/plenum return system. Space heating may be provided either by hot water baseboard heaters supplied by a gas boiler or by terminal electric resistance coils in the ducts. The simple air pressure activated VAV boxes have a minimum turn down ratio of 30%. Reheating is significantly minimized but not eliminated.

The central DX cooler has separate supply and return fans, with variable inlet vanes and constant speed motor on the supply fan. The evaporator is installed inside the building (in a penthouse or mechanical space), while the condenser may be roof or ground mounted. Low leakage (i.e., less than 3%) dampers are provided. In appropriate climates, an economizer (sensible or enthalpy) is included. Otherwise, the specified outside air fraction is maintained. The DX cooler has the ASHRAE 90–75 (1980) minimum cooling COP of 2.5. The gas boiler has a seasonal efficiency of 65%, while the electric resistance coils are 100% efficient.

A central controller should be provided to maintain a minimum space temperature of 55°F during unoccupied hours and also to prevent cooling during such times. During occupied hours the space is maintained at 72°F for heating, 74°F for cooling. Since the supply air quantity is varied, a fixed cold/hot deck temperature is employed to enhance reliability and simplicity of operation.

HHP

Hydronic Heat Pump Loop

C. Hydronic Heat Pump Loop

This system involves a series of small, packaged water-to-air heat pumps that are connected through a common hydronic (hot water) loop. It is sometimes termed the "California" or "runaround" heat pump concept.

Each unit, which has cooling capacities in the one to three ton range, serves an individual zone and operates independently of all other units. Thus, some units may be in a cooling mode (sunny side or interior zone in winter) while others are in a heating mode (north side in winter). Those in a cooling mode extract heat from the space and inject it into the hydronic loop. Those units in a heating mode extract heat from the hydronic loop and pump it into the space. This load redistribution function is a highly efficient operating strategy.

The hydronic loop water temperature is typically maintained in the 70–90°F range in order to permit effective heat pump operation. A backup boiler (usually electric because of the typically low heating loads) and cooling tower are provided to maintain loop temperatures within the designed limits. The individual heat pumps have the ASHRAE 90–75 (1980) minimum cooling COP of 2.5 and heating COP of 2.8 at 80°F.

Each individual unit has its own circulating fan that draws in room air, conditions it, and then circulates it back into the space. This involves minimal ductwork, if any. A separate ducted ventilation system is provided to supply the required outside air. Economizer operation is not possible because this outside air ventilation system is assumed not to be sized large enough to accommodate the large volume fresh air flow associated with economizer operation.

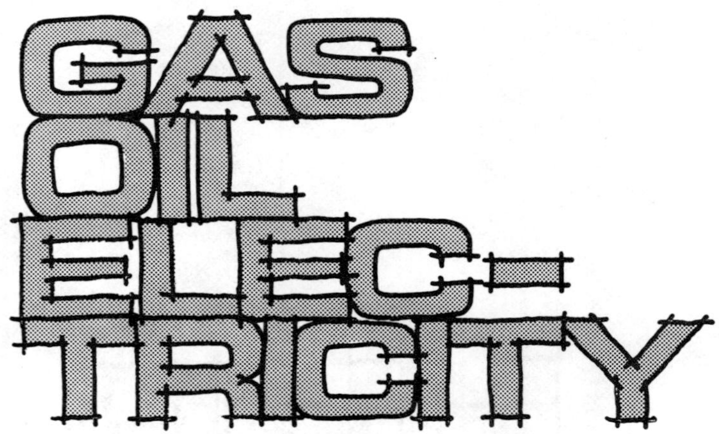

FUEL CHOICES

Gas and electricity are listed as the fuel choices. The data for gas may be used for installations that use oil as a fuel. Conversion factors are given to develop oil quantities and costs from the gas data. These factors consider the differences in plant efficiency as well as fuel unit differences.

When considering electricity it is important to look at all the possible rates that may be available to your project. Different rates can result in large differences in utility cost for a given system. For example, a hydronic heat pump system is an efficient and flexible system for an all-electric building with good levels of electricity consumption. It does, however, have substantial peak electrical demand. If your general utility rate has high demand charges, the hydronic heat pump system may not be very attractive. If, however, there is a heating rate with low demand charges and if the electric heating portion of the system can be wired and metered separately, the hydronic heat pump may prove to be quite cost-effective.

DATA LIBRARY

The energy use data in the Data Library have been developed by the extensive use of computer simulations and analysis. Two state-of-the-art procedures have been used: DOE–2.1b, the first version of DOE-2 to include daylighting routines; and the ESE program (by Control Data Corporation), the first available automated calculation method based on the work of ASHRAE Technical Committee 4.7. All daylighting simulations were performed on DOE–2.1b. All other building performance calculations were performed either on DOE–2.1b or on ESE, depending on the complexity of the problem under study.

Over 1,500 runs have been performed in the development of the energy use data for base case buildings and for strategy sets. This work has been reviewed by a Technical Advisory Panel and has been checked against similar simulation efforts being conducted by ASHRAE and by the California Energy Commission.

The research basis for the project and all detailed analyses and calculations underlying this handbook are contained in reports to Battelle/Pacific Northwest Laboratories.

The Data Library is organized by building case number, and each base case analyzed has a unique case number. All data pertaining to a building case are grouped together. The first column indicates whether the data are for the base case building or for a related strategy set applied to the base building. Positive numbers indicate an increase in first cost over the base case. Negative numbers indicate a decrease in first cost over the base case.

The third column displays the energy form of the tabulated data. The commonly known quantities such as total gas consumption, total electric consumption, and total peak electrical demand have been tabulated. For all-electric buildings, additional data have been listed in order to allow calculations of energy costs using more complex rate schedules. These specific components of energy use are:

- Electric consumption for space and water heating
- Remaining electric consumption (not used for heating)
- Peak electrical demand for space and water heating
- Peak electrical demand for remaining load (not heating)

These data are used where the HVAC systems could be wired and submetered for separate measurement of heating energy. Systems that fall into this category are:

- Constant Volume/Electric (CV/El)
- Variable Air Volume/Electric (VAV/El)
- Multi-Zone/Electric (MZ/El)
- Rooftop/Electric (RT/El).

The fourth column lists units of measurement, which are:

- *Therms* of gas
- *Kilowatt hours* of electric consumption
- *Kilowatts* of peak electrical demand

The next 12 columns present monthly data *(per square foot)* for each energy category. The final column presents annual data. It should be noted that both annual gas and annual electric consumption are the sum of the monthly values. Annual peak electrical demand is the highest of all the 12 monthly peak demand figures.

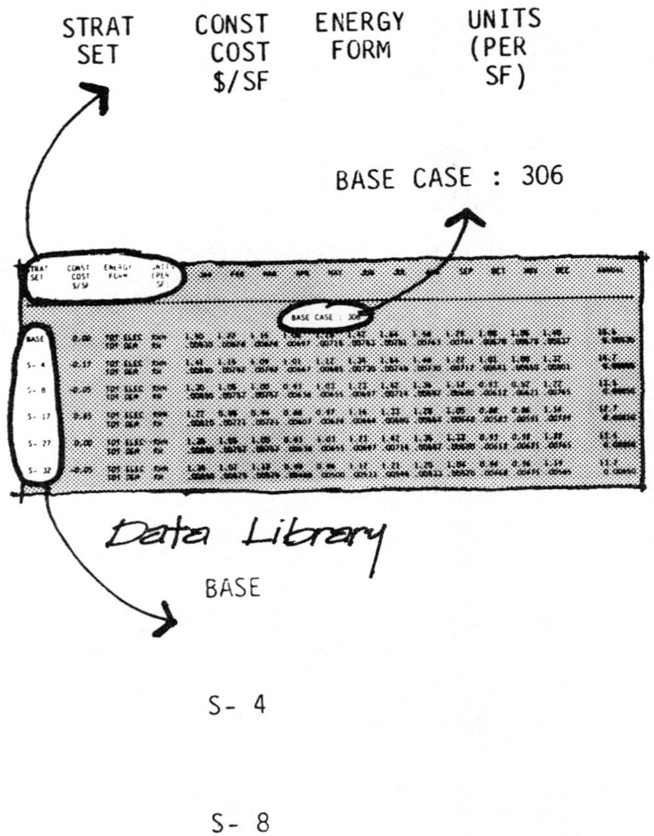

COST DEVELOPMENT

The user of this handbook may choose from among three methods for developing annual energy costs, depending on the complexity of the local utility rate schedule and the complexity of the design project. Worksheets are provided for the three different methods:

A. User analysis—annual basis
B. User analysis—monthly basis
C. Utility analysis—detailed monthly basis

A table characterizing utility rates is given in Appendix A to aid in selection of one of the three methods of calculation.

For any of the worksheets, it is important to use the actual square footage of *your* building design rather than the nominal square footage for the small, medium, or large building category. This will assure that the utility rates calculated apply to your specific design.

Both in Worksheet A and Worksheet B, per square foot energy units are built up to a whole building quantity, costed at the utility rate, and then broken down to a per square foot basis again. This seems an unnecessary complexity of multiplication and division, but the reason this procedure is followed is to recognize that utility rates vary with the actual magnitude of energy used in the building. For the category of large buildings, for example, the range in size is from 25,000 sq. ft. to 50,000 sq. ft. Though energy usage per square foot (with a given set of building characteristics) may be even across this range, the utility rate that would apply to the amount of energy used by a 25,000 sq. ft. building might be quite different from that applying to the amount of energy used by a 50,000 sq. ft. building. This method recognizes those differences.

Following is a discussion of the details of each process:

A. User Analysis—Annual Basis.

This method will be practical when utility rates are relatively uniform. In the case of electricity units, both demand and consumption are relatively uniform throughout the year. In the case of gas rates, there are only irrelevant differences between winter and summer rates. In method A, the user should have confidence that average rates applied to annual data will give representative results. An experienced user may find it possible to use average or weighted average annual numbers to represent rates that have some variation over the year. A separate worksheet A must be completed for each base case and for each strategy set selected for analysis.

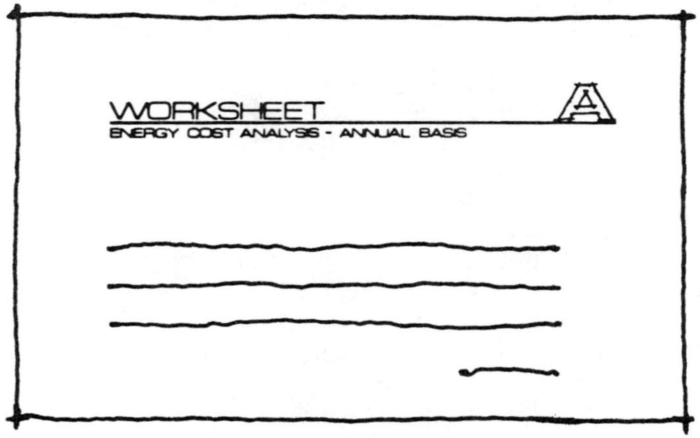

Method A is generally not capable of utilizing specific component energy data: electric consumption for space and water heating, remaining electric consumption, peak electric demand for space and water heating, and peak electric demand for remaining load. These data are helpful with rate schedules that emphasize heating in all-electric buildings.

B. User Analysis—Monthly Basis.

This method will be practical when electric demand charges are relatively uniform throughout the year (with no ratchet clause) and consumption rates for both gas and electricity have only seasonal changes. With method B, as with A, it is practical to deal only with the three common energy quantities: gas consumption, electrical consumption, and total peak electrical demand. Each energy form requires three entries: energy use per sq. ft. (from the data library) on the first line; your actual building square footage on the second line; and the monthly utility rates for your project on the fourth line. Performing the indicated multiplications, additions, and divisions yields the cost of each energy form and the total energy cost for the project. A separate worksheet B must be completed for each base case and for each strategy set selected for analysis.

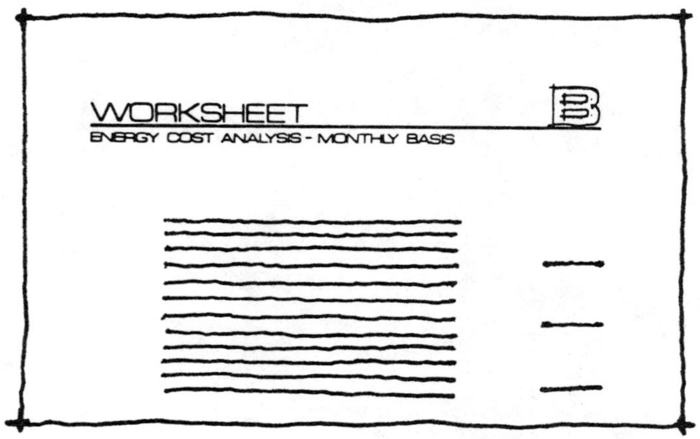

C. Utility Analysis—Detailed Monthly Basis.

This method is simpler for the user of the handbook than the previous two, but it does involve a response from your local utility. This method requires the proper identification of building characteristics and strategy sets to be studied and the preparation of a mail-off package to all applicable utilities. It allows complex rate schedules to be applied by the utility. This method

may wish to go back to the decision tree and select other options for evaluation.

Ultimately, the user should select one of these options for his Level 2 base case. Such a decision will depend to some degree on the energy costs data just developed. But the user should definitely not overlook other factors that will play a part in this decision. Energy costs must be weighed in conjunction with first cost. These and other economic parameters should fulfill the owner's economic objective, whether it be one of quick return on investment expected by speculative builders or one of long-term savings and reliability desired by corporate builders. Program requirements will also be a big factor. Decreases in glazing will save first costs and energy costs but must be weighed against issues such as aesthetics and marketability. The quality of the built environment, its comfort, its effect on productivity, and its compliance with local building codes are all issues that must be weighed.

Once a Level 2 case has been selected, proceed to Level 3 for the selection of strategy sets to improve your base case building energy performance.

LEVEL 3: Select Strategy Sets

This chapter describes 17 basic strategies suitable for small office buildings in general and explains how these basic strategies are combined into strategy sets suitable for each building case selected in Level 2. Information supplementing the Level 3 analysis worksheets (found in Section I) is presented at the end of this chapter.

Level 3 savings are achieved by applying specific strategy sets to the building case number identified at Level 2. Several criteria were used to select strategies and to compose Level 3 strategy sets. Following is a list of the principal criteria which guided this process:

1. There should be at least one strategy set with no added construction costs.
2. Sets should be composed to provide increased levels of energy saving and costliness.
3. Sets should address the major components of energy use:
   a) Lighting
   b) Electrical demand
   c) Cooling
   d) Heating (cold climate)
4. There should be a separate daylighting strategy set.
5. Sets should be tailored to building and heating, ventilating, and air conditioning (HVAC) system characteristics.
6. Sets should be tailored to climate characteristics (for example, use evaporative coolers in hot/arid climates; use pulse boilers in cold).
7. Strategies must be reliable (operable windows have questionable reliability because of the need for consistent occupant control).
8. Strategies must not compromise comfort/productivity.
9. It must be possible to simulate the strategy set by computer.
10. Strategies must be achievable with common construction practice.

These criteria were first applied to a list of over 80 possible conservation strategies (see Appendix B). Out of this screening process, 17 basic strategies were selected. Each of these strategies is described briefly later in this chapter. Again using the criteria listed above, these 17 basic strategies were combined into 56 strategy sets suitable for all the possible "base cases" defined in Level 2. Although the list of 56 possible strategy sets seems long (see Table A), they have been grouped on the Level 3 worksheets into six groups (or options), which may be summarized as follows:

- **Option 1: Simplest lighting improvements**
  - Generally involves installing high efficiency ballasts.
  - Savings from HVAC downsizing more than offset any increase in first cost.
  - Downsizing applies to all but smallest HVAC equipment.
- **Option 2: Modest lighting and mechanical improvements**
  - Generally involves installing high efficiency ballasts and lamps plus HVAC COP improvements.
  - Downsizing mechanical systems is limited to Option 1 levels only, as a designer cannot always count on high efficiency lamps being present.
  - Special cases: COP improvements are not used for cooling equipment in climates where cooling load is small; non-uniform lighting is offered as an option.
- **Option 3: Major lighting and mechanical improvements**
  - Generally involves installing the best practical lighting system using high efficiency ballasts, lamps, and fixtures, plus Option 2 strategies.
  - Downsizing is limited to credit for high efficiency ballasts and fixtures but not for lamps.
  - Special casė: high efficiency boilers are recommended for medium buildings in cold climate.
- **Option 4: Daylight dimming controls and lighting improvements**
  - Generally involves installing one or more of the following:
    1. perimeter daylight dimmers and high efficiency ballasts

   2. high efficiency lamps in the core of some buildings
   3. high efficiency lamps and high efficiency fixtures in the core of some buildings
  - Also includes COP improvements
  - HVAC downsizing for high efficiency ballasts and fixtures only, not for daylighting.

- **Option 5: Skylighting, daylight dimming controls and lighting improvements**
  - Generally involves installing skylights in outermost core area, plus perimeter daylighting dimmers, high efficiency ballasts, and COP improvements.
  - Applies only to single-story buildings with significant core areas (medium building).
  - No downsizing credit is allowed for perimeter or skylight daylighting.
- **Option 6: Evaporative cooling and lighting improvements**
  - Generally involves installing evaporative coolers on small or medium one-story buildings in hot/arid climate only.
  - Includes installing high efficiency ballasts and lamps.
  - Downsized direct expansion cooler is used as backup.

A complete list of the 56 possible strategy sets is given in table A. The allocation of the 17 basic strategies to each strategy set is shown in Matrix B. Following these exhibits, section C provides detailed descriptions of each of the 17 basic strategies. Section D presents a description of the interactive effects of combining strategies into sets. Finally, Section E presents important supplementary information on the computation of simple payback and the selection of one strategy set from several options.

After you have chosen a strategy set using one of the worksheets provided in section I, the following steps are recommended for gaining a clear understanding of that strategy set and its proper application to the base case you have chosen (Level 2):

1. Identify the strategy set in Table A.
2. Identify the individual strategies that comprise the set from Matrix B.

3. Review the descriptions of all basic strategies that apply in Section C.
4. Review the discussion of combined effects that apply in Section D.
5. Review the discussion of cost based strategy selection in Section E.

A. DESCRIPTIVE LIST OF STRATEGY SETS

1. High eff. ballasts over all bldg.
2. High eff. ballasts over all bldg. + tinted glazing.
3. High eff. ballasts over all bldg. + HVAC downsized for ballasts.
4. High eff. ballasts over all bldg. + HVAC downsized for ballasts + tinted glazing + HVAC downsized for tinted glass.

5. High eff. ballasts + high eff. fluorescent lamps over all bldg.
6. High eff. ballasts and lamps over all building + HVAC downsized for ballasts.
7. High eff. ballasts and lamps over all bldg. + HVAC downsized for ballasts + cooling COP improvement.
8. High eff. ballasts and lamps over all bldg. + HVAC downsized for ballasts + cooling COP improvement + tinted glazing + HVAC downsized for tinted glass.
9. High eff. ballasts and lamps over all bldg. + HVAC downsized for ballasts + heat pump COP improvement.
10. High eff. ballasts and lamps over all bldg. + cooling COP improvement.
11. High eff. ballasts and lamps over all bldg. + cooling COP improvement + tinted glazing.
12. High eff. ballasts and lamps over all bldg. + heat pump COP improvement.
13. High eff. ballasts and lamps over all bldg. + heat pump COP improvement + tinted glazing.

14. High eff. ballasts and lamps + high eff. fixtures over all bldg.

15. High eff. ballasts, lamps and fixtures over all bldg. + HVAC downsized for ballasts and fixtures.
16. High eff. ballasts, lamps and fixtures over all bldg. + HVAC downsized for ballasts and fixtures + cooling COP improvement.
17. High eff. ballasts, lamps and fixtures over all bldg. + HVAC downsized for ballasts and fixtures + cooling COP improvement + tinted glazing + HVAC downsized for tinted glass.
18. High eff. ballasts, lamps and fixtures over all bldg. + HVAC downsized for ballasts and fixtures + heat pump COP improvement.
19. High eff. ballasts, lamps and fixtures over all bldg. + HVAC downsized for ballasts and fixtures + high eff. boilers.
20. High eff. ballasts, lamps and fixtures over all bldg. + cooling COP improvement.
21. High eff. ballasts, lamps and fixtures over all bldg. + cooling COP improvement + tinted glazing.
22. High eff. ballasts, lamps and fixtures over all bldg. + heat pump COP improvement.
23. High eff. ballasts, lamps and fixtures over all bldg. + heat pump COP improvement + tinted glazing.

24. Non-uniform lighting + high eff. ballasts over all bldg.
60. Non-uniform lighting + high eff. ballasts over all bldg. + HVAC downsized for ballasts.
25. Non-uniform lighting + high eff. ballasts over all bldg. + cooling COP improvement.
61. Non-uniform lighting + high eff. ballasts over all bldg. + HVAC downsized for ballasts + cooling COP improvement.
26. Non-uniform lighting + high eff. ballasts over all bldg. + cooling COP improvement + tinted glazing.
27. Non-uniform lighting + high eff. ballasts over all bldg. + HVAC downsized for ballasts + cooling COP improvement + tinted glazing + HVAC downsized for tinted glass.
28. Non-uniform lighting + high eff. ballasts over all bldg. + heat pump COP improvement.

62. Non-uniform lighting+high eff. ballasts over all bldg.+HVAC downsized for ballasts+heat pump COP improvement.
29. Non-uniform lighting+high eff. ballasts over all bldg.+heat pump COP improvement+tinted glazing.

30. Perimeter daylighting dimmers+high eff. ballasts over all bldg.
31. Perimeter daylighting dimmers+high eff. ballasts over all bldg.+HVAC downsized for ballasts.
32. Perimeter daylighting dimmers+high eff. ballasts over all bldg.+HVAC downsized for ballasts+cooling COP improvement.
33. Perimeter daylighting dimmers+high eff. ballasts over all bldg.+HVAC downsized for ballasts+heat pump COP improvement.
34. Perimeter daylighting dimmers+high eff. ballasts over all bldg.+HVAC downsized for ballasts+high eff. boilers.
35. Perimeter daylighting dimmers+high eff. ballasts over all bldg.+cooling COP improvement.
36. Perimeter daylighting dimmers+high eff. ballasts over all bldg.+heat pump COP improvement.

37. Perimeter daylighting dimmers+high eff. ballasts over all bldg.+high eff. lamps in core.

38. Perimeter daylighting dimmers+high eff. ballasts over all bldg.+high eff. lamps and fixtures in core.
39. Perimeter daylighting dimmers+high eff. ballasts over all bldg.+high eff. lamps and fixtures in core+HVAC downsized for ballasts and fixtures.
40. Perimeter daylighting dimmers+high eff. ballasts over all bldg.+high eff. lamps and fixtures in core+HVAC downsized for ballasts and fixtures+cooling COP improvement.
41. Perimeter daylighting dimmers+high eff. ballasts over all bldg.+high eff. lamps and fixtures in core+HVAC downsized for ballasts and fixtures+heat pump COP improvement.

42. Perimeter daylighting dimmers+high eff. ballasts over all bldg.+high eff. lamps and fixtures in core+HVAC downsized for ballasts and fixtures+high eff. boilers.
43. Perimeter daylighting dimmers+high eff. ballasts over all bldg.+high eff. lamps and fixtures in core+heat pump COP improvement.

44. Perimeter daylighting dimmers+high eff. ballasts over all bldg.+high eff. lamps in core+HVAC downsized for ballasts.
45. Perimeter daylighting dimmers+high eff. ballasts over all bldg.+high eff. lamps in core+HVAC downsized for ballasts+cooling COP improvement.
46. Perimeter daylighting dimmers+high eff. ballasts over all bldg.+high eff. lamps in core+HVAC downsized for ballasts+heat pump COP improvement.
47. Perimeter daylighting dimmers+high eff. ballasts over all bldg.+high eff. lamps in core+HVAC downsized for ballasts+high eff. boilers.
48. Perimeter daylighting dimmers+high eff. ballasts over all bldg.+high eff. lamps in core+heat pump COP improvement.

49. Perimeter daylighting dimmers+skylights with dimmers+high eff. ballasts over all bldg.+HVAC downsized for ballasts.
50. Perimeter daylighting dimmers+skylights with dimmers+high eff. ballasts over all bldg.+HVAC downsized for ballasts+cooling COP improvement.
51. Perimeter daylighting dimmers+skylights with dimmers+high eff. ballasts over all bldg.+HVAC downsized for ballasts+heat pump COP improvement.
52. Perimeter daylighting dimmers+skylights with dimmers+high eff. ballasts over all bldg.+HVAC downsized for ballasts+high eff. boilers.

53. High eff. ballasts and lamps over all bldg.+HVAC downsized for ballasts+evaporative cooler.

B. MATRIX OF STRATEGIES VS. STRATEGY SETS

		1	2	3	4	5	6	7	8	9	10	11	12	13	14	15	16	17	18	19	20	21	22	23
1	BALLASTS	●	●	●	●	●	●	●	●	●	●	●	●	●	●	●	●	●	●	●	●	●	●	●
2	LAMPS					●	●	●	●	●	●	●	●	●	●	●	●	●	●	●	●	●	●	●
3	LAMPS (CORE)																							
4	FIXTURES														●	●	●	●	●	●	●	●	●	●
5	FIXTURES (CORE)																							
6	NON-UNIFORM LIGHTING																							
7	DAY-LIGHTING																							
8	SKY-LIGHTS																							
9	DEMAND LIMITER																							
10	TINTED DOUBLE GLAZING		●		●				●			●		●				●				●		●
11	DOWNSIZE - BALLASTS			●	●		●	●	●	●														
12	DOWNSIZE - BALLASTS + FIXTURES															●	●	●	●	●				
13	DNSZE.-BAL. (PER.); BAL+ FIXT. (CORE)																							
14	DOWNSIZE - TINTED DOUB. GLAZING				●				●									●						
15	COOLING COP							●	●		●	●					●	●			●	●		
16	HEAT PUMP COP									●			●	●					●				●	●
17	BOILERS																			●				
18	EVAP. COOLING																							

		24	60	25	61	26	27	28	62	29	30	31	32	33	34	35	36	37	38	39	40	41	42	43	44	45	46	47	48	49	50	51	52	53
1	BALLASTS	●	●	●	●	●	●	●	●	●	●	●	●	●	●	●	●	●	●	●	●	●	●	●	●	●	●	●	●	●	●	●	●	●
2	LAMPS																																	●
3	LAMPS (CORE)																	●	●	●	●	●	●	●	●	●	●	●	●					
4	FIXTURES																																	
5	FIXTURES (CORE)																		●	●	●	●	●	●										
6	NON-UNIFORM LIGHTING	●	●	●	●	●	●	●	●	●																								
7	DAY-LIGHTING										●	●	●	●	●	●	●	●	●	●	●	●	●	●	●	●	●	●	●	●	●	●	●	
8	SKY-LIGHTS																													●	●	●	●	
9	DEMAND LIMITER																																	
10	TINTED DOUBLE GLAZING					●	●			●																								
11	DOWNSIZE - BALLASTS		●		●		●		●			●	●	●	●										●	●	●	●		●	●	●	●	●
12	DOWNSIZE - BALLASTS + FIXTURES																																	
13	DNSZE.-BAL. (PER.); BAL+ FIXT. (CORE)																			●	●	●	●											
14	DOWNSIZE - TINTED DOUB. GLAZING						●																											
15	COOLING COP			●	●	●	●						●			●					●					●					●			
16	HEAT PUMP COP							●	●	●				●			●					●		●			●		●			●		
17	BOILERS														●								●					●					●	
18	EVAP. COOLING																																	●

Note: The vertical bands identify strategy sets that have a common basic set composition. These are described in D.—Combined Effects.

C. DETAILED DESCRIPTIONS

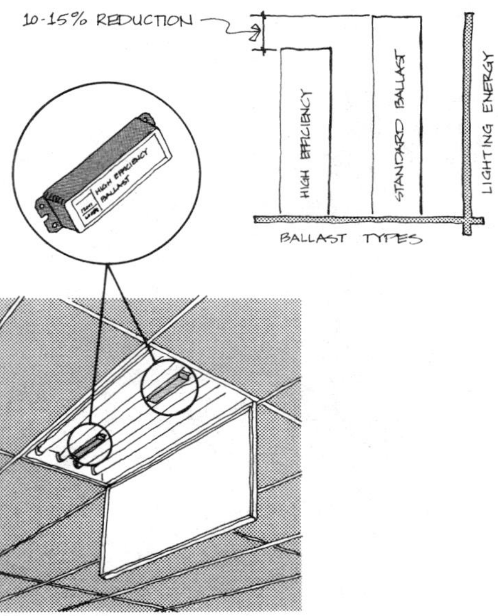

1. HIGH EFFICIENCY BALLASTS OVER ALL BUILDING

Substitute high efficiency ballasts for conventional fluorescent ballasts. High efficiency ballasts are improved versions of conventional electromagnetic ballasts that require between 8 and 16 watts less input than conventional ballasts to operate two standard fluorescent lamps. High efficiency ballasts, because of their lower power input, reduce heat gain from lights by 20%. Other ballast/lamp combinations show similar reduction in input wattage.

These ballasts are widely available for two-lamp installations (a four-lamp fixture would require two ballasts), and some manufacturers have sizes for one and three-lamp installations. Many fixtures may now be ordered from the manufacturer with high efficiency ballasts already installed. Three-lamp fixtures may also be tandem wired using two-lamp ballasts. To do this, a pair of fixtures (containing 6 lamps) is wired as follows: 1 ballast for 2 lamps in the first fixture, 1 ballast for 2 lamps in the second fixture, and 1 ballast to the single lamps remaining in each of the fixtures.

2. HIGH EFFICIENCY FLUORESCENT LAMPS— OVER ALL BUILDING

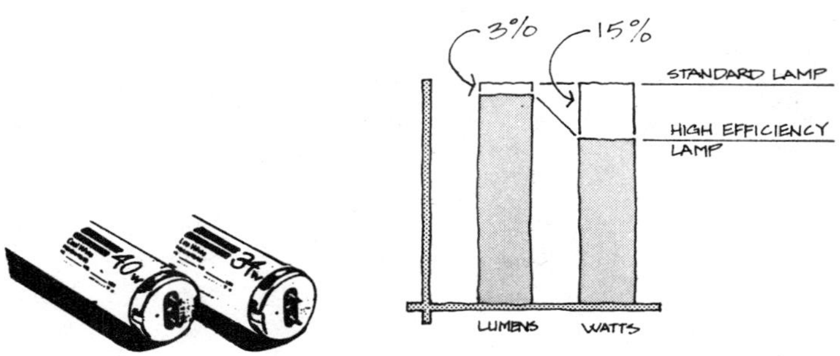

Substitute high efficiency fluorescent lamps for standard lamps. High efficiency fluorescent lamps are lamps that provide approximately 96% of the rated output of a standard 40 watt lamp with a nominal 34 watt input. They are made in warm, cool, and high color rendition phosphors and are available from several major manufacturers. Comparable lamps are available in a number of common lamp sizes.

In this strategy, high efficiency lamps are used in fixtures over all the building, both perimeter and core. They are not used with dimming systems as they do not dim reliably.

3. HIGH EFFICIENCY FLUORESCENT LAMPS— NON-DAYLIT CORE

Substitute the high efficiency fluorescent lamps of strategy 2 for standard lamps in the core area of the building where lights are not controlled by daylight dimmers. High efficiency lamps are not well suited to dimming and thus are not reliable in the perimeter daylit area when dimmers are used. This strategy is particularly attractive in buildings that have large interior areas—generally the one story medium building and two story large building.

4. HIGH EFFICIENCY FLUORESCENT FIXTURES— OVER ALL BUILDING

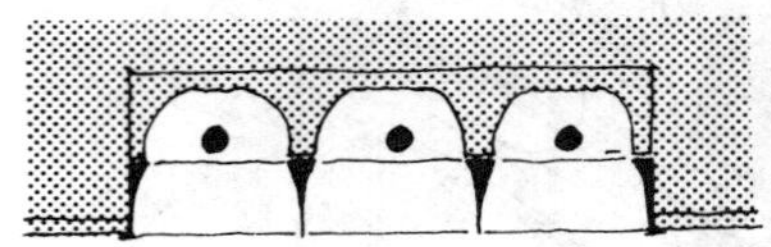

Substitute high efficiency fluorescent fixtures for conventional fixtures. These fixtures are high efficiency 2 ft. × 4 ft. parabolic (no lens) troffers with three lamps, tandem wired (i.e., 3 two-lamp ballasts per two fixtures). When installed in a standard layout of 80 sq. ft./fixture, this installation will provide an illumination level of approximately 50 fc in open offices with a reduction of one lamp per fixture. High efficiency parabolic fixtures are available from numerous manufacturers.

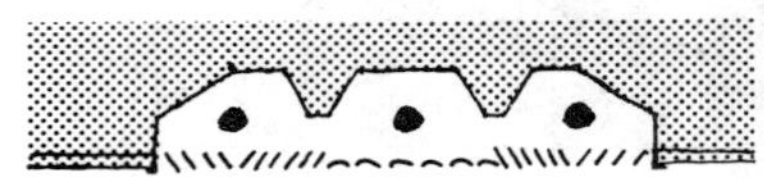

Approximately similar results may be achieved by using an advanced design prismatic lens with a linear fresnel lens pattern. Such lenses are designed to provide good light distribution from three lamps in a conventional fixture, thus achieving energy reduction similar to the high efficiency parabolic fixtures. It is recommended that an actual installation of such lenses be viewed to assure that the light pattern and ceiling appearance are acceptable, as there is some striping effect when viewing the lens.

5. HIGH EFFICIENCY FLUORESCENT FIXTURES— NON-DAYLIT CORE

Substitute the high efficiency fluorescent fixtures of strategy 4 for conventional fixtures in the core area of the building where lights are not controlled

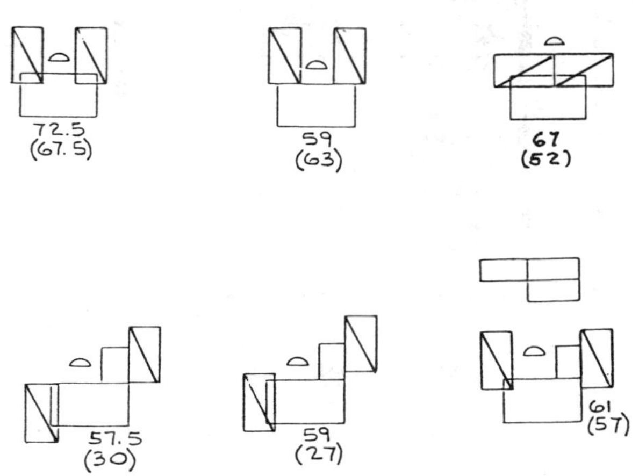

Ceiling-mounted Task Lighting Showing Six Alternate Arrangements

Number shown by each desk is footcandles at work surface. Number in parentheses is ESI for pencil task at 25% viewing angle; however, the ESI approaches the footcandle value for most common office reading materials. The average footcandle level for all six locations is calculated at 63 footcandles. Troffers are two-lamp fluorescent energy-conserving type, with 2,500 mean lumens per lamp, maintained.

by daylight dimmers. This strategy applies to buildings with a high percentage of core area. This core area is not reached by daylighting and can thus benefit from high efficiency fixtures in addition to other measures.

6. NON-UNIFORM LIGHTING

With a non-uniform but more efficient distribution of fixtures, sufficient light may be provided in offices using a lesser total amount of equipment than a conventional uniform layout.

More design time and costs may be required with non-uniform lighting to achieve proper light level distribution. (These costs are not included in the installed per-square-foot cost figures.) This additional time and cost would be incurred not only when the space is first occupied, but also whenever it is remodeled. On the other hand, the quality of the lighting in a non-uniform scheme should be better than uniform layouts and may result in higher worker satisfaction and productivity.

Equipment

The non-uniform lighting design uses 2 ft. × 4 ft. two-lamp standard fixtures.

Application

For typical office environments, layout should provide one lamp on each side of a task to achieve 50–60 fc on the work surface. (See attached diagram for possible layouts.) The lighting level drops off to 30 fc in areas surrounding the task and down to an average of 10 fc in corridors, bathrooms, and other core areas.

This layout should result in fewer fixtures than conventional uniform designs. For speculative buildings, where task locations are generally not known at the time of design, the designer may specify a quantity of fixtures (about 75% of the quantity that is required for the regular grid pattern) with 8 ft. pigtails pre-wired to the fixtures. These fixtures are delivered to the jobsite on pallets and are layed-in and wired up to the flexible electrical distribution system in the ceiling once tenant requirements are known.

The HVAC equipment should not be downsized for the reduced lighting if there is a possibility of fixtures being added during subsequent occupancy changes.

7. DAYLIGHT DIMMERS

Application

Install daylight dimmers around the perimeter of the building in all areas with windows. The dimmers should control the perimeter fluorescent fixtures to a depth of 15 ft. from the window wall. Dimmers may control individual fixtures or groups of fixtures. The alternatives to consider are cost vs. space layout flexibility. A common middle ground would be to have one dimmer/sensor combination control two or four fixtures. The typical layout illustrated here shows a 15 ft. deep perimeter daylighting zone with daylight sensors located 3 ft. and 10 ft. in from the windows, each controlling two rows of fixtures. It is important to properly adjust the dimmers to achieve adequate but not excessive use of the fluorescent lights.

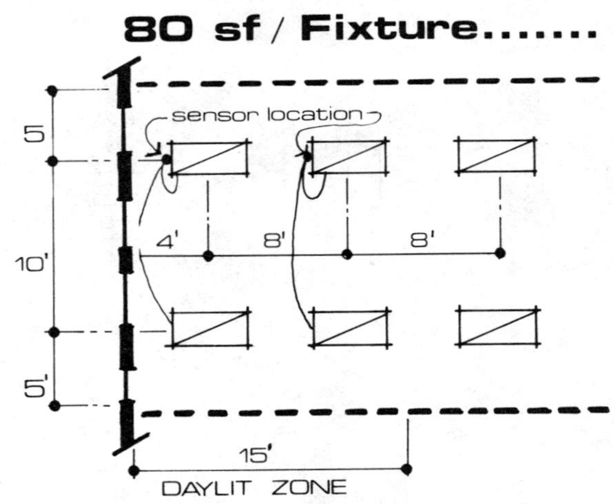

Equipment

Daylight dimmers automatically dim fluorescent luminaires to maintain a set illumination level (such as 50 fc) through a mixture of daylight and artificial light. They should be ones that are commercially available to dim standard or high efficiency ballasts (not electronic ballasts). These types of dimmers generally do not require the use of special dimming ballasts. The dimmers should, at a minimum, provide continuous dimming down to 25% of original light output (at 30% power). To prevent visual disruption for building occupants, it is felt that smooth/continuous dimming is preferable to step switching controls.

In determining the effectiveness of the daylight control dimmers, it was assumed that the usual type of office window coverings would be used: either blinds or drapes. It was further assumed that these window coverings would be open whenever glare from direct sunlight was not a problem.

The savings determined for this strategy are based on a building that has full exposure to daylight: i.e., a free-standing structure. If the actual building will have reduced daylight access because of shading by adjacent structures, terrain, or foliage, the savings will be less. Furthermore, the base case building has an aspect ratio of 1.5:1 and is oriented with the long axis E-W. This is a desirable orientation if site choice is possible. A more elongated aspect ratio is not particularly beneficial until such extremes are reached that the daylight zone (approximately 15 ft. depth from windows) covers most of the floor area. With small buildings this may be practical, but on larger buildings the extremely elongated shape that results cannot be considered a generally useful office plan shape. It should be noted that daylight dimming is an effective means of reducing electrical demand, as lights are typically dimmed when demand is greatest: hot sunny summer days.

8. SKYLIGHTS WITH DAYLIGHT DIMMERS

Add skylights and daylight dimmers on fluorescent lights in a zone just inside the perimeter daylit zone to extend the benefits of daylighting. The strategy is keyed to a medium one-story building that has a high percentage of core area. Such a building can receive a significant lighting benefit from skylights in the areas that perimeter daylighting does not reach.

There are a number of challenges that a skylight installation presents to the designer:

- *Summer cooling.* With a horizontal skylight, the higher sun in summer increases the solar load on the skylight.
- *Control of direct sunlight.* As with windows, direct sunlight may present an annoyingly brilliant light source if not controlled.
- *Intereference with use of the plenum.* To be most effective, a skylight should have a widely sloped light well from aperture to ceiling. This spreads the useful light and provides a tapering light transition to the surrounding ceiling area. Such sloped light wells take much space from the plenum area, which may limit duct and wiring runs and make partition intersections with the ceiling more difficult.
- *Cost.*

- *Aesthetics.* This can be a positive advantage for a skylight. Natural light can be most attractive in the office environment and can provide spaces that command rental premiums.

Equipment

Skylights are standard double glazed plastic domed units with a clear exterior layer and a transluscent interior layer. Three foot square units are the size that has been analyzed, but this dimension is not critical.

Daylight dimmers are the same as described in strategy number 7.

Application

The skylights should total approximately 3½ % of the roof area over the skylit zone and be spaced to illuminate a zone just inside the 15 ft. deep perimeter daylit zone. The illustration shows a representative layout and cross-section. The skylight well should slope away from the unit at approximately 45°.

The dimmers associated with the skylit zone should control pairs of fixtures to provide full coverage of the zone. For a 3 ft. × 3 ft. horizontal skylight in a building with an 8 ft. 6 inch ceiling, a roof height of 12 ft., and a light well sloped at 45°, a typical skylit zone would be 15 ft. wide.

In determining the effectiveness of the daylight control dimmers, it was assumed that the usual type of office window coverings would be used: either blinds or drapes. It was further assumed that these window coverings would be open whenever glare from direct sunlight was not a problem.

The savings determined for this strategy set are based on a building that has full exposure to daylight: i.e., a free-standing structure. If the actual building will have reduced daylight access because of shading by adjacent structures, terrain, or foliage, the savings will be less. It should be noted that daylight dimming is an effective means of reducing electrical demand, as the period of greatest dimmer action occurs when demand is greatest: hot sunny summer days.

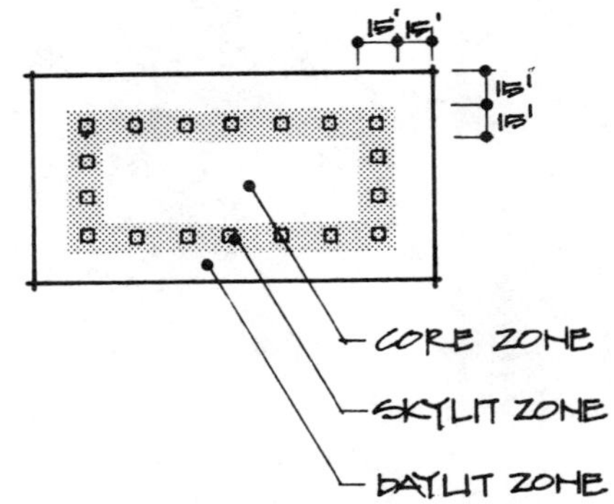

SKYLIT PLAN LAYOUT
1 STORY • MEDIUM

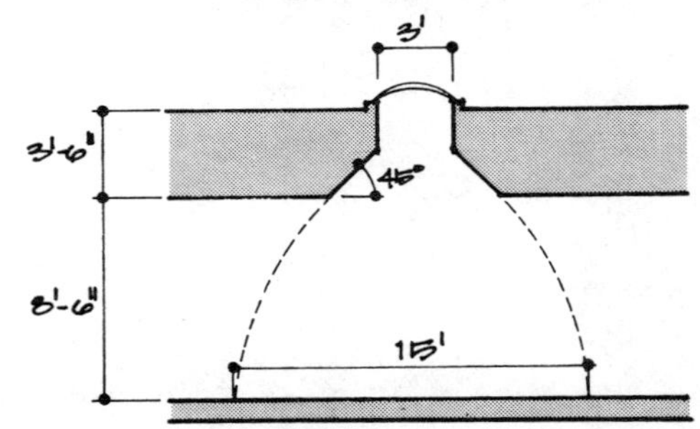

SKYLIGHT SECTION

9. TINTED DOUBLE GLAZING

Replace clear double glazing (shading coefficient=0.81) with double glazing with one light tinted (shading coefficient=0.55) for reduced cooling load and glare control.

10. DOWNSIZE HVAC SYSTEM FOR HIGH EFFICIENCY BALLASTS

HVAC systems, including plant, ducts, and fans, should be sized to meet the reduced cooling load resulting from the use of high efficiency ballasts over all the building. High efficiency ballasts operate with less power input, which reduces the heat gain from the lighting system by 20% (0.35 watts/sq. ft.). This represents a significant reduction in internal gains, thus shifting the building's balance point temperature upward. This has the net effect of reducing peak and total annual cooling load and slightly increasing annual heating load. As most system components (except heating components) are sized to meet peak cooling requirements, they may be downsized (typically up to 5%) because of the reduced cooling load. This results in a reduction in system cost. Conversely, heating loads will increase slightly, but the costs associated with this increase are usually very small.

This strategy is applicable to all buildings/climates/HVAC systems except those systems employing packaged unitary equipment too small to be downsized (e.g., the Packaged Terminal Air Conditioner and heat pump systems in the small building and the hydronic heat pump system in the large building.)

11. DOWNSIZE HVAC SYSTEM FOR HIGH EFFICIENCY BALLASTS AND HIGH EFFICIENCY FIXTURES

HVAC systems, including plant, ducts, and fans, should be sized to meet the reduced cooling load resulting from the use of high efficiency ballasts and high efficiency fixtures over all the building. Note that it is not considered practical to downsize HVAC systems for the installation of high efficiency fluorescent lamps, as it is relatively easy for these to be replaced with standard lamps at a future date, thus increasing peak cooling loads.

High efficiency ballasts operate at lower temperatures, which typically reduces the heat from the lights by 20% (0.35 watts/ sq. ft.). High efficiency fixtures reduce the number of fluorescent lamps per fixture from four to three. The heat load from all fixtures affected by the change of ballasts and fixtures may thus be reduced by 25%. Together these two strategies represent a significant reduction in the internal gains portion of the load, which results in shifting the building's balance point temperature upward. This has the net effect of reducing peak and annual cooling loads and slightly increasing annual heating loads. As HVAC systems are typically sized to meet peak cooling loads, the reduction in peak cooling capacity permitted by these measures allows system downsizing (typically up to 10%) and an associated reduction in system cost. Although heating loads do increase slightly, the costs associated with this increase are usually very small.

This strategy is applicable to all buildings/climates/HVAC systems except for those systems employing packaged unitary equipment too small to be downsized (e.g., the PTAC and heat pump systems in the small building and the hydronic heat pump system in the medium and large buildings).

12. DOWNSIZE HVAC SYSTEM FOR HIGH EFFICIENCY BALLASTS OVER ALL BUILDING AND HIGH EFFICIENCY FIXTURES IN NON-DAYLIT CORE

HVAC systems, including plant, ducts, and fans, are sized to meet the reduced cooling load resulting from the use of high efficiency ballasts over all the building and high efficiency fixtures in the non-daylit core. Note that it is not considered practical to downsize HVAC systems for the installation of high efficiency fluorescent lamps, as it is relatively easy for these to be replaced with standard lamps at a future date, thus increasing cooling peak loads.

HVAC system downsizing should be based on the load reduction described in strategy number 10 for the daylit perimeter and for the load reduction described in strategy number 11 for the non-daylit core.

This strategy is applicable to all buildings/climates/HVAC systems except for those systems employing small packaged unitary equipment that cannot be downsized by an increment as small as 5–10% (e.g., the PTAC and heat pump systems on the small building and the hydronic heat pump system in the medium and large buildings).

13. DOWNSIZE HVAC SYSTEM FOR TINTED GLAZING

Revise the design of the HVAC system for a lower cooling load. This is possible because of the reduction in solar load resulting from the change from double clear to double tinted glass. Shading coefficient changes from 0.81 to 0.55.

14. COOLING COP IMPROVEMENT

Upgrade the COP of the HVAC system cooling equipment from the base level (the ASHRAE–90A 1980 minimum standards) to the ASHRAE–90A 1984 minimum standards. This provides an average performance improvement of 10–15%.

15. HEAT PUMP COP IMPROVEMENT

Upgrade the COP of the HVAC system heat pump from the base level (the ASHRAE 90A–1980 minimum standards) to the ASHRAE 90A–1984 minimum standards. This provides an average performance improvement of 10-15%.

16. HIGH EFFICIENCY BOILERS

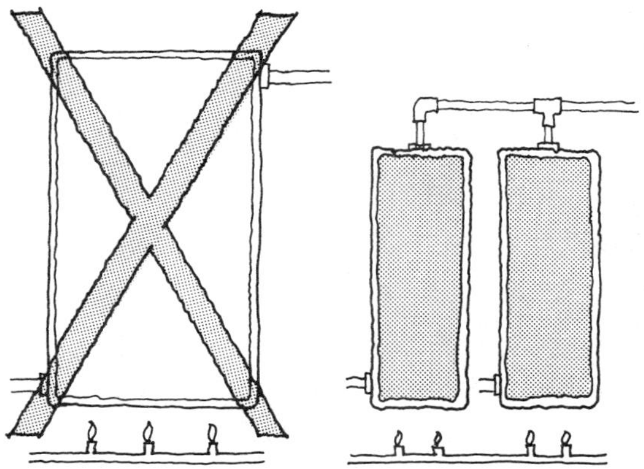

Equipment

Substitute high efficiency gas or oil boilers for conventional boilers. High efficiency (85% or better annual efficiency) boilers are currently available in sizes ranging up to 150,000 BTUH. The high efficiency is obtained through a "pulse" or recuperative combustion process. The recommended application, a medium building in a cold climate, has a design heating load that would probably require multiple units connected in parallel. This would generally be the case in any of the buildings in the medium range. A small factory-supplied controller is required to stage the boilers.

Application

In areas where gas (or oil) is used for space heating and the heating cost represents a large portion of the total bill, high efficiency boilers may be a cost effective option. This typically is true for colder climates with moderate to expensive gas (or oil) rates.

17. EVAPORATIVE COOLERS

Application

In hot/dry climates, a substantial air cooling effect may be achieved through the use of water. This may be accomplished both by using the (normally) low water temperature (relative to the air) in a sensible heat exchanger and also by evaporating water directly into the dry ventilation air. The first process involves variable cooling, while the second process is an adiabatic process. A space cooling system that combines these processes can be highly energy efficient: Energy Efficiency Ratios (EER) of up to 45 are possible. However, the process does consume substantial quantities of water and may require more maintenance than a conventional Direct Expansion (DX) cooler.

It should be noted that in almost all locations it is not possible to obtain satisfactory space cooling at all times by the use of water alone. Periods of high humidity or atmospheric pollution (dust, smog, etc.) may preclude its use. Accordingly, in order to maintain the reliable comfort necessary for office buildings, a conventional DX backup is usually needed. Furthermore, water is consumed in this process so a dependable supply should be available before considering use of an evaporative cooler.

Equipment

There are a variety of means of using well or city supplied water to achieve space cooling. The system recommended for office buildings is the indirect/direct evaporative method. As indicated above, such a system would generally be backed up with a conventional DX cooler. This is, in effect, a three step cooling process. The first, or indirect, stage involves drawing out-

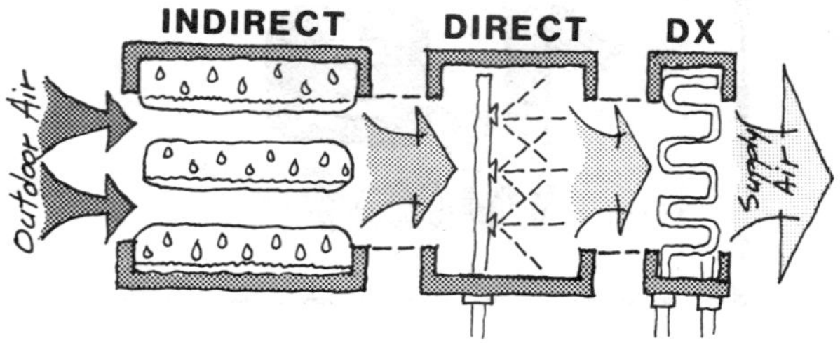

side makeup air through the inside of a heat exchanger such as tubes. The outside of the tubes is then wetted, and the dry ambient air is blown through them. The resulting evaporation of water off the tubes results in the sensible cooling of the makeup air inside the tubes.

The second stage of this system involves the direct evaporation of water into the outside air stream. This is generally accomplished by drawing the dry (but sensibly cooled) outside air and/or return air across a wetted membrane. An adiabatic cooling effect is obtained. The third stage involves conventional cooling of the supply air by a DX (or possibly chilled water) coil in the airstream.

The first stage of this cooling system is used alone when the outside air is dry and the load is not too great. Normally this would involve the use of 100% outside air. This is similar to an economizer cycle but is useful over wider temperature conditions. When additional cooling is required, the second stage would be activated. Depending upon conditions, the percent outside air may be reduced (down to the minimum required for ventilation) and return air used for the remainder of the supply air. Appropriate ducting and dampers are required to achieve these operating modes.

The third and final stage, the DX coil, may be activated at those times when additional cooling is required. The DX coil is normally placed in the supply airstream just downstream of the evaporative unit. (However, it may also be placed upstream of the evaporative unit, making it the actual second stage). Space heating may be provided either by a duct coil downstream of this cooling equipment or by baseboard radiation units.

D. COMBINED EFFECTS

SETS 1 THROUGH 4

The basis of these sets is the use of high efficiency ballasts with various other strategies.

Application of the high efficiency ballasts is based on a lighting layout with one fixture/80 sq. ft. that achieves a power level of approximately 1.9 watts/sq. ft. and an illumination level of 60 fc or greater in open offices. Fixtures are assumed to be the same as the base building: standard 2 ft. × 4 ft. prismatic troffers with four fluorescent lamps. High efficiency ballasts should be included on all fluorescent fixtures in both perimeter and core zones of the building.

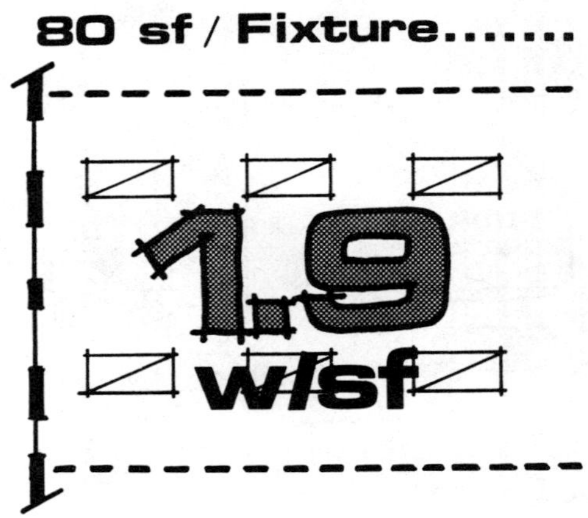

SETS 5 THROUGH 13 AND 53

The basis of these sets is the combined use of high efficiency ballasts and high efficiency lamps with various other strategies.

Application of the high efficiency ballasts and lamps is based on a lighting layout with one fixture/80 sq. ft. that achieves a power level of approximately 1.6 watts/sq. ft. and an illumination level of approximately 50 fc in open offices. Fixtures are assumed to be the same as the base building: standard 2 ft. × 4 ft. prismatic troffers with four fluorescent lamps. High efficiency ballasts and high efficiency lamps should be included on all fluorescent fixtures in both perimeter and core zones of the building.

Note that HVAC system downsizing, if used in these strategy sets, should be on the basis of the reduced cooling load from the high efficiency ballasts alone, not for any cooling load reduction resulting from the high efficiency lamps. It is felt that, because the ballasts are permanently installed, their effect is relatively reliable. Lamp installation, on the other hand, may be changed easily by occupants or maintenance personnel.

SETS 14 THROUGH 23

The basis of these sets is the combined use of high efficiency ballasts, high efficiency lamps, and high efficiency fixtures with various other strategies.

Application of the high efficiency ballasts, lamps, and fixtures is based on a lighting layout with one fixture/80 sq. ft. that achieves a power level of approximately 1.25 watts/sq. ft. and an illumination level of approx-

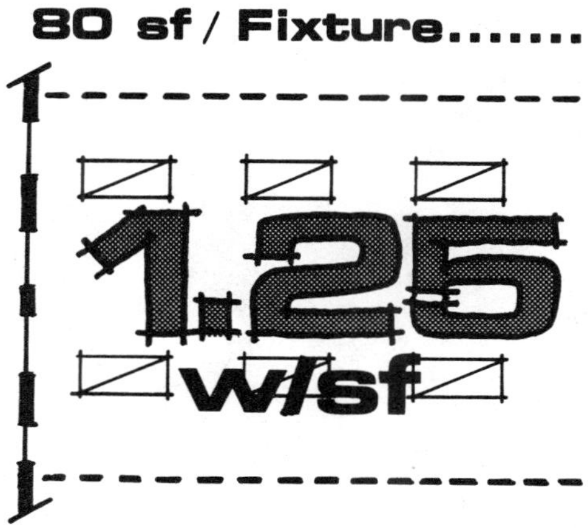

imately 50 fc in open offices. High efficiency ballasts, lamps, and fixtures should be used in both perimeter and core zones of the building.

Note that HVAC system downsizing, if used in these strategy sets, should be on the basis of the reduced cooling load from the high efficiency ballasts and high efficiency fixture but not for any cooling load reduction resulting from the high efficiency lamps. It is felt that, because the ballasts and fixtures are permanently installed, their effect is relatively reliable. Lamp installation, on the other hand, may be changed easily by occupants or maintenance personnel.

SETS 24 THROUGH 29 AND 60 THROUGH 62

The basis of these sets is the combined use of non-uniform lighting and high efficiency ballasts with various other strategies.

Application of the non-uniform lighting and high efficiency ballasts is based on a lighting layout using 75% of the number of fixtures of a conventional grid. Further, this design is based on the use of 2-lamp fixtures instead of conventional 4-lamp fixtures and achieves a power level of approximately 1.6 watts/sq. ft. with a non-uniform illumination level.

It is intended that a non-uniform lighting with efficient ballast combination be compared to the other lighting strategies embodied in strategy sets 1 through 23. The lighting energy savings with non-uniform lighting is almost the same as with high-efficiency ballasts and lamps (strategy sets 5 through 13),but the installed per-square-foot cost for lighting equipment is less. For comparative purposes, these strategy sets do not include high efficiency lamps, but the non-uniform design will benefit from their use just as more conventional layouts do.

SETS 30 THROUGH 52

The basis of these sets is the combined use of daylight dimmers and high efficiency ballasts with various other strategies. Be sure to check with the dimmer manufacturer to assure that the dimmer is recommended for use with the high efficiency ballast that is chosen, as not all ballasts and dimmers are compatible.

High efficiency ballasts are included because they require less electrical energy and allow HVAC equipment downsizing on all but small systems. The potential for occupant intervention to either change or disable dimmer controls is great enough that energy use reductions resulting from dimming should not be included in calculations aimed at downsizing HVAC equipment. Were these shortcomings to be overcome, equipment might be downsized for cooling by up to 20% in some climates.

ADDITIONAL COMBINED EFFECT, SETS 30 THROUGH 36

Application of the high efficiency ballasts and daylight dimmers is based on a lighting layout with one fixture/80 sq. ft. that achieves a power level of approximately 1.9 watts/sq. ft. and an illumination level of 60 fc or greater in open offices. Fixtures are assumed to be the same as the base building: standard 2 ft. × 4 ft. prismatic troffers with four fluorescent lamps. High efficiency ballasts should be included on all fluorescent fixtures, including those with daylight dimmers, in both perimeter and core zones.

ADDITIONAL COMBINED EFFECT, SETS 37 AND 44 THROUGH 48

The basis of these sets is the addition of high efficiency lamps in the non-daylit core to the basic combination of daylight dimmers and high efficiency ballasts with various other strategies. This results in two different design power levels.

In the daylit perimeter, the design and power level of 1.9 watts/sq. ft. is the same as described for strategy sets 30 through 36. In the non-daylit core, the design and power level of 1.6 watts/sq. ft. is the same as described for strategy sets 5 through 13.

Note that HVAC system downsizing, if used in these strategy sets, should be on the basis of the reduced cooling load from the high efficiency ballasts alone and not for any cooling load reduction resulting from the high efficiency lamps. It is felt that, because the ballasts are permanently installed, their effect is relatively reliable. Lamp installation, on the other hand, may be changed easily by occupants or maintenance personnel.

ADDITIONAL COMBINED EFFECT, SETS 38 THROUGH 43

The basis of these sets is the addition of high efficiency fixtures and high efficiency lamps in the non-daylit core to the basic combination of daylight dimmers and high efficiency ballasts with various other strategies. This results in two different design power levels.

In the daylit perimeter, the design and power level of 1.9 watts/sq. ft. is the same as described for strategy sets 30 through 36 except that the effect is confined to the perimeter area. In the non-daylit core, the design and power level of 1.25 watts/sq. ft. is the same as described for strategy sets 14 through 23 except that the effect is confined to the core area.

Note that HVAC system downsizing, if used in these strategy sets, should be on the basis of the reduced cooling load from the high efficiency ballasts in the perimeter zone and from the high efficiency ballasts and high efficiency fixtures in the core zone. No downsizing should be done for any cooling load reduction resulting from the high efficiency lamps. It is felt that, because the ballasts and fixtures are permanently installed, their effect is relatively reliable. Lamp installation, on the other hand, may be changed easily by occupants or maintenance personnel.

ADDITIONAL COMBINED EFFECT, SETS 49 THROUGH 52

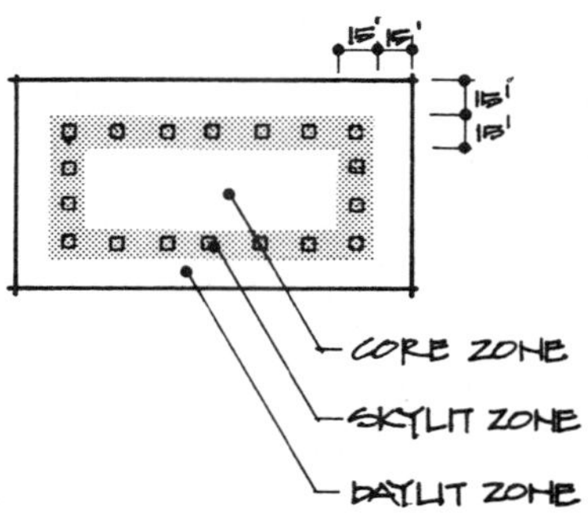

The basis of these sets is the addition of skylights with daylight dimmers in the skylit zone to the basic combination of perimeter daylight dimmers and overall high efficiency ballasts with various other strategies.

The design and power level of 1.9 watts/sq. ft. is the same as described for strategy sets 30 through 36.

E. COST BASED STRATEGY SELECTION

The process for filling out the Level 3 analysis worksheet is explained in some detail in the instruction sheet that accompanies the worksheet itself. This section supplements the basic instruction sheet with added information that is useful in finally selecting a strategy set.

BASE CONSTRUCTION COSTS—$ / SF (Step 3)

The base construction cost for each strategy set has been completed for you and is given in the second column for each case in the Data Library. Note that this cost estimate is based on standard cost estimating techniques. Construction costs are based on non-union labor. It is felt that the majority of buildings of this size would be built in non-central city areas where non-union labor is most common. The costs presented are representative of average quality levels for HVAC systems and should be adjusted if significantly higher or lower quality/cost systems are planned.

The Base Construction Cost is the *national average* cost for applying the strategy set and may be adjusted to reflect costs in your area. This is done most simply by multiplying the Base Construction Cost by the "Location Factor" for your area given in one of the standard cost estimating handbooks (as shown in step 4). The "Location Factor" is typically a decimal slightly larger or slightly smaller than one. You may want to adjust this factor yourself, adjusting it upward or downward based on your own judgment of costs in your area or likely costs on the project under analysis.

Finally, the Base Construction Cost needs to be adjusted to reflect cost changes since 1983, when this book was written. Lacking specific information for your area, you could reasonably use the cumulative change in the Consumer Price Index (CPI) since 1983 as a simple cost escalator. Applying these two adjustments yields the "Adjusted Construction Cost" estimates—item D on the worksheet.

Note that in some cases, applying a strategy set actually *reduces* the first cost of your building. In this case, the Base Construction Cost is given a *negative* value in the Data Library. In all cases, if the Base Construction Cost is negative, the Adjusted Construction Cost will also be negative.

COMPUTING SIMPLE PAYBACK—D÷C (Step 12)

In computing simple payback, two situations can occur. In the first case, there is some added cost for applying the strategy set, so the Adjusted Construction Cost is a positive value. Energy cost savings, item C, are computed by subtracting energy use after applying the strategy set (B) from base case energy use (A). "Simple Payback," or the number of years it will take to "pay back" the original first cost investment for the strategy set with annual savings, is computed by simple division (D÷C). The answer is a positive number indicating the number of years to payback—usually as a decimal (1.2 years, .6 years, etc.)

In the second case, there is either no cost for applying the strategy set or the first cost of the building actually declines from applying the set. In either of these cases, the Adjusted Construction Cost will be a negative number. Simple payback will, therefore, also be negative.

If simple payback is computed to be a negative number, payback can be thought of as "instantaneous." Degrees of negative payback are not really useful for comparison. If a strategy set pays for itself immediately, it is generally not useful to debate just "how immediately."

COMPARING STRATEGY SETS: POSITIVE PAYBACK VERSUS INSTANT PAYBACK

At first glance, one might assume that a strategy set with instant payback will always be better than a strategy set that takes a year or two of savings to pay back. This is not always true. The choice depends on the financial sophistication of your client.

The benefit of "simple payback" analysis is that it is simple. It is an easy way to compare costs and benefits. Unfortunately, it can be misleading in one key way. Payback analysis only considers annual energy savings up to the point where "payback" occurs. In fact, a client will continue to receive savings annually, in most cases for the life of the building. If annual savings are substantial, the benefit of many years of such savings can be very significant.

In selecting a strategy set, review the magnitude of annual energy savings as well as simple payback. If annual savings are large, it may be worth accepting a slightly longer payback in exchange for a large benefit over a long term. This is especially true for the daylighting strategy sets. In nearly all cases, both the first costs and annual savings of these sets are higher than the simpler strategy sets. As a result, paybacks may run in the range of 2–3 years or less. Although significantly longer than the "instant" payback strategies, the magnitude of total savings over the life of the building will also far exceed that of the simpler strategy sets.

If your client is willing, you have the basic information necessary (first cost estimates and annual savings) to do more sophisticated financial analysis, such as discounted cash flow or internal rate of return. Any such analysis, using the client's own financial criteria, will more clearly demonstrate the long term value of increased annual energy savings. Examples of such analysis are not included here because they are highly client specific. It is worth investigating, however, if your client is willing to consider the more comprehensive strategy sets.

LEVEL 4: . . . Add More

The preceding section, Level 3, identified a large number of strategy sets that have the potential to significantly reduce building energy use and building/operational cost. These strategy sets were selected primarily on the basis of their practicality, cost effectiveness, predictability, and ability to have a significant impact on total building energy use. There exist a number of additional strategies that may be of interest but were not included in Level 3, either because they did not meet one or another of the Level 3 criteria, or for reasons such as:

- Implementation would require highly subjective, non energy related decision (e.g., making use of reflective glazing).
- Strategy is appropriate, but does not address the major energy problem (minor energy impact).
- Strategy is not capable of being simulated with the ESE/DOE-2 computer programs used in this research.
- Strategy requires occupant or operator interface to achieve energy reductions.
- Strategy impacts comfort or amenity level.
- Implementation would require skill levels not typically found in small commercial building contractor crews.

These strategies may be attractive for reasons such as high cost effectiveness, increased comfort, maintenance ease, or ability to allow individual choice in energy efficient operation.

Two groups of additional strategies have been developed:

GROUP A: Those that should be selected on the basis of technical judgment

1. Increase thermostat deadband for occupied periods from 2°F to 8°F (70°F heating—78°F cooling)

2. Design for efficient space planning
3. Optimize building orientation
4. Minimize building height to reduce impact of glazed area
5. Reduce shading coefficient to minimally acceptable levels (not with daylighting)
6. Reduce glazing to minimally acceptable levels (not with daylighting)
7. Landscape to minimize wind induced infiltration effect
8. Design for passive heating (particularly for small buildings)
9. Increase roof insulation
10. Increase wall insulation
11. Increase roof reflectivity
12. Increase manual light switching
13. Install operable windows (best for buildings without an economizer cycle)
14. Install electronic ignition, flue dampers, and powered burners for gas or oil boilers
15. Install appropriate energy monitoring and control systems (medium and large buildings)
16. Locate outdoor cooling equipment in shade (north or east side of building)
17. Install night window insulation (cold climate)
18. Design for building mass with nighttime ventilation (Hot/Arid climate)

GROUP B: Those whose value may be determined by simple calculations

19. High efficiency motors and pumps
20. Multiple speed compressors
21. Variable or two-speed circulating fan
22. Use electric resistance heaters if electric demand rates are low, the ratio of electric consumption rate to (net) gas rate is low, and/or the total heating load is relatively low
23. Desuperheater or stand-alone heat pump for DHW

24. Point-of-use electric domestic hot water heaters
25. External shading (overhangs, fins, trees, etc.)
26. Solar domestic hot water heater
27. Chilled water/ice storage (Hot/Humid and Hot/Arid climates)
28. Hot water storage (Cold climate)
29. Demand limiter
30. Absorption chillers/heaters
31. Cogeneration

A discussion of these strategies follows.

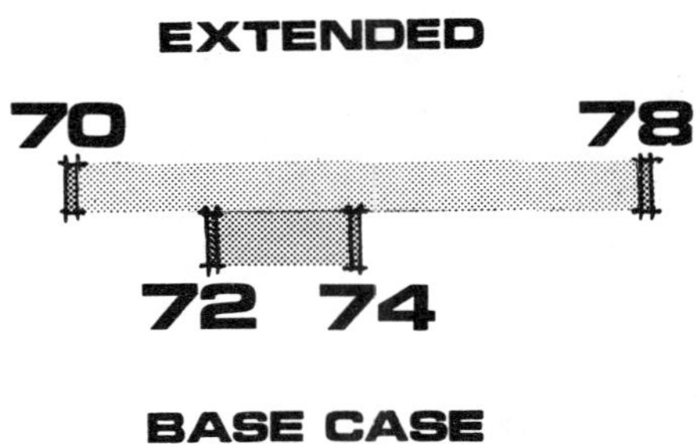

1. Increase Thermostat Deadband from 2°F to 8°F. The thermostat set temperatures that are the basis of the analysis work for the handbook are 72°F for heating and 74°F for cooling during occupied hours. Changing these set temperatures to 70°F for heating and 78°F for cooling may result in major energy savings. Analysis done on typical buildings from the handbook with only a 6°F deadband indicate building energy use reductions ranging from 6% to 16% depending on climate region and HVAC system type. The strategy is simple to accomplish and costs almost nothing to implement. This often makes it cost effective.

A potentially adverse aspect of the strategy should be recognized: building occupant productivity may be reduced. If so, this can easily outweigh any energy cost savings achieved.

2. Design for Efficient Space Planning. Clearly, one way to improve energy cost effectiveness is to distribute energy costs over a larger productive building population. If business objectives can be met while increasing the population of the building, all energy costs will have less relative impact.

Skillful and thorough space planning can significantly increase densities. The research for this handbook assumed a population density of 200 sq. ft./person. If this can be decreased to 180 or 150 sq. ft./person and still maintain acceptable productivity and amenity, significant cost gains will be realized.

Higher densities will mean increased internal loads resulting from people and ventilation. This will increase cooling costs and decrease heating costs, but the change is so small that practically no cost impact will be seen.

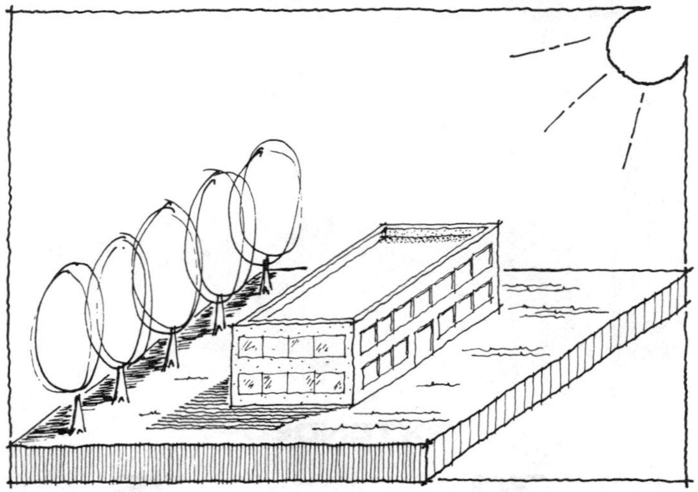

3. Optimize Building Orientation. If building proportions are quite elongate, orientation may have an effect on energy costs. Simulations done during this research project of a four-story 40% glazed building in the Temperate and Hot/Arid climate regions with a 4:1 aspect ratio indicated that changing the building orientation from a long axis north-south position to a long axis east-west position resulted in a reduction in energy use of approximately 5% to 7%. It is fair to conclude that a long axis east-west orientation is preferred.

The effect of orientation is, however, quite weak compared to many other strategies. The impact cited above was for a nearly maximum representation of the effect. Most buildings will have less than a 4:1 aspect ratio and will have glazing around their entire perimeters. If these buildings come up to the performance of Level 2, the effect of orientation changes will be very slight.

4. Minimize Building Height. Decreasing the surface area of a building gives it less exposure to the exterior environment and, thus, a lowered heating and cooling load. One method of achieving this decrease is to lower building height, generally by decreasing floor to floor height. The primary benefits of this strategy, in addition to thermal benefits, are lower construction costs and lower wind loads on the structure.

5. Reduce Glass Shading Coefficient (SC). One way to reduce cooling load caused by solar radiation is to reduce the shading coefficient of the glass to the minimally acceptable level. Glass may be obtained with a shading coefficient as low as .15, which has only 8% light transmittance. More reasonable levels would probably be required to allow good vision, one of the principal reasons for providing glass. Local experience and market forces will probably also exert an influence on SC choices. This strategy is most feasible in climate regions with very high solar radiation and cooling load.

It should be kept in mind that SC, like insulation thickness, decreases in effectiveness as greater levels of shading are achieved. The basic buildings used in the handbook represent a good balance of thermal performance and amenity. Decreasing the SC beyond the values of strategy Levels 1 through 3 will have nowhere near the effect that has already been achieved by Level 1 through 3 strategies.

Reducing SC is, of course, at cross purposes with effective daylighting. Again, the base case buildings (Level 2) achieve a good balance of thermal performance and daylight effectiveness. Slightly higher SC may improve daylighting performance, but there will be a cooling penalty that will counter some of the savings from daylighting.

6. Reduce Glazing to Minimally Acceptable Levels. This strategy should not be used if daylighting is employed in the building. If daylighting is used, the optimal glazing percentage is near 30% of the wall area. Once the Level 1 insulation values have been reached, heat gains and losses and solar gains through the glass area of the building are the major external environmental loads, and they may be reduced by cutting back on glass area.

There are some factors to consider in reducing glazing area below the ranges in Level 2. View and amenity will be reduced, and in the small building it is doubtful that codes or market conditions would allow reduction much below the bottom of the range noted. As with many other strategies, a point of diminishing returns may be reached as one goes below the lower end of the Level 2 ranges. At the 10% glass level, very little of the building utility cost is attributable to the glazed area, and further reductions in glass percentage will have little total impact on costs.

7. Landscape to Minimize Wind Induced Infiltration. For office buildings of less than 50,000 sq. ft., it is possible to have site situations where adjacent trees may be tall enough to form effective wind breaks. If wind patterns during the heating season are relatively stable (and this is by no means always the case—check with your local weather bureau) and your office building can be properly positioned downstream of a substantial wind break, lower wind pressures will impinge on the building. This will reduce infiltration losses.

One useful reference to consult for designing relative to wind breaks is *Plants, People and Environmental Quality*, G.O. Robinette (Washington, D.C., National Park Service in collaboration with the American Society of Landscape Architects Foundation, 1972).

8. Design for Passive Heating. This strategy is of primary usefulness in the cold climate region on buildings where heating represents a substantial portion of the energy cost. Generally, office buildings at the small end of our size scale would be most appropriate candidates for passive heating.

In terms of heating effect and occupancy period, there would appear to be a good match between solar availability and the occupied hours of an office building. However, because much of the building's heating requirement is at night (when internal heat gain is low), some heating carryover would be desirable. A passively heated building would probably benefit from night window insulation.

Before considering passive heating, the designer should study the interactive effects of the entry of strong sunlight into the building. Glare and local overheating will have to be controlled. It is desirable to conduct the heat gain from the south side around the building to meet heat requirements on the east, west and north faces.

An example of passive heating combined with daylighting is shown in the adjacent sketch. Here, a west facing atrium has widely spaced glass walls that act as a plenum on the outside of the building. Sunlight may be modulated by the vertical louvers, and solar heated air can circulate vertically. In this design, the warm air is drawn off by the HVAC return air system and can be used to satisfy building heating needs. The warm air may be vented directly to the outside in the summer.

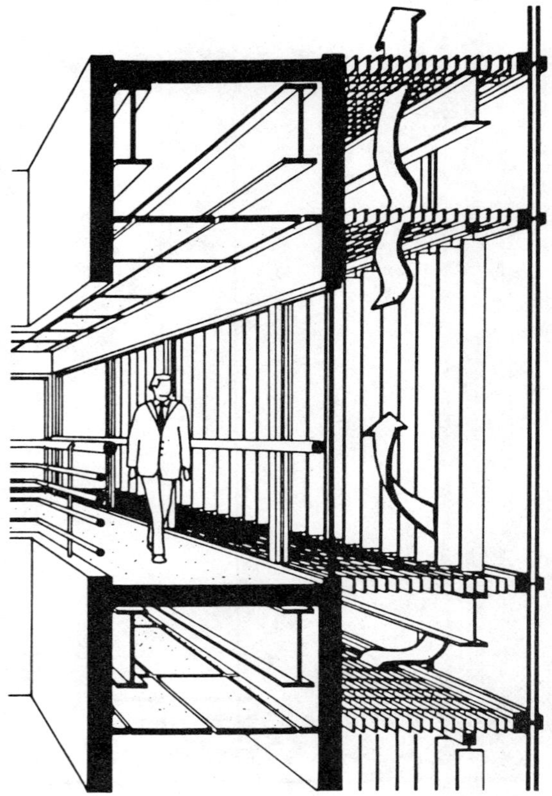

9. Increase Roof Insulation. The roof insulation levels recommended in Level 1 are felt to be the most cost effective design that will meet the requirements of ASHRAE Standard 90–75 for a given climate region. They are keyed to the heating degree days for the reference city used for the computer simulations. In the cold region, the degree days were for Madison, Wisconsin.

The cost benefit of one inch additional insulation was analyzed for the Madison climate data and found to have a payback in excess of 50 years. If you are in a location that has significantly greater heating degree days than Madison (7863), additional roof insulation may be worthwhile. In all other locations, it cannot be recommended.

10. Increase Wall Insulation. For reasons very similar to those discussed under roof insulation, additional wall insulation would be recommended in the cold climate region when local heating degree days are substantially greater than for the design climate, Madison, Wisconsin (7863).

In hot climates, insulation in walls above a modest level actually has a detrimental effect. It reduces the beneficial re-radiation from the walls that can help cool a building when outdoor temperatures are a bit lower than room temperature but not yet to a level that requires building heating. This beneficial heat loss from a building that almost exclusively needs cooling can aid in lowering operating costs.

11. Increase Roof Reflectivity. In climates with large cooling requirements and strong solar radiation, a high roof reflectivity will be beneficial in reducing roof temperatures and, thus, the building cooling load. Light color roof surfaces are common. The principal concern with regard to thermal effect is whether the aging process, including dirt deposition and material deterioration, will significantly reduce the reflectivity of the surface. Careful selection of material for durability and effectiveness and a roof design that enhances long-term performance (such as a shape that washes off and drains well) are important to satisfactory long-term performance.

12. Increase Manual Switching. By providing sufficient light switches it is possible for building occupants to manually control lights and reduce unnecessary energy use.

It is desirable to separately switch areas that have different functions and, thus, different periods of use. Further, it is desirable to separately switch lights near windows so that benefit may be taken of useful daylight.

There are recommendations that partial switching be provided for every 100 sq. ft. of office space to allow light reduction with a fine degree of individual control. Finally, it is recommended that switching be provided for a low level of lighting rather than full office illumination for night-time cleaning.

13. Install Operable Windows. In all areas but those where humidity is a constant problem, operable windows can provide beneficial cooling effects. Computer simulations have shown the economizer cycle to be effective in most climates. This probably represents the best effect to be achieved from operable windows, as it uses outdoor air for cooling whenever temperature and humidity are suitable. Operable windows will fall short of this performance for numerous reasons:

- They rely on occupants to sense the appropriateness of outdoor conditions and tailor window use to these conditions.
- Window ventilation, unlike the HVAC system, has difficulty reaching the full depth of an office structure and, thus, serving all zones.
- Natural ventilation is often curtailed because of annoyance factors such as: wind velocity, rain, dirt, noise or insects.
- Their use may conflict with automatically controlled HVAC systems, causing inefficient operation.

Operable windows probably make the most sense on small buildings where there is the best chance of cross ventilation and with those HVAC systems that cannot use an economizer, notably the small PTAC and heat pump.

14. Install Electronic Ignition, Flue Dampers, and Powered Burners. Electronic ignition may be used to replace a pilot on gas boilers. A typical pilot can consume 6,000 cubic feet of gas a year while the burner is off. An electronic ignition device, which allows gas flow and creates a spark only when the boiler is on, can eliminate this waste. Four to six per cent improvements are common.

Flue dampers are remotely controlled (usually electromagnetically) circular plates, mounted in the flue, that close down when the boiler stops

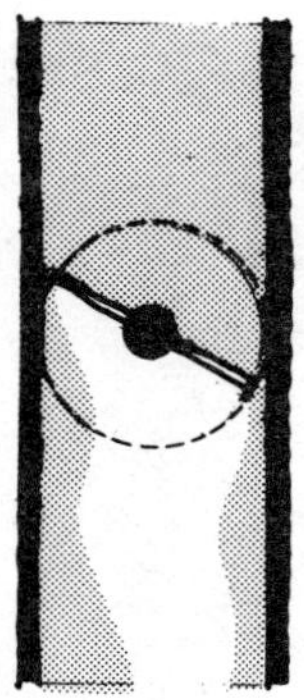

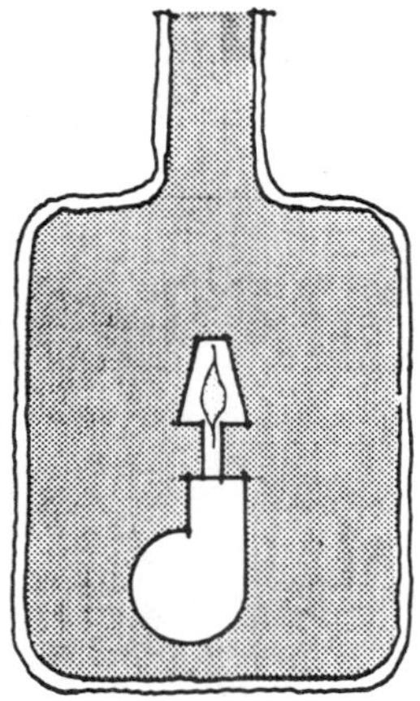

firing. This prevents a draft from occurring by natural convection from the boiler, up the flue, and to the outside. Wind induced drafts through the boiler are also reduced. The boiler's thermal capacitance is thus retained and efficiency improved. Up to fourteen per cent improvements are possible. Because hazardous combustion fumes could escape into the structure should the flue damper fail or activate at the wrong time, only UL approved equipment should be used, and care should be observed when installing.

Powered burners with flame retention heads are another means of improving a boiler's efficiency. A standard gas burner is open to the atmosphere. In a powered burner, the blower draws the proper amount of air for the combustion process. This means less excess air and, hence, greater efficiency. The flame retention burner, which requires a pressure blower, has a combustion head in the shape of a cone. This provides greater turbulence and confines the flame to a smaller zone. In addition to providing greater efficiency, the hood restricts air flow and reduces naturally induced drafts when the boiler is off.

15. Install Energy Management and Control (EMC) Systems. Energy management and control systems can be very effective in reducing building energy consumption, including electrical demand. They can be used to control a wide variety of building conditions and subsystems, including temperature, HVAC, and lighting. These systems typically control, on a time scheduled basis, the various building subsystems in order to obtain acceptable comfort at minimal energy cost. They are generally programmable, so that a generic processor can be employed for a variety of situations, and operational instructions can be modified as conditions warrant. An example of a typical set of functions for an EMC system is as follows:

- Optimal start-stop of equipment
- Duty cycling (demand limiting) of electricity using equipment
- Resetting discharge temperatures
- Providing an alarm for failure of critical functions, such as discharge temperature, dirty filters, or equipment failures

Additional functions, such as energy use accounting and security, can readily be integrated into an EMC system. Design of an EMC system is very building specific.

EMC system equipment includes a central controller/data processor, various sensors, controllers, input/output and recording devices, cabling, and conduit. Although their cost per function has decreased in the past few years, these systems are nevertheless often too expensive for smaller buildings. This strategy is thus most appropriate for buildings of 25,000 sq. ft. and more.

16. Locate Outdoor Cooling Equipment in Shade. If site and building conditions permit, location of DX equipment, air intakes, and other cooling equipment components in a shaded location will make their operation more efficient. The elimination of direct solar radiation and the probable lower ambient air temperatures can only help cooling equipment operation. This can be a no-cost strategy.

17. Install Night Window Insulation. Insulation of windows at night during heating periods holds potential for attractive energy savings. In cold climates, even buildings with double glazing that use nighttime temperature setback to 55°F experience a major portion of their heating requirement during unoccupied overnight periods. The reason for this is that there is very little internal heat generated during this period: lights are off and occupants are absent; outdoor temperatures are generally the lowest of the 24 hour period; and there is no solar heating effect. During this time, the glazed area of the building is one of the major areas of heat loss, even if double glazing is employed.

To cut this heat loss, computer simulations have shown that movable insulation of modest quality (R-3 or R-4) can have a substantial beneficial effect. The strategy is reported to be cost effective. As with any movable insulation, the reliability of use is a major concern. If the actuation of the movable insulation is manual and is left to the occupants, reliability of effect would have to be questioned. On the other hand, automated devices, which may promise reliable effect, are more costly to install. Further concerns with window insulation are the tightness of edge seal and the possibility of heat loss and condensation on the glass if the edge seal is not air tight. These movable devices also should be durable and easy to maintain.

18. Design for High Mass Buildings with Nighttime Ventilation. Buildings designed with high mass construction and appropriate ventilation systems may offer opportunities for reducing cooling energy costs. In such buildings, nighttime air would be circulated through the building to withdraw heat absorbed into the mass during the day. Continued flushing may reduce the mass temperature to the point where it will act as a heat sink for some of the daytime internal heat load.

This is a difficult design problem. Methods of designing buildings for this type of heat exchange and methods of predicting the effectiveness of such a design are currently being developed. At this time, local experience is probably the most reliable guide both for design configuration and probable effectiveness.

A few pertinent factors are as follows:

- There must be a day-night temperature range during the cooling season that drops low enough to provide the necessary cooling effect. This pattern will need to be relatively persistent over the whole of the cooling season for it to be worthwhile considering.
- The quality of the air used for nighttime flushing must be satisfactory. Factors to consider are humidity, dust, etc.
- The quantity of air required for nighttime flushing may be larger than that required during the day for ventilation purposes. The building and HVAC system will have to be designed to handle these larger volumes of air.
- The building design must contain sufficient mass to provide an adequate heat sink for the heat quantities and exchange rates of the design load conditions.
- Substantial portions of the mass may need to be directly exposed to the nighttime flushing air. This may require special design of either the structural mass or the air conduits or both.
- The building should be designed with the insulation (if any) outside of the mass. This will mean the use of a protective cladding system

outside of the insulation unless it is sufficiently durable and fire resistive to meet requirements.

- Designers should consider the acoustic effect of large amounts of exposed mass in occupied spaces.

19. High Efficiency Pumps and Motors. High efficiency pumps and motors can reduce purchased electrical energy consumption simply through their greater efficiency. Energy savings vary depending upon the specific models chosen, with 10 per cent being a reasonable goal. This high efficiency equipment also tends to operate at a lower temperature, thus increasing insulation, lubrication, and bearing life and decreasing maintenance.

In addition to selecting high efficiency pumps and motors, it is perhaps even more important to correctly size the equipment to the load. Equipment that is grossly oversized or that often operates at partial load tends to be inefficient.

Power factor is another consideration related to partial load operation. An AC induction motor that operates partially loaded or unloaded much of the time will have a poor power factor. This results in low efficiency due to excessively high current draw. If the building's net power factor is bad enough (usually less than 85%, which is unusual for an office building), then the utility may even assess a power factor penalty. Power factor controllers are available to reduce this problem and increase efficiency. These devices monitor the motor's load and control voltage accordingly. When the load is low they will reduce voltage, thereby reducing current draw and increasing efficiency.

20. Multiple Speed Compressors. Multiple speed compressors improve the efficiency of a DX cooler by providing a modulated refrigeration control that is more responsive to actual conditions. Single speed units are designed to operate optimally at a given set of indoor/outdoor conditions. However, when less than peak load is required, the efficiency of the system decreases and cycling increases. A multiple speed compressor allows the cooling equipment to more closely follow the load, reduce cycling, and, hence, improve efficiency.

21. Variable or Two Speed Circulating Fans. Fan energy in a VAV system can represent a significant portion of total HVAC system energy consumption. According to the ideal fan laws, airflow is directly proportional to motor speed, horsepower is proportional to the cube of the speed, and, thus, significant savings are possible at part load conditions.

A variety of means exist for regulating airflow, each with its own cost and performance benefits. Traditional means of modulating VAV capacity include discharge damper control and the somewhat more efficient variable inlet vane control (which was assumed for this handbook). These approaches use a "constant speed" motor. More efficient, but more expensive, approaches include AC adjustable frequency drive, mechanically adjustable pulleys, and forward curved fans in conjunction with discharge dampers or, optimally, inlet vanes. The cost/performance of each concept must be evaluated for a given application.

22. Electric Resistance Heating. Electric resistance heating equipment (including the required electrical hookup) is less expensive than a gas hydronic system but is generally more expensive to operate. Depending on a building's specific load profile and the utility rate structure, electric resistance heaters may be more cost effective than the gas boiler otherwise assumed in this handbook. Generally, in order for this to be true, the electric demand rates (that are time coincident with heating demand) must be low, the ratio of electric consumption rate to (net) gas rate must be low, and/or the total heating load must be low. Situations in which the ratio of average to peak heating load is low may also tend to favor electric resistance heating systems. The energy economics of this approach are very site specific.

23. Stand-Alone Heat Pump or Desuperheater for Service Hot Water. Alternate means of producing service hot water include the use of a stand-alone heat pump or desuperheater. These devices offer annual energy savings of up to 50 per cent or more, depending upon operating conditions.

A stand-alone hot water heat pump is a unit that draws heat from the surrounding air and transfers it into the water. It is used for hot water heating exclusively. It uses electricity to power a small compressor and has a condensing coil at the water interface and an evaporating coil at the air interface. Normally, such units are installed indoors and ideally utilize "waste heat" from a boiler or other heat producing device. When installed indoors they aid in space cooling during the cooling season, but unless they truly utilize waste heat, water heater operation results in an additional heating load in the heating season. Alternatively, in mild climates, some models may be installed outdoors. Electric resistance coils are generally used as a backup energy source.

Hot water heat pumps come in sizes small enough that they could be installed in a semi-point-of-use pattern; i.e., one per floor. This could be an attractive alternative when they would operate off of the core zone of the building, which often requires cooling for the entire year.

Desuperheaters are essentially simple heat exchangers that are installed at the hot gas discharge of a compressor (and before the reversing valve in a heat pump system). They are placed in contact with the service water, either through direct immersion or as a separate device piped in parallel with the tank. The hot refrigerant from the compressor gives up some of its heat to the water, thus heating it. This results in a net increase in cooling capacity but represents a decrease in capacity when applied to a heat pump in the heating mode. Such devices are often relatively inexpensive and can be quite effective. However, it is important to recognize that they are only operative when the compressor is on. Thus, a 100 per cent backup must be provided. Desuperheaters also introduce additional pressure drop into the overall refrigerant pressure drop.

Each stand-alone heat pump and desuperheater application must be engineered and a cost/benefit analysis performed in order to establish technical and economic viability.

24. Point-of-Use Hot Water Heaters. Service hot water load in an office

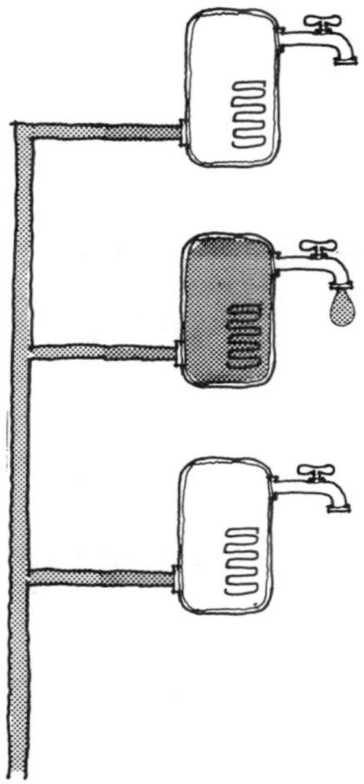

building (that does not have a kitchen or other process load) is often small and intermittent. The line losses between a central heater and the load often represent more than half of the gross load. This loss can be largely eliminated through the use of small distributed electric hot water heaters at each point of use. In this approach, the cost of the distributed electric heaters, associated power wiring, high efficiency, and resulting operating cost must be traded off against the cost of a central (gas or electric) fired hot water heater, the hot water piping and insulation, low efficiency, and resulting operating cost.

25. External Shading. External shading is a powerful method for reducing cooling loads resulting from solar gain. Shading may be provided by fixed or movable overhangs, fins, screens, louvers, etc., or by trees, shrubs, vines, or other plant materials. There are numerous references that give guidance on the design of shading devices.

The basic Level 2 buildings described in the handbook mainly have tinted glass with a moderate level shading coefficient (.55). In these buildings, the amount of utility cost attributable to solar gain is rather small, and savings that can be achieved by adding window shading devices are not large. Furthermore, though shading of south facing windows is not difficult, east and west facing glass presents a major challenge to the designer who is trying to achieve effective sun control.

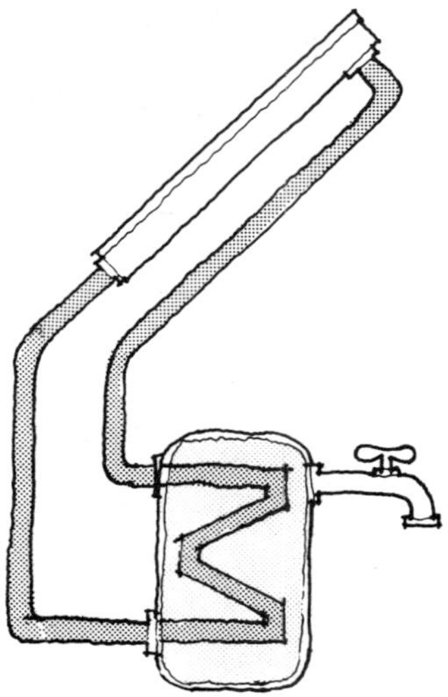

26. Solar Domestic Hot Water (DHW) Heater. The domestic hot water loads in office buildings that do not have cafeterias or other large hot water uses is very small. Packaged solar DHW heaters are available that can easily fulfill these hot water requirements. A solar DHW system is a centralized system and suffers from the same problem of large distribution losses as conventional centralized systems. Cost effectiveness and maintenance requirements should be analyzed carefully before such an installation is undertaken to be sure they are consistent with project objectives.

27. Chilled Water or Ice Storage. Some utilities may impose significant demand penalties or time-of-day rates in an effort to reduce their peak capacity problems. Depending upon the specific rate schedule, it may be desirable to avoid a demand penalty and/or take advantage of low night-time rates through the use of chilled water or ice storage for cooling. In this concept, "coolth"—either cold water, ice, or a slurry mixture—is generated through conventional means during periods of low rates and/or low demand. This stored cooling capacity can then be used later during periods of peak cooling demand, thus avoiding purchasing electricity at such times.

In addition to displacing the purchase of electricity, this approach can also allow downsizing of the DX capacity. This results in a cost credit. However, the required storage tank and plumbing (and possibly ice generation equipment) can be expensive and more than negate this DX cost credit. A significant amount of floor space is also often required for storage tanks large enough to be practical. Large size results from the practical limitation of about a 20°F temperature differential. Ice storage systems require less room for a given capacity but are more complex than chilled water.

Chilled water or ice storage systems are generally most applicable to hydronically based cooling systems. Chilled water and/or ice storage systems generally increase a building's first cost. Depending on utility rate structure, the purchased energy savings may result in an acceptable payback. This is most typically true for large buildings (100,000 sq. ft. and more) in hot climates. Under any conditions, a thorough engineering cost/benefit analysis is required for each specific building.

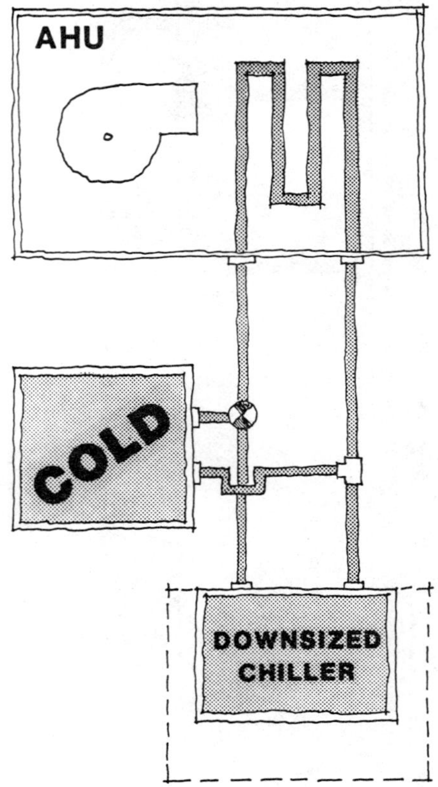

28. Hot Water Storage. High demand charges and/or heavily weighted time-of-day rates may suggest the use of hot water storage as a means of time shifting the purchase of electricity. This would generally be true only for all-electric buildings that experience heating loads at hours during which the utility assesses some form of cost penalty. The availability of hot storage would avoid or reduce consumption during such times. Conversely, hot storage may also be useful for reclaiming excess building heat generated during daytime operating hours and using it for space heating at night. In order to use hot water storage, the heating (and possibly, cooling) system should be hydronically based.

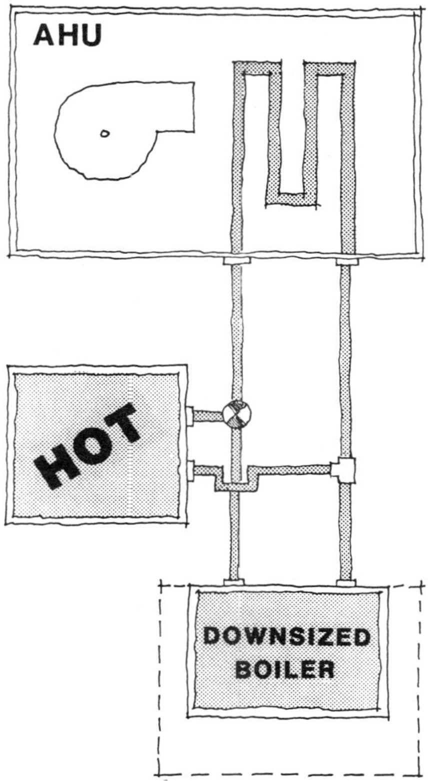

Heat may be stored at temperatures up to 212°F or beyond if pressurized storage is used. Thus, depending on use temperatures, large temperature differentials (100°F) are possible. This reduces the required volume of storage for a given capacity (relative to cold storage). Nevertheless, storage tanks and piping are expensive and do take up signficant floor space. Accordingly, each application must be carefully engineered and analyzed in detail to determine technical/cost viability. Large buildings (100,000 sq. ft.) in colder climates typically represent the best candidate applications.

29. Demand Limiter. There are light dimming systems available that can be controlled by the level of building electrical demand. They are generally operated by a microprocessor type demand controller. Such a dimming system might be worth investigating where an Energy Management and Control (EMC) system is going to be installed.

The total building fluorescent lighting/dimming control system is a set of dimmers connected to all (practical) fluorescent lights in the building. The dimmers must have peak demand reducing controls that can be set to initiate dimming at a predetermined level of electrical demand and dim the connected lights to a predetermined level (normally a 10% to 20% reduction of light output).

The dimmers are most efficient when installed in 277 V lighting system and, for cost reasons, may practically be limited to 277 V installations. For peak demand reduction, the dimming control is useful only to reduce summer peaks, because dimming lights reduces their heat output and, thus, the cooling load. In the winter, this lost heat would have to be made up by additional heating. The amount of dimming is modest, up to 20% maximum, and is considered to be imperceptible to occupants (illumination may drop from 50 fc to 40 fc).

This strategy would be most practical to consider where demand charges are very high. Further, it is not as effective as daylight dimmers and should be considered only when daylighting is not considered practical.

30. Absorption Chiller-Heaters. In the past few years, packaged absorption chillers and heaters have become available for space heating, cooling, and service hot water applications. Such units typically employ a double effect, lithium bromide based absorption cycle to obtain cooling COPs of about 1.0. Cooling is provided in the form of chilled water. Such units may also be operated as a boiler to generate hot water for space heating or service water. Depending on the design of the specific unit, simultaneous heating and cooling operation may or may not be possible. These units generally can be fired with gas, propane, or oil.

In addition to providing heating and cooling in a single prepackaged/pre-engineered system, absorption chiller/heaters can offer a variety of other benefits to a building owner. With respect to a conventional DX chiller and gas boiler, potential advantages include:

- Reduced floor space.
- Comparable equipment cost.
- Lower operating cost, depending on the relative rates of electricity to gas.
- Reduction in electrical demand.
- Better part load performance.

These benefits are dependent upon the specific situation. For example, relative operating costs depend on both fuel rates and conversion efficiency. For a DX chiller operating at an effective COP of 2.5, the utility gas rate would have to be more than 40 percent of the utility (marginal) electric consumption rate (in equivalent energy units) in order to provide an equivalent cost of delivered cooling. The absorption unit will also have greater parasitic energy costs (due to a larger cooling tower) than a conventional DX system, and, thus, the actual ratio will be lower. However, the absorption chiller/heater will offer some relief in demand charges if such are applicable.

Absorption technology is well established, but these packaged combined cooling/heating units are relatively new. They are currently available in sizes ranging from a few to over a thousand tons cooling capacity, and they may be applicable to all building sizes included in this handbook. When selecting such a system, the designer should insure that the chosen equipment will meet all cooling and heat loads, wherever such loads occur. Relative cooling to heating capacity and simultaneous operation may be issues. The economics of each application is specific to the building's heating and cooling load magnitude/profile and relative fuel cost.

31. Cogeneration. Prepackaged cogeneration units have recently become available in sizes ranging from 25 kw to 840 kw and are thus applicable to the building sizes considered in this handbook. These units usually consist of a diesel engine coupled to an electrical generator with a heat recovery system to reclaim the exhaust heat. Typically, about one third of the energy content of the input fuel (such as oil, propane, or gas) is converted to electricity, and about half is recovered as useful heat for space heating or process heating applications. Absorption refrigeration is generally available as an option.

The economics of such a cogeneration system relative to a conventional heating and/or cooling system is a complex issue. Capital installation costs for equivalent systems, operating costs (dependent on both efficiency and consumption rates for electric and fuel), and maintenance/lifetime costs are involved. In addition to displacing utility supplied electricity, a cogeneration system might also sell energy back to a utility, obtain a "capacity credit" from a utility, and offer a user some reduction in demand charges. Conversely, use of a cogeneration system may cause the utility to increase the rates for the remaining electricity purchased by a user. In addition, unless a backup system is available, it may be necessary to operate a cogeneration system at undesirable times in order to meet space heating and/or cooling loads. Typical office building load profiles may not represent the best match for such a system.

	DESIGN	CONSTRUCTION	TURNOVER	BUILDING OPERATION
DESIGN TEAM (ARCHITECTS/ ENGINEERS)	• Document all design assumptions (1) • Estimate energy use by sub-system (2) **& For Large Buildings:** • Document all systems and control strategies (3) • Specify contents of O&M Manuals (4) • Specify HVAC balancing requirements (5) • Specify equipment protection and cleaning (6) • Specify on-site instruction (7) • Specify access doors, valves, test ports (8) • Specify piping system ID (9) • Specify price quotations on all long-term maintenance contracts (10)	• Prepare "as-built" drawings (11) • Observe construction process for compliance (12)	• Conduct "punch list" inspection (15) • Visually check for air leakage (15) • If possible, participate in on-site training session(s) (7)	
CONTRACTOR & SUB-CONTRACTOR	**For Small Buildings Only:** • Document all systems and control strategies (3) • Specify contents of O&M Manuals (4) • Specify HVAC balancing requirements (5) • Specify equipment protection and cleaning (6) • Specify on-site instruction (7) • Specify access doors, valves, test ports (8) • Specify piping system ID (9) • Specify price quotations on all long-term maintenance contracts (10)	• Prepare O&M Manual (4) • Perform final calibration of all controls (5) • Hire independent balancing agent or require of subs (5) • Protect and clean equipment (6) • Provide all access doors, valves, and test ports (8) • Place piping system ID tags (9) • Prepare "as-built" drawings (11) • Use HVAC system to "flush" construction odors (13)	• Deliver O&M Manual (4) • Conduct on-site instruction and training (7)	• Honor warranty and call back as required (16)
EQUIPMENT MANUFACT.		• Support calibration of controls (5) • Provide all special tools required (14)	• Support O&M Manual preparation (4) • Support on-site instruction and training (7)	
OWNER & BLDG. STAFF	• Check all key design assumptions (1)	• Periodically check compliance on-site (12)	• Attend on-site instruction and training sessions (7)	• Monitor energy consumption against estimates (2) • Periodically test equipment efficiency (17) • Periodically sample and test water (18) • Inspect and calibrate controls (19) • Monitor occupant complains (20)

LEVEL 5: Finish Right

A CHECKLIST TO INSURE SUCCESSFUL TURNOVER AND PROPER BUILDING OPERATION

The best designed building in the world will not save energy unless its turnover and operation are in accordance with the intentions of the design. In effect, the effort spent in achieving a good design (Levels 1–4) could largely be wasted unless the process is "finished right" (Level 5) by planning an orderly turnover process and good building operation. The consequences of slighting or ignoring this step are all too well known to owners: uncomfortable and, therefore, unproductive employees; increased fuel and electrical costs; high maintenance costs; and, in extreme cases, even safety hazards.

Turnover and building operation need not be difficult. There are two key factors for success:

- Plan for turnover and operation during design and continue to check key items throughout construction and occupancy.
- Include all key players (owner, staff, design team, contractors, subcontractors, and manufacturers) and be sure that each takes responsibility for specific tasks at the proper time.

A checklist including many of the key steps that must occur for successful turnover and building operation is shown on facing page. The checklist is arranged as a matrix to indicate both when a step should occur (design, construction, turnover, operations) and who should be responsible (design team, contractor and subcontractors, equipment manufacturer, or building owner and staff). The matrix clearly illustrates the shift in major responsibilities over time: the design team during design, the contractor and subcontractors during construction, and the building owner or staff during operation.

It is interesting to note, however, that all players have responsibilities in other phases of the process and that all are needed during turnover. It is also very important to note how many steps occur *before* turnover. If turnover and building operation are to be successful, the majority of work must be done before turnover occurs. Unfortunately, this is not typical practice today and, as a result, the potential of many buildings to be energy efficient is never met.

Steps listed on the matrix are purposely brief in order to make the whole checklist visible in a single figure. The rest of this chapter discusses each of these steps in detail. Each step on the checklist is keyed by number to the following text. If one or more steps clearly relate to one another, they are given the same reference number and are discussed in the same section of the text. As always, the basic guidance suggested here should be expanded upon and adapted to each building, based on good professional judgment.

1. DOCUMENT ALL DESIGN STRATEGY ASSUMPTIONS

In developing building and systems design concepts, the A/E must make assumptions regarding the manner in which a building will be operated. These assumptions include temperature set points for HVAC controls; minimum and planned ventilation rates; planned occupancy schedules and variations for cleaning and maintenance; and operation of any special systems or building components, especially passive components or components requiring user interaction. Preliminary discussions of these assumptions among the building owner, the operating staff, and the design team are necessary. Such discussions will insure not only that the designer's initial assumptions are based on the owner's requirements but also that these assumptions can be implemented by the operating staff. The importance of documenting these assumptions and passing them on to the operating staff cannot be overemphasized, as the efficiency of actual building performance will depend on the degree to which operating conditions match design assumptions.

Documentation can be made part of the "as-built" drawings or part of the O&M manual prepared by the contractor. It is important, however, that the documentation be completed during the design phase, while the assumptions are still fresh in everyone's mind.

2. ESTIMATE ANNUAL ENERGY CONSUMPTION BY SUBSYSTEM

During design, the design engineer should prepare a system-by-system breakdown of estimated annual energy consumption. Depending on the building size, type of systems used, and the fee available for energy analyses, these estimates may range from fairly crude to fairly accurate predictions. This breakdown will be used later by the building operating staff in several important ways:

A. Operators can review assumptions used in the analyses to determine such items as daytime and night-setback thermostat settings or lighting system switching arrangements for cleaning activities.
B. Operators can review energy use estimates to identify systems that are causing highly excessive energy consumption due to a control malfunction (e.g., failure of outside air dampers to close during unoccupied hours or a preheat coil control malfunction, which could cause continuous overheating of ventilation air that must be cooled back down to the proper supply air temperature).
C. Operators can compare actual monthly energy costs with design estimates as another indicator of poor system performance. Once problems are identified, the necessary response will depend on the complexity of the problem and the experience of the operating staff.

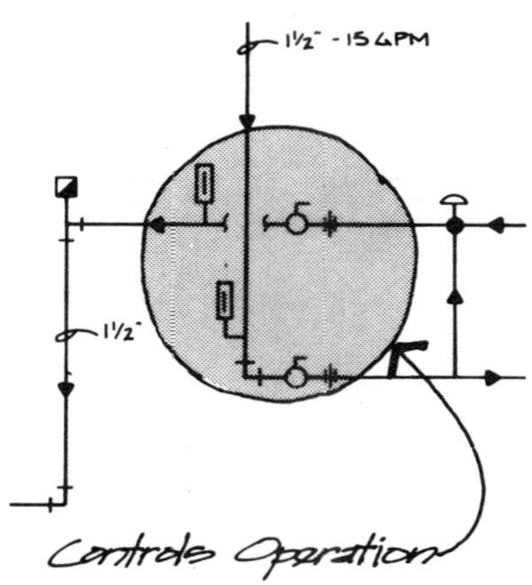

3. DOCUMENT ALL SYSTEMS AND CONTROL STRATEGIES

The design engineer should prepare schematic diagrams illustrating the major components of each building subsystem, including the proper location of each valve or control device. These diagrams should be accompanied by written descriptions of all control sequences in each mode of operation. Together, these documents will allow the building operating staff to understand how each system is intended to function and to determine if all system components are installed and operating properly. These diagrams are often placed in glass-enclosed frames and mounted in the mechanical room or maintenance manager's office so that they become permanent records for future reference. They should, at least, be made part of the O&M manuals.

4. SPECIFY CONTENTS OF CONTRACTOR-SUPPLIED OPERATING AND MAINTENANCE MANUALS

The design team's specifications writer should provide detailed instructions regarding the content and format of all contractor-supplied O&M manuals. Otherwise, the building operating staff may simply be provided with a set of equipment "catalog cuts" containing little or no useful information. Depending on the type of equipment provided, the O&M manual should contain the following items to aid the operating/maintenance staff in the performance of their duties:

A. Contractor/Subcontractor's name, address, and telephone number
B. Index
C. General description of each system and catalog cuts of equipment
D. Step-by-step equipment start-up, shutdown, and operating procedures under both fully occupied and partially occupied conditions
E. Spare parts list, including local supplier's name and address
F. Routine preventive maintenance and lubrication procedures
G. Wiring and control diagrams for each piece of equipment, showing "as-installed" conditions (see also 3 and 12)
H. Performance curves for all pumps, fans, refrigeration machines and similar equipment.

As an alternative, the O&M manuals can be prepared by the A/E design team. This option must be made clear in the original contract for A/E services, as this is not part of a normal A/E contract.

5. SPECIFY HVAC SYSTEM BALANCING REQUIREMENTS

For all large buildings, and often for smaller buildings, it is important to insure that all water and air distribution systems are properly balanced. Otherwise, thermal comfort conditions designed for individual spaces may never be realized.

For larger buildings, typical specifications usually require that the contractor who installs a particular subsystem be responsible for hiring an independent agency to complete the balancing in accordance with the procedures specified, with a final report provided to the owner and signed by a professional engineer. Standard balancing procedures developed by various trade organizations (such as the Sheet Metal and Air Conditioning Contractors National Association) are often referenced in the construction specifications.

Having an independent agency verify that a particular subsystem is properly balanced may cost on the order of 5% to 10% of the subsystem construction cost. While this is money well spent in larger buildings, it is not absolutely essential for small buildings. One method of reducing first costs is to place responsibility for balancing with each subsystem contractor, through the general contractor, as described below:

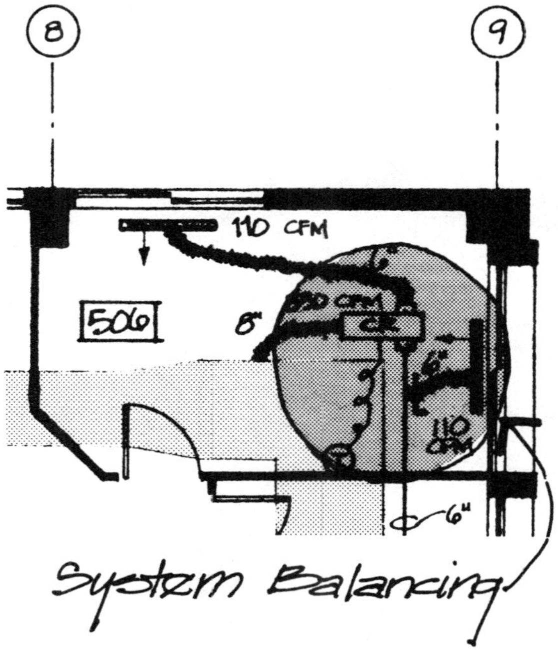

A. The contractor who installs a particular subsystem is responsible for balancing that subsystem to achieve the thermal comfort conditions that the system was designed to provide. Balancing verification by an independent agency is not mandatory and, therefore, is not included in the contractor's price.

B. If, after initial occupancy, the owner is not satisfied with the subsystem performance, an independent balancing agency is hired directly by the owner to document both actual comfort conditions and the adequacy of the original system balancing.

C. If the independent agency verifies that the system is balanced and operating as described in the construction documents, the contractor who installed the system is relieved of any responsibility and the owner pays for the independent balancing report. However, if it is found that the system was not installed or balanced properly, the contractor becomes responsible for the cost of the independent balancing report and any modifications required to make the system perform as originally designed.

This type of agreement benefits both the owner and the contractor, as long as the contractor installs and balances his systems properly. The owner is protected if the contractor does not perform adequately, and the contractor is protected if the cause of the owner's complaint is a design error or the result of another trade's installation.

6. SPECIFY CONTRACTOR REQUIREMENTS FOR CLEANING, PROTECTING, AND ADJUSTING EQUIPMENT PRIOR TO BUILDING OCCUPANCY

Normally, construction specifications require the contractor to be responsible for protecting any equipment installed early from damage resulting from subsequent construction operations. The contractor is also responsible for final cleaning of all equipment upon completion of construction and making any adjustments required for proper system operation. Unfortunately, these requirements are often overlooked as the construction period draws to a close and contractors become anxious to "get on with the next job." Therefore, the owner or his representatives must pay particular attention to see that these responsibilities are fulfilled prior to final payment.

Specific areas of concern that must be addressed during the construction period include the following:

A. Protect equipment, piping and ductwork from construction activity, damage or dirt. Vertical sections of piping systems and ductwork should be capped to avoid construction dirt accumulating within them prior to final system hook-ups.

B. Replace air handling equipment filters. Accumulation of construction dirt on installed equipment filters can degrade equipment performance and cause dirt "streaking" of finished ceilings. Periodic filter replacement *during* the construction phase may be necessary if installed systems are used to provide temporary heat or ventilation during the construction period. Care should be taken that no equipment is operated without filters.

C. Clean, flush, and sterilize domestic (potable) water piping systems.
D. Clean, flush, and protect from corrosion and freezing all piping systems that distribute non-potable liquids.
E. Clean, repaint, or refurbish all equipment that is damaged or soiled during the construction period.
F. Adjust all balancing valves, air vents, automatic controls, and similar devices in accordance with specfied balancing procedures (as discussed in 5 above).

7. SPECIFY ON-SITE INSTRUCTION AND TRAINING

Normally, the contractor who installs a piece of equipment is responsible for providing adequate on-site instruction and training to the building operating staff prior to the owner's final acceptance of the work. In some cases, it is wise to require that the contractor provide the services of a factory representative to conduct these instruction sessions, as a contractor may not be qualified to instruct others on operational procedures for complex equipment (such as energy management systems and chillers). This is rarely done on small projects and should only cover complex equipment. These efforts should be coordinated with the preparation and delivery of the O&M manuals (as described in 4 above).

If at all possible, some members of the design team should participate in the on-site training sessions. Not only is this a good opportunity for the design team to convey the overall strategy of the building for energy efficiency while meeting its other requirements; it is also an excellent time to re-focus on key operating assumptions, such as:

A. Temperature set points for HVAC controls
B. Minimum and planned ventilation rates
C. Planned occupancy schedules and variations for cleaning and maintenance
D. Operation of any special systems or building components (especially passive components or components requiring user interaction).

8. SPECIFY ADEQUATE ACCESS DOORS, MAINTENANCE VALVES, AND TEST PORTS

In order for a piece of equipment to be maintained properly, the maintenance staff must have easy access to it. This applies to small items (valves, dampers, etc.), as well as to larger pieces of equipment. Maintenance of equipment that is difficult to reach tends to be delayed or forgotten by the building staff. It is the responsibility of the design team to see that adequate access is specified; the responsibility of the contractor (and subcontractors) to see that all access doors are installed; and the responsibility of the owner (or his representative) to check on both.

During construction, many key items such as in-line pipe thermometers, pressure gauges, and sight glasses may be omitted from the installation in an effort to reduce the first cost of mechanical systems. In some cases, even service shutoff valves are omitted, preventing the building maintenance staff from servicing individual pieces of equipment without shutting down (and possibly draining) an entire piping system.

In order to permit smooth, inexpensive, and long-term operation and maintenance of building systems, adequate service shutoff valves and instrumentation must be specified and installed to facilitate systems maintenance. Instruments such as thermometers and pressure gauges are necessary to determine whether systems or components are performing as designed, to fine-tune systems after building occupancy, and to troubleshoot malfunctioning systems. These devices can be very expensive to install after the fact, often requiring system shutdown, time delays, and loss of building comfort control during the installation process. Even if only a minimum of these devices can be included in the original installation, it is wise to provide a number of additional test ports (e.g., capped thermometer pipe wells) at critical points within the systems so that troubleshooting devices can be temporarily installed without shutting down the entire system. The location of these test ports should be documented as carefully as all other system components.

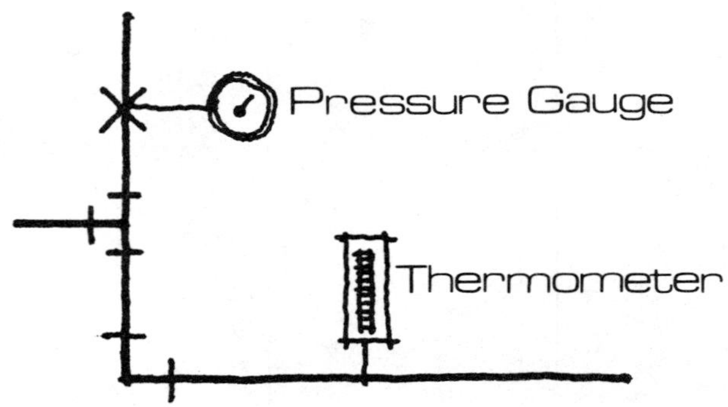

Instrumentation......

9. SPECIFY EQUIPMENT AND PIPING SYSTEM IDENTIFICATION

In large buildings the quality, number, and location of piping system identification tags and flow direction markers are typically spelled out in the construction specifications. Brass or plastic identification tags correspond to the diagrams and written descriptions of each system (see 3 above). The purpose of these identification markers is to help members of the various construction trades, balancing agencies, building inspectors, and operating staff to more easily understand each piping system and its many components. Obviously, for larger buildings or more complex subsystems, it is more important that a good identification system be provided to facilitate inspections, troubleshooting, and future modifications.

10. SPECIFY PRICE QUOTATIONS ON ALL LONG-TERM MAINTENANCE CONTRACTS

If an automated energy management system or similar "high technology" system is to be provided for the building, selection criteria must include the cost of long-term maintenance. Too many building owners have purchased an expensive system with a nominal one-year maintenance service contract, only to find out later that the actual costs of a long-term maintenance contract are much higher. In some cases, the owner may also be locked into a specific arrangement because other service companies are either not available in the local area or are not qualified to service the proprietary equipment included in the system. To avoid such unhappy surprises, price quotations for long-term service contracts should be requested and evaluated for competitive systems before final system selection is made.

11. PREPARE "AS-BUILT" OR RECORD DRAWINGS

If an architect is engaged to provide construction documents, the development of "as-built" drawings is a simple procedure. Subcontractors may simply mark up sepias of original drawings to show changes that occurred during construction. This process does require close supervision by the A/E team to make sure that all changes are recorded. However, if the owner works directly with the contractor, the quality and extent of record drawings to be furnished by the contractor should be established as part of the negotiation process. If inadequate drawings are provided, the trouble-shooting, system fine-tuning, or modifications required after construction is complete may be extremely difficult to accomplish or prohibitively expensive for the owner. Because of the complexity and interaction of mechanical/electrical systems (even in small, "simple" buildings), it is extremely important to long-term building operation that the as-built conditions be thoroughly documented for future reference (see also 2 above).

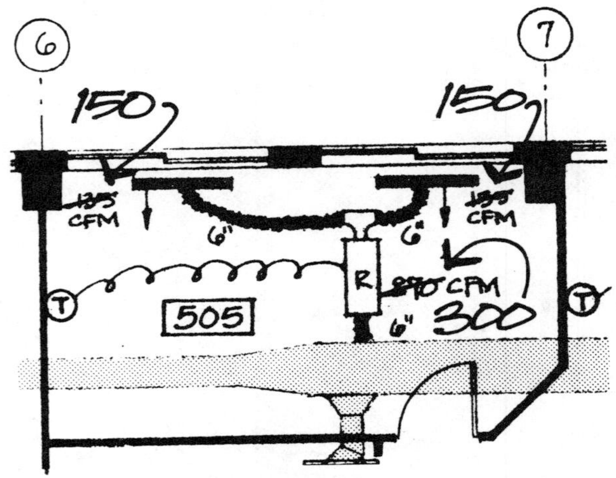

"As-Built" Drawings....

12. OBSERVE THE CONSTRUCTION PROCESS FOR COMPLIANCE

On-site inspection throughout the construction process is traditionally part of the design team's responsibility. There are, however, a number of areas that merit special attention if the design team is attempting to achieve an energy efficient building:

A. Carefully check the integrity of the building envelope. Air leakage (infiltration) is a common reason that buildings do not achieve their potential for energy efficiency. Areas to be given close inspection are the continuity of insulation and the integrity of air/ vapor barriers. Both of these tend to receive less than careful attention from most contractors.

B. Check ductwork for air tightness and proper insulation where necessary. Again, this is a common area for less than perfect workmanship. Excessive leakage will make it difficult to balance the HVAC system and, should leakage occur in the wrong areas, could lead to excessive energy consumption.

C. Check all lighting equipment. Make sure that the right ballasts and fixtures are supplied. At times, high efficiency ballasts are specified but not delivered. Also check to see that fixtures are installed in the proper places. A lighting strategy that mixes high efficiency and regular fixtures can be confusing to or ignored by installers. Also, check to see that specified lamps are actually installed. If the owner is not represented by an architect or other member of the design team, it is essential that the owner or the owner's representative on-site check all of the items above.

13. USE HVAC SYSTEM TO "FLUSH" BUILDING OF CONSTRUCTION ODORS

If the building HVAC system includes an economizer cooling cycle, use it to purge the building of construction fumes. One of the last operations in the construction phase is usually the installation of carpeting, which often requires the use of particularly strong-smelling adhesives. In some cases, the economizer cycle may have to be run 24-hours a day for several days prior to initial building occupancy to purge these odors from the building.

14. MANUFACTURERS SHOULD PROVIDE ALL SPECIAL TOOLS REQUIRED

In some cases, special tools are required to service or maintain a particular piece of equipment. For example, some boilers require a special tool to remove boiler tubes for cleaning or replacement. Any such tools should be provided to the maintenance staff and included as part of the equipment first cost. Responsibility for supplying any spare parts, such as fuses, belts, etc., should be clearly assigned to either the manufacturer or contractor and be included in the cost estimate.

15. CONDUCT A "PUNCH LIST" INSPECTION

As part of the normal turnover process, the design team, the contractor, and the owner will jointly inspect the building for items to be completed before the owner accepts the building and occupies it. It is important to use this event as an opportunity to double-check those items that will have significant impact on energy use. Items to include are:

A. A visual check for air leakage (around doors and windows)
B. Operation of the HVAC system within the comfort ranges specified for both heating *and* cooling in *all* areas of the building
C. Delivery of the O&M manuals and the holding of good quality training sessions
D. Adequate system identification (tagging)
E. Clean equipment, including filters, following construction
F. Delivery of any special tools required by the building maintenance staff.

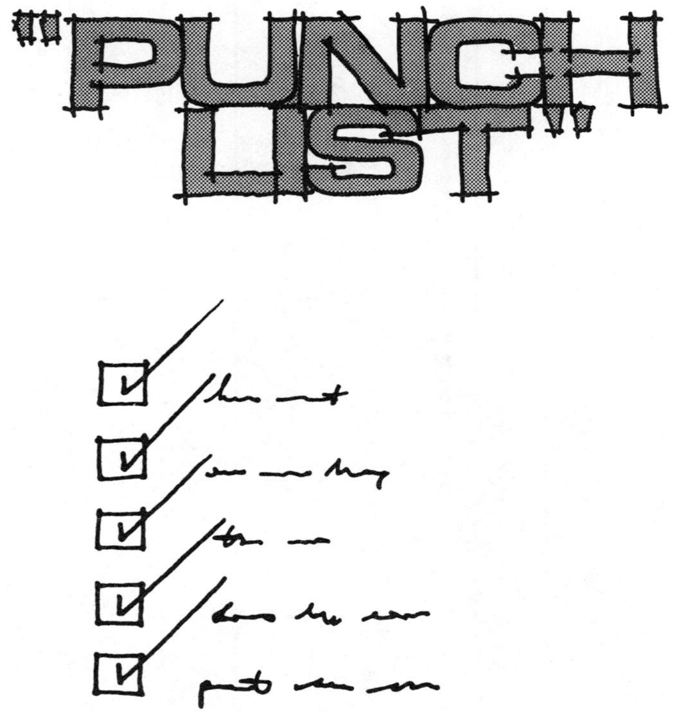

16. CALL BACK CONTRACTOR AND SUBCONTRACTORS AS REQUIRED TO MEET WARRANTY

If careful attention has been paid to the items listed above, it is hoped that neither the contractor nor subcontractors will have to be called back in order to meet their obligations under the original contracts. Nevertheless, it is important for the owner to work with the maintenance staff to see that energy related systems are working as expected before the contractor and subcontractors are released from all obligation. In this regard, it will be useful to check actual energy consumption (and costs) during the first year and compare the results to anticipated consumption. It should be expected that the building will use more energy than anticipated until systems are "fine-tuned" for best operation. Whether this fine-tuning should be done by the building staff or by calling back contractors depends on the experience of the building staff, the complexity of the system (or systems) involved, and contractual agreements in place. In any event, prior to the end of any warranty period the building owner should have an understanding of how the building subsystems are operating and be satisfied that this operation is within "reasonable" limits of the design.

17. PERIODICALLY TEST EQUIPMENT EFFICIENCY

After the building is turned over to the owner, the maintenance staff should take steps to see that major fuel-consuming equipment (most notably gas-fired boilers) is tested periodically to determine its actual operating efficiency. Combustion efficiency testing of gas boilers at full and part-load conditions should be conducted at the beginning of each heating season. In many cases, a simple adjustment of controls will return the boiler to its original operating efficiency. The fuel cost savings over a heating season will often more than offset the cost of the efficiency test.

18. PERIODICALLY SAMPLE AND TEST WATER SYSTEM

Depending on the type of mechanical systems within the building, various types of water treatment may be required to avoid premature deterioration of the distribution system piping. The addition of water treatment chemicals may be required to minimize corrosion and scale build-up. The type and amount of water treatment and the frequency of testing should be specified in the original design. Solar system piping and systems using rooftop air-cooled condensing units with indoor water-cooled compressors should be checked periodically to verify that the proper proportion of ethylene glycol or other freeze-protection additives is being maintained. These tests and maintenance procedures will normally be done by the building maintenance staff or by a chemical treatment contractor. Following the specified frequency of testing is critical to the proper operation of these systems.

19. PERIODICALLY INSPECT AND CALIBRATE SYSTEM CONTROLS

Usually, control components represent the weakest link in the building's network of environmental control systems. Even the most energy efficient heating system can perform poorly if a malfunction in the control system results in the equipment receiving erroneous control signals. All controls, even solid state digital controls, "drift" off their setpoint, sometimes within a period of only a few hours. Pneumatic controls are notorious offenders in this regard. Periodic inspection and calibration of these components should be carried out by the maintenance staff. Records should be kept on those controls that tend to drift from their setpoints, the degree of drift, and the date of recalibration, so that particularly troublesome items can be identified and inspected more frequently or be replaced.

20. MONITOR OCCUPANT COMPLAINTS

After the building is occupied, the operating staff should maintain a log of all occupant complaints regarding the environmental control systems. Pertinent information that might offer clues for future troubleshooting include:

A. Nature of complaint
B. Date and time complaint was received
C. Temperature control zone or room in which the problem occurred
D. Sky condition (clear vs. cloudy), outside air temperature, and relative humidity at the time the complaint was received
E. Any unusual climatic conditions that may have occurred during the previous night
F. Corrective action taken and results obtained.

By carefully logging these items over a period of time, a trail of clues may develop that will be extremely useful in identifying problems. Without such clues, the time and expense involved in investigations can be greatly increased.

SECTION III

ALTERNATIVE PROCESS

Introduction

The Design Selector described in Sections I and II is intended to address the majority of small office design situations and to offer a reasonable, but limited, number of alternatives to each. The result will be more energy efficient designs at little or no increase in first cost. This section is for those who want to:

- Increase annual energy savings
- Try innovative design solutions
- Solve uncommon or difficult design problems

These situations are beyond those faced in the majority of design situations and, therefore, require more than the simplified approach outlined in Section I.

Unfortunately, because the range of possible problems and energy conscious solutions is so broad, this section cannot propose to give "answers," as was done in Section I. Rather, this section offers a process: a process that the authors believe will help any designer arrive at a better energy conscious design solution. This section is divided into two parts. The first part is a narrative description of the proposed process, with illustrations of key points. The second part is a collection of lessons learned, offered by practitioners who have done the type of analysis described here.

A Process for Solving the Right Energy Problem

The purpose of this part of Section III is to describe an approach for developing innovative energy efficient building designs. It is only one of several different approaches, but one that the authors feel can be followed successfully by any design professional. The process is divided into seven steps:

1. Use basic research to identify opportunities.
2. Define a base case building.
3. Compute base case energy use and energy cost. Then define goals.
4. Analyze base case energy use patterns for added opportunities.
5. Modify the design to improve energy performance.
6. Estimate first costs and annual energy savings. Then test financial feasibility.
7. Refine the design using design tools.

Following is a brief description of each step.

1. USE BASIC RESEARCH TO IDENTIFY OPPORTUNITIES.

The environment surrounding a building is both the cause of much of its energy use and a source of many opportunities for reducing energy use. A careful site investigation can be used to better understand all of the environmental factors impinging on a building, so that both problems and opportunities can be more precisely defined. In addition, information must be collected about the building program, the financial criteria of the client, and local utility rates.

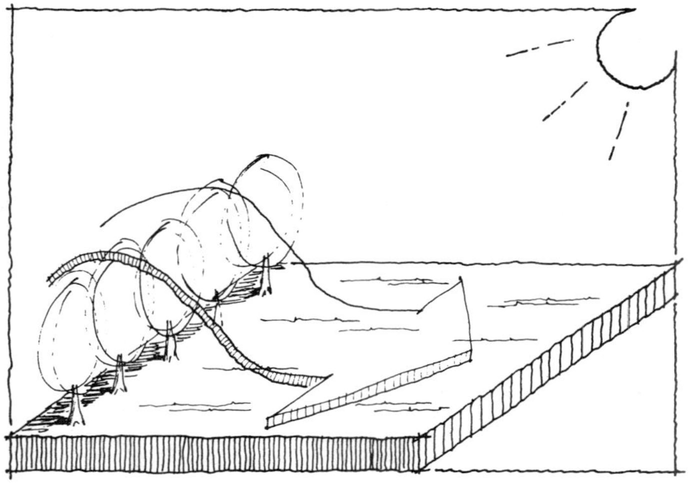

Site Specific Environmental Data

Begin by collecting each of the following:

- Temperature (dry bulb ranges, diurnal cycle, degree days, and dew points)
- Insolation (monthly and hourly direct and diffuse sunlight, and solar access)
- Humidity (wet bulb)
- Earth Temperature (if appropriate)
- Wind Patterns (direction and velocity)

  Two good sources of climatological information are your local weather bureau and the National Oceanic and Atmospheric Administration (NOAA). The Solar Energy Research Institute (SERI) is also in the process of publishing a number of studies on insolation data and cloud cover. The *ASHRAE Handbook of Fundamentals* is another excellent source of these data.

Client's Building Program

In addition to collecting standard architectural program information, such as building size, code and zoning requirements, etc., identify any unusual conditions that may influence energy requirements. For example, a large number of computers or other equipment that generates heat may present an opportunity to store heat for use during unoccupied periods. Chilled water storage may permit equipment to be operated on a longer cycle, resulting in reduced capacity and lower demand requirements. Extended office occupancy will require longer operation of mechanical and lighting systems.

Client Specific Financial Data

In order to assist a client in making decisions among various options, it is essential to understand how a client is likely to view the options presented.

In addition to the basic project budget, it is important to discuss the client's willingness (or not) to spend additional capital in order to reduce annual energy bills. Both the total amount that could be spent up front and the number of years of savings necessary to return the original investment (payback) are useful figures to know. If the client uses more sophisticated financial analysis, agree on precisely what numbers will be necessary for this analysis and who will be responsible for doing the financial analysis.

Utility Rate Data

As was clearly illustrated in the simplified design approach, the purpose of an energy analysis is to save money, not simply Btus. As part of the basic information gathering phase, either utility rate schedules should be collected or communications should be established with the utilities that will serve the site. This is necessary so that estimated energy bills and savings resulting from energy design analysis can be computed accurately and included in the financial analysis. Using average costs of consumption and demand may often cover up real opportunities for saving money within the rate schedule. For example, some rate schedules reward electrical use during non-peak periods with very preferential prices; this may present opportunities for mechanical system options that reduce demand and have lower electrical consumption costs.

Once all of this information has been collected, it is often useful to display it in graphic form. Visual inspection of the resulting graphics will generate ideas to be tested later in the analysis. This basic file of data will also be useful in applying any analytical method or design tool chosen for the subsequent energy analysis.

2. DEFINE A BASE CASE BUILDING.

It is difficult to do energy analysis in the abstract. For this reason, the authors believe it is best to tie energy analysis directly to something designers

already understand: a building design with shape, orientation, transparency, and systems. Therefore, the process we are proposing here starts from the assumption of *some* building (really, *any* building) and proceeds to a more energy efficient design by modification to the original building—called for convenience the "base case."

The purpose of the base case is to be a "straw man." It gives a first look at how and when a building is likely to use energy on the given site. There is no "right" or "wrong" base case. It is simply a starting point that will likely be revised—often significantly—by the time the desired energy conserving design is finally achieved.

For most simple energy computation tools, a base case can be defined by specifying only a handful of major variables:

- Shape—the number of stories and the basic dimensions of the building
- Orientation—the compass direction of each main building face
- Transparency—the amount of glass relative to the amount of wall exposed to the environment in each orientation
- HVAC System type, control system, and operating schedule
- Lighting type, quantity, and operating schedule
- Number of occupants or occupants per square foot.

If the designer has a preferred building configuration in mind or if many of the key variables are set by the building program, this is a reasonable base case building to start with. If the designer's choices are largely unlimited or if there is no preferred configuration, then it is reasonable to select specifications that are roughly near the mid-point of the available range of options, so that there is freedom to revise in either direction once the base case energy analysis has been completed. The level of specification will depend to a great extent on the reasons why the designer chose to go through this process in the first place. If the goal is increased energy cost savings, the base case will probably already be a reasonably efficient building. A reasonable base case could also be developed using the selector process in Section I. If the designer is attempting an innovative solutions or has a special design problem, these issues could be a major constraint on the base case.

3. COMPUTE BASE CASE ENERGY USE AND ENERGY COST. THEN DEFINE GOALS.

There are many calculation tools that can be used to compute base case energy use. It is reasonable to use a rather simple tool at this point—one that will give answers at the same level of accuracy as the rest of the schematic design process. In the best of all worlds, one would choose a tool that could easily address simultaneously both architectural and mechanical system variables. Unfortunately, there are few computation tools available that even attempt this. It is important to include mechanical and lighting system efficiencies in the best available form to get an accurate concept of the characteristic energy requirements.

One important goal in tool selection should be its ability to give the design team answers to "what if" questions quickly and at low cost, so that the impulse to try alternative designs is enhanced rather than discouraged. Another important consideration is the form of output from the calculation process.

The output should show energy use or cost by end use and should be readily usable by the design team without additional time consuming analysis or graphic work. The descriptions of available design tools found in the Appendix are intended to help the reader in choosing an appropriate computation method. Once a tool is chosen, use base case specifications to compute energy use and cost for the building. Depending on the tool chosen, these data will be reported hourly, daily, monthly, quarterly, or annually. Regardless of the tool chosen and its output, take time to draw up two annual energy use/cost summaries like the ones illustrated in the adjacent figure and the figure on the facing page. The plot that normalizes to 100% provides an easy comparison of:

- Load (energy use in Btus, excluding COPs)
- Consumption (energy use in Btus, including equipment COPs)
- Cost (energy use in $, reflecting site utility rates).

All three figures are shown as a percentage of the annual total so that they can be displayed on the same graph. This method of display is useful

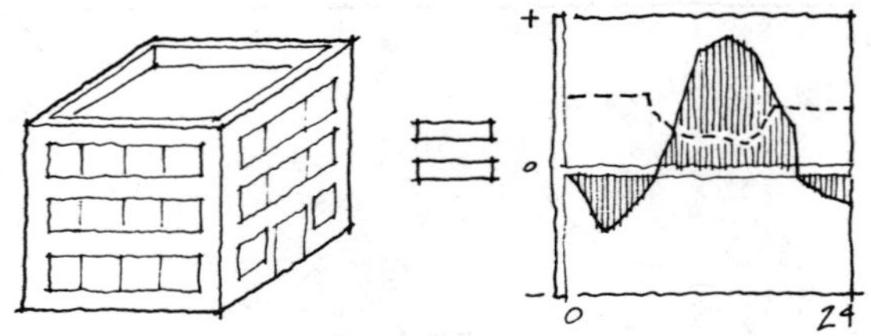

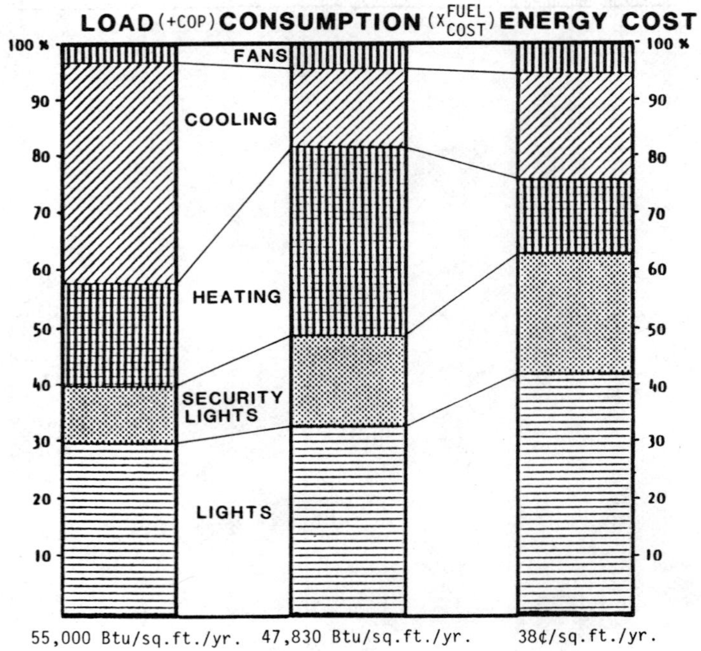

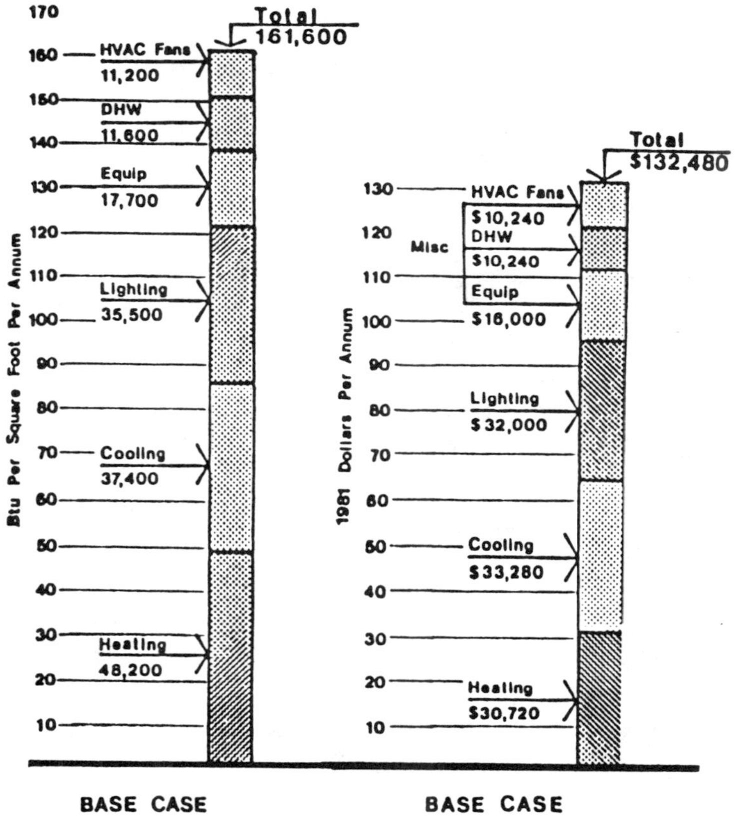

because it clearly illustrates the importance of setting design priorities from a cost point of view rather than from an energy use point of view. As shown in the illustration, if load was used to give design direction, cooling would be the most important design problem to be solved in this building. However, from a cost point of view, cooling ranks second or third. In fact, lighting is the most important design issue.

It is also useful to plot both annual energy consumption and annual energy cost as simple bar graphs. These bar graphs are useful not only as the basis of comparison for future iterations of this analysis, but also to compare this analysis with others. When preparing the bar graph of cost, be sure to include a bar (or portion of a bar) for demand. Demand is a major factor of cost and deserves separate attention along with each component of consumption. If possible, it would be desirable to display demand by end use. Unfortunately, this calculation is extremely difficult, which means that demand is typically displayed as a single number (or bar) for a given building.

Once the base case energy analysis is complete, it is useful to compare it to other buildings and historic data on the building type. The base case buildings presented in Section I conform to ASHRAE 90–75 standards and are a useful basis of comparison, as are those buildings with strategies already applied. Additional comparisons can also be made to actual buildings. Case studies of energy conscious designs are increasingly available.

With the above information in hand, it is useful to pause and review the situation before proceeding. The implications of the original "base case" set of assumptions are now clear, and they have been compared to what is both theoretically possible (the simulations) and actually possible (measured performance in a real building). This is a good point in the process to try to set some goals for the remaining energy analysis. Two important goals should be set: (1) the desired level of energy cost reduction and (2) the possible level of first cost increase, if any. These goals should be discussed with the client before proceeding so that the analysis does not presume to accomplish more (or less!) than the client expects. This is also a good time to revisit basic assumptions or limitations that the client may have set down in the original building program. If any specification is clearly causing energy problems, it should be discussed early to determine the extent to which it might be made more flexible.

4. ANALYZE BASE CASE ENERGY PATTERNS FOR ADDED OPPORTUNITIES.

This step focuses on analysis of the information gained in the previous step. Depending on the tools chosen for base case analysis, output from Step (3) can vary significantly. As suggested in the adjacent figure, there tends to be a distinct difference between analysis tools that can address the thermal loads and consumption in the designs and tools that can address lighting. To date, most simplified analysis tools have done only thermal analysis. Simplified tools for lighting design (including daylighting) are either still under development or have become available only recently. Most designers have used physical models with much greater success in analyzing lighting problems.

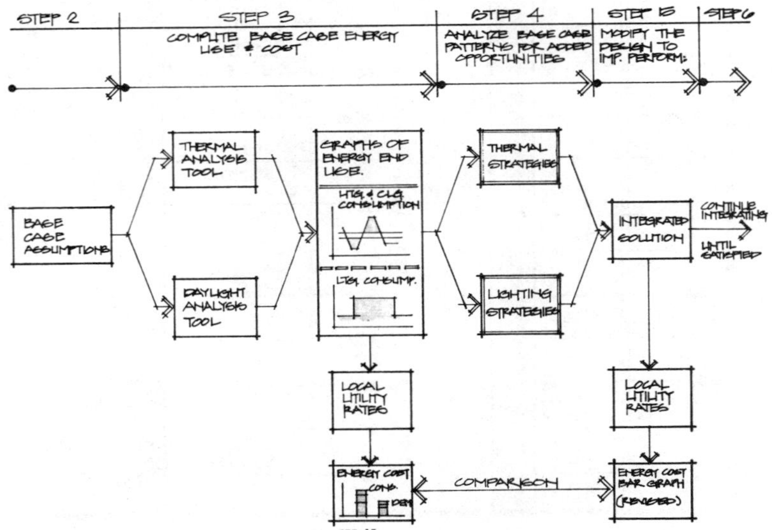

III-15

The difference in available tools means that the analysis process in this step will generally follow two separate tracks: lighting analysis and thermal analysis. The advice available to design teams follows this dichotomy. Unfortunately, no simplified analysis tool developed to date can approach both lighting and thermal issues in an integrated fashion. This integration is still completely on the shoulders of the design team.

Analyzing Thermal Energy Use Patterns

Thermal analysis tools typically display energy use over time. Energy use can either be displayed as a net of all end uses (as in *Energy Graphics*) or as individual end uses (as in color keyed-bar graphs or the energy nomographs). The time period is usually either a typical day in each season (or month) or the heating and cooling design days. Because we have selected no single analysis tool to discuss here, it is difficult to describe precisely what the output will look like for the tool you choose.

With these plots of use over time in hand, carefully examine them for opportunities—either to reduce energy use, to offset gains and losses, to reduce net energy use, or to shift energy use to less costly times of the day.

There are basically four circumstances that can occur if you are reviewing a plot of new energy use over a typical or design day:

- All heat loss
- A mixture of loss and gain
- All gain
- No gain or loss

In the final case (no gain or loss), the base case you have designed is very good and will probably not benefit from further analysis. In the other three cases, you have an opportunity to improve energy performance.

Next, analyze the observed opportunities in order to translate them into general directives that can be carried out in the subsequent design. For example:

- To reduce heat loss, either reduce envelope and ventilation losses or increase internal heat gains or solar gains to offset the loss.
- To reduce heat gains, do the reverse: either reduce internal and solar gains or offset the gains with increased envelope and ventilation losses.
- To reduce a mixture of gain and loss, review all possible strategies for decreasing both gain and loss and carefully analyze when each occurs. Opportunities can then be identified for eliminating the problem altogether or offsetting it in a way that will reduce net energy cost.

A number of books are available that categorize this process of reading use patterns and selecting design strategies. *Energy Graphics*, the *AIA Level II Workshop* (derived from *Energy Graphics*), and *Energy Nomographs* are three good sources. (See references at the end of this section.)

Analyze Lighting Energy Use Patterns

Because the tools used to analyze lighting use in a proposed building vary so greatly, it is difficult to address them as a group. In general, the analysis must first address the amount of light needed and the period for which it is needed. Then, daylighting resources must be investigated (the Solar

Energy Research Institute and Lawrence Berkeley Laboratories have recently done a great deal of work in this area). Finally, design options that take advantage of the daylight resource while not significantly deteriorating the thermal performance of the building must be investigated. As was mentioned above, one of the best and lowest cost current design techniques uses physical models to find optimum lighting solutions, while simultaneously doing thermal analysis with a simplified tool to monitor the thermal effects of various lighting options.

The integration, or balancing, of lighting and thermal performance is one of the greatest challenges to the design team and one for which there is still limited low cost design-tool help. There is now a large and growing number of buildings in which integration of lighting and thermal performance has been tried. The Passive Solar Commercial Buildings Program has carefully cataloged many of these case studies. There are also others regularly reported in the literature.

One common example of the integration process proceeds as follows. The design team notes from the thermal analysis that there is an excessive cooling load during occupied hours and that this is causing a serious cost problem, both from electrical consumption costs and demand charges as indicated on the base case cost plots. Heating, on the other hand, is very cheap.

Using models and guidance from the literature, the team develops a daylighting solution that reduces the amount of artificial lighting required during daylight occupied hours. They also use automatic controllers to insure that lights will be turned off when they are not needed. The result is interesting. The reduction in lighting cuts electrical costs both for the lights themselves and for the part of the cooling load caused by the heat-of-lights. Roughly 80% of the electrical power consumed for lighting ends up in the building as heat. Mechanical heating, however, *increases* during certain parts of the day because of the lost heat-of-lights. The reductions in electrical cost so far outweigh the increase in heating costs that the net effect is positive.

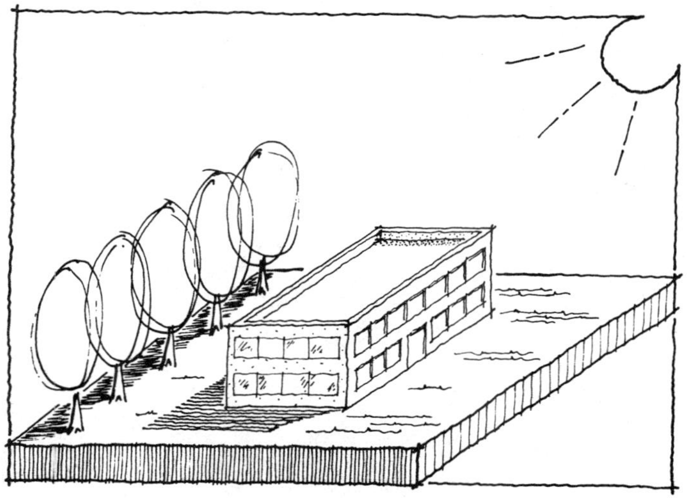

5. MODIFY THE DESIGN TO IMPROVE ENERGY PERFORMANCE.

Once the base case analysis is complete, the next major phase of the process is to modify the base case in order to improve energy performance. This is accomplished by modifying the original set of base case assumptions and then re-running the analysis tool. If the tool is a manual method, the "re-run" is by hand calculations.

This modification process can be very directed or it can be by trial and error, depending in large measure on the experience of the design team. As the body of literature on workable energy solutions grows, it will become easier for the general practitioner to come to this modification process with a list of good ideas to try. Lacking that information or experience, trial and error can be a very useful teacher in this step of the process.

It is important to try more than one iteration on the base case. Few design projects are simple enough to permit a correct, simultaneous assessment on the first try of all variables. In fact, it usually takes several iterations until all reasonable variations have been tried. Because of the need to try several iterations, it is important to keep the base case analysis simple. Complex multi-zone analysis is not likely to be useful too early in the analysis process.

It is also important to keep careful records throughout this process. These records serve as a valuable reference for ideas tried and rejected. They also serve later in design development when it is necessary to recall precisely what was done to suggest a particular strategy. It can be extremely frustrating not to be able to identify the set of assumptions that led to a key design decision.

There is no "right" number of runs necessary to complete this step. The process is limited only by the patience (and the fee) of the design team. As a general rule-of-thumb, the process is complete when total energy consumption cannot be significantly reduced or when any particular end use cannot be further reduced without an offsetting increase in another end use.

6. ESTIMATE FIRST COSTS AND ANNUAL ENERGY SAVINGS. THEN TEST FINANCIAL FEASIBILITY.

Any recommendations to improve energy performance of a building must make financial sense if they are to have any hope of being accepted. It is essential that the design team present their energy analysis in financial terms that are meaningful to the client. The purpose of this step in the proposed process is to address financial feasibility.

If the analysis tool you have been using also computes annual energy bills, you are a step ahead here. If not, which is generally the case, convert the annual energy use of your modified design into the annual energy bill using the utility rate schedule for your area. It is important to seek the assistance of your utility's Commercial Rate Application Specialist if the rate schedule in your area is complex or if your design depends heavily on details such as time-of-use rate changes. A comparison to the base case energy bill (computed in Step 3) should yield annual energy savings for your modified design. This figure will be key in any financial calculations aimed at judging financial feasibility.

Next, it is important to have a comparison of likely first costs for both the base case and the proposed modified design. These estimates will identify the first costs (if any) of the proposed energy efficient design. This is the other key figure necessary for any financial analysis. Although it may seem somewhat time-consuming, it is recommended that whole building cost estimates be done rather than component cost analysis. For one thing, it is virtually impossible to isolate some of the costs associated with energy efficient design on a component basis. For example, when does glass cease to be part of an owner's requirement for view and start to be part of a solar heat collector? For another, component analysis can hide trade-offs that might be made between energy related building systems and other building systems or components. Comparing the "bottom line" of the base case to the modified design for the whole building will avoid both of the problems discussed above.

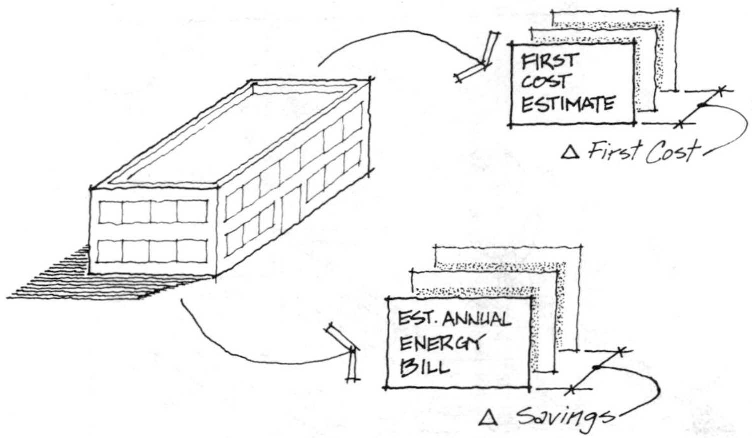

The level of financial analysis required will depend on the client. If possible, the original building program should spell out the level of analysis to be done, who does it, and with what assumptions. In addition to the overall budget for the building, it is useful to know whether or not the client would allow an increase in first cost if it could be justified by annual energy savings. If so, the speed with which savings must equal costs, or some similar measure of financial performance, must be given. It is also important to know if the client will be flexible on fee in order to achieve better energy performance.

Barring a more complex analysis, the following simple formula can be used to illustrate most key points in a discussion of the financial viability of an energy efficient project:

$$\frac{\left(\begin{array}{l}\text{first cost}\\ \text{revised}\\ \text{design}\end{array} + \begin{array}{l}\text{revised}\\ \text{design}\\ \text{fee}\end{array}\right) - \left(\begin{array}{l}\text{first cost}\\ \text{base case}\end{array} + \begin{array}{l}\text{standard}\\ \text{design}\\ \text{fee}\end{array}\right)}{\left(\begin{array}{l}\text{base case}\\ \text{energy bill}\end{array} - \begin{array}{l}\text{revised design}\\ \text{annual energy bill}\end{array}\right)} = \begin{array}{l}\text{simple}\\ \text{payback}\end{array}$$

Simple payback is a very crude and limited financial measure. It does a poor job of accounting for a client's required return-on-investment, and it completely ignores the value of annual energy savings that continue to occur after payback is achieved. Nevertheless, it can be a simple tool for discussing necessary issues if no other analysis is required.

If more complex financial analysis is required, the six numbers in the equation above are the key numbers necessary to fit any analysis.

7. REFINE THE DESIGN USING DESIGN TOOLS.

Simplified analysis tools used to make basic decisions and to give general guidance to the design process must eventually be augmented by tools that can give more precise answers as they are needed during design development. The ideal design tool would be one that could give simple answers based on limited assumptions and then give more precise answers as more and more precise specifications become available. Unfortunately, the only design tools approaching this level of sophistication are the mainframe computer programs such as DOE-2 and BLAST, which are generally too costly to use on buildings under 50,000 sq. ft.

For buildings of this size, the design team will have to select a limited number of options to be investigated and then select an appropriate design tool to do each analysis. A list of design tools and a description of the capabilities of each is given in the Appendix following this section. Integration of individual analyses from different tools is a real problem that must be dealt with by the design team. This is a rapidly changing environment at the moment, with new and better tools constantly replacing older ones. As always, a healthy measure of professional judgment is the best means to insure reasonable answers and reasonable integration of answers to key design questions.

More and more case studies that document attempts of design teams to use various tools are becoming available in the literature. The Building Energy Design Tool Development Council is another source of information on the state-of-the-art.

Lessons Learned

The process of designing energy efficient buildings is not without its problems and pitfalls. This part of Section III presents several pieces of "good advice" about how to avoid problems in doing an energy analysis. These items have been collected over the past several years from both architects and engineers and are presented here to help others avoid potentially painful and costly learning experiences.

SECURE YOUR CLIENT'S COMMITMENT TO ENERGY CONSCIOUS DESIGN BEFORE YOU START

The process of energy conscious design requires a good deal of client and design team interaction. How much energy efficiency is to be achieved and at what first cost are decisions that will ultimately involve the client in the design process. It is unlikely that your client will have the patience to participate in achieving an energy conscious design unless a commitment has been made up front to do so.

A client commitment to energy conscious design can be achieved in many ways, not the least of which is through sound financial analysis. A designer wishing to secure such a commitment can use general reading, case studies of similar buildings, or generic building analysis to make clients aware of the potential cost and benefits of energy conscious design. On larger projects, where energy costs are likely to be a significant issue, many firms prepare preliminary estimates of potential energy savings based either on previous experience or on analysis using simple design tools and reasonable assumptions.

DO NOT RELY ON INTUITION TO IDENTIFY THE MOST IMPORTANT ENERGY PROBLEM IN YOUR PROJECT

Analysis, not intuition, is the only way most designers can be sure they are actually solving the right energy problem. In spite of all that has been written recently on energy issues, two informal polls (reported in a paper by Hart, *et al.*, *ASHRAE Proceedings 1982*, v. 88, p. 2) showed that fewer than one architect or engineer in ten could correctly guess, unaided by an analysis tool, the annual utility bill of a small commercial office building. Only one in five could guess the most significant energy problem. These rather low scores clearly indicate that intuition alone is not yet a suitable or reliable analysis tool.

ALWAYS START AN ENERGY ANALYSIS DURING PRE-DESIGN

Because of the sequential nature of the design process, major decisions made early in schematic design cannot easily be changed during design development. However, studies have shown that many of these decisions, such as number of stories, amount of glass, HVAC system type, sources of energy (fuels), etc., can have a major impact on subsequent annual energy costs. As with overall construction cost, where most opportunities for cost control occur in the schematic design stage, it is generally agreed that half or more of the potential energy that can be saved by building design and equipment selection will have been achieved (or not) by the end of schematic design. Only equipment selection carries as much energy impact as decisions made during schematic design. To achieve maximum energy savings, energy analysis must begin in the earliest stages of the design sequence. Design development is too late in the sequence to change most of the key variables that determine a building's annual energy bill.

KEY DESIGN TEAM MEMBERS SHOULD INTERACT SIMULTANEOUSLY

Strategies to improve energy performance can simultaneously affect virtually *all* disciplines on the design team. Given that the design fee is probably not large enough to afford a great deal of group meeting time, it is desirable from an energy point of view that at least the key members of the design team meet together throughout pre-design and schematic design. Key energy players are the architect, mechanical engineer, energy analyst, and lighting designer, if there is one on the job. Interacting simultaneously offers a number of advantages:

- All team members hear the estimated energy effect of a design option and all can simultaneously judge its significance.
- Synergy among team members can build on a good idea from all points of view
- Group decision making informs all participants, reducing the likelihood that any key decision must change later for unforeseen reasons.

The team should work together at least through schematic design. By this time, most key building decisions will be in place with agreed upon energy consequences. This also limits the total time taken up in team meetings.

CONSIDER CAPITAL COST FROM THE BEGINNING OF THE ENERGY ANALYSIS

In the end, the goal of an energy analysis is to recommend a building design that will reduce annual energy bills at a cost, if any, that is acceptable within the financial requirements of the client. A good energy analysis constantly trades off annual savings *in dollars* against first cost.

For this reason, it is essential that capital cost estimates begin with the energy analysis and proceed throughout the process. A common mistake over the past few years has been to expend significant design time trying to solve every energy problem in a proposed building, only to find out later that many elegant solutions had to be abandoned because they were too costly relative to the amount of money that could actually be saved on utility bills.

APPENDICES AND GLOSSARY

APPENDIX A:
Worksheets

This appendix contains the following pages to assist in developing your building's energy costs:

- "Utility Rate Characteristics" (guidelines to assist in finding the right worksheet to use with your utility rate structure)
- Instructions for using Worksheet A
- "Worksheet A" (annual analysis method)
- Instructions for using Worksheet B
- "Worksheet B" (monthly analysis method)
- Instructions for using Worksheet C
- "Worksheet C" (form letter requesting utility assistance)

UTILITY RATE CHARACTERISTICS

Select Worksheet A, B, or C for energy cost analysis with the aid of these guidelines (note that more than one utility rate may apply to a project). You may wish to contact your utility to clarify any of the guidelines below.

A Annual Analysis

IF UTILITY RATES:

1. Have relatively uniform demand *and* consumption charges throughout the year.

(You do the analysis)

B Monthly Analysis

IF UTILITY RATES:

1. Have relatively uniform demand charges throughout the year with no “ratchet” charges.
2. Have consumption rates with only seasonal changes.

(You do the analysis)

C Form Letter To Utility

IF UTILITY RATES:

1. Have complicated “ratchet” clauses for demand.
2. Have consumption rates that depend on level of demand.
3. Have seasonal changes in demand and consumption rates.
4. Have separate heating and general service rates that require many calculations

   and

   Your HVAC system can be wired and metered to use these rates—(including CV/elec., VAV/elec., MZ/elec., or RT-DX/elec.).

(Utility does the analysis)

INSTRUCTIONS FOR USING **WORKSHEET A**

Use one "Worksheet A" per base case (level 2) or strategy set option (level 3) investigated.

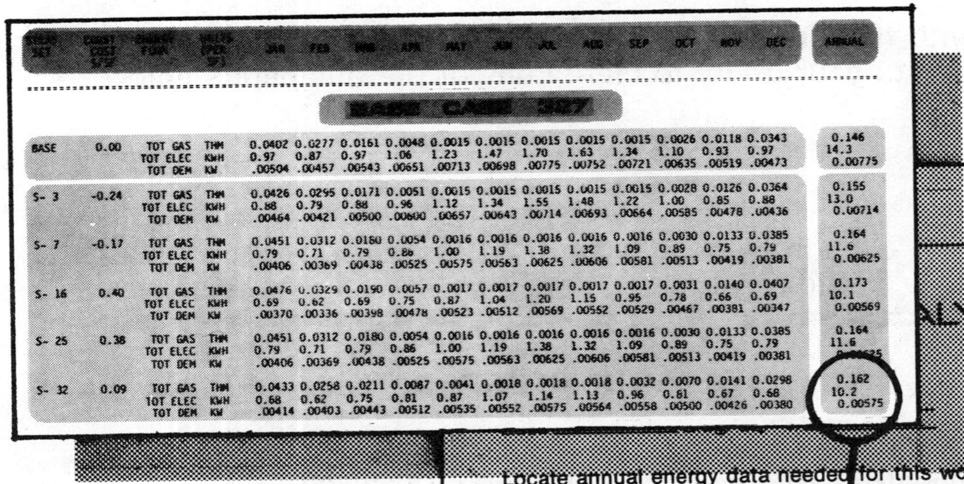

				JAN	FEB	MAR	APR	MAY	JUN	JUL	AUG	SEP	OCT	NOV	DEC	ANNUAL
BASE	0.00	TOT GAS	THM	0.0402	0.0277	0.0161	0.0048	0.0015	0.0015	0.0015	0.0015	0.0015	0.0026	0.0118	0.0343	0.146
		TOT ELEC	KWH	0.97	0.87	0.97	1.06	1.23	1.47	1.70	1.63	1.34	1.10	0.93	0.97	14.3
		TOT DEM	KW	.00504	.00457	.00543	.00651	.00713	.00698	.00775	.00752	.00721	.00635	.00519	.00473	0.00775
S- 3	-0.24	TOT GAS	THM	0.0426	0.0295	0.0171	0.0051	0.0015	0.0015	0.0015	0.0015	0.0015	0.0028	0.0126	0.0364	0.155
		TOT ELEC	KWH	0.88	0.79	0.88	0.96	1.12	1.34	1.55	1.48	1.22	1.00	0.85	0.88	13.0
		TOT DEM	KW	.00464	.00421	.00500	.00600	.00657	.00643	.00714	.00693	.00664	.00585	.00478	.00436	0.00714
S- 7	-0.17	TOT GAS	THM	0.0451	0.0312	0.0180	0.0054	0.0016	0.0016	0.0016	0.0016	0.0016	0.0030	0.0133	0.0385	0.164
		TOT ELEC	KWH	0.79	0.71	0.79	0.86	1.00	1.19	1.38	1.32	1.09	0.89	0.75	0.79	11.6
		TOT DEM	KW	.00406	.00369	.00438	.00525	.00575	.00563	.00625	.00606	.00581	.00513	.00419	.00381	0.00625
S- 16	0.40	TOT GAS	THM	0.0476	0.0329	0.0190	0.0057	0.0017	0.0017	0.0017	0.0017	0.0017	0.0031	0.0140	0.0407	0.173
		TOT ELEC	KWH	0.69	0.62	0.69	0.75	0.87	1.04	1.20	1.15	0.95	0.78	0.66	0.69	10.1
		TOT DEM	KW	.00370	.00336	.00398	.00478	.00523	.00512	.00569	.00552	.00529	.00467	.00381	.00347	0.00569
S- 25	0.38	TOT GAS	THM	0.0451	0.0312	0.0180	0.0054	0.0016	0.0016	0.0016	0.0016	0.0016	0.0030	0.0133	0.0385	0.164
		TOT ELEC	KWH	0.79	0.71	0.79	0.86	1.00	1.19	1.38	1.32	1.09	0.89	0.75	0.79	11.6
		TOT DEM	KW	.00406	.00369	.00438	.00525	.00575	.00563	.00625	.00606	.00581	.00513	.00419	.00381	0.00625
S- 32	0.09	TOT GAS	THM	0.0433	0.0258	0.0211	0.0087	0.0041	0.0018	0.0018	0.0018	0.0032	0.0070	0.0141	0.0298	0.162
		TOT ELEC	KWH	0.68	0.62	0.75	0.81	0.87	1.07	1.14	1.13	0.96	0.81	0.67	0.68	10.2
		TOT DEM	KW	.00414	.00403	.00443	.00512	.00535	.00552	.00575	.00564	.00558	.00500	.00426	.00380	0.00575

Locate annual energy use data needed for this worksheet in the **Data Library** section of the climate being investigated. Proceed to calculate your building energy costs using the square footage of your building and applicable **annual** local utility rates.

The annual utility rate should be obtained from your local utility company. To use this worksheet, your rate structure should have relatively uniform demand **and** consumption charges throughout the year.

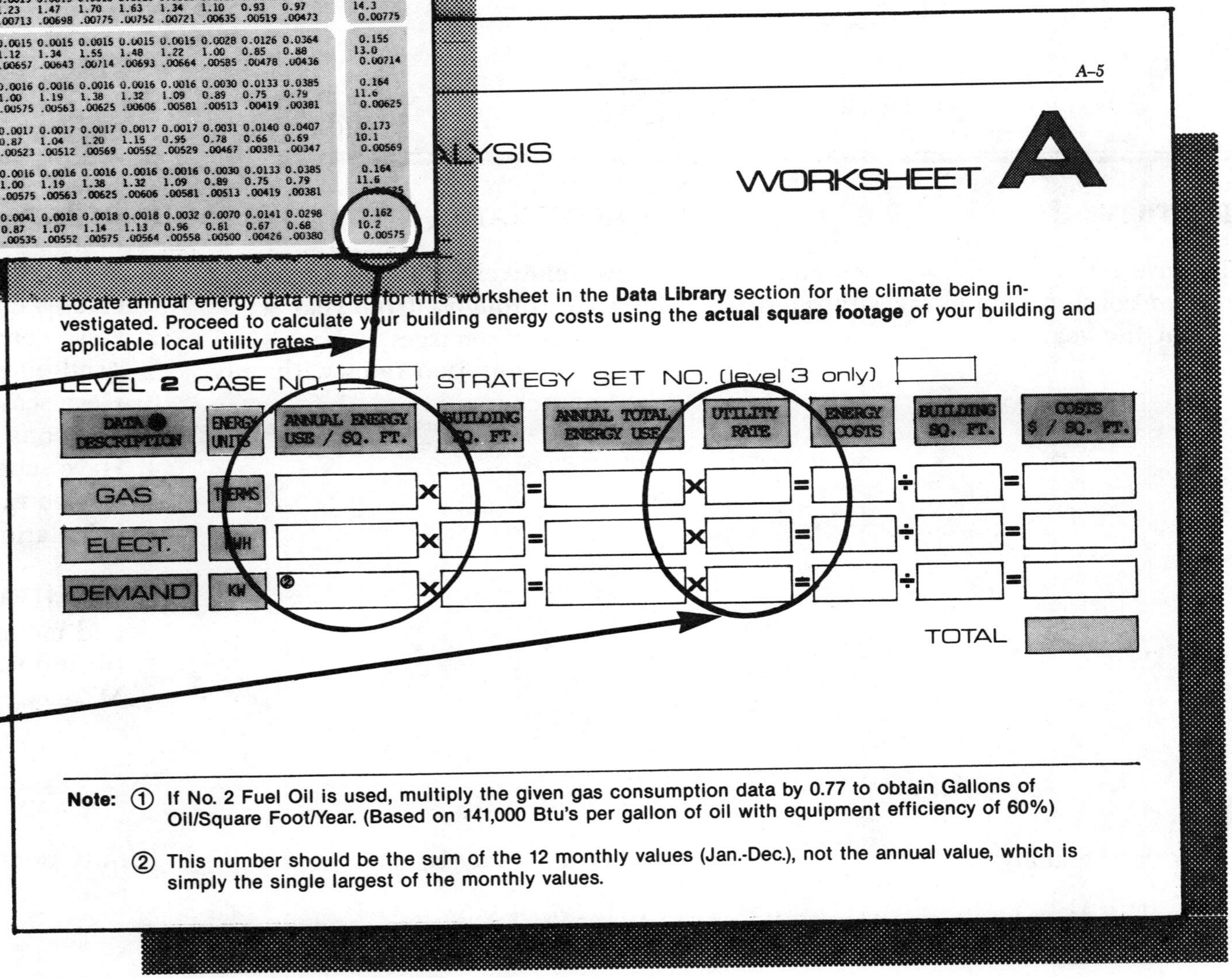

A–5

…ALYSIS

WORKSHEET **A**

Locate annual energy data needed for this worksheet in the **Data Library** section for the climate being investigated. Proceed to calculate your building energy costs using the **actual square footage** of your building and applicable local utility rates.

LEVEL **2** CASE NO. ______ STRATEGY SET NO. (level 3 only) ______

DATA ① DESCRIPTION	ENERGY UNITS	ANNUAL ENERGY USE / SQ. FT.		BUILDING SQ. FT.		ANNUAL TOTAL ENERGY USE		UTILITY RATE		ENERGY COSTS		BUILDING SQ. FT.		COSTS $ / SQ. FT.
GAS	TERMS		×		=		×		=		÷		=	
ELECT.	KWH		×		=		×		=		÷		=	
DEMAND	KW	②	×		=		×		=		÷		=	
												TOTAL		

Note: ① If No. 2 Fuel Oil is used, multiply the given gas consumption data by 0.77 to obtain Gallons of Oil/Square Foot/Year. (Based on 141,000 Btu's per gallon of oil with equipment efficiency of 60%)

② This number should be the sum of the 12 monthly values (Jan.-Dec.), not the annual value, which is simply the single largest of the monthly values.

ENERGY COST ANALYSIS ANNUAL BASIS

WORKSHEET A

Project ______________________
date __________ bldg. sq. ft. __________

Locate annual energy data needed for this worksheet in the **Data Library** section for the climate being investigated. Proceed to calculate your building energy costs using the **actual square footage** of your building and applicable local utility rates.

LEVEL **2** CASE NO. [ ] STRATEGY SET NO. (level 3 only) [ ]

DATA ① DESCRIPTION	ENERGY UNITS	ANNUAL ENERGY USE / SQ. FT.		BUILDING SQ. FT.		ANNUAL TOTAL ENERGY USE		UTILITY RATE		ENERGY COSTS		BUILDING SQ. FT.		COSTS $ / SQ. FT.
GAS	THERMS		×		=		×		=		÷		=	
ELECT.	KWH		×		=		×		=		÷		=	
DEMAND	KW	②	×		=		×		=		÷		=	
												TOTAL		

Note: ① If No. 2 Fuel Oil is used, multiply the given gas consumption data by 0.77 to obtain Gallons of Oil/Square Foot/Year. (Based on 141,000 Btu's per gallon of oil with equipment efficiency of 60%)

② This number should be the sum of the 12 monthly values (Jan.-Dec.), not the annual value, which is simply the single largest of the monthly values.

INSTRUCTIONS FOR USING **WORKSHEET B**

Use one "Worksheet B" per base case (level 2) or strategy set option (level 3) investigated.

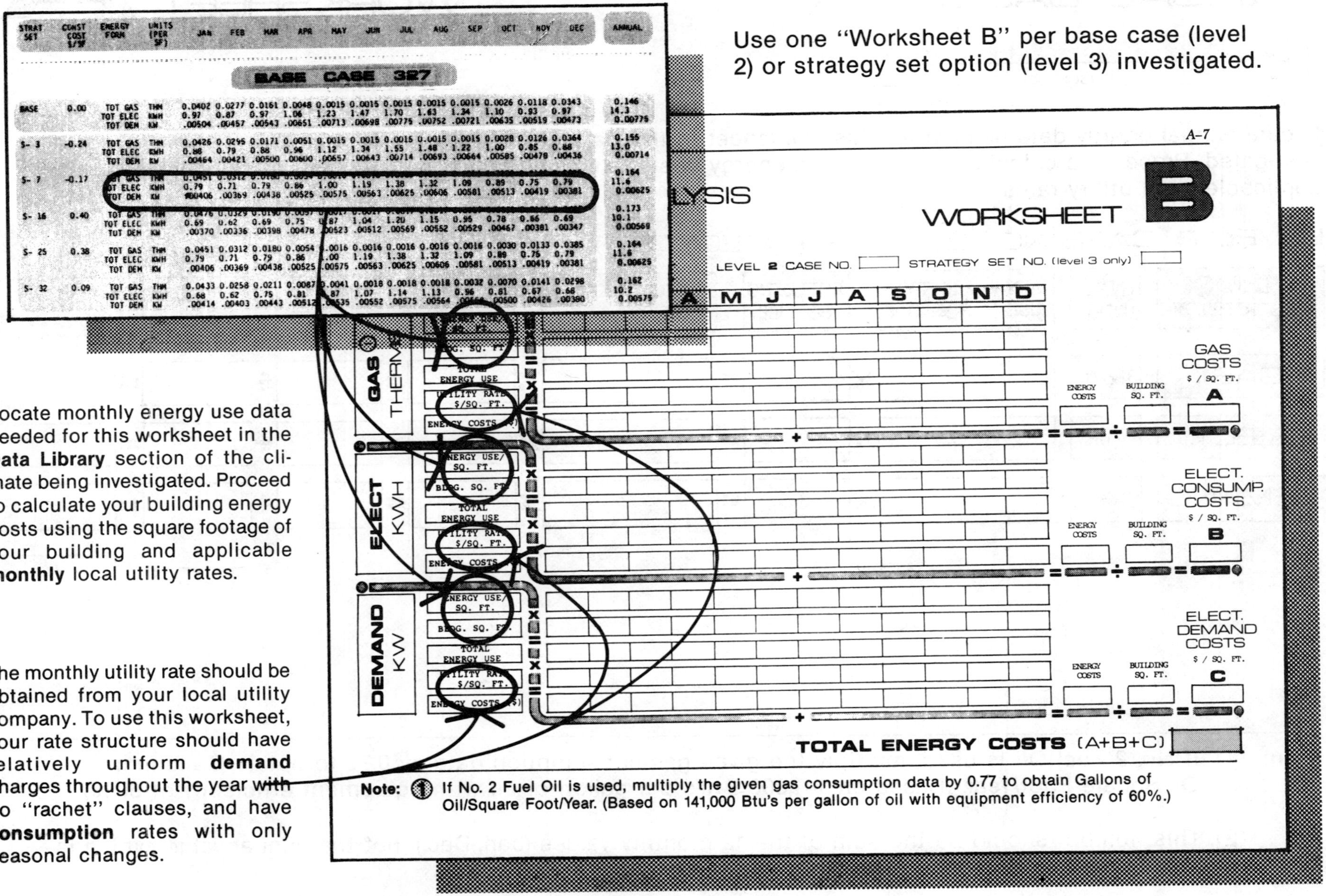

Locate monthly energy use data needed for this worksheet in the **Data Library** section of the climate being investigated. Proceed to calculate your building energy costs using the square footage of your building and applicable **monthly** local utility rates.

The monthly utility rate should be obtained from your local utility company. To use this worksheet, your rate structure should have relatively uniform **demand** charges throughout the year with no "rachet" clauses, and have **consumption** rates with only seasonal changes.

ENERGY COST ANALYSIS MONTHLY BASIS

WORKSHEET B

Project ____________________

date ____________ bldg. sq. ft. ____________

LEVEL 2 CASE NO. [ ] STRATEGY SET NO. (level 3 only) [ ]

		J	F	M	A	M	J	J	A	S	O	N	D
GAS ① THERMS	ENERGY USE/ SQ. FT.												
×	BLDG. SQ. FT.												
=	TOTAL ENERGY USE												
×	UTILITY RATE $/SQ. FT.												
=	ENERGY COSTS ($)												
ELECT KWH	ENERGY USE/ SQ. FT.												
×	BLDG. SQ. FT.												
=	TOTAL ENERGY USE												
×	UTILITY RATE $/SQ. FT.												
=	ENERGY COSTS ($)												
DEMAND KW	ENERGY USE/ SQ. FT.												
×	BLDG. SQ. FT.												
=	TOTAL ENERGY USE												
×	UTILITY RATE $/SQ. FT.												
=	ENERGY COSTS ($)												

	ENERGY COSTS (+ of monthly energy costs)	÷ BUILDING SQ. FT.	= $ / SQ. FT.
GAS COSTS			A
ELECT. CONSUMP. COSTS			B
ELECT. DEMAND COSTS			C

TOTAL ENERGY COSTS (A+B+C) [ ]

Note ① If No. 2 Fuel Oil is used, multiply the given gas consumption data by 0.77 to obtain Gallons of Oil/Square Foot/Year. (Based on 141,000 Btu's per gallon of oil with equipment efficiency of 60%.)

INSTRUCTIONS FOR USING **WORKSHEET C** (REQUEST TO UTILITY)

"Worksheet C" is a form letter that may be sent by the user to a utility to obtain the necessary information regarding energy costs. This letter should be sent with the appropriate level 1, 2, and when applicable, level 3 forms to give your utility an idea of the building characteristics and systems employed. In all cases, you must mark the appropriate sections of **energy use data** and provide your **building gross square footage** for use by the utility in their calculations.

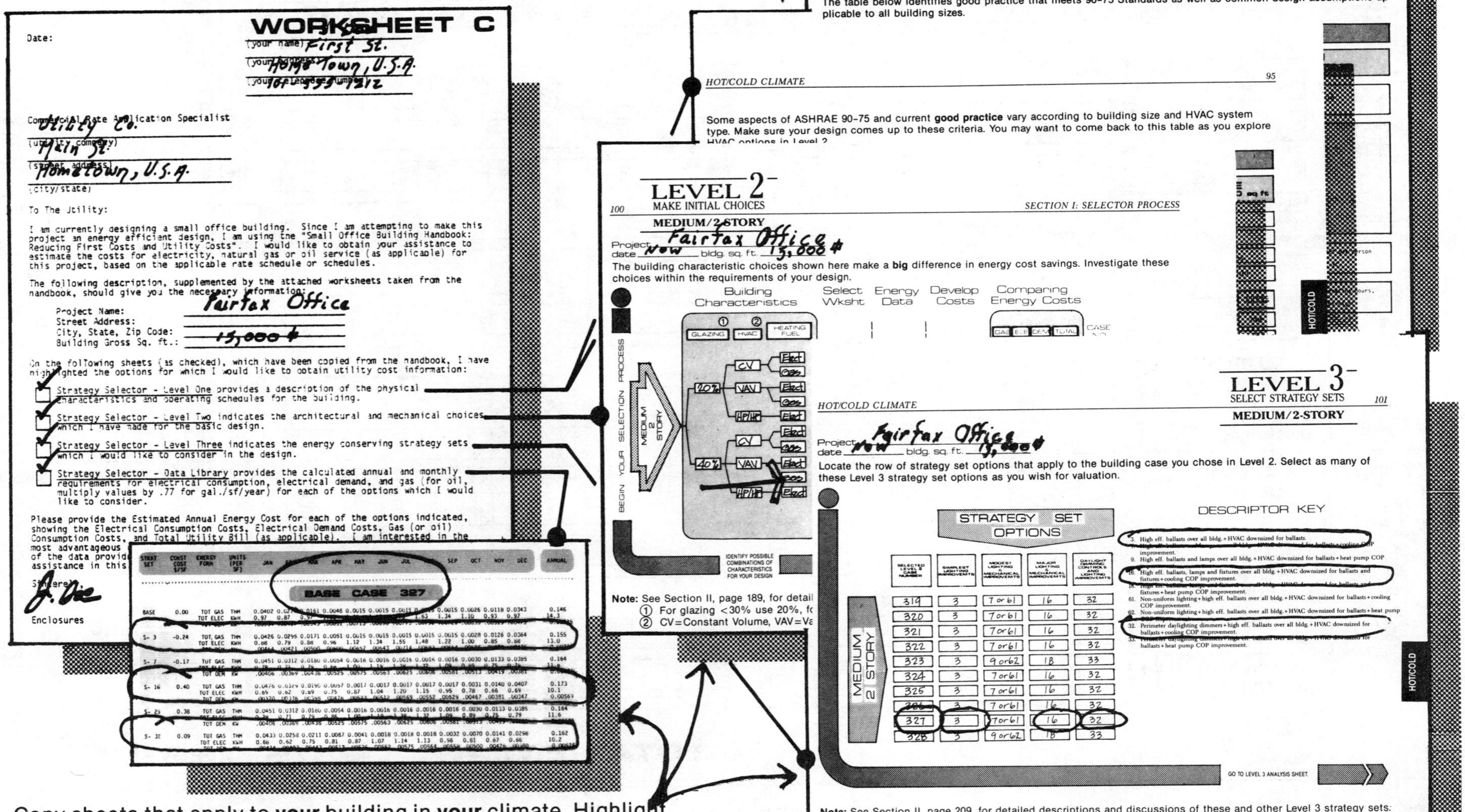

Copy sheets that apply to **your** building in **your** climate. Highlight or mark as appropriate all options you wish to have analyzed.

REQUEST TO UTILITY

WORKSHEET C

Date:

____
(your name)
____
(your address)
____
(your telephone number)
____

Commercial Rate Application Specialist

____
(utility company)
____
(street address)
____
(city/state)

To The Utility:

I am currently designing a small office building. Since I am attempting to make this project an energy efficient design, I am using the "Small Office Building Handbook: Reducing First Costs and Utility Costs". I would like to obtain your assistance to estimate the costs for electricity, natural gas or oil service (as applicable) for this project, based on the applicable rate schedule or schedules.

The following description, supplemented by the attached worksheets taken from the handbook, should give you the necessary information:

Project Name: ______________________
Street Address: ______________________
City, State, Zip Code: ______________________
Building Gross Sq. ft.: ______________________

On the following sheets (as checked), which have been copied from the handbook, I have highlighted the options for which I would like to obtain utility cost information:

- ☐ Strategy Selector - Level One provides a description of the physical characteristics and operating schedules for the building.
- ☐ Strategy Selector - Level Two indicates the architectural and mechanical choices which I have made for the basic design.
- ☐ Strategy Selector - Level Three indicates the energy conserving strategy sets which I would like to consider in the design.
- ☐ Strategy Selector - Data Library provides the calculated annual and monthly requirements for electrical consumption, electrical demand, and gas (for oil, multiply values by .77 for gal./sf/year) for each of the options which I would like to consider.

Please provide the Estimated Annual Energy Cost for each of the options indicated, showing the Electrical Consumption Costs, Electrical Demand Costs, Gas (or oil) Consumption Costs, and Total Utility Bill (as applicable). I am interested in the most advantageous utility rates for the project and would appreciate your assessment of the data provided in relation to your various applicable rates. Thank you for your assistance in this matter; please call me if you have any questions.

Sincerely,

Enclosures

APPENDIX B:
Strategies Examined

In preparing this handbook, over 80 individual energy conservation strategies were identified and investigated to see whether or not they could be recommended with confidence as cost effective strategies for reducing energy costs in small office buildings. The table on the following four pages summarizes the results of this investigation.

Those strategies selected as reasonable for one or more small office building application are allocated to one of five sections in the main text:

Level 1—Start Right
Level 2—Initial Choices
Level 3—Strategy Sets
Level 4—. . . Add More
Level 5—Finish Right

Those strategies that may be appropriate for small offices but require analysis beyond the scope of this handbook are recommended for review under the "Alternative Process" described in Section III.

Finally, those strategies not included in the handbook are indicated. A brief explanation of why each was omitted follows the table.

STRATEGIES EXAMINED

	Strategy	1—START RIGHT	2—INITIAL CHOICES	3—STRATEGY SETS	4—. . . ADD MORE	5—FINISH RIGHT	III—ALTERNATIVE PROCESS	NOT INCLUDED
Building Shape/ Siting	1. Modify aspect ratio to reduce thermal load							●
	2. Modify aspect ratio/shape to maximize daylighting contribution		●					
	3. Minimize building height to reduce thermal loads				●			
	4. Optimize building orientation to reduce peak A/C load				●			
Building Shell —Roof	5. Increase roof insulation				●			
	6. Increase roof mass							●
	7. Increase roof reflectivity				●			
	8. Ventilate roof cavity							●
	9. Horizontal skylights			●				
	10. Clerestories/roof monitors						●	
Building Shell —Walls	11. Increase wall insulation				●			
	12. Increase mass with external insulation				●			
	13. Reduce % glazing, increase % glazing				●			
	14. Reduce shading coefficient (e.g., reflective glass)	●						
	15. Increase shading (e.g., exterior shading)				●			
	16. Increase glazing insulation (single to double)	●						
	17. Redistribute glazing to maximize passive contribution						●	
	18. Interior window management device (e.g., venetian blinds)	●						
Building Shell —Infiltration	19. Improve construction quality					●		
	20. Airlock entry	●						
	21. Infiltration barriers (air/vapor barrier)	●						
	22. Operable windows for "economizer" cooling				●			
	23. Operable windows for fresh air requirements only				●			

STRATEGIES EXAMINED

	Strategy	1—START RIGHT	2—INITIAL CHOICES	3—STRATEGY SETS	4—. . . ADD MORE	5—FINISH RIGHT	III—ALTERNATIVE PROCESS	NOT INCLUDED
HVAC Concepts	24. Delete interior core area zone							●
	25. Separate heating and cooling systems		●					
	26. Powered economizer cooling	●						
	27. Powered VAV box for perimeter heating							●
	28. VAV cooling plenum induction box							●
	29. Night ventilation with mass				●			
	30. Split HVAC (CV interior/VAV perimeter)		●					
	31. Convert base HVAC to other type			●				
Heating Plant	32. Electronic ignition burners				●			
	33. Powered burners				●			
	34. Flue damper				●			
	35. Modular boiler sizing			●				
	36. Pulse/condensing boiler			●				
	37. Electric resistance boiler				●			
	38. Boiler hot water reset	●						
	39. Hot water storage				●			
Cooling Plant	40. Variable speed compressor				●			
	41. Evaporative cooling (direct/indirect)			●				
	42. Improved equipment Coefficient of Performance (COP)			●				
	43. Freon-water heat recovery				●			
	44. Heat recovery chiller							●
	45. Ice/cold water storage				●			

STRATEGIES EXAMINED

	Strategy	1—START RIGHT	2—INITIAL CHOICES	3—STRATEGY SETS	4—. . . ADD MORE	5—FINISH RIGHT	III—ALTERNATIVE PROCESS	NOT INCLUDED
Mechanical Distribution System	46. Variable speed fan control—two speed				●			
	47. Variable speed fan control—continuous (invertor, etc.)				●			
	48. Low leakage dampers	●						
	49. Oversize ducts/undersize coils to reduce fan total static pressure							●
	50. Air to air heat recovery							●
	51. Runaround heat recovery		●					
	52. High efficiency fan motors				●			
	53. High efficiency pump motors				●			
Controls	54. Microprocessor demand limiting/load shedding				●			
	55. Night setback	●						
	56. Deadband thermostats (heating/cooling)				●			
	57. Performance feedback metering				●			
	58. Electronic Direct Digital Controls (DDC)							●

STRATEGIES EXAMINED

	Strategy	1—START RIGHT	2—INITIAL CHOICES	3—STRATEGY SETS	4—. . . ADD MORE	5—FINISH RIGHT	III—ALTERNATIVE PROCESS	NOT INCLUDED
Lighting	59. Reduce lighting level provided (non-uniform lighting)			●				
	60. High efficiency—sources			●				
	61. High efficiency fixtures—parabolic			●				
	62. High efficiency fixtures—task/ambient							●
	63. High efficiency fluorescent lamps			●				
	64. High efficiency ballasts			●				
	65. Auto (stepped) lighting levels (occupied—cleaning—emergency)							●
	66. Increased manual switching capability				●			
	67. Combine auto/manual lighting controls							●
	68. Photocell daylighting controls, zoned on/off							●
	69. Photocell daylighting controls, zoned dimming			●				
	70. Photocell daylighting controls, individual fixture dimming			●				
	71. Occupancy sensors							●
	72. Demand limiting on lights				●			
Additional	73. Point of use Domestic Hot Water ((DHW) heaters				●			
	74. Efficient space planning				●			
	75. Landscape to reduce air infiltration				●			
	76. Passive heating				●			
	77. Shade outdoor cooling equipment				●			
	78. HVAC system downsizing			●				
	79. Solar Domestic Hot Water (DHW) heaters				●			
	80. Absorption heater/chillers				●			
	81. Cogeneration				●			

STRATEGIES NOT INCLUDED

1. **Modify building aspect ratio to reduce thermal load.** Project investigations have shown that making the building more elongate (4:1 aspect ratio rather than 1.5:1) actually increases energy use.
6. **Increase roof mass.** Structural considerations and standard building practices in small commercial buildings make this costly in comparison to light weight roofs. This strategy is covered to some extent in Level 4, item number 18: "Building Mass with Night Ventilation."
8. **Ventilate roof cavity.** This strategy is principally of interest in the Hot/Arid climate. It requires a roof space with an air tight ceiling and insulation at the ceiling level. (This is very like residential construction). Any ceiling ducts must be insulated. No plenum is practical, and surface mounted lights must be used. All heat from light must go to the space, none to the plenum. A residential style duct design may be required.

   The relative impact of this strategy on total building energy use is questionable, particularly with a well insulated roof.
24. **Delete interior core area zone.** The energy impact is believed to be negligible.
27. **Fan powered VAV box for perimeter heating.** Small energy impact is expected due to limited heating requirements.
28. **VAV cooling plenum induction boxes.** Small energy impact is expected due to limited heating requirements.
44. **Heat recovery chiller.** Not appropriate for small buildings.
49. **Reduce fan total static pressure by oversizing ducts and undersizing coil.** High cost and minor impact on small systems.
50. **Air—air heat recovery.** Quantities of outside air intake and building air exhaust are too low to justify the cost of such heat exchangers.
58. **Electronic direct digital control (DDC).** A control concept that currently is too advanced for small office buildings.
62. **High efficiency fixtures—task/ambient.** A design strategy that,

though effective, is very dependent on the program that the user of the office develops. Thought to be inappropriate at the overall building design level addressed by the handbook.

65. **Automatic (stepped) lighting levels.** Such controls automatically provide decreased levels of illumination for night cleaning and for security lighting. It is felt that a separate system for this purpose alone is not warranted. This type of control is offered by two of the Level 4 strategies: Item number 12, "Increase Manual Light Switching" and item number 15, "Energy Monitoring and Control Systems."
67. **Combine automatic and manual lighting controls.** Energy impact depends on occupant actions. The effect of this strategy is difficult to predict.
68. **Photocell daylighting controls, zoned on/off.** Because of the annoyance caused by on/off switching of lights, continuous dimmers are considered preferable.
71. **Occupancy sensors.** Such controls do eliminate unneeded light when occupants are not present. Many people object to having their lights out when they are not present, as the lights act as an indicator that they are on the job, just not in the office. Also, in an office environment where people come and go, there may be considerable annoyance from lights continually being turned off and on. This strategy cannot be easily modeled.

APPENDIX C:
Design Tools

This appendix is provided to aid designers in the selection of energy design tools for analysis of small commercial buildings. It is not intended as a comparison of design tools to determine "the best." It is meant to inform designers of the variety of tools available, how they differ, and which are most appropriate for use during a particular stage of the design process.

With an ever-increasing number of design tools becoming available, it is important that the design tool survey be easily updated. Therefore, a blank copy of the form (Figure 1) developed to perform the original survey is included to allow the user of this handbook to evaluate any new tools. To make proper use of this form, the designer should contact someone involved in the development of the design tool and someone who thoroughly understands and has used the complete tool. Contact can be made in person, by telephone, or by mailing the form to that individual for completion. Every item on this form is important, and every effort should be made to obtain all the information. This may require more than one contact, but it will insure a complete understanding of the tool and its usefulness to the designer.

The remainder of this appendix is a compilation of 20 survey forms for various design tools. These tables allow quick reference to General Information, Applications, and Input and Output Data. The design tools listed here have been limited to those that calculate several components of building energy consumption. Numerous analysis tools have been developed for specific building components (i.e., shading or daylighting). Although these tools are very helpful in portions of analysis work, they are not included because of their narrowness.

The first table (Figure 2) describes the tool generically, what type of hardware is required, and who the intended user is. This page will help designers select tools based on equipment requirements and initial investment costs. For example, a designer who has a micro-computer in his office may not wish to invest money in a new hand-held calculator just to get a design tool that operates on the calculator.

The second table (Figure 3) details the specific capabilities of the design tools. Through its use, a selection can be made of tools that are capable of addressing specific types of energy problems required for a full building analysis. As can be seen, relatively few tools are capable of analyzing all the factors considered in the strategy selector process. Relatively few, for example, analyze daylighting.

The third table (Figure 4) provides information on the input and output data required to operate the design tools. This helps the designer rate the various tools in two ways. First, a tool that requires an abundance of information to perform an analysis is probably not useful in the early design phases. However, a tool that requires very little input may not address all of the facets of the energy consumption of a completed design as well as a tool that requires more data. Assessing these factors will aid the selection of the appropriate tool in the appropriate design phase.

By reviewing these three tables and adding information on additional tools, a designer will be better prepared to select tools for the design of energy efficient commercial buildings.

Figure 1. Survey Form.

General

TOOL NAME: ______ AVAILABLE THROUGH: ______
DEVELOPED BY: ______
PHONE NO.: ______
SUPPORTED BY: ______
DATE DEVELOPED: ______
DATE OF LAST REVISION: ______
PHONE NO.: ______
BRIEF DESCRIPTION: ______

PLEASE ATTACH ANY VALIDATION OR TESTING REPORTS.

Tool Hardware & Available Forms

- ☐ MAIN FRAME COMPUTER
  - ☐ CARD DECK
  - ☐ TAPE
  - ☐ TIME SHARING
  - ☐ LISTING - HARD COPY
  - (COMPLETE SECTIONS 1, 2, 3)
- ☐ MICRO-COMPUTER
  - ☐ DISC
  - ☐ TAPE
  - ☐ LISTING
  - ☐ RECALL ONLY MEMORY INTEGRATED CIRCUIT
  - (COMPLETE SECTIONS 1, 2, 4)
- ☐ HAND CALCULATOR
  - ☐ MAGNETIC CARD
  - ☐ LISTING
  - ☐ RECALL ONLY MEMORY
  - ☐ INTEGRATED CIRCUIT
  - (COMPLETE SECTIONS 1, 2, 5)
- ☐ GRAPHIC OR MANUAL
  - ☐ TEMPLATES, CHARTS, TABLES
  - ☐ BOOK
  - ☐ DEVICE
  - (COMPLETE SECTIONS 1, 2, 6)

Intended Use

INTENDED FOR USE BY
☐ ARCHITECT ☐ ENGINEER ☐ TECHNICIAN ☐ RESEARCH ANALYST ☐ BUILDER

PHASE FOR WHICH DESIGN TOOL WAS DEVELOPED
☐ PRE-DESIGN ☐ SCHEMATICS ☐ DESIGN DEVEL. ☐ POST DESIGN SERVICES ☐ RESEARCH

BUILDING TYPE WHICH CAN BE ANALYZED
☐ RESIDENTIAL/SMALL COMMERCIAL ☐ LARGE COMMERCIAL ☐ BOTH

Major & Minor Energy Components Addressed By Tool

- ☐ HEATING
  - ☐ LOADS
  - ☐ SPACE TEMPS.
  - ☐ HVAC SYSTEMS
  - ☐ PASSIVE SOLAR
  - ☐ ACTIVE SOLAR
  - ☐ SHADING
  - ☐ SYSTEM DESIGN
  - ☐ ECONOMICS
  - ☐ UNDERGROUND LOADS
  - ☐ MASS
- ☐ COOLING
  - ☐ LOADS
  - ☐ SPACE TEMPS.
  - ☐ HVAC SYSTEMS
  - ☐ PASSIVE CLNG.
  - ☐ SHADING
  - ☐ SYSTEM DESIGN
  - ☐ ECONOMICS
  - ☐ UNDERGROUND LOADS
  - ☐ SLOPED GLAZING
  - ☐ MASS
- ☐ LIGHTING
  - ☐ LOADS
  - ☐ FC LUX LEVELS
  - ☐ SYSTEM DESIGN
  - ☐ ECONOMICS
  - ☐ DAYLIGHTING FC LUX LEVELS
  - ☐ ARTIFICIAL LTNG. REDUCTION
- ☐ DHW
  - ☐ LOADS
  - ☐ SOLAR ACTIVE
  - ☐ SOLAR PASSIVE
  - ☐ ECONOMICS
- ☐ MISCELLANEOUS
  - ☐ FANS
  - ☐ PUMPS
  - ☐ MISC. ELECTRICAL
  - ☐ ELEV. & ESCALATOR

Input Data Required

	DOES NOT ACCOMMODATE	DEFAULT VALUES SUPPLIED	INPUT REQUIRED
PRE-DESIGN AND SITE ANALYSIS DATA			
LOCATION - ASSOCIATED WEATHER DATA	☐	☐	☐
BUILDING TYPE AND SCHEDULE	☐	☐	☐
OCCUPANCY RATES	☐	☐	☐
BUILDING AREA	☐	☐	☐
SPACE TEMPERATURES	☐	☐	☐
LOCAL ENERGY COSTS	☐	☐	☐
GENERIC BUILDING SHAPE DUE TO SITE RESTRICTIONS	☐	☐	☐
LOCAL CODE REQUIREMENTS (VENTIL., INSUL., ETC.)	☐	☐	☐
LIGHTING REQUIREMENTS	☐	☐	☐
SCHEMATIC DESIGN DATA ENTATIONS			
BUILDING SURFACE AREAS	☐	☐	☐
GLAZING AREAS & ORIENTATIONS	☐	☐	☐
ZONING	☐	☐	☐
ROOM SHAPES	☐	☐	☐
OPERATING SCHEDULES & PROFILES	☐	☐	☐
ARCHITECTURAL DESIGN DEVELOPMENT DATA			
BUILDING MATERIALS & ASSOCIATED DATA (R, α, ϵ, ETC.)	☐	☐	☐
BUILDING MASS DATA	☐	☐	☐
SHADING COEFFICIENTS & DAYLIGHT TRANSMISSION	☐	☐	☐
INTERIOR SURFACE DATA	☐	☐	☐

(continued)

ENGINEERING DESIGN DEVELOPMENT DATA

MECHANICAL SYSTEM DESIGN	☐	☐	☐
MECHANICAL SYSTEM CONTROL	☐	☐	☐
ELECTRICAL SYSTEM DESIGN	☐	☐	☐
ELECTRICAL SYSTEM CONTROL	☐	☐	☐
LIGHTING SYSTEM DESIGN	☐	☐	☐
LIGHTING SYSTEM CONTROL	☐	☐	☐

Weather Data

TEMPERATURE DATA: ☐ HOURLY TAPE ☐ DAILY ☐ MONTHLY DEGREE DAYS ☐ ANNUAL DEGREE DAYS
☐ AVE. MONTHLY MIN. AND MAX. ☐ AVE. MONTHLY TEMP. ☐ OTHER ______

SOLAR DATA: ☐ HOURLY TAPE ☐ MONTHLY AVE. DAILY TOTAL ☐ MONTHLY
☐ MONTHLY AVE. PERCENT POSS. SUNSHINE ☐ OTHER ______

SOLAR ORIENS. CALC: ☐ ANY ORIEN AND SLOPED ☐ HORIZ. AND 4 VERTICAL CARDINAL DIREC.
☐ SLOPED FACING SOUTH ☐ OTHER ______

Calculation Procedures

LANGUAGE: ☐ FORTRAN ☐ BASIC ☐ MACHINE LANGUAGE ☐ OTHER ______ ☐ GRAPHS, CHARTS & SIMPLE CALC.
USER TYPE: ☐ INTERACTIVE ☐ INTERACTIVE GRAPHIC ☐ PREPARE FILE ☐ HAND CALCULATION
UNITS OF CALCULATION: ☐ SI UNITS ☐ ENGLISH ☐ BOTH
CHECK ALL APPROPRIATE BOXES:

HEAT TRANSFER:	☐ FINITE DIFFERENCE	☐ RESPONSE FACTOR	☐ STEADY STATE
SOLAR COMP. CALCULATED:	☐ DIFFUSE/DIRECT/RE-RADIATED	☐ DIFFUSE/DIRECT	☐ TOTAL
INTEGRATION:	☐ SIMPLE EULER	☐ IMPLICIT	☐ OTHER
SHADING:	☐ ANY SOLAR OBSTRUCTION	☐ OVERHANG ONLY	☐ NO SHADING
MOVABLE SHADING:	☐ DAILY & SEASONAL SWITCHING	☐ SEASONAL SWITCHING	☐ NOT CALCULATED
MASS EFFECT IS CALCULATED:	☐ TRANSIENT HEAT FLOW	☐ TIME CONSTANT FACTORS	☐ ASSUME NO MASS AFFECT
ROOM TEMP. BASED ON:	☐ SURFACE & AIR	☐ AIR ONLY	☐ NOT CALCULATED
INSIDE TEMPERATURE:	☐ INPUT SCHEDULE BY USER	☐ FIXED BY TOOL	☐ VARIED BY TOOL
U-VALUES:	☐ CHANGE W/WIND SPEED	☐ REMAIN CONSTANT	☐ MOVABLE INSULATION
INFILTRATION:	☐ AIR CHANGE PER HOUR	☐ CRACK METHOD	☐ VARIES W/WIND SPEED
INTERNAL LOADS INCLUDE:	☐ SENSIBLE & LATENT SEPARATE	☐ SENS. & LAT. TOTAL	☐ SENSIBLE ONLY
VENTILATION:	☐ SENSIBLE	☐ LATENT	☐ VARIES BY SCHEDULE OR COMMAND
DAYLIGHT COEFFICIENTS:	☐ SKY, REFL. & DIRECT	☐ SKY & REFL.	☐ SKY ONLY
ZONES PER RUN:	☐ > 25 ☐ 10 - 25	☐ 2 - 10	☐ 1 ONLY
SYSTEM MODELING:	☐ SYSTEM EFFIC. INPUT	☐ SYSTEM OPTIMIZING	☐ COMPONENT SENSITIVITY
ECONOMIC ANALYSIS:	☐ ANNUAL COST	☐ SIMPLE PAYBACK	☐ LIFE CYCLE COSTING

Output

LOAD DETERMINANTS: ☐ COMPONENT ☐ ZONE ☐ BUILDING
LOADS OUTPUT BY: ☐ HOUR ☐ DAY ☐ MONTH ☐ SEASON ☐ YEAR
TEMPERATURES: ☐ AIR ☐ SURFACE ☐ GRAPHIC PLOT
FUEL USE BY: ☐ MONTHLY CONSUMPTION ☐ ANNUAL CONSUMPTION ☐ SYSTEM COMPONENTS
☐ MONTHLY PEAK DEMAND ☐ ANNUAL PEAK DEMAND ☐ ENERGY SYSTEMS
☐ OTHER ______ ☐ OTHER ______ ☐ TOTAL BUILDING ONLY

Hardware

MANUFACTURER AND MODEL NUMBER: ______
RANDOM ACCESS MEMORY (RAM) REQUIRED: ______ K
DOES THIS TOOL REQUIRE A PRINTER? ☐ YES ☐ NO ☐ OPTIONAL
DISC DRIVES REQUIRED: NOT REQUIRED ☐ 1 ☐ 2 ☐ OPTIONAL
TAPE DRIVES REQUIRED: NOT REQUIRED ☐ 1 ☐ 2 ☐ OPTIONAL
COMPATIBLE WITH A CPM SYSTEM? ☐ NO ☐ YES, TYPE ______

Costs

FIRST COST:
MICRO-COMPUTER: ______
SOFTWARE: ROM IC ______ DISC ______ TAPE ______ LISTING ______
SUPPORT INFORMATION: USER'S GUIDE ______ DATA MANUAL ______ OTHER ______
TIME TO INPUT AND DEBUG: ______ MAN-DAYS ______ MAN-HOURS
RUN COST/TIME:
TYPICAL* INPUT SET-UP TIME: ______ MAN-DAYS ______ MAN-HOURS
TYPICAL* RUN TIME: ______ HRS. ______ MIN.
TYPICAL* PRINT TIME: ______ HRS. ______ MIN.

*FOR THIS FORM, ASSUME "TYPICAL" TO BE A SINGLE-ZONED 100 SQUARE METER RESIDENCE WITH ALL OUTPUTS CHECKED ( ) IN SECTION 2.

RUN COST TIME	BEGINNER	INTERMEDIATE	EXPERIENCED
TYPICAL INITIAL SET-UP TIME:	______	______	______
TYPICAL* INPUT TIME:	______	______	______
TYPICAL* RUN TIME:			
TYPICAL* PRINT TIME:			

Figure 1. (Continued).

Figure 2.

Program Code Name		Latest Version	Intended Use: Architect	Intended Use: Engineer	Intended Use: Technician	Intended Use: Research Analyst	Intended Use: Builder	Data Entry: Interactive	Data Entry: Prepare File	Data Entry: Units: English	Data Entry: Units: SI	Availability: System Compatibility	Random Access Memory (RAM) Reqd. K	Peripherals	1982 Software Purchase Price (U.S. $)	Users Manual	Can Be Purchased From
AXCESS	MF	1/81	●	●		●			●	●		UCC Time-Share			100-900/run	●	Vancar Pace Eng. Services, Inc. 215/885-5900
BLAST 3.0	MF	1/80	●	●		●		●		●		IBM, CDC			1,437	●	NTIS Nat. Energy Software Center 312/972-7250
DOE-2.1A,B	MF	1/80	●	●		●		●	●	●		IBM, CDC			1,660	●	NTIS Nat. Energy Software Center 312/972-7250
ESE	MF		●	●		●				●						●	Control Data Corp. 301/468-8009
EMPS-2.1	MF	1/82	●	●		●			●	●		IBM, UNIVAC				●	Electric Power Research Institute (EPRI) 415/855-2168
ASEAM	MC	7/83	●	●	●	●		●	●	●		APPLE II Plus	48	P	50	●	W. S. Flemming & Assoc., Inc. 202/543-7212
E20-II	MC		●	●	●							TRS 80		OP		●	Local Carrier Representative
F CHART PLUS	MC	1/82	●	●	●			●		●	●	TRS 80 III	48	P	400	●	F Chart Software 608/263-1590
MICROPAS	MC		●	●	●	●		●		●		CPM/IBM (POS)	64/128	P	895/1,095	●	Enercomp 916/753-3400
SASEP	MC	6/83	●	●	●		●	●		●		TRS 80 III APPLE II IBM	48	OP	1,000	●	SUD Associates, Inc. 919/493-5277
SEA	MC	2/84	●	●	●		●	●		●		CPM/IBM (POS)	64	OP	995	●	FKAI Assoc., Inc. 617/996-4499
SOLITE 1	MC	1/82	●	●				●		●		APPLE II	48	OP		●	SOLITE 714/278-7069
STENET	MC	1/81	●	●				●		●	●	HP85	32		6 + DISC	●	Hewlett Packard Users Group 503/757-2000, EXT. 3885
QUICKLITE 1	PC	1/81	●	●					●	●		TI-59			Free		Solar Age, Aug. 1983, pp. 37-47
TEANET III	PC	1/81	●	●					●	●	●	TI-59			90	●	Total Environmental Action, Inc. 603/827-3374
ASHRAE DEGREE DAYS	MS	1/84	●	●			●		●	●							ASHRAE Handbook 1984 Systems, Chapter 42
SEET (BUEHRER METHOD)	MS	1/82	●	●				●		●					15 Members 18 Nonmembers		AIA Publications 202/626-7332
CARRIER SYSTEMS DESIGN MANUAL PART 1	MS			●	●			●		●							Local Carrier Representative
ENERGY GRAPHICS	MS		●	●	●		●		●	●					8.		Booz•Allen & Hamilton, Inc. 301/951-2000
ENERGY NOMOGRAPHS	MS	2/84	●	●				●		●							AIA Publications 202/626-7332

Program Code Name	Keyword	Design Phase: Pre-Design	Design Phase: Schematic Design	Design Phase: Design Development	Design Phase: Post Design	Design Phase: Research	Applications – Heating: Loads	Heating: Space Temperatures	Heating: HVAC Systems	Heating: Passive Solar	Heating: Active Solar	Heating: Shading	Heating: Underground	Heating: Mass	Cooling: Loads	Cooling: Space Temperatures	Cooling: HVAC Systems	Cooling: Passive Solar	Cooling: Active Solar	Cooling: Shading	Cooling: Underground	Cooling: Mass	Lighting: Loads	Lighting: Daylighting FC(LUX) Levels	Lighting: Artificial Lighting FC(LUX) Levels	DHW: Loads	DHW: Passive Solar	DHW: Active Solar	Misc.: Fans	Misc.: Pumps	Misc.: Misc. Electrical	Misc.: Elevator & Escalator	Economics	Number Of Bldg. Zones Per Run	Algorithm
AXCESS	MF	●	●	●	●	●	●	●	●	●	●	●	●	●	●	●	●	●	●	●	●	●	●			●		●	●	●	●	●		180	STEADY STATE
BLAST	MF		●	●	●	●	●	●	●	●	●	●		●	●	●	●	●	●	●		●	●			●		●	●	●	●	●	●	>25	OTHER*
DOE-2	MF		●	●	●	●	●	●	●	●	●	●		●	●	●	●	●	●	●		●	●	●	●	●	●	●	●	●	●	●	●	>25	OTHER*
EMSP 2.1	MF		●	●			●	●	●	●		●	●	●	●	●	●			●	●	●	●	●		●		●	●	●	●	●	●	2-10	THERMAL NETWORK
ESE	MF			●			●		●						●		●						●			●			●	●	●			10-25	ASHRAE TC4.7
ASEAM	MC		●	●	●		●	●	●					●	●	●	●					●	●						●	●	●			10	ASHRAE TC4.7
E20-II PROGRAMS	MC	●	●	●			●	●	●						●	●	●						●			●			●	●			●	150	OTHER*
F-CHART PLUS	MC	●	●				●			●	●	●		●	●			●	●	●		●				●		●					●	1	UNUTILIZ-ABILITY
MICROPAS	MC		●	●	●		●	●	●	●		●		●	●	●	●	●		●		●	●			●			●	●	●	●		3	
SASEP (ISHSUD)	MC	●	●	●			●		●						●		●																	1	ASHRAE BIN DATA
SEA (FKAI)	MC	●	●				●		●						●		●						●			●			●	●			●	2-50	ASHRAE TC4.7
SOLITE 1	MC	●	●	●																			●	●	●								●	2-10	OTHER*
STENET	MC		●	●			●	●	●	●		●		●	●	●	●	●		●		●												10-25	THERMAL NETWORK
QUICKLITE-1	PC	●	●																				●	●	●										
TEANET III	PC		●				●	●		●		●		●	●	●		●		●		●												2-7	THERMAL NETWORK
ASHRAE DEGREE DAYS	MN	●	●				●																											1	ASHRAE DEGREE DAYS
SEET			●				●		●				●		●		●						●		●	●			●	●	●			1	STEADY STATE
CARRIER MANUAL OF AIR CONDITIONING	MN		●	●											●		●																		STEADY STATE
ENERGY GRAPHICS	MN	●	●				●			●					●																			1	OTHER*
ENERGY NOMO-GRAPHS	MN		●				●		●			●			●		●			●			●	●	●	●		●	●	●	●			1	STEADY STATE

* INCOMPLETE INFORMATION

Figure 3.

Input Data

Program Code Name	Keywords	Pre-Design									Schematic					Design Devel.				Eng. Design Development					
		Location (Weather Data)	Bldg. Type & Schedule	Occupancy Rates	Bldg. Floor Area	Space Temperatures	Local Energy Costs	Bldg. Shape & Orient	Lighting Requirements	Code Requirements	Surface Areas	Glazing Areas & Orients	Zoning	Room Shapes	Operating Schedules	Bldg. Materials	Mass Data	Shading Coeff & Daylight Trans	Interior Surface Data	Mechanical Systems	Mechanical Controls	Electrical Systems	Electrical Controls	Lighting Systems	Lighting Controls
AXCESS	MF	●	●	●	●	●	0	●	0	0	●	●	●	●	●	●	●	●	●	●	●	●	●	●	●
BLAST	MF	●	●	●	●	●	●	●	●	0	●	●	●	●	●	●	●	●	●	●	●	●	●	●	●
DOE-2	MF	●	●	●	●	●	●	●	●	0	●	●	●	●	●	●	●	●	●	●	●	●	●	●	●
EMSP 2	MF	●	●	●	●	●	●	0	●	0	●	●	●	0	●	●	●	●	●	●	●	●	●	●	●
ESE (CDC)	MF	●	●	●	●	●	●	●	0	●	●	●	●	0	●	●	0	●	0	●	●			●	0
ASEAM (WSFLEMMING)	MC	●	●	●	●	●		●	●		●	●	●		●	●		●	●	●	●	●	●	●	●
E20-II PROGRAMS	MC	●	●	●	●	●	●	0	0	0	●	●	●	0	0	●	●	●	0	●	●	0	0	0	0
F-CHART PLUS	MC	●	●	●	●	●	0	0	0	0	●	●	0	0	0	●	0	●	●	0	0	0	0	0	0
MICROPAS	MC	●																							
SASEP (ISHSUD)	MC	●	●	●	●	●	0		0	0	●	●	●	0	●	●	●	●	●	●	0	●	0	0	0
SEA (FKAI)	MC	●	●	●	●	●	●	0	0	0	●	●	●	0	●	●	●	●	0	●	●	0	0	0	0
SOLITE 1	MC	●	●	0	0	0	●	●	●		●	●	●	●	●	●	●	●	●	0	0	0	0	0	0
STENET	MC	●	●	●	●	●	0	0	0	0	●	●	●	0	0	●	●	0	0	0	0	0	0	0	0
QUICKLITE-1	PC	0	0	0	0	0	0	0	0	0	0	0	0	●	0	0	0	●	0	0	0	0	0	0	0
TEANET III	PC	●	●	●	●	●	0	0	0	0	●	●	●	●	0	●	●	0	0	0	0	0	0	0	0
ASHRAE DEGREE DAYS	MN	●	0	0	●	0	0	0	0	0	●	●	0	0	0	●	0	0	0	0	0	0	0	0	0
SEET	MN	●	●	●	●	●	●	●	0	●	●	●			●	●	0	●	0	●				●	0
CARRIER MANUAL OF AIR CONDITIONING	MN																								
ENERGY GRAPHICS	MN	●	●	●	●	●	●	●	0	●	●	●	●		●	●		●							
ENERGY NOMOGRAPHS	MN	●	●	●	●	●	●	●	●	●	●	●		●	●	●	0	●	●	●	●	0	0	●	●

Input Data (continued): Weather

Program Code Name	Temperature								Solar		
	Hourly Tape	Daily	Monthly	Annual	Monthly Degree Days	Annual Degree Days	Ave. Monthly Min. & Max	Average Monthly	Hourly Tape	Typical Daily Profile	Monthly Ave. Day & Total
AXCESS	●								●		
BLAST	●								●		
DOE-2	●								●		
EMSP 2	●								●		
ESE (CDC)	●								●		
ASEAM (WSFLEMMING)		●								●	
E20-II PROGRAMS							●				●
F-CHART PLUS					●		●				●
MICROPAS											
SASEP (ISHSUD)	●								●		
SEA (FKAI)	●								●		
SOLITE 1											
STENET							●			●	
QUICKLITE-1											
TEANET III							●			●	
ASHRAE DEGREE DAYS					●						
SEET											
CARRIER MANUAL OF AIR CONDITIONING											
ENERGY GRAPHICS		●								●	
ENERGY NOMOGRAPHS	●								●		

Output Data

Program Code Name	Load Determinants			Load Output					Temps.			Fuel Use							
	Component	Zone	Building	Hour	Day	Month	Season	Year	Air	Surface	Graphic Plot	Monthly Consumption	Annual Consumption	Monthly Peak Demand	Annual Peak Demand	System Components	Hourly Profile	Energy Systems	Total Building Only
AXCESS		●	●	●	●	●	●	●	●	0	0	●	●	●	●	●	●	●	0
BLAST	●	●	●	●	●	●		●	●	●		●	●	●	●	●	●	●	
DOE-2	●	●	●	●	●	●		●	●	●	●	●	●	●	●	●	●	●	
EMSP 2			●			●		●	●			●	●					●	●
ESE (CDC)	●					●						●	●	●	●	●			
ASEAM (WSFLEMMING)																			
E20-II PROGRAMS			●	●	●	●		●				●	●						
F-CHART PLUS		●	●			●		●											
MICROPAS																			
SASEP (ISHSUD)			●			●		●				●	●						●
SEA (FKAI)			●			●		●				●	●						
SOLITE 1																			
STENET		●	●		●														
QUICKLITE-1																			
TEANET III			●	●	●				●	●									
ASHRAE DEGREE DAYS			●			●		●											
SEET			●					●					●		●				
CARRIER MANUAL OF AIR CONDITIONING																			
ENERGY GRAPHICS		●	●				●												
ENERGY NOMOGRAPHS	●							●					●						

●: REQUIRED/GIVEN INFORMATION; 0: DOES NOT ACCOMMODATE; BLANK: INFORMATION NOT AVAILABLE

Figure 4.

Glossary

ASHRAE Standard 90–75. Developed by the American Society of Heating, Refrigerating, and Air-Conditioning Engineers (ASHRAE), Standard 90 is a guideline officially titled *Energy Conservation in New Building Design*. Standard 90–75 is the version approved in 1975, although portions have been revised in 1977 and 1980.

The standard includes energy performance requirements for all new building systems, including the exterior envelope, HVAC systems, service water heating, and lighting. Standard 90 is the technical base for building energy codes in at least 45 states.

ASHRAE TC 4.7 Procedure. A method developed by Technical Committee 4.7 of ASHRAE for calculating the energy use of buildings. For calculating the effects of climatic conditions, the TC 4.7 procedure uses a modified temperature bin method rather than an hour-by-hour simulation. Although the calculation procedure can be completed manually, it is the technical basis of several energy analysis computer programs. One of these is the Energy Systems Evaluation (ESE) program used in the development of this handbook. Other programs based on the TC 4.7 procedure are listed in Appendix C, Design Tools.

Balance Point (Cooling). The ambient temperature at which the internal gains of a conditioned building just equal its heating load and the maximum acceptable interior temperature is maintained.

Balance Point (Heating). The ambient temperature at which the internal gains of a conditioned building just equal its heating load and the minimum acceptable interior temperature is maintained.

Cold Deck. The cooling portion of an air distribution system. It includes the cooling coil and, if applicable, associated cooling duct work (e.g., dual duct system).

Constant Volume System. An HVAC system in which the air flow to building spaces is constant. The temperature of the air entering the space modulates in response to a demand for heating or cooling. See Section II, Level 2, for details of the constant volume systems in this handbook.

Core Zone. In this handbook, the core zone or area is defined as all interior building areas more than 20 feet from an exterior wall. The conditioning needs of the core zone typically are not affected by outdoor weather conditions; therefore, because of internal gains, these areas usually have fairly constant cooling loads during occupied hours.

DOE, Version 2.1b. DOE, version 2.1b, is an hour-by-hour simulation computer program for calculating the energy use of buildings. It simulates the response of building envelopes and systems to variations in climatic and operating conditions. DOE 2.1b was developed initially by Lawrence Berkeley Laboratory under the sponsorship of the U.S. Department of Energy and the State of California.

Desuperheater. A heat exchanger in which the heat of refrigerant gas is used for domestic water heating. In this process, the refrigerant gas is initially at a temperature above its boiling point (superheated), and as it gives up its heat to the water, it is cooled down to a temperature just above or at its boiling point. Cold water passing through the heat exchanger is therefore heated by the superheat of the gas.

Direct Expansion (DX) Coil. A heat exchange coil in which liquid refrigerant undergoes a pressure reduction and evaporates (and expands) by absorbing heat from air passing over the coil. This direct cooling of air by refrigerant expansion is in contrast with an indirect process in which the refrigeration system is used to chill water or another medium, which is then used in a heat exchanger to cool the airstream.

Dual Heat Pump. As used in this handbook, an HVAC system by which the core and perimeter areas of a building are separate HVAC zones served by separate heat pump systems. More details can be found in Section II, Level 2.

Duty Cycling. A control strategy under which mechanical equipment is cycled on a preprogrammed basis during periods of heavy electrical demand within a building. The equipment cycling serves to reduce the peak demand of the building.

Economizer Cycle (Enthalpy-Controlled or Sensibly-Controlled). An economizer cycle is a control sequence of an air supply system that modulates the quantity of outdoor air supplied to provide a cooling effect. An economizer cycle may be designed to either reduce or completely eliminate the use of refrigeration energy for cooling.

The control system for an economizer must be able to sense whether the outdoor air at a given time would have a beneficial cooling effect if introduced into the space. Typically, the system is either sensibly-controlled, which means that it senses the dry bulb temperature of the outdoor air, or it is enthalpy-controlled, which means that it senses the total heat content of the air and, hence, senses its humidity as well as dry bulb temperature.

Evaporative Cooler. An air conditioning device using the evaporation of water to produce a cooling effect. More details are given in Section II, Level 3.

Exfiltration. Air flow outward through a wall, door, window, or other building component.

Fixed Deck Temperature. A control scheme in which the leaving air temperature of a heating or cooling coil is fixed for any load condition.

Global Radiation. The total amount of radiation from a source (typically the sun) on a horizontal surface, including direct, reflected, and diffuse radiation.

Insolation. Radiation from the sun.

Hot Deck. The heating portion of an air distribution system. It includes the heating coil and, if applicable, associated heating duct work (e.g., dual duct systems).

Humidity Ratio. Ratio of the mass of water vapor to the mass of dry air in a sample of air.

Hydronic Heat Pump System. An HVAC system in which each zone is served by a separate water-to-air heat pump. The heat pumps are connected by a hydronic loop that acts simultaneously as a heat source for some pumps and a heat sink for others. More details on the systems used in this handbook can be found in Section II, Level 2.

Multi-Zone System. An HVAC system that makes use of zone controlled space conditioning. In this handbook, the term multi-zone refers in particular to a three-deck multi-zone system. More details on this system can be found in Section II, Level 2.

Non-Uniform Lighting. A lighting scheme in which lighting fixtures are distributed according to the location of work areas, not spread uniformly throughout the interior space. Further details are given in Section II, Level 3.

OTTV (Overall Thermal Transfer Value). A term used in ASHRAE Standard 90 to connote the maximum thermal transfer permissible into the building through its walls or roof. The transfer value is determined by equations given in Standard 90 and accounts for thermal transfer caused by solar heat gain and outdoor-indoor temperature differences.

Open Plenum Return. An air return system in which the space over a dropped ceiling acts as a return air plenum. Conditioned air passes through return grills or through specially designed light fixtures into the open ceiling space and then flows to the air handler. An open plenum return eliminates the need for a ducted return system.

Outdoor Air Temperature Reset. A control scheme by which the temperature of the heating system supply (air or water) is set according to the outdoor air temperature. In this way, the heating system operates no hotter than is needed to meet the building heating load, and system heat distribution losses are minimized.

Packaged Terminal Air Conditioning (PTAC) System. A heating and cooling system integrated into one assembly and intended to serve an individual room or zone.

Parabolic Troffer. A fluorescent light fixture with its reflector having a parabolic cross section. The reflector distributes the available light more evenly and efficiently within the lighted space than a conventional flat reflector.

Prismatic Lens. A clear plastic cover for a fluorescent light fixture that has a surface of small prismatic faces. The faces refract the light from the fixture and thus serve to spread the light more evenly over a wider area than bare lamps would.

Reheat. A space heating scheme in which air supplied from a central conditioning system is heated locally as it enters a building zone according to the temperature in that zone. Since air from the central system may have been cooled to meet cooling loads in some building areas, the air may undergo reheat before entering an area needing heating or relatively little cooling.

Seasonal Efficiency. The overall efficiency of a system or of a piece of space conditioning equipment over the course of an entire heating or cooling season. This efficiency takes into account reductions in efficiency because of equipment cycling, part-load operation, defrosting cycles, and other variations from steady-state, peak-load operation.

Shading Coefficient. A fraction that characterizes a window's ability to exclude solar heat gain. It is defined as the following ratio:

$$\frac{\text{Solar Heat Gain of Fenestration}}{\text{Solar Heat Gain of Clear Double-Strength Glass}}$$

To determine the shading coefficient, solar heat gains are measured under the same solar conditions. More details can be found in 1981 ASHRAE *Fundamentals* at 27.17.

Temperature Deadband. A range of temperature in a building space within which no heating or cooling is provided. For example, the thermostat for a space may be set for the cooling system to be on at temperatures above 75°F and for the heating system to be on at temperatures below 70°F. The range 70–75°F will be the deadband, as no heating or cooling will be done while the space temperature is in this range.

Temperature Ramp. A control strategy for setting the supply air temperature of a heating or cooling system. The supply air temperature is varied as a function of some control variable, such as outside air temperature. The control variable determines the supply temperature according to a relationship that, when plotted, takes the form of a ramp.

An example is shown in which the supply temperature increases with a decrease in outdoor air temperature.

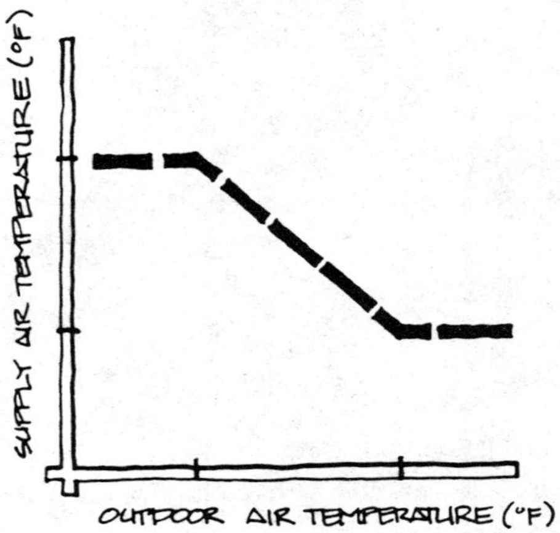

Test Reference Year Data. A reference year of hourly weather data for a particular location. Further details are given in Section II under climatic zone description.

Variable Air Volume (VAV) System. An HVAC system in which the conditioning of each building zone is varied primarily by modulating the air flow to that zone. Control of each zone's air flow is achieved with a thermostat and a variable volume damper in each zone. See Section II, Level 2, for details of the VAV system used in this handbook.

Variable Inlet Vanes. Controllable air guide vanes at the inlet of a constant speed supply air fan. They can be manually or automatically varied to adjust fan delivery efficiency by effectively varying delivered cfm and, hence, horsepower, over a wide operating range. Such vanes are commonly used in VAV systems with a static pressure sensor to maintain constant pressure loss through the system. Typically, a VAV box restricts/increases flow, which increases/decreases static pressure. Then the vanes change position to lower/increase cfm and balance static pressure.

Zone Control. An HVAC layout and control scheme in which each area within a building with distinct heating and/or cooling requirements is conditioned and controlled separately.